ADVANCES IN X-RAY ANALYSIS

Volume 10

ADVANCES IN X-RAY ANALYSIS

Volume 10

Edited by

John B. Newkirk and Gavin R. Mallett

**Proceedings of the Fifteenth Annual Conference on
Applications of X-Ray Analysis
Held August 10-12, 1966**

Sponsored by
University of Denver
Denver Research Institute

PLENUM PRESS
NEW YORK

Plenum Press
A Division of Plenum Publishing Corporation
227 West 17 Street, New York, N. Y. 10011

ISBN-13: 978-1-4684-7837-2   e-ISBN-13: 978-1-4684-7835-8
DOI: 10.1007/978-1-4684-7835-8
Library of Congress Catalog Card Number 58-35928

PREFACE

The featured subject of the 1966 Denver X-Ray Conference was *X-Ray Diffraction Topography and Dynamical X-Ray Phenomena*. One of the chairmen of the featured sessions, Professor R. A. Young, made the following remarks at the conclusion of his session. We think they are quite appropriate to the occasion and with his permission we reproduce them here.

"... It is a source of great pleasure to me to have been an observer of X-ray diffraction topography from what I would consider its infancy in 1958 until now, when it is at least in robust adolescence, if not in young adulthood. Looking at the program we have had, I think that we must all be very happy that we took the effort to come here today. This has been an excellent program, and in it you can see how far topography has come in just a few short years. From 1958, when we had dramatic pictures which at last allowed X-ray crystallographers to compete with electron microscopists for management's attention, we have come to a point where we have such amazing sensitivity that a region within a crystal having a misalignment of only a few hundredths of a second of arc can be detected; we have X-ray moiré fringe techniques and X-ray interferometers; we are able to determine the signs of strains and of dislocations; we now have a rather good understanding of most of the geometric aspects of topography; we have complicated X-ray topographic patterns which we think we understand, at least in part; and we are well on the way to the quantitative interpretation of all of the intensity contrast phenomena in our patterns. Much has indeed been accomplished. It is a source of great personal gratification to me to be a part of today's activities, to see the culmination of these last eight years, and to look forward to the bright promise of at least the next eight."

These words express what many of us feel about the growth and present status of X-ray diffraction topography and related X-ray methods. We join Professor Young in giving due recognition to this promising new research tool.

As co-chairmen of the Conference, we wish to express our thanks to all of the technical contributors, including the authors of the papers, as well as those who made the ensuing discussions so meaningful. In particular, we thank the individual session chairmen under whose skillful guidance each subject was developed for the maximum benefit of the participants.

The chairmen for the various sessions were:

Topography and Dynamical X-Ray Phenomena (I)
> John B. Newkirk, *University of Denver, Denver, Colorado*

Topography and Dynamical X-Ray Phenomena (II)
> R. A. Young, *Georgia Institute of Technology, Atlanta, Georgia*

Structure of Solids
> G. R. Mallett, *University of Denver, DRI, Denver, Colorado*

Stress Analysis
> Karl E. Beu, *Goodyear Atomic Corporation, Piketon, Ohio*

Fluorescence Analysis
> Merlyn L. Salmon, *Fluo-X-Spec Laboratory, Denver, Colorado*

Diffraction Techniques
>Deane K. Smith, *Lawrence Radiation Laboratory, Livermore, California*

Electron Microprobe
>K. F. J. Heinrich, *National Bureau of Standards, Washington, D.C.*

Long Wavelength X-Rays
>L. F. Vassamillet, *Mellon Institute, Pittsburgh, Pennsylvania*

Finally, we wish to acknowledge with thanks the capable cooperation of several members of the University of Denver Research Institute staff whose unstinting efforts made possible a useful and smoothly running conference. Mrs. Mildred Cain efficiently transcribed the discussions; Mrs. Elaine Mason gathered and systematized the discussions offered by each participant. Our special thanks go to Mrs. Ilene Blattner who worked diligently as conference secretary and kept the records and various papers in order. We are grateful also to Mr. Frank Rivera who did so much to keep the conference running smoothly by supervising the audiovisual equipment and personnel.

>*J. B. Newkirk*
>and
>*G. R. Mallett*

CONTENTS

X-Ray Diffraction Topography.................................... 1
   U. K. Bonse, M. Hart, and J. B. Newkirk

Contrast of Dislocation Images in X-Ray Transmission Topography............ 9
   A. Authier

The Asymmetric Bragg Reflection and Its Application in Double Diffractometry. 32
   M. Renninger

Experimental Determination of the Integrated Contribution of Temperature
Diffuse Scattering in X-Ray Reflections.................................. 42
   M. Renninger

The X-Ray Diffraction Image of a Stacking Fault........................ 46
   Norio Kato, Katsuhisa Usami, and Takeshi Katagawa

Dynamical Theory for Simultaneous X-Ray Diffraction...................... 67
   P. Penning

Measuring Techniques of Parallel-Beam-Diffraction Micrography.............. 80
   H. Barth

Some Recent Applications of X-Ray Topography.......................... 91
   A. R. Lang

The Dilemma of Anomalous X-Ray Reflections........................... 108
   James F. McGee and Veli I. Olli

X-Ray Diffraction Microscopy of Planar Diffused Junction Structures.......... 118
   J. K. Howard and G. H. Schwuttke

Experimental Procedures in X-Ray Diffraction Topography.................. 134
   Stanley B. Austerman and John B. Newkirk

X-Ray Diffraction Contrast from Impurity Precipitates in CdS Single Crystals.. 153
   Jun-ichi Chikawa

Lang X-Ray Topographic Studies of Ruby Grown by Different Methods........ 159
   Roger F. Belt

The Analysis of Berg–Barrett Skew Reflections and Their Applications in the
Observation of Process-Induced Imperfections in (111) Silicon Wafers........ 173
   E. M. Juleff, A. G. Lapierre, III, and R. G. Wolfson

The Effect of Small Additions of Magnesium on the Preprecipitation Behavior
of Al–Zn Alloys.. 185
   Robert W. Gould

Analysis of High Angle Diffuse Scattering from Small Platelets................ 204
   A. D. Thomas, Jr., and Gerald L. Liedl

X-Ray Diffraction Study of Ordering in Two Sigma Phases.................. 213
   R. W. Spor, H. Claus, and Paul A. Beck

A Study of the Unusual Line Structure in Powder Patterns of Pyrolytically
Deposited Boron Compounds and Other Materials........................ 221
   Robert L. Prickett, R. L. Hough, and Duane Earley

The Expansion upon Cooling of Thick Cu_2O Films Grown on Copper Substrates 234
   T. F. Swank and K. R. Lawless

New Results on the Iron–Nickel Equilibrium Diagram—The Gamma/Gamma-
Plus-Alpha Boundary... 240
   N. I. Ananthanarayanan and R. J. Peavler

Lattice Constant and Crystallite Size of Condensed Gold Vapor.............. 250
    Frank G. Karioris, Jerome J. Woyci, and Richard R. Buckrey
Line Shape Analysis of Deformed Cu–Ge Alloys........................... 265
    M. Ahlers and L. F. Vassamillet
Anomalous Residual Stresses.. 273
    R. E. Ricklefs and W. P. Evans
Experimental Factors Concerning X-Ray Residual Stress Measurements in High-
    Strength Aluminum Alloys.. 284
    Michael E. Hilley, James J. Wert, and Robert S. Goodrich
X-Ray Measurement of Residual Stresses in Titanium Alloy Sheet............. 295
    David N. Braski and Dick M. Royster
The Application of X-Ray Diffraction Techniques to the Study of Wear........ 311
    T. F. J. Quinn
X-Ray Analysis of Fatigue Damage in Copper............................. 328
    Roy G. Baggerly, Regis M. N. Pelloux, and William F. Flanagan
Improvement of Accuracy in Representation of Conventional Pole Figures....... 342
    K. Aoki, S. Hayami, and M. Matsuo
Precision Lattice Parameter Determination at Liquid Helium Temperatures by
    Double-Scanning Diffractometry..................................... 354
    Hubert W. King and Carolyn M. Preece
Numerical Control X-Ray Powder Diffractometry........................... 366
    R. W. Rex
The Effects of Electronic Structure and Interatomic Bonding on the Soft X-Ray
    Al K Emission Spectrum from Aluminum Binary Systems.................. 374
    David W. Fischer and William L. Baun
Multilayer Soap Film Structures... 389
    R. C. Ehlert and R. A. Mattson
Production Efficiencies of X-Ray Emission Spectra by Proton Bombardment..... 399
    A. A. Sterk, C. Marks, and W. P. Saylor
Electron Microprobe Analyses and X-Ray Diffraction Study of $SrSi_2$.......... 409
    G. M. Faulring and E. S. Malizie
The Application of the Electron Microprobe in the Analysis of Nuclear Fuel
    Meltdown Experiments.. 422
    S. J. Stachura and L. Cooper
Preparation of Electron Probe Microanalyzer Standards Using a Rapid Quench
    Method... 431
    J. I. Goldstein, F. J. Majeske, and H. Yakowitz
Quantitative Microprobe Analysis by Means of Target Current Measurements... 447
    J. W. Colby, W. N. Wise, and D. K. Conley
A Comparison of Four Slit Apertures for Selected-Area Analysis with the X-Ray
    Secondary-Emission Spectrometer.................................... 462
    Eugene P. Bertin
The Analysis of the Light Elements in Ferrotitanium Ores and Residues of
    Widely Varying Composition by X-Ray Spectrography................... 474
    Benjamin S. Sanderson and James A. Yeck
A Glass Fusion Method for X-Ray Fluorescence Analysis................... 489
    J. O. Larson, R. A. Winkler, and J. C. Guffy
Quantitative X-Ray Emission Analysis of Magnesium Through Fluorine with
    X-Ray and Electron Excitation...................................... 494
    F. Bernstein and R. A. Mattson

A Method of Liquid Analyses Providing Increased Sensitivity for Light Elements 506
    D. W. Beard and E. M. Proctor
The Demountable Tube in Light Element Fluorescence Analysis............... 520
    M. A. Short and M. J. H. Ruscoe
Characteristics of Flow Proportional Counters for X-Rays.................... 534
    N. Spielberg
Author Index.. 547
Subject Index.. 555

X-RAY DIFFRACTION TOPOGRAPHY

U. K. Bonse

University of Münster
Münster, Germany

M. Hart

University of Bristol
Bristol, England

and

J. B. Newkirk

University of Denver
Denver, Colorado

ABSTRACT

In recent years, a number of experimental X-ray diffraction techniques have been developed by which a topographical display of the microscopical defects in a crystal can be obtained. This brief review of the most useful of these techniques is intended to summarize the elements of the various methods and to compare their respective features and limitations. Contrary to microradiographic methods, in which image contrast is due entirely to variations in X-ray *absorption* from point to point in the specimen, X-ray diffraction topography is concerned with point-to-point variations in the directions or the intensities of X-rays that have been *diffracted* by crystals. From these variations the defect structure of the crystal may be examined. Methods that mainly measure local variations in the *direction* of the diffracted beam are useful for the detection of gross misorientations such as sub-grains or grains (methods of Guinier and Tennevin,[1] Schulz,[2] Weissmann[3]). *Intensity* mapping methods are chiefly concerned with individual defects such as dislocations, stacking faults, etc. In both groups there are experimental arrangements with both Laue-case (transmission) and Bragg-case (back reflection) geometry.

METHODS

The Schulz Method[2]

A bundle of white X-rays divergent from a point source is diffracted by the crystal and recorded on a film (Figure 1). The use of white radiation ensures that there is no significant change in diffracted intensity due to different incident angles within the bundle. However, misorientations in the crystal cause gaps or overlap regions in the image. With dimensions shown, typical rotations of $2\theta''$ (seconds of arc) can be detected.

Guinier and Tennevin used a similar technique in transmission. In a polycrystal or a heavily distorted single crystal, many images from different grains or different reflections may occur at the same time. Weissmann reduced the number of spots by using crystal monochromatized radiation. Furthermore, in order to identify the particular reflections

1

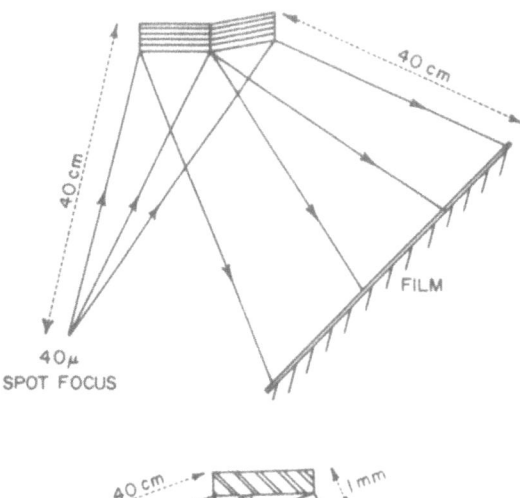

Figure 1. Typical arrangement for the Schulz technique.[2]

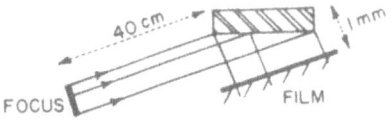

Figure 2. The Berg–Barrett arrangement.[5]

which occur, he traced the direction of the diffracted beams with films at varying distances from the specimen.

The Berg–Barrett Method[4–6]

The crystal is set to Bragg-reflect the characteristic radiation from a line focus (Figure 2). By placing the plate close to the crystal, geometric resolutions of about 1 μ are realized. Across the diffracted beam, variations of the integral reflection power due to the imperfections in the crystal occur. Newkirk[7] showed that single dislocations could be resolved by this technique and their Burgers vector experimentally determined.

The corresponding Laue-case arrangement has been developed by Barth and Hosemann.[8]

The Double-Crystal Method[9,10]

X-rays from a line focus are Bragg-reflected from a perfect reference crystal, then from the specimen, and finally are recorded on a film (Figure 3). The reference crystal and specimen consist of the same kind of material so that exactly *the same spacing* of reflecting planes can be used in both crystals. As a result, the shape of the rocking curve is principally independent of the spectral distribution $\Delta\lambda$ of the radiation used (dispersion is eliminated), and becomes 10 to 100 times narrower than any spectral line. This makes the method very suitable for measuring small tilts. When the specimen is slightly misset from the exact parallel position, i.e., on the flank of the rocking curve, tilts of $\lesssim 0.1''$ and corresponding strains of $|\Delta d/d| \lesssim 10^{-8}$–$10^{-9}$ result in detectable changes of reflected intensity. Deformations of this magnitude occur at distances of up to 50 or 100 μ from the cores of single dislocations, depending upon the material examined. The double-crystal arrangement may also be used with the specimen set for transmission.[11] When imaging larger tilts—for instance, those that occur between subgrains or strained regions of a deformed crystal (> 10 sec of arc)—only partial elimination of the dispersion by using a "monochromatizing" crystal with different spacing of reflecting planes,[12] or simply the use of characteristic radiation (Berg–Barrett technique), is usually sufficient for good contrast between regions with different orientations.

The Lang Method[13]

In this method, a ribbon X-ray beam is collimated to an angular divergence sufficiently small that only one characteristic wavelength is diffracted by the crystal (Figure 4). A stationary opaque screen allows only the diffracted beam to reach the photographic plate. A large area/volume of crystal can be surveyed by scanning both the crystal and photographic plate in synchronism past the incident beam. The line X-ray source in the method of Barth and Hosemann and the traversing movement in the Lang method achieve the same result, namely, a horizontally extended field of view. The drawback of the more complicated traverse apparatus, however, is compensated by the following advantages of the Lang technique: lower background; less trouble with simultaneous reflections because a strongly collimated beam is used; and a higher resolution, since the photographic plate may be placed closer to the specimen.

Many different types of lattice defects have been observed by this technique.

The Anomalous Transmission Method[14]

In a perfect crystal, incident and diffracted waves form the entity of the X-ray wavefield, which is a standing-wave pattern fitting onto the set of reflecting planes. Depending on the position of the antinodes of this pattern, the energy transport of the wavefield occurs with anomalously high (antinodes coincident with planes) or anomalously low (antinodes between the planes—the Borrmann effect) absorption. Therefore, a perfect crystal set at the precise Bragg angle is capable of transmitting interfering X-rays at crystal thicknesses which would absorb almost all the energy of a noninterfering X-ray

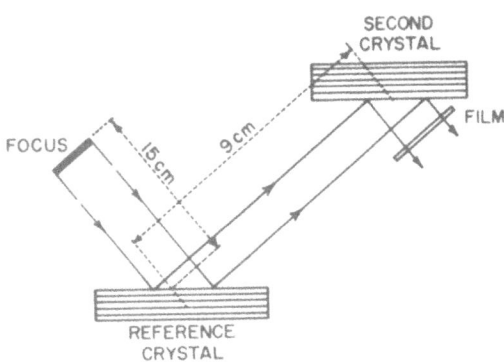

Figure 3. The double-crystal geometry.[10]

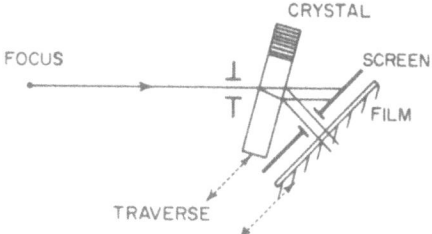

Figure 4. The Lang scanning method.[13]

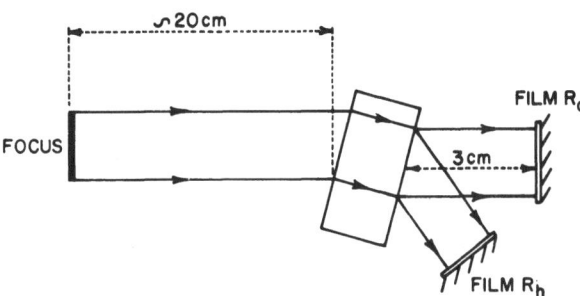

Figure 5. The Borrmann arrangement for anomalous transmission topography.[14]

beam (Figure 5). If $\mu_0 t \geq 20$, where μ_0 is the normal absorption coefficient and t is the thickness of the specimen, both the transmitted diffracted beam R_h and the transmitted direct beam R_0 have similar intensities and either can be used to obtain diffraction topographs. Lattice defects cause a local reduction in the anomalous transmission and are therefore seen as *shadows*.

CAUSES OF CONTRAST[15,16]

With respect to the local amount of distortion produced by the lattice defects, the crystal may be divided into *dynamical* regions, where the primary wave $\mathbf{K}_0$ and the diffracted wave $\mathbf{K}_h$ are still coherently bound to each other as a wavefield, and *kinematical* regions, where this coupling of $\mathbf{K}_0$ and $\mathbf{K}_h$ is destroyed because of too large lattice distortions. Dynamical regions still behave locally like perfect crystals; taken over longer distances, a continuous change of orientation and spacing of reflecting planes occurs. The wavefield can adapt[17] itself to these changes by a continuous variation of the direction of its energy flow $\mathbf{j}$ and, correspondingly, of the ratio I_h/I_0 of the intensities of its components $\mathbf{K}_h$ and $\mathbf{K}_0$. Dynamical regions have locally the same narrow reflection curve as a perfect crystal (<10 sec of arc wide) and especially the same integral reflection power, i.e., the same area under the reflection curve.

Homogeneous Dilation and Tilt Contrast

In reflection techniques, i.e., where the image is formed by the beam split off at the surface, the most straightforward cause for the contrast of dynamical regions is that these regions may simply be more or less misset from the diffraction peak given by the Bragg equation, $\lambda = 2d \sin \theta$. Missetting can be due to either a change of orientation θ or of lattice parameter d, or of both simultaneously. How large a missetting has to be before it causes a detectable intensity change depends on the "precision" of the diffraction condition, i.e., the width of the rocking curve (double-crystal arrangement) or of the λ-spread (Berg–Barrett, etc.). In any case, the contrast should be proportional to

$$\frac{1}{W}\left(\frac{\Delta d}{d} \tan \theta \pm \Delta \theta \right)$$

where W is the width of either the rocking curve or the λ-spread. Owing to the extreme narrowness of the double-crystal rocking curve, this missetting is the main cause of contrast with the double-crystal technique. It is worth noting that the sign of the contrast depends on the sign of the deformation.

"Extinction Contrast" or Direct Image

In kinematical regions, the local reflection curve is generally lower and wider, combined with an increase of integral (total) reflection power, compared with the perfect crystal. Cores of dislocations, up to about 2μ diameter, are typical kinematical regions. What has commonly been called "extinction contrast" is exactly this increase of integral reflection power of kinematical regions. However, an increase of diffracted intensity will be observed only if the reflection curve is utilized to its full width, i.e., only if the divergence of the incident beam is wide enough. This is in general the case for the Berg–Barrett, Barth–Hosemann, and Lang methods, however, not with the double-crystal technique. Therefore the two former methods yield *enhanced* intensity from kinematical regions, and the latter, decreased intensity (if the image of the kinematical region plays a role here at all).

The kinematical image contrast is also often referred to as the "direct" image. The sign of the contrast is independent of the sign of the deformation.

Dynamical Image

The change which the entering wavefield has experienced after traversing distorted dynamical regions also contributes to the image contrast, particularly in transmission methods. These effects are important for large values of $1/\chi_{rh}$ (χ_{rh} is the Fourier coefficient of order h of electric susceptibility χ_r, which is proportional to the structure factor F_h).

In reflection techniques, wavefield beams can be bent back to the entrance surface, where they add to the beam which is reflected at the surface itself.[18]

In transmission methods, besides this bending of beam paths, the shift of energy between $\mathbf{K}_0$ and $\mathbf{K}_h$ is responsible for considerable contrast effects in the $\mathbf{K}_0$ and $\mathbf{K}_h$ waves when they leave the crystal at its exit surface.[16] Moreover, the Borrmann contrast may also be counted in this category since it is due to breaking up or continuous modification of wavefield beams with anomalously low absorption into highly absorbed wavefields or single waves.

Conclusion

In the Berg–Barrett technique and in transmission techniques for $\mu_0 t \lesssim 1$, extinction contrast predominates. In the range $1 \lesssim \mu_0 t \leq 10$, both direct and dynamical images are visible. With $\mu_0 t \gtrsim 10$, practically only Borrmann contrast is observed.

Contrast in the double-crystal arrangement is mainly due to homogeneous dilations and tilts, and—sometimes—to wavefield beams which have been bent back to the surface.

The general condition for the visibility of a defect is that it produces a sufficiently large disturbance of the set of reflecting planes used. From this it follows, in the isotropic case, that dislocations can, in principle, vanish only if $\mathbf{g} \cdot \mathbf{b} = 0$ and $\mathbf{g} \cdot \mathbf{n} = 0$, where $\mathbf{g}$ is the diffraction vector, $\mathbf{b}$ is the Burgers vector, and $\mathbf{n}$ is the vector normal to the slip plane. Since, for mixed dislocations, both equations cannot be satisfied simultaneously, *mixed* dislocations can never vanish completely. In the anisotropic case, dislocations can only vanish if their direction is normal to a mirror plane of elastic symmetry.

A particular cause of contrast not mentioned so far is due to the interaction of two wavefields (Pendellösung fringes) in wedge-shaped portions of the crystal. Since stacking faults and dislocations may, in effect, generate wedges in crystals, they are often observed by these fringes.

Table I. Features of Several X-Ray Topographic Methods

Technique	Schulz, Guinier and Tennevin[1,2]	Berg-Barrett[4,5,6,7]	Double crystal[9,10,11,15]¶	Wide-beam transmission (Barth-Hosemann)[8]		Scanning transmission (Lang)[13]	
Apparatus	Simple	Simple	Complicated	$\mu_0 t > 10$ Simple	$\mu_0 t < 1$ Simple	$\mu_0 t \sim 3$ Complicated	$\mu_0 t < 1$ Complicated
Exposure time	10–25 hr	Short (~ 1 hr)	Short (~ 1 hr)	Long (~ 10 hr)	Short (~ 1 hr)	10–30 hr	2–10 hr
Defect for which technique is most suited with kind of contrast*	Grain misorientation, subgrains (1)	Subgrains (1) Dislocations (2)	Subgrains (1) Dislocations (1) Stacking faults	Dislocations (3)	Subgrains (1) Dislocations (2) Stacking faults	Dislocations (2) (3)	Subgrains (1) Dislocations (2) Stacking faults
Best geometric resolution	50 μ	1 μ	1 μ	1 μ	1 μ	1 μ	1 μ
Sensitivity to deformations	Low	Low	High	High	Low	High	Low
Sensitive to the *sense* of deformations	Tilts: Yes Inhomogeneous deform.: No	Subgrains: Yes Disloc.: No	Yes	Yes	No	Yes	No
Thickness t of specimen contributing to topograph†	Schulz: ≤ 5 μ G.–T. 50 → 1000 μ	≤ 5 μ	≤ 5 μ (back refl.) ≤ 300 μ (transm.)	1 → 5 mm	0 → 2 mm	0.1 → 5 mm	0 → 2 mm
Dislocation image width‡	—	1 → 5 μ	Up to 150 μ	$\gtrsim 50$ μ	~ 5 μ	Up to 150 μ	1 → 10 μ
Upper limit of dislocation density (lines/cm²)	—	5×10^8	10^5	5×10^3	5×10^6	5×10^3	5×10^6

* (1) Homogeneous dilation and tilt contrast, (2) extinction contrast, (3) dynamical contrast.
† This is determined in the Bragg case by the extinction depth, and in the Laue case, by the value of μ_0 (the absorption coefficient) for the material and the value of t imposed by the technique.
‡ Based on the assumption that this is determined by normal image overlap.
¶ References 9, 10, and 15 deal with back reflection; reference 11 with transmission.

COMPARISON AND APPLICATION OF THE
DIFFERENT TECHNIQUES

Table I gives a comparative survey of the different methods described above and their main fields of application. It has to be kept in mind, however, that the limits between different categories are somewhat diffuse, depending on the special experimental conditions under which a particular method is employed. Observations have been made of dislocations, stacking faults, low- and high-angle grain boundaries, twin boundaries, magnetic domains, and precipitation and segregation of impurities. The materials so far studied by various workers include diamond, silicon, germanium, InSb, NaCl, AgCl, SiC, MgO, Al_2O_3, BeO, aluminum, copper, Fe-5%Si, calcite, quartz, and ice.

The methods are given below in order of their decreasing sensitivity to lattice perturbation for the same thickness of crystal: double crystal; anomalous transmission; Lang; Berg–Barrett, Barth–Hosemann; and Guinier–Tennevin, Schulz.

In some cases, the transmission techniques may exceed in sensitivity the double-crystal reflection method because a larger volume of crystal is sampled. There are, however, applications where high sensitivity is not desired, such as subgrain delineation, slip-band survey, and any investigation of heavily strained regions. For these applications, the less sensitive methods are more suitable. For stacking faults, dislocations, twins, and lamellar structures the transmission techniques are generally more informative than the Bragg-case techniques.

Webb[19] and Azaroff[20] discussed some of the aspects outlined here a few years ago. A fairly comprehensive list of literature up to 1961 may be found in those references.

ACKNOWLEDGMENT

This brief review is reprinted from *Encyclopaedic Dictionary of Physics*, Pergamon Press, Oxford, London, New York, Paris.

REFERENCES

1. A. Guinier and J. Tennevin, "Sur Deux Variantes de la Methode de Laue et leurs Applications," *Acta Cryst.* **2**: 133, 1949.
2. L. G. Schulz, "Method of Using a Fine-Focus X-Ray Tube for Examining the Surface of Single Crystals," *J. Metals* **6**: 1082, 1954.
3. S. Weissmann, "Method for the Study of Lattice Inhomogeneities Combining X-Ray Microscopy and Diffraction Analysis," *J. Appl. Phys.* **27**: 389, 1956.
4. W. Berg, "An X-Ray Method for Study of Lattice Disturbances of Crystals," *Naturw.* **19**: 391, 1931.
5. C. S. Barrett, "New Microscopy and its Potentialities," *Trans. AIME* **161**: 15, 1945.
6. R. W. K. Honeycombe, "Simple Method of X-Ray Microscopy and its Application to Study of Deformed Metals," *J. Inst. Metals* **80**: 39, 1951.
7. J. B. Newkirk, "Observation of Dislocations and Other Imperfections by X-Ray Extinction Contrast," *Trans. AIME* **215**: 483, 1959.
8. H. Barth and R. Hosemann, "Use of a Parallel Beam Transmission Method for the X-Ray Examination of Crystal Structure," *Z. Naturforsch.* **13a**: 792, 1958.
9. U. Bonse and E. Kappler, "X-Ray Photography of the Distortion Planes of Single Dislocations in Germanium Single Crystals," *Z. Naturforsch.* **13a**: 348, 1958.
10. U. Bonse, "Zur röntgenographischen Bestimmung des Typs ein zelner versetzungen in Einkristallen," and "X-Ray Picture of the Field of Lattice Distortions Around Single Dislocations," in: *Direct Observation of Imperfections in Crystals*, edited by J. B. Newkirk and J. H. Wernick, Interscience Publishers, New York, 1962, pp. 431–460.
11. M. Kohra, M. Yoshimatsu, and J. Shimizu, "X-Ray Observation of Lattice Defects Using a Crystal Monochromator," *ibid.*, pp. 461–470.
12. J. Intrater and S. Weissman, "An X-Ray Diffraction Method for the Study of Substructure of Crystals," *Acta Cryst.* **7**: 729, 1954.

13. A. R. Lang, "Direct Observation of Individual Dislocations by X-Ray Diffraction," *J. Appl. Phys.* **29**: 597, 1958; **30**: 1748, 1959.

14. G. Borrmann, W. Hartwig, and H. Irmler, "Shadow of a Dislocation Line in X-Ray Diagrams," *Z. Naturforsch.* **13a**: 423, 1958.

15. U. Bonse, "Zum Kontrast an Versetzungen im Röntgenbild," *Z. Physik* **177**: 543, 1964.

16. M. Hart, Ph.D. Thesis, Bristol, 1963; M. Hart and A. R. Lang, 6th. Intern, Cong. Cryst., Rome, 1963, Paper 12.4.

17. P. Penning and D. Polder, "Anomalous Transmission of X-Rays in Elastically Deformed Crystals," *Philips Res. Repts.* **16**: 419, 1961.

18. U. Bonse, "Starke Ablenkung von Röntgen-Wellenfeldstrahlen in elastich gebogenen Kristallen," *Z. Physik* **177**: 529, 1963.

19. W. Webb, "X-Ray Diffraction Topography," J. B. Newkirk and J. H. Wernick, *op. cit.*, pp. 29–76.

20. L. V. Azaroff, "X-Ray Diffraction Studies of Crystal Perfection," *Progr. Solid State Chem.* **1**: 347, 1964.

DISCUSSION

R. F. Belt (Airtron): Could you suggest some explanation for the intensity variations that can be seen along the horizontal plane of the Borrmann photographs?

J. B. Newkirk: It is possible that this effect is due to variations in intensity of the X-ray source, because the X-rays used in this method are parallel to each other and therefore come from specific places on the X-ray source.

M. Renninger (Krist. Institut der Universität): It should be mentioned that the method usually ascribed exclusively to Bonse and Kappler was developed independently by Bond and Andrus earlier, *Am. Mineralogist* **37**: 622, 1952.

Written comment by R. D. Deslattes: In the review of topographic procedures presented by Professor Newkirk, two crystal techniques were described only in the form originally used by Bonse and Kappler. In this form, the technique appears to be limited in generality by the requirement of a perfect reference crystal of the species under investigation. As we pointed out at this conference two years ago (Deslattes, Torgesen, Paretzkin, and Horton, *Advan. X-Ray Anal.* **8**: 315–324, 1965, cf. *J. Appl. Phys.* **37**: 541–548, 1966), a significant increase in generality of application can be obtained at little cost in sensitivity by use of reference crystals differing slightly in grating spacing from the specimen of interest. A concurrent advantage of this "mismatching" is that neither beam collimation nor scanning is required to produce unique imaging in large area topographs with high strain sensitivity. We have shown, in fact, (Paretzkin, Deslattes, Meeting of the International Union of Crystallography, Moscow, July 1966), that with a set of only six reference crystals (various cuts of silicon), a specimen with $1 \text{ Å} < d < 2 \text{ Å}$ can be topographed with a dispersive contribution to the total rocking-curve width of no larger than about five seconds of arc. A paper in which these remarks are more fully detailed has been prepared for publication and will be submitted to *Acta Cryst.*

CONTRAST OF DISLOCATION IMAGES IN X-RAY TRANSMISSION TOPOGRAPHY

A. Authier

*Laboratoire de Minéralogie–Cristallographie
Paris, France*

ABSTRACT

A brief review of the different topographic methods is given in order to attempt a classification of the various types of contrast observed. Depending on whether the image is obtained by a reflection or transmission method, is simple or integrated, dynamic, direct, or intermediary, the contrast and image width vary greatly. A detailed analysis is given in the case of transmission section topographs.

INTRODUCTION

It is possible to show experimentally that three different types of image arise simultaneously. The *direct* image is due to the reflection of the direct beam by the perturbed region around the dislocation line. This image gives a direct view of the distortion field. When the width of the rocking curve for the perfect crystal is narrow enough, it appears as a *double* line; we have interpreted this result. The *dynamic* image is the shadow of the dislocation line in the fan of wave-fields. It is due to the disappearance of the Borrmann effect because of the deformation; it can also be interpreted in terms of a Fresnel diffraction effect (diffraction by a line). The converse effect, amplification of the Fresnel diffraction by a slit in a crystalline medium at the Bragg setting, has been described by C. Malgrange and the author.

The third type of image, or *intermediary* image, is due to an interesting effect; when a wave-field enters a highly distorted region, it gives rise to a second wave-field, of which the tie-point belongs to the other branch of the dispersion surface (interbranch scattering). The effect has been predicted by G. Borrmann and A. R. Lang and studied theoretically by P. Penning. We have shown that it takes place in the case of impurity layers in natural calcite and of dislocation lines in various materials. The wave-fields created at the dislocation line or the impurity layer give rise to Pendellösung fringes which are visible both on section and traverse topographs. We have compared them to the fringes obtained when there is a twin lamella across a crystalline slab. In this latter case, each wave-field splits into its two components when it enters a twin lamella which is out of the Bragg-setting, and they will in turn induce new wave-fields when they reenter the parent crystal. These new wave-fields give rise to Pendellösung fringes which we have studied theoretically and experimentally. There is good agreement between the properties of these fringes and those observed in the shadow of a dislocation line.

CLASSIFICATION OF DISLOCATION IMAGES

Since 1958, a great many topographic methods have been developed which permit observation of single dislocations: Berg–Barrett (Newkirk[1]), Lang,[2,3] anomalous transmission (Borrmann[4]), double-crystal setting by reflection,[5,6] double-crystal setting by

transmission [7,8] and parallel beam technique.[9,10] A study of these methods and the principles on which they are based are given in the paper by Newkirk, Bonse, and Hart.[11]
As they have pointed out, there are various reasons for the contrast and it is not easy to
make a simple classification. We first note that two different things may be required from
a topograph: that the images of dislocations be as narrow as possible, i.e., that the
method have the highest *spatial* resolving power, and that the smallest misorientations
be observed, i.e., that the method have the highest *angular* resolving power.

Obviously the same type of image will not fill both requirements. Two different
experimental settings are in general required, although in transmission topography both
types of image may occur simultaneously.

To characterize a given type of image, it is suitable to use three criteria:

1. The technique is one of *reflection* or of *transmission*; interpretations of the contrast
   are always different in both cases.
2. The image is *direct*, *dynamic*, or *intermediary*: Direct means reflection of the direct
   beam by the deformed region as if it were a small mosaic crystal.[12] The kinematical, or geometric, theory of diffraction applies. Dynamic means that there is
   coupling of the reflected and incident waves into wave-fields,[12] and that an extension of the dynamical theory to the perturbed-crystal case should be used. Intermediary means that there has been creation of new wave-fields from the wave-
   fields incident on the defect (interbranch scattering).
3. The image is *simple* or *integrated*: Simple means that the angular divergence of the
   beam falling on a given point of the defect is small. Integrated means that this
   divergence is, on the contrary, large and that the intensity recorded on the photographic plate is an integrated intensity. For example, this is always the case in
   traverse topographs.

Depending on which criterion is found, it is possible to predict the contrast of dislocation images. In this paper, we shall discuss in more detail the images obtained in transmission topography.

IMAGE FORMATION IN X-RAY TRANSMISSION TOPOGRAPHY

There are two main types of transmission topography: Lang's method, corresponding
to small values of μt, where μ is the absorption coefficient and t the thickness of the
sample; and anomalous transmission methods, corresponding to high values of μt.
Actually, it is better to consider values of μt between 2 and 4 and to extrapolate to the
two extreme cases. It is also more convenient to study first what happens when the
crystal is immobile, the X-ray beam comes from a point focus, and is limited by a fine slit
(section topographs); and then to study the influence of a translation applied to the crystal
(traverse topographs) or that of a long focus in the high μt range (parallel beam method).

Figure 1 depicts the well-known experimental setup for section topographs. The
influence of a dislocation can be described schematically as being threefold:

1. The dislocation cuts the direct beam. The region around the dislocation line
   reflects intensity from this direct beam, giving rise to the direct image.
2. The dislocation line cutting the paths of wave-fields propagating within the
   Borrmann fan ABC casts a shadow, giving rise to the dynamic image.
3. The wave-fields intercepted by the dislocation line decouple into their incident
   and reflected wave components which, on reentering good crystal, excite new
   wave-fields. These give rise to a third type of image, the intermediary image. This
   is equivalent to interbranch scattering, or transfer of energy from one branch to the
   other. As the path of the wave-fields incident on the dislocation line becomes

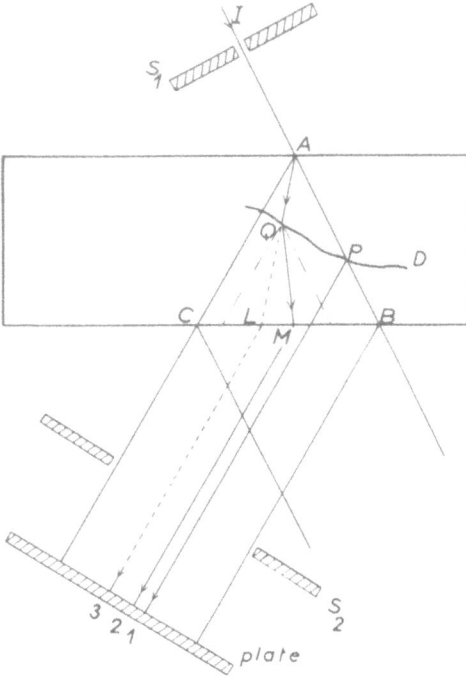

Figure 1. Image formation in transmission topography. Point P of the dislocation line (D) reflects the direct beam (I) to give the direct image (1). Point Q casts a shadow in the direction of the incident wave-fields AQ, giving rise to the dynamic image (3). New wave-fields created at Q propagate along QM, giving rise to the intermediary image (2).

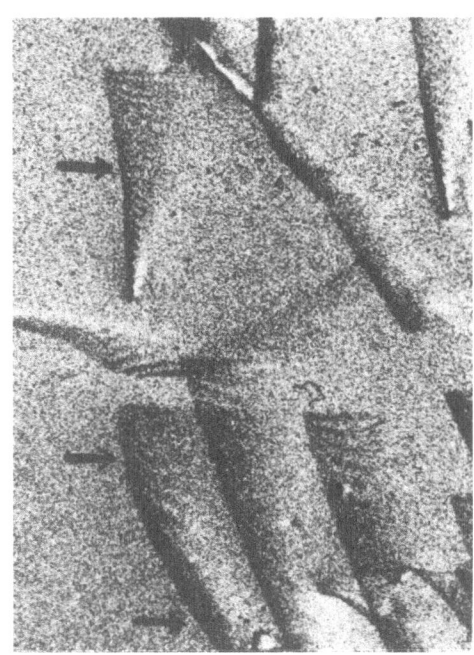

Figure 2. Traverse topograph of dislocations in silicon: reflection $2\bar{2}0$; crystal face (111); thickness 0.8 mm; Mo K_α radiation. The arrow shows the positions of the direct images visible in Figure 3a. (180×, reduced 10% for reproduction.)

closer to that of the direct beam (when Q moves into P), the intermediary image gradually merges into the direct image.

When crystal and photographic plate are traversed, the direct images form projections of the defects in the reflected direction. They are the interesting features of Lang-type topographs, or *projection topographs*.[3] The dynamic images will give more or less diffuse images which are the only ones visible in the high μt range. The intermediary images give rise to fringes in the shadow of the direct images if the dislocation line is not parallel to the crystal faces; when μt is very high, these are only visible when the dislocation line lies very close to the surface.

Figures 2, 3a, and 4a give examples of the three types of image. Figure 2 is a traverse topograph of silicon (0.8 mm thick, reflection $2\bar{2}0$, Mo K_α). One sees the thin black direct images, the white diffuse dynamic images, and, in between, the fringes in the shadow of the dislocation images. Figure 3a is a section topograph of the same region. The direct image here is a black point, the dynamic image a thick white line, and the intermediary image a series of black fringes. Figure 4a is a schematic drawing showing the formation of the images. The top part of the drawing is a projection on the incidence plane, the lower part a projection on the photographic plate placed normal to the reflected direction.

In the next three sections we shall discuss in more detail the properties of these three types of image.

DIRECT IMAGES

Theoretical Considerations

A small angular fraction only of the beam incident on the crystal is near enough the Bragg angle to give rise to wave-fields propagating within the Borrmann fan ABC (Figure 1).

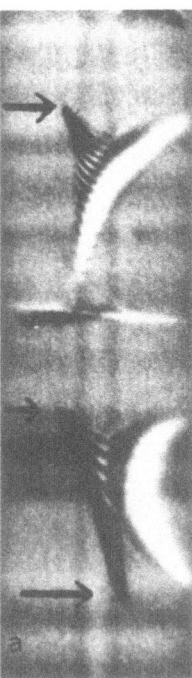
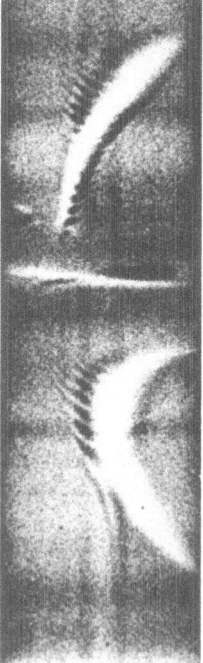

Figure 3. Section topograph of the same region as in Figure 2. The direct images are shown by arrows. (a) $2\bar{2}0$ reflection; (b) $\bar{2}20$ reflection. ($180\times$, reduced 10% for reproduction.)

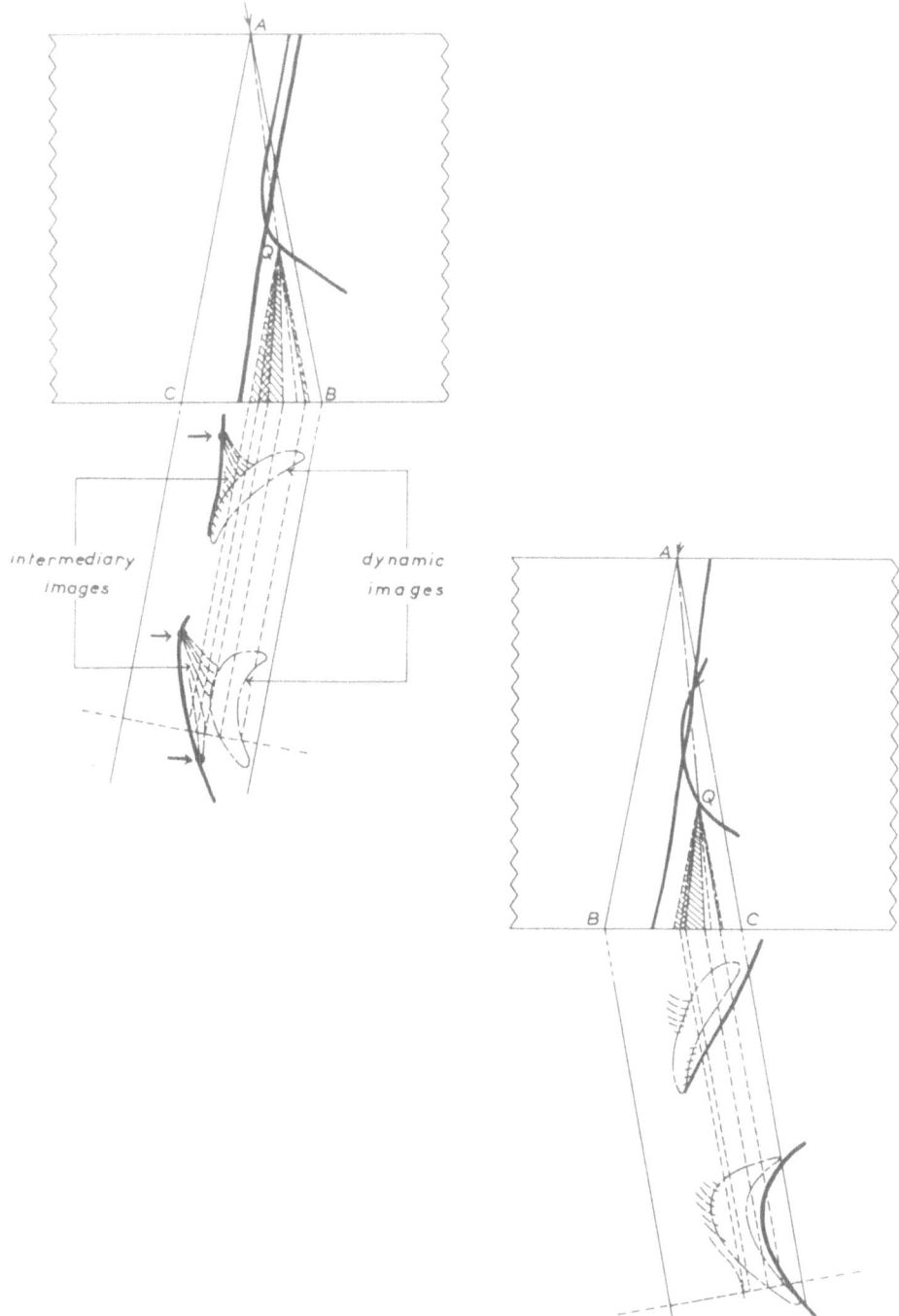

Figure 4. Construction of the images for two of the dislocations visible in Figure 3. (a) $2\bar{2}0$ reflection—the direct images (thick black lines) have been drawn from the traverse topograph; their positions in the section topograph are indicated by arrows. (b) $\bar{2}20$ reflection—the direct beam does not cut the dislocation lines; there are no direct images; the intermediary images are weak.

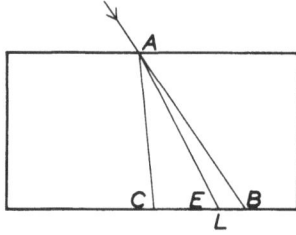

Figure 5

The major part propagates along AB, undergoing normal attenuation. It is made of "rays" with a large departure from Bragg's law with regard to the perfect crystal, i.e., a Fourier analysis of the direct beam AB would show that its plane-wave components with a departure from Bragg's law less than once or twice the width of the rocking curve for the perfect crystal have very low intensity.

There is of course no abrupt transition between the "fan" and the direct beam. A possible definition might be that a wave-field belongs to the direct beam if the intensity ratios $|R_h{}^2|$ of its reflected and incident components are less than 1.5% and 5%, respectively. We shall see that as far as the gross interpretation of the contrast of direct images is concerned, this definition is satisfactory.

It is possible to show that

$$|R_h{}^2| = \frac{\gamma_0}{\gamma_h} \cdot \frac{1 - Y}{1 + Y}$$

where $Y = EL/EB$, E is the middle of BC, and L is the emergence point of a given wave-field from the crystal (Figure 5); γ_0 and γ_h are the cosines of the angles of the incident and reflected directions with the normal to the entrance surface, respectively. Y is related to a parameter proportional to the departure from Bragg's law[13] by

$$Y = \pm \frac{\eta}{\sqrt{1 + \eta^2}}$$

The width of the reflecting range corresponds to $\Delta\eta = 2$ (absorption need not be taken into account here), and intensity ratios of 5% and 1.5% therefore correspond to values of Y equal to 0.90 and 0.97 and to departures from Bragg's law equal to once or twice the width of the rocking curve, respectively.

The direct beam, thus defined, is very intense in low absorbing material. When it enters a deformed region, it will satisfy the Bragg condition for this deformed region provided the effective misorientation is larger than once or twice the width of the rocking curve for the perfect crystal and less than the total divergence of the direct beam. This last limitation is unimportant in the case of a dislocation line for which the misorientation is bigger the closer one is to the core. It is the first limitation which leads to an explanation of the width and the contrast of the images. If the divergence of the direct beam were not bigger than the width of the rocking curve, there would be no direct image.

Figures 6a and 7 give equal effective misorientation curves around an edge dislocation line parallel to the crystal face with Burgers vectors parallel and perpendicular to the crystal face, respectively (see Appendix).

From what we have seen above, it is the region inside these curves, drawn for an effective misorientation equal to x times the width of the rocking curve ($x \sim 1$ or 2), which reflects the direct beam and contributes to the formation of the direct image. Because of the divergence of the direct beam, it is reasonable to assume that the intensity

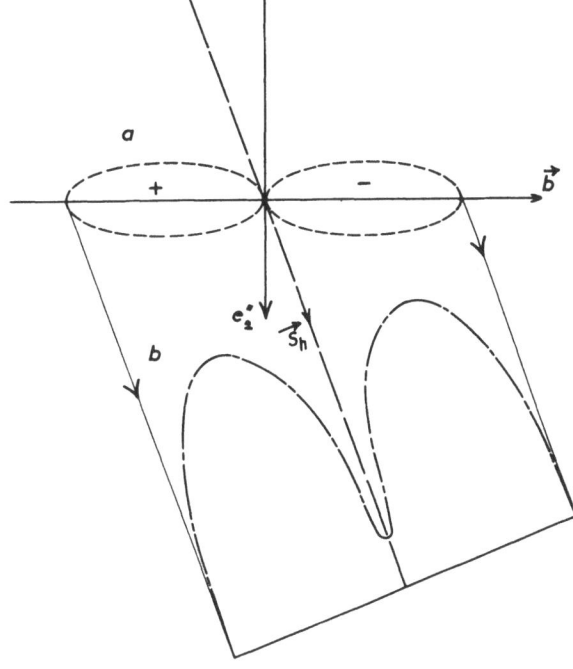

Figure 6. Principle of the formation of the direct image. (a) Equal effective misorientation curve around an edge dislocation in mica with Burgers vector parallel to the crystal face; $\vec{b}$ indicates direction of the Burgers vector; $\vec{e}_2$ shows the direction of the normal to the entrance surface and, $\vec{s}_h$ the reflected direction. (b) Corresponding intensity distribution in the reflected direction.

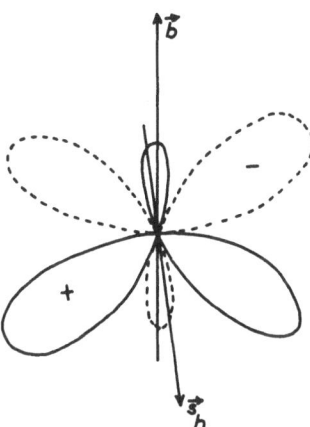

Figure 7. Equal effective misorientations in the plane of incidence around an edge dislocation in mica; Burgers vector $\vec{b}$ normal to the crystal faces.

reflected by these regions is an integrated intensity and that the direct image is an integrated image. This is true in all cases for a traverse topograph. Since these regions reflect as "mosaic crystals" imbedded in an X-ray beam, the integrated intensity on the photographic plate is proportional to the *volume* of the mosaic crystal crossed by the reflected beam. Figure 6b shows the profile we should expect for the direct image of an edge dislocation with Burgers vector parallel to the crystal faces.* The direct image should thus have a double contrast on traverse as well as on section topographs.

It can be shown that the distance L_1 between the two maxima is equal to the maximum distance from the dislocation core for which the effective misorientation is equal to $x\delta$

* A similar study of the contrast of direct images has been given by J. Chikawa (*J. Appl. Phys.* **36**: 3496, 1965).

where δ is the width of the reflecting range for the perfect crystal. From equations A-6 and A-7 (see Appendix),

$$L_1 = \frac{1}{2\pi \cos\theta}\bigg[b_v \cos\beta \cos(\alpha + \theta) \cos(\alpha + 2\theta)$$

$$+ b_c \sin\beta \frac{2\cos(\alpha + \theta)\cos(\alpha + 2\theta) + (1 - 2\nu)\cos\theta}{2(1 - \nu)}\bigg]\frac{1}{x\delta}$$

where α is the angle between the reflecting planes and the normal to the crystal faces; β is the angle between the dislocation line and the incidence plane; ν is Poisson's ratio; and b_v and b_c are the screw and edge components of the Burgers vector, respectively.

In the case of traverse topographs, the distance L_1 should be measured along the section of the image by a plane parallel to the plane of incidence.

We shall not discuss here the more complicated case of Figure 7, corresponding to a Burgers vector perpendicular to the crystal face. The image on a section topograph should be a four- or six-fold rosette, which we have observed in the case of triglycine sulfate; on a traverse topograph the image simply has double contrast.

The width δ of the rocking curve is given by

$$\delta = \frac{2|C|}{\sin 2\theta}\sqrt{\frac{\gamma_0}{\gamma_h}\chi_h\chi_{\bar{h}}}$$

with:

$$\chi_h = -\frac{e^2}{mc^2} \cdot \frac{\lambda F_h}{\pi V}$$

where e^2/mc^2 is the classical radius of the electron, λ is the wavelength, F_h is the structure factor, V is the unit cell volume, and $C = 1$ or $\cos 2\theta$, depending on the polarization direction.

The width δ and the resolving power of the Lang method increase with increasing wavelength, as observed by Lang and Polcarova.[14]

Experimental Evidence

Qualitative Study of the Double Contrast. The contrast of direct images is usually simple, but is sometimes double as has been noted by Authier and Pétroff[15] and by Lang.[16] We shall show that when the width δ is small enough, L_1 becomes larger, and the two parts of the image can be separated and the double contrast appears. The value of δ may be decreased by using either a shorter wavelength or a different reflection. The latter case is illustrated by Figure 8 in the case of silicon (double contrast for 333, $\delta = 2.6 \times 10^{-6}$; single contrast for 111, $\delta = 14 \times 10^{-6}$, Mo K_α radiation) and by Figure 9 in the case of mica[17] (double contrast for 400, $\delta = 2.26 \times 10^{-6}$; single contrast for 200, $\delta = 3.63 \times 10^{-6}$). In both cases, the Burgers vector is parallel to the crystal face and has the same indices ($\frac{1}{2}[1\bar{1}0]$). For practical purposes, it may be noted that in silicon the {111} topographs have the highest resolving power.

Figure 10 compares a section and a traverse pattern of two dislocations in mica, showing the double contrast (reflection 060, $\delta = 4.8 \times 10^{-6}$).

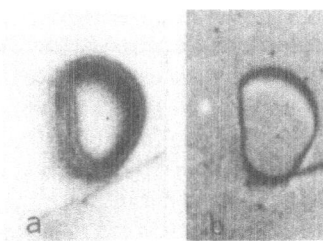

Figure 8. Traverse topographs of a dislocation loop in silicon: Mo K_α; thickness 500 μ. (a) 333; (b) 111. (60 ×).

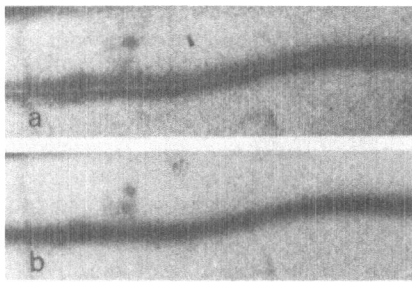

Figure 9. Traverse topographs of a dislocation line in mica: Mo K_α; thickness 100 μ. (a) 400; (b) 200. (55 ×).

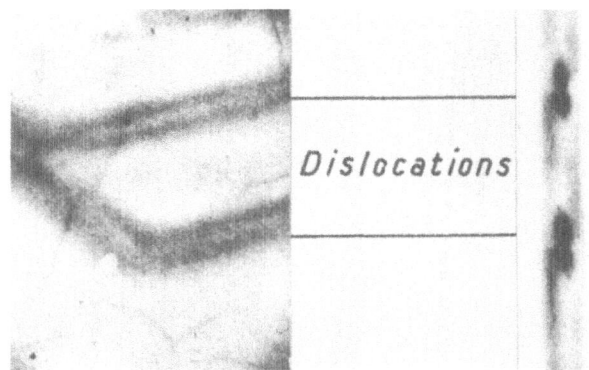

Figure 10. Traverse and section patterns in mica: Mo K_α; 060; thickness 100 μ. (100 ×).

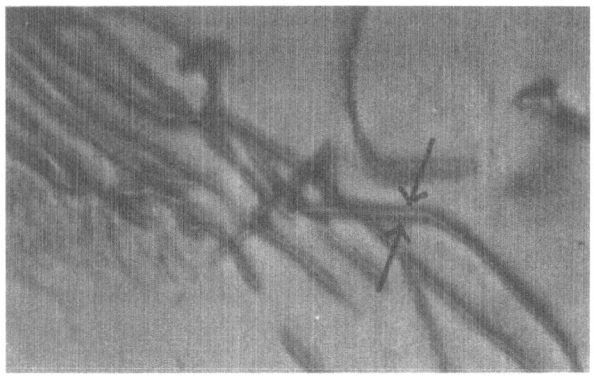

Figure 11. Traverse topograph of triglycine sulfate: 200: Mo K_α; thickness 300 μ. (18 ×).

Sign of Misorientations. The two regions giving rise to the two sides of the image correspond to misorientations of opposite sign, as Figure 6 shows. It is possible to check this with a technique first used by Chikawa.[18]

The experiment was performed by Petroff on a dislocation in triglycine sulfate. He used a straight portion running parallel to the crystal faces. It is shown between two arrows on Figure 11, which is a 200 topograph with Mo K_α radiation ($\delta = 3 \times 10^{-6}$). Three rocking curves were then carefully recorded for three different positions of the crystal in front of a slit $10\,\mu \times 25\,\mu$, such that the incident beam would hit one after the other the two maxima of the dislocation image, $55\,\mu$ apart, and a dislocation-free region. The angular position of the peak was accurately determined for each rocking curve by taking the middle point of successive chords. Petroff was thus able to show that the effective misorientation is of opposite sign for each maximum and equal to $\pm 5 \times 10^{-6}$ rad, which is what one expects from the effective strain field $30\,\mu$ away from the dislocation core.

Quantitative Study. A quantitative comparison between the calculated strain field and the observed profile of dislocation images has been made for mica.[17] To calculate the strain field around a dislocation line in this crystal, it is necessary to take anisotropy into account. General expressions have been derived for this by H. Schlangenotto[19] and R. Siems[20] using the theory of Eshelby, Read, and Shockley.[21] In the appendix we give the value of the effective misorientation for mica.

Figure 12 compares (a) a microphotometer recording of the dislocation image in

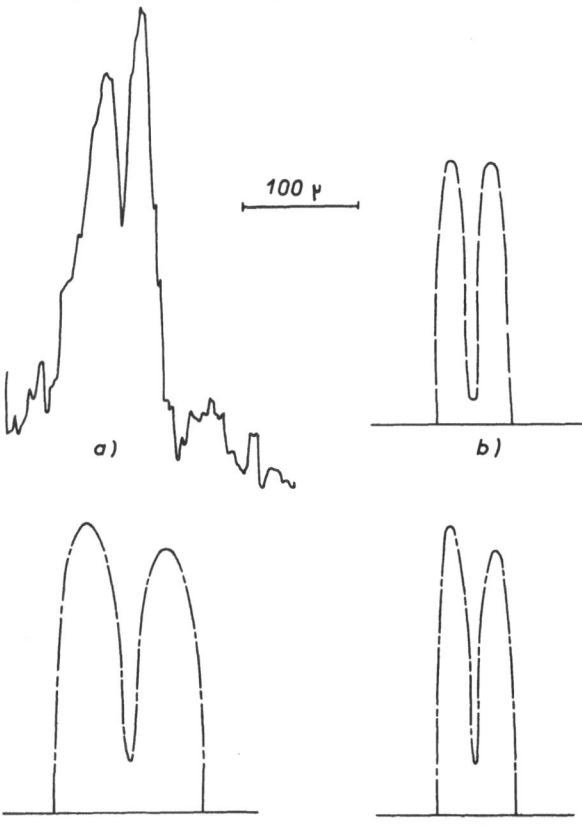

Figure 12. Comparison of experimental and theoretical profiles of the direct image of a dislocation in mica. (a) Experimental; (b) theoretical —isotropic case $x = 1$; (c) theoretical—anisotropic case $x = 1$; (d) theoretical—anisotropic case $x = 2$.

Table I. Theoretical and Experimental Dislocation Image Widths

Reflection	λ, A	$\delta \times 10^6$, rad	L_1 (theoretical)	L_2 (experimental)	$\dfrac{L_2}{L_1}$	Contrast
200	0.71	3.63	42.1	42.5	1.01	single
200	1.54	7.60	19.9	20	1.005	single
400	0.71	2.26	66.8	67.5	1.01	double
400	1.54	4.10	35.6	42.5	1.19	single
060	0.71	4.78	35.1	56	1.60	double
060	1.54	9.00	18.7	28	1.50	single
20$\bar{2}$	0.71	5.90	25	20	0.80	single
131	0.71	5.80	41	35	0.85	single
131	1.54	12.00	21	19	0.90	single

Figure 9a (200, Mo K_α); (b) the theoretical profile, assuming isotropic elasticity and $x = 1$; (c) the theoretical profile, assuming *anisotropic* elasticity and $x = 1$; and (d) theoretical profile, assuming anisotropic elasticity and $x = 2$.

It can be seen that the best agreement between the experimental and theoretical curves is obtained for curve (c), but the experimental shape is somewhat different from the theoretical one. This may be due to experimental broadening, to the existence of the other two types of images, and, of course, to the roughness of the theory.

The most convenient experimental quantity to compare with the parameter L_1 defined above is the width L_2 of the profiles. They are compared in Table I. L_1 was calculated for $x = 1$ and assuming anisotropic stresses.

The ratio L_2/L_1 gives an experimental order of magnitude for x. We can see that for a given reflecting plane it is independent of the wavelength and the order of the reflection, which justifies our assumption that the width of the image is inversely proportional to the

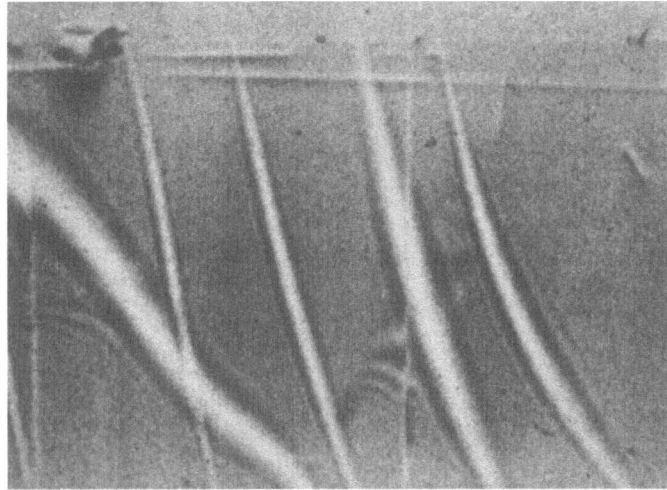

Figure 13. Section topograph of silicon: 220; Ag K_α; thickness 4 mm.
(60 ×, reduced 10% for reproduction.)

width of the rocking curve. The value of L_2/L_1 depends on the reflecting plane, however, and oscillates between 0.8 and 1.6. It is therefore reasonable to assume that the region contributing to the direct image is that for which the effective misorientation is of the order of one or two. This is in agreement with the results of Lang and Polcarova.[14]

DYNAMIC IMAGES

Experimental Study: Section Patterns (Simple Images)

Dynamic images were first observed in silicon by Borrmann, Hartwig, and Irmler,[4] and their contrast studied by Borrmann,[22] Authier,[23] and Ishii.[24]

They are best visible when μt is greater than 1 or 2, that is, either in a highly absorbing crystal or in a relatively thick, low-absorbing material. They usually present a black–white–black contrast[22] and sometimes subsidiary fringes[24] which are visible in Figure 13. Their contrast is the same in hkl and $\bar{h}\bar{k}\bar{l}$ reflections,[23] as can be seen in Figures 3a and 3b. When they correspond to dislocations lying close to the surface of the crystal, their contrast is black-white;[23] it is reversed in the direct-beam section topograph and in the $\bar{h}\bar{k}\bar{l}$ reflections. This effect may easily be explained by Penning and Polder theory.[25]

As is shown schematically in Figures 1 and 4, dynamic images may be considered as shadows cast by the dislocation along the path of the wave-fields incident at each point. Knowing both the direct and dynamic images, it is possible to reconstruct the position of the dislocation line within the crystal without using a stereopair.[23]

The width of the dynamical images may be considered as having a triple origin:

1. The width $2L_1$ of the region surrounding the dislocation line and within which the effective misorientation is a fraction of the width of the rocking curve. This width is evaluated in the same way as the direct image, but is much larger of course.

2. The divergence of the wave-fields intercepted by this width $2L_1$. It is largest when point Q of Figure 1 lies in the middle of the fan or near the entrance surface. It can be seen in Figure 13 that the image is narrower near both sides of the section and it broadens considerably when the dislocation is near the entrance surface.[23,29]

3. Fraunhofer diffraction. The influence of the dislocation line may be considered as that of a linear screen. Fresnel and Fraunhofer diffraction effects are magnified by the crystal which acts as an angular amplifier:[26] the angular spread in path directions is much bigger than that of the wave vectors. The type of diffraction depends on the value of a parameter:

$$w = e' \sqrt{\left(\frac{1}{r_1} + \frac{1}{r_2}\right) \frac{2}{A\lambda} \frac{\cos\alpha}{\cos\theta}}$$

where r_1 and r_2 are the distances of the dislocation line from entrance and exit surfaces, e' is the projection of the width $2L_1$ in the direction normal to the propagation direction, α is the angle between the path and the reflecting planes, θ is the glancing angle of the wave vector on the reflecting planes, and

$$A = \frac{d\alpha}{d\theta}$$

is the amplification ratio.

When w is smaller than 1 or 2 (thick crystal, high amplification ratio, long wavelength, small effective misorientation), Fraunhofer diffraction takes place and there is broadening. The subsidiary maxima are best visible in this case.

The width also depends on the orientation of the dislocation line with respect to the incidence plane corresponding to point Q of Figure 1. Since the main broadening occurs in the incidence plane, the dislocation image is quite thin if the dislocation happens to lie in an incidence plane. An example is shown in Figure 19.

Experimental Study: Traverse Patterns and Parallel Beam Method (Integrated Images)

When crystal and photographic plate are traversed or a long focus line is used, the dynamic image becomes diffuse, since in this case the shadow is cast in all directions within the 2θ-fan drawn from Q. In low-absorbing, relatively thick crystals, it is not very visible and does not decrease the visibility of the direct images. On the contrary, for high-absorbing material, the direction of energy flow is close to that of the reflecting planes (Borrmann effect) and the image is not broad. However, its angular divergence is seldom smaller than a degree, except for extremely high values of μt, and it is only when the dislocation lies less than a few tenths of a millimeter from the exit surface that the resolving power may be compared to that of the direct images in Lang's method. These methods are nevertheless very useful and the only possible ones for high values of μt. They have been used with great success by many authors: for example, Meier with germanium,[27] Hart with indium antimonide,[28] and Young with copper.[29]

Theoretical Study

Two origins may be attributed to the dynamical images:

1. Curvature of wave-fields propagating in the lightly distorted region away from the dislocation line. The paths of the X-rays near a dislocation line have been thus calculated by Kambe using Penning and Polder theory;[30]

2. Creation of new wave-fields due to the decoupling of the wave-fields incident on the more distorted regions nearer the dislocation (interbranch scattering). These new wave-fields take up intensity from the incident wave-fields and propagate in a different direction, giving rise to the intermediary image discussed in the next section. In the high μt case, when the only propagation direction is practically that of the reflecting plane, the incident wave-fields correspond to branch 1 of the dispersion surface, the newly created wave-fields, to branch 2 and are absorbed out. In both cases, the result is depletion of intensity in the direction of the incident wave-fields. We have interpreted it by saying the dislocation casts a shadow.

It is possible, using a very rough theory, to explain thus the black–white–black contrast of the images.[23]

Actually, both effects occur simultaneously. They are taken into account in the more general theories. This has been done by D. Taupin in the case of an incident plane wave[31] and by F. Dumesnil when the dislocation cuts the Borrmann fan.[32] Figures 14a and b show the results obtained by Dumesnil for two positions of point Q within the fan. The calculations were made using Takagi's equations[33] on a 7094 IBM computer for silicon, 220 reflection and Mo K_α. The dislocation is parallel to the crystal faces and lies 300 μ below the entrance surface. In both figures, the normal Kato–Pendellösung fringes may be observed. In Figure 14a, the dislocation line cuts the incidence plane halfway between the direct beam and the median line of the Borrmann fan. Both the dynamic white image and the new wave-fields giving rise to the intermediary image are seen. In Figure 14b, the dislocation line cuts the incidence plane very near the direct beam. The dynamical image

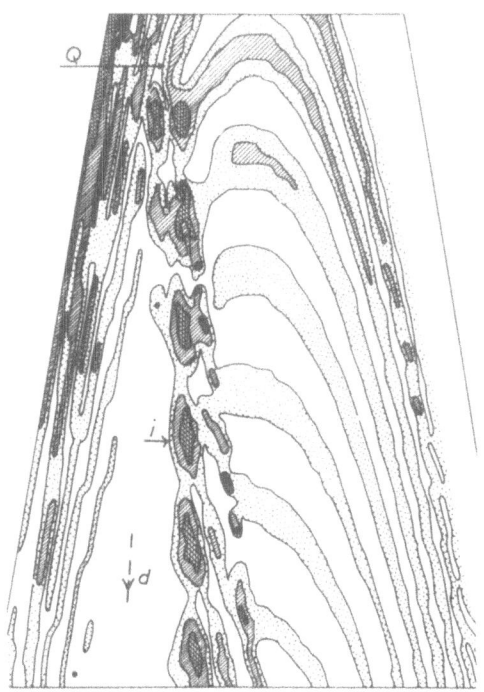

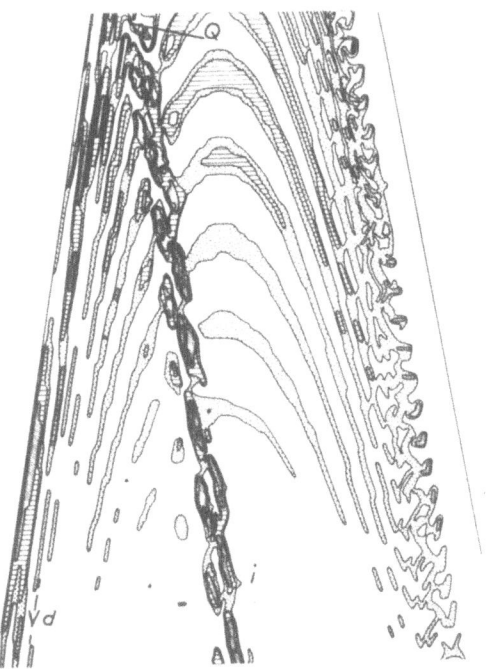

Figure 14. Calculated distribution of intensity in the incidence plane for a screw dislocation in silicon: Mo K_α; 220; Q is the point where the dislocation line cuts the incidence plane, d the dynamic image, and i the intermediary image.

is very narrow and is barely seen, but the intermediary image is very strong; it is practically a direct image.

The results show that the more general theories do take into account the creation of new wave-fields in highly distorted regions and give a satisfactory interpretation of dynamic images.

INTERMEDIARY IMAGE

The decoupling of a wave-field into its two components—the incident and reflected waves—when it crosses a highly distorted region and the subsequent creation of new wave-fields—or interbranch scattering—were predicted by Borrmann,[34] Penning,[35] and Lang.[16] It was taken into account specifically in the new theory developed by Penning.[36] Actually, as shown by the results of the calculations mentioned at the end of the last section, it is also automatically taken into account in the more general theories, such as Taupin's and Takagi's, although it is difficult to see it directly from their equations.

The creation of new wave-fields can best be understood in the case of a twin lamella or a stacking fault. This has been studied independently by N. Kato, K. Usami, and T. Katagawa[37] and by A. Authier and M. Sauvage.[38] Figure 15 illustrates the path of the new wave-fields. Let B_1C_1 be the intersection of a fault plane with the incidence plane. Two wave-fields propagate along Ap, with tie-points P_1 and P'_2. They decouple upon crossing the fault plane and each wave creates two wave-fields in the second half of the crystal. Their tie-points are at the intersection of the dispersion surface with the normal to the fault plane: P_1 and P'_2 are reexcited and two new tie-points P'_1 and P_2 are excited. The corresponding wave-fields propagate along pm. They interfere and all the wave-fields created along B_1C_1 form a new set of Pendellösung fringes which are represented schematically on the figure. Their characteristic shape may be seen on a section pattern when

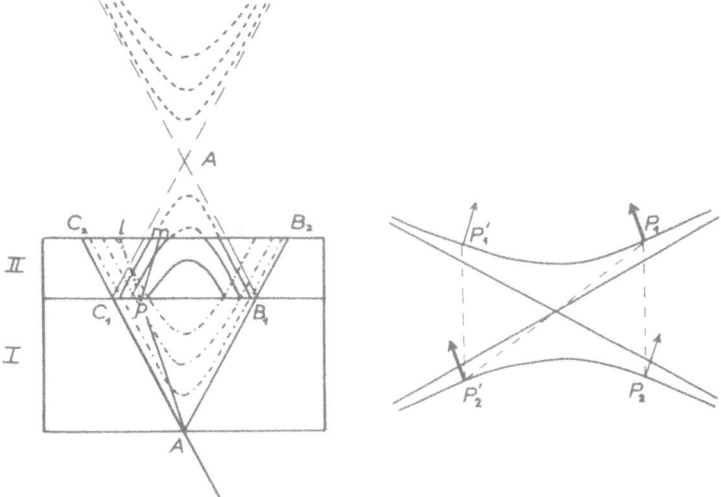

Figure 15. Energy flow in a crystal containing a fault plane B_1C_1. P_1, P'_2 are the tie points of the wave-fields propagating along Ap, and P_2, P'_1 are the tie points of the new wave-fields propagating along pm. Solid lines represent interference fringes between the new wave-fields; and dashed lines the interference fringes between the normal wave-fields.

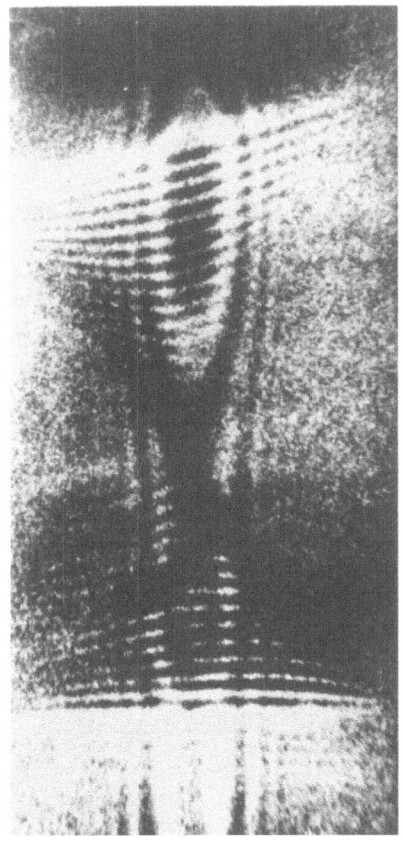

Figure 16. Image of a stacking fault in dolomite. 100; Mo K_α; thickness 1 mm. (230 ×, reduced 10% for reproduction.)

the fault plane makes a wedge with the exit surface of the crystal. An example is given in Figure 16 for a stacking fault in dolomite.

The decoupling of wave-fields also occurs when the distortion is sufficiently high. Let us call the diameter of the region surrounding a dislocation line, and in which the deformation is large enough, the *effective width*. The wave-fields incident on the effective width have a divergence which depends on this width and on the position of the dislocation line in the Borrmann fan. Their tie-points build up a certain domain on the dispersion surface. The tie-points of the newly excited wave-fields lie on the opposite branch of the dispersion surface and occupy a domain of similar size. Diffraction effects should also be taken into account and result in a certain broadening. However, the divergence of the new beam, giving rise to the intermediary image, is of the same order of magnitude as that of the incident wave-fields. If the dislocation lies very close to the entrance surface, this domain is very large and practically the whole dispersion surface may be excited. When this is not the case, and if the dislocation does not happen to lie in the median line of the fan, the incident and new wave-fields do not overlap, and the *dynamic and intermediary images are separated*, as can be observed on Figures 3a and 17a. When μt is small, the new wave-fields correspond to conjugate domains of the dispersion surface (around P'_1 and P_2 in Figure 15) and interfere. If the excited domain covered the whole dispersion surface, these interference fringes would form a complete series of Kato–Pendellösung fringes. Since, in fact, only two small conjugate regions of the dispersion surface are excited, the fringes correspond to a small portion of the Kato hyperbolae. Figures 17b and 18 enable one to interpret the intermediary images in Figures 17a and 3a, respectively.

The intensity of the intermediary image is highest when the dislocation lies in the direct-beam side of the fan (AEB in Figure 5), since it is due to the reflection of wave-fields for which the intensity of the incident wave component is very high. On the other hand, when the dislocation lies in the other side, the intermediary image has little intensity, since the reflected wave component of the incident wave-field on the dislocation is very weak. This can be observed by comparing Figures 3a and 3b, taken at the same point on the crystal but with hkl and $\bar{h}\bar{k}\bar{l}$ reflections, respectively. The intermediary image is very intense in the first case, where it merges into the direct image. It is very weak in the second case because the dislocations do not cut the direct beam and there is no direct image (see Figure 4b). The intermediary images in traverse patterns are for this reason always observed either as overlapping the dynamic image or between direct and dynamic images.

When the dislocation cuts the fan near the entrance surface, the whole dispersion surface is excited and the fringes of the intermediary image cover the whole width of the

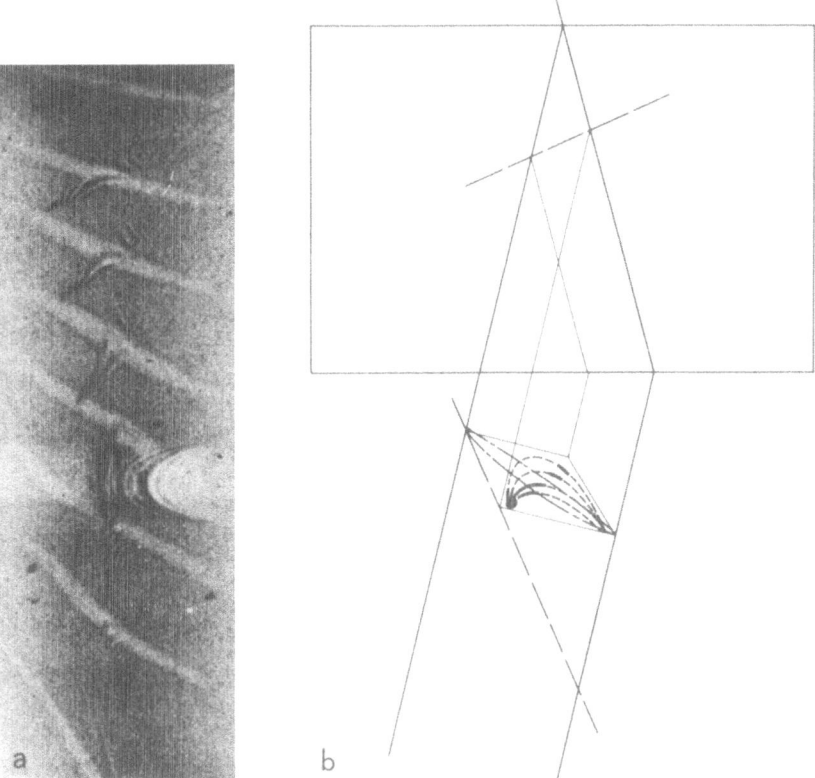

Figure 17. Dynamic and intermediary images in silicon: (a) Section topograph; thickness 4 mm; 220; Ag $K\alpha$; the fringes are the intermediary images (37 ×, reduced 10% for reproduction). (b) Diagram showing that the fringes are part of a series of hyperbolic fringes. The interrupted lines are the projections of the dislocation line on the incidence plane (top part of diagram) and the photographic plate (lower part).

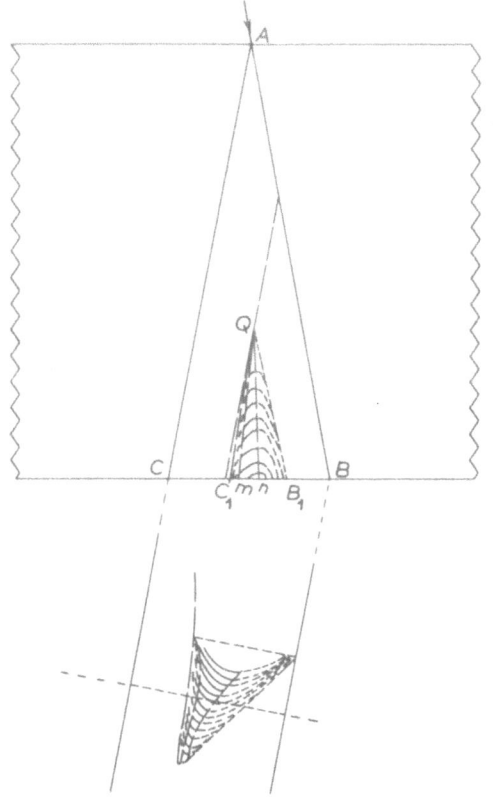

Figure 18. Formation of the intermediary image of the upper dislocation in Figures 3 and 4. The lower part is a projection on the photographic plate. The upper part is a projection on the incidence plane cutting the photographic plate along the broken line. The newly created wave-fields at Q propagate within the triangle Qmn; their interference fringes (solid lines) are part of a set of Kato hyperbolae (broken lines).

Figure 19. Section pattern in silicon: 220, Mo K_α; thickness 0.8 mm. Three dislocations are observed. One (1) is parallel to the plane of incidence; its direct image lies on the dynamic image. The second (2) makes a 30° angle with the incidence plane and lies close to the exit surface; the first fringes of the intermediary image are practically superposed on the dynamic image. The third (3) makes a 60° angle with the incidence plane; it cuts the entrance surface on the right-hand side of the section. The fringes due to the new wave-fields cross the whole section. As the dislocation sinks inside the crystal (toward the left), the widths of the dynamic and intermediary image decrease (180 ×).

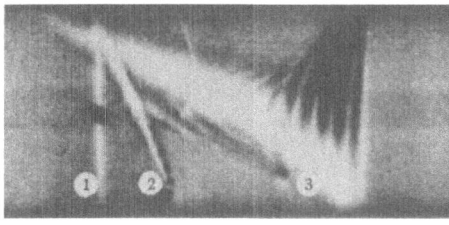

section. This can be seen in the case of the dislocation in Figure 19: as the depth of the dislocation increases, the widths of the dynamic and intermediary images diminish.

In a traverse pattern, the intermediary image will give rise to fringes if the dislocation is not parallel to the crystal face, or to a single fringe along the direct image if it lies parallel to the exit surface. Figure 20 shows examples of the various types of contrast to be expected.

When μt is very high, the intermediary image may be seen in the vicinity of the crystal faces. Let us consider, for example, a dislocation lying close to the exit surface. It receives wave-fields corresponding to tie-points near the summit of branch 1 of the dispersion

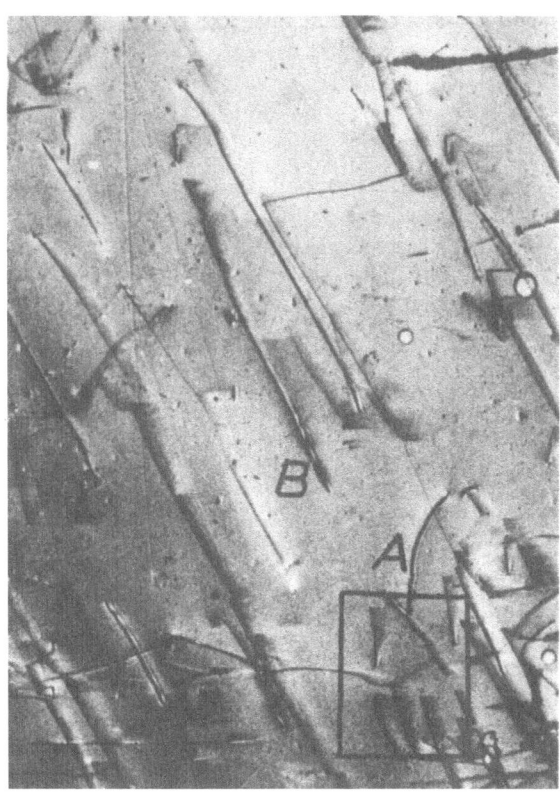

Figure 20. Traverse topograph of silicon: 220; Mo K_α. The framed region is reproduced in Figure 2 at a higher magnification. The dislocation in A lies parallel and close to the exit surface; the intermediary image appears as a continuous black fringe along the direct image. The dislocation in B makes a small angle with the exit surface; the intermediary image appears as periodic black streaks (21 ×).

surface. They will excite tie-points around the summit of branch 2. The corresponding wave-fields will not be absorbed out since they are created near the exit surface and they will interfere with the incident wave-fields giving rise to fringes.

CONCLUSION

Yet other phenomena may play a role in the contrast of dislocation images. For instance, N. Kato has shown that the Pendellösung fringes are deformed in the presence of a defect.[39,40] This leads to a modification of the usual equal thickness fringes in traverse topographs, as has been observed by Authier and Lang[41] and to equal deformation fringes around groups of dislocations, as has been shown by Pétroff and the author.[42]

As this review shows, the contrast phenomena in transmission topography are quite complex but are beginning to be fairly well understood.

APPENDIX: THE EFFECTIVE STRAIN-FIELD
AROUND A DISLOCATION LINE

Isotropic Case

The effective misorientation around a defect is[43]

$$\delta(\Delta\theta) = -\frac{1}{k \sin 2\theta} \frac{\partial(\mathbf{h} \cdot \mathbf{u})}{\partial s_h} = -\left(\frac{\partial}{\partial x'_1} - \tan\theta \frac{\partial}{\partial x'_3}\right)\mathbf{u} \cdot \mathbf{e}'_3 \qquad (A\text{-}1)$$

where **u** is the atomic displacement, **h** is the reciprocal lattice vector, $\mathbf{e}'_3$ is a unit vector in the direction of **h**, ∂/∂_{s_h} represents a partial derivative in the reflected direction, x'_1 and x'_3 are coordinates in a system related to the reflecting planes by

$$\mathbf{e}'_1 = \frac{\mathbf{s}_0 + \mathbf{s}_h}{2\cos\theta} \qquad \mathbf{e}'_3 = \frac{\mathbf{s}_0 - \mathbf{s}_h}{2\sin\theta} \qquad \mathbf{e}'_2 = \mathbf{e}'_3 \times \mathbf{e}'_1 \tag{A-2}$$

$\mathbf{s}_0$ and $\mathbf{s}_h$ are unit vectors in the incident and reflected directions, respectively, θ is the Bragg angle, and k is the wave number.

The displacement vector is usually given with respect to a coordinate system related to the dislocation line. Let $\mathbf{e}_1$, $\mathbf{e}_2$, $\mathbf{e}_3$ be unit vectors defined by $\mathbf{e}_3$, parallel to the dislocation line; $\mathbf{e}_1$ parallel to the edge component of the Burgers vector; and $\mathbf{e}_2 = \mathbf{e}_3 \times \mathbf{e}_1$.

Let $A_i^{\ j}$ be the matrix elements

$$A_i^{\ j} = \mathbf{e}_i \cdot \mathbf{e}'_j \tag{A-3}$$

Then

$$\partial(\Delta\theta) = -\left[\left(A_1^{\ 1} - \tan\theta A_3^{\ 1}\right)\left(A_3^{\ 1}\frac{\partial u^1}{\partial x_1} + A_3^{\ 2}\frac{\partial u^2}{\partial x_1} + A_3^{\ 3}\frac{\partial u^3}{\partial x_1}\right)\right.$$
$$\left. + \left(A_1^{\ 2} - \tan\theta A_3^{\ 2}\right)\left(A_3^{\ 1}\frac{\partial u^1}{\partial x_2} + A_3^{\ 2}\frac{\partial u^2}{\partial x_2} + A_3^{\ 3}\frac{\partial u^3}{\partial x_2}\right)\right] \tag{A-4}$$

In the isotropic case, one has the well known expressions:

$$\left.\begin{aligned}
u_1 &= \frac{b_c}{2\pi}\left[\arctan\frac{x_2}{x_1} + \frac{x_1 x_2}{2(1-\nu)(x_1^2 + x_2^2)}\right] \\
u_2 &= -\frac{b_c}{8\pi(1-\nu)}\left[2(1-2\nu)L\frac{r}{r_0} + \frac{x_1^2 - x_2^2}{x_1^2 + x_2^2}\right] \\
u_3 &= \frac{b_v}{2\pi}\arctan\frac{x_2}{x_1}
\end{aligned}\right\} \tag{A-5}$$

where b_c and b_v are the edge and screw components, respectively.

In contrast problems, it is usually interesting to study the distribution of the effective misorientation in the incidence plane. Let us introduce a third coordinate system: $\mathbf{e}''_1$, parallel to the intersection of the incidence plane and the surface of the crystal; $\mathbf{e}''_2$, parallel to the normal to the surface drawn towards the inside of the crystal; and $\mathbf{e}''_3 = \mathbf{e}''_1 \times \mathbf{e}''_2$. Furthermore, let x''_{1D} and x''_{2D} be the coordinates of the point where the dislocation line cuts the incidence plane.

Strains Due to the Screw Components. Let ϵ be the angle of the dislocation line with the crystal surface, β, the angle between the projection of the dislocation line on the crystal surface and the incidence plane, and, α, the angle between the lattice planes and the normal to the surface of the crystal. Then

$$\left.\begin{aligned}
\delta(\Delta\theta) = &-\frac{b_v}{2\pi}[\sin\epsilon\sin(\alpha+\theta) - \cos\epsilon\cos\beta\cos(\alpha+\theta)] \\
&\times \frac{\cos(\alpha+2\theta)(x''_1 - x''_{1D}) + \sin(\alpha+2\theta)(x''_2 - x''_{2D})}{\sin^2\beta(x''_1 - x''_{1D})^2 + [\cos\beta\sin\epsilon(x''_1 - x''_{1D}) + \cos\epsilon(x''_2 - x''_{2D})]^2}
\end{aligned}\right\} \tag{A-6}$$

A symmetrical reflection corresponds to $\alpha - \theta = 0$.

Strains Due to the Edge Components. We now give the expressions for the case where the dislocation line lies parallel to the crystal faces:

$$
\begin{aligned}
\delta(\Delta\theta) = \frac{b}{2\pi}\Bigg\{ &\left[A_3{}^1(A_1{}^2 - \tan\theta\, A_3{}^2)\left(1 + \frac{x_1{}^2 - x_2{}^2}{2(1-\nu)(x_1{}^2 + x_2{}^2)}\right) \right. \\
&\left. - \frac{A_3{}^2}{2(1-\nu)}\left((1-2\nu)(A_1{}^1 - \tan\theta\, A_3{}^1) - (A_1{}^2 - \tan\theta\, A_3{}^2)\frac{x^1 x^2}{x_1{}^2 + x_2{}^2}\right) \right] \\
&\times \frac{x_1}{x_1{}^2 + x_2{}^2} + \left\{ -A_3{}^1(A_1{}^1 - A_3{}^1\tan\theta)\left(1 + \frac{x_1{}^2 - x_2{}^2}{2(1-\nu)(x_1{}^2 + x_2{}^2)}\right) \right. \\
&- \frac{A_3{}^2}{2(1-\nu)}\Bigg[(1-2\nu)(A_1{}^2 - \tan\theta\, A_3{}^2) \\
&\left. \left. - (A_1{}^1 - \tan\theta\, A_3{}^1)\frac{x^1 x^2}{x_1{}^2 + x_2{}^2}\right]\right\} \frac{x_2}{x_1{}^2 + x_2{}^2} \Bigg\}
\end{aligned}
\tag{A-7}
$$

with

$$
\begin{aligned}
A_3{}^1 &= -\cos(\alpha+\theta)\cos\omega\sin\beta - \sin\omega\sin(\alpha+\theta) \\
A_1{}^2 &= \sin(\alpha+\theta)\sin\omega\sin\beta + \cos\omega\cos(\alpha+\theta) \\
A_1{}^1 &= \sin(\alpha+\theta)\cos\omega\sin\beta - \sin\omega\cos(\alpha+\theta) \\
A_3{}^2 &= -\cos(\alpha+\theta)\sin\omega\sin\beta + \sin(\alpha+\theta)\cos\omega \\
x_1 &= -\cos\omega\sin\beta(x''_1 - x''_{1D}) - \sin\omega(x''_2 - x''_{2D}) \\
x_2 &= -\sin\omega\sin\beta(x''_1 - x''_{1D}) + \cos\omega(x''_2 - x''_{2D})
\end{aligned}
$$

where β is the angle between the dislocation line and the incidence plane and ω is the angle between the Burgers vector and the crystal surface.

Anisotropic Case

The displacement vector around a dislocation in mica is given (after Siems)[20] by

$$
\begin{aligned}
u_1 &= \frac{b_1}{2\pi}\left(B_1 \arctan C_1 \frac{x_2}{x_1} + D_1 \arctan E_1 \frac{x_2}{x_1}\right) \\
u_2 &= \frac{b_1}{2\pi}\left(B_2 L\sqrt{x_1{}^2 + C_2 x_2{}^2} + D_2 L\sqrt{x_1{}^2 + E_2 x_2{}^2}\right) \\
u_3 &= \frac{b_3}{2\pi} \arctan C_3 \frac{x_2}{x_1}
\end{aligned}
\tag{A-8}
$$

In the case of the dislocation in Figure 9:

$$
b_1 = -2.59\ \text{Å} \qquad \text{and} \qquad b_3 = 4.49\ \text{Å}
$$

The coefficients have been calculated from the elastic coefficients of mica:[44]

$$
\begin{aligned}
B_1 &= 1.034 & B_2 &= 0.139 \\
C_1 &= 3.677 & C &= 13.53_2 & C_3 &= 2.357
\end{aligned}
$$

$$D_1 = -0.0386 \qquad D_2 = -0.508$$
$$E_1 = 0.493 \qquad E_2 = 0.243$$

The effective misorientation is found to be equal to

$$
\begin{aligned}
\partial(\Delta\theta) = \frac{-1}{2\pi}\Bigg\{ & \sin\beta\frac{\sin(\alpha+2\theta)}{\cos\theta} \\
&\times\left[\sin\beta\cos(\alpha+\theta)b_1\left(\frac{B_1C_1}{\sin^2\beta\,x''_1{}^2 + C''_1{}^2x''_2{}^2} + \frac{D_1E_1}{\sin^2\beta x''_1{}^2 + E_1{}^2x''_2{}^2}\right)x''_2\right. \\
&+ \sin\beta\sin(\alpha+\theta)b_1\left(\frac{B_2}{\sin^2\beta\,x''_1{}^2 + C_2x''_2{}^2} + \frac{D_2}{\sin^2\beta\,x''_1{}^2 + E_2x''_2{}^2}\right)x''_2 \\
&+ \left.\cos\beta\cos(\alpha+\theta)b_3C_3\frac{x''_2}{\sin^2\beta\,x''_1{}^2 + C_3{}^2x''_2{}^2} + \frac{\cos(\alpha+2\theta)}{\cos\theta}\right]\times \\
&\left[-\sin^2\beta\cos(\alpha+\theta)b_1\left(\frac{B_1C_1}{\sin^2\beta\,x''_1{}^2 + C_1{}^2x''_2{}^2} + \frac{D_1E_1}{\sin^2\beta x''_1{}^2 + E_1{}^2x''_2{}^2}\right)x''_1\right. \\
&+ \sin(\alpha+\theta)\left(\frac{B_2C_2}{\sin^2\beta x''_1{}^2 + C_2x''_2{}^2} + \frac{D_2E_2}{\sin^2\beta\,x''_1{}^2 + E_2x''_2{}^2}\right)x''_2 \\
&- \left.\left.\cos\beta\sin\beta\cos(\alpha+\theta)b_3C_3\frac{x''_1}{\sin^2\beta\,x''_1{}^2 + C_3{}^2x''_2{}^2}\right]\right\}
\end{aligned}
\tag{A-9}
$$

ACKNOWLEDGMENTS

Topographs 8 and 20 are due to M. Simon, 9 and 11 to M. Willaime, 10 to M. Pétroff, and 16 to M. Zarka. I would like to thank Dr. Siems for showing me his calculations prior to publication.

REFERENCES

1. J. B. Newkirk, *Trans. AIME* **215**: 483, 1959.
2. A. R. Lang, *J. Appl. Phys.* **29**: 597, 1958.
3. A. R. Lang, *Acta Cryst.* **12**: 249, 1959.
4. G. Borrmann, W. Hartwig, and H. Irmler, *Z. Naturforsch.* **13a**: 423, 1958.
5. U. Bonse and E. Kappler, *Z. Naturforsch.* **13a**: 348, 1958.
6. M. Renninger, *Phys. Letters* **1**: 104 and 106, 1962.
7. K. Kohra, M. Yoshimatsu, and J. Shimzu, *Direct Observations of Imperfections in Crystals*, Interscience Publishers, New York, 1962, p. 461.
8. A. Authier, *J. Phys. Radium* **21**: 655, 1960.
9. H. Barth and R. Hosemann, *Z. Naturforsch.* **13a**: 792, 1958.
10. V. Gerold and F. Meier, *Z. Physik* **155**: 387, 1959.
11. J. B. Newkirk, U. Bonse, and M. Hart, this volume, pp. 1–8.
12. A. Authier, Informal Meeting on Dynamical Theory, Munich, 1962.
13. A. Authier, *Bull. Soc. Franc. Mineral. Crist.* **84**: 51, 1961.
14. A. R. Lang and M. Polcarova, *Proc. Roy. Soc. (London)* **A285**: 297, 1965.
15. A. Authier and J. F. Petroff, *Compt. Rend.* **258**: 4238, 1964.
16. A. R. Lang, *Z. Naturforsch.* **20a**: 636, 1965.
17. C. Willaime and A. Authier, *Bull. Soc. Franc. Mineral. Crist.* **89**: 279, 1966.
18. J. Chikawa, *Appl. Physics Letters* **4**: 154, 1964.
19. H. Schlangenotto, *Z. Physik* **171**: 537, 1963.
20. R. Siems, private communication, 1965.

21. J. D. Eshelby, W. T. Read, and W. Shockley, *Acta Met.* **1**: 251, 1953.
22. G. Borrmann, *Physik. Bl.* **15**: 508, 1959.
23. A. Authier, *Bull. Soc. Franc. Mineral. Crist.* **84**: 115, 1961.
24. Z. Ishii, *J. Phys. Soc. Japan* **17**: 838, 1962.
25. P. Penning and D. Polder, *Philips Res. Rpt.* **16**: 419, 1961.
26. A. Authier and C. Malgrange, *Compt. Rend.* **262**: 429, 1966.
27. F. Meier, *Z. Physik* **168**: 10 and 29, 1962.
28. M. Hart, Ph.D. Thesis, Bristol University, 1963.
29. F. Young, *Advances in X-Ray Analysis, Vol.* 9, Plenum Press, New York, 1965, pp. 1–13.
30. K. Kambe, *Z. Naturforsch.* **18a**: 1010, 1963.
31. D. Taupin, *Bull. Soc. Franc. Mineral. Crist.* **87**: 469, 1964.
32. C. Malgrange and F. Dumesnil, 7th Int. Congr. Cryst., Moscow, 1966.
33. S. Takagi, *Acta Cryst.* **15**: 1311, 1962.
34. G. Borrmann, private communication, 1962.
35. P. Penning, *Colloque de l'Association Française de Cristallographie*, Nancy, 1964.
36. P. Penning, Thesis, Delft, 1966.
37. N. Kato, K. Usami, and T. Katagawa, this volume, pp. 46–66.
38. A. Authier and M. Sauvage, *J. Phys. Radium* **27**: C3-137, 1966.
39. N. Kato, *J. Phys. Soc. Japan* **19**: 971, 1964.
40. N. Kato and Y. Ando, *J. Phys. Soc. Japan* **21**: 964, 1966.
41. A. Authier and A. R. Lang, *J. Appl. Phys.* **35**: 1956, 1964.
42. J. F. Petroff and A. Authier, *Phys. Status Solidi* **17**: K3, 1966.
43. A. Authier, *J. Phys. Radium* **27**: 57, 1966.
44. K. S. Alexandrov and T. V. Ruzhova, *Iz. Akad. Nauk SSSR Ser. Geofiz.* **12**: 1799, 1961.

THE ASYMMETRIC BRAGG REFLECTION AND ITS APPLICATION IN DOUBLE DIFFRACTOMETRY

M. Renninger

*Kristallographisches Institut der Universität
Marburg, Germany*

ABSTRACT

Guided by the dispersion surface of dynamical theory, we easily obtain evidence of the fact that in asymmetrical surface reflections (Bragg case) of X-rays at a perfect crystal, the angular widths of the incident and reflected beams are different from the same widths in the symmetrical Bragg case. The beam with the smaller glancing angle to the surface is angularly expanded, the other one, contracted (though in cross section, the former is contracted and the latter, expanded). Consequently, if the incident glancing angle is the smaller one (striping incidence, V-reflection), the angular width of the reflected beam is smaller than in the symmetric case (S-reflection), and that of the reflected section out of the primary beam is greater, and vice versa (striping emergence, R-reflection). At striping incidence, a contraction occurs, whereas at striping emergence, expansion of the angular reflection range occurs. These facts offer a number of possibilities for applying asymmetrical reflections to the well-known technique of the two crystal diffractometer —the author proposes the term "diffractometer" instead of the widely used "spectrometer"—in the $(n, -n)$-position. These possibilities extend the efficiency of the double diffractometer with multiple effects, and some of them are given below: (1) Realization of rocking curves of extremely small angular width in the $(n^V, -n^R)$-position, and consequently increased sensibility of that width to lattice distortions. (2) Use of the extremely small angular width (or more strictly, the sharp θ–λ-coordination) of the beam emerging from a first crystal. This is in V-reflection for scanning the intrinsic diffraction pattern of the second crystal used in S- or V-reflection [$n^V, -n^S$)- or $(n^V, -n^V)$-position]. (3) Increased steepness of the sides of the rocking curves using $(n^V, -n^R)$ or $(n^V, -n^S)$-positions which allows double diffractometric topography (Bond–Andrus, Bonse–Kappler) of increased angular resolving power. (4) Repeated V-reflections (eventually within a single crystal provided with a suitable groove) leading to further angular contraction of the resulting reflected beam and therefore further increased angular resolving power [$(n^V, -n^V, n^R, -n^R)$- and $(n^V, -n^V, +n^S)$-positions]. Here the tails of the rocking curve are also suppressed by multiple reflection (Bonse–Hart).

According to dynamic theory, the diffraction pattern of a plane monochromatic X-ray wave reflected at the face of a perfect crystal has the form shown in Figure 1. The dependence of the coefficient of reflection $R = P/P_0$ on the direction of incidence is shown. For a nonabsorbing perfect crystal this is the well known Darwin–Ewald "top hat" ("Zylinderhut") curve, with total reflection (ordinate = 1) extending over some few seconds of arc. If absorption is taken into account, the curve is modified, according to Prins and Kohler, to the typical shape shown by some examples in solid lines in the figure, with characteristic asymmetry and reduced height.

How these diffraction curves come about may be understood from the dynamic theory by looking at the dispersion surface in reciprocal space. The dispersion surface is

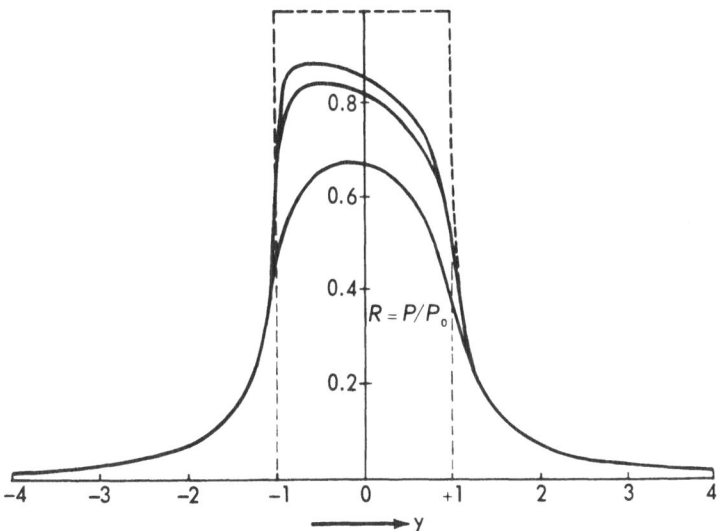

Figure 1. Diffraction pattern for surface reflection according to dynamic theory. Broken line: nonabsorbing crystal (Darwin–Ewald); solid line: absorption taken into account (Prins–Kohler). Some different examples [Si (333); Cu K_α; $\beta = 0$, ± 0.6, ± 0.9].

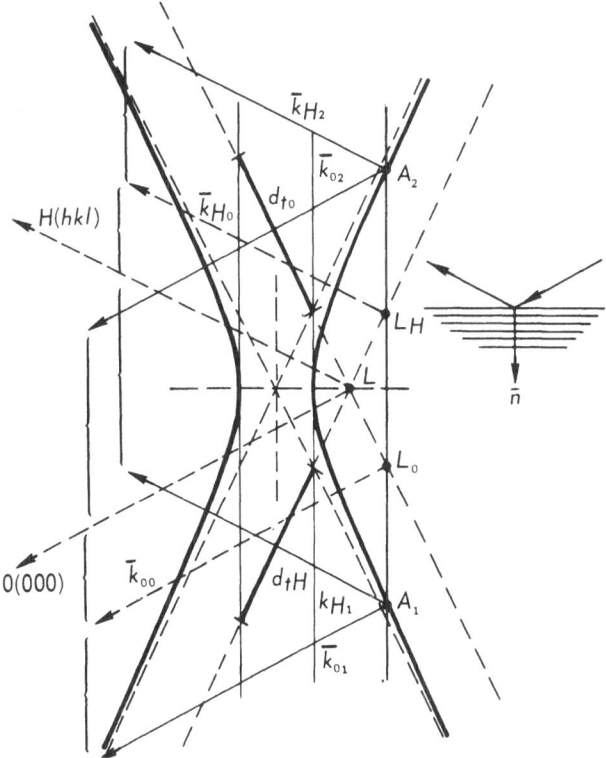

Figure 2. Dispersion surface and angular range of total reflection for symmetric Bragg reflection.

the geometrical locus of all possible wave points—i.e., origins of wave vectors—directed to all points of the reciprocal lattice which take part in the interference: for example, 0(000) and H(hkl) (Figure 2).

We restrict our attention to one direction of polarization: that one for which the electric vector is normal to the plane of reflection (here the plane of the drawing). The planes drawn in broken lines (in reality these are spheres with radii of kilometers in the scale of the drawing) are the "Laue" spheres; they are loci of origins L_0 of wave vectors of waves in external space, called "excitation points" ("Anregungspunkte") by Ewald, and the Laue sphere around 0(000) is called by him the "excitation sphere" ("Anregungs-kugel"). Motion of an excitation point L_0 on the excitation sphere through a distance d_0 corresponds to a rotation of the related wave, with wave-vector $\bar{k}_{00} = \bar{s}_0/\lambda'$, by an amount d_0/k_{00}, i.e., $d_0 \cdot \lambda$.

Now we assume a wave with vector k_{00} and origin at L_0, incident on a crystal face parallel to the reflecting lattice planes, as indicated on the right side of Figure 2. This is the *symmetrical* "Bragg case." We seek the direction and velocity of the waves excited in the interior of the crystal by this incident wave, i.e., we seek their wave points on the dispersion surface. We find them (A_1 and A_2) simply as points of intersection of the dispersion surface and a line parallel to the normal of the crystal face through L_0. To each of the wave points A_1 and A_2 there corresponds a pair of wave vectors toward 0(000) and H(hkl). Movement of L_0 on the Laue sphere, i.e., rotation of the incident wave direction, corresponds to a motion of the wave points on the dispersion surface; in the Bragg case, these wave points always lie on the same branch of the hyperboloid. However, within a region d_{t0} of movement of L_0, lying between the intersection points of the Laue sphere with tangents to the dispersion surface parallel to the face normal, there exist no real wave points. The angular range corresponding to that region d_{t0} is the range of total reflection of the Darwin–Ewald curve.

Simultaneous with the motion of L_0 on the Laue sphere around 0(000), a point L_H moves with the same speed and in the opposite direction on the Laue sphere around H(hkl). L_0 and L_H indicate the directions of waves in external space: L_0, that of the incident wave and L_H, that of the emergent wave. To the region d_{t0} of the incident wave we now have a corresponding region d_{tH} for the reflected wave. This means that the angular range of total reflection is the same for both waves.

But the situation is quite different if we now consider asymmetrical reflections (asymmetric Bragg case), i.e., reflections at crystal faces which form an angle $\varphi \neq 0$ with the reflecting lattice planes (Figure 3). If we make the same construction as before, we see that now the angular range over which no real wave points exist—i.e., the range of total reflection—is different for the incident and the reflected wave; the former is expanded, the latter contracted compared with the symmetrical case. The reverse is true if the angle φ has the opposite sign. We define the sign of φ as positive if it diminishes the angle between the incident beam and the surface, compared with the symmetric case (glancing incidence). Thus, in Figure 3, φ is positive.

We can make the general statement: *that* wave which has the *smaller* glancing angle at the crystal face has the *greater* angular width, and vice versa. At glancing *incidence*, *contraction* of the angular range of reflection takes place, and at glancing *emergence*, *expansion* occurs; note that this behavior of angular range of reflection is just opposite to that of the beam cross section.

Quantitative formulation of all these matters is easily derived from dynamic theory. We will give only the results. There are two equally convenient means for formulation of the results. Both facilitate representation of variations with the degree of asymmetry, not only of the angular range of reflection in which we are presently interested, but also of

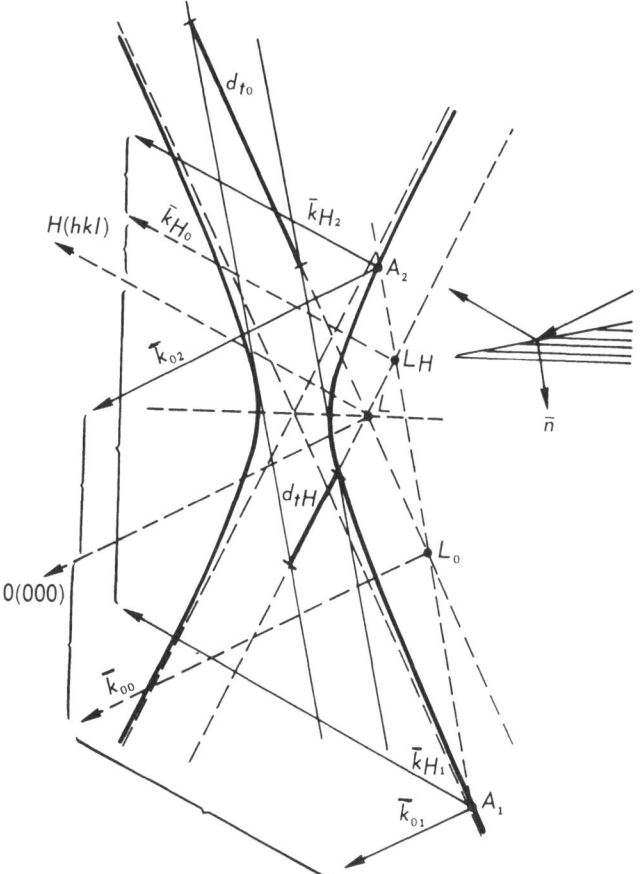

Figure 3. Dispersion surface and angular range of total reflection
for asymmetric Bragg reflection.

other quantities such as the deviation from Bragg's law, and the integrated reflection. One of these variations, introduced 30 years ago by the author,[1] employs the ratio

$$\beta = \tan \varphi / \tan \theta$$

as a variable for the degree of asymmetry. The other, introduced by Zachariasen,[2] uses $b = \gamma_0/\gamma_H$, the ratio of the direction cosines of the primary and the reflected beam, which ratio is always negative in the Bragg case. β and b are related very simply:

$$\beta = \frac{b + 1}{b - 1}, \qquad b = \frac{\beta + 1}{\beta - 1}$$

Figure 4 and Table I show the characteristic variation of various quantities: against β on the left, against $-b$ on the right. We see that some properties are more simply expressed in β and others in b. An advantage of the expression in β may be seen in the symmetry of the abscissa scale relative to the symmetric case.

We are now particularly interested in the last of the tabulated quantities, which is drawn in solid lines in Figure 4—namely, the angular width of the reflection, which we have already considered qualitatively. The formulae are exact, of course, only for the non-

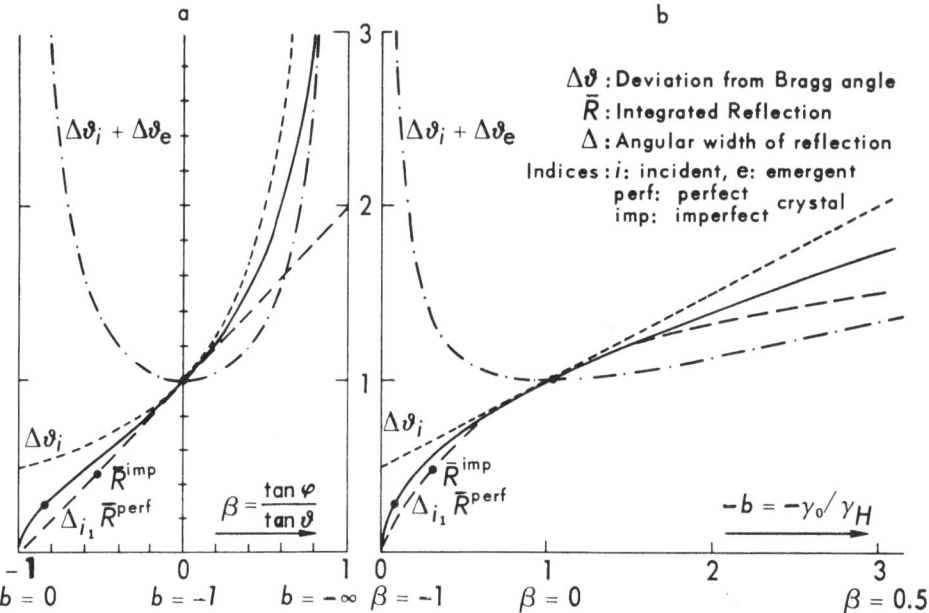

Figure 4. Variation of different quantities with degree of reflection asymmetry, related to the same
quantities for symmetric reflection. (a) Against $\beta = \tan \varphi / \tan \vartheta$; (b) against $b = \Upsilon_0 / \Upsilon_H$
– – – – – – deviation $\Delta \vartheta_i$ of incident beam from Bragg angle ϑ_B;
– · – · – deviation $\Delta \vartheta_i + \Delta \vartheta_e$ of total deflection $\vartheta_i + \vartheta_e$ from double Bragg angle $2 \vartheta_B$;
– – – – integrated reflection $\bar{R}^{imp}$ of imperfect crystal;
—————— angular width Δ_i of the diffraction pattern and integrated reflection $\bar{R}^{perf}$ for a nonabsorbing perfect crystal.

Table I. Variation of Different Quantities with β and with b
related to their values at symmetrical reflection ($\beta = 0, b = -1$)

Deviation from Bragg angle	$\dfrac{\Delta \vartheta_i}{\Delta \vartheta_{sym}} = \dfrac{1}{1 - \beta} = \dfrac{1}{2}(1 - b)$
Deviation of total deflection from double Bragg angle	$\dfrac{\Delta \vartheta_i + \Delta \vartheta_e}{2 \Delta \vartheta_{sym}} = \dfrac{1}{1 - \beta^2} = - \dfrac{(b - 1)^2}{4b}$
Integrated reflection of imperfect crystal	$\dfrac{\bar{R}^{imp}}{\bar{R}_{sym}{}^{imp}} = 1 + \beta = \dfrac{2b}{b - 1}$
Angular width of diffraction pattern and integrated reflection of (nonabsorbing perfect crystal)	$\dfrac{\Delta_i}{\Delta_{sym}} = \dfrac{\bar{R}^{perf}}{\bar{R}_{sym}{}^{perf}} = \sqrt{\dfrac{1 + \beta}{1 - \beta}} = \sqrt{-b}$

absorbing perfect crystal. They must be modified somewhat, the stronger the influence of
absorption. A general expression which includes the influence of absorption is not possible.
However, the differences become significant only for $|\beta|$ near 1, i.e., nearly parallel
incidence or emergence, and they are of equal size for both signs of β [for example, for
Si (333) the difference of the angular width compared with the curve for nonabsorbing
case is less than 5% up to $\beta = \pm 0.9$].

Some examples of experimental results are next presented. The first of these, Figure 5, is an instructive series of experimental diffraction patterns obtained by R. Bubakova[3] (1962). These measurements followed a short communication from the author[4] on the effects of asymmetrical reflection, and, for the purpose of a further test of the theoretical predictions, involved a triple diffractometric method described earlier.[5]

In this technique, a beam of very small angular and spectral width is produced by double diffractometric $(n, +n)$ reflection from a pair of crystals having an order of angular sharpness much higher than that of the pattern to be tested. The sharp beam thus produced is then used to scan the reflection under study, that is, Ge (111) in Figure 5. The solid curve is obtained at symmetric reflection, both of the others at asymmetric, $\varphi = \pm 4.5°$ (1/3 of θ), the right one by reflection at glancing incidence, and the left one at glancing emergence. The difference of the angular widths of all three patterns is evident.

In the above example, the asymmetric reflection was not itself employed in a double diffractometric arrangement. We shall consider what is to be expected for $(n, -n)$ double reflection if one or both of the individual reflections are asymmetric.* In normal use of the double diffractometric method, the first crystal is adjusted in the reflecting position for the desired order of interference, the second crystal is turned slowly through that same position, and the reflected power is recorded. The rocking curve obtained in this manner is a convolution of the angular distribution of *emergence* of the *first* crystal with the angular distribution of *incidence* of the *second* crystal; the spectral distribution has no effect on it. A number of different combinations of asymmetric and symmetric reflections of both crystals are possible. Some of them will be discussed. In the following, superscripts will be assigned as follows to characterize the various types of reflections: n^S for symmetric reflections, n^V and n^R for glancing incidence and emergence, respectively.

1. $(n^V, -n^R)$ position. Both angular distributions in convolution are small compared with that of the symmetric reflection. The resulting rocking curve is smaller than that for $(n^S, -n^S)$ reflection, but it is also symmetric. Figure 6 shows a series of rocking curves obtained at this setting with Ge (111) and different degrees of asymmetry.

2. $(n^V, -n^S)$ position. Here, the sharpened angular width of the beam emerging from the first crystal is suitable for scanning the intrinsic diffraction pattern of the second crystal in the S (or V) position (convolution of a narrow peak with an essentially wider peak). The result is the same as in the earlier triple crystal method, but the difference is that here the scanning device to test the diffraction pattern of the sample is not an extremely

* Independently of the author's first communication,[4] mentioned earlier, the same idea (of applying asymmetric reflection for double diffractometric work) has been published with preliminary experimental results by K. Kohra.[6]

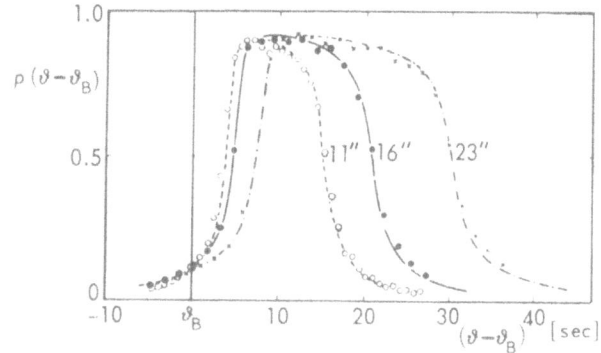

Figure 5. Diffraction patterns of Ge (111) and Cu $K\alpha$ ($\varphi = 0$ and $\pm 4\frac{1}{2}°$), recorded by triple crystal reflection (R. Bubakova, 1962). Monochromatic parallel primary beam produced by Si [(333), +(333)] double reflection.

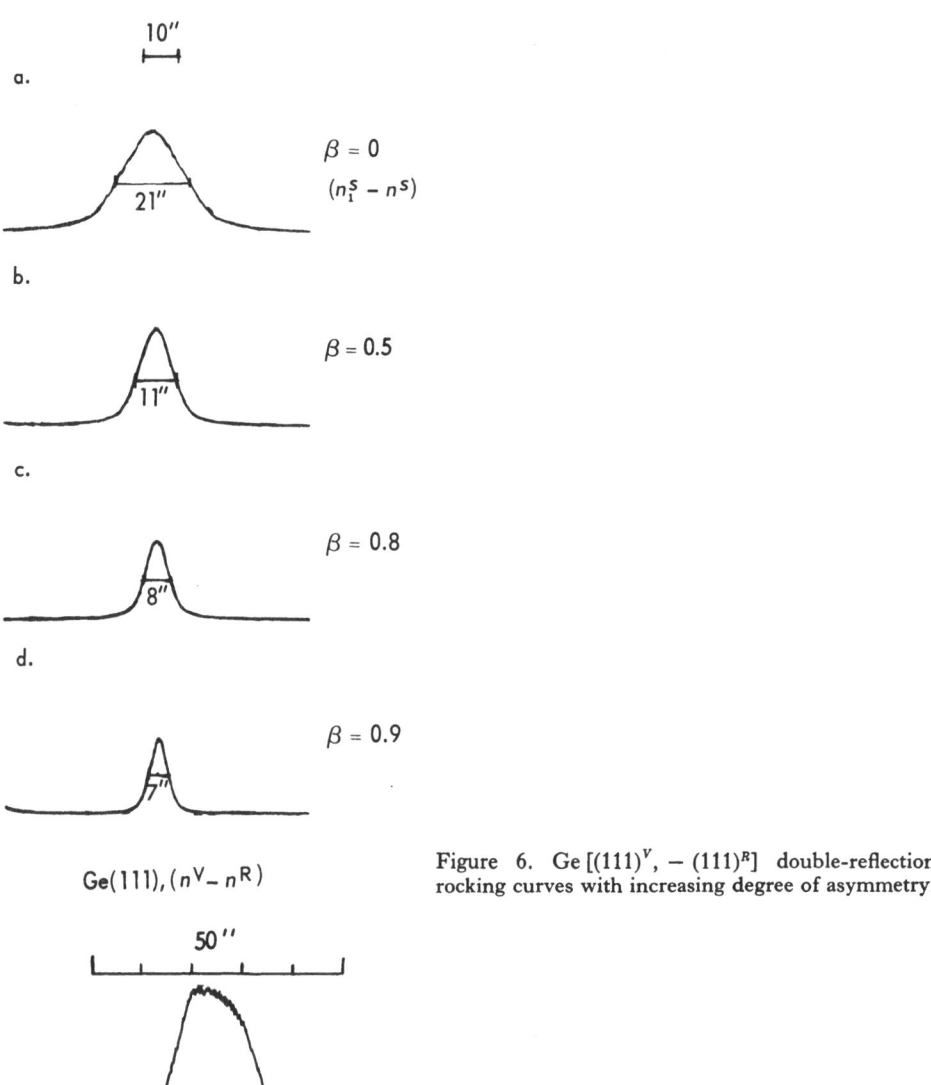

a.

10″

b.

21″

$\beta = 0$

$(n_1^S - n^S)$

c.

11″

$\beta = 0.5$

d.

8″

$\beta = 0.8$

7″

$\beta = 0.9$

Ge(111), $(n^V - n^R)$

Figure 6. Ge $[(111)^V, - (111)^R]$ double-reflection rocking curves with increasing degree of asymmetry.

50″

Ge(111), $n_1^V - n_1^S$, $\beta = 0.9$

Figure 7. Ge $[(111)^V, - (111)^S]$ double reflection with $\beta = 0.9$ for the first crystal.

parallel and monochromatic beam, but only a beam of extremely sharp coordination of direction and wavelength; and merely by the well known functioning of the $(n, -n)$ double reflection a simultaneous reflection of that whole wavelength range is effected.

Examples of patterns obtained in this setting are shown in Figures 7 and 8. The first one is again of Ge (111), with the first crystal in the V position and angle φ giving the smallest width at the $(n^V, -n^R)$ reflection in Figure 6, and the second crystal having reflecting face parallel to (111). In this way a curve is obtained similar to that of Figure 5 as measured by Bubakova[3] and the author;[4] but in this case, only two, not three, crystals are used, and the intensity is enormously increased (by a factor in excess of 10^3). Figure 8

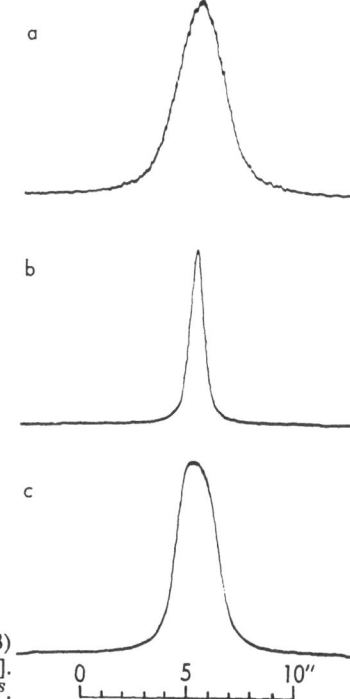

Figure 8. $(n, -n)$ double-reflection rocking curves of Si (333) and (511) respectively [(333) and (511) are of the same order]. (a) $(333)^S$, $-(333)^S$, (b) $(511)^V$, $-(511)^R$, (c) $(511)^V$, $-(333)^S$.

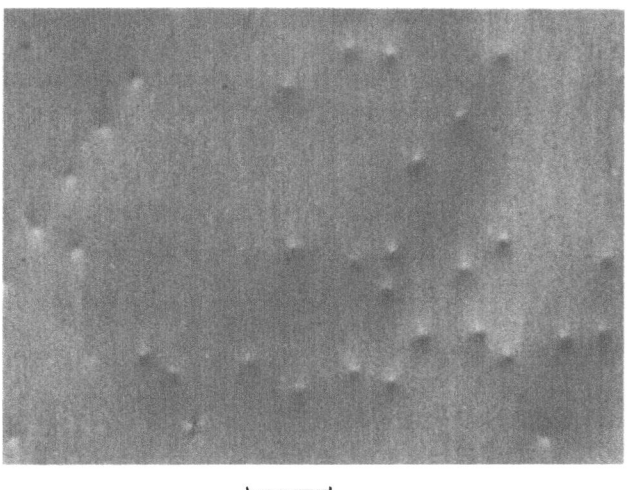

1mm

Figure 9. Example of topograph obtained by $[(511)^V, -(333)^S]$ reflection with the double diffractometer. The contrasts are produced by the screw component of the dislocations. Exposure time, 7 min (Renninger[10]).

illustrates the same arrangement for the much narrower silicon (333) reflection. In (a) and (b) of that figure, $(n^S, -n^S)$ and $(n^V, -n^R)$ curves are shown for comparison. The pattern of Figure 8c again reveals the main characteristics of the intrinsic diffraction pattern of

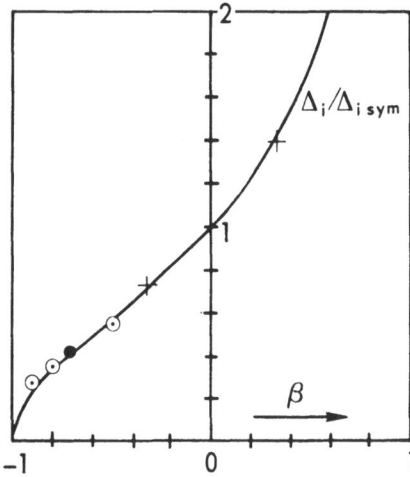

Figure 10. Theoretical variation of the angular width Δ_i with β, repeated from Figure 4, with measuring points entered, obtained from Figures 5, 6, and 8: $+$ Ge (111) (Bubakova); $\odot$ Ge (111) and $\bullet$ Si (333) (Renninger).

silicon (333), the angular width of which is eight times smaller than that of Ge (111). To obtain the Si (333) pattern by the three-crystal method would be very difficult.

3. Using the same position because of the extreme steepness of the rocking curves of Figure 8c or similar curves, double diffraction topographs may be obtained, following the methods of Bond and Andrus[7] and Bonse and Kappler,[8] but with greatly increased angular resolution (the author[9,10]). In this way, lattice distortions of the order of 10^{-7} or of 0.02 seconds of arc are perceptible in silicon. The distortion fields of single dislocations are perceptible at a distance of 0.2 mm from the dislocation cores. Figure 9 illustrates such observations with one example.

4. Repeated V reflections $(n^V, n^V, n^V, \ldots)$, eventually within one single crystal provided with a suitable groove, may lead to further sharpness of the beam emerging from the final reflection. Thus, an even greater increase in angular resolution may be attainable. Experiments to explore this possibility are being conducted.

In Figure 10, for a further experimental test, the theoretical curve of variation of the angular width with β was repeated, and experimental points obtained from the Figures 5, 6, and 8 entered.

ACKNOWLEDGMENT

The author is greatly indebted to the "Deutsche Forschungsgemeinschaft" for liberal support.

REFERENCES

1. M. Renninger, "Überlegungen zur Interferenztheorie," *Z. Krist.* **97**: 95–106, 1937.
2. W. H. Zachariasen, *Theory of X-Ray Diffraction in Crystals*, John Wiley & Sons, Inc., New York, 1945.
3. R. Bubakova, "Diffraction Pattern of Germanium (111)-Asymmetrical Bragg Case," *Czech. J. Phys.* **B12**: 776–783, 1962.
4. M. Renninger, "Asymmetriche Bragg-Reflexion am Idealkristall zur Erhöhung des Doppelspektrometer-Auflösungsvermögens," *Z. Naturforsch.* **16a**: 1110–1111, 1961.
5. M. Renninger, "Messungen zur Röntgenstahl-Optik des Idealkristalls, (I) Bestätigung der Darwin–Ewald–Prins–Kohler-Kurve," *Acta Cryst.* **8**: 597–606, 1955.
6. K. Kohra, "An Application of Asymmetric Reflection for Obtaining X-Ray Beams of Extremely Narrow Angular Spread," *J. Phys. Soc. Japan* **17**: 589–590, 1962.
7. W. L. Bond and J. Andrus, "Structural Imperfections in Quartz Crystals," *Am. Mineralogist* **37**: 622–632, 1952.

8. U. Bonse and E. Kappler, "X-Ray Recording of Distortion Field Round Isolated Dislocations in Germanium Single Crystals," *Z. Naturforsch.* **13a**: 348–349, 1958.

9. M. Renninger, "Doppeldiffraktometrische Transmissions-Topographie," *Z. Naturforsch.* **19a**: 783–787, 1964.

10. M. Renninger, "Beiträge zur doppeldiffraktometrischen Kristall-Topographie mit Röntgenstrahlen: I. Methodik und Ergebnisse typischer Art," *Z. Angew. Phys.* **19**: 20–33, 1965; "II. Ein Si-Kristallstab mit bemerkenswerten Verlauf von Versetzungen und Punktdefekt-Konzetrationsschwankungen," *Z. Angew. Phys.* **19**: 34–35, 1965.

DISCUSSION

N. Spielberg (Philips Laboratories): How does the decay of the wings of the asymmetric reflections compare with that of the symmetric reflections?

M. Renninger: That is a matter which is under exploration now. It is of special interest to me, but I have not yet come to an answer.

EXPERIMENTAL DETERMINATION OF THE INTEGRATED CONTRIBUTION OF TEMPERATURE DIFFUSE SCATTERING IN X-RAY REFLECTIONS

M. Renninger

Kristallographisches Institut der Universität
Marburg, Germany

ABSTRACT

A double diffractometric method which allows clean separation of the thermal diffuse contribution from X-ray reflection peaks is demonstrated.

Twenty five years ago, the author[1]—and almost simultaneously, Witte and Wölfel[2]—made absolute measurements of the integrated X-ray intensities of rocksalt reflections. Despite the fact that the author's measurements were in reflection and those of Witte and Wölfel were in transmission, the agreement of the derived F-values between these was very good. However, both lie much higher than all earlier measured and theoretically expected values. The exponent M of the temperature factor, derived by the author from his structure factors, was lower than expected, on the basis of the characteristic temperature: 281° as determined from specific heat measurements, or 303° from elastic constants. The characteristic temperature derived from the X-ray measurements was 319°. This result has been justifiably questioned subsequently (Blackman,[3] Chipman and Paskin[4]), and it has been conjectured that the values of the integrated X-ray reflections obtained were too high because diffuse thermal maxima were superimposed on the principal interference peaks. Quantitative expression has been given to that conjecture in a theoretical paper of Nanny Nilsson,[5] in which correction factors $\alpha = (\bar{R} - \bar{R}_0)/\bar{R}_0$ to the X-ray intensities are calculated. Development of a direct experimental basis for such corrections is the aim of the work reported here.

Several factors render proper analysis of the principal and thermal-diffuse reflections from a rocking curve difficult, in spite of the fact that the angular width of both peaks is very different. The demarcation is blurred by the angular width of the primary beam, by the natural slopes of the X-ray spectral lines, and by the misorientation within the crystal (mosaic texture). These difficulties can be overcome through the use of crystals of the highest possible perfection and the use of the two-crystal diffractometer technique with crystals in the $(n, -n)$-position. Rocking curves are then obtainable as shown in Figures 1 and 2. These curves allow unambiguous separation of the principal and diffuse contributions. These figures present examples of automatic records; Figure 1 shows such rocking curves made with NaCl (820) with Mo K_α radiation. Crystals of sodium chloride with large perfect domains are apparently not available as yet, and hence the angular width of the principal reflection is spread over a range of 10′ (minutes of arc). Nevertheless the separation of the principal and thermal peaks is uniquely possible, as can be seen in the upper curve of Figure 1 whose ordinate scale (abscissa scale kept unchanged) is 25 times

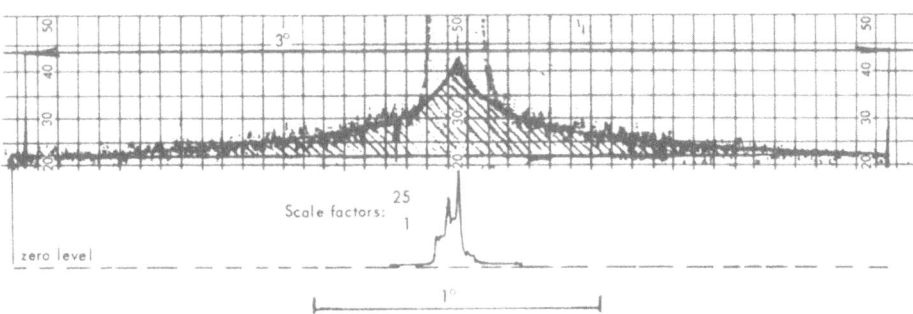

Figure 1. $(n, -n)$ rocking curve for NaCl (820) with Mo K_α.

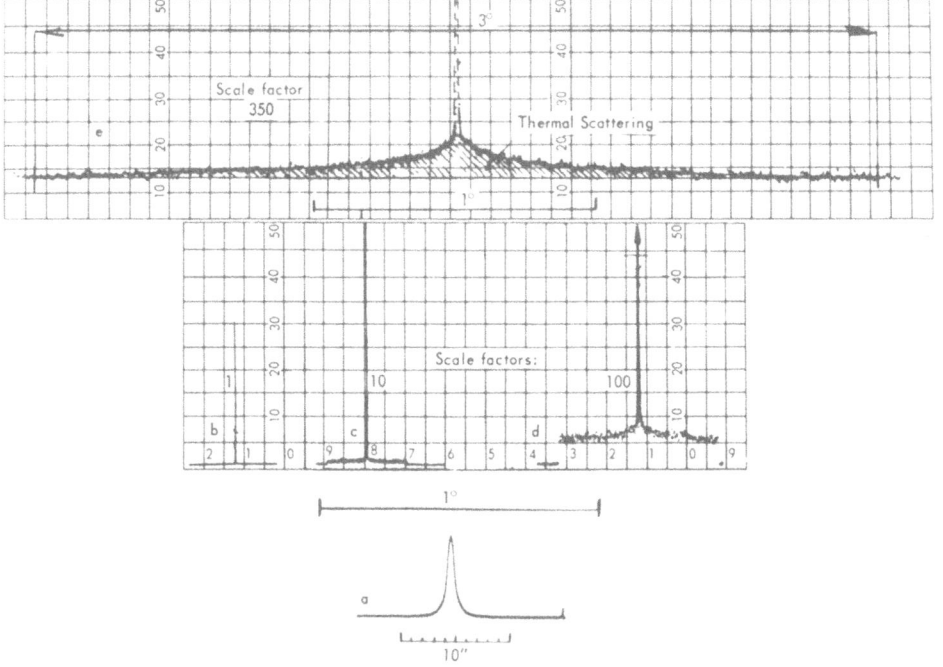

Figure 2. The $(n, -n)$ rocking curve for Si (822) using Mo K_α radiation. (a) Ordinate scale factor unity, extended abscissa scale; (b)–(e) ordinate scale factor increases from unity to 350 as indicated.

greater than that of the lower one. Figure 2 shows chart records for the (822) reflection of a perfect silicon crystal. Here, both scales, the abscissa and the ordinate, are very much increased in curve e compared with curve a (180 times for the abscissa, 350 times for the ordinate), with all the diagrams of this figure showing the same curve with increasing scale factor for the ordinate and two different abscissa scales. The width at half maximum for the principal peak is only 0.7″, and therefore its distinctness from the thermal peak is still more evident.

Figure 3 shows transformed drawings of the same silicon rocking curve in logarithmic ordinate scale and two different abscissa scales.

Measurements have been made at only a few reflections: (820) and (10, 2, 0) of

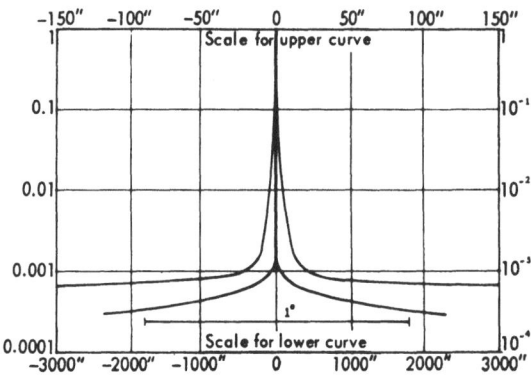

Figure 3. Transformed drawings of the Si (822) rocking curve of Figure 2 in logarithmic ordinate scale.

Table I. Contribution of Thermal Diffuse Scattering (correction factors $\alpha = \bar{R} - \bar{R}_0/\bar{R}_0$) to NaCl X-Ray Reflections Integrated Over an Angular Range $\pm 1.5°$*

Order	Experimental	Theoretical (Nilsson)
(820)	$0.20 \pm 10\%$	0.183
(10, 2, 0)	$0.31 \pm 20\%$	0.320

* The range for which the theoretical values of α are calculated by Nilsson.

sodium chloride and (822) of silicon. The purpose was just to test the method and to check Nilsson's correction factors at some points.*

Among precautions to be observed in the quantitative evaluation, one is mentioned in detail. As is well known $(n, -n)$ rocking curves are square convolutions of the diffraction patterns of the crystals. It is easy to see that the diffuse contributions appear in double amount within the square convolution: double that in the single diffraction pattern. At each angular position of the second crystal two components are reflected: one, principally by the first crystal and weakly by the second, and the other, the reverse of this. Therefore, the diffuse contributions obtained from the recorded rocking curves must be halved.

The diffuse contributions have been measured in two ways: absolutely, i.e., with reference to the incident beam of the second crystal; and, in the case of sodium chloride, also relatively, i.e., compared with the principal reflection from an imperfect (ground) crystal. The evaluation is based on the assumption that, because of its weakness and angular width, the diffuse scattering must be regarded as kinematically reflected, as from an imperfect crystal. Its integrated intensity should be the same for perfect and imperfect crystals.

The results are in good agreement with the correction factors calculated by Nilsson, as can be seen in Table I. Thus, Nilsson's calculations and the value of the characteristic temperature, 302°, derived by her from them can be regarded as experimentally confirmed.

* High-order reflections—those for which a large thermal contribution could be expected—were chosen, but only *those* which are *first-orders* at the reflecting lattice planes, in order to avoid trouble from the contribution of low-order reflections of larger wavelength.

REFERENCES

1. M. Renninger, *Acta Cryst.* **5**: 711, 1952.
2. H. Witte and E. Wölfel, *Z. Physik. Chem.* **3**: 296, 1955.
3. M. Blackman, *Acta Cryst.* **9**: 734, 1956.
4. D. R. Chipman and A. Paskin, *J. Appl. Phys.* **30**: 1992, 1998, 1959.
5. Nanny Nilsson, *Arkiv. Fysik.* **12**: 247, 1957.

DISCUSSION

N. Spielberg (Philips Laboratories): In these studies, did you use the asymmetrically cut crystals?

M. Renninger: No. The angular width of the principal peak is small enough even at symmetrical reflection to be separated properly from the thermic peak.

THE X-RAY DIFFRACTION IMAGE OF A STACKING FAULT

Norio Kato, Katsuhisa Usami,* and Takeshi Katagawa

Nagoya University
Nagoya, Japan

ABSTRACT

The spherical wave theory of X-ray Pendellösung fringes in perfect crystals (N. Kato, *Acta Cryst.* **14**: 526, 627, 1961) is extended to the case of crystals including a single stacking fault in an arbitrary way. Kelvin's stationary phase method is used extensively. The stationary phase condition gives us the trajectories of X-ray beams in the crystal. The phase and the amplitude along each trajectory are obtained by straightforward calculation. Based on this crystal wave field, the section patterns in X-ray diffraction topographs are obtained both for the direct and the Bragg-reflected waves. Characteristic fringe-patterns are expected. Through the image of a fault plane in a single section pattern, the geometrical configuration inside the crystal and the magnitude of the fault vector can be determined. Traverse patterns are also discussed. The fault image based on the plane wave theory (Whelan and Hirsch, *Phil. Mag.* **2**: 1121, 1303, 1957) is also reformulated in the most general Laue case without the use of *ad hoc* assumptions on the shape of the dispersion surface.

INTRODUCTION

The diffraction image of a single stacking fault was first observed by Whelan and Hirsch in thin metal films by means of electron microscopy.[1] The image is made of characteristic parallel fringes. In X-ray diffraction topographs, Kohra and Yoshimatsu observed similar patterns in silicon single crystals and interpreted them as the images of stacking faults.[2] Their observations were carried out by using Lang's traverse technique,[3,4] the patterns obtained by this technique being called traverse patterns. Before that, in the first observation of Pendellösung fringes in X-ray cases, Kato and Lang also had observed similar but more complicated fringes in quartz.[5] They interpreted them as the diffraction contrast due to plate-like defects.

In X-ray cases, through the experiments of Kato and Lang and the theoretical work of Kato,[6-8] it is certain that the diffraction pattern of section-type is more fundamental than the traverse pattern, since the traverse pattern is a superposition of the section patterns. Recently, Yoshimatsu[9] and Homma[10] have obtained independently the section patterns for parallel-sided quartz crystals in which the above-mentioned fringes were observed in the traverse patterns.† Yoshimatsu pointed out also that the fringes appeared at the boundary of a faintly coloured part included in the matrix. The section patterns obtained by Yoshimatsu and Homma are not identical and it seems that there are many variations in this kind of pattern. Nevertheless, the fact that two sets of hook-shaped fringes appear in the region of hourglass shapes is characteristic of this type of pattern.

Under these circumstances it is desirable to study theoretically the diffraction contrast

---

* Present address: Hitachi Central Research Laboratory, Kokubunji, Tokyo, Japan.
† Authier and Sauvage also obtained similar fringes in calcite (private communication).

due to a stacking fault and other conceivable plate-like defects. Actually, in electron cases, Whelan and Hirsch have presented the dynamical theory on the diffraction contrast of a single stacking fault based on the plane-wave theory for wedge-shaped crystals.[11,12] Although the plane-wave theory is very adequate in electron cases, a spherical-wave theory must be used in X-ray cases, particularly for understanding the section patterns. Since a stacking fault is the simplest case of plate-like defects, it is very natural to start with a single stacking fault for understanding plate-like defects.

This paper is a development of the spherical-wave theory[7] for perfect crystals to the case of the crystal including a single stacking fault. The basic approach, therefore, is essentially the same as that used in the previous theory. In the following section, the plane-wave theory is reformulated for the most general case of a geometrical configuration regarding the fault plane, the net plane, and the crystal surfaces, without adding any *ad hoc* approximation to the two-beam approximation used for perfect crystals in the conventional dynamical theory. Next, we shall consider the trajectories of X-ray beams in the crystal using the stationary phase condition. Finally, the wave field and the intensity field are calculated; absorption is not included. Based on this treatment, some numerical results are given as an example for a symmetrical Laue case. Although the result obtained here is not identical to the observed patterns, the characteristic features of observed fringe systems and the intensity distribution can be explained properly.

In order to minimize the complexity of the notation, it is summarized in Appendix I. Principal formulas are presented in the text, but the subsidiary ones are listed in the other four Appendices. Mathematical operations for obtaining them are described in Appendix VI.

GENERAL CONSIDERATION OF THE CRYSTAL WAVE FIELD

As shown in Figure 1, the crystal is assumed to be divided by a single stacking fault into two regions, denoted by I and II. At the fault plane, region II is displaced by a vector **u** with respect to region I. We shall consider the crystal wave field produced by an incident beam of X-rays. In the case of X-ray diffraction, the incident wave is to be considered as a spherical wave.[6,7] Because a spherical wave is represented by a superposition of plane waves as follows:

$$\frac{\exp(iKr)}{4\pi r} = \frac{i}{8\pi^2} \iint_{-\infty}^{\infty} \frac{1}{K_z} \exp[i(\mathbf{K}\cdot\mathbf{r})]\, dK_x\, dK_y \tag{1}$$

where **r** is a position vector and K_x, K_y, and K_z are the x, y, and z components of the wave vector **K**, respectively, then the crystal wave field can be written in the form

$$\varphi_{0,g} = \frac{i}{8\pi^2} \iint_{-\infty}^{\infty} \frac{1}{K_z} d_{0,g}(\mathbf{K}, \mathbf{r}) \exp[i(\mathbf{K}\cdot\mathbf{r})]\, dK_x\, dK_y \tag{2}$$

where the subscripts 0 and g indicate the direct and Bragg-reflected waves, respectively. In the integrand, the expression $d_{0,g}(\mathbf{K}, \mathbf{r}) \exp[i(\mathbf{K} \cdot \mathbf{r})]$ is the crystal wave field created by a plane wave $\exp[i(\mathbf{K} \cdot \mathbf{r})]$. The functions $d_{0,g}(\mathbf{K}, \mathbf{r})$ in regions I and II, therefore, are obtained by the conventional dynamical theory of crystal diffraction in which the incident beam is assumed to be a plane wave.

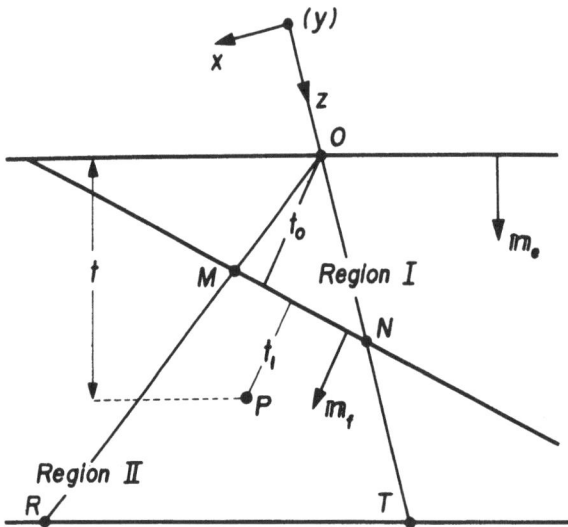

Figure 1. The crystal including a single stacking fault. O: Entrance point. P: Observation point. The lines ONT and OMR are the direct and the Bragg-reflected beams satisfying the Bragg condition in a kinematical sense. MN: Trace of the fault plane. Other notations are described in Appendix I.

The Wave Field Due to a Plane Wave

The wave field in region I is the same as the wave field expected in perfect crystals.[7] As shown in the right half of Figure 2, two dispersion points, D^1 and D^2, are excited on two branches of the dispersion surface. They are connected with the dispersion point of the incident plane wave by the condition of tangential continuity on the entrance surface. As a result, four plane waves are excited, the corresponding wave vectors being $\mathbf{k}_0^1 = \overrightarrow{D^1O}$, $\mathbf{k}_0^2 = \overrightarrow{D^2O}$, $\mathbf{k}_g^1 = \overrightarrow{D^1G}$, and $\mathbf{k}_g^2 = \overrightarrow{D^2G}$. Thus, the wave fields in region I are represented in the form

$$d_0 \exp[i(\mathbf{K}\cdot\mathbf{r})] = \sum_i d(^i_0) \exp[i(\mathbf{k}_0^i \cdot \mathbf{r})] \tag{3a}$$

$$d_g \exp[i(\mathbf{K}\cdot\mathbf{r})] = \sum_i d(^i_g) \exp[i(\mathbf{k}_g^i \cdot \mathbf{r})] \tag{3b}$$

where the superscript i runs over (1) and (2). The amplitudes $d(^i_0)$ and $d(^i_g)$ are given by

$$d(^i_0) = C_0^i \exp[i(\mathbf{K} - \mathbf{k}_0^i \cdot \mathbf{r}_e)] \tag{4a}$$

$$d(^i_g) = C_g^i \exp[i(\mathbf{K} - \mathbf{k}_0^i \cdot \mathbf{r}_e)] \tag{4b}$$

where the vector $\mathbf{r}_e$ indicates a position on the entrance surface. The wave vectors $\mathbf{k}_0^i$ and the amplitudes* C_0^i and C_g^i are represented in terms of a few parameters specifying the departure from the Bragg condition, the structure factors, and others, and the results are listed in Appendix II.

In region II, each of the four plane waves of region I excites four plane waves, just as the incident wave does in region I. In all, therefore, sixteen waves are excited. Two of them, however, have the same wave vector, so that they are reduced to eight plane waves. In fact, it is easily seen that only four dispersion points are excited, each point representing

---

* Although the quantities C_0^i and C_g^i are not exactly the amplitudes, we shall call them amplitudes, since no confusion is expected.

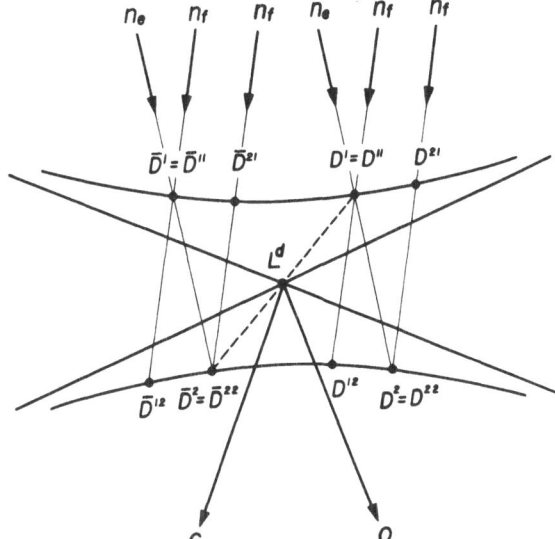

Figure 2. The dispersion surface of the crystal waves. Two groups of dispersion points are illustrated. Each group is excited from an incident plane wave. The wave points $D^1 = D^{11}$ and $\bar{D}^2 = \bar{D}^{22}$, $D^2 = D^{22}$ and $\bar{D}^1 = \bar{D}^{11}$, D^{12} and $\bar{D}^{21}$, and D^{21} and $\bar{D}^{12}$ are conjugate with each other, respectively, with respect to the dynamical Laue point, L^d.

a pair of the direct- and the Bragg-reflected waves. The dispersion points can be constructed from D^1 and D^2 through the tangential continuity of the wave vector with respect to the fault plane. The situation is illustrated in Figure 2. Here, two points, D^{11} and D^{22}, are identical to D^1 and D^2, respectively, and others are denoted by D^{12} and D^{21}. The wave fields are represented as follows:

$$d_0 \exp[i(\mathbf{K}\cdot\mathbf{r})] = \sum_{ij} \{d(^{ij}_{00}) \exp[i(\mathbf{k}^{ij}_{00}\cdot\mathbf{r})] + d(^{ij}_{g0}) \exp[i(\mathbf{k}^{ij}_{g0}\cdot\mathbf{r})]\} \qquad (5a)$$

$$d_g \exp[i(\mathbf{K}\cdot\mathbf{r})] = \sum_{ij} \{d(^{ij}_{0g}) \exp[i(\mathbf{k}^{ij}_{0g}\cdot\mathbf{r})] + d(^{ij}_{gg}) \exp[i(\mathbf{k}^{ij}_{gg}\cdot\mathbf{r})]\} \qquad (5b)$$

where the superscripts i and j specify the branch of the dispersion surface. The wave with amplitude $d(^{ij}_{0g})$, for example, is a plane wave which passes through regions I and II in the states specified by $(^i_0)$ and (^j_g), respectively. The expression for the amplitudes $d(^{ij}_{00})$ etc., are given as follows:

$$d(^{ij}_{00}) = C^{ij}_{00} \exp i\{(\mathbf{K} - \mathbf{k}^i_0\cdot\mathbf{r}_e) + (\mathbf{k}^i_0 - \mathbf{k}^{ij}_{00}\cdot\mathbf{r}_f)\} \qquad (6a)$$

$$d(^{ij}_{g0}) = C^{ij}_{g0} \exp i\{(\mathbf{K} - \mathbf{k}^i_0\cdot\mathbf{r}_e) + (\mathbf{k}^i_0 - \mathbf{k}^{ij}_{g0}\cdot\mathbf{r}_f)\} \qquad (6b)$$

$$d(^{ij}_{0g}) = C^{ij}_{0g} \exp i\{(\mathbf{K} - \mathbf{k}^i_0\cdot\mathbf{r}_e) + (\mathbf{k}^i_g - \mathbf{k}^{ij}_{0g}\cdot\mathbf{r}_f)\} \qquad (7a)$$

$$d(^{ij}_{gg}) = C^{ij}_{gg} \exp i\{(\mathbf{K} - \mathbf{k}^i_0\cdot\mathbf{r}_e) + (\mathbf{k}^i_g - \mathbf{k}^{ij}_{gg}\cdot\mathbf{r}_f)\} \qquad (7b)$$

where $\mathbf{r}_f$ is a position vector indicating the fault plane. The detailed expressions of the wave vectors $\mathbf{k}^{ij}_{00}$ etc., and the amplitudes C^{ij}_{00} etc., are shown in Appendix III.

The formulae listed in Appendix III are essentially the same as those obtained by Whelan and Hirsch in the case of electron diffraction.[11] Their results, however, can be used only for symmetrical Laue cases and for electrons of high energy, since they are based on *ad hoc* assumptions for the shape of the dispersion surface. In X-ray cases, we need to have more general and rigorous formulae. The results in Appendix III are free from any approximation and limitation with respect to the configuration of the fault plane, the net plane, and the crystal surfaces.

The Wave Field Due to a Spherical Wave

Here, the Bragg-reflected wave φ_g is treated in detail. The wave field in region II is denoted by φ_g^{II} in particular. The direct wave field φ_0 can be treated in an analogous way to the Bragg-reflected wave. As shown in Appendix III, the wave vectors $\mathbf{k}_{0g}^{ij}$ and $\mathbf{k}_{gg}^{ij}$ are the same. We shall denote them by $\mathbf{k}_g^{ij}$. Two plane waves, $d(_{0g}^{ij}) \exp[i(\mathbf{k}_{0g}^{ij} \cdot \mathbf{r})]$ and $d(_{gg}^{ij}) \exp[i(\mathbf{k}_{gg}^{ij} \cdot \mathbf{r})]$, therefore, can be reduced to a single plane wave. Thus, the wave field φ_g^{II} can be written in the form

$$\varphi_g^{II} = \sum_{ij} \varphi_g^{ij} \tag{8}$$

where

$$\varphi_g^{ij} = \frac{i}{8\pi^2} \iint\limits_{-\infty}^{\infty} \frac{1}{K_z} \{d(_{0g}^{ij}) + d(_{gg}^{ij})\} \exp[i(\mathbf{k}_g^{ij} \cdot \mathbf{r})] \, dK_x \, dK_y \tag{9}$$

This can easily be obtained by substitution from equation (5b) to equation (2). For the later purpose, we classify the wave field φ_g^{II} into two parts:

$$\varphi_g^{II} = A_g + B_g \tag{10}$$

where

$$A_g = \varphi_g^{11} + \varphi_g^{22} \tag{11a}$$

$$B_g = \varphi_g^{12} + \varphi_g^{21} \tag{11b}$$

The wave field A_g consists of the waves which pass through the fault plane without a jumping of the dispersion point, whereas the wave B_g suffers a jumping at the fault plane.

So far, the x, y, and z axes have not been specified. Here, we shall take them as shown in Figure 1. The plane of illustration is the reflection plane determined by the directions of the direct beam $\mathbf{K}_0^B$ and the Bragg-reflected beam $\mathbf{K}_g^B = \mathbf{K}_0^B + 2\pi\mathbf{g}$ which satisfy the Bragg condition exactly in a kinematical sense, the vector $\mathbf{g}$ being the reciprocal lattice vector concerned. The integration with respect to K_y can, then, be carried out by the stationary phase method, independently, upon crystal diffraction. The method is exactly the same as in the case of perfect crystals.[7] After integrating, the wave fields can be written in the form

$$\varphi_g^{ij} = D \int\limits_{-\infty}^{\infty} F(s) \exp[iG(s)] \, ds \tag{12}*$$

where the integral variable s is defined by equation (I.1) and the factor D can be given by equation (I.4).† The functional forms of the amplitudes $F(s)$ and the phases $G(s)$ are listed in Appendix IV for every pair of i and j.

The integral (12) can also be obtained by the stationary phase method.[7] The stationary condition of the phase $G(s)$ is

$$\frac{d}{ds}G(s) = 0 \tag{13}$$

---

* The suffix g and the superscripts i and j are omitted here for simplicity.
† The equation numbers preceded by Roman numerals refer to equations in the Appendices.

This gives us the trajectory of the beam. The wave along the trajectory has the form,

$$\varphi_g^{ij} = D[F(s)\sqrt{2\pi/|G''(s)|}] \exp i\{[G(s)] + \text{sign}[G''(s)]\pi/4\} \tag{14}$$

where G'' implies the second derivative of the function $G(s)$ and $[\ ]$ denotes the value of the relevant quantity for a particular s-value, $[s]$, determined by equation (13).[7,8] The detailed expressions of the amplitude $[F(s)\sqrt{2\pi/|G''|}]$, the phase $[G(s)]$, and the sign of $[G''(s)]$ are given in Appendix V.

THE TRAJECTORIES OF X-RAY BEAMS

The stationary condition (13) which determines the trajectories becomes, in region I,

$$\pm \frac{s}{\sqrt{s^2 + \beta^2}}\xi_1 - \xi_2 = 0 \tag{15}$$

where ξ_1 and ξ_2 are defined by equations (IV.1a and b) and β is a constant defined by equation (I.3). The double signs correspond to branch (1) and (2) respectively. Equation (15) represents a set of straight lines, including parameter s, all of which start at the entrance point O, namely the point specified by $\xi_1 = \xi_2 = 0$.

When $|s|$ is large enough, the trajectories tend to two lines $\xi_1 + \xi_2 = 0$ and $\xi_1 - \xi_2 = 0$. Through the definitions of ξ_1 and ξ_2, it can be shown that

$$\xi_1 + \xi_2 = x_0 \tag{16a}$$

$$\xi_1 - \xi_2 = (\gamma_0/\gamma_g)x_g \tag{16b}$$

Here, x_0 and x_g mean the perpendiculars to the lines $\mathbf{K}_0^B$ and $\mathbf{K}_g^B$ from the observation point, respectively. Thus, the lines $\xi_1 \pm \xi_2 = 0$ imply the lines OT and OR in Figure 3. The wave fields are limited within the region bounded by these lines. Since the trajectory concerned here is the trajectory of the wave bundle of the Bloch wave, it includes the direct and the Bragg-reflected waves.

In region II we shall consider the wave fields A_g and B_g separately. The trajectories of beams corresponding to the waves A_g are given by the same equation as equation (15). This is easily seen from the functional forms of G. These beams, therefore, can be considered as the continuation of the beams in region I without the deflection at the fault plane.

Similarly, the trajectories of the beams corresponding to the wave fields B_g are given by the stationary condition (13) as follows:

$$\pm \frac{s}{\sqrt{s^2 + \beta^2}}\eta_1 - \eta_2 = 0 \tag{17}$$

where η_1 and η_2 are defined by equations (IV.2a and b). In contrast with the beams of the wave field A_g, the beams concerned are deflected at the fault plane. For large values of $|s|$, the trajectories tend to the lines $\eta_1 + \eta_2 = 0$ and $\eta_1 - \eta_2 = 0$. From the definitions of η_1 and η_2, we can show that

$$\eta_1 - \eta_2 = -(\gamma_0/\gamma_g)(\gamma_g'/\gamma_0')x_0' \tag{18a}$$

$$\eta_1 + \eta_2 = -(\gamma_0'/\gamma_g')x_g' \tag{18b}$$

where x_0' and x_g' are the perpendiculars to the lines MT' and NR' from the observation point, respectively. Thus, the lines $\eta_1 \pm \eta_2 = 0$ imply the lines NR' and MT', respectively. An important fact is that a set of lines given by equation (17) passes through

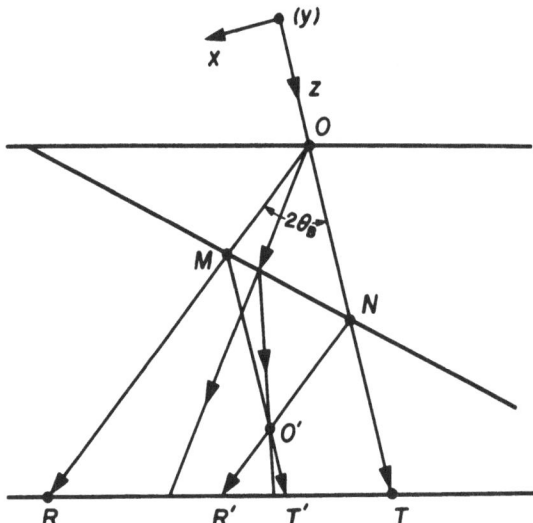

Figure 3. The trajectories of X-ray beams in the crystal. The point O' is the focussing point of the trajectories of the B-type waves. The wave of type A does not deflect at the fault plane.

the intersection O' of the lines MT' and NR', because the intersection is given by the conditions $\eta_1 = \eta_2 = 0$; equation (17) then holds, irrespective of s-values. Moreover, we know that the wave field B_g is confined within two triangular regions MNO' and $R'T'O'$.

THE INTENSITY FIELD OF THE BRAGG-REFLECTED WAVE

As shown in equation (10), the wave field φ_g^{II} is composed of two wave fields, A_g and B_g. Their concrete expressions are obtained by substituting the formulas listed in Appendix V into equations (11a and b). The results are given as follows:

$$A_g = 2\sqrt{2\pi}iD\left(\frac{\Delta_1\xi_1 + \Delta_2\xi_2}{\Gamma_1\xi_1 + \Gamma_2\xi_2}\right)\frac{\cos(\beta\sqrt{\xi_1^2 - \xi_2^2} - \pi/4)}{(\beta\sqrt{\xi_1^2 - \xi_2^2})^{\frac{1}{2}}} \tag{19a}$$

$$B_g = \pm\, 2\sqrt{2\pi}iD\left(\frac{\Delta(\eta_1 - \eta_2)}{\Gamma_1\eta_1 + \Gamma_2\eta_2}\right)\frac{\cos(\beta\sqrt{\eta_1^2 - \eta_2^2} - \pi/4)}{(\beta\sqrt{\eta_1^2 - \eta_2^2})^{\frac{1}{2}}} \tag{19b}$$

where Δ_1, Δ_2, and Δ are given by equations (IV.3a, b, and c), respectively, and Γ_1 and Γ_2 are defined by equations (IV.4a and b), respectively. Here, the double signs correspond to the regions MNO' ($\eta_1 > 0$) and $R'T'O'$ ($\eta_1 < 0$), respectively. The intensity field is therefore made up of the intensity fields $|A_g|^2$ and $|B_g|^2$ and the interference term $2Re(A_g^* B_g)$. We shall consider the behavior of $|A_g|^2$ and $|B_g|^2$ separately.

The Intensity Field $|A_g|^2$

The intensity field $|A_g|^2$ is given by

$$|A_g|^2 = 8\pi^2|D|^2\left[1 - \frac{4(x_0/\gamma_0')(x_g/\gamma_g')}{(x_0/\gamma_0' + x_g/\gamma_g')^2}\sin^2\frac{\delta}{2}\right]\frac{\cos^2(\beta\sqrt{(\gamma_0/\gamma_g)}x_0x_g - \pi/4)}{\beta\sqrt{(\gamma_0/\gamma_g)}x_0x_g} \tag{20}$$

in terms of the perpendiculars x_0 and x_g. Through the $(\cos)^2$ term, a fringe-like distribution of intensity is expected. The fringe contours are hyperbolas which are the same in form as those expected in perfect crystals. The intensity distribution, however, is different from that of the perfect crystal by the factor denoted by the square bracket [ ]. The difference becomes maximum on the line $x_0/\gamma_0' = x_g/\gamma_g'$, which is actually the line OO' in Figure 3. The line OO' does not lie on the net plane, in general.

The Intensity Field $|B_g|^2$

Here, we are concerned with the wave field $|B_g|^2$ in the triangular regions NMO' and $T'R'O'$. The intensity field can be written as

$$|B_g|^2 = 8\pi^2|D|^2 \frac{4(x_0'/\gamma_0')^2}{(x_0'/\gamma_0' + x_g'/\gamma_g')^2} \sin^2\frac{\delta}{2} \frac{\cos^2(\beta\sqrt{(\gamma_0/\gamma_g)}x_0'x_g' - \pi/4)}{\beta\sqrt{(\gamma_0/\gamma_g)}x_0'x_g'} \quad (21)$$

in terms of the perpendiculars x_0' and x_g'. The fringe-like contours for $|B_g|^2 = 0$ are also of the form of hyperbola whose asymptotes are the lines NR' and MT' in Figure 3. It should be noticed that the behavior of the $(\cos)^2$ terms in both intensity fields $|A_g|^2$ and $|B_g|^2$ is exactly the same with respect to the corresponding asymptotes of the hyperbolic fringes. The intensity distribution of $|B_g|^2$ is very asymmetric with respect to the asymptotes NR' and MT'. It increases when the observation point P is close to the line NR' due to the factor $(x_0'/\gamma_0')^2$.

In Figure 5, the fringe systems are illustrated in a particular case which will be discussed later in more detail. Here, the broken lines correspond to the intensity field $|A_g|^2$ and the full lines correspond to the field $|B_g|^2$.

THE INTENSITY FIELD OF THE DIRECT WAVE

The wave field of the direct wave is also divided into two parts, A_0 and B_0, in the same way as the Bragg-reflected wave. The intensity field, therefore, includes three terms $|A_0|^2$, $|B_0|^2$, and the interference term $2Re(A_0^*B_0)$. The expressions for $|A_0|^2$ and $|B_0|^2$ can be given as follows after a straightforward calculation:

$$|A_0|^2 = 8\pi^2|D|^2 \left[1 - \frac{4(x_0/\gamma_0')(x_g/\gamma_g')}{(x_0/\gamma_0' + x_g/\gamma_g')^2} \sin^2\frac{\delta}{2} \right] \left(\frac{\gamma_0}{\gamma_g} \frac{x_g}{x_0} \right) \frac{\cos^2(\beta\sqrt{(\gamma_0/\gamma_g)}x_0x_g - \frac{3}{4}\pi)}{\beta\sqrt{(\gamma_0/\gamma_g)}x_0x_g}$$

$$(22)$$

$$|B_0|^2 = 8\pi^2|D|^2 \frac{4(x_0'/\gamma_0')(x_g'/\gamma_g')}{(x_0'/\gamma_0' + x_g'/\gamma_g')^2} \left(\sin^2\frac{\delta}{2} \right) \left(\frac{\gamma_0}{\gamma_g} \frac{\gamma_g'}{\gamma_0'} \right) \frac{\cos^2(\beta\sqrt{(\gamma_0/\gamma_g)}x_0'x_g' - \frac{3}{4}\pi)}{\beta\sqrt{(\gamma_0/\gamma_g)}x_0'x_g'} \quad (23)$$

The expression (22) without the factor denoted by the square bracket [ ] is the same as the expression for the direct wave in the perfect crystal, obtained by the method of stationary phase.[7]

THE INTENSITY DISTRIBUTION ON THE EXIT SURFACE

In this section we shall consider the intensity distribution on the exit surface. The section patterns due to the direct and the Bragg-reflected waves are merely the projections of the corresponding intensity distributions on the exit surface in the directions of the direct and the Bragg-reflected beams, respectively.

Numerical Example

For the sake of simplicity, we will concern ourselves with a symmetrical Laue case of the parallel-sided crystal in which the fault plane lies obliquely to the crystal surfaces but perpendicular to the net plane, as shown in Figure 4. In this particular case, by definition, $\gamma_0 = \gamma_g = \cos \theta_B$ and $\gamma_0' = \gamma_g' = \cos \theta_B \cos \phi$, where ϕ is the angle between the crystal surface and the fault plane. A position on the exit surface is specified by the coordinates u and v. Throughout the following, the phase angle δ is assumed to be $\pm 120°$. The distance of Pendellösung fringes along the net plane is taken to be $(2/25)T_0$, T_0 being the total thickness of the crystal.

The fringe systems within this crystal are illustrated in Figure 5. The corresponding intensity fields on the exit surface are shown schematically in Figures 6a and b for the

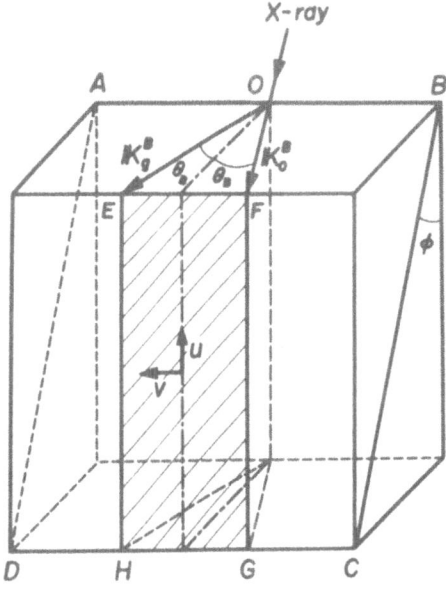

Figure 4. The geometrical relations among the crystal surfaces, the net plane, and the fault plane $ABCD$ in the case for which numerical calculation was carried out. The intensity field on the exit surface is limited within the region $EFGH$.

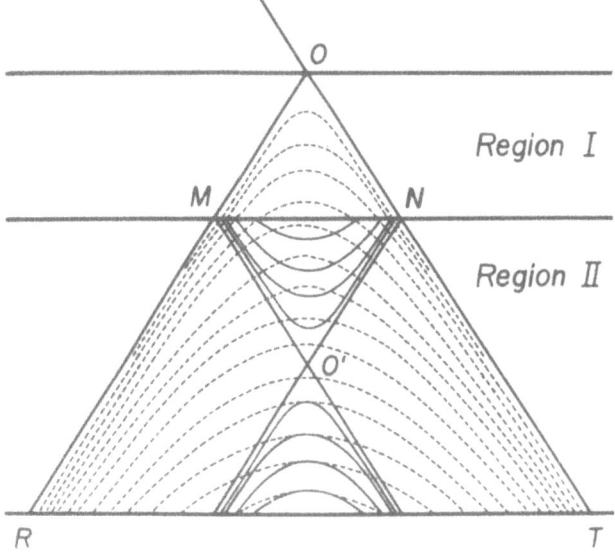

Figure 5. The fringe systems in the crystal. The broken and the full lines correspond to the fringes due to waves of type A_g and B_g, respectively.

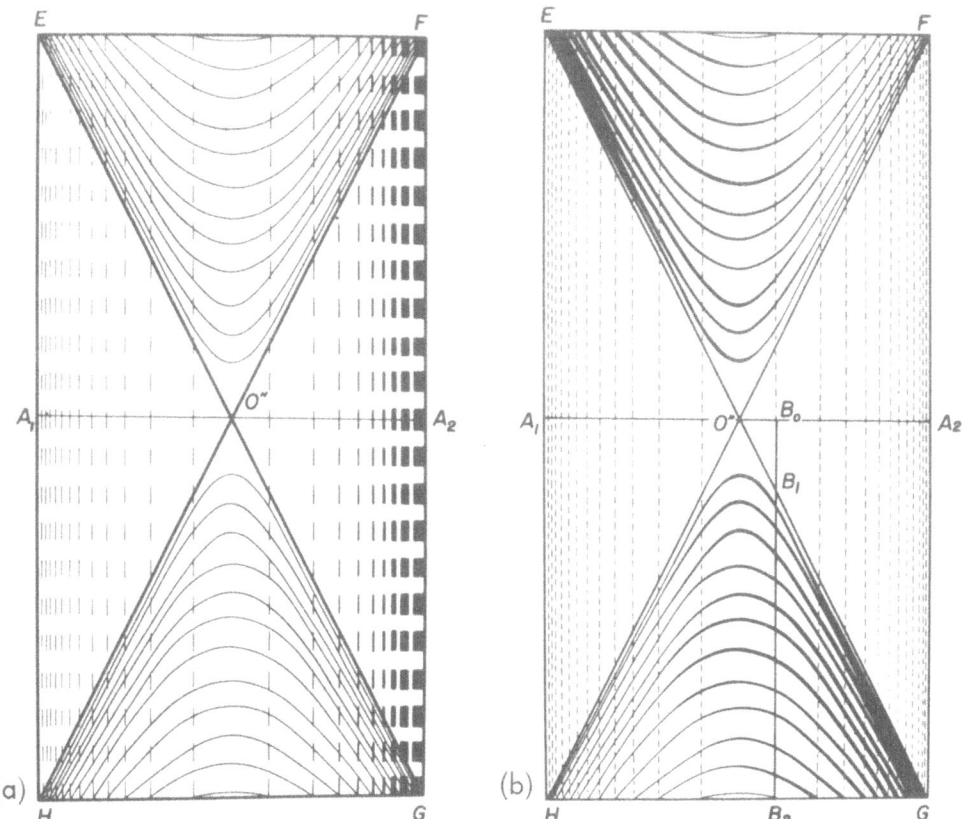

Figure 6. Schematic illustration of the intensity distribution on the exit surface. (a) Direct wave. (b) Bragg-reflected wave. The region $EFGH$ is identical to the region $EFGH$ in Figure 4.

direct and the Bragg-reflected waves. The intensity field consists of two types of intensity field corresponding to the waves A and B mentioned above,* and their interference term. The straight fringes correspond to the fringe system of $|A|^2$ and the hyperbolic ones correspond to the fringe system of $|B|^2$. Their asymptotes cross at the center O'' of each section pattern. The intensity distribution along the horizontal line A_1A_2 corresponds to the case that the fault plane is located in the middle of the crystal.

The intensity field of the Bragg-reflected wave along the line A_1A_2 is shown in Figure 7. Here, the distance is normalized by the factor $T_0 \tan \theta_B$—the half-width of the X-ray field on the exit surface. The intensity distribution along the vertical line $B_0B_1B_2$ of Figure 6b is also illustrated in Figure 8, where the distance is normalized by the factor $\frac{1}{2}T_0/\tan \phi$—the half-height of the stacking-fault image. Between B_0 and B_1 the intensity is constant; in the region between B_1 and B_2 it oscillates. The amplitude of the oscillation does not decrease monotonically since the interference term $2Re(A_g^*B_g)$ is overlapped. This can be seen more clearly in the intensity distribution along a particular fringe. One of the examples is shown in Figure 9.

* We shall omit the subscripts 0 and g when we are concerned with both direct and Bragg-reflected waves.

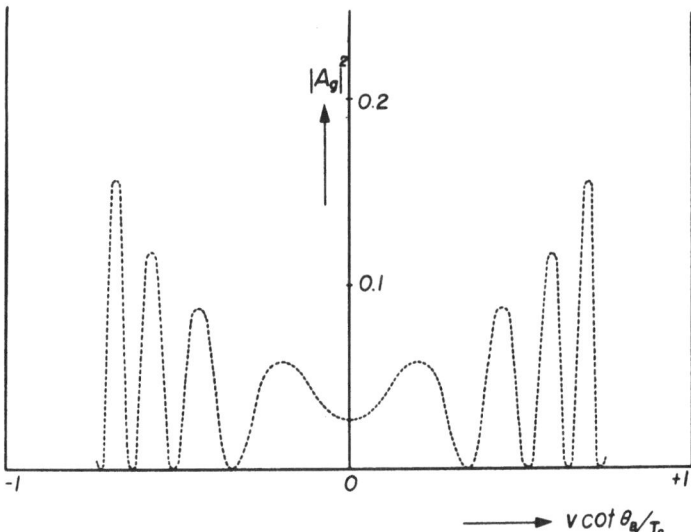

Figure 7. The intensity distribution along the line A_1A_2 of Figure 6b.

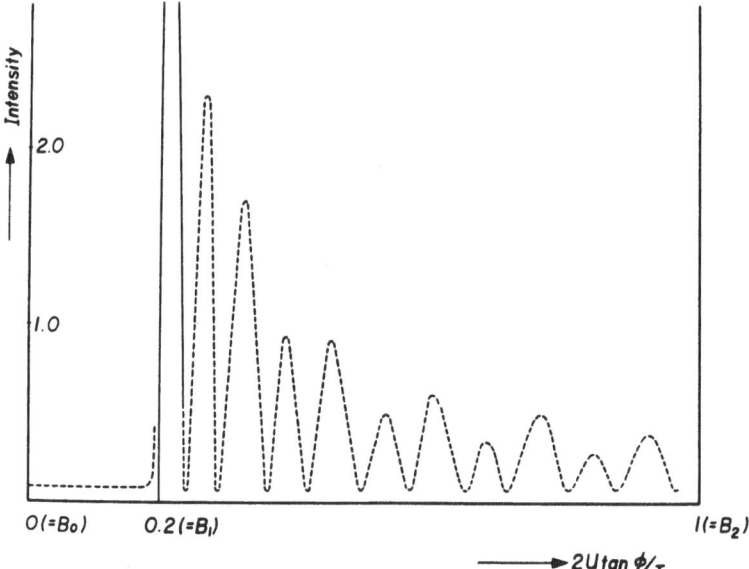

Figure 8. The intensity distribution along the line $B_0B_1B_2$ of Figure 6b.

Consideration of General Cases

In more general cases, the geometry and the intensity distribution of the fringe systems are similar to the above case in a topological sense. This is clearly seen from the fact that the expressions of the intensity fields can be represented as the functions of x_0 and x_g or x'_0 and x'_g, modified by the factors γ_0, γ_g, γ'_0, and γ'_g. Through elementary geometrical consideration we will see the following rules:

1. No significant difference of the fringe form is expected between the symmetrical and nonsymmetrical Laue cases as in the case of perfect crystals. Because the form of

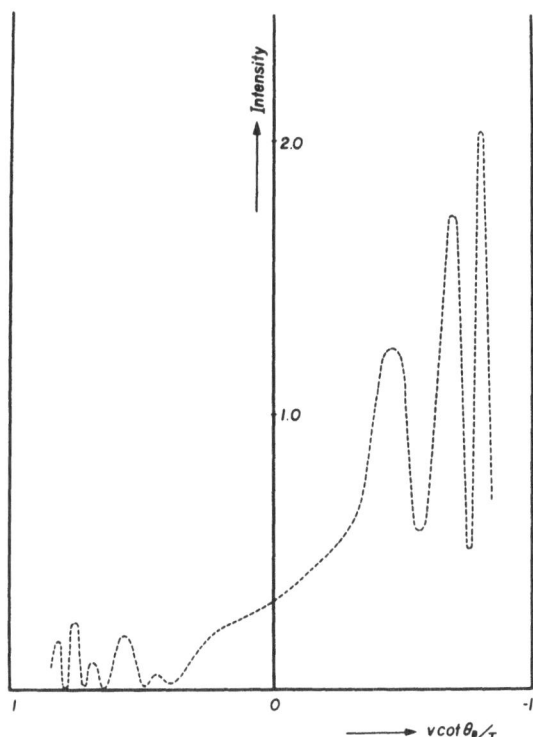

Figure 9. The intensity distribution along a hyperbolic fringe of Figure 6b.

the fringes $|A_g|^2$ in the crystal is a hyperbola, the two intersections of a particular fringe and the exit surface appear at the same distance from the edges of the section pattern in any case. The same is true for the fringe system of $|B_g|^2$ with respect to the asymptotic lines $EO''G$ and $FO''H$ in the particular case of Figure 6b.

2. If the fault plane lies arbitrarily in the crystal, the image of the fault is of trapezoidal form in general. However, one end is always rectangular as shown in Figure 10a. The rectangular end corresponds to the position where the fault plane passes through the entrance point. The fringe system due to the intensity field $|B_g|^2$ is again of hyperbolic form, the asymptotes being the diagonals of the fault image.

3. In a very particular case, where the fault plane is perpendicular to the plane determined by the lines $\mathbf{K}_0^B$ and $\mathbf{K}_g^B$, both fringe systems of the waves A and B are parallel to the edges of the section pattern as shown in Figure 10b.

4. The center of the hyperbolic fringes is closer to one of the edges corresponding to the lines $\mathbf{K}_0^B$ and $\mathbf{K}_g^B$. For example, if the intersection of $\mathbf{K}_0^B$ with the fault plane is closer to the exit surface compared with the intersection of $\mathbf{K}_g^B$ and the fault plane, the center is closer to the edge corresponding to $\mathbf{K}_0^B$.

5. The above rule can be generalized to the particular case mentioned in (3).

The intensity distributions in the general case are anticipated easily from the simplest case described above. The characteristic features are summarized as follows:

6. In the case of the Bragg-reflected wave, the straight fringes due to the wave A_g are symmetrical with respect to the central line parallel to the edge of the section pattern. On the other hand, the hyperbolic fringes due to the wave field B_g are very asymmetric. The intensity of the region which is close to the asymptote corresponding to the line $NO'R'$ in Figure 3 is higher than that of the region close to the other asymptote.

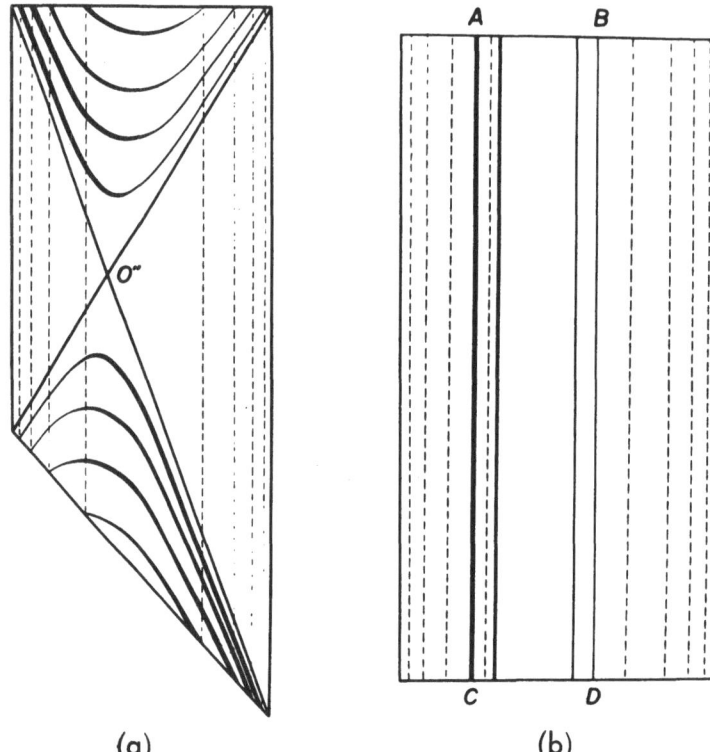

(a)   (b)

Figure 10. Schematic illustration of the intensity distribution of the
Bragg-reflected wave on the exit surface. (a) General cases, (b) special
case in which the fault plane is perpendicular to the planes illustrated in
Figures 1 and 3.

7. In the case of the direct wave, the hyperbolic fringes due to the wave B_0 are
symmetrical with respect to the asymptotes, whereas the straight fringes due to the
wave A_0 are asymmetric with respect to the central line. The intensity of the region close
to the direct beam $\mathbf{K}_0^B$ is stronger than that close to the Bragg-reflected beam. The situation
is similar to that in perfect crystals.

DISCUSSION AND CONCLUSIONS

Interpretation of the Wave Field in Terms of the Dispersion Surface

As discussed earlier (see Figure 2), a single plane wave excites two Bloch waves
specified by two dispersion points D^1 and D^2 in region I. In region II, they excite four
Bloch waves, the corresponding dispersion points being $D^{11} = D^1$, $D^{22} = D^2$, D^{12}, and
D^{21}. According to the concept of energy flow,[13,14,15,16] they propagate in the directions
normal to the dispersion surface at these dispersion points. Thus, we may obtain the
trajectories of the crystal waves as described above. It is easily seen that the trajectory
of a Bloch wave which suffers a jumping of the dispersion point changes the direction at
the fault plane.

Since the incident wave should be considered as a spherical wave, the whole of the dispersion surface is excited and the plane waves specified by dispersion points are coherent with each other. The conjugate dispersion points of region I, for example, D^1 and $\bar{D}^2$, again excite the conjugate points D^{11} and $\bar{D}^{22}$ or D^{12} and $\bar{D}^{21}$ in region II, respectively. Thus, two pairs of conjugate Bloch waves are excited in region II and each pair propagates in the same direction. Since the wave vectors are slightly different for the Bloch waves of each pair, we may expect interference fringes among the paired Bloch waves.

Geometry of the Fault Plane

From rules (1) through (7) on the geometry and the intensity distribution of the fault image, we can predict the geometrical configuration of the fault plane inside the crystal. In particular, rule (2) is very useful for predicting which end of the fault image corresponds to the fault plane closer to the entrance surface. This prediction can be checked also by the intensity asymmetry of the hyperbolic fringes due to $|B_g|^2$, explained in rule (6). Next, rule (4) can be used to predict the side of the fault plane closer to the entrance surface.

Magnitude of the Phase Displacement δ

The intensity distribution along the horizontal line in the central part of the section pattern ($A_1 O'' A_2$ in the case of Figure 6b) indicates the intensity $|A_g|^2$ given by equation (20). As this equation shows, the amplitude of the oscillating intensity is different from that of the perfect crystal. Thus, by comparing the intensity distributions of the perfect region and the central part of the fault image, we can determine the phase displacement δ. The intensity distribution for the perfect crystal is obtained experimentally from the part of the section pattern above or below the fault image. The sign of δ, however, can not be distinguished on the basis of the present theory.

Traverse Pattern

As discussed in the case of perfect crystals,[7] the intensity of the traverse pattern corresponding to a point P on the exit surface can be obtained by integrating spatially the intensity of the Bragg-reflected wave at the entrance surface for the case when the X-ray source is virtually located at point P (Figure 2 in Reference 7). This is generally proved by using the reciprocity theorem in optics for the case of Bragg-reflection.[17] The spatial integration of the Bragg-reflected intensity produced by a point source is a function of the total thickness T_0 of the crystal, if the crystal is perfect. Obviously, the integrated intensity is constant for the parallel-sided crystal. If the fault plane cuts the triangular fan of the intensity field produced by a point source of X-rays, the spatial integration turns out to be a function of the total thickness T_0 and the perpendicular distance t_1 from the point P to the fault plane. Thus, we may expect parallel fringes as the fault image in the traverse pattern. A fringe contour indicates the intersection of the exit surface and a plane of equal distance from the fault plane.

Comparison with Experimental Results

So far, we have not yet obtained section patterns which are exactly the same as those predicted by the present theory. Nevertheless, the section patterns obtained by Yoshimatsu[9] and Homma[10] in quartz are very similar to the patterns shown in Figure 6b. Characteristic hyperbolic fringes and the intensity asymmetry described in rule (6) are noticed. The trapezoidal images (Figure 10a) and the type of images as shown in Figure 10b are frequently observed. It should be pointed out, however, that there are many variations in the section patterns of this type regarding the details of the geometry and the intensity distribution. It seems very desirable to compare the observations with the theoretical predictions obtained here in order to elucidate the physical nature of plate-like defects in real crystals.

ACKNOWLEDGMENT

This work was partly supported by the Scientific Research Fund of the Ministry of Education, Japan.

REFERENCES

1. M. J. Whelan, P. B. Hirsch, R. W. Horne, and W. Bollmann, "Dislocation and Stacking Faults in Stainless Steel," *Proc. Roy. Soc.* **240**: 524, 1957.
2. K. Kohra and M. Yoshimatsu, "X-Ray Observations of Lattice Defects—in Particular, Stacking Faults in the Neighbourhood of a Twin Boundary," *J. Phys. Soc. Japan* **17**: 1041, 1962.
3. A. R. Lang, "Direct Observation of Individual Dislocations by X-Ray Diffraction," *J. Appl. Phys.* **29**: 597, 1958.
4. A. R. Lang, "The Projection Topograph: A New Method in X-Ray Diffraction Microradiography," *Acta Cryst.* **12**: 249, 1959.
5. N. Kato and A. R. Lang, "A Study of Pendellösung Fringes in X-Ray Diffraction," *Acta Cryst.* **12**: 787, 1959.
6. N. Kato, "A Theoretical Study of Pendellösung Fringes. Part I, General Considerations," *Acta Cryst.* **14**: 526, 1961.
7. N. Kato, "A Theoretical Study of Pendellösung Fringes, Part II, Detailed Discussion Based upon a Spherical Wave Theory," *Acta Cryst.* **14**: 627, 1961.
8. N. Kato, "Wave Optical Theory of Diffraction in Single Crystals," in: G. N. Ramachandran, *Crystallography and Crystal Perfection*, Academic Press, London, 1963, p. 153. [A correction should be made to equation (14) of this review article (see S. Homma, Y. Ando and N. Kato, "Absolute Positions of Pendellösung Fringes in X-Ray Cases," *J. Phys. Soc. Japan* **21**: 1160, 1966).]
9. M. Yoshimatsu, "A New Type of X-Ray Pendellösung Fringe Observed in a Quartz Single Crystal," *Japanese J. Appl. Phys.* **4**: 619, 1965.
10. S. Homma, unpublished.
11. M. J. Whelan and P. B. Hirsch, "Electron Diffraction from Crystals Containing Stacking Faults (I)," *Phil. Mag.* **2**: 1121, 1957.
12. M. J. Whelan and P. B. Hirsch, "Electron Diffraction from Crystals Containing Stacking Faults (II)," *Phil. Mag.* **2**: 1303, 1957.
13. M. von Laue, "Die Energieströmung bei Röntgenstrahl-Interferenzen in Kristallen," *Acta Cryst.* **5**: 619, 1952.
14. N. Kato, "Dynamical Theory of Electron Diffraction for a Finite Polyhedral Crystal. (I) Extension of Bethe's Theory," *J. Phys. Soc. Japan* **7**: 397, 1952.
15. P. P. Ewald, "Group Velocity and Phase Velocity in X-Ray Crystal Optics," *Acta Cryst.* **11**: 888, 1958.
16. N. Kato, "The Flow of X-Ray and Material Waves in Ideally Perfect Single Crystals," *Acta Cryst.* **11**: 885, 1958.
17. M. von Laue, "Die Fluoreszenzröntgenstrahlung von Einkristallen," *Ann. Physik* **23**: 705, 1935.
18. W. H. Zachariasen, *Theory of X-ray Diffraction in Crystals*, John Wiley, New York, 1945, Section III, p. 111.

APPENDIX I

Notation

ψ_0, ψ_g = The zeroth and g-th order Fourier coefficients of the polarizability of the crystal for X-rays, respectively

$\sin \chi$ = The polarization factor of X-rays, 1 or $\cos 2\theta_B$

θ_B = The Bragg angle

$\mathbf{n}_e, \mathbf{n}_f$ = The unit normals of the entrance surface and the fault plane, respectively

$\mathbf{K}_0^B$ = The wave vector of the incident beam which satisfies the Bragg condition exactly in a kinematical sense

$\mathbf{K}_g^B = \mathbf{K}_0^B + 2\pi\mathbf{g}$ = The wave vector of the Bragg-reflected beam for $\mathbf{K}_0^B$

K = The magnitude of the wave vector of the incident wave

$$\gamma_0 = \cos(\mathbf{K}_0^B, \mathbf{n}_e)$$
$$\gamma_0' = \cos(\mathbf{K}_0^B, \mathbf{n}_f)$$
$$\gamma_g = \cos(\mathbf{K}_g^B, \mathbf{n}_e)$$
$$\gamma_g' = \cos(\mathbf{K}_g^B, \mathbf{n}_f)$$

s = The parameter specifying the departure from the Bragg condition in a dynamical sense

$$= -K_x - \frac{K\psi_0}{2\sin 2\theta_B}\left(\frac{\gamma_g}{\gamma_0} - 1\right) \tag{I.1}$$

$$\alpha = (\sin 2\theta_B)/2\gamma_g \tag{I.2}$$

$$\beta = K\sin\chi\,\sqrt{\psi_g\psi_{\bar{g}}}\left(\sqrt{\gamma_g/\gamma_0}\big/\sin 2\theta_B\right) \tag{I.3}$$

$\mathbf{r}$ = The position vector of the observation point

$\mathbf{r}_e$ = The position vector of the entrance surface

$\mathbf{r}_f$ = The position vector of the fault plane

$\mathbf{u}$ = The displacement of crystal II with respect to crystal I at the fault plane

x, y, z = The rectangular coordinates of $\mathbf{r}$

$\delta = -2\pi(\mathbf{g}\cdot\mathbf{u})$ = The phase displacement in the Fourier amplitude on the fault plane

$t = (\mathbf{r} - \mathbf{r}_e\cdot\mathbf{n}_e)$ = The depth of $\mathbf{r}$ with respect to the entrance surface

$t_1 = (\mathbf{r} - \mathbf{r}_f\cdot\mathbf{n}_f)$ = The depth of $\mathbf{r}$ with respect to the fault plane

D = The constant factor appearing in the expressions of the crystal wave fields

$$= \frac{i}{8\pi^2}\left(\frac{2\pi}{Kz}\right)^{\frac{1}{2}}\left(\frac{K\psi_g\sin\chi}{\sin 2\theta_B}\right)\exp i\left(Kz - \frac{\pi}{4} + P\right)\exp 2\pi i(\mathbf{g}\cdot\mathbf{r})$$

P = The phase independent of the parameter s

$$= \frac{K\psi_0}{2\gamma_0}t - \frac{K\psi_0}{2\sin 2\theta_B}\left(\frac{\gamma_g}{\gamma_0} - 1\right)x \tag{I.5}$$

APPENDIX II

Wave Vectors and Amplitudes in Region I

(1) *Wave Vectors:* $\quad \mathbf{k}_0^i = \mathbf{K} + A^i\mathbf{n}_e \qquad (i = 1, 2) \tag{II.1a}$

$$\mathbf{k}_g^i = \mathbf{k}_0^i + 2\pi\mathbf{g} \tag{II.1b}$$

$$A^1 = \frac{K}{2\gamma_0}\psi_0 + \alpha(s + \sqrt{s^2 + \beta^2}) \tag{II.2a}$$

$$A^2 = \frac{K}{2\gamma_0}\psi_0 + \alpha(s - \sqrt{s^2 + \beta^2}) \tag{II.2b}$$

(2) *Amplitudes:* C_0^i and C_g, and amplitude ratios c^1 and c^2.

$$C_0^1 = \frac{c^2}{c^2 - c^1} \tag{II.3a}$$

$$C_0^2 = \frac{c^1}{c^1 - c^2} \tag{II.3b}$$

$$C_g^1 = \frac{c^1 c^2}{c^2 - c^1} \tag{II.3c}$$

$$C_g^2 = \frac{c^1 c^2}{c^1 - c^2} \tag{II.3d}$$

$$c^1 = \frac{\sin 2\theta_B}{K\psi_{\bar{g}} \sin \chi}(s + \sqrt{s^2 + \beta^2})\frac{\gamma_0}{\gamma_g} \tag{II.4a}$$

$$c^2 = \frac{\sin 2\theta_B}{K\psi_{\bar{g}} \sin \chi}(s - \sqrt{s^2 + \beta^2})\frac{\gamma_0}{\gamma_g} \tag{II.4b}$$

The derivation will be seen in any text book on dynamical theory.[18]

APPENDIX III

Wave Vectors and Amplitudes in Region II

(1) *Wave Vectors:*

$$\mathbf{k}_{00}^{ij} = \mathbf{k}_0^i + A^{ij}\mathbf{n}_f \tag{III.1a}$$

$$\mathbf{k}_{0g}^{ij} = \mathbf{k}_{00}^{ij} + 2\pi\mathbf{g} = \mathbf{k}_{gg}^{ij} \tag{III.1b}$$

$$\mathbf{k}_{gg}^{ij} = \mathbf{k}_g^i + A^{ij}\mathbf{n}_f \tag{III.1c}$$

$$\mathbf{k}_{g0}^{ij} = \mathbf{k}_{gg}^{ij} - 2\pi\mathbf{g} = \mathbf{k}_{00}^{ij} \tag{III.1d}$$

$$A^{11} = A^{22} = 0 \tag{III.2a}$$

$$A^{12} = \alpha\left[(s - \sqrt{s^2 + \beta^2})\frac{\gamma_g}{\gamma_g'} - (s + \sqrt{s^2 + \beta^2})\frac{\gamma_0}{\gamma_0'}\right] \tag{III.2b}$$

$$A^{21} = \alpha\left[(s + \sqrt{s^2 + \beta^2})\frac{\gamma_g}{\gamma_g'} - (s - \sqrt{s^2 + \beta^2})\frac{\gamma_0}{\gamma_0'}\right] \tag{III.2c}$$

(2) *Amplitudes:* C_{0g}^{ij}, etc.

I \ II	$\binom{1}{0}$	$\binom{2}{0}$	$\binom{1}{g}$	$\binom{2}{g}$
$\binom{1}{0}$	$\dfrac{c^{12}}{c^{12} - c^{11}}C_0^1$	$\dfrac{c^{11}}{c^{11} - c^{12}}C_0^1$	$\dfrac{c^{11}c^{12}}{c^{12} - c^{11}}C_0^1$	$\dfrac{c^{11}c^{12}}{c^{11} - c^{12}}C_0^1$
$\binom{2}{0}$	$\dfrac{c^{22}}{c^{22} - c^{21}}C_0^2$	$\dfrac{c^{21}}{c^{21} - c^{22}}C_0^2$	$\dfrac{c^{21}c^{22}}{c^{22} - c^{21}}C_0^2$	$\dfrac{c^{21}c^{22}}{c^{21} - c^{22}}C_0^2$
$\binom{1}{g}$	$\dfrac{1}{c^{11} - c^{12}}C_g^1$	$\dfrac{1}{c^{12} - c^{11}}C_g^1$	$\dfrac{c^{11}}{c^{11} - c^{12}}C_g^1$	$\dfrac{c^{12}}{c^{12} - c^{11}}C_g^1$
$\binom{2}{g}$	$\dfrac{1}{c^{21} - c^{22}}C_g^2$	$\dfrac{1}{c^{22} - c^{21}}C_g^2$	$\dfrac{c^{21}}{c^{21} - c^{22}}C_g^2$	$\dfrac{c^{22}}{c^{22} - c^{21}}C_g^2$

$$\tag{III.3}$$

The table should be read as follows: the states of the waves in region I are specified by column I and the states in region II are specified by row II. For example, C_{0g}^{ij} is shown in $\binom{i}{0}$ row and $\binom{j}{g}$ column.

(3) **The Amplitude Ratios:**

$$c^{11} = c^1 e^{i\delta} \tag{III.4a}$$

$$c^{22} = c^2 e^{i\delta} \tag{III.4b}$$

$$c^{12} = c^2 \frac{\gamma_0' \, \gamma_g}{\gamma_0 \, \gamma_g'} e^{i\delta} \tag{III.4c}$$

$$c^{21} = c^1 \frac{\gamma_0 \, \gamma_g'}{\gamma_0' \, \gamma_g} e^{i\delta} \tag{III.4d}$$

APPENDIX IV

Crystal Wave Fields (a)

$$\varphi_g^{ij} = D \cdot \int_{-\infty}^{\infty} F(s) \exp iG(s) \, ds$$

(i,j)	$F(s)$	$G(s)$	wave field
$(1,1)$	$\dfrac{\Delta_1\sqrt{s^2+\beta^2}+\Delta_2 s}{(\Gamma_1\sqrt{s^2+\beta^2}+\Gamma_2 s)\sqrt{s^2+\beta^2}}$	$\xi_1\sqrt{s^2+\beta^2}-\xi_2 s$	$\left.\begin{array}{c} \\ \\ \end{array}\right\}A_g$
$(2,2)$	$\dfrac{-\Delta_1\sqrt{s^2+\beta^2}+\Delta_2 s}{(\Gamma_1\sqrt{s^2+\beta^2}-\Gamma_2 s)\sqrt{s^2+\beta^2}}$	$-\xi_1\sqrt{s^2+\beta^2}-\xi_2 s$	
$(1,2)$	$\dfrac{\Delta(\sqrt{s^2+\beta^2}-s)}{(\Gamma_1\sqrt{s^2+\beta^2}+\Gamma_2 s)\sqrt{s^2+\beta^2}}$	$\eta_1\sqrt{s^2+\beta^2}-\eta_2 s$	$\left.\begin{array}{c} \\ \\ \end{array}\right\}B_g$
$(2,1)$	$\dfrac{\Delta(-\sqrt{s^2+\beta^2}-s)}{(\Gamma_1\sqrt{s^2+\beta^2}-\Gamma_2 s)\sqrt{s^2+\beta^2}}$	$-\eta_1\sqrt{s^2+\beta^2}-\eta_2 s$	

The notations used are defined as follows:

$$\xi_1 = \alpha t \tag{IV.1a}$$

$$\xi_2 = x - \alpha t \tag{IV.1b}$$

$$\eta_1 = \alpha t - \alpha t_1\left(\frac{\gamma_0}{\gamma_0'}+\frac{\gamma_g}{\gamma_g'}\right) \tag{IV.2a}$$

$$\eta_2 = x - \alpha t + \alpha t_1\left(\frac{\gamma_0}{\gamma_0'}-\frac{\gamma_g}{\gamma_g'}\right) \tag{IV.2b}$$

$$\Delta_1 = \frac{\gamma_0}{\gamma_0'}+\frac{\gamma_g}{\gamma_g'}e^{i\delta} \tag{IV.3a}$$

$$\Delta_2 = \frac{\gamma_0}{\gamma_0'} - \frac{\gamma_g}{\gamma_g'}e^{i\delta} \tag{IV.3b}$$

$$\Delta = \frac{\gamma_g}{\gamma_g'}(1 - e^{i\delta}) \tag{IV.3c}$$

$$\Gamma_1 = \frac{\gamma_0}{\gamma_0'} + \frac{\gamma_g}{\gamma_g'} \tag{IV.4a}$$

$$\Gamma_2 = \frac{\gamma_0}{\gamma_0'} - \frac{\gamma_g}{\gamma_g'} \tag{IV.4b}$$

APPENDIX V

Crystal Wave Fields (b)

$$\varphi_g^{ij} = D \cdot \left[F(s)\sqrt{\frac{2\pi}{|G''(s)|}} \right] \exp i\left\{ [G(s)] + \text{sign}[G''(s)]\frac{\pi}{4} \right\}$$

(i,j)	$[s]$	$[G(s)]$	$\text{sign}[G''(s)]$	$\left[F(s)\sqrt{\dfrac{2\pi}{|G''(s)|}}\right]$	wave field
$(1,1)$	$\dfrac{\beta\xi_2}{\sqrt{\xi_1^2 - \xi_2^2}}$	$\beta\sqrt{\xi_1^2 - \xi_2^2}$	$+$	$\dfrac{\sqrt{2\pi}(\Delta_1\xi_1 + \Delta_2\xi_2)}{\Gamma_1\xi_1 + \Gamma_2\xi_2} \dfrac{1}{(\beta\sqrt{\xi_1^2 - \xi_2^2})^{\frac{1}{2}}}$	A_g
$(2,2)$	$\dfrac{-\beta\xi_2}{\sqrt{\xi_1^2 - \xi_2^2}}$	$-\beta\sqrt{\xi_1^2 - \xi_2^2}$	$-$	$-\dfrac{\sqrt{2\pi}(\Delta_1\xi_1 + \Delta_2\xi_2)}{\Gamma_1\xi_1 + \Gamma_2\xi_2} \dfrac{1}{(\beta\sqrt{\xi_1^2 - \xi_2^2})^{\frac{1}{2}}}$	
$(1,2)$	$\dfrac{\beta\eta_2}{\sqrt{\eta_1^2 - \eta_2^2}}$	$\beta\sqrt{\eta_1^2 - \eta_2^2}$	$+$	$\dfrac{\sqrt{2\pi}\Delta(\eta_1 - \eta_2)}{\Gamma_1\eta_1 + \Gamma_2\eta_2} \dfrac{1}{(\beta\sqrt{\eta_1^2 - \eta_2^2})^{\frac{1}{2}}}$	B_g
$(2,1)$	$\dfrac{-\beta\eta_2}{\sqrt{\eta_1^2 - \eta_2^2}}$	$-\beta\sqrt{\eta_1^2 - \eta_2^2}$	$-$	$-\dfrac{\sqrt{2\pi}\Delta(\eta_1 - \eta_2)}{\Gamma_1\eta_1 + \Gamma_2\eta_2} \dfrac{1}{(\beta\sqrt{\eta_1^2 - \eta_2^2})^{\frac{1}{2}}}$	$(\eta_1 > 0)$
$(1,2)$	$\dfrac{-\beta\eta_2}{\sqrt{\eta_1^2 - \eta_2^2}}$	$-\beta\sqrt{\eta_1^2 - \eta_2^2}$	$-$	$\dfrac{\sqrt{2\pi}\Delta(\eta_1 - \eta_2)}{\Gamma_1\eta_1 + \Gamma_2\eta_2} \dfrac{1}{(\beta\sqrt{\eta_1^2 - \eta_2^2})^{\frac{1}{2}}}$	B_g
$(2,1)$	$\dfrac{\beta\eta_2}{\sqrt{\eta_1^2 - \eta_2^2}}$	$\beta\sqrt{\eta_1^2 - \eta_2^2}$	$+$	$-\dfrac{\sqrt{2\pi}\Delta(\eta_1 - \eta_2)}{\Gamma_1\eta_1 + \Gamma_2\eta_2} \dfrac{1}{(\beta\sqrt{\eta_1^2 - \eta_2^2})^{\frac{1}{2}}}$	$(\eta_1 < 0)$

APPENDIX VI

Wave Vectors and Amplitudes

(1) *Wave Vectors.* Under two-beam approximations, the fundamental equation determining the crystal waves are represented by

$$\begin{bmatrix} (\mathbf{k}_0^i)^2 - K^2(1 + \psi_0) & K^2\psi_{\bar{g}}\sin\chi \\ K^2\psi_g\sin\chi & (\mathbf{k}_g^i)^2 - K^2(1 + \psi_0) \end{bmatrix} \begin{bmatrix} d(_0^i) \\ d(_g^i) \end{bmatrix} = 0 \tag{VI.1}$$

The dispersion equation is given by this secular equation. By the tangential continuity of wave vectors on the entrance surface, the wave vectors in region I are represented by equations (II.1a and b). Substituting these into the dispersion equation, we have

$$(2\gamma_0 A^i - K\psi_0)(2\gamma_g A^i - K\psi_0 + 2K_z \sin 2\theta_B) = K^2 \psi_g \psi_{\bar{g}}; \sin^2\chi \qquad (\text{VI.2})$$

neglecting small quantities of higher order than $(A^i)^2$ and $(K_z)^2$. From this, A^i is given by equations (II.2a and b).

In region II, the wave vectors are represented by equations (III.1a, b, c, and d) from the tangential continuity of the wave vectors at the fault plane. Substituting these again into the dispersion equation, we have

$$(2\gamma_0 A^i + 2\gamma_0' A^{ij} - K\psi_0)(2\gamma_g A^i + 2\gamma_g' A^{ij} - K\psi_0 + 2K_z \sin 2\theta_B) = K^2 \psi_g \psi_{\bar{g}}; \sin^2\chi$$
$$(\text{VI.3})$$

under similar approximations to those used in deriving equation (VI.2). Combining (VI.2) and (VI.3) we obtain A^{ij} as follows:

$$A^{ij} = 0 \qquad (\text{VI.4a})$$

or

$$A^{ij} = -(\gamma_0/\gamma_0')(\alpha s \pm \alpha\sqrt{s^2 + \beta^2}) - (\gamma_g/\gamma_g')(-\alpha s \pm \alpha\sqrt{s^2 + \beta^2}) \quad (\text{VI.4b})$$

where the upper and the lower indices correspond to the cases in which the superscript i implies the branches (1) and (2), respectively. Solution (VI.4a) corresponds to the case where the dispersion point in region II is identical to the dispersion point in region I, whereas solution (VI.4b) implies the case where the dispersion point jumps at the fault plane.

(2) *Amplitudes.* The amplitudes $d\binom{i}{0}$ and $d\binom{i}{g}$ for unit amplitude of the incident wave are obtained from the boundary conditions on the entrance surface, which are given by

$$d\binom{1}{0} \exp i(\mathbf{k}_0^1 \cdot \mathbf{r}_e) + d\binom{2}{0} \exp i(\mathbf{k}_0^2 \cdot \mathbf{r}_e) = \exp i(\mathbf{K} \cdot \mathbf{r}_e) \qquad (\text{VI.5a})$$

$$d\binom{1}{g} \exp i(\mathbf{k}_g^1 \cdot \mathbf{r}_e) + d\binom{2}{g} \exp i(\mathbf{k}_g^2 \cdot \mathbf{r}_e) = 0 \qquad (\text{VI.5b})$$

The ratio c^i of the amplitude $d\binom{i}{g}$ to the amplitude $d\binom{i}{0}$ is determined by equation (VI.1). The results are given in equations (II.4a and b). Substituting equations (4a and b) into the above, we see that

$$C_0^1 + C_0^2 = 1 \qquad (\text{VI.6a})$$

$$C_g^1 + C^2 = 0 \qquad (\text{VI.6b})$$

The latter can be rewritten as

$$c^1 C_0^1 + c^2 C_0^2 = 0 \qquad (\text{VI.6b}')$$

From equations (VI.6a and b′) we obtain C_0^i and C_g^i immediately, as shown in equations (II.3a, b, c, and d).

In a similar way, the amplitudes $d\binom{ij}{00}$, $d\binom{ij}{0g}$, $d\binom{ij}{g0}$, and $d\binom{ij}{gg}$ are obtained from the boundary conditions at the fault plane. The conditions are represented by

$$d(^{i1}_{00}) \exp i(\mathbf{k}^{i1}_{00} \cdot \mathbf{r}_f) + d(^{i2}_{00}) \exp i(\mathbf{k}^{i2}_{00} \cdot \mathbf{r}_f) = d(^i_0) \exp i(\mathbf{k}^i_0 \cdot \mathbf{r}_f) \quad \text{(VI.7a)}$$

$$d(^{i1}_{0g}) \exp i(\mathbf{k}^{i1}_{0g} \cdot \mathbf{r}_f) + d(^{i2}_{0g}) \exp i(\mathbf{k}^{i2}_{0g} \cdot \mathbf{r}_f) = 0 \quad \text{(VI.7b)}$$

$$d(^{i1}_{g0}) \exp i(\mathbf{k}^{i1}_{g0} \cdot \mathbf{r}_f) + d(^{i2}_{g0}) \exp i(\mathbf{k}^{i2}_{g0} \cdot \mathbf{r}_f) = 0 \quad \text{(VI.7c)}$$

$$d(^{i1}_{gg}) \exp i(\mathbf{k}^{i1}_{gg} \cdot \mathbf{r}_f) + d(^{i2}_{gg}) \exp i(\mathbf{k}^{i2}_{gg} \cdot \mathbf{r}_f) = d(^i_g) \exp i(\mathbf{k}^i_g \cdot \mathbf{r}_f) \quad \text{(VI.7d)}$$

Substituting equations (4a, b), (6a, b), and (7a, b) into the above, we have

$$C^{i1}_{00} + C^{i2}_{00} = C^i_0 \quad \text{(VI.8a)}$$

$$C^{i1}_{0g} + C^{i2}_{0g} = 0 \quad \text{(VI.8b)}$$

$$C^{i1}_{g0} + C^{i2}_{g0} = 0 \quad \text{(VI.8c)}$$

$$C^{i1}_{gg} + C^{i2}_{gg} = C^i_g \quad \text{(VI.8d)}$$

In deriving these equations, the relations (III.1b and d) are used. Now, we define the amplitude ratios

$$c^{ij} = d(^{ij}_{0g})/d(^{ij}_{00}) = d(^{ij}_{gg})/d(^{ij}_{g0}). \quad \text{(VI.9)}$$

Using the above, equations (VI.8b and d) can be rewritten as

$$c^{i1}C^{i1}_{00} + c^{i2}C^{i2}_{00} = 0 \quad \text{(VI.8b')}$$

$$c^{i1}C^{i1}_{g0} + c^{i2}C^{i2}_{g0} = C^i_g \quad \text{(VI.8d')}$$

From equations (VI.8a, b', c, and d'), the quantities C^{ij}_{00} etc. are obtained as listed in (III.3).

From the fundamental equation in region II, which is similar to equation (VI.1), the amplitude ratios c^{ij} are given by

$$c^{ij} = \frac{(\mathbf{k}^{ij}_{00})^2 - K^2(1 + \psi_0)}{K^2 \psi_{\bar{g}} e^{-i\delta} \sin \chi} \quad \text{(VI.9)}$$

Now, in region II, the wave vector $\mathbf{k}^i_0$ should be replaced by $\mathbf{k}^{ij}_{00}$ and the $\bar{g}$-th Fourier coefficient must be multiplied by the phase factor $e^{-i\delta}$. From equation (VI.9), and the wave vector $\mathbf{k}^{ij}_{00}$ given by equation (III.1a) and equations (III.2a, b, and c), we obtain

$$c^{ij} = \frac{\sin 2\theta_B}{K \psi_{\bar{g}} \sin \chi}(s \pm \sqrt{s^2 + \beta^2}) \frac{\gamma_0}{\gamma_g} e^{i\delta} \quad (i = j) \quad \text{(VI.10a)}$$

$$c^{ij} = \frac{\sin 2\theta_B}{K \psi_{\bar{g}} \sin \chi}(s \mp \sqrt{s^2 + \beta^2}) \frac{\gamma'_0}{\gamma'_g} e^{i\delta} \quad (i \neq j) \quad \text{(VI.10b)}$$

where the upper and the lower indices correspond to the cases in which the superscript i specifies the branches (1) and (2), respectively. With the use of the expressions for c^i [equations (II.4a and b)], the amplitude ratios c^{ij} can be written as in equations (III.4a, b, c, and d).

DISCUSSION

P. Penning (Philips Research Laboratories): Is absorption included in these calculations?
E. Sickafus: No.

DYNAMICAL THEORY FOR SIMULTANEOUS X-RAY DIFFRACTION

P. Penning

Philips Research Laboratories
Eindhoven, The Netherlands

ABSTRACT

Complete dynamical solutions for three coupled plane-wave components in crystal structures with inversion symmetry have been found. After reviewing briefly the dynamical solutions for wave fields with two coupled plane-wave components, the results for the three-beam case are discussed in qualitative terms. Attention is paid to singular points and lines on the ω-surface, and to the attenuation of the mode-intensity because of absorption. The most surprising result is that in the case one of the reflections is forbidden (Umweganregung) the absorption is reduced in comparison with the adjoining two-beam cases. Experimental data are in reasonable agreement with the theory. Quantitative data are presented for a few three-beam cases of simultaneous diffraction of Cu $K\alpha$ radiation in germanium.

INTRODUCTION

X-ray diffraction is an interference phenomenon. An incident, monochromatic plane-parallel wave is scattered by the atoms by a very small amount. It is easily calculated that one single layer of atoms reflects the incident wave for a fraction of the order of 10^{-5}. A reasonable reflected intensity is obtained only when thousands of layers contribute. Because of this large number, the phase of all scattered waves has to be the same within a high degree of accuracy, resulting in a very narrow range of directions of incidence for which diffraction with a reasonable amplitude takes place. Apart from this small divergence the direction for constructive interference is given in the well-known Bragg equation. In the dynamical theory it is usually given in a different form (see Fig. 1):

$$\mathbf{k}_2 = \mathbf{k}_1 + 2\mathbf{b}_{12} \tag{1}$$

The vectors $\mathbf{k}_1$ and $\mathbf{k}_2$ represent the wave vectors of the waves in vacuum that satisfy the Bragg condition exactly, and $2\mathbf{b}_{12}$ the reciprocal-lattice vector for the set of planes that reflect. With d the spacing of these planes, $|2\mathbf{b}_{12}| = 2\pi/d$. The reflecting planes are perpendicular to $\mathbf{b}_{12}$. A large number of vectors $\mathbf{k}_1$ satisfy equation (1), *viz.* all those lying on a circular cone with $\mathbf{b}_{12}$ as axis. It is conceivable that some of these vectors also satisfy the Bragg condition for another set of reflecting planes, characterized by the reciprocal-lattice vector $2\mathbf{b}_{13}$. Then a third wave is generated because of diffraction, with wave vector $\mathbf{k}_3$:

$$\mathbf{k}_3 = \mathbf{k}_1 + 2\mathbf{b}_{13}$$

In this situation there is simultaneous diffraction on two sets of reflecting planes. The situation is sketched in Figure 2. Since both vectors $2\mathbf{b}_{12}$ and $2\mathbf{b}_{13}$ are reciprocal-lattice vectors, the vector $2\mathbf{b}_{23}$, connecting $\mathbf{k}_2$ and $\mathbf{k}_3$, is one also, indicating that the diffracted

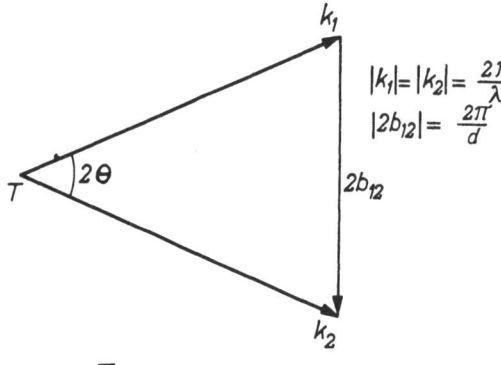

$$|k_1| = |k_2| = \frac{2\pi}{\lambda}$$
$$|2b_{12}| = \frac{2\pi}{d}$$

Figure 1. An incident wave with wave vector $\mathbf{k}_1$ satisfies exactly the Bragg condition for a set of reflecting planes, characterized by the reciprocal-lattice vector $2\mathbf{b}_{12}$, if $|\mathbf{k}_2| = |\mathbf{k}_1 + 2\mathbf{b}_{12}|$ is exactly equal to $|\mathbf{k}_1|$.

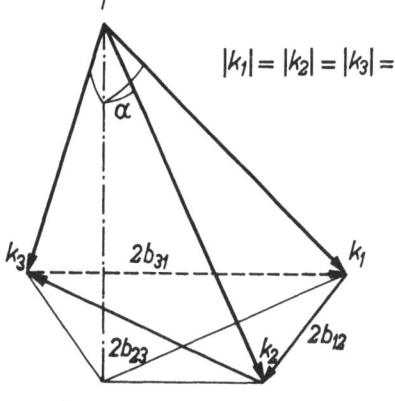

$$|k_1| = |k_2| = |k_3| = \frac{2\pi}{\lambda}$$

Figure 2. The condition for simultaneous diffraction on two sets of reflecting planes, characterized by the reciprocal-lattice vectors $2\mathbf{b}_{12}$ and $2\mathbf{b}_{31}$, respectively, is satisfied exactly if $|\mathbf{k}_2|$ and $|\mathbf{k}_3|$ are equal to $|\mathbf{k}_1|$. Note that the waves with wave vectors $\mathbf{k}_2$ and $\mathbf{k}_3$ satisfy exactly the Bragg condition for reflection on a third set of planes, characterized by $2\mathbf{b}_{23}$.

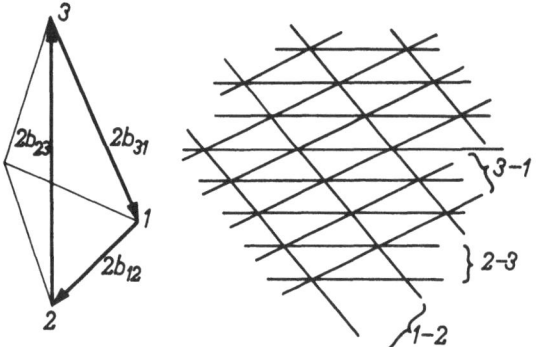

Figure 3. In the three-beam case three sets of reflecting planes play a part. On the left-hand side, the relation between the three reciprocal-lattice vectors is shown. Their sum is zero. On the right-hand side, the orientation and spacing of the planes is shown in a projection on the plane through the reciprocal-lattice vectors. The reflecting planes have lines normal to the plane of drawing in common.

waves 2 and 3 satisfy exactly the Bragg condition for a third set of planes. The Miller indices of this third set are easily found from the condition that the sum of the Miller indices for all three must be zero, for example, (111) (111) (200). For an arbitrary case the reciprocal-lattice vectors and the orientation of the corresponding reflecting planes are shown in Fig. 3. The three reflecting planes have in common the line perpendicular to the plane through $2\mathbf{b}_{12}$, $2\mathbf{b}_{23}$, and $2\mathbf{b}_{31}$. Before considering a particular case one must first determine whether one deals with a three-beam case or whether more beams play a part. Quite often, once having satisfied the Bragg condition for three sets of planes, the Bragg condition is satisfied automatically for more sets of reflecting planes.

Although we have obtained the dynamical solutions for the general three-beam case, we shall present here only the qualitative considerations and a few numerical results for simultaneous diffraction of Cu K_α radiation in germanium. The derivation will be published elsewhere.[1]

To understand the results of the three-beam case, the two-beam case is first briefly reviewed.

TWO-BEAM CASE[2,3,4]

If there is interaction with only one set of reflecting planes, the wave field consists of two predominant plane-wave components. The wave vectors of these two waves are rigidly coupled. If we denote the wave vector of the component traveling almost parallel to $\mathbf{k}_1$* by $\mathbf{k}$, the wave vector of the other component is given by $\mathbf{k}+2\mathbf{b}_{12}$. Therefore, it is sufficient to characterize the entire wave field with only one wave vector. The dynamical theory tries to find solutions for $\mathbf{k}$ with a given vacuum wavelength λ. Excluding absorption and exponentially damped modes because of extinction, we find that there is a limited number of solutions. For any allowed value of $\mathbf{k}$ there is one specific wave field, the *mode of propagation*. The allowed values of $\mathbf{k}$ together form the *ω-surface*; in this case it consists of four branches. Figure 4 gives a cross section of the ω-surface with the plane through $\mathbf{k}_1$ and $\mathbf{k}_2$. The full ω-surface is obtained by rotating the figure around a line parallel to $\mathbf{b}_{12}$ through point T of Figure 1. The multiple-beam cases that are generally present are excluded. The large magnification in scale with respect to Figure 1, is in agreement with the remark in the introduction that diffraction takes place only over a very small angle.

The mathematical treatment of the two-beam case is simplified greatly because two principal directions of polarization can be distinguished (σ and π). In the σ-polarization, the dielectric displacement vector $\mathbf{D}$ lies in the reflecting planes (perpendicular to the plane of drawing and parallel for both waves). In the π-polarization, $\mathbf{D}$ lies in the plane of drawing. A σ-polarized mode does not interact with a mode in π-polarization.

The amplitudes of the two plane-wave components have the following properties: In the first place, they are either in phase or in antiphase at the reflecting planes. Secondly, their ratio ξ, indicated in Figure 4, depends on the deviation from the exact Bragg angle. Assuming a crystal structure with inversion symmetry, the ratio is $+1$ (in phase) or -1 (in antiphase) if the Bragg condition is satisfied exactly. Far off the Bragg angle it is

_____

* The possible choice $\mathbf{k}_2$ works equally well.

Figure 4. Cross section of the ω-surface in the two-beam case with an arbitrary plane through point T of Figure 1 parallel to $\mathbf{b}_{12}$. It is a large magnification of the region around the end-point of vector $\mathbf{k}_1$. The full ω-surface, except regions of multiple diffraction, is obtained by rotating the figure around an axis parallel to $\mathbf{b}_{12}$ through point T, lying far outside the drawing to the left. For modes in σ- and π-polarization, the dielectric displacement vector is perpendicular and parallel to the plane of drawing respectively. The parameter ξ is the ratio in amplitude of the plane-wave components traveling parallel to $\mathbf{k}_2$ and $\mathbf{k}_1$.

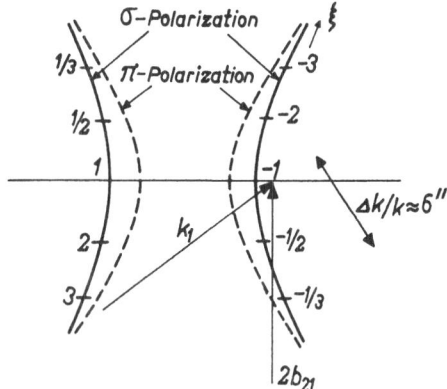

either zero (wave with wave vector $\mathbf{k}_1$ predominant) or infinite (wave with wave vector $\mathbf{k}_2$ predominant). It is interesting to consider the magnitude squared of the total dielectric displacement as a function of position for the case of equal amplitude and σ-polarization (Figure 5). When they are in antiphase the total amplitude is zero at the reflecting planes. When they are in phase the amplitude squared at the reflecting planes is maximum and twice the average value. In the π-polarization, the dielectric displacement vectors for the two components are not parallel and cannot cancel, even for $\xi = -1$, thus leaving a finite amplitude at the reflecting planes. Moving $\mathbf{k}$ out of the Bragg angle results in a decreasing value of maximum over minimum. Far off the Bragg angle, $|\mathbf{D}|^2$ is independent of place.

Let us now introduce absorbing matter, located near the nucleus of the atoms. When all atoms lie in the reflecting planes, as is the case for the (220)-reflection in germanium, the absorption is very low for modes in the σ-polarization with $\xi = -1$. Because the absorbing matter is spread out a little and the atoms vibrate, the absorption is not zero, but very close to it. If we consider a crystal slab with surfaces perpendicular to the reflecting planes (symmetrical Laue case) the relative decrease in intensity of a mode per unit thickness K is given by

$$\frac{K(\cos\theta)}{\mu_0} = 1 - \epsilon_{12}$$

where θ is the Bragg angle, μ_0 is the absorption coefficient if no diffraction takes place,

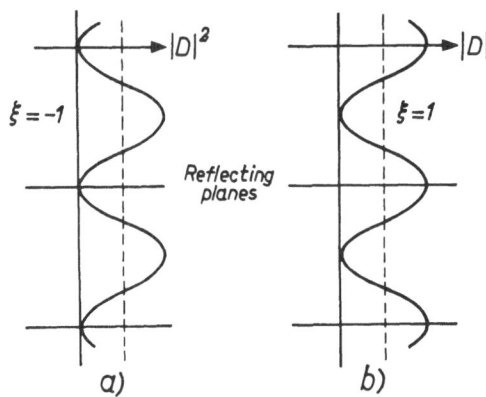

Figure 5. The local value of the total amplitude squared of the dielectric displacement for modes of propagation in the σ-polarization with $\xi = \pm 1$. For $\xi = -1$, the amplitude squared is zero in the reflecting planes. For $\xi = +1$, it is maximum there and twice the average value indicated by the dashed line.

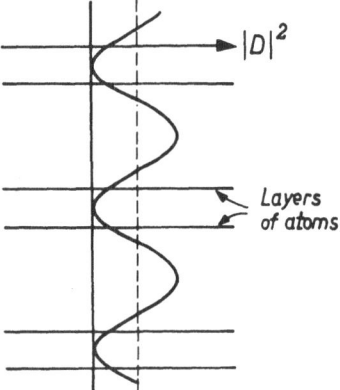

Figure 6. Similar to Figure 5a but now for the (111)-reflection in the diamond lattice where the atoms are arranged in double layers.

and ϵ_{12} is a parameter characteristic for the reflection. Okkerse[5] found experimentally $1 - \epsilon_{220}$ to be 0.0408. The attenuation coefficient K is, hence, much smaller than the decrease in intensity because of normal absorption—the Borrmann effect.

Not all reflections in the diamond lattice show such a strong Borrmann effect. In the case where the Miller indices are odd we have a situation similar to the one shown in Figure 6, corresponding to a (111)-reflection. Note that now the atoms do not lie in the reflecting planes (passing through the inversion center) but in double layers. For the mode in σ-polarization with $\xi = -1$, there are still nodes with zero amplitude at the reflecting planes. In the layers of atoms, the amplitude squared is finite, although below the average. Hence the absorption is reduced ($1 - \epsilon \approx 0.3$), but not so strongly as in the case of the diffraction on planes with even Miller indices.

THREE-BEAM CASE, THREE NONZERO STRUCTURE FACTORS, NO ABSORPTION

In the three-beam case there are three predominant plane-wave components, each with its own direction of polarization. The main difficulty in treating this case is caused by the fact that separation in main directions of polarization is only possible in very special situations. James in his paper on the dynamical theory[3] treats such a situation, *viz.* the case where the three wave vectors lie in one plane. The main directions of polarization are now perpendicular to and in the plane. In the general case, there are three plane-wave components, each with two directions of polarization and hence six relevant amplitudes. The dynamical theory leads to a secular problem, involving the solution of a six by six determinant. We found a way to solve the secular problem explicitly for crystal structures with inversion symmetry, where all structure factors are real parameters. In this paper only some relevant results will be discussed. For the derivation of the results, the reader is referred to a forthcoming paper.[1]

The interesting region is located closely around $\mathbf{k}_1$,* within a sphere with radius a few times $10^{-4}/\lambda$. A three-dimensional model of the ω-surface has been built in the laboratory, but unfortunately it is not easy to make a good drawing of the six branches that brings out all its characteristics. The best thing to do is to make a suitable cross section for a simple case. As an example, we take the case of simultaneous reflection of Cu K_α radiation on three (220)-planes in germanium, and as plane of section the plane through $\mathbf{k}_1$ perpendicular to $\mathbf{b}_{23}$ and hence parallel to the set of planes 2-3 (see Figure 7). First consider the asymptotic values of $\mathbf{k}$ in this plane, far away from the point where the three-beam case prevails. One possible solution is a mode of propagation with only one predominant plane-wave component, *viz.* the one that travels parallel to $\mathbf{k}_1$. This wave suffers no diffraction, since it is far off the Bragg angle for both sets of planes 1-2 and 1-3. Its $\mathbf{k}$-values lie on a straight line perpendicular to $\mathbf{k}_1$: the asymptote E in Figure 7. It is a double asymptote, since the polarization of the wave is irrelevant. Other limiting possibilities follow from the observation that for all modes in the plane section, the waves traveling parallel to $\mathbf{k}_2$ and $\mathbf{k}_3$, respectively, satisfy exactly the Bragg condition for the set of planes 2-3. Far from the region of simultaneous diffraction, modes consisting of waves 2 and 3 must exist. From Figure 4 it follows that in the plane where the Bragg condition is exactly satisfied, the ω-surface gives four lines—two for the σ-polarization and two for the π-polarization. These four lines appear in the cross section as the four asymptotes A, D, B, and C. The first two correspond to modes in the σ-polarization, the last two to modes in the π-polarization.

* Identical solutions lie in the regions around $\mathbf{k}_2$ and $\mathbf{k}_3$.

The solution for the three-beam case must indicate how these six limiting values are connected through the region of simultaneous diffraction. The result is shown in Figure 7 also. Several characteristics can be seen. In point P, three branches intersect. The asymptote D is part of the ω-surface. In Q, two branches intersect. It must be remarked that Saccocio and Zajac,[6] who treated this highly symmetrical case, also found the solutions in P and Q, together with the composition of the wave fields in these points.

The general case with three different sets of reflecting planes, and with the cross section still perpendicular to $\mathbf{b}_{23}$, is very similar. Only the point Q with two branches intersecting does not lie in the cross section but somewhere else on the ω-surface. There, the two branches have one point in common. In the immediate vicinity, the ω-surface

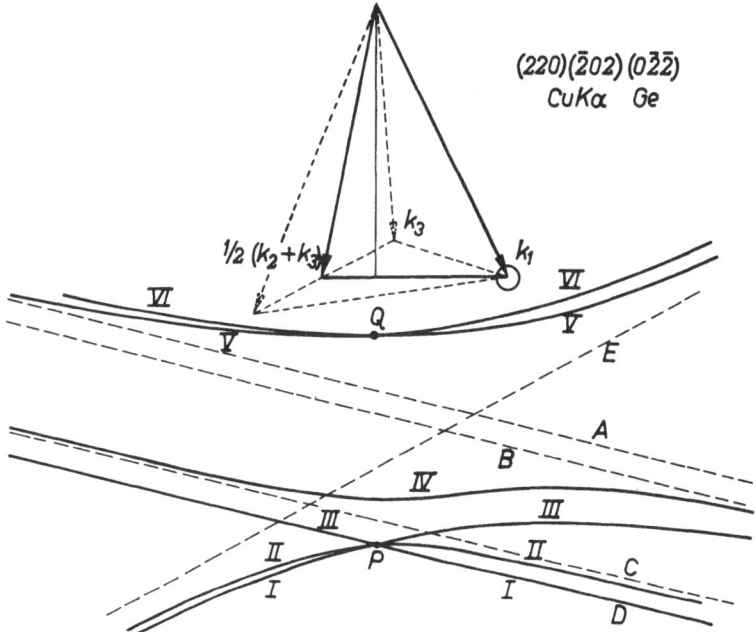

Figure 7. Cross section of the ω-surface for simultaneous diffraction of Cu K_α radiation on three different sets of (220)-planes in germanium, with a plane through $\mathbf{k}_1$ perpendicular to $\mathbf{b}_{23}$. Note that there appears in this cross section a point with threefold degeneracy P, a point of twofold degeneracy Q, and a straight line D.

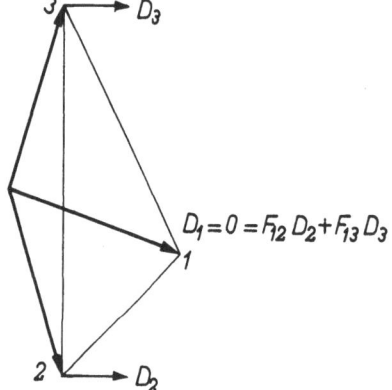

Figure 8. For modes of propagation with wave vectors on the line D of Figure 7, the plane-wave components 2 and 3 form a two-beam pair in the σ-polarization: $\mathbf{D}_2$ and $\mathbf{D}_3$ are normal to the plane through $\mathbf{k}_2$ and $\mathbf{k}_3$. Although the waves 2 and 3 are scattered both in the direction 1, with amplitudes proportional to the appropriate structure factor F, the total amplitude of wave 1 is zero because the scattered waves just cancel.

degenerates into a cone. The straight line D is still present. The modes of propagation on this line correspond to a two-beam pair in the σ-polarization with waves traveling parallel to $\mathbf{k}_2$ and $\mathbf{k}_3$, while the wave traveling parallel to $\mathbf{k}_1$ is absent (see Figure 8). That such modes are possible throughout the entire region of simultaneous diffraction can be concluded from the following argument. We imagine that the two-beam pair is present with amplitudes $\mathbf{D}_2$ and $\mathbf{D}_3$, respectively. In the region of simultaneous diffraction, both waves 2 and 3 are scattered in direction 1 because the Bragg condition for the sets of planes 1-2 and 1-3 is satisfied here. According to the kinematical theory, the amplitude of the scattered wavelets is proportional to the structure factor for the reflection F and the incident amplitude. The total amplitude of the wavelets scattered in direction 1 is therefore proportional to $F_{12}\mathbf{D}_2 + F_{13}\mathbf{D}_3$. Hence $\mathbf{D}_1$ is equal to zero if $F_{12}\mathbf{D}_2 = -F_{13}\mathbf{D}_3$. Since the structure factors are different in the general case, the value of ξ for the two-beam pair is not equal to -1. Accordingly, the wave vector must lie a fixed distance from the plane through the end-point of $\mathbf{k}_1$ perpendicular to $\mathbf{b}_{23}$, where the Bragg condition is satisfied exactly (see Figure 4). Two other solutions very similar to the one found now are obtained by cyclic interchange of the indices. One group consists of a two-beam pair in the σ-polarization of waves 1 and 3, with the wave traveling parallel to $\mathbf{k}_2$ absent. Their wave vectors lie in a plane perpendicular to $\mathbf{b}_{13}$. The third group has a σ-polarized two-beam pair of waves 1 and 2, with wave 3 absent. Their wave vectors lie on a straight line in the plane perpendicular to $\mathbf{b}_{12}$. The three straight lines intersect in one point that apparently shows threefold degeneracy. In this point, any combination of the three modes mentioned above is a possible solution.

ABSORPTION, THREE NONZERO STRUCTURE FACTORS

Absorption is introduced as a minor correction in the wave behavior by adding small imaginary parts to the real structure factors. For simplicity, only plane-parallel crystal slabs are considered, with surfaces normal to the three sets of reflecting planes. The relative decrease in intensity of a single mode per unit thickness is denoted by K. It is convenient to use as a parameter, $K(\cos \alpha)/\mu_0$, with α defined as the angle between the wave vector of any plane-wave component and the surface normal (see Figure 2). The value of this parameter is unity far off any Bragg angle. In Figure 9, results are shown for the symmetrical case of three equal structure factors. The calculations are based on diffraction of Cu K_α radiation on (220)-planes in germanium. Only the modes with $\mathbf{k}$-vectors given in Figure 7 are considered. Each branch in that figure gives its own curve here. The Roman numerals correspond in the two figures. In the point of twofold degeneracy the absorption is high and more than twice that without diffraction. The upper curve consists of two lines crossing over on the vertical axis, but the difference is so small that it could not be brought out in the figure. The lowest absorption occurs along the straight line D—which is not surprising since the modes consist here of a two-beam mode in the σ-polarization exactly at the Bragg angle. The value of $K(\cos \alpha)/\mu_0$ is equal to the minimum in attenuation for a single reflection on (220)-planes. In the point of three-fold degeneracy, three such modes are possible and the three curves coincide accordingly.

In less symmetrical cases, the three modes on the three different straight lines have unequal absorption coefficients and all three are higher than the minimum value obtainable in the respective two-beam cases. The absolute minimum has to be determined for each three-beam case separately. It seems probable that the lowest value is obtained in one of the limiting two-beam cases and that the minimum value in the point of three-fold degeneracy is equal to or exceeds the absolute minimum. Experimental evidence obtained by Borrmann and Hartwig[7] indicates that the difference, if present, is small.

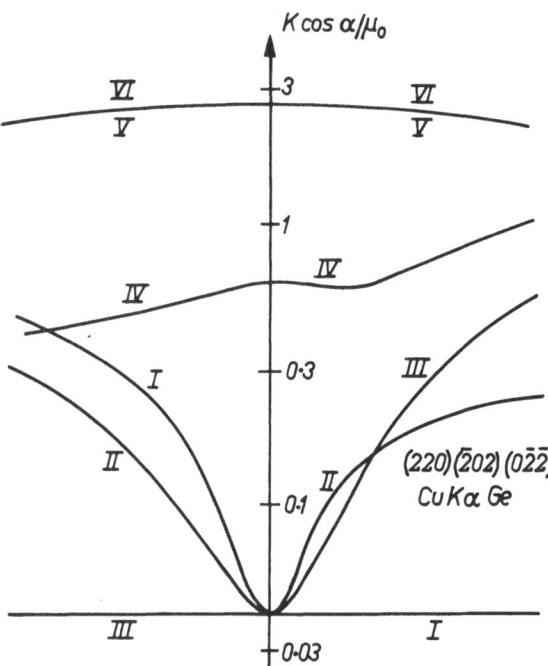

Figure 9. The attenuation coefficient K for the modes shown in Figure 7. The two curves V and VI lie so close to each other that they have been drawn as one line. The Roman numerals refer to the numbering of the branches in Figure 7. The data used to calculate the curves were obtained from Okkerse.[5] Figures 7 and 9 have the same horizontal axis.

THREE-BEAM CASE, ONE ZERO STRUCTURE FACTOR

Let us now consider the case that one structure factor is zero. Renninger[8] has shown that this situation is particularly interesting, because of the possibility of "Umweganregung"—the generation of a forbidden reflection. In the diamond lattice it is important from a dynamical point of view, because the "Umweganregung" must go there via reflections that show a weak Borrmann effect. For example, the forbidden (200)-reflection can be obtained via the reflections (111) and (111). Borrmann and Hartwig[7] investigated this particular case experimentally and observed a much more pronounced Borrmann effect in the three-beam case.

The ω-surface is much simpler now. The points of twofold and threefold degeneracy disappear. Instead, a straight line of twofold degeneracy is obtained. On this line, two of the six branches intersect. We choose F_{23} to be zero. The modes with $\mathbf{k}$-vectors on this straight line have no plane-wave component that travels in direction 1. With the presence of the two other plane-wave components, this is only possible if the wavelets scattered from these 2 waves into direction 1 just cancel. Hence $F_{12}\mathbf{D}_2 + F_{13}\mathbf{D}_3$ must be parallel to $\mathbf{k}_1$. The restriction that $\mathbf{D}_2$ and $\mathbf{D}_3$ form together a mode of propagation is no longer present, since $F_{23} = 0$ means that they are uncoupled. The direction of polarization for $\mathbf{D}_2$ may be chosen arbitrarily, indicating that the solution of the straight line is twofold degenerated.

Since Borrmann and Hartwig[7] obtained some numerical results in the case of simultaneous reflection of Cu K_α radiation on (111) and (111)-planes in germanium, with the third (200)-reflection forbidden, it is worthwhile to treat this case as an example. Figure 10 shows the cross section of the ω-surface with the plane of symmetry through $\mathbf{k}_1$ perpendicular to the reciprocal-lattice vector of the (200)-planes. The twofold-degenerated line lies in this plane and replaces the four parallel asymptotes in Figure 7. The other four branches are hyperbolas with, as asymptotes, the twofold degenerate line and the line

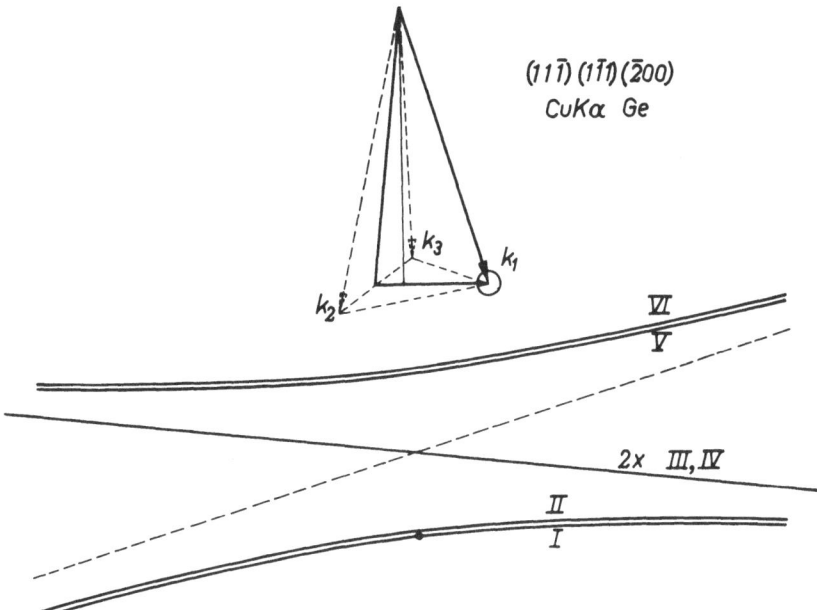

Figure 10. Similar to Figure 7 but now for simultaneous diffraction on $(11\bar{1})$-, $(1\bar{1}1)$-, and $(\bar{2}00)$-planes. The structure factor for the $(\bar{2}00)$-reflection is zero. The reciprocal-lattice vector for this reflection connects $\mathbf{k}_2$ and $\mathbf{k}_3$. Note the differences with Figure 7: the points of twofold and threefold degeneracy have disappeared, the straight line is double now.

perpendicular to $\mathbf{k}_1$. This last asymptote corresponds to modes with only one plane-wave component, *viz.* the one traveling parallel to $\mathbf{k}_1$.

Figure 11 gives the results of calculations on the absorption for the modes shown in Figure 10; the horizontal axis is the same as in the latter figure. The difference in absorption for the two upper branches is so small that it could not be shown in the figure. For the modes on the twofold-degenerated line, the absorption is normal, as if no diffraction takes place. For the modes on the lower branches the absorption is reduced. The most important result is that it drops well below the minimum value that is obtained in the limiting two-beam case with a (111)-reflection only. This result is in good qualitative agreement with Borrmann and Hartwig's measurements. Their rough estimate of the minimum value for $K(\cos \alpha)/\mu_0$ is 0.13, whereas we calculate 0.05. The difference can be partly explained by the fact that there are two modes with a rather weak absorption, whereas Borrmann and Hartwig assumed that there is only one. More important, however, is the effect of a decreasing "window" with increasing thickness, an effect neglected by the authors although they used a divergent incident beam. In Figure 12 are plotted lines of equal attenuation for modes on the branch with the lowest absorption. The projection of the wave vectors is on the plane through the three reciprocal-lattice vectors. For infinite thickness only the mode with minimum attenuation coefficient survives. But for thinner crystals, modes with a slightly higher absorption coefficient may reach the back-surface of the crystal with a reasonable amplitude. The transmitted intensity I therefore increases more strongly than expected from the minimum absorption coefficient alone, since a larger and larger part of the divergent incident beam contributes to I. The curve of log I *vs.* the thickness t must be curved upward. Determination of K from

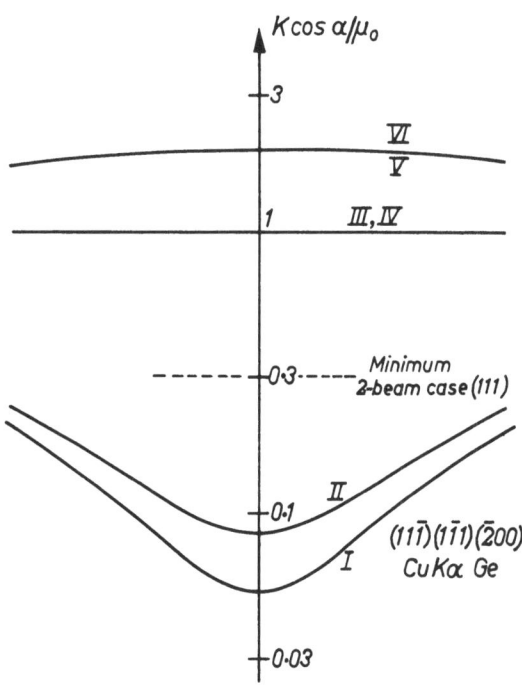

Figure 11. The attenuation coefficient for the modes shown in Figure 10. The horizontal axis is the same as in that Figure. The Roman numerals refer to the numbering of branches in Figure 10. Note that the minimum in attenuation is much lower than for reflection on (111)-planes alone.

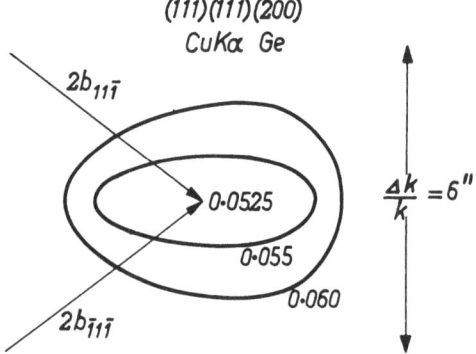

Figure 12. Lines of equal attenuation for modes on the branch with least absorption. The projection is on the plane through the reciprocal-lattice vectors. The numbers refer to the value of $K(\cos\alpha)/\mu_0$. The divergence in **k**-vector of the modes that suffer weak absorption is very small.

two points only, as was done by Borrmann and Hartwig, necessarily results in too high values.* Accurate determination of the minimum value of K is only possible by comparing the experimental results with calculated integrated intensities, taking into account the relation between K and the precise orientation of **k**, and the fact that two branches may contribute.

It is also possible to calculate the intensities as distributed over the three plane-wave components. For this special case and the mode that suffers least absorption, the intensities in the waves 1, 2, and 3 are in the proportion 2 : 1 : 1. For infinite thickness, the emerging beam traveling parallel to $\mathbf{k}_1$ is expected to be twice as intense as the emerging

---

* Okkerse,[5] while dealing with a two-beam case, took this "window" correction into account, because he did not calculate the attenuation coefficient from the curved plot log I vs. t, but from the straight line log $(I\sqrt{t})$ vs. t.

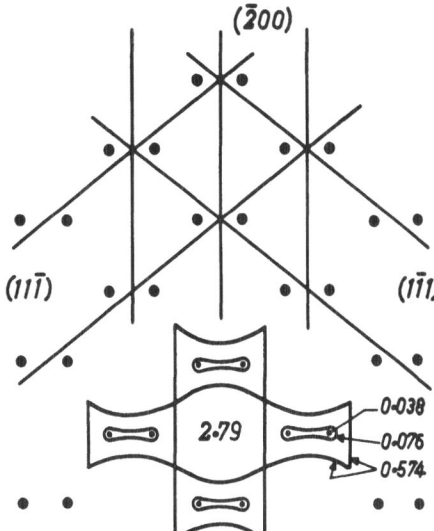

Figure 13. The projection of the diamond lattice on a (011)-plane and the lines of equal total amplitudes for the least damped mode excited by simultaneous diffraction of Cu K_α radiation on (11$\bar{1}$)-, (1$\bar{1}$1)-, and ($\bar{2}$00)-planes. Numbers refer to the ratio in the local and the averaged amplitude squared.

Table I. Minimum Value of the Attenuation Coefficient in Several Two- and Three-Beam Cases with One Structure Factor Equal to Zero

The attenuation coefficient K is the relative decrease in intensity per unit thickness in a crystal slab with surfaces normal to the reflecting planes. It is expressed in units of the absorption coefficient μ_0, when no diffraction takes place. The angle α is defined in Figure 2.

	Reflecting planes			$K(\cos \alpha)/\mu_0$
Two-beam cases	(220)			0.0408_5
	(111)			0.304
	(113)			0.332
Three-beam cases with one forbidden reflection	(1$\bar{1}$1)	(11$\bar{1}$)	($\bar{2}$00)	0.053
	(113)	(1$\bar{1}$3)	($\bar{2}$00)	0.092
	(113)	(11$\bar{1}$)	($\bar{2}\bar{2}\bar{2}$)	0.089
	(311)	($\bar{1}$13)	($\bar{2}\bar{2}\bar{2}$)	0.168
	($\bar{1}$3$\bar{1}$)	(1$\bar{1}$3)	(0$\bar{2}$4)	0.198

beams traveling parallel to $\mathbf{k}_2$ and $\mathbf{k}_3$ respectively. This is also in qualitative agreement with Borrmann and Hartwig's observations. The small discrepancies are probably the result of neglecting the "window" effect. It must be kept in mind that the ratios mentioned above depend also on the precise orientation of $\mathbf{k}$.

In Figure 13 it is demonstrated how this low absorption is brought about. The projection of the diamond lattice on a (011)-plane is shown, together with the orientation of the reflecting planes. In the lower part are drawn lines of equal value for the total amplitude squared. The numbers refer to the ratio in the local and the averaged value of the amplitude squared. It is minimum in the rows of atoms perpendicular to the plane of projection. Although the minimum value is not zero, it is very small (0.038). The regions with high electric fields are far away from the atoms. It is evident that such an amplitude distribution leads to a small absorption coefficient.

In Table I the results are given of calculations on other three-beam cases with one structure factor equal to zero, together with data concerning two-beam cases for comparison. The minimum values of $K(\cos \alpha)/\mu_0$ are given.

The following ratios in structure factors were used: $F(111)/F(000) = \pm 0.593$ and $F(113)/F(000) = \pm 0.483$. They are based on calculations by Berghuis *et al.*[9] The sign depends on the sign of the Miller indices.

ACKNOWLEDGMENT

The author is greatly indebted to Professor D. Polder. In interesting discussions the problem was brought to an elegant solution.

REFERENCES

1. P. Penning, to be published in *Philips Res. Repts.*; see also P. P. Ewald, *Z. Krist.* **A97**: 1, 1937.
2. B. W. Batterman and H. Cole, "Dynamical Diffraction of X-Rays by Perfect Crystals", *Rev. Mod. Phys.* **36**: 681, 1964.
3. R. W. James, "The Dynamical Theory of X-Ray Diffraction," *Solid State Physics* **15**: 55, 1963; G. Mayer, *Z. Krist.* **66**: 585, 1928; see also E. Lamla, *Ann. Physik* **36**: 194, 1939.
4. M. von Laue, *Röntgenstrahlinterferenzen*, Akademische Verlagsgesellschaft, Frankfurt a. M., 1960, 3rd ed.
5. B. Okkerse, "Anomalous transmission of X-rays in germanium. Part I; The imaginary part of the atomic scattering factor", *Philips Res. Repts.* **17**: 464, 1962.
6. E. J. Saccocio and A. Zajac, "Simultaneous Diffraction of X-rays and the Borrmann Effect", *Phys. Rev.* **139A**: 255, 1965.
7. G. Borrmann and W. Hartwig, "Die Absorption der Röntgenstrahlen im Dreistrahlfall der Interferenz", *Z. Krist.* **121**: 401, 1965.
8. M. Renninger, "Umweganregung, eine bisher unbeachtete Wechselwirkungserscheinung bei Raumgitterinterferenzen", *Z. Physik* **106**: 141, 1937.
9. J. Berghuis, G. M. Haanappel, M. Potters, B. O. Loopstra, C. H. MacGillavry, and A. L. Veenendaal, "New Calculations of Atomic Scattering Factors", *Acta Cryst.* **8**: 478, 1955.

DISCUSSION

H. K. Herglotz (E. I. duPont de Nemours): When you talk about absorption, do you always mean the so-called true absorption, the process causing ionization of inner atomic shells?

P. Penning: The absorption is introduced as a correction to the wave-field behavior in non-absorbing crystals by means of a complex dielectric constant. For strong reflections and for wavelengths not very close to an absorption edge, this method should be correct, independent of the cause of absorption.

H. K. Herglotz: Are Kossel lines explainable by your theory?

P. Penning: Yes.

D. K. Smith (Lawrence Radiation Lab.): What is the effect of asymmetry, taking germanium as the symmetric example, and then considering gallium arsenide, which would be an asymmetric set with almost the same intensities? Also, I might suggest that you consider a photographic stereo pair of the model of the dispersion surfaces that you will include in the manuscript. I think it would be very worthwhile in stereo.

P. Penning: As far as your first point is concerned, I didn't investigate this crystal structure without the inversion center in detail. There are changes, but I think that when the asymmetry is not too strong, the effect will not be very strong either. It is, of course, possible to make stereo pair photographs; however, whenever you have a model, you want to rotate it in your hands to get a more complete picture. The stereo pair will not help you in that, of course. Perhaps if you come to me afterward, I can show you the model and you can see how complicated the ω-surface is.

M. Renninger (Krist. Institut der Universität): Do you have an explanation of the fact reported by Borrmann that interaction between three different (220)-reflections does not give strongly decreased absorption?

P. Penning: The absorption in the case of three {220}-planes is shown in Figure 9. In the minimum, the absorption is equal to the minimum absorption coefficient for the two-beam (220)-reflections. So, the only difference is that now three modes are contributing instead of one, as in the two-beam case.

MEASURING TECHNIQUES OF PARALLEL-BEAM-DIFFRACTION MICROGRAPHY

H. Barth*

ABSTRACT

Examination of the substructure of crystalline solids by diffractographic methods has recently developed into an independent field of work alongside atomic-structure analysis. Diffraction micrography with electrons is characterized by the small size of the distortion fields and the high resolving power in small crystal ranges (1000–3000Å thickness). X-ray diffraction micrography is characterized by great reciprocation between wave field and distortion field and by undisturbed preparation and undisturbed testing in the large crystal ranges (up to 15 cm²). There are two groups of examination methods for diffraction micrography with X-rays: (1) Examination with a finely limited, polychromatic or monochromatic X-ray source and moving sample, according to A. R. Lang *et al.* (2) Examination with a parallel-ray beam of polychromatic or monochromatic X-rays with fixed sample, in accordance with Berg–Barrett *et al.* For the examination of coarse defects in single-crystalline and polycrystalline matter, the parallel-beam method offers a wide scope for studies in the physics and applications of single-crystalline and polycrystalline solids. This paper therefore includes a summary of the methods using collimation systems and grating diaphragms. Measuring techniques and results are illustrated with the help of reflection and transmission pictures on various crystals. The various methods and refined measuring technique of the parallel-beam method enable the following to be defined: (1) Localization of crystallites from 20 μ diam. upward in a surface up to 15 cm². (2) Determination of the faces of averted crystallites from 20 μ diam. upward in crystal surfaces. (3) Angle of avertence of crystallites or curvature angles of net faces from 1′ to 3° in crystal surfaces up to 15 cm². (4) Subangle grain boundaries, slip bands, and dislocation lines; also distortion fields (from 20 μ upward) resulting from mechanical, thermal, and radiation damage.

INTRODUCTION

Thirty-five years ago, W. Berg published details of a method by which the surface of monocrystals can be examined through diffraction of monochromatic X-rays. The measuring technique of micrography of crystal surfaces with X-rays has been improved several times during the last three decades[1-7] in order to make even more precise statements about the substructure of monocrystals. The exploration of the substructure of crystals is not only of importance to science, but to technology as well. The ever-growing application of crystals in different areas of technology requires simple methods for the nondestructive analysis of monocrystal substructure. The range of application of the procedure named after Berg and Barrett also extends to solids with almost ideal structure, up to perfect mosaic structure, and to polycrystalline substructure. In the

* Fritz-Haber-Institut der Max-Planck-Gesellschaft, Berlin-Dahlem, Germany. Home address: 1 Berlin 45, Ringstrasse 85.

following, the physics and the known technical variations of the parallel-beam-method according to Berg–Barrett will be explained in brief.

Figure 1 shows a schematic representation of the parallel-beam method in a horizontal section. The monochromatic X-rays emitted by a line focus are reflected at the net plane of an ideal monocrystal (1) under the reflection angle ϑ as a parallel beam and are registered by a photoplate (5) that can be placed at right angles to the reflected beam or parallel to the crystal surface. The reflecting crystal surface is larger, the longer the line focus and the smaller the reflection angle ϑ. On low index net planes, crystal surfaces up to 15 cm^2 can be examined in one picture.

Since the reflection of monochromatic X-rays at a net plane runs on a cone slope, the X-rays are divergent in the vertical section. If a cross grating is set into the reflected beam at (4) behind an ideal monocrystal with plane surface, we obtain on the photoplate (5) a shadow picture as shown in Figure 2a. From the X-ray picture of the cross

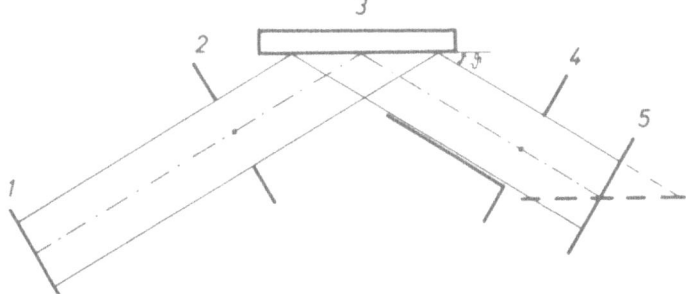

Figure 1. Schematic diagram showing the paths of rays in the parallel-beam method. (1) Line focus, (2) diaphragms or gratings, (3) monocrystal, (4) diaphragms or gratings, (5) photoplate.

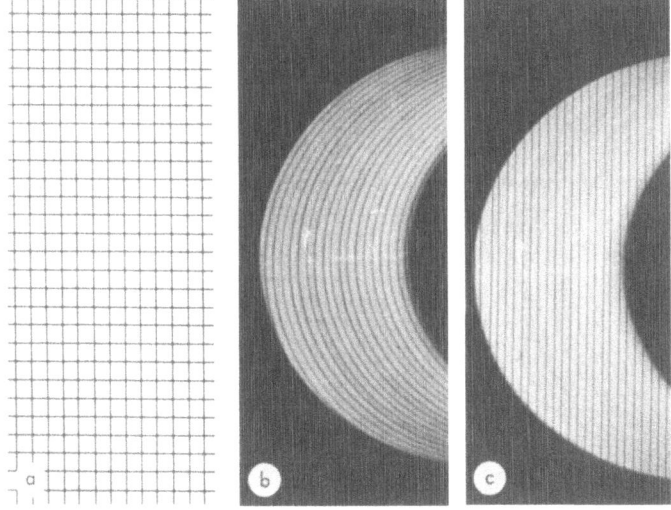

Figure 2. Shadow pictures of grating diaphragms. (a) Monocrystal with plane surface and cross-grating at (4), (b) monocrystal with cylindrical surface and line grating at (2), (c) monocrystal with cylindrical surface and line grating at (4).

grating, we note that defects in the crystal surface too can be stigmatically imaged with X-rays.

If an ideal monocrystal with a cylindrical surface is used, a line grating is imaged in different ways, according to whether it is placed into the parallel beam before the crystal at (2) or behind the crystal at (4). Figures 2b and 2c show shadow pictures of a line grating in front of and behind a cylindrical monocrystal. Only when placing the line grating in the reflected parallel beam, does the image not depend upon the form of the crystal surface.

If defects occur in a crystal surface, the relative angle of divergence between two microcrystals, or the curvature angle of a net face, can be measured by means of the relative deviation of the shadow lines. Since the monochromatic X-rays are divergently emitted by the line focus, a slit diaphragm at (2) or at (4) can be projected into different directions, as shown in Figure 3. Misalignment angles between 1′ and 3° can be measured. The surface of misaligned crystallites can be measured down to a diameter of 50 μ.

The X-rays emitted by a line focus are not only divergent, but also polychromatic. The characteristic X-ray lines are K_{α_1}, K_{α_2}, and K_{β}, where the proportion of the integrated intensities in the same order is approximately 1 : 2 : 4. In addition, a polychromatic Bremsstrahlung appears whose integrated intensity is much lower than the integral intensities of the X-ray lines. Figure 4 shows schematically the path of rays of a diaphragm at (2), pictured through the parallel beam of the X-ray lines $K_{\alpha_{1,2}}$ and K_{β}. For each X-ray line, there exists a parallel bundle of rays and therefore a multiple projection of a diaphragm or of a crystal defect. These multiple projections are made with different integrated intensities and can therefore be distinguished.

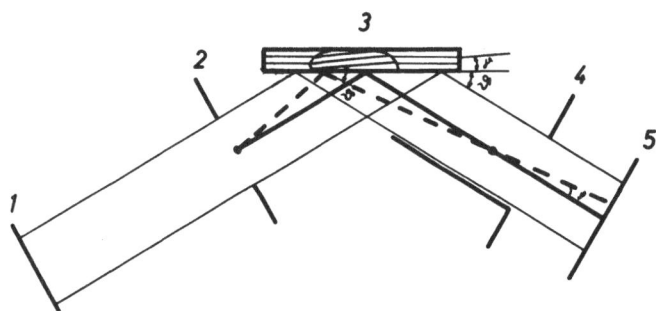

Figure 3. Path of rays during reflection of monochromatic X-rays by a misaligned crystallite.

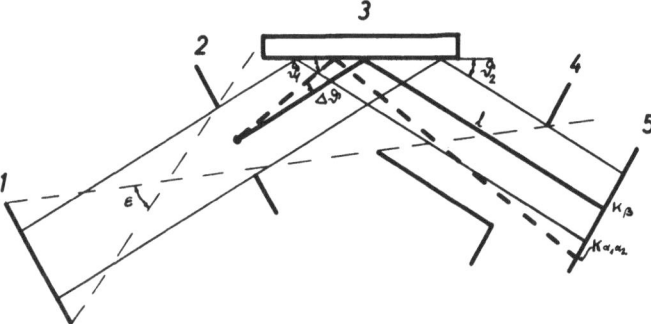

Figure 4. Paths of rays during reflection by an ideal monocrystal with the X-ray lines $K_{\alpha_{1,2}}$ and K_{β}.

The X-ray lines themselves are not monochromatic either and have a relatively large spectral intensity distribution which causes a projection which is not sharp. The blur of the image increases with the distance l between crystal (3) and photoplate (5). In examining large crystal surfaces on low indicated net planes, *i.e.*, at a small reflection angle ϑ, the distance l is limited by the primary X-ray beam with the opening angle ϵ, even with functional diaphragm design. In Figure 4, the primary X-ray beam runs between the thin dotted lines. The distance l can be shortened only, and a small reflection angle ϑ and a short distance between line focus (1) and crystal (3) given, if the opening angle ϵ of the primary beam is being decreased.

PHOTOGRAPHING TECHNIQUE WITH MOVING COLLIMATORS

A reduction of the opening angle ϵ, while still using the whole beam width emitted by the line focus, can be gained with a collimator (6) set up as shown in Figure 5 between the line focus (1) and the crystal (3). By selecting the opening angle of the collimator, the beam reflected by an ideal crystal becomes monochromatic. The half-width of a spectral intensity distribution let through by the collimator and reflected by the crystal is

$$\Delta\vartheta = \frac{\Delta\lambda}{\lambda}t_s\vartheta \sim \frac{\epsilon}{2}$$

If one wishes to work with a bundle of rays of one X-ray line, *e.g.*, only with $K_{\alpha_{1,2}}$ or K_β, or only with K_{α_1}, the opening angle of the collimator

$$\epsilon \leqslant 2\Delta\vartheta$$

has to be smaller than the reflection angle difference, *e.g.*, between $K_{\alpha_{1,2}}$ and K_β. From Table I we can see which opening angles ϵ are necessary for collimators in order to separate the parallel rays of $K_{\alpha_{1,2}}$ and K_β at a certain net plane and with certain X-rays. In order to separate the parallel rays of K_{α_1} and K_{α_2}, collimators with very small opening angle are necessary. If one wishes to work with the parallel beam of one X-ray line only, it is easier to set K_β. It is possible to produce, for example, a parallel beam with a spectral half-width of 0.39 XE at the net plane 2 3 $\bar{5}$ 4 of an ideal quartz monocrystal with Cu K_{α_1} at a reflection angle of $\vartheta = 77°$ and a collimator with an opening angle of 8.6′. This is a subarea of the natural spectral intensity distribution of Cu K_{α_1}. The adaptation to a certain problem requires a number of collimators with different opening angles, *e.g.*, $\epsilon_1 = 8.6′$; $\epsilon_2 = 18.9′$; $\epsilon_3 = 36.2′$; $\epsilon_4 = 70.5′$; $\epsilon_5 = 106.6′$; $\epsilon_6 = 143.6′$; $\epsilon_7 = 163′$; $\epsilon_8 = 204′$.

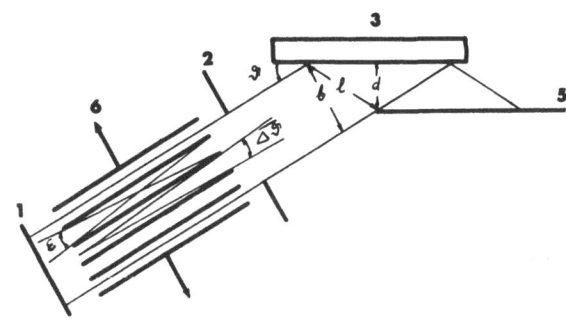

Figure 5. Schematic diagram of the parallel-beam method for reflection pictures with a collimator between line-focus and crystal. (1) Line focus, (2) diaphragms or gratings, (3) mono-crystal, (5) photoplate, (6) collimator.

Figure 6 shows shadow images of a wire set up in front of a quartz monocrystal at (2). In Figure 6a, no collimator is placed in the path of rays. One sees the shadow of the wire through the parallel beam of $K_{\alpha_{1,2}}$ and K_β. In 6b, a moving collimator (6) with $\epsilon_3 = 36.2'$ is placed in the path of the rays. The wire is projected only through the parallel beam of $K_{\alpha_{1,2}}$. If with the same collimator the reflection angle of K_β is set, the wire is projected through the parallel beam of K_β only, as shown in Figure 6c.

Table I. Reflection Angle Difference of K-Lines of Some Elementaries at Certain Quartz Net Planes

Quartz	$d(\text{Å})$	X-ray	$\vartheta,$ K_{α_1}	$2\Delta\vartheta,$ $K_{\alpha_1}\!-\!K_\beta$	$2\Delta\vartheta,$ $K_{\alpha_2}\!-\!K_{\alpha_1}$
10$\bar{1}$0	4.260	Cr	15°36′	170′	3.3′
10$\bar{1}$1	3.340	Cu	13°21′	159′	4.0′
32$\bar{5}$0	0.975	Mo	21°17′	290′	16.0′
23$\bar{5}$4	0.790	Ag	20°21′	282′	20.0′

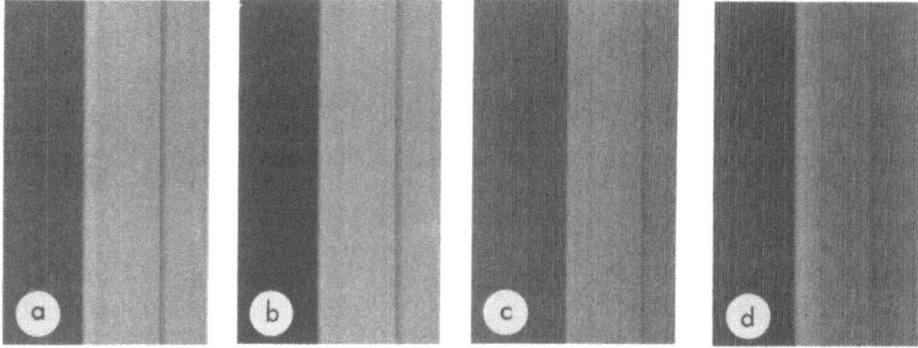

Figure 6. Pictures with parallel beams of different X-ray lines. (a) $K_{\alpha_{1,2}}$ and K_β; (b) $K_{\alpha_{1,2}}$, collimator $\epsilon_3 = 36.2'$; (c) K_β, collimator $\epsilon_3 = 36.2'$; (d) K_{α_1}, collimator $\epsilon = 4.3'$.

Figure 7. Parallel-beam-diffraction micrograph for reflection pictures. (1) X-ray tube with line focus, (2) exchangeable diaphragms, (3) crystal, (5) photoplate, (6) exchangeable collimators.

In 6d, a wire projection through the parallel beam of K_{α_1} can be seen. As mentioned before, it is not necessary to work with K_{α_1} or K_{α_2} alone.

From Figure 5 it can be seen that the distance l between the crystal surface (3) and the photoplate (5) depends upon the effective width b of the parallel beam:

$$l = \frac{b}{\sin 2\vartheta}$$

The blur of the image grows less, the smaller the examined crystal area. Since the distortion fields of the lattice irregularities have a width of $10\,\mu$ to $50\,\mu$, according to the kind of crystal examined, a distance d between 5 mm and 10 mm is sufficient.

The photoplate can be adjusted parallel to the crystal surface and at right angles to the crystal. The parallel-beam-diffraction micrograph* with a setup for reflection pictures is shown in Figure 7. The instrument can be attached to all common X-ray tube hoods with line focus. The collimators (6) are moved back and forward by a synchronous motor and are easily replaceable. Diaphragms and gratings can be set up at spot (2). The crystal is attached to a microgoniometer. The photoplate (5) is in a frame that can be moved at right angles to the surface and close up to the crystal surface. The adjustment of the crystal is done from the outside with the aid of a fluorescent screen or a counter tube. The radiation protection travels with the adjustment movement. Slide plates of the Perutz Company, size 5×5 cm, have proven useful as photo material.

RESULTS

By using collimators and grating diaphragms, a variation of the photographing technique according to the kind of problem is possible. In the following, the photographing technique and the resulting statements will be described with the aid of some examples.

In Figure 8, parallel beam pictures with Cu $K_{\alpha_{1,2}}$ radiation at net plane (100) of a copper monocrystal are shown. The picture 8a has been made with a moving collimator of $\epsilon_4 = 70.5'$ opening angle and a plate distance of $d = 9$ mm in 2 min. The net face is very much curved. Areas of the net face are misoriented by more than $35'$ and therefore have not been imaged. The bright lines that run somewhat irregularly and vertical to each other across the picture are slip bands in the directions [010] and [001], as have been pictured with electrons elsewhere. The form and density of the slip bands are different in each crystal. The influence of the lattice distortion through these slip bands on the integrated intensity of thermal neutron reflections is being examined currently.

Figure 8b has been made on the same crystal and in the same setup, but with a fixed collimator having an opening angle of $\epsilon_5 = 106.6'$. Not only do the slip bands become visible, but the differing curvature of the net face appears as well. Size and course of the curvature angle γ of the net face can be measured on the basis of the shadow lines of the collimator's lamellas. The photographing technique of 8a and 8b can be combined, too, by moving the collimator during half the exposure time and leaving it fixed during the other half.

Figure 8c has been made on the same crystal and in the same setup as 8a and 8b. After the exposure, though, the collimator was taken away and a fine slit was set into the beam. Shadow pictures of this slit have been made from different positions on the same photoplate. Statements about the angles of crystallite misalignment can be made on the basis of the shadow lines of the slit. This picturing technique has been discussed already in detail by G. Wadewitz[8]. In parallel beam imaging of crystals with polychromatic substructure, only crystallites with net face positions in the area of $\pm\epsilon/4$ will

---

* Emka Company, 1 Berlin 15, Bayerische Strasse 7.

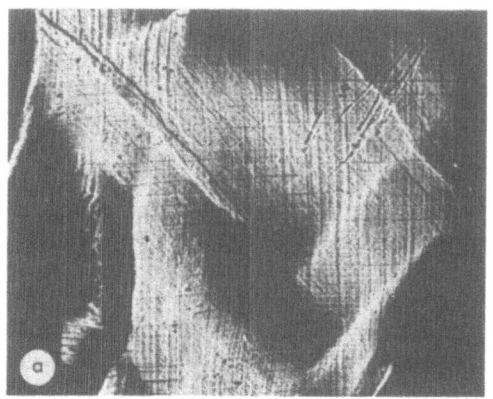

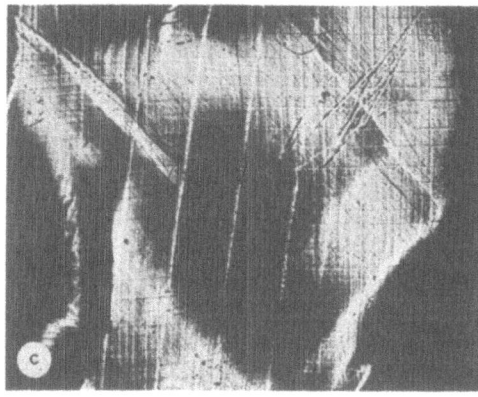

Figure 8. Parallel-beam pictures with Cu $K_{\alpha_{1,2}}$ at the (100) plane of a copper monocrystal. Magnified 2.8 times. (a) Collimator with $\epsilon_4 = 70.5'$ moved, $d = 9$ mm; (b) collimator with $\epsilon_5 = 106.6'$ fixed; (c) collimator with $\epsilon_5 = 106.6'$ moved. Collimator taken away and with slit exposed additionally in different positions.

reflect when a collimator is used. These pictures thus show crystallites with almost identical net plane position.

In Figure 9, parallel beam pictures on a crystallite of a stilbene crystal are shown. The crystallites in this crystal are averted from each other by $\pm 6°$. Figure 9a has been made with a moving collimator $\epsilon_2 = 18.9'$ and a line grating in the reflected beam.

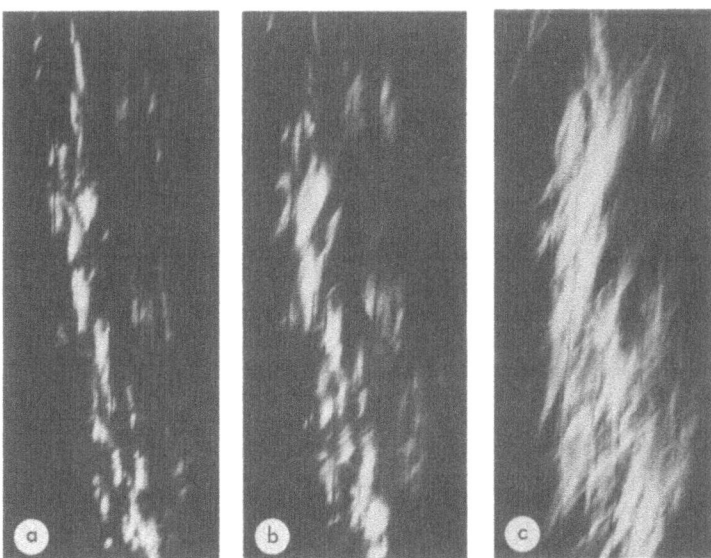

Figure 9. Parallel-beam pictures on a crystallite of a stilbene crystal. Magnified 5.0 times. (a) Collimator with $\epsilon_2 = 18.9'$, $l = 55$ mm; (b) collimator with $\epsilon_3 = 36.2'$; (c) collimator with $\epsilon_8 = 204.0'$.

The distance of the photoplate from the crystal is $l = 55$ mm. We recognize that the crystallite, too, consists of different subareas. Figure 9b has been made with the same arrangement, but with a collimator $\epsilon_3 = 36.2'$. Crystallites with a greater angle of misalignment become visible, corresponding to the greater opening angle. By the shadow lines of the line grating the angle of misalignment can be measured from one subarea to the other. Figure 9c has been made with a collimator $\epsilon_8 = 204'$. One recognizes that these subareas are not only misaligned with respect to each other, but that they are curved as well. This type of photographing technique is especially suited for examination of the crystallite structure in polycrystalline solids. While the crystal in Figure 9 remained in the same position and the collimators were exchanged, it is also possible to use only one collimator and to vary the crystal's position.

Figure 10 shows a parallel beam picture at a (0002) net plane of a beryllium monocrystal, intended to be used as analyzer crystal for a neutron spectrometer. 10a has been made with a moving collimator $\epsilon_4 = 70.5'$. The whole radiated area reflects. We recognize bright lines, parallel to each other, caused by lattice irregularities during crystal growth. In addition, the beryllium monocrystal has a network of small lattice distortions, the cause of which is not known to me. In Figure 10b, a collimator with an opening angle $\epsilon_1 = 8.6'$ has been moved. Only part of the radiated crystal surface reflects. If the position of the crystal surface is changed, the collimator remaining the same, a neighbouring area reflects, as can be seen in 10c. This beryllium monocrystal consists of areas that are misaligned in certain directions by $5'$ to $6'$ each. Size and relative position of the areas can be determined with this photographing technique.

Transmission pictures on thin monocrystal plates can also be made with the same arrangements. Since different authors[9-17] have described double crystal schemes for observing dislocation lines, I want to be content with the reference that almost

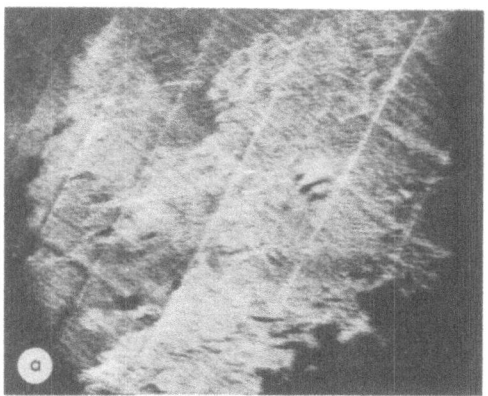

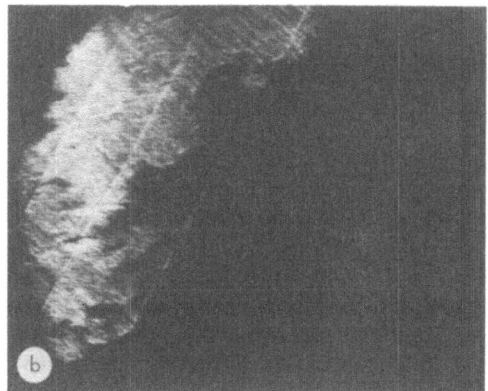

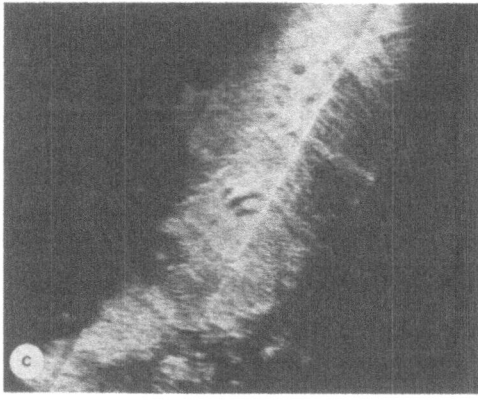

Figure 10. Parallel-beam pictures at the (0002) net plane of a beryllium-mono-crystal. Magnified 1.3 times. (a) Collimator moved with $\epsilon_4 = 70.5'$, $d = 13$ mm; (b) collimator moved with $\epsilon_1 = 8.6'$; (c) crystal position turned by $6'$ compared with position at 10(b).

all ideal crystals in transmission can be examined with the parallel beam method, and that contact pictures can also be made.

In Figure 11, a schematic drawing of the path of a ray during transmission picture taking can be seen. The radiation emitted by the line focus (1) and collimated afterward is being reflected and monochromated by an asymmetrically polished ideal monocrystal.

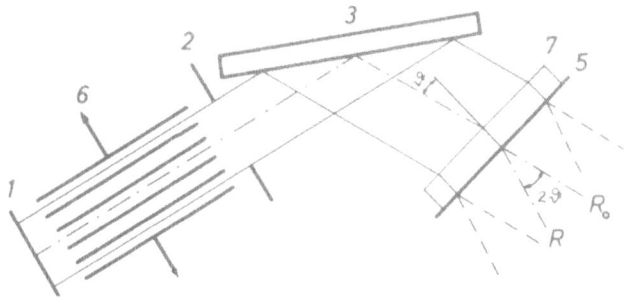

Figure 11. Schematic diagram of the path of rays of the parallel-beam method for transmission pictures with a collimator between focus and monochromator crystal. (1) Line focus, (2) diaphragm, (3) monochromator crystal, (5) photoplate, (6) collimator, (7) crystal plate.

Figure 12. Parallel-beam-diffraction micrograph for transmission pictures.

The crystal plate (7) is being adjusted to the parallel- or antiparallel-reflection position with a counter tube. With appropriate selection of crystal plate thickness, contact pictures can also be made. The pictures are of the same quality as the pictures made with the diffraction micrograph, according to A. R. Lang.

Figure 12 shows a picture of the transmission setup in the parallel diffraction-micrograph. All adjustments can be made here too from the outside with radiation protection.

CONCLUSION

The parallel-beam method with the measuring techniques described here is applicable to many problems of solid state physics and solid state technology that could not be presented here. This technique is especially useful for the nondestructive radiation damage in solids.

ACKNOWLEDGMENTS

I wish to thank my wife for the idea of using moving collimators. I am also indebted to Dr. Marius Kratzenstein for his interest and support for the work.

REFERENCES

1. W. Berg, "Über eine röntgenographische Methode zur Untersuchung von Gitterstörungen an Kristallen," *Naturwiss.* **19**: 391, 1931.
2. W. Berg, "About the History of Load and Deformed Crystals," *Z. Krist.* **89**: 286, 1934.
3. C. S. Barrett, *Trans. AIME* **161**: 15, 1945.
4. A. Guinier and J. Tennevin, "Sur Deux Variantes de la Méthode de Laue et Leurs Applications," *Acta Cryst.* **2**: 133, 1949.
5. H. Barth, "Monochromatisierung einer Röntgen-Spektrallinie," *Z. Naturforsch.* **13a**: 680, 1958.
6. H. Barth, "Analyse der Realstruktur von Einkristallen," *Z. Elektrochem.* **63**: 908, 1959.
7. H. Barth, "Gefügeanalyse von Einkristallen mit der Parallelstrahlmethode," *Fortschr. Mineral* **38**: 53, 1960.
8. H. Wadewitz, "Zur Bestimmung der Subkornwinkel mittels Berg–Barrett–Technik," Manuscript, 1963.
9. W. L. Bond and J. Andrus, "Structural Imperfections in Quartz Crystals," *Am-Mineralogist* **37**: 622–632, 1952.
10. U. Bonse, "Zur röntgenographischen Bestimmung des Typs einzelner Versetzungen in Einkristallen," *Z. Physik* **153**: 278, 1958.
11. J. B. Newkirk, "Method for the Detection of Dislocations in Silicon by X-ray Extinction Contrast," *Phys. Rev.* **110**: 1465, 1958.
12. A. R. Lang, "Direct Observation of Individual Dislocations by X-Ray Diffraction," *J. Appl. Phys.* **29**: 597, 1958.
13. H. Barth und R. Hosemann, "Anwendung der Parallelstrahlmethode im Durchstrahlungsfall zur Prüfung des Kristallinneren mit Röntgen-Strahlen," *Z. Naturforsch.* **13a**: 792, 1958.
14. A. R. Lang, "The Projection Topograph: A New Method in X-Ray Diffraction Microradiography," *Acta Cryst.* **12**: 249, 1959.
15. V. Gerold und F. Meier, "Der röntgenographische Nachweis von Versetzungen in Germanium," *Z. Phys.* **155**: 387–394, 1959.
16. M. Renninger, "Eingefrorene und reversible Gitterverzerrungen," *Phys. Letters* **1**: 106, 1962.
17. M. Renninger, "Netzebenen-Interferonetrie," *Phys. Letters* **1**: 104, 1962.

SOME RECENT APPLICATIONS OF
X-RAY TOPOGRAPHY

A. R. Lang

H. H. Wills Physics Laboratory
University of Bristol, England

ABSTRACT

Dislocations, Inclusions, and Precipitates. Impurity precipitated after growth and foreign particles accidentally included during growth both produce intense local strain fields which give rise to diffraction contrast effects resembling those seen in electron microscope images of precipitates. The relationship between the dislocation configuration and these localized strain centers can show whether the latter arise from inclusions or precipitates. Precipitates will generally be found strung along the grown-in dislocations, decorating them. On the other hand, inclusions often generate dislocations by lattice closure errors; such dislocations then fan out from the inclusion in the general direction of advance of the growth interface.

Twin Boundaries and Fault Surfaces. The cases when the twins have parallel lattices, such as in Brazilian and Dauphiné twinning in quartz, are interesting. When the crystals on either side of the twin boundary are both Bragg reflecting, the twin boundary may exhibit 'stacking fault' type fringes. From an analysis of the variation of visibility of these fringes in different reflections, the fault vector at the twin boundary and its variation with boundary orientation may be found. In quartz, other types of fault surfaces producing fringe contrast may lie parallel to growth horizons or they may mark growth sector boundaries. In synthetic quartz, they can also mark cell boundaries under conditions of cellular growth.

Internal Magnetic Domain Structures. In plates of Fe + 3% Si roughly parallel to (110), a variety of previously undetected domain structures has been discovered and analyzed. Diffraction contrast is produced by 90° domain walls but not by 180° walls. The 90° walls produce strong diffraction contrast even though the magnetostriction of silicon–iron is only about 10^{-5}. In plates parallel to (112), the main lamination pattern below the complex pattern of surface closures can be revealed and, in favorable cases, interpreted.

X-ray Moiré Patterns. The most direct method of observing X-ray moiré patterns—by topography of one crystal closely superimposed upon another—involves considerable theoretical complexities and produces a variety of curious diffraction patterns. However, it shows promise of providing a means for the comparison of lattice spacings to about one part in 10^7 and for mapping strain fields very sensitively.

INTRODUCTION

The aim of this paper is to illustrate some current activities in the practical use of X-ray topography for detecting and displaying lattice defects, and at the same time to demonstrate the interplay of this work with investigations of the diffraction behavior of nearly perfect crystals. X-ray topographic study of single crystals in which individual dislocations can be resolved remains a leading activity, and it is especially directed toward investigations of the growth history of crystals. The ability to sample a large crystal volume and present on a single topographic record the variation in degree and type of imperfection over a distance in the crystal corresponding to a substantial fraction, if not all, of its period of growth is an asset of the X-ray topographic method which in some measure counterbalances its inferior resolution compared with the electron microscope. Grown-in defects in diamonds have been extensively studied[1-6], and natural and synthetic quartz, and the amethyst variety of quartz[7] have been investigated to a considerable degree. In the cases of both quartz and diamond, added interest arises from the availability of both natural and synthetic specimens; it is always instructive to compare the products of the laboratory with those of nature. The nearly perfect crystals upon which the bulk of X-ray topographic work has been done have all had quite simple structures, such as the diamond structure, the face-centered cube (e.g., aluminium[8]) and the body-centered cube (e.g., iron[9]). In the more complex structure of quartz we might expect to find diffraction evidence for a wider variety of lattice imperfections than those present in the simplest structures. Such variety has indeed been found. In the earliest X-ray topographic studies of quartz various sheet defects or 'fault surfaces' were revealed.[10,11] Such fault surfaces may be of general occurrence in complex structures; the recent observations by Yoshimatsu[12] on fault surfaces in *ADP* crystals are suggestive.

Observations on fault surfaces in alpha quartz will be reported below. The type of fault surface most amenable to exact study is the boundary between twins whose lattices are parallel. The two common types of twinning in quartz—those following the Brazil and Dauphiné laws—fall in this category. The experiments described here illustrate the timeliness of the new theoretical work of Kato, Usami, and Katagawa[13] reported elsewhere in this volume. They also show the need for development of the theory to take into account X-ray absorption.

Two aspects of X-ray topographic study that do not directly involve dislocations will also be discussed. These are the study of internal magnetic domain structures and of X-ray moiré fringes, both of which have implications outside the field of lattice imperfections.

The question may be asked, how far is it possible to use X-ray topography merely as a tool for detecting and identifying lattice imperfections without reference to complicated diffraction theory? The answer is that one can proceed satisfactorily quite a useful distance in many directions. Examples are the study of the sizes and shapes and deformations of subgrains in single crystals, the locating of surface damage and the verification of its satisfactory removal, and the counting of dislocations and even the determination of their Burgers vectors. However, in the last-mentioned work one soon runs into diffraction phenomena which require some understanding of the processes of production of diffraction contrast if a proper relation of the image to the nature of lattice defect is to be made, as some examples in this paper will show. Indeed, it is worth emphasizing that due regard should be taken of the theoretical work on diffraction by perfect and nearly perfect crystals. It is certainly necessary if X-ray topographic experiments are to be designed and executed to give the maximum information yield. The most significant theoretical papers for transmission X-ray topography are those by Kato and his colleagues, and a summary of work done up to 1962 will be found in a useful review by Kato.[14]

DISLOCATIONS, INCLUSIONS, AND PRECIPITATES

A common dislocation configuration found in crystals is one in which dislocations radiate from a central point within the crystal and run outward to the crystal faces.

It is frequently seen in natural diamonds.[1] The interpretation of the pattern—that the dislocations were generated at the crystal nucleus and were subsequently grown into the crystal—is doubtless correct, but it leaves unsettled the question whether the nucleation was homogeneous or heterogeneous. Rapid initial growth under the conditions of supersaturation that attend homogeneous nucleation could lead to the introduction of dislocations, especially if the initial growth were dendritic. However, if the topographs show a concentration of strain at the nucleus greater than that attributable to the dislocations, then it is likely that a foreign body is present there, and that nucleation was heterogeneous, the dislocations being generated by initial imperfect growth on this body or by lattice closure errors arising in the course of its envelopment. Evidence on the origin of dislocations drawn from X-ray topographic studies of a wide variety of crystal species, but of necessity restricted to those specimens which are of sufficient purity and perfection of crystal matrix to allow individual dislocation images to be recognized, indicates that the dislocations in these crystals are generated by lattice closure errors at foreign bodies incorporated during growth rather than by some mechanism involving growth accidents in the course of crystallization of pure material.

The configurations adopted by grown-in dislocations, their relationship with inclusions, and the effects of impurity segregation at dislocations after growth, are well illustrated by natural and synthetic quartz; some recent observations on these materials will now be described. The record of the growth history of quartz crystals that appears on X-ray topographs includes not only the configuration of grown-in dislocations but also the disposition of fault surfaces, the shapes of twins, and the presence of growth stratifications which delineate the surfaces upon which material was crystallizing during the stages of growth of the crystal.

The structure and twinning of quartz are described in a standard work.[15] Note that there has been some confusion in the indexing of planes of quartz in Miller–Bravais axes. In certain X-ray crystallographic work, the plane which is structurally and morpho-

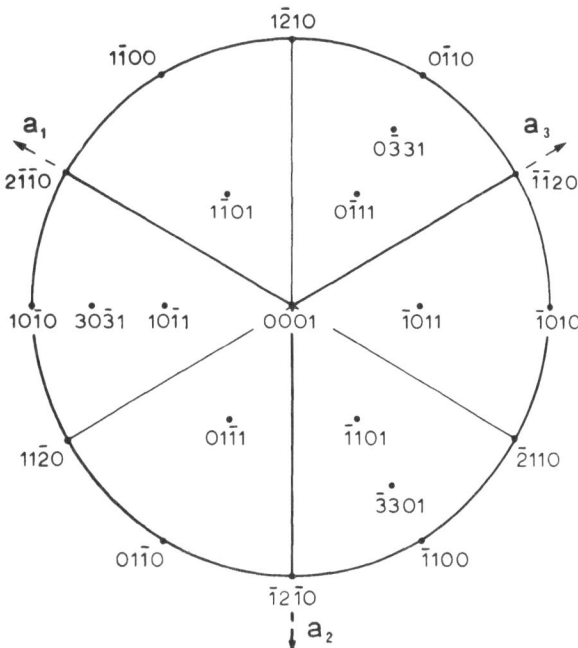

Figure 1. Stereographic projection on to the basal plane of quartz showing poles of reflecting planes most used in X-ray topographic surveys.

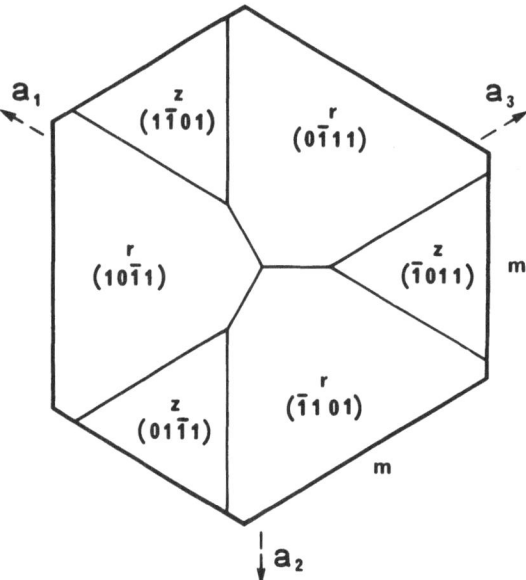

Figure 2. Idealized division into major and minor rhombohedral growth sectors in a quartz slice cut parallel to the basal plane.

logically recognized as the major rhombohedron is given the index of the minor rhombohedron. The history of this confusion, and a mnemonic for the correct orientation of the Miller–Bravais axes with respect to the structure have been stated;[16] the correct orientation is adopted here.

Figure 1 is a stereographic projection on the basal plane of quartz which indicates the planes chiefly used in X-ray topographic studies, in some cases in several orders of reflection. Most of the work on natural quartz has been done on plates about 15 mm square and 1 mm thick cut in the BT orientation. This lies between $(10\bar{1}1)$ and (0001), being 11° off $(10\bar{1}1)$. Suppose a slice were cut from a quartz prism parallel to the basal plane. Then, assuming growth had occurred only on the major and minor rhombohedral faces, the slice would be divided into areas which had grown on these faces in the way shown ideally in Figure 2. These divisions are called *growth sectors*, and it will be seen that the r–z growth sector boundaries are parallel to $\{10\bar{1}0\}$ and the r–r sector boundaries are parallel to $\{11\bar{2}0\}$. Growth stratifications parallel to either r or z faces will cut the slice in lines parallel to the sides of the hexagon. When topographs are taken using some of the various planes indicated on Figure 1, the specimen may perforce be viewed from directions making large angles with the c-axis, and the geometry of Figure 2 will be appreciably distorted in consequence. Nevertheless, it is usually a straightforward matter to determine the orientations of growth layers and growth sector boundaries. Possibly section topographs[17] may be needed to supplement the projection topographs.[18]

Figures 3 and 4 show projection topographs of parts of plates cut in the BT orientation. On Figure 3 there can be seen the diffraction images of strings of centers of intense localized strain. These strings lie on surfaces parallel to the r and z planes. Traces of these planes are more directly indicated by the pattern of growth stratifications revealed in Figure 4. In both Figures 3 and 4 it is easy to identify the trace of $(\bar{1}011)$ which is vertical and that of $(\bar{1}101)$ which slopes upward to the right. (On these figures there also appear images of twin boundaries and growth sector boundaries; these will be described below.)

The X-ray diffraction contrast arises from strain gradients in the specimen. A

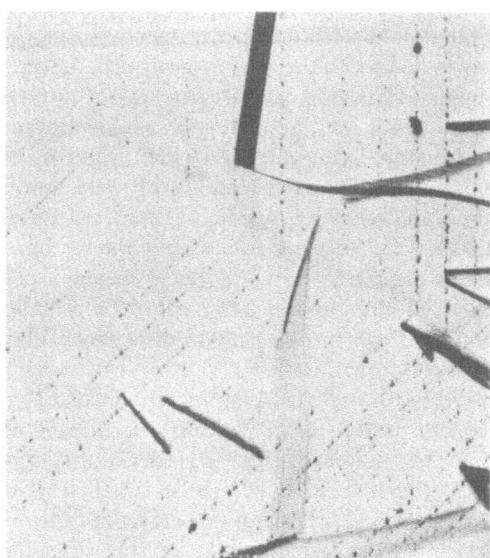

Figure 3. X-ray topograph of natural quartz plate about 1 mm thick. Field edge length 8 mm; reflection $\bar{2}4\bar{2}0$, diffraction vector vertical; Mo K_α radiation; fault fringes at Dauphiné twin boundaries.

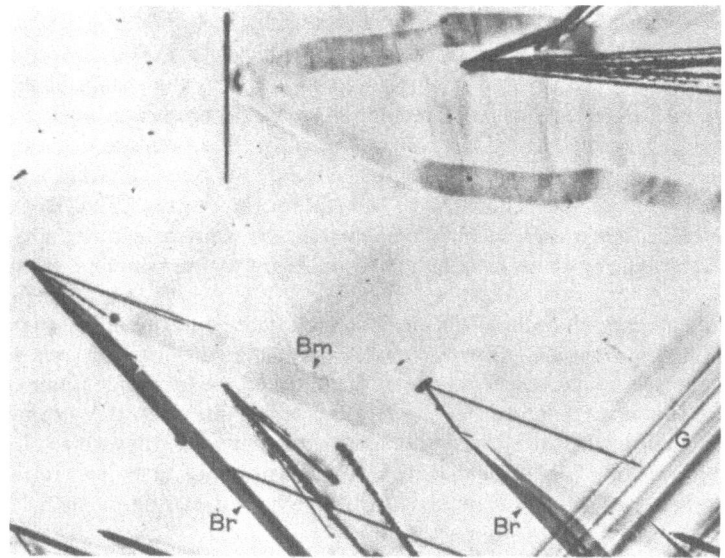

Figure 4. Topograph of natural quartz plate, thickness about 1 mm. Field area 0.9 mm by 1.1 mm; reflection $10\bar{1}\bar{1}$, Ag K_α radiation. Projection of diffraction vector is horizontal, direction $[\bar{1}2\bar{1}0]$ is vertical on topograph (orientation similar to Figure 3). Growth layers parallel to $(\bar{1}101)$ marked (G); Brazil twin boundaries (Br) are parallel to $(0\bar{1}11)$ and (Bm) parallel to $(0\bar{1}10)$.

variability from point to point of the identity, mode of incorporation in the structure, or quantity of impurity will cause changes in cell dimensions; but lattice curvature is more effective than a gradient of dilation in producing diffraction contrast. Experiments show that the diffraction contrast arises chiefly from lattice tilts where the growth layers intersect the free surface of the specimen and stresses are locally relieved. The constraints of the situation confine displacements to the plane normal to the line of intersection of the growth layer with the specimen surface; reflections from this plane will hence give zero or minimal diffraction contrast from the growth layer. The contrast in other reflections will depend both upon the angle between the diffraction vector and the growth layer, and upon the structure factor of the reflection. Interbranch scattering, and hence a locally intensified reflection, occurs more readily with weak and high-order reflections.

When slices of quartz of 1 mm thickness or more are examined topographically, it is usual to find some dislocations which originate at points within the slice, and examples of this can be seen in Figures 3, 4 and 5. These dislocations may occur singly or in small bundles. In the latter case, the members of the bundle usually diverge gently from each other. The general direction of the individual (or the bundle) is normal to the local growth surface. It is believed that all these dislocations originate by lattice closure errors at inclusions incorporated during growth. This idea is supported by the frequent occurrence of the diffraction image of a local strain concentration at the point of origin. However, there is evidence that in quartz, as in diamond, not all inclusions which make themselves visible by their strain fields do generate dislocations. The strain may arise wholly or in part from different coefficients of thermal expansion of the inclusion and the matrix, the crystal when topographed being at a temperature much different from that at which it crystallized. There is also the possibility that the strong localized strain-field images seen now in the topographs (for example in Figure 3) may be due to growth of a precipitate nucleated on an inclusion which by itself would produce much less diffraction contrast. However, the fact that dislocations radiate from some points on the growth horizons which are delineated by the strings of strain centers shows that some imperfections were present there at the time of growth. Sometimes a strain center seen isolated or at the point of origin of one or more dislocations shows in all reflections the "butterfly" type of image, i.e., two lobes of blackening separated by a *line of no contrast* perpendicular to the diffraction vector. This is indicative of dominantly radial strain and is most probably a post-growth feature; it is associated chiefly with the images of precipitates, both in X-ray topography and electron microscopy.[19]

There is one large bundle of dislocations in Figure 4, in the upper right corner of the field. Most of these are grown-in nearly normal to the local growth surface, the z face (1011) which is also nearly normal to the plane of the specimen. Some dislocations cross the growth sector boundary between the z sector and the (0111) r growth sector along the top margin of the field, and change direction when they do so. In the lower part of Figure 4 there is a wedge-shaped Brazil twin. This, together with several dislocations, is nucleated at a strain center near the left-hand margin of the field. Figure 5 is another reflection from the center bottom area of Figure 4, at greater magnification. Note the fine-structure of some dislocation lines, and that some dislocation images, intensely visible in Figure 4, have vanished in Figure 5 (center, and above left thereof), save for a line of faint blobs of darkening (a dynamical diffraction effect).

Dislocations observed in X-ray topographs of both natural and synthetic crystals may be described as "clean" or "dirty." For X-ray topographic purposes, a clean dislocation is one whose visibility rules are not sensibly modified by the strains introduced by precipitation upon it. Many dislocations in natural quartz are obviously dirty. With some of these the images of decorating precipitates remain visible when the dislocation is viewed using a reflection in which it should be invisible; an example appears in Figure 4 above and to the left of the letter *G*. All dislocations observed so far in synthetic quartz are fairly straight (e.g., Figure 6). In natural quartz, some dislocations are exactly

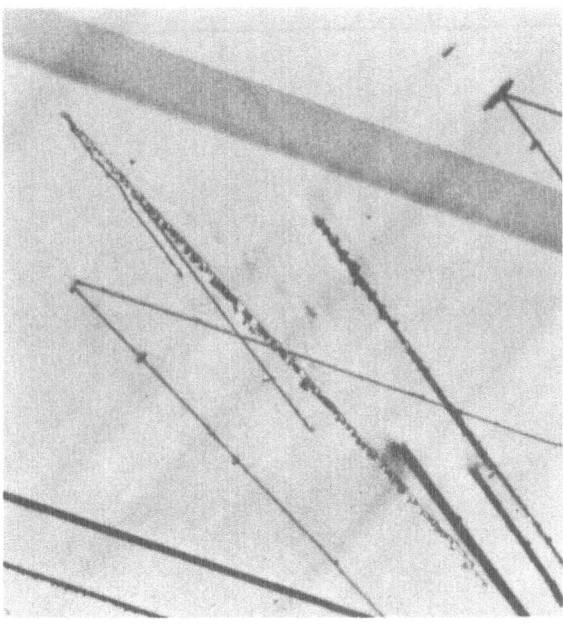

Figure 5. Detail of dislocation configuration in specimen of Figure 4. Reflection $\bar{1}101$, Mo K_α radiation; field edge length 2.5 mm.

straight within the precision of topographic measurement; others may be quite tightly coiled. It is remarkable that straight dislocations and strongly coiled ones may occur close together in a crystal; Figure 5 presents examples of this juxtaposition. Coiled dislocations generally appear to be dirty and it is difficult to determine their Burgers vectors, so the relationship between straightness, direction of dislocation line, and Burgers vector is not yet clear. The majority of the dislocations seen in Figures 3 and 4 have as Burgers vector, **b**, the lattice translation **c** plus one of the lattice translations **a** in the basal plane. The two long, highly coiled dislocations that have their origin in the upper left corner of Figure 5 and run down to the bottom right appear to have their Burgers vectors parallel to **c**.

When investigating the character of dislocations, it is important to bear in mind how the diffraction conditions control the appearance of the dislocation image, particularly with respect to its width and its profile. A much simplified approach to the problem of the widths of dislocation images, but one which does not agree at all badly with measurements, relates this width to the diameter of the volume in the crystal within which the crystal is misoriented by an amount about equal to the angular width at half-maximum intensity of the perfect-crystal reflection curve.[2,9,10,20,21] Incorporating the relation between this angular range and the interplanar spacing and the extinction distance ξ_g, one obtains an expression for the image width W of a pure screw dislocation

$$W \approx (2\pi)^{-1}\xi_g \mathbf{g} \cdot \mathbf{b} \tag{1}$$

The image of a pure edge dislocation, viewed in the reflection from the plane containing the line and normal to **b**, is about 1.75 times the value given by equation (1). With weak reflections, for which ξ_g is large, large values of W are apparent. An example is shown in Figure 6 which is part of a slice cut normal to the basal plane of a synthetic quartz crystal, seen in the 0003 reflection. The much greater width of the dislocation images compared with Figure 5 is clear. The uneven background intensity in Figure

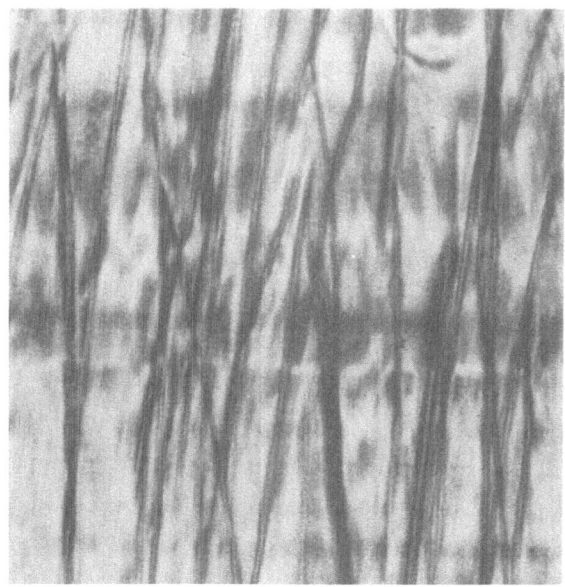

Figure 6. Dislocations and growth layers in synthetic quartz grown on a seed plate parallel to (0001). Specimen thickness about 1 mm; field edge length 5 mm; reflection 0003, Ag $K\alpha$ radiation; direction [0001] is vertical, growth layers are parallel to (0001).

6 arises in part from growth layering parallel to (0001) and in part from the faint images of many dislocations which are, according to the simple $\mathbf{g} \cdot \mathbf{b} = 0$ criterion, "invisible" in this reflection. Image widths calculated from equation (1) agree well with measurements on the specimen of Figures 4 and 5, but the widths in Figure 6, wide though they are, amount to only about two-thirds the prediction from equation (1). The discrepancy can be attributed to the high background intensity in Figure 6.

Notable is the bimodal intensity profile of the dislocations in Figure 6. This is seen also on a much finer scale when ξ_g is small, in cases when $\mathbf{g} \cdot \mathbf{b} = 2n$, n being an integer. When ξ_g is large, and W correspondingly large, double images appear for all values of $\mathbf{g} \cdot \mathbf{b}$. They can be reasonably understood by considering that contours, on the Bragg plane, of equal misorientation with respect to the setting for Bragg reflection are of figure-of-eight shape, with their lobes respectively on either side of a ray passing through the dislocation line. Thus there is a minimum thickness of misoriented crystal volume to be traversed, and hence a minimum of enhancement of diffracted intensity, for rays passing through the center of the image. Various aspects of double images of dislocation images have been studied by several workers.[22-24]

TWIN BOUNDARIES AND FAULT SURFACES

When planar defects of stacking-fault type, which exhibit on X-ray topographs a fringe pattern analogous to the fringes they produce in electron micrographs,[25] occur in the diamond structure, as they do in small areal extent in natural diamond[6] and both small[26] and large[27] in silicon, their analysis by diffraction contrast is much simplified by the small choice of fault vectors to which it is permissible to restrict attention. With defects such as are found in quartz,[10,11] on the other hand, the fault vectors are quite unknown, and it is the task of X-ray topography to determine them as precisely as possible. Since the X-ray extinction distance in a typical quartz reflection is about 50 μ, and, for X-ray optical purposes, a layer of material with anomalous interplanar spacings will act effectively as a single interface provided the layer is substantially less thick than the extinction distance, it is clear that X-ray topography cannot show directly the structure of fault surfaces on a submicron scale, or indeed resolve them separately when they occur closer together than about 1 μ. However, the thickness of distorted material at a twin boundary is only of atomic dimensions. Moreover, it appears probable that some of the other observed fault surfaces may arise from sheets of impurity again only of atomic dimensions thick. With such defects, the concept of a single interface, and associated with it a fault vector which measures the displacement of the Bravais lattice on one side of the fault surface with respect to that on the other, is quite realistic. The fault surfaces involved in the present discussion do not disturb the parallelism of the lattice on one side of the fault with respect to that on the other, as long as points sufficiently remote from the surface are considered. The conditions characterizing the fault are thus that lattice-parallelism is maintained but lattice-coincidence is not. The fault surfaces so characterized include, besides Dauphiné and Brazil twin boundaries, the boundaries between growth sectors, various surfaces in natural quartz not obviously related to either growth layers or growth sector boundaries and whose nature is not yet properly understood, and boundaries between cells in regions of cellular growth in synthetic quartz. In addition there should be included cases of surfaces coincident with growth layers at which a change in interplanar spacing due to change in impurity concentration occurs sufficiently abruptly to cause considerable "interbranch scattering" of the X-ray waves and hence a clearly visible fringe pattern. In the last-mentioned case it may not be easy to tell from the projection topograph whether the pattern of bands represents the images of a set of stratifications which individually scatter only weakly, or whether it is a set of fringes due to strong scattering at a single stratum. The section topograph should make the situation clear, for in the latter case there will be evident on either side of the fault surface a set of the "hook-shaped" fringes which occur in section patterns of wedge-shaped crystals[11] and which, upon translation of the specimen, give rise to the observed equal-thickness fringes in wedge-shaped volumes.

The section topograph will also make clear another significant distinction—that between fault surfaces which are and which are not attended by long-range strains. In the former case the Pendellösung fringe pattern will be perturbed,[28-31] often in a remarkable way. Fortunately, the twin boundaries in rather perfect crystals such as those of Figures 3, 4, and 5 are quite free from long-range strain. Compare the section pattern of a Brazil-twin boundary shown in Figure 7a with that arising from a different type of fault surface, Figure 7b. Figure 7a shows the hour-glass pattern described in the calculations of Kato et al.[13] superimposed upon the usual Pendellösung fringes which run parallel to traces of the specimen surfaces. At the fault surface cut by the section shown in Figure 7b, on the other hand, the Pendellösung fringe pattern corresponding to the full specimen thickness is shifted to a higher order of interference just as it is at the strong dislocation strain-field also included in the field. One can still see, however, a set of hook-shaped fringes on either side of the fault surface. It is only in the absence of long-range strains that it appears feasible to make an accurate determination of the fault vector, for then we will expect to find certain reflections in which the fault surface has zero or vanishingly small visibility. For these reflections, $\mathbf{g} \cdot \mathbf{f} = n$ ($\mathbf{g}$ is the reciprocal lattice vector, in magnitude equal to the reciprocal of the interplanar spacing, and n is zero or a positive or negative integer).

Fault surfaces which coincide with growth sector boundaries show more strongly in crystals which also show pronounced images of growth stratifications. For example, in the top part of Figure 4 one sees quite clearly the fault fringes on the images of the sector boundaries between the (1101) r sector and the (1011) z sector, and between the latter sector and, right at the top margin of the field, the (0111) r sector. In Figure 3,

Figure 7. Section topographs cutting fault surfaces. Specimen thickness about 1 mm, width of section topograph about $\frac{1}{2}$ mm; reflection 30$\bar{3}$1, Mo K_α radiation. (a) Section through Brazil twin boundary: no long-range strains; (b) section through growth sector boundary, long-range strains present. Note also long-range strains associated with dislocation in lower part of section.

on the other hand, in which banding parallel to growth layers is not evident, the one growth sector boundary in the field—that between the (1101) r sector and the (1011) z sector—which crosses the field from left to right, is only faintly visible. In no case so far have fault fringes been found at r–r sector boundaries.

A few of the findings relating to twin boundaries will now be described. In most cases the material on either side of a Dauphiné twin boundary will reflect with different intensity. If the reflection on one side is $hkil$, then that on the other is $\bar{h}\bar{k}\bar{i}l$. This follows from the Dauphiné twin law, which is a rotation of 180° about the c-axis. The reflections $hkil$ and $\bar{h}\bar{k}\bar{i}l$ have the same interplanar spacing, and so both simultaneously make the Bragg angle with the incident beam, but they are not in general structurally equivalent. Indeed in some cases they have greatly differing structure factors, and the resulting great difference in intensity reflected from volumes related by the Dauphiné twin law provides a sensitive X-ray topographic method for picking out small volumes of Dauphiné-twinned material.[32] However, for investigating the fault vectors at Dauphiné twin boundaries it is preferable to use reflections, such as those from {112l} planes, which reflect equally strongly on either side of the twin boundary (as long as Friedel's law is obeyed). With such reflections, one sees that Dauphiné twin boundary fault vectors depend upon twin-boundary orientation. This is well illustrated by the Dauphiné twin boundaries in Figure 3. Consider the two segments of boundary which run roughly vertically on the topograph. Measurements show that the top segment is roughly normal to the basal plane; it shows intense diffraction contrast. The orientation of the lower segment, where it shows a wide image on the topograph, is roughly parallel to the basal plane, and it then shows weak contrast. Another observation is that Dauphiné twin boundaries are stepped on a fine scale, down to the topographic resolution limit of about 1 μ. For example, where the twin boundaries run horizontally on Figure 3, to the upper right-hand margin, it will be seen that they have a "staircase" structure, with different contrast produced at the "treads" and "risers." The standard method of delineating Dauphiné twin boundaries is by etching polished crystal surfaces. The etch pits are generally several microns in diameter. Hence, fine structure of the twin boundary on the micron scale is not revealed by this method.

Material on either side of a Brazil twin boundary will reflect equally strongly in all reflections, unless experimental conditions are chosen to produce a significant departure from Friedel's law.[32] This is because the twin law is a reflection, turning a left-handed structure into a right-handed one, and the members of an enantiomorphic pair cannot be distinguished as long as Friedel's law is obeyed. The reflection is in the planes $\{11\bar{2}0\}$ which are normal to the twofold axes of the structure. The equality of structure factor moduli on either side of a Brazil twin boundary implies that more reflections are usable for determining the twin boundary fault vector than in the case of Dauphiné twinning. It also helps greatly that Brazil twin boundaries can be planar, with no X-ray topographically-detectable steps or deviations in orientation, over areas of several square millimeters. They usually lie parallel to low-index planes. For example, the segments of boundary shown in Figures 4 and 5 are accurately parallel to $(0\bar{1}11)$ and $(01\bar{1}0)$. As in the case of Dauphiné twinning, it is found that the fault vector is a function of orientation of the composition plane. Thus, in the $10\bar{1}1$ reflection (Figure 4), the twin boundary parallel to $(0\bar{1}11)$ shows strongly, but that parallel to $(01\bar{1}0)$ only weakly. In Figure 5, on the other hand, the twin boundary parallel to $(01\bar{1}0)$, which runs across the top of the field, shows quite strongly, whereas that parallel to $(0\bar{1}11)$ which is lower in the picture—running from the left edge to the bottom edge—is quite faint. Although X-ray topography is much slower than electron microscopy, it is much surer in that it is certain that the topograph is produced with only one Bragg reflection excited, and there can be no doubt which plane is Bragg reflecting. One is not troubled by the systematic interactions between various orders of reflection from the same plane that occur in electron diffraction. Indeed, the separate examination of several orders of reflection from a given plane, and the comparison of fault fringe visibility in them, is the chief X-ray topographic method for determining fault vectors, and is capable of measuring them to a percent or so of the lattice translation.

The assessment of fault fringe visibility may, of course, be hindered by the simultaneous presence of images of other imperfections. When twin boundaries transect growth layers, a complex fringe pattern is generated. Figure 8 shows part of the $1\bar{1}00$ reflection from the region in Figure 4 where a pair of Brazil twin boundaries parallel to

Figure 8. Detail of fringe pattern at intersection of Brazil twin boundaries with growth layers. Specimen thickness about 1 mm; field edge length 2 mm; reflection $1\bar{1}00$, Mo K_α radiation.

(0111) join onto a single boundary parallel to (0110). The latter boundary has near maximum visibility, the phase jump at it being about 0.9 π. In this field, diffraction contrast due to growth stratifications is locally quite strong. Qualitatively, the pattern can be described as the superimposition of extra scattering at the fault surface upon the extra scattering from individual growth layers. Thus the nearly vertical dark bands crossing the twin boundary image just to the right of the field center are the projections of the intersection of the twin boundary with the strongly-scattering growth stratifications. Two dislocation images run across the field from the left margin and outcrop at the specimen surface. Their direction is {2111} or close to it. The upper dislocation image is very straight, and double. The lower image is single and locally coiled and decorated. Both dislocations perturb the images of the growth stratifications where they transect them, showing that there is an appreciable "dynamical" contribution to the images of the stratifications, involving interference between waves of both branches of the dispersion surface, which is sensitive to the dislocation strain-fields.

INTERNAL MAGNETIC DOMAIN STRUCTURES

The easy directions of magnetization in single crystals of iron are the cube directions. As is well known, large single crystals of iron are found to be divided into magnetic domains. Their configuration is determined by the externally applied field and, of particular importance in the cases now to be considered, the need to avoid the presence of free magnetic poles at the specimen surface because of the high magnetostatic energy such poles involve. At the magnetic domain boundaries, the direction of magnetization may reverse, producing a '180° wall,' or it may turn through 90° to follow another cube axis, producing a '90° wall.' Now a single crystal of iron which is also a single magnetic domain, with the magnetization along [001], say, is not cubic, but slightly tetragonal because of magnetostriction. At room temperature, its axial ratio would exceed unity by 2×10^{-5}. If the single crystal is divided up into magnetic domains by 180° walls only, all the 'tetragonal' domains have the same axis of extension and fit together perfectly. When 90° walls are present the situation is different. Consider a 90° wall parallel to (011) at which the direction of magnetization changes from [001] to [010]. A drawing with the tetragonality due to magnetostriction much exaggerated quickly shows that planes in the [100] zone which cross the boundary have a kink there. If there is a network of 90° walls in the specimen then there will be curvature as well as kinks in the lattice planes because of the deformations required to fit all the domains together. The strain fields produced at the boundaries are not dissimilar in type and magnitude to those encountered when passing a dislocation at a distance of some microns. Thus it is not surprising to find that 90° walls are visible on X-ray topographs and produce diffraction images possessing quite a lot in common with the images of dislocations. In the usual experimental arrangement, the image arises from interbranch scattering at and close to the 90° boundary. The different interplanar spacing in the bulk material on either side of the boundary changes the Bragg angle by a fraction of a second of arc only, and so no "volume contrast" is produced.

The iron specimens have been examined in the form of single crystals of the alloy Fe + 3.5%Si grown from the melt, annealed, and cut into discs which were then mechanically and chemically polished. Diffraction contrast from internal domain structures has been studied in specimens 3 μ thick to over a 100 μ thick. The X-ray topographic method can provide information not obtainable by other means.[33] The colloid method and the optical Faraday rotation method show only the outcrops of domain boundaries at the crystal surfaces and indeed X-ray topography has demonstrated that some domain

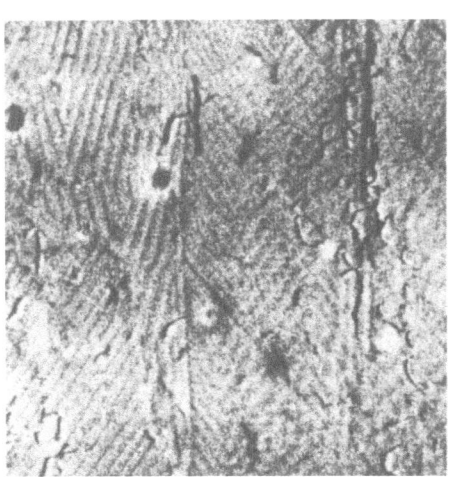

Figure 9. Magnetic domain structures in iron–silicon alloy. Plate parallel to (112), thickness about 30 μ; field edge length 1 mm. Stripe pattern due to domain laminations, dislocation images also present.

boundaries that touch specimen surfaces are not detected by the colloid method.[34] Transmission electron microscopy—Lorentz microscopy—of thin films has high resolution but is limited to films not more than about 1000 Å thick.

Some observations made in the course of X-ray topography of Fe + 3.5% alloy are as follows. In plates parallel to (100) which have two easy directions of magnetization in the plane of the plate, 90° walls cutting the plate and those involved in the 'fir-tree' type of magnetic closure structure can be mapped. In plates parallel to (110) a variety of complex closure structures has been found in parts of the specimens which taper. Many of these structures are internal, though they may touch one or other of the specimen surfaces. In the case of specimen plates parallel to (112), there is no easy direction of magnetization parallel to the surface, and the whole specimen is divided into laminations in which domains of different magnetization directions are sandwiched together in such sequence as will minimize the magnetostatic energy due to free poles on the specimen surface. The scale of division gets finer as the surface is approached and may, at the surface, be at the limit of resolution of the colloid technique (about 1 μ). Such fine structures are not resolvable by X-ray topography. What the X-ray topographs can show by penetrating the full specimen thickness is the underlying basic lamination pattern into which the specimen is divided. Even this basic pattern is generally very complicated (Figure 9 shows an example), but occasionally in thinner specimens (10–25 μ thick) the lamination pattern can be satisfactorily interpreted. X-ray topographic experiments can of course be made repetitively and magnetic domain movements observed.[35] Two serious questions amenable to investigation through X-ray topographic study of domain structures are: first, how far are minimum energy structures achieved in the observed configuration of magnetic domains; and second, how are the actual observed structures influenced by the presence of lattice imperfections. At last a sensitive experimental method is available to enable one to say when an observed magnetic structure can be attributed to local stress concentrations, and when it cannot.

X-RAY MOIRÉ PATTERNS

Where two periodic media overlap, moiré patterns can be produced when light, electrons, or X-rays are transmitted through them. The same basic geometry applies in the formation of the patterns with all radiations—only the scale of the periodicities is changed. Moiré fringes are quite frequently observed in the electron microscope when two thin crystals overlap and both are Bragg reflecting. The conditions for their production are not very critical since the angular range of Bragg reflection is quite large, approximately 10^{-2} of a radian, and the high resolution of the electron microscope enables closely-spaced moiré fringes to be observed, as are produced when the difference of interplanar spacing or of orientation of the two crystals is relatively large. In electron microscope moiré patterns, dislocations and stacking faults in one or other of the overlapping crystals can be revealed by the dislocation or displacement of the moiré fringe pattern. Moiré fringes in X-ray topographs of directly overlapping, simultaneously reflecting crystals have been observed only recently.[36,37] The fringes observed in the X-ray interferometer[38] are also essentially moiré fringes formed between the first and third crystal in this device, the symmetry of the X-ray beam paths making it possible to have a large spatial separation between the two crystals whose periodicities interact to form the moiré fringe system.

The basic analysis of X-ray and electron microscope moiré patterns produced by a pair of Bragg-reflecting crystals is as follows. Let us call the crystal through which the radiation first passes A, and the second, B. In a fixed coordinate system let $\mathbf{g}_A$ be the reciprocal lattice vector of the active Bragg reflection in crystal A and $\mathbf{g}_B$ the vector in the crystal B. (The magnitudes of these vectors are the reciprocals of the interplanar spacings.) Then the reciprocal lattice vector $\mathbf{G}$ of the moiré fringe system (in the same chosen coordinates) is given simply by $\mathbf{G} = \mathbf{g}_A - \mathbf{g}_B$. The period D of the moiré fringe system is given by $D = |\mathbf{G}|^{-1}$. Interpretation of moiré patterns is assisted by considering separately the contributions of the misorientation of B relative to A and of the change of interplanar spacing of B relative to A. These are the 'rotation' and 'compression' components in the difference vector $\mathbf{G}$. In the pure 'rotation' moiré, one is concerned with the case when there is no difference in interplanar spacing, but where $\mathbf{g}_A$ makes a small angle ϵ with $\mathbf{g}_B$. Then $\mathbf{G}$ is perpendicular to $\mathbf{g}_A$ and $\mathbf{g}_B$, the moiré fringes are perpendicular to the traces of the Bragg planes, and the moiré fringe spacing D is given by d/ϵ, where d is the interplanar spacing of the Bragg planes. In the pure 'compression' moiré, $\mathbf{g}_A$ and $\mathbf{g}_B$ have the same direction but different lengths. The difference vector $\mathbf{G}$ is thus parallel to $\mathbf{g}_A$ (and $\mathbf{g}_B$) and the moiré fringes run parallel to the traces of the Bragg planes. The spacing D of the compression moiré is given by $d_A d_B/(d_A - d_B)$, where d_A and d_B are the interplanar spacings of the operating Bragg planes in crystals A and B, respectively.

The above simple analysis indicates one of the restrictions on the range of difference vectors $\mathbf{G}$ within which X-ray moiré patterns may be observed. Consider the geometry of the experimental way in which X-ray reflections are observed, with the reciprocal lattice vector touching the Ewald sphere, and with the incident beam and the reciprocal lattice vector together defining the plane of incidence in which the diffracted beam also lies. It will be seen that in the case of the pure rotation moiré, as long as ϵ is small, not only can $\mathbf{g}_A$ and $\mathbf{g}_B$ be made to touch the Ewald sphere, but also $\mathbf{g}_B$ when the incident beam on B is the diffracted beam from A. This condition must be fulfilled if both crystals are to simultaneously reflect with equally strong intensity and give a moiré fringe pattern of maximum visibility. The condition can be attained in practice by setting $\mathbf{g}_A$ to touch the Ewald sphere and then rotating the crystal about $\mathbf{g}_A$ until $\mathbf{g}_B$ also touches the sphere. When $\mathbf{g}_A$ and $\mathbf{g}_B$ are of different lengths, it is not possible to satisfy the Bragg condition exactly both for crystals A and B, and the maximum visibility of moiré fringes will not be obtained.

Some of the complexities of X-ray moiré fringes are illustrated in Figure 10. The fringes are formed by the relative displacement between the two sides of a crack in

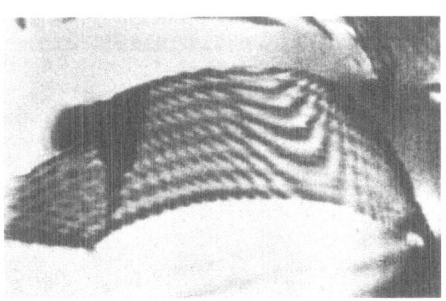

Figure 10. X-ray moiré fringes at crack in quartz plate 1 mm thick. Field width 2 mm; reflection 10$\bar{1}$0, Ag K_α radiation; projection of diffraction vector nearly horizontal.

quartz. The crack surface is roughly cylindrical and it cuts both surfaces of the plate which is about 1 mm thick. The concave edge of the crack image in Figure 10 is nearer the observer. Near its left end the crack branches. Since the crack divides the crystal into two overlapping wedges, Pendellösung fringes which run as thickness contours of these wedge-shaped volumes, and hence as contours of the distance of the crack from the specimen surface, are produced. These are the fringes which run parallel to the curve of intersection of the crack with the specimen surfaces. The moiré fringes are the more highly curved fringes, trending horizontally somewhat to the right of center of the crack image, and nearly vertically at either end of the crack. Since the projection of the diffraction vector is nearly horizontal on the topograph, it follows that where the moiré fringes run nearly horizontally they are dominantly a rotation moiré, and where vertically, then compression is dominant. As indicated in the discussion in the previous paragraph, maximum fringe visibility is achieved in the pure rotation case. Then it is possible for both crystals A and B to satisfy the Bragg condition equally well and when they do so the period of the thickness contours is halved, as may be seen in the topographs. This is because the situation giving rise to the moiré fringes is equivalent to a stacking fault with a spatially periodic fault vector—and in stacking faults with exact satisfaction of the Bragg condition the period of depth contours is halved.[25] Note that the moiré fringes provide a method for measuring the displacement field at the crack directly in units of interplanar spacing, just by counting fringes. The great advantage of X-ray moiré fringes over those seen in the electron microscope is the large moiré-fringe spacing observable in the X-ray case, say a spacing of the order of 1 mm. If one is studying a compression moiré with a fringe spacing of 1 mm and the Bragg spacing is 1 Å, then the relative difference in interplanar spacing being measured is only one part in 10^7, about a couple of orders of magnitude smaller than that measurable by conventional X-ray diffraction methods.

REFERENCES

1. A. R. Lang, "Dislocations in Diamond and the Origin of Trigons," *Proc. Roy. Soc. (London)* **A278**: 234, 1964.
2. F. C. Frank and A. R. Lang, "X-Ray Topography of Diamond," in *Physical Properties of Diamond*, ed. R. Berman, Clarendon Press, Oxford, England, 1965, Chapter III, p. 69.
3. M. Takagi and A. R. Lang, "X-Ray Bragg Reflection, 'Spike' Reflection, and Ultra-violet Absorption Topography of Diamond," *Proc. Roy. Soc. (London)* **A281**: 310, 1964.
4. Y. Kamiya and A. R. Lang, "On the Structure of Coated Diamonds," *Phil. Mag.* **11**: 347, 1965.
5. Y. Kamiya and A. R. Lang, "X-Ray Diffraction and Absorption Topography of Synthetic Diamonds," *J. Appl. Phys.* **36**: 579, 1965.

6. B. Lawn, Y. Kamiya, and A. R. Lang, "An X-Ray Topographic Study of Planar Growth Defects in a Natural Diamond," *Phil. Mag.* **12**: 177, 1965.
7. H. H. Schlössin and A. R. Lang, "A Study of Repeated Twinning, Lattice Imperfection, and Impurity Distribution in Amethyst," *Phil. Mag.* **12**: 283, 1965.
8. A. Authier, C. B. Rogers, and A. R. Lang, "On the Macroscopic Distribution of Dislocations in Single Crystals of High-Purity Recrystallized Aluminium," *Phil. Mag.* **12**: 547, 1965.
9. A. R. Lang and M. Polcarová, "X-Ray Topographic Studies of Dislocations in Iron-Silicon Alloy Single Crystals," *Proc. Roy. Soc. (London)* **A285**: 297, 1965.
10. A. R. Lang, "Studies of Individual Dislocations in Crystals by X-Ray Diffraction Microradiography," *J. Appl. Phys.* **30**: 1748, 1959.
11. N. Kato and A. R. Lang, "A Study of Pendellösung Fringes in X-Ray Diffraction," *Acta Cryst.* **12**: 787, 1959.
12. M. Yoshimatsu, "Some Observations of Imperfections in *ADP* Single Crystals by X-Ray Diffraction Micrography," *Japan. J. Appl. Phys.* **5**: 29, 1965.
13. N. Kato, K. Usami, and T. Katagawa, "The X-Ray Diffraction Image of a Stacking Fault," *Advances in X-Ray Analysis, Vol. 10*, ed. G. R. Mallett and J. B. Newkirk, Plenum Press, New York, 1967, pp. 46–66.
14. N. Kato, "Wave-Optical Theory of Diffraction in Single Crystals," in *Crystallography and Crystal Perfection*, ed. G. N. Ramachandran, Academic Press, New York and London, 1963, p. 153.
15. C. Frondel, *Dana's System of Mineralogy, Vol. III. Silica Minerals*, 7th ed., Wiley, New York and London, 1962.
16. A. R. Lang, "The Orientation of the Miller–Bravais Axes of Alpha Quartz," *Acta Cryst.* **19**: 290, 1965.
17. A. R. Lang, "A Method for the Examination of Crystal Sections Using Penetrating Characteristic X-Radiation," *Acta Met.* **5**: 358, 1957.
18. A. R. Lang, "The Projection Topograph: A New Method in X-Ray Diffraction Microradiography," *Acta Cryst.* **12**: 249, 1959.
19. V. A. Phillips and J. D. Livingston, "Direct Observation of Coherency Strains in a Copper–Cobalt Alloy," *Phil. Mag.* **7**: 969, 1962.
20. M. Wilkens and F. Meier, "Zur Kontrastbreite röntgenographisch abgebildeter Versetzungen," *Z. Naturforsch.* **18a**: 26, 1963.
21. M. Hart, "Dynamical X-Ray Diffraction in the Strain Fields of Individual Dislocations," *Thesis, University of Bristol*, 1963.
22. A. R. Lang, "X-Ray Topographic Determination of the Sense of Pure Screw Dislocations," *Z. Naturforsch.* **20a**: 636, 1965.
23. J.-I. Chikawa, "X-Ray Topographic Observation of Dislocation Contrast in Thin Cadmium Sulfide Crystals," *J. Appl. Phys.* **36**: 3496, 1965.
24. A. Authier and M. Sauvage, "Études Topographiques de défauts dans les Cristaux. Contraste et applications," *Seventh International Congress of Crystallography*, Moscow, July 1966, paper 2.2.
25. M. J. Whelan and P. B. Hirsch, "Electron Diffraction from Crystals Containing Stacking Faults," *Phil. Mag.* **2**: 1121, 1303, 1957.
26. G. H. Schwuttke and V. Sils, "X-Ray Analysis of Stacking Fault Structures in Epitaxially Grown Silicon," *J. Appl. Phys.* **34**: 3127, 1963.
27. M. Yoshimatsu, K. Kohra, and I. Shimizu, "X-Ray Observation of Lattice Defects using a Crystal Monochromator," in *Direct Observation of Imperfections in Crystals*, ed. J. B. Newkirk and J. H. Wernick, Interscience, New York and London, 1962, p. 461.
28. N. Kato, "Pendellösung Fringes in Distorted Crystals, Parts I, II and III," *J. Phys. Soc. Japan* **18**: 1785, 1963; **19**: 67, 971, 1964.
29. M. Hart, "Observations of Pendellösung Fringes in Elastically Deformed Crystals," *Appl. Phys. Letters* **7**: 96, 1965.
30. N. Kato and Y. Ando, "Contraction of Pendellösung Fringes in Distorted Crystals," *J. Phys. Soc. Japan* **21**: 964, 1966.
31. Y. Ando and N. Kato, "X-Ray Diffraction Patterns of an Elastically Distorted Crystal," *Acta Cryst.* **21**: 284, 1966.
32. A. R. Lang, "Mapping Dauphiné and Brazil Twins in Quartz by X-Ray Topography," *Appl. Phys. Letters* **7**: 168, 1965.
33. D. J. Craik and R. S. Tebble, *"Ferromagnetism and Ferromagnetic Domains,"* North-Holland, Amsterdam, 1965.
34. F. C. Frank, J. Kaczér, A. R. Lang, and M. Polcarová, in preparation.

35. M. Polcarová and A. R. Lang, "X-Ray Topographic Studies of Magnetic Domain Configurations and Movements," *Appl. Phys. Letters* **1**: 13, 1962.
36. A. R. Lang and V. F. Miuscov, "Angstrom-Scale Displacements Revealed by X-Ray Moiré Topographs," *Appl. Phys. Letters* **7**: 214, 1965.
37. J. -I. Chikawa, "X-Ray Observation of Moiré Patterns with Superposed CdS Crystals," *Appl. Phys. Letters* **7**: 193, 1965.
38. U. Bonse and M. Hart, "An X-Ray Interferometer," *Appl. Phys. Letters* **6**: 155, 1965.

DISCUSSION

G. A. Walker (3-M Company): Would it be possible to study ferroelectric domains by this means where the polarization reverses only in a few unit cells?

J. B. Newkirk: As far as I know, this hasn't been reported in the literature, although it certainly is reasonable that there should be a topographical effect due to ferroelectric strains.

THE DILEMMA OF ANOMALOUS
X-RAY REFLECTIONS

James F. McGee

*Saint Louis University
St. Louis, Missouri*

and

Veli I. Olli

*University of Turku
Turku, Finland*

ABSTRACT

That X-rays focused by reflection from concave surfaces exhibit anomalous image broadening was shown by Ehrenberg using a line source. Even flat non-focusing surfaces of many different materials produced an anomalous striated reflection. According to Ehrenberg, the common cause for the observed broadening in the focused image and the striated pattern in the nonfocused reflection is a periodic surface structure. In later years, Eliot, using a point source of X-rays reflected from a concave surface, observed a striated pattern at a position well beyond the focal plane. It is to be noted that Eliot's experiment combined elements of Ehrenberg's two experiments—a focusing surface and observations made far from any focal plane, real or virtual. Eliot observed his striations with fused silica surfaces but not with obsidian. Recently, Yoneda, using X-rays collimated by a Soller slit and incident on a plane surface, observed an anomalous line reflection not previously reported.

Various experiments performed by the present authors have attempted to duplicate some of the above situations as closely as possible, either directly with X-rays or in an analogous manner with visible light. It will be shown that satisfactory explanations can be formulated from previously unsuspected diffraction phenomena, recently confirmed by experiment as well as by a simple experimental oversight. In spite of the dilemma presented, either horn eliminates all previous limitations due to surface conditions, so that the tolerable surface roughness is once again determined by the well-known Rayleigh criterion rather than by any periodic surface irregularities introduced through the polishing process or by nature.

EXPERIMENTAL METHODS

Anyone considering the use of the total reflective property of X-rays in the design of an X-ray microscope or X-ray telescope will soon encounter the recent literature on the subject of anomalous reflection of X-rays.[1-4] While the first three authors dispose of the anomaly reported by Yoneda,[4] there remain earlier anomalies reported by Ehrenberg[5] in connection with his experiments on the reflection of X-rays from optical flats and concave reflectors.

An uncoated optical flat M_1, rendered concave by straining in a specially designed clamp and shown in Figure 1, was used by Ehrenberg[5,6] to focus X-rays emanating from a fine slit S_1 illuminated with X-rays from a G.E. CA-type X-ray diffraction tube.

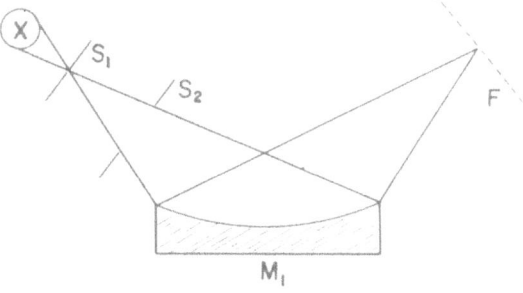

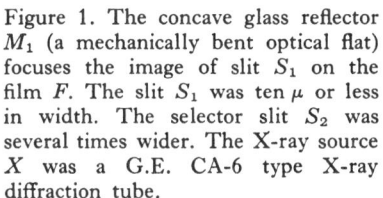

Figure 1. The concave glass reflector M_1 (a mechanically bent optical flat) focuses the image of slit S_1 on the film F. The slit S_1 was ten μ or less in width. The selector slit S_2 was several times wider. The X-ray source X was a G.E. CA-6 type X-ray diffraction tube.

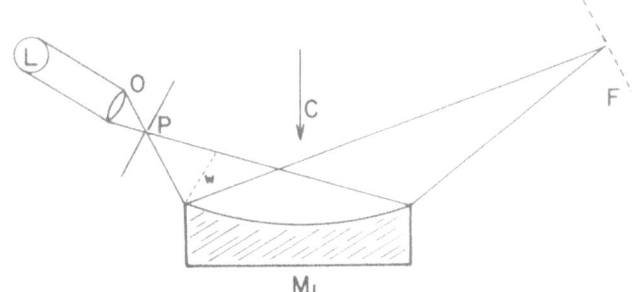

Figure 2. A laser analog experiment in which light from a single mode operating gas laser L is focused by objective lense O through pinhole P to form a diffraction limited cone of coherent radiation incident upon a concave glass reflector M_1 of 3-m radius of curvature.

The "selector" slit S_2 was much wider than S_1. A two-second exposure yielded a fine line image in accordance with the mirror equation. However, upon increasing the exposure time to 20, 200, 2000, and 20,000 seconds, respectively, the line image widened considerably. Ehrenberg accounted for the presence of this "stray light in the neighborhood of the focus" by postulating an imperfect reflecting surface with "hills and dales."

The essential elements of Ehrenberg's experiment were recently repeated using 6328 Å light from a gas laser instead of X-rays. The physical arrangement is shown in Figure 2 where a microscope objective O is used to focus the laser light L in the plane of the pinhole P. The resulting diffraction-limited cone of light is incident on a spherical glass reflector M_1 with a radius of curvature of approximately 3 m.

The image appearing on the screen is the usual astigmatic line image of a point source formed at grazing incidence. Examination with a low-power magnifier shows, however, that it is broadened unsymmetrically by diffraction, as may be seen in Figure 3. It has been shown by Hesser and McGee[7] from electromagnetic wave theory that this type of diffraction pattern should be formed at grazing incidence by a concave reflecting surface whenever a large amount of spherical aberration is present. It is clear that a rough surface is not necessary for the production of an image which will broaden with an increase of exposure time.

Conceivably, a short exposure would show only the first few fringes of high intensity. Longer exposures would finally reveal the entire width of the diffraction pattern. However the fringes shown in Figure 3 would not be resolved if X-rays were used instead of the visible laser beam. To confirm the latter, we repeated this part of Ehrenberg's experiment. The result of a four-hour exposure is shown in Figure 4. The reflected

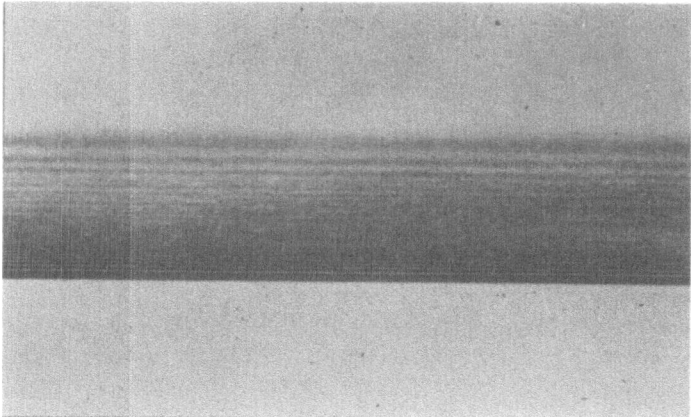

Figure 3. Image of pinhole P of Figure 2 broadened by diffraction in the presence of spherical aberration. The image is about 3 cm in width with broad fringes at top and narrow sharp fringes at bottom. The first bright fringe corresponds to the position of the image according to geometrical optics.

Figure 4. X-ray image of slit S_1 of Figure 1 formed with 1.54 Å radiation and a reflector M_1 of 3-m radius of curvature. The X-ray image at the top is broadened by diffraction. The exposure time was 4 hr so that the central beam is over-exposed and shows strong halation. The weaker nonmeridonal rays serve to outline the slit on the right and left sides of the picture. They also give the position of the image for low exposure. That it spreads upward (increasing angle of reflection) is in agreement with Figure 3.

beam is on top. The broad lower beam includes the direct beam, the refracted continuous, and refracted $K\,\alpha\beta$ unresolved, plus a large amount of halation. A shorter exposure of two seconds yielded a very fine line-image. Its equivalent may be seen in the tapering of the image on the right and left sides of Figure 4, which corresponds to a short exposure time, since this part of the image is formed by weak nonmeridonal X-rays. Thus, a single long-exposure-time photograph shows simultaneously the effects of short and long exposure. The broadening of the X-ray image of Figure 4 in the direction of increasing angle is seen to be in agreement with the diffraction pattern of Figure 3.

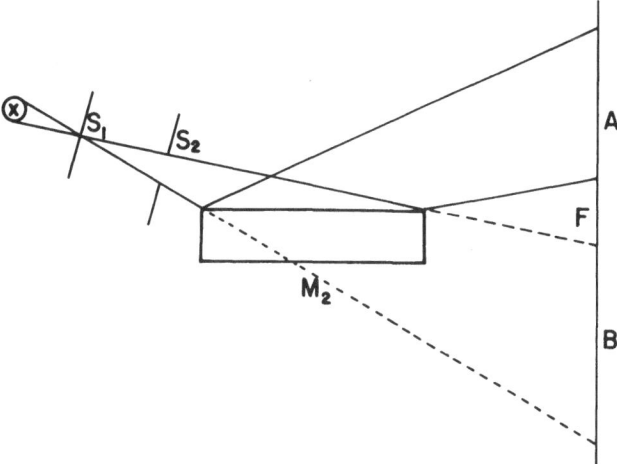

Figure 5. Experimental arrangement used by Ehrenberg. The reflector M_2 is an optical glass flat. Film F at position A revealed a striated pattern consisting of approximately 25 lines, each 10 μ in width. When M_2 was removed and the X-rays recorded on film F at position B, a homogeneous distribution was noted.

In Ehrenberg's second experiment, represented in Figure 5, X-rays diverging from a "narrow" slit S_1 were reflected from an *optical flat* M_2 of unspecified quality. A photographic plate F placed in position A to receive the X-rays after reflection did not show the uniform darkening expected from a homogeneous divergent beam. Instead, anomalous bands or striations not unlike those of an interference pattern appeared. When the flat reflector was removed and the divergent beam recorded directly upon the photographic plate as in film position B of Figure 5, a more uniform blackening was evident. The latter would seem to substantiate the flat surface as the cause of the striations. The theory formulated by Ehrenberg to explain this result postulated that the optical flat surface had imperfections in the form of hills and dales. The "dales," as concavities in the flat reflector, obey the lens law and thus cause a point source to be imaged as a line according to the known optical properties of image formation at angles of grazing incidence. A reverse type calculation by Ehrenberg yielded widths and depths which were in apparent agreement with the results for defects in polished glass and metal surfaces found by Tolansky[8] and others by interferometric methods. Ehrenberg did not give details of these calculations.

Later, Buteux,[9] an associate of Ehrenberg, investigated reflections from a large variety of reflecting surfaces including steel, calcite, coated plastic, gypsum, alkalide halides, etc. The striated pattern was reproduced in each case with only small differences. Is it coincidental that so many plane surfaces produced in a variety of ways by polishing, cleaving, and casting should all have imperfections of the type required for the focusing of X-rays into a striated pattern? Or does it mean that the phenomenon is completely independent of surface details and can be explained by optical phenomena associated with geometrically perfect surfaces? The latter view is held by the present authors.

The laser arrangement of Figure 2 was again used but with an *optical flat* by analogy with Ehrenberg's second experiment. The picture, Figure 6, was taken with the axis of the beam making a very small angle with the surface of the optical flat. The observed pattern is the typical Lloyd's mirror interference pattern with the wider fringes occurring

Figure 6. Lloyd type fringes recorded at position A in the laser analogue of X-ray experiment shown in Figure 5 where slit $S_{1,2}$ and X-ray source X have been replaced by laser L, objective lense O, and pinhole P of Figure 2 with beam axis approximately parallel to the surface of the optical flat.

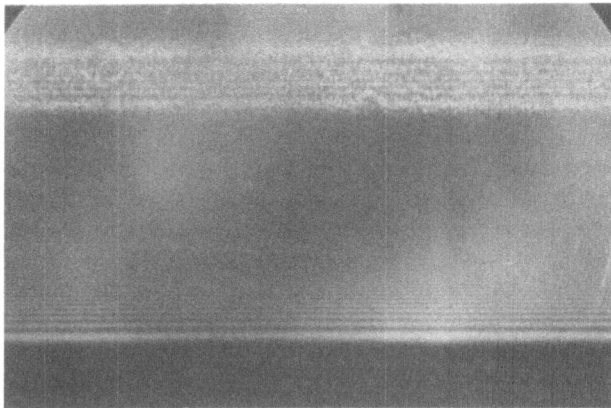

Figure 7. Newly observed phenomenon when optical flat M_2 used in making Figure 6 is rotated about an axis normal to the plane of Figure 2 where M_2 replaces M_1.

near the edge of the reflector. As the angle is increased, it may be noted in Figure 7 that a new phenomenon is superimposed on the Lloyd's pattern. With further increase of angle, the new fringe system moves bodily upward, away from the geometrical shadow edge.

If a curtain or aperture stop C is now lowered toward the optical flat as shown in Figure 2, it is first noticed that the Lloyd's system of fringes which appeared as background in the previous pictures is now limited, as seen in Figure 8, to where it appears totally under the new striated phenomenon shown in the previous figure. Further lowering of the aperture stop results in almost complete disappearance of the background fringe system. The new fringe system revealed in Figure 9 bears a striking resemblance to those observed by Ehrenberg. However, they are more uniform and of better quality. Their position is that corresponding to the reflected position of the incident beam. The fringes are similar in appearance to the diffraction fringes of a narrow slit. Since the *flat* is oriented at a small angle to the beam axis, one would possibly expect the appearance of fringes corresponding to diffraction by a narrow obstacle, the latter being the projected width w (Figure 2) of the reflector M_1 which has been replaced by M_2 of Figure 5. However, because of the presence of the reflecting surface, one should expect the complementary diffraction pattern of a slit rather than a narrow obstacle, the width w being determined by the projected length of the reflecting flat in image space instead of object space. The subsidiary experiment of actually inserting a separate narrow slit into the direct laser beam and observing it was performed. Better intensity agreement occurred when the edges were non-coplanar by an amount less than the length of the mirror.

While the diffraction pattern described above bears a strong resemblance to Ehrenberg's, it is highly questionable whether he could have produced its equivalent in the X-ray region. Lack of parallel spacings casts doubt on his X-ray pattern's being a true interference or diffraction effect. A simpler explanation of his result will be discussed later.

Since the publication of Ehrenberg's work, a third type of experiment has been described as supporting his conclusion. Reference is made to the arrangement of Eliot's[10] experiment in which radiation from a point source of 1.54 Å X-rays is incident on a concave reflecting surface at small angles of grazing incidence. Several significant differences in experimental arrangement should be noted. Eliot records the X-ray pattern at a position well beyond the focal plane. In retrospect, one might interpret Eliot's arrangement as an Ehrenberg optical-flat experiment using, however, a concave

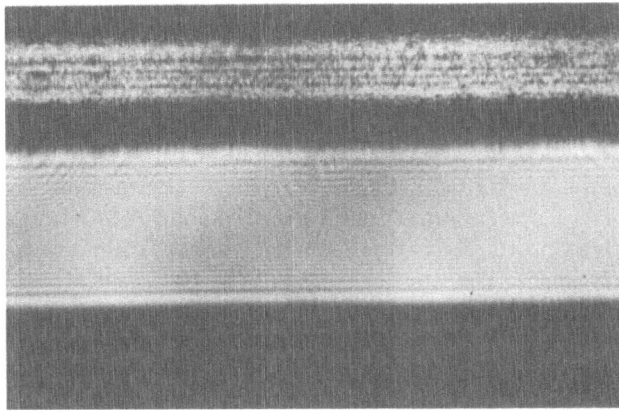

Figure 8. Effect of lowering curtain C toward the surface of the reflector M_2.

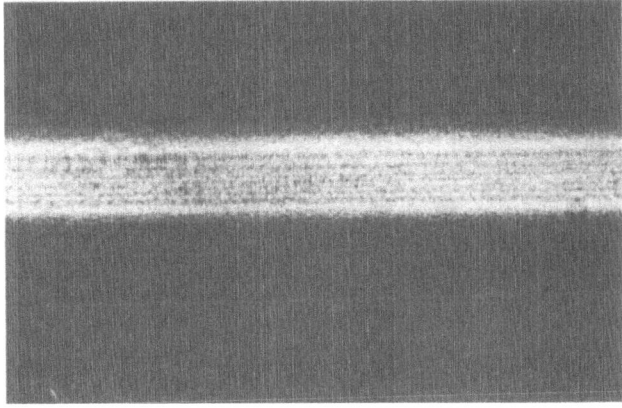

Figure 9. Effect of further lowering curtain C toward the surface of M_2. Since the direct beam has been cut off completely, it cannot interfere with the reflected beam in the production of Lloyd type fringes. The remaining band of fringes are in angular position corresponding to reflection of the incident beam.

reflector of large radius of curvature and with the observations made well beyond the focal plane to avoid any confusion with the focused image of Ehrenberg's first type of experiment. A second difference is that Eliot had an aperture stop similar to C of Figure 2 mounted quite close to the surface of the concave reflector for the express purpose of blocking the direct beam from the photographic plate. Since Eliot used a point source X-ray tube, the curtain or aperture stop close to the reflector surface results in collimation of the X-ray beam to one of very small divergence, essentially parallel for all practical purposes. That a parallel beam of radiation incident on a concave reflector can give rise to an interference pattern also somewhat similar to that observed by Ehrenberg is shown by the following laser experiment.

In this third and final laser experiment, an uncollimated laser beam was made to graze the entire surface of a concave reflector. The adjustment was fairly critical. It was noted that the whole surface must be illuminated in order to produce the effect shown in Figure 10. The observed fringes are not wide as in the other pictures because of the lack of divergence of the laser beam. Eliot offers his fringes as evidence in support of Ehrenberg's surface defect theory. This laser experiment shows that a fringe system well beyond the focal plane is to be expected with a perfectly smooth concave reflector. It is to be recalled that Eliot did not observe any fringe system when he used an obsidian reflector. This is odd, in view of Ehrenberg's experience. As previously stated, both he and his associate observed a fringe system with a number of widely different materials. Perhaps Eliot's lack of success with obsidian is more indicative of a failure to illuminate the entire length of surface as was required with the laser illumination. Failure to satisfy the latter condition yields a reflected beam but no fringes.

Aside from all arguments involving diffraction or interference, it is now felt that the results of Ehrenberg's second experiment can best be explained by what he ruled out early in his paper, namely, structure in the X-ray tube target. Repeating the experiment but using a structureless target is an obvious suggestion. This was Eliot's approach using a point source X-ray tube of the Cosslett–Nixon type. That he should record fringes provided the entire concave surface is illuminated is shown by the third laser experiment. The resemblance of Eliot's fringes to Ehrenberg's is very tenuous.

Structure in the anode of the X-ray tube does not necessarily mean tool or die marks in the target material but rather variations in X-ray intensities from various points of the anode surface because of a smaller or larger density of electrons striking the anode.

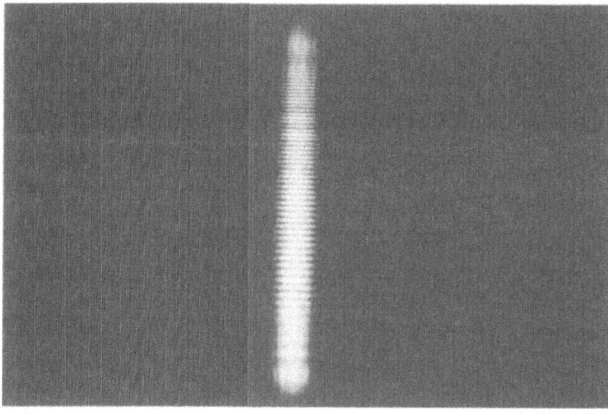

Figure 10. Interference type pattern using experimental arrangement of Figure 2 where M_1 is again a concave glass reflector but the film F is placed well beyond the focal plane. The objective lens O and pinhole P have been removed from the system.

Conceivably, the simple electrostatic focusing system used in a modern X-ray diffraction tube focuses a poor electron image of the filament onto the anode. That such is the case is easily shown by putting a small pinhole on the X-ray port and recording the pin-hole image of the X-ray anode on a film placed a few centimeters away. The result is shown in Figure 11. It is evident that the electron-focusing system does not produce a uniform density of electrons across the face of the anode. The high electron density caused by hotter portions of the filament is clearly evident. Incidentally, the small, black, triangular-shaped areas represent damaged areas of the target where material has been removed and a pit results. X-rays produced in the bottom of the pit cannot escape in the direction of observation of the X-rays, so black areas are produced on the print. In Figure 11, the parallel dense lines are due to the fact that the filament is wound in a helix. The lines are not exactly parallel because the filament winding varies of spacing due to loss of material by evaporation, which in turn results in a variation in electron density along the length of the helical winding. If the pinhole is replaced by a slit system as in Figure 1, the structure takes on the appearance shown in Figure 12, which could easily be mistaken for an interference or diffraction effect. It is notable that the CA-6 tube used by Ehrenberg had a helical filament wound from 0.005 in. of tungsten wire. The helix of the CA-6 has a length of 0.730 in., an ID of 0.020 in. and an OD of 0.030 in. The most interesting specification is that the helix has a total

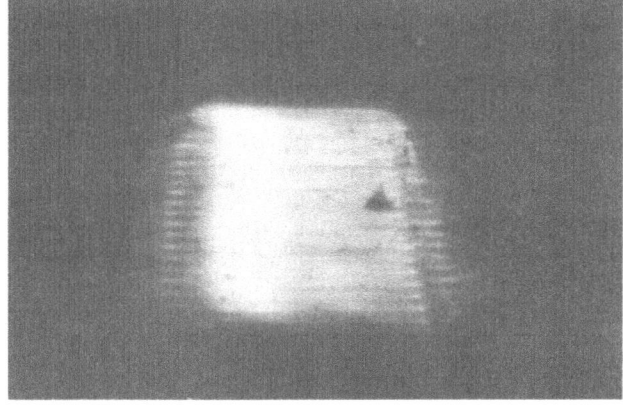

Figure 11. X-ray pinhole camera picture of anode of X-ray tube showing image of filament. About 25 turns may be counted. Large triangular mark is a pit in the anode. Dots distributed over the entire picture are smaller damaged areas.

Figure 12. X-ray slit picture of the anode transmitted through slits S_1 and S_2 of Figure 5. Same picture observed at position A or B.

of 28 turns. When one subtracts a turn at each end as being too cool to contribute to the electron image, the number of effective turns is down to 26. Ehrenberg's striated image showed by his count some 25 fringes which are not parallel or constant in intensity along their length. The experiment which resulted in Figures 11 and 12 was performed with a Philips tube of similar design. Information on the G.E. CA-6 was supplied by the X-ray Department of General Electric Company.

DISCUSSION

The propagation of the filament image by reflection from the optical flat does not alter its intensity distribution significantly. This distribution of intensity was observed in position A of Figure 5. How then did Ehrenberg's image obtained for position B of Figure 5 with the optical flat removed give an apparently uniform distribution when, according to the above, the image obtained after the X-rays pass through two slits would show the structure of Figure 12 with or without reflection? Careful reading of Ehrenberg's paper and examination of his apparatus will reveal that the second slit S_2 was removed as a unit and replaced. The two slits actually determine a certain area of the X-ray-tube anode which is emitting X-rays that get through the system. A slight error in replacing the "selector" slit would cause the apparatus to view a slightly different part of the anode. Pinhole camera pictures reveal that outside the dense image of the filament there exists a region of weak but fairly uniform intensity formed by aberrant electrons from the filament. Actual experiment has revealed that a slight variation of one slit can cause the image of the filament to move out of the field revealing a weak but more uniform region of X-rays. Did Ehrenberg, in removing the mirror to take the picture corresponding to position A in Figure 5, also disturb slit S_2? Was S_2 moved or disturbed when the film was placed in position B? Information on exposure times in position A and B are not available in Ehrenberg's paper. His photographic reproduction for position B is very grainy and may conceal structure of the type discussed above. If such is the case, no dilemma exists and the structural details of his observed fringes can be explained as a slit-image of the filament structure.

ACKNOWLEDGMENT

It is a sincere pleasure for the senior author to acknowledge the use of the facilities of the Institute of Physics, University of Turku, Turku, Finland and the gracious hospitality of its director, Professor Martti H. Kantola during the school year 1964–65.

REFERENCES

1. A. N. Nigam, "Origin of Anomalous Surface Reflection of X-Rays," *Phys. Rev.* **138**: A1189–A1191, 1965.
2. O. J. Guentert, "Study of the Anomalous Surface Reflection of X-Rays," *J. Appl. Phys.* **36**: 1361–66, 1965.
3. B. S. Warren and J. S. Clarke, "Interpretation of the Anomalous Surface Reflection of X-Rays," *J. Appl. Phys.* **36**: 324–325, 1965.
4. Y. Yoneda, "Anomalous Surface Reflection of X-Rays," *Phys. Rev.* **131**: 2010–13, 1963.
5. W. Ehrenberg, "X-Ray Optics: The Production of Converging Beams by Total Reflection," *J. Opt. Soc. Am.* **39**: 741–746, 1949.
6. W. Ehrenberg, "X-Ray Optics: Imperfections of Optical Flats and Their Effect on the Reflection of X-Rays," *J. Opt. Soc. Am.* **39**: 746–751, 1949.
7. D. R. Hesser and J. F. McGee, "Diffraction Theory of Image Formation by Total Reflection," *J. Opt. Soc. Am.* **53**: 525 1963.

8. S. Tolansky and W. L. Wilcock, "Interference Studies of Diamond Faces. A Crossed Fringe Technique," *Proc. Roy. Soc. (London) Ser. A* **A191**: 182, 1947.

9. R. H. Buteux, "Examination of Surfaces by X-Ray Reflection," *J. Opt. Soc. Am.* **43**: 618, 1953.

10. S. B. Eliot, "Effects of Polishing Imperfections on Specular Reflection of X-Rays," in: *Third International Symposium on X-Ray Optics and X-Ray Microanalysis*, ed. H. H. Patee, V. E. Cosslett, and Arne Engstrom, Academic Press, New York, 1963.

X-RAY DIFFRACTION MICROSCOPY OF PLANAR DIFFUSED JUNCTION STRUCTURES

J. K. Howard and G. H. Schwuttke*

International Business Machines Corporation
Hopewell Junction, New York

ABSTRACT

Basic processing steps utilized in the fabrication of planar silicon devices include (a) substrate preparation, (b) epitaxial deposition, (c) thermal growth of silicon dioxide over the entire silicon wafer surface, (d) device pattern formation by photoetching techniques, and (e) diffusion of n-type or p-type elements through the holes to produce localized regions of desired conductivity. A detailed study of the various diffraction phenomena associated with such structures is presented. X-ray topographs of planar transistors show distinct contrast features such as excess diffraction intensity along the silicon oxide/silicon boundary and/or excess diffraction intensity inside the device area. The diffraction phenomena are discussed in terms of reversible elastic deformations, frozen-in lattice deformations, strain fields and imperfections generated by the various processing steps. A technique is presented to measure the sign of the elastic deformations. The phenomenon of stress-jumping across semiconductors interfaces is described, and finally the implications of stress–strain relations on junction performance are stated.

INTRODUCTION

The increasing interest in semiconductor component reliability and fabrication costs necessitates the implementation of in-process methods of device evaluation. An important facet of this inspection should be a determination of the expected device performance based on certain indicators discovered in the fabrication process. If the potential failures are rejected upon detection, a considerable cost reduction could be realized.

As crystal imperfection is a dominant factor that can degrade device quality, X-ray diffraction microscopy is an effective method to relate crystal faults to device failure. Since low energy X-ray irradiation of silicon has little or no effect on its crystal structure, each phase of device processing can be monitored for the formation of crystal flaws. Reflection and transmission methods have been employed to record the defect patterns.[1-5] These methods complement each other as each probes different aspects of crystal perfection.

This paper describes various diffraction contrast phenomena observed during the X-ray inspection of silicon diffused junction devices. The influence of a slight elastic strain, which results from the deposition of thin films on silicon, on diffraction contrast is explored. New findings resulting from this investigation, such as stress-jumping across semiconductors interfaces, are applied to interpret complex crystal deformations observed after certain device processing steps. The observed contrast effects are catalogued to

* Work sponsored in part under AFCRL Contract No. 19(628)-5059.

serve as diagnostic charts for future monolithic device process control, and the implications of localized strains on junction performances are stated.

EXPERIMENTAL

X-Ray Measurements

X-ray diffraction microscopy has been utilized to inspect the defect character of various monocrystalline materials.[1–6] Crystal discontinuities can be investigated with various diffraction geometries; the specific experimental arrangement depends on the intent of the evaluation. Transmission methods (Laue geometry) are used to characterize the bulk crystal perfection.[1,2,5,6] Imperfections at the crystal surface can be detected preferentially by employing a reflection technique (Bragg geometry). If large area surfaces are to be displayed, the scanning–reflection method can be used.[4] The scanning oscillator technique (SOT) was frequently employed to compensate for extended strain fields induced in normal device fabrication processes.[2] This same strain renders standard methods ineffectual if large-area crystal recordings are desired. (A more complete discussion of these methods can be found in the list of references.[2,4])

The SOT topographs were registered on Ilford G-5 (50 μ) nuclear plates; the exposure time varied from 5–15 hr depending on the crystal deformation. Ilford G-5 (25 μ) plates were employed for the reflection method; 3–7 hr exposure times were realized.

Material Preparation and Processing Steps Utilized in Planar Diffused Junction Devices

Figure 1, a simplified flow chart, depicts the fabrication of planar diffused junction devices, either diode or transistor structures. The elemental processes observed in the fabrication of planar diodes or transistors consist of oxidation, opening of windows in the oxide, and diffusion into these apertures. The X-ray results reported here are applicable to integrated device structures as well as to a single diode or transistor.

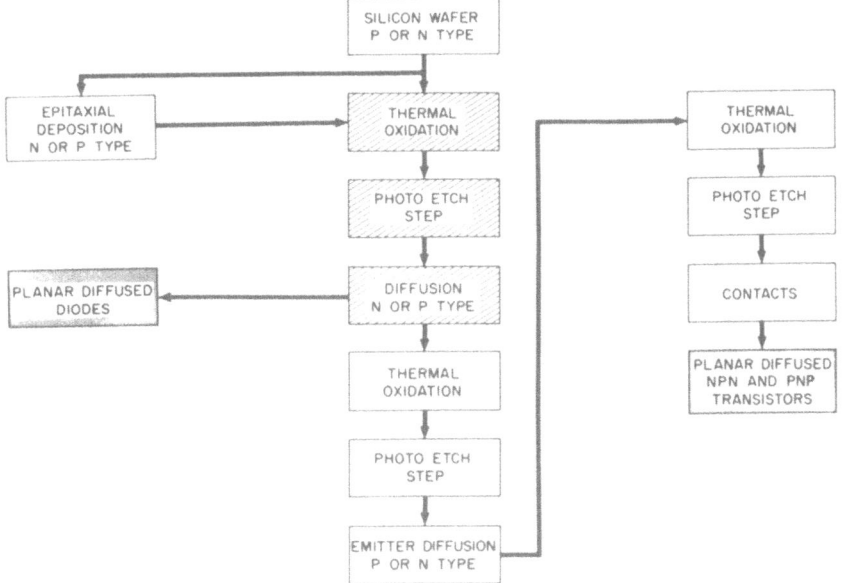

Figure 1. Flow chart for planar diffused junction structures.

Table I. Typical Crystal Faults Due to Device Processing

Process	Type of deformation	Determinant factor	Diffraction effect	Stress	Remarks
Oxidation and window formation	Elastic strain	Mismatch in the coefficient of thermal expansion between SiO_2 and Si	Reflection-dependent contrast at window edge	Tension or compression	Contrast is eliminated if oxide is removed
Diffusion	1. Frozen-in strain	Concentration gradient	Reflection-dependent contrast at window edge	Tension or compression	Contrast is not eliminated if oxide is removed
	2. Diffusion-induced dislocations	Lattice contraction strain due to substitutional boron or phosphorus	Entire diffused volume is faulted	—	Dislocations are contained in the diffused volume
	3. Interfacial dislocations	Concentration gradient at window edge	Dislocations radiate outward into undiffused area	Tension or compression	Various degrees of residual strain contrast evident at window boundary

Substrate Preparation. Silicon monocrystals are employed for planar device fabrication. Wafers are usually cut parallel to the (111) plane with an ID saw to minimize damage.[7] The sawed slice is typically 20 mils in thickness. Wafer processing consists of machine lapping the sawed slice in 25 μ SiC grit. This procedure removes the saw damage and promotes a uniform damaged layer which can be easily removed by etching. After lapping, the slice thickness is about 16 mils. The slice is then etch-polished to remove the residual damage; the resulting thickness varies from 8–10 mils.

Thermal Oxidation. Oxidation of silicon surfaces is accomplished by a number of different processes, such as anodic oxidation, pyrolytic oxidation, or oxidation by reacting an oxidant such as oxygen or steam with the silicon surface at a high temperature ($\sim 1000°C$). The oxide thickness is normally a few thousand angstroms. The thickness depends primarily on the ambient, reaction time, and temperature.[8,9] The oxide layer performs the dual function of protecting the silicon and serving as an impurity mask during diffusion.[10]

Photo Resist. The photo-resist method permits apertures of selected size and shape to be formed in the oxide layer. For that purpose, the oxide layer is coated with a light sensitive organic compound (typically, polyvinyl alcohol), thus simulating a photographic plate. A stencil containing the desired pattern is photographically reproduced on the photo-resist emulsion with a high-intensity ultraviolet light. The unexposed area is easily removed in deionized water; however, the exposed pattern is polymerized and etch resistant.[11] A reactive etch such as HF is then used to remove the oxide and form the diffusion windows.

Masked Diffusion. Since boron and phosphorus have the property of much slower diffusion in SiO_2 than in silicon, only the conductivity and type of the silicon below the window is modulated by the diffusion.

Planar diffused P^+N and N^+P diode structures were fabricated according to the flow chart. The boron diffusion involved a two-step process consisting of a capsule deposition cycle at $\sim 1100°C$, followed by an oxidation drive-in cycle[12] at $1150°C$. PH_3 was employed as source material for the single-cycle phosphorus diffusion. Surface concentrations were about 10^{20} atoms/cc for boron and 10^{21} atoms/cc for the phosphorus. The junction depths varied from 0.5–9.5 μ and were determined by standard angle-lap and stain techniques.

DIFFRACTION CONTRAST PHENOMENA ASSOCIATED WITH DEVICE PROCESSING

Causal Relationship

Crystal flaws introduced in device processing were analyzed in terms of contrast variations as seen in X-ray topographs recorded by different reflections.[5] This contrast evidence, coupled with the physical understanding of the device processing steps, permits an interpretation of the defect structure in regard to a cause and effect relationship. The format for this investigation is depicted in Table I.

Oxidation and Window Formation. After the oxide layer is grown and windows are cut with photo-resist techniques, transmission SOT topographs reveal strong diffraction contrast at the SiO_2/Si boundary.[2,13] The strain contrast is primarily activated by the mismatch in expansion coefficients. Since the thermal coefficient of expansion of silicon differs from its oxide,* the substrate will contract more than the film upon cooling

---

* Different oxides may differ in their coefficients of expansion relative to silicon. For thermally grown oxide, $\alpha_{Si} > \alpha_{SiO}$.

($\alpha_{Si} > \alpha_{SiO_2}$). Thus the slice shape is nearly parabolic and convex to the film surface.[14] The compressive film stress generates a tensile stress at the substrate/film interface. When windows are etched into the oxide, the local stress is partially relieved; however, a strain gradient develops normal to the window edge.[2] The absence of diffraction contrast after removal of the oxide confirms the elastic nature of the strain.

A pyrolytically grown oxide layer (2250 Å; $\alpha_{Si} > \alpha_{SiO_2}$) was deposited at 735°C onto a (111) silicon crystal (1 mm thick); the substrate was dislocation-free. The crystallographic orientation of the triangle is shown in Figure 2. The illustration defines the triangle sides A, B, and C; identification vectors $\mathbf{R}_A$, $\mathbf{R}_B$, and $\mathbf{R}_C$ are assigned orthogonal to the window edges in the direction of $SiO_2 \rightarrow Si$. These vectors are colinear to the actual strain gradient in the crystal. Based on a planar lattice displacement,* the contrast dependence is summarized in Table II. Figures 3a, b, and c show the corresponding SOT topographs for this window. Mo K_α radiation was employed; thus μt was near unity (μ is

_____

* The invariance of Young's modulus and Poisson's ratio permits a treatment of two dimensional stress problems in cubic materials as if the material were isotropic, if the stress plane is the (111) plane.

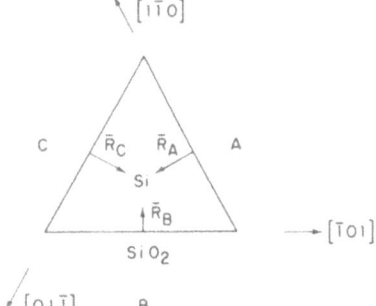

Figure 2. Crystallographic orientation of window in oxide film.

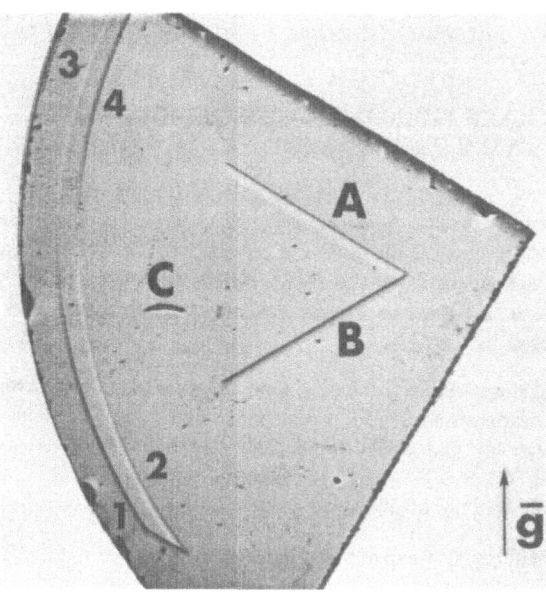

Figure 3a. The (0$\bar{2}$2) topograph of a window formed in a pyrolytic oxide film. Note that cos ∢$(\mathbf{g}, \mathbf{R}_C) = 0$ implies minimal edge contrast at C.

the linear absorption coefficient and t is the slice thickness). The topograph was recorded with the oxide layer facing the X-ray source.

The triangle images (Figures 3a, b, and c) are delineated by dark (strong contrast) lines; light (weak contrast) images are observed adjacent to the dark line. Note that the relative intensity at the window edges varies with the sign and magnitude of cos $\sphericalangle(\mathbf{g}, \mathbf{R})$, where $\mathbf{g}$ is the diffraction vector and $\mathbf{R}$ is the edge identification vector. The negative sign implies the existence of the weak contrast image. The $(0\bar{2}2)$ topograph (Figure 3a) depicts *lower* intensity for edge A (cos $\sphericalangle(\mathbf{g}, \mathbf{R}_A) = -\sqrt{3/2}$) than for edge

$$B \text{ (cos } \sphericalangle(\mathbf{g}, \mathbf{R}_B) = +\sqrt{3/2})$$

Side C is out of contrast as cos $\sphericalangle(\mathbf{g}, \mathbf{R}_C) = 0$.

From Table II, it can be seen that the absolute magnitude of the angle between $\mathbf{g}$ and $\mathbf{R}$, $|\sphericalangle(\mathbf{g}, \mathbf{R})|$, as measured from $\mathbf{g}$, can also serve as a guide to the eclipse of the window edge. The edge contrast is inversely proportional to the absolute magnitude of that angle ($\alpha_{\text{substrate}} > \alpha_{\text{film}}$ and $\sphericalangle(\mathbf{g}, \mathbf{R}) \neq 0$). Maximum contrast is realized if $\sphericalangle(\mathbf{g}, \mathbf{R}) = 0$; therefore, cos $\sphericalangle(\mathbf{g}, \mathbf{R}) = 1$. This effect is shown in Figure 3a. Note the change in peripheral contrast along the semicircular-shaped window in Figure 3a. The contrast varies along this mask boundary because the angle $|\sphericalangle(\mathbf{g}, \mathbf{R})|$ is constantly changing. The lower left side of the image (indicated by 1 in Figure 3a) shows maximum contrast on the left of the cut; here, $|\sphericalangle(\mathbf{g}, \mathbf{R})|$ is less than for the opposite side (2 in Figure 3a). This phenomenon is reversed at the top left of the image (3 and 4 in Figure 3a).

The scanning–reflection method[4] is utilized to obtain topographs of the surface layer; Cu K_α radiation is employed. The (333) topograph (Figure 4) shows good contrast at each window edge. The (422) and (511) reflection topographs also display a similar image. The contrast criterion, cos $\sphericalangle(\mathbf{g}, \mathbf{R})$, predicts minimal contrast at the window edges since cos $\sphericalangle(\mathbf{g}, \mathbf{R}) = 0$ for any $\mathbf{R}$ in the (111) plane. This anomaly is understood if axial strain components are considered. From the X-ray measurements, it must be inferred that the strain field enclosing the window is essentially isotropic in the (111) plane but has a component perpendicular to (111).

The diffraction effects exhibited by the pyrolytic SiO_2/Si interface are typical for most oxide layers grown by different methods ($\alpha_{Si} > \alpha_{SiO_2}$).

Diffusion. The type of crystal faults introduced in the masked-diffusion process are numerous and complex. The faults can be classified in three different groups: (1) frozen-in strains, (2) dislocations generated in the diffusion plane, and (3) dislocations that propagate into the undiffused material. These stress-relief mechanisms occur

Table II. Relationship Between Operating Reflection and the Contrast at the Window Edge

| Identification vector | Contrast at window edge diffraction vector $\mathbf{g}$ | | | | | |
| | $[0\bar{1}1]$ | | $[\bar{1}01]$ | | $[2\bar{1}\bar{1}]$ | |
$\mathbf{R}$	$\|\sphericalangle(\mathbf{g}, \mathbf{R})\|$	cos $\sphericalangle(\mathbf{g}, \mathbf{R})$	$\|\sphericalangle(\mathbf{g}, \mathbf{R})\|$	cos $\sphericalangle(\mathbf{g}, \mathbf{R})$	$\|\sphericalangle(\mathbf{g}, \mathbf{R})\|$	cos $\sphericalangle(\mathbf{g}, \mathbf{R})$
$\mathbf{R}_A$ $[11\bar{2}]$	150°	$-\sqrt{\frac{3}{2}}$	150°	$-\sqrt{\frac{3}{2}}$	60°	$\frac{1}{2}$
$\mathbf{R}_B$ $[1\bar{2}1]$	60°	$+\sqrt{\frac{3}{2}}$	90°	0	60°	$\frac{1}{2}$
$\mathbf{R}_C$ $[\bar{2}11]$	90°	0	60°	$+\sqrt{\frac{3}{2}}$	180°	-1

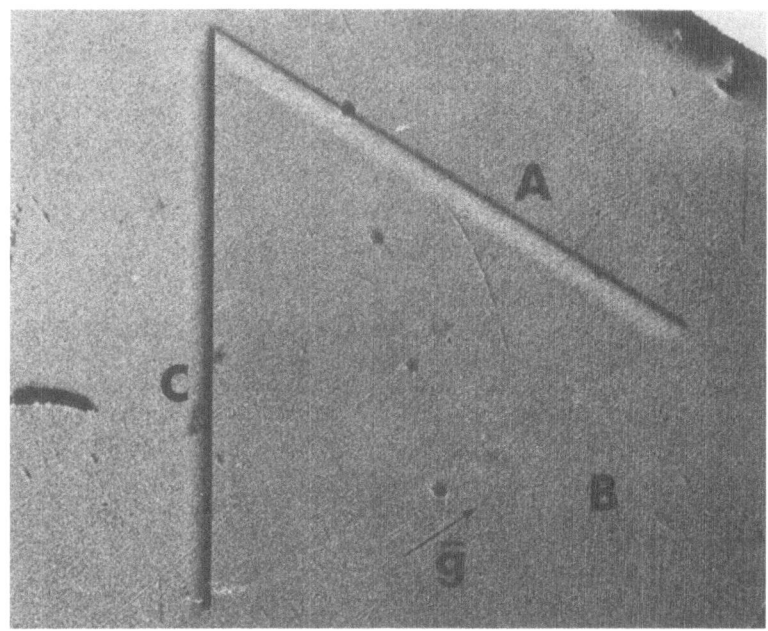

Figure 3b. The ($\bar{2}$02) topograph shows maximum contrast at C; i.e.,
cos $\not\prec(\mathbf{g}, \mathbf{R}_C) = + \sqrt{\tfrac{3}{2}}$.

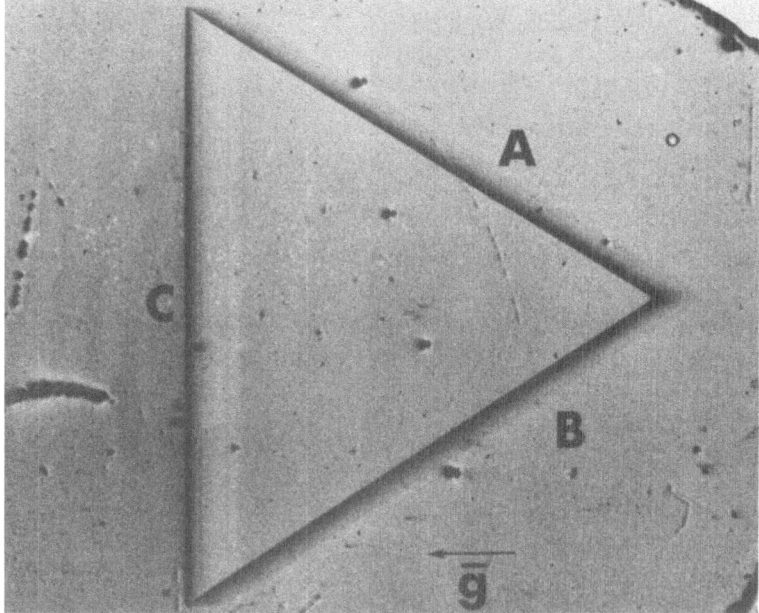

Figure 3c. The (4$\bar{2}\bar{2}$) topograph reveals strong contrast at each window edge.

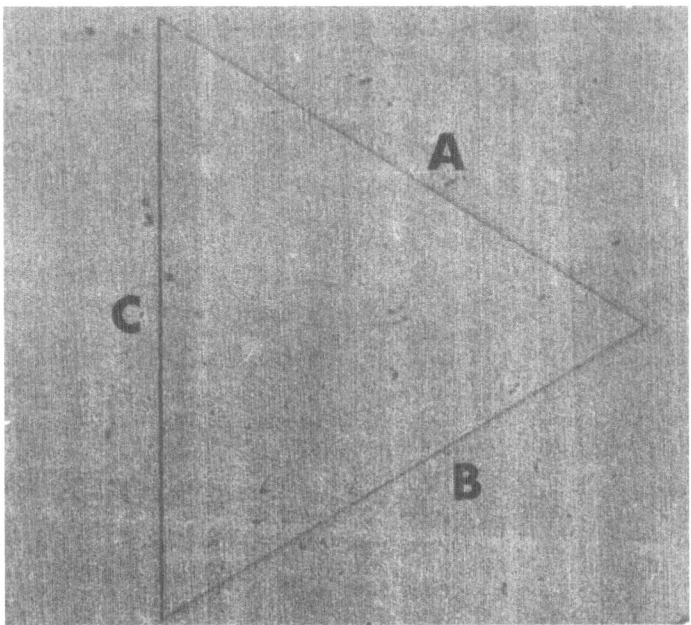

Figure 4. The (333) reflection topograph of this window also depicts strong contrast at each edge even though cos $\not< (\mathbf{g}, \mathbf{R}) = 0$.

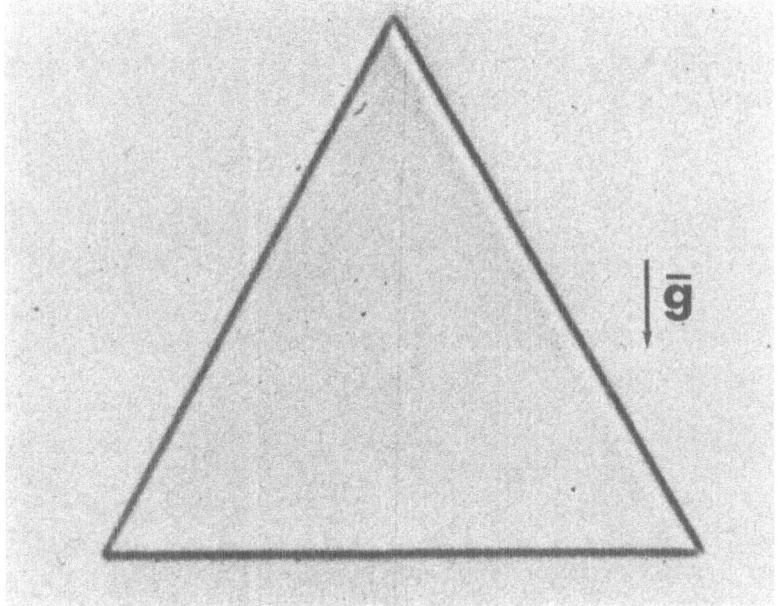

Figure 5. A {220} topograph after masked phosphorus diffusion ($C_0 \sim 10^{20}$ atoms/cc; $X_f = 0.5\ \mu$) and oxide removal. The triangle sides (5 mm) are aligned in $\langle 112 \rangle$ directions.

simultaneously, but differentiation is needed to assign a cause and effect relationship. The pertinent features of these crystal flaws are outlined in Table I.

If the concentration of the diffusant (phosphorus) is below 10^{20} atoms/cc, a locked-in strain defines the diffused/undiffused boundary in the X-ray topographs. The strain contrast is not reversible because removal of the oxide layer does not eliminate the image. Figure 5 displays a window after a low C_0 phosphorus diffusion and oxide removal. Here, the residual strain is still apparent at the window edges, the strain is isotropic in the (111) plane, and the edge contrast varies with $\cos \measuredangle(\mathbf{g}, \mathbf{R})$. The sharp concentration gradient at the interface is responsible for the strain contrast.

Diffusion-induced dislocations were first analyzed through X-ray diffraction microscopy.[15] These results were later confirmed through electron microscopy.[16,17] When the diffusant concentration is greater than $\sim 10^{20}$ atoms/cc (phosphorus), the stress generated in the diffused lattice is relaxed through spontaneous formation of dislocations. SOT topographs, after phosphorus diffusion ($C_0 = 5 \times 10^{20}$ atoms/cc; $X_j = 5.5 \mu$) depict strong contrast in the diffused area (Figure 6a). Dislocation densities in excess of 10^8 cm^{-2} are responsible for the fault contrast. This was verified by transmission electron microscopy[18] as shown in Figure 6b. The dislocations created by lattice contraction are contained in the diffused volume. A scanning–reflection topograph (Figure 7) reveals a window after boron diffusion. The traces of dislocation slip in $\langle 110 \rangle$ directions are evident; dislocations at the edge are bent upward to penetrate the surface at the diffused/undiffused junction (arrow). A more complete analysis of this effect will be reported.[18]

The propagation of dislocations into undiffused material has been reported by several investigators. These dislocations were found below the junction after blanket diffusion[19] and underneath the oxide mask following a localized diffusion.[20,21] In general, it is believed that the dislocations relieve the strain in the diffused volume created by substitutional

Figure 6a. A {220} topograph following a deep phosphorus diffusion (5.5 μ). The entire diffused volume is faulted.

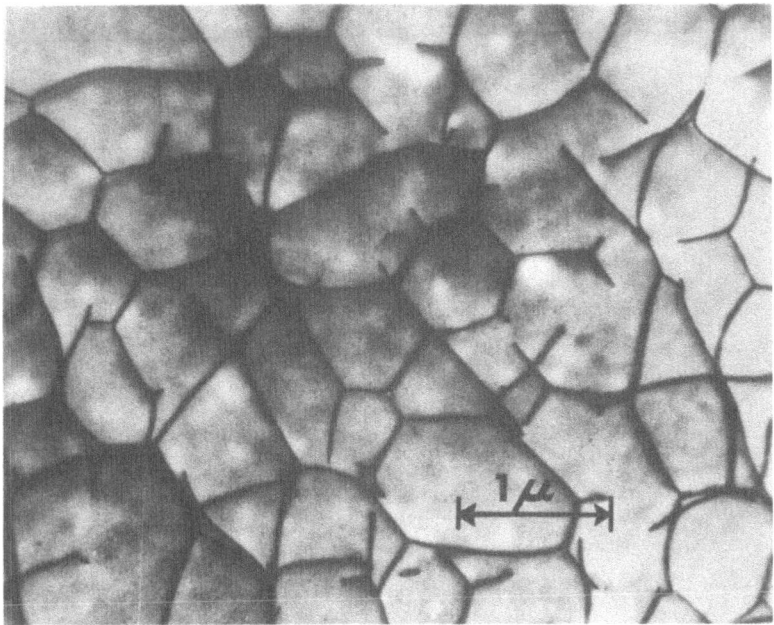

Figure 6b. Transmission electron micrograph (48,000 ×) reveals a hexagonal array of dislocations; these dislocations ($>10^8$ cm^{-2}) are responsible for the contrast in Figure 6a.

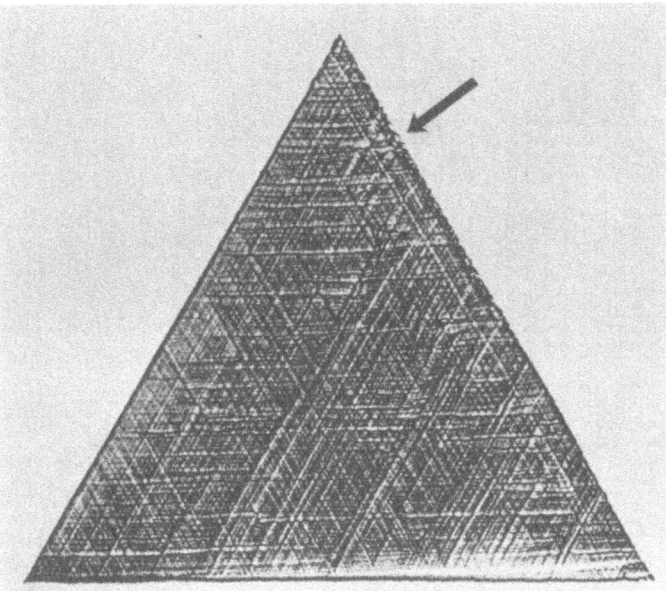

Figure 7. The (115) reflection topograph (Cu $K\alpha$) of a boron diffused area; dislocation slip in $\langle 110 \rangle$ directions is apparent. The dislocations are bent back to intercept the surface at the junction edges (arrow).

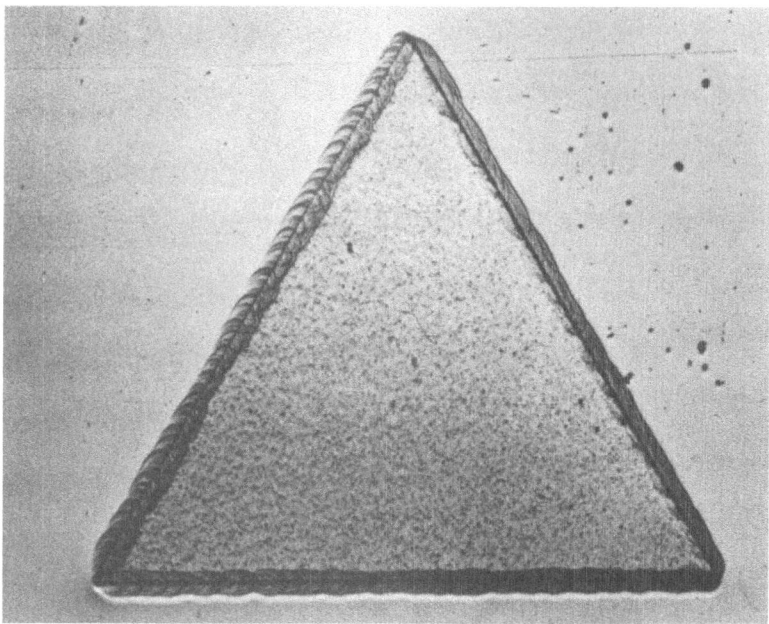

Figure 8. Interfacial dislocations created by residual strain at the diffused/undiffused boundary. These dislocations propagate into the undiffused material.

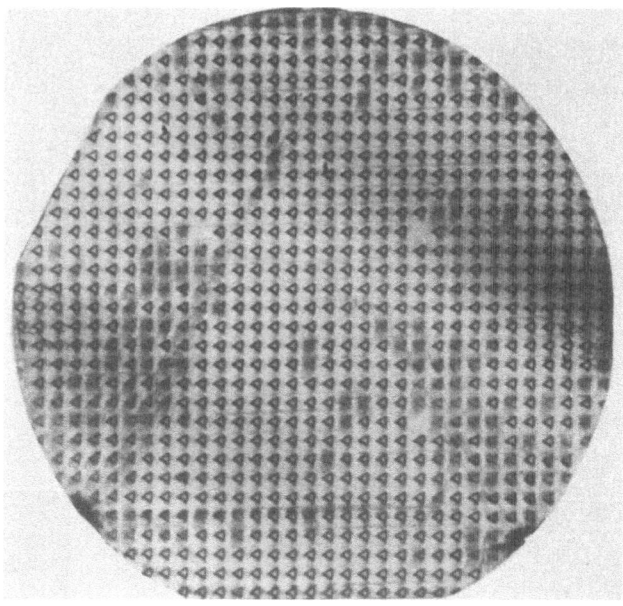

Figure 9. Dislocations formed at the diffused/undiffused boundary are seen to coalesce when normal device geometrics are employed.

boron or phosphorus. However, evidence has been presented which clearly explains the origin of these dislocations.[22] An SOT topograph (Figure 8) reveals a triangular window (5-mm sides) following phosphorus diffusion. The junction is delineated by strong contrast lines; profuse dislocation arrays project into the undiffused silicon. The dark line contrast was encountered after shallow phosphorus diffusions and was shown to indicate residual strain. The interfacial dislocations were created to relieve the stress at the window edge; the stress was activated by the concentration gradient across the junction. An effect of these dislocations is depicted in Figure 9. The normal device structure (boron diffusion; $C_0 = 5 \times 10^{20}$; $X_j = 6.7\ \mu$) yields closely spaced windows ($\sim$ 17-mil sides); the coalescence of these dislocations is evident in certain regions of the slice.

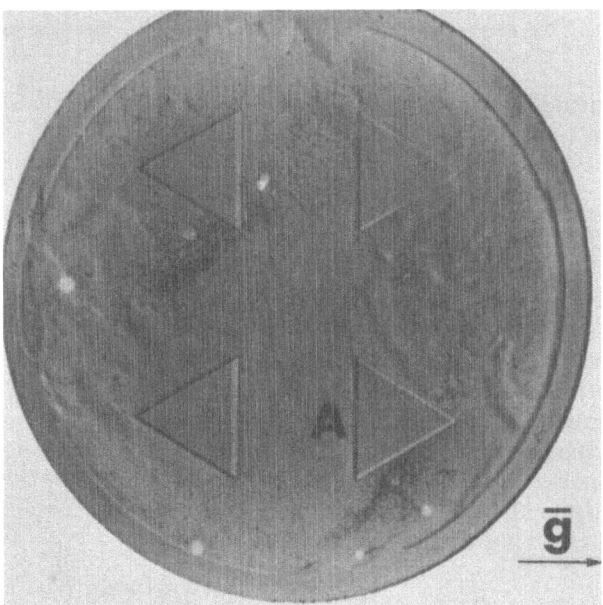

Figure 10. The ($\bar{2}20$) topograph of windows cut into a compressive Molybdenum film; the silicon substrate is 1 mm thick.

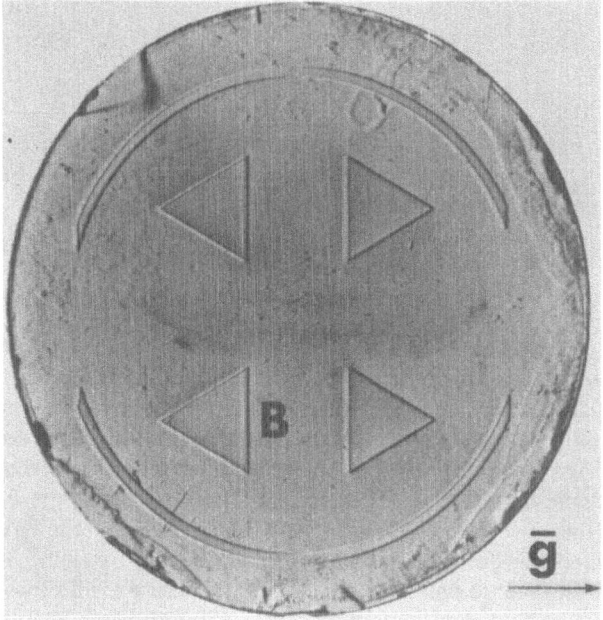

Figure 11. The ($\bar{2}20$) topograph of windows formed in a tensile Molybdenum film. The relative contrast is reversed with respect to Figure 10.

Table III. Evaluation of Tensile or Compressive Stresses Through
the Contrast Condition

Maximum edge contrast	$\mathbf{g}$	$\mathbf{R}$	$\angle(\mathbf{g}, \mathbf{R})$	$\cos \angle(\mathbf{g}, \mathbf{R})$	Film stress
A (Figure 10)	$[\bar{1}10]$	$[\bar{1}10]$	$0°$	$+1$	Compressive
B (Figure 11)	$[\bar{1}10]$	$[1\bar{1}0]$	$180°$	-1	Tensile

Diffraction Contrast at Windows Forming in Tensile or Compressive Films on Silicon

X-ray diffraction contrast characteristics of window edges as described in the preceding sections can be used to differentiate between compressive or tensile properties of films deposited on silicon. Molybdenum films can be grown on silicon with either compressive or tensile properties.[23] The stress situation is determined by the growth conditions of the molybdenum film. For our experiments, tensile and compressive molybdenum films were prepared about 0.8 μ thick on 1-mm-thick silicon substrates. The film stress was measured and calculated[14] prior to the X-ray examination. After the stress measurements, triangular windows (5-mm-sides) were etched into the film using photo-resist techniques. In Figure 10, the $(\bar{2}20)$ topograph of the compressive film, the wrinkled texture is indicative of buckling stresses.[23,24] The dark–light image relationship observed in the compressive film (Figure 10) is transposed in the $(\bar{2}20)$ topograph of the tensile film (Figure 11). If $\mathbf{R}_i$ indicates the edge with maximum contrast, then the compressive film is defined by the sign of the contrast criterion:

$$\cos \angle(\mathbf{g}, \mathbf{R}_i) > 0$$

where

$$-\pi/2 < \angle(\mathbf{g}, \mathbf{R}_i) < \pi/2.$$

A tensile film is identified when

$$\cos \angle(\mathbf{g}, \mathbf{R}_i) < 0$$

or $\pi/2 < \angle(\mathbf{g}, \mathbf{R}_i) < 3\pi/2$ [the angle $(\mathbf{g}, \mathbf{R}_i)$ is measured from $\mathbf{g}$]. These relations were applied to the compressive and tensile films depicted in Figure 10 (side A) and Figure 11 (side B). The results are summarized in Table III.

DISCUSSION AND CONCLUSION

Anisotropic elastic stress can greatly affect the properties of shallow p–n junctions. In the elastic range, completely reversible resistance decreases by several orders of magnitude and considerable reduction of breakdown voltage have been observed.[25,26] Rindner and Braun report also that qualitatively similar, but irreversible, changes in the junction characteristic are obtained after the elastic limit is exceeded and stress relief occurs by plastic deformation. The mechanical and electrical anisotropics of semiconductor junctions, as reported by Rindner and Braun, are not yet fully explained; but various mechanisms have been considered, in particular, internal field emission, stress-controlled carrier generation–recombination statistics, and stress-controlled ionization rates.[27,28] A different stress effect resulting also in the lowering of breakdown voltage of silicon p–n junctions was reported by A. Goetzberger and R. H. Finch.[27] They observed a reversible reduction of breakdown voltage under mechanical stress whereby the junction resistance remains unchanged. They indicate that this effect is related to a decrease in the band gap. From a comparison of hydrostatic pressure measurements with those made by uniaxial stress, they conclude that shear stresses are very important in the lowering of the breakdown voltage.

A detailed account of stress distributions around planar semiconductor devices has been presented in this paper. Obviously, the strain and stress relations are complicated and very complex, and determined by the process procedures. A hole cut into an oxide film to make a diffusion window is a discontinuity in an otherwise homogeneous stress distribution. Linear elastic theory shows that at all such discontinuities the stress is increased. Mathematically, the stress approaches infinity; practically, material adjustment will limit these stresses, but stress pile-up will occur. Such a local stress enforcement is the underlying cause for the enhanced diffraction contrast along window edges cut into oxide films deposited on single crystals. Subsequently, oxidation and diffusion procedures change and determine magnitude and sign of the final stress situation.

The X-ray technique described is a rather elegant tool making it possible to map the stress buildup step-by-step and obtain qualitative information about its magnitude; but, most important of all, the sign of the stress can be determined. We found that the mechanism leading to the formation of dislocations outside the diffused area (Figure 8) is activated when the stress in the silicon jumps from tensile to compressive across the boundary undiffused/diffused silicon. Dislocations of this type have a strong influence on junction characteristics and are deleterious to the device.[28] The segregation of impurities along border lines of oxide windows, as reported previously,[2] is also the result of local stress enhancement enforced through stress-jumping. The mechanism of stress-jumping activated across semiconductor interfaces is a new fundamental failure mechanism.

This paper has not attempted to explain the various diffraction phenomena associated with diffused junction structures. Rather, we have used these phenomena to characterize the existing stress situation in the silicon device. The X-ray phenomena displayed in the topographs must be accounted for by the dynamical theory of X-ray diffraction, in particular by the theories describing the propagation of X-rays in slightly deformed crystals as formulated by Penning and Polder[29] and also by Bonse.[30]

In conclusion, we would like to say that the origin and action of the various described stress and strain effects and their correlation with the semiconductor junction as well as their scientific and technical implications pose many more interesting questions for future research.

ACKNOWLEDGMENT

The authors wish to acknowledge the assistance of C. Hoogendoorn and T. A. Hansen in obtaining some of the SOT topographs presented in this paper.

REFERENCES

1. A. R. Lang, "Studies of Individual Dislocations by X-Ray Diffraction Microradiography," *J. Appl. Phys.* **30**: 1748, 1959.
2. G. H. Schwuttke, "A New X-Ray Diffraction Microscopy Technique for the Study of Imperfections in Semiconductor Crystals," *J. Appl. Phys.* **36**: 2712, 1965.
3. J. B. Newkirk, "Method for the Detection of Dislocations in Silicon by X-Ray Extinction Contrast," *Phys. Rev.* **110**: 1465, 1958.
4. J. K. Howard and R. D. Dobrott, "A New Scanning–Reflection X-Ray Topographic Method," *Appl. Phys. Letters* **7**: 101, 1965.

5. A. F. Jenkinson and A. R. Lang, "X-Ray Diffraction Topographic Studies of Dislocations in Floating-Zone Silicon," in: *Direct Observation of Imperfections in Crystals*, Interscience Publishers, New York, 1962, p. 471.

6. G. H. Schwuttke, "Semiconductor Junction Properties As Influenced by Crystallographic Imperfections," Contract No. AF19 (628)-5059, Scientific Report No. 1, Feb. 15, 1965–Feb. 15, 1966.

7. J. W. Faust, Jr., "Factors That Influence the Damaged Layer Caused by Abrasion on Silicon and Germanium," *Electrochem. Tech.* 2: 339, 1964.

8. B. D. Joyce and J. A. Baldrey, "Selective Epitaxial Deposition of Silicon," *Nature* 195: 485, 1962.

9. H. C. Evitts, H. W. Cooper, and S. S. Flaschen, "Rates of Formation of Thermal Oxides of Silicon," *J. Electrochem. Soc.* 111: 688, 1964.

10. C. J. Frosch and L. Derick, "Surface Protection and Selective Masking during Diffusion in Silicone," *J. Electrochem. Soc.* 104: 547, 1957.

11. R. M. Warner and J. N. Fordemwalt, *Integrated Circuits: Design Principles and Fabrication*, McGraw-Hill, New York, 1965, p. 298.

12. W. J. Armstrong and M. C. Duffy, "A Closed Tube Technique for Diffusing Impurities into Silicon," to be published by *Electrochem. Tech.*

13. E. S. Meieran and I. A. Blech, "X-Ray Extinction Contrast Topography of Silicon Strained by Thin Surface Films," *J. Appl. Phys.* 36: 3162, 1965.

14. R. Glang, R. A. Holmwood, and R. L. Rosenfeld," Determination of Stress in Films on Single Crystalline Silicon Substrates," *Rev. Sci. Instr.* 36: 7, 1965.

15. G. H. Schwuttke and H. J. Queisser, "X-Ray Observations of Diffusion-Induced Dislocations in Silicon," *J. Appl. Phys.* 33: 1540, 1962.

16. J. Washburn, G. Thomas, and H. J. Queisser, "Diffusion-Induced Dislocations in Silicon," *J. Appl. Phys.* 35: 1906, 1964.

17. M. L. Joshi and F. Wilhelm, "Diffusion-Induced Imperfections in Silicon, *J. Electrochem. Soc.* 112: 185, 1965.

18. F. Wilhelm and G. H. Schwuttke, to be published in *Bull. Am. Phys. Soc.*

19. S. Prussin, "Generation and Distribution of Dislocations by Solute Diffusion," *J. Appl. Phys.* 32: 1876, 1961.

20. Y. Sato and H. Arata, "Distribution of Dislocations near the Junction Formed by Diffusion of Phosphorus in Silicon," *Japan. J. Appl. Phys.* 3: 511, 1964.

21. I. A. Blech, E. S. Meieran, and H. Sello, "X-Ray Topography of Diffusion-Generated Dislocations in Silicon," *Appl. Phys. Letters* 7: 176, 1965.

22. G. H. Schwuttke and J. M. Fairfield, "Dislocations in Silicon due to Localized Diffusion," *J. Appl. Phys.* 37: 4394, 1966.

23. R. Glang, R. A. Holmwood, and P. C. Furois, "Bias Sputtering of Molybdenum Films," IBM Fast Fishkill Technical Rpt. 22.173.

24. R. A. Holmwood and R. Glang, "Vacuum Deposited Molybdenum Films," *J. Electrochem. Soc.* 112: 827, 1965.

25. W. Rindner, "Resistance of Elastically Deformed Shallow p–n Junctions (I)," *J. Appl. Phys.* 33: 2479, 1962.

26. W. Rindner and I. Braun, "Resistance of Elastically Deformed p–n Junctions (II)," *J. Appl. Phys.* 34: 1958, 1963.

27. A. Goetzberger and R. H. Finch, "Lowering of Breakdown Voltage of Silicon p–n Junctions by Stress," *J. Appl. Phys.* 35: 1851, 1964.

28. J. M. Fairfield and G. H. Schwuttke, "The Influence of Crystallographic Defects on Device Performance," *J. Electrochem. Soc.* 113: 1229, 1966.

29. P. Penning and D. Polder, "Anomalous Transmission of X-Rays in Elastically Deformed Crystals," *Philips Res. Repts.* 16: 419, 1961.

30. U. Bonse, "Zum Kontrast an Versetzungen im Roentgenbild," *Z. Physik* 177: 543, 1964.

DISCUSSION

H. W. Pickett (General Electric, X-Ray): What is the angular range and period of the oscillating crystal in the SOT procedure?

J. K. Howard: I'm not sure what the angular frequency is. The angular width depends on the deformation itself. It really depends on how intense is the elastic strain introduced by the diffusion, or, say, the process of deformation. It can be of the order of seconds, or it can be much larger. Really, we have no calibration on this.

H. W. Pickett: You adjust the device according to what you need?

J. K. Howard: That is right.

EXPERIMENTAL PROCEDURES IN X-RAY DIFFRACTION TOPOGRAPHY

Stanley B. Austerman

Atomics International
Canoga Park, California

and

J. B. Newkirk

University of Denver
Denver, Colorado

ABSTRACT

The quality and interpretation of diffraction topographic images are strongly dependent on the detailed laboratory techniques that are used in making them. In this paper practical instructions are given for the preparation of Berg–Barrett and Lang topographs. It is hoped that these suggestions will enable the novice in the field of X-ray topography to produce high quality images and to interpret them with a minimum of learning time. The topics treated include adjustment of the critical conditions for attaining highest resolution, choice of radiation and the specific *hkl* planes to be used, conditions limiting the size of the image, cause and avoidance of image distortion, choice of photographic emulsions, plate processing for best contrast and resolution, photomicrographs of the original image, and plate preservation.

INTRODUCTION

The value of X-ray diffraction topography as a powerful tool for investigating defects in nearly perfect crystals is evident from the many papers dealing with this topic that have been published during recent years. These publications deal, by and large, with theoretical and interpretive aspects of topography, and the practical laboratory techniques of diffraction topography usually have been ignored. In view of the rapidly broadening use of diffraction topography, the authors believe that discussion of recommended topographic techniques is timely, and will be useful to those laboratories attempting to establish an efficient capability in diffraction topography with minimum wasted time. This paper, therefore, is intended as a guide for the novice who otherwise might experience many fruitless, frustrating hours in the laboratory. It is acknowledged that many of the details described here will not appear important to the experienced topographer.

The general principles of X-ray diffraction topography, as well as the underlying theory of kinematical and dynamical X-ray diffraction effects have been amply described in previous publications.[1-7] A discussion of equipment variables and diffraction effects will be undertaken here only to the extent necessary for the primary objectives. Although

134

the discussions will center on the Berg–Barrett and Lang techniques, many of the details presented may be applicable to the many closely related variations of topographic techniques.

The Berg–Barrett (B–B) and Lang techniques are complementary in many ways. By virtue of simplicity of the experimental setup, laboratory capability for the B–B techniques can be rapidly established. On the other hand, the Lang technique requires more complex and precise apparatus and, therefore, requires more attention and longer lead-time for equipment to be either purchased or constructed.

For reasons that will be evident in the following sections, the Berg–Barrett method has its most general application to investigating the *surface* region of the specimen, using relatively "soft" X-rays and large diffraction angles. With more refined geometrical configurations, the Lang method lends itself to investigation of the *volume* of the specimen (for which $\mu t < 1$), using "hard" X-rays and relatively small diffraction angles.

The quality and usefulness of X-ray diffraction topographs find expression in terms of resolution and contrast in image details, and in freedom from artifacts associated with scratches, dust particles, and emulsion distortions. Topograph quality and interpretive significance depend on careful attention to three general areas of technique: camera design, manipulation of specimen and photographic plate, and processing of the plate. To the extent that these areas of technique are accessible to control by the topographer, they will be discussed in the following sections. Although this information represents the "state of the art" as it is practiced currently in the authors' laboratories, undoubtedly the techniques will be continually refined and reduced to standard practice as studies in the field of X-ray diffraction topography progress.

BERG–BARRETT TOPOGRAPHY

Camera Design

The camera originally prescribed by Barrett[6] has been modified by the present authors to provide more freedom of motion of the specimens and the photographic plate* (see Figure 1). Many different designs are usable with good results so long as the operator is able to control the orientation of the plate and the specimen, and to limit the boundaries of the incident X-ray beam. All of the degrees of freedom indicated in Figure 1 have been needed in actual practice. The specimen holder itself can be a standard eucentric goniometer of the type commonly used by crystallographers. The ball and socket joint in the plate holder assembly can be adapted from an inexpensive photographer's camera tripod head. An adjustable slit assembly is mounted at the end of $\frac{1}{2}$-in.-diameter brass tube approximately a foot long; this is not a critical dimension, as will be discussed later. The size and angular divergence of the X-ray beam at the specimen are controlled by the size of opening between slits and the projected size of the X-ray source. The camera is shown at a side port of a standard diffraction X-ray tube, for reasons to be discussed. The takeoff angle should be chosen as a compromise between high X-ray intensity ($\sim 6°$) and high image resolution ($\sim 2°$), depending on requirements. Use of the camera is most convenient if the mounting track is horizontal and the tube is tilted to the appropriate takeoff angle. The plate holder assembly should be backed by a sheet of lead at least $\frac{1}{8}$-in. thick and the camera itself should be on the opposite side of a thick shield of plastic from the operator, to intercept the incoherent rays scattered by the specimen and camera parts.

As an aid to aligning the crystal specimen, a 1- × 3-in. strip of fluorescent screen†

---

* The term "plate" will be used to designate an X-ray sensitive photographic emulsion supported on a glass plate backing.

† Such as DuPont CB-2.

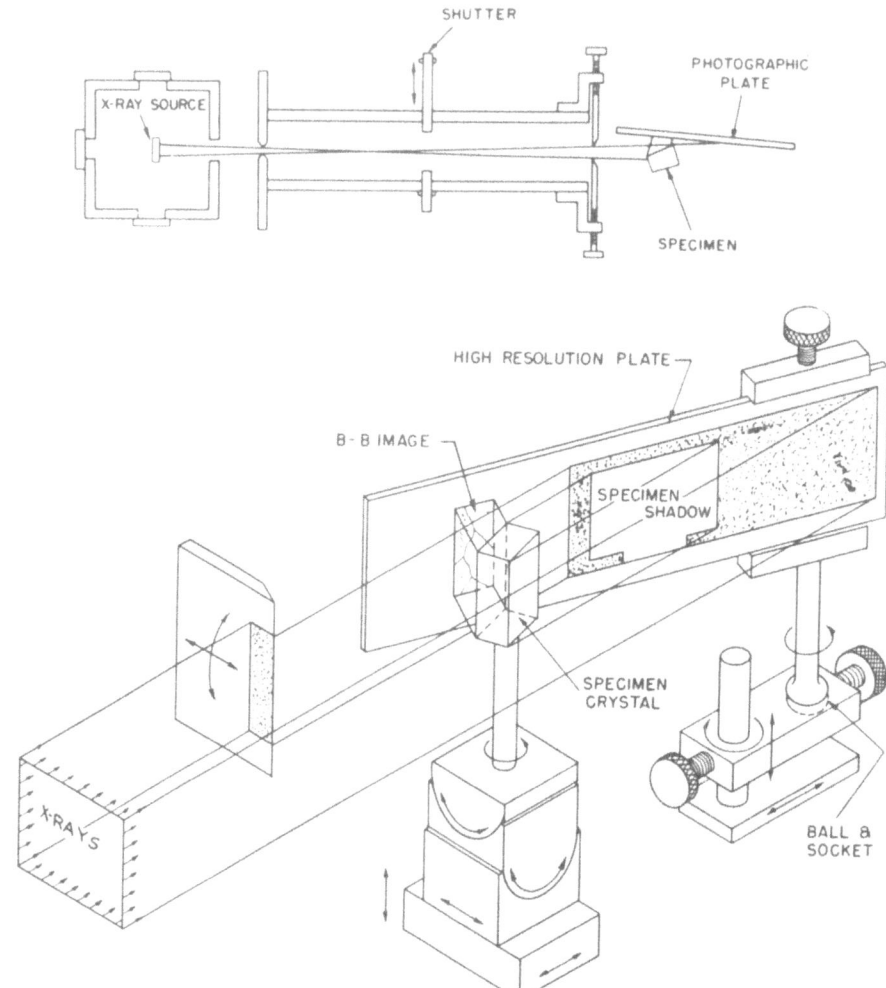

Figure 1. Diagrams of component and diffraction geometry for reflection Berg–Barrett X-ray diffraction topography.

may be placed in the plate holder. Alternatively, a large sheet of fluorescent screen may be mounted independently behind the specimen relative to the observer's position. The large screen is preferred since a broader field of reflections can be seen, and there is a lower background of illumination due to scattered X-rays and occasional visible specimen fluorescence. The room is darkened and an unfiltered beam of X-rays is allowed to fall on the specimen.* When the eyes are dark-adapted, Laue spots will be seen moving systematically across the screen when the specimen is rotated manually about its vertical axis. By adjusting the settings of the two horizontal arcs, it is possible to bring a major crystallographic axis into coincidence with the vertical rotation axis. One set of Laue spots will then move along a horizontal line which is at the same level as the specimen itself. Occasionally one of these moving spots will be seen to flash out with sudden and unmistakably high intensity. This

* Extreme caution is needed at this stage to avoid X-ray exposure to the fingers and hands.

occurs when the Bragg conditions are satisfied for the K_α or K_β wavelength. It is usually easy to deduce the indices of a spot from the 2θ angle of the diffracted beam, the X-ray wavelength, and the specimen crystallography.

Having thus set the crystal to diffract from a desired set of planes, the beam-limiting knife edge (Figure 1) is adjusted so that the edge of the beam coincides as closely as possible with the vertical edge of the crystal nearest the screen. This is a critical adjustment since the plate will later be brought as close to the specimen as possible for the exposure. Any error in setting the beam knife edge will either cut off useful radiation from the specimen or flood the latent image on the plate with the intense direct beam.

With the crystal oriented as desired and the beam limited on both sides of the crystal, the fluorescent screen is replaced by a photographic plate.* This may be done safely under a yellow safelite used for contact printing papers. The plate should be held nearly parallel to the incident beam so that the plate-to-specimen distance will be small over as large an area as possible. With a suitably shaped specimen, it is possible to bring the plate to within a few tenths of a millimeter of the specimen edge without interfering with the diffracted image of its surface. Care must be taken, however, to avoid jostling the specimen out of alignment during plate placement. With the safelite properly located over the specimen and plate, it is possible to obtain a reflection of the specimen on the emulsion surface of the high resolution plate. The gap between the specimen and its reflection can be used as a critical gauge of proximity. It is obvious that this critical alignment makes it impossible to wrap the plate in light-tight material.

The requirement of maintaining separation between the primary beam and the diffracted image on the plate, as well as several other geometrical requirements, lead to certain restrictions on the use of B–B topography. Because of these requirements, it is most convenient for the diffracted X-ray beam to emerge more or less perpendicularly to the primary beam. This geometry is most readily accomplished with the longer X-ray wavelengths (Cu to Cr, K_α and K_β). Since these X-rays are attenuated relatively strongly in most specimen materials, the Berg–Barrett method (given the above restrictions) is most adaptable to *reflection* topography of shallow surface regions.

The size of the area of the specimen which can diffract a given wavelength depends directly on the projected horizontal length of the X-ray source. X-rays issuing from a point on the target can diffract only from a line arc on the specimen surface which satisfies the Bragg relationship, as shown in Figure 2. Thus, a single point on the X-ray source can suitably irradiate the entire specimen height. However, if the specimen has a larger width, projected parallel to the X-ray beam, than the length of the X-ray source, part of the surface will not be bathed in X-rays at the proper angle to diffract. Consequently, it is preferable to use an X-ray source having a broad horizontal dimension. However, it is undesirable to flood the specimen with a beam that is broader than necessary, because attendant incoherent X-ray scattering contributes to build-up of detail-obscuring background. The distance between the target and the specimen determines the band curvature.

A light-tight enclosure over the B–B unit is desirable, since it will allow operation of the unit without interfering with other use of the laboratory. If the X-ray tube and B–B camera are mounted on a level table surface, a removable, lightweight cardboard box, painted black inside, can be placed over the tube and camera during exposure. A rail mounted on the table top and matched to the edge of the box provides adequate light-trapping to prevent fogging of the enclosed plate.

---

* Some fine grain emulsions, being transparent, are difficult to distinguish from the glass on the opposite side of the plate. A quick way to find the emulsion side is by pressing a slightly moist thumb against one corner of the plate. The emulsion will soften and feel quite different from the bare glass.

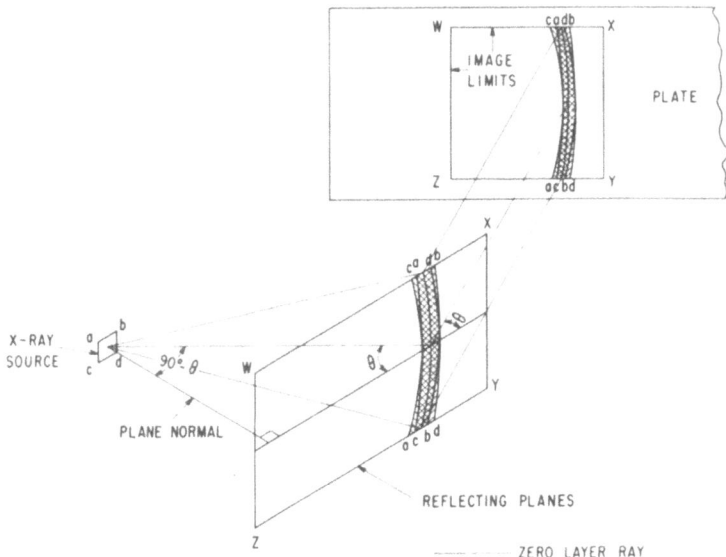

Figure 2. Diagram of diffraction in divergent beam. The amount of the specimen surface which can diffract depends upon the angle between the beam and the surface and upon the effective size of the X-ray source, but not upon the distance from source to specimen.

Image Resolution

Vertical divergence of the incident radiation is the most serious cause of poor resolution. Any X-radiation falling upon a point P (see Figure 3) will be diffracted if the Bragg conditions are satisfied. Thus, X-rays of wavelength λ, which lie upon a cone of semiapex angle $(90° - \theta)$ and are incident upon P as shown, will be diffracted. Since rays satisfying these conditions can originate from any point of an arc on the X-ray target, the image of point P is also an arc. For maximum resolution, the projected height of the X-ray source should be as small as possible. This is obtained by viewing the target focal spot from the side and by using a small takeoff angle (at the consequent loss of intensity). Because of the Bragg restriction, horizontal divergence is not important in reducing resolution.

Illumination of a broad specimen with a horizontally wide X-ray source is at the expense of horizontal beam definition at the specimen. The broad source casts a divergent penumbra shadow of the knife edge, which may interfere with positioning the plate very close to the specimen without allowing the primary beam and the diffraction image to partially overlap. Use of vertical soller slits help to reduce horizontal divergence, although they should not be so restrictive as to fragment the overall image. (An interesting refinement, reported at this conference, is the use of moving slits.[8])

The most direct way to achieve high resolution is to keep the ratio of D' to D as low as practicable. This is best accomplished by reducing D' since extending D lowers the incident intensity seriously. When D'/D is of the order of 0.005, a small decrease in D' brings a great improvement in resolution, whereas a large increase in D gives only slightly better resolution at considerable loss of diffracted intensity. Since the plate and the specimen surface generally cannot be parallel, D'/D will vary across the image.

The resolution at $D'/D = 0.005$ is much better than can be recorded on standard X-ray film or even on standard fine-grained spectroscopic plates (*e.g.*, Eastman V–O

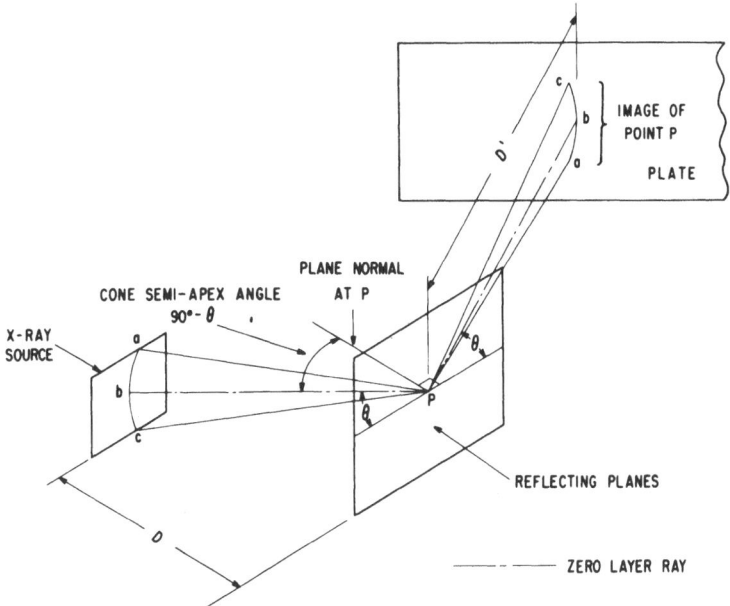

Figure 3. Diagram of resolution-related component spacings. Radiation permitting Bragg diffraction to a point P on the specimen can issue from an arc on the X-ray source area. Thus, the image of point P is an arc. Resolution is lost as the plate-to-specimen distance D' is increased.

plates). However, the finest grain emulsions, having resolving power greater than 1000 lines/mm, can record the detail in the best B–B image we have obtained to date. For the ultimate in resolution, it is necessary for the diffracted rays to pass through the emulsion normal to the plane of the plate rather than at an oblique angle. Finally, mechanical vibration of the camera components must be eliminated during the exposure. This is done by keeping all parts rigid and firmly clamped. Vibration of the plate itself is prevented by allowing the unclamped end of the plate to rest lightly upon the top of the goniometer which holds the specimen.

Reflections Available from a Specified Surface

It often happens that a reflection from a certain set of crystallographic planes cannot be obtained from a prepared specimen surface. This may be due to one of the following circumstances:

1. The diffracted rays are directed into the crystal where they are absorbed instead of being reflected away from the surface.
2. In order to bring the desired planes to the Bragg angle, the surface to be examined is in the shadow of the rest of the crystal and, therefore, is not illuminated by X-rays.

One may predict by means of stereographic projections (Figure 4) what reflections are available for making B–B topographs of a given specimen surface. Let the normal to the specimen surface be represented by the center of a Wulff stereographic net and let circles, as in Figure 4, be drawn around the center, with radii equal, respectively, to the number of degrees in the Bragg angle for the various reflections. It can be shown that if a pole representing a specific set of diffracting planes lies inside the circle which has as its radius the Bragg angle for that set (or a higher-order reflection thereof), then a zero-layer B–B topograph of the surface may be made using those planes. It is evident that zero-layer diffraction can occur only when the pole of the diffracting plane lies on the equator and touches the appropriate circle. For reflections off the zero layer the scheme shown in Figure 4b applies.* The poles of all planes which make the Bragg angle θ with the beam will lie on a circle $(90° - \theta)$ from the incident beam. This circle will be called the *reflecting circle*. Only the 111 reflecting circles are shown here. Since, in this illustration, the Bragg conditions are met whenever a {111} pole touches the reflecting circle, any {111} pole lying to the right of the reflecting circle represents a set of planes by which a B–B image could be made (by rotating the crystal about A–A toward the X-ray beam) as long as the total rotation does not exceed 2θ. Thus, a *limiting circle* also may be drawn as indicated. Planes whose poles lie to the right of the shaded area are excluded due to condition (a) above; those whose pole lie to the left are excluded due to condition (b). Only planes whose poles can be made to touch the reflecting circle on the equator, without having to rotate the crystal about the beam as an axis, will give zero-layer reflections.

Image Distortion

Usually the B–B diffraction image is a distorted facsimile of the specimen surface. This distortion is associated with the image projection geometry between the specimen surface and the plate. If two points in the image lie on a line which is parallel with the primary beam (the plate being parallel with the beam), their distance apart (b) is related to the separation (a) of the corresponding points on the specimen surface by:

$$b = \frac{a \sin(2\theta - \phi)}{\sin 2\theta}$$

where ϕ is the angle between the specimen surface and the primary beam. Since there is no image distortion when $\sin 2\theta = \sin(2\theta - \phi)$ it is possible by the appropriate choices of specimen, X-radiation, reflection, and plate orientation, to adjust the distortion within limits. If the crystallographic orientation of the specimen is known with respect to the surface, the angle ϕ may be easily derived from the stereographic plot in (Figure 4a).

Rectification of the image distortion is not possible by simple photographic methods since these introduce unwanted keystone. effects. It is simpler to superpose a reference grid upon the photomicrograph of the original specimen surface and another, drawn with the proper distortion for the topograph in question, upon the corresponding diffraction topograph. Corresponding areas can then be conveniently described by their identical coordinates.

Emulsions and Exposure Times

All of the ultrafine grain emulsions are very slow. By direct comparison, little difference in characteristics has been found between the Eastman GH-649 and the Eastman High Resolution plates which are recommended for use in B–B topography. A rough approximation to the needed exposure time may be reached on the basis of the brightness of the B–B image as seen on the fluorescent screen. An exposure of $\frac{1}{2}$ hr for the finest grain emulsion will serve as an initial trial until experience is gained.

---

* This construction was suggested by Mr. Sam Leber of the General Electric Laboratory, Nela Park, Ohio.

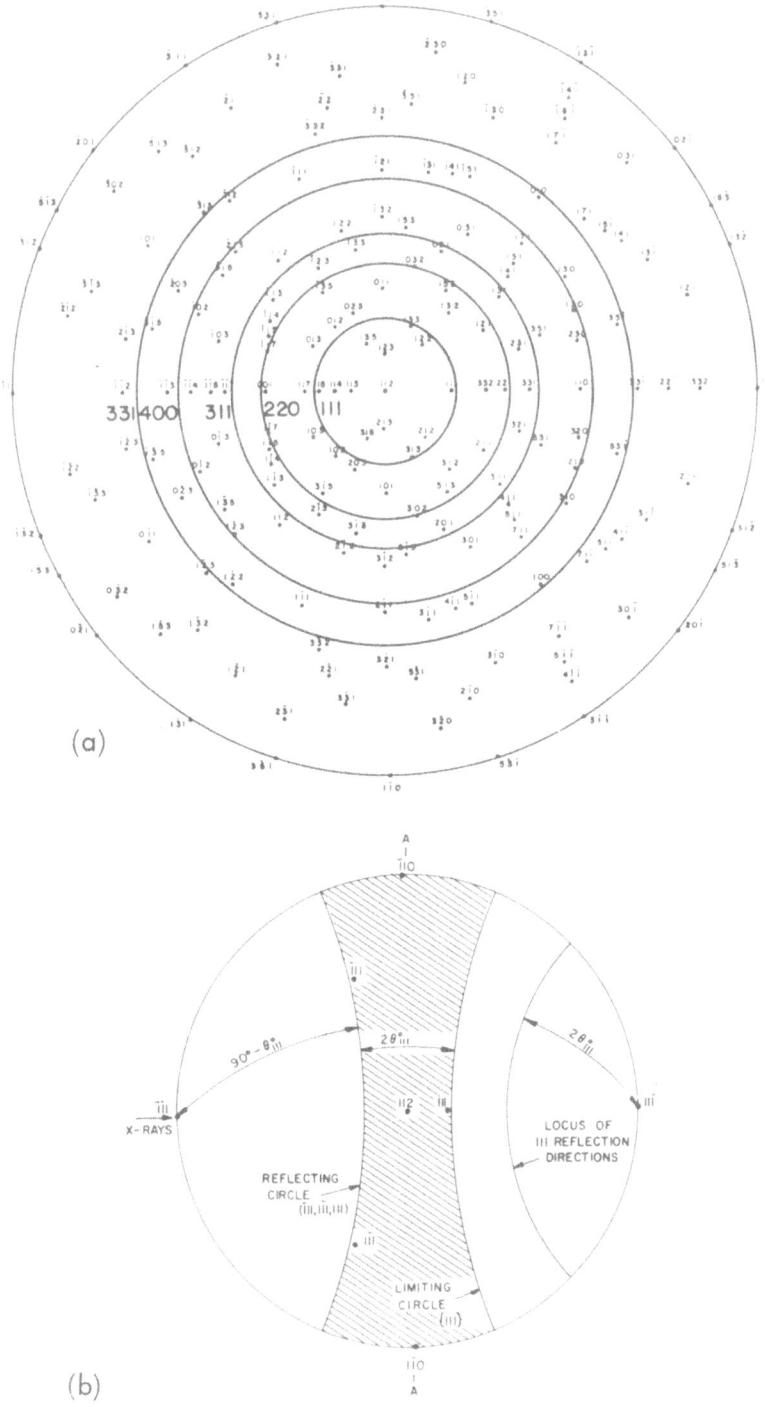

(a)

(b)

Figure 4. Schemes for determining possible reflections for diffraction topographs of a (112) surface of a silicon crystal using Cr K_α radiation. (a) For producing zero layer reflections by rotation about the center and then about the vertical axis A–A. (b) For predicting all possible reflections, whether on or off the zero layer. See text for explanation.

LANG TOPOGRAPHY

Geometry

The geometry and mechanical operation of the Lang technique of X-ray diffraction topography were first prescribed by A. R. Lang.[9] The essential elements for this technique are shown in Figure 5. The fine X-ray source and the collimating slit define a narrow beam of X-rays incident on the specimen. The direct beam transmitted through the sample is stopped by the (receiving) slit structure and the diffracted beam passes through the slit to strike the photographic plate.

The most unique feature of this method is that the specimen and the plate are carried on a common platform (carriage) and are translated together, as shown in Figure 5. By virtue of this feature, there is a 1 : 1 projection correlation between specimen details and the recorded image. Also due to this geometry, there is no inherent limit to the specimen length in the direction of traverse; the limit is established only by the lengths of carriage travel and the plate. Since the primary beam is blocked from the plate (unlike the B–B method), the entire plate is available for recording the diffracted image. Also, this makes it feasible to perform *transmission* topography at small diffraction angles.

The X-ray source should be a fine-focus X-ray target. Commercial units are available under the general designation "microfocus" X-ray generator.* Typically, the

* Jarrell-Ash Company, Waltham, Mass.; Engis Equipment Company, Morton Grove, Illinois; and Rigaku Denki Co., Ltd., Tokyo, Japan.

1. FINE SOURCE OF X-RAYS
2. COLLIMATING SLIT
3. SPECIMEN
4. RECEIVING SLIT
5. PLATE CASSETTE
6. X-RAY DETECTOR

TRAVERSE

Figure 5. Diagram of important components for Lang topography.

Figure 6. Photograph of Lang camera on microfocus X-ray generator. Plate cassette is not shown. Note the special cabinet-top extension necessary to accommodate the length of the camera.

recommended focal spot size on the target is 1.4 × 0.1 mm. When viewed end-on at 3° takeoff angle, the projected X-ray source size is ∼100 μ × 100 μ; this size is optimum for maximum X-ray intensity with small size, and is recommended for normal use with the Lang technique.

The "camera" is a fairly complex and precisely made piece of equipment. Although it can be custom designed and built for special applications, the availability of commercial units* eases the task of establishing laboratory capability. A typical commercial Lang camera is shown in Figure 6. The unit shown contains the mechanical components required to collimate the beam, critically align the sample, align the slit and X-ray detector, and to slowly translate the specimen and plate between pre-set travel limits.

The primary X-ray beam incident on the specimen is defined by the size of the X-ray tube focal spot and by the collimating slit. The horizontal divergence of the beam typically may be on the order of two minutes of arc. The vertical divergence, however, is far greater in order to bathe the entire specimen height. The vertical dimension at the collimating slit is limited by sliding lead stops that are positioned to illuminate *only* the entire specimen height. The vertical divergence may be as much as two degrees.

Intensity Monitoring

A very practical feature of the removable cassette (plate holder) is its relative transparency to X-rays. The front and back openings can be covered with material† that has negligible absorption of X-rays, is light tight, and contributes no detectable texture to the topograph. Consequently, the primary reduction in transmitted beam intensity is due to absorption in the photographic emulsion and supporting glass plate. It is feasible, therefore, to put an X-ray detector behind the plate and to observe the transmitted X-ray intensity during final crystal alignment or continuously during traverse and exposure.

Several advantages are gained by monitoring the X-ray intensity during the topographic exposure. In the first place, the crystal can be given final alignment for maximum intensity after the film cassette has been placed on the carriage. Secondly, continuous monitoring with strip-chart recording will reveal gradual (if any) loss of diffraction intensity during exposure, also indicating drift of specimen alignment. In case of alignment drift, the specimen can be realigned at any time during the exposure, using intensity as a criterion for best alignment. Of course, if the crystal position shifts too far, the topographic image at initiation and termination of exposure will not be superimposed precisely, leading to "smearing" of image details and loss of resolution.

A qualitative evaluation of crystalline perfection can be gained quickly from the variations in intensity while scanning the crystal from one end to the other. If the specimen is relatively free of defects, the intensity will be more or less proportional to the height of the specimen where it intercepts the X-ray beam. If, however, the specimen contains highly defective regions that locally give rise to strong diffraction intensity the non-uniformity of defect distribution will be revealed by wide and rapid variations in the monitored X-ray intensity during specimen scan. An example of intensity distribution for a specimen with relatively perfect and imperfect regions is shown in Figure 7 which reproduces a strip-chart recording of the monitored intensity.

Specimen Mounting and Alignment

The procedure for mounting a specimen at preselected angular positions presently is somewhat inadequate. It would be highly desirable to be able to mount the specimen

* Jarrell–Ash Company, Waltham, Mass.; Rigaku Denki Co., Ltd., Tokyo, Japan; and Crystallogenics, Inc., P.O. Box 634, Belmont, Calif.
† Nylon Sheet Supronyl Black, American Hoechst Corporation, Mountainside, New Jersey.

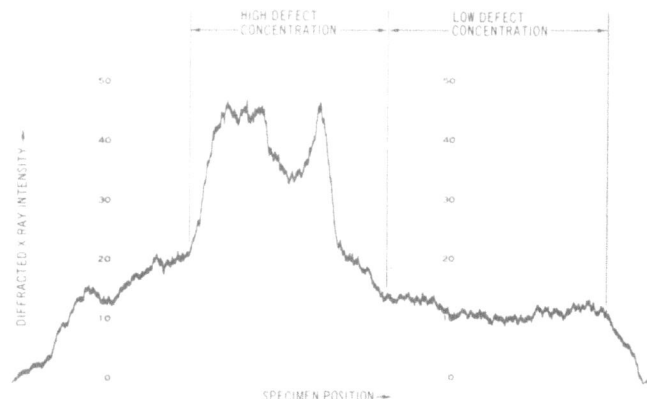

Figure 7. Recorder tracing of diffracted X-ray intensity from a specimen containing regions of differing defect concentration.

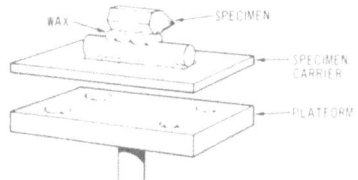

Figure 8. Sketch of specimen carrier.

on a goniometer head, align it with an optical goniometer or by Laue X-ray diffraction techniques, and then transfer the mounted and aligned specimen to the Lang camera so that the desired family of lattice planes are at least close to the Bragg angle relative to the incident primary beam. No such arrangements have been made by the instrument manufacturers. In lieu of such provisions, the following procedures have been used for visual alignment of the specimen.

The specimen is initially mounted on a carrier, which for convenience is made of plastic components, with a bit of wax for attachment as shown in Figure 8. If the specimen has identifiable crystallographic faces, it usually is possible to align the specimen visually to within a few degrees of the intended position relative to the forward edge of the carrier. If useful faces are lacking, then the specimen can be aligned from Laue X-ray diffraction patterns. In either case, with the specimen aligned relative to the edge of the carrier, alignment of the edge relative to the Lang camera axes serves the purpose of simultaneously aligning (approximately) the specimen. The specimen and carrier are attached to a platform, already mounted on a goniometer head, with small dabs of wax as shown. The advantage of this procedure is that for visual alignment on the Lang camera, it is far easier to sight on the reference edge than on the specimen itself. This procedure is generally adequate to place the desired family of diffraction planes in approximately the proper position for diffraction. Since adjustment to diffraction position is critical, final adjustment can be done only with the precise angle-position devices built into the camera itself.

A minor annoyance arises with mechanical relaxation of the wax used to attach the specimen to the carrier. When the wax ball is kneaded to proper shape and then impressed from opposite sides by the specimen and the carrier, a system of internal stresses is established. These stresses slowly relax by creep deformation over a period of time that may last a number of hours. Since the specimen position tends to drift with the wax relaxation, it is preferable not to attempt critical specimen alignment until the wax has fully relaxed.

Alignment of a small specimen to the Bragg angle is a relatively straightforward affair. When a large specimen is used, however, extra complications in alignment arise. This is primarily associated with the vertical divergence of the X-ray beam from the point source. If the specimen is poorly aligned, one part of the specimen may be bathed by X-rays incident at the Bragg angle, while elsewhere the angle of incidence between the diffracting planes and the X-ray beam is off the Bragg angle sufficiently to seriously reduce diffraction intensity. Under these conditions, when the entire specimen is rotated about the vertical axis to gain good diffraction intensity from the second area, intensity is lost from the first. As a result, the *apparent* rocking curve breadth is broader than it should be for proper alignment, and resolution of the $\alpha_1 - \alpha_2$ doublet is considerably reduced. A practical test for good specimen alignment, therefore, involves observation of the doublet resolution and of total diffraction intensity. If the entire specimen goes in and out of reflection at the same time, both resolution and intensity will be at a maximum. With particularly large specimens or awkward orientations, the best alignment of the supporting goniometer head can be sought in a systematic procedure.

Cassette Placement

A relatively large specimen-to-plate separation is necessary in the Lang technique in order to accommodate the slit assembly between the cassette and the specimen. It should be noted that, since the carriage usually does not travel parallel to the slit assembly, clearance between the slit and the specimen or the cassette constantly changes; the specimen-cassette spacing must also allow for these varying clearances. In spite of this rather large spacing, image resolution is not adversely affected because, with the essentially point source of X-rays, there is relatively little divergence of X-rays diffracted from any given feature in the specimen. Therefore, resolution is not as dependent on close specimen-plate separation as in the Berg–Barrett method.

Still in the interest of maximum image resolution, the plate must be placed perpendicular to the diffracted beam. Resolution is lost if the direction of observation (such as during photography) is not closely parallel to the direction of X-rays through the plate during prior exposure. This is true especially for thick emulsions. For example, an error of five degrees in placing the plate normal to the X-ray beam will cause *apparent* broadening of image details of nearly five microns in 50 μ emulsion, unless the plate is similarly (and inconveniently) tilted during subsequent observation. In general, it should not be difficult to orient the plate within a couple of degrees of the correct position by visual sighting, and this should be sufficient for most topographic studies.

Vibration

As with the Berg–Barrett method, any vibration can cause broadening of image details and loss of resolution. The Lang cameras are designed to operate without creating vibration. It is important that any other laboratory source of vibration be eliminated.

Traverse Sectioning

The geometrical overlapping of regions, in the specimen, that are bathed in the X-ray beam, and "visible" through the slits leads to effects in the topograph that can be used advantageously if properly understood, or misinterpreted if not recognized. The geometrical bases for discussing these effects are shown in Figure 9. The primary X-ray

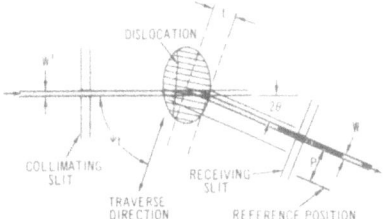

Figure 9. Diagram and slit geometry for topographic sectioning.

beam is incident on the specimen at the Bragg angle for a particular family of lattice planes. Since the external shape of the specimen is not important here, the specimen is represented as a simple shape. X-rays are diffracted from the primary beam along its entire length through the specimen. The width W of the slit, however, governs the width of the diffracted beam that reaches the plate or X-ray detector. The region of the crystal that is "visible" to the topograph, therefore, is the region of overlapping of the primary and slit-limited diffraction beams. Since the primary beam and the slit-limited diffracted beam each have significant width (W' and W, respectively), the region of overlap forms a rhombohedral prism. If the specimen is traversed through the primary beam, the region of visibility will scan a section through the specimen as represented by the dashed lines in Figure 2. The thickness of the section, being the projection of the prism edges in the traverse direction, is related to the beam widths by

$$t = \frac{1}{\sin 2\theta}[W' \cos(2\theta + \psi_t - 90°) + W \cos(90° - \psi_t)]$$

If the traverse direction is approximately normal to the diffracting planes, as usually is the case, the expression above can be reduced, with simplifying approximations, to

$$t \simeq \csc 2\theta(W' + W)$$

It should be noted that the topographic image will be dominated by the details near the midplane of the section. This is because during traversal, a point along the midplane of the section travels a longer distance through the "visibility" prism than does a point near the tip of the rhomb, and, therefore, contributes to the image exposure for a correspondingly longer period of time. It is evident from the figure that the viewed section is parallel to the traverse direction. It can be seen that the direction and thickness of the section can be chosen by appropriate settings of the traverse angle (ψ_t) and slit width (W).

This type of section, which we will term "traverse section," can be very useful to examination of intricate defect structures. With a wide slit, the image of most or all the specimen can be recorded on a single topograph. If the defect structures are complex, the overlapping details in the topography may be impossible to evaluate. In this case, a narrow slit can be used to investigate the details in a thin section. By moving the slit to various positions P, a parallel series of section topographs can be prepared, facilitating both resolution and three-dimensional tracing of individual defect structures even in moderately complex defect arrays.

Stationary Sectioning

Another type of sectioning procedure can be useful in resolving details in complex defect arrays; this type of section can be designated as a stationary "section," because its essential feature is that the specimen does not move during the topograph exposure. The slit should be opened wide enough to accept the entire diffracted beam. By moving the specimen to a series of positions, a corresponding series of topographs can be prepared, also facilitating resolution and three-dimensional mapping of defect structures. The reader will recognize that in an ordinary wide-slit scanning topograph, the image details in the stationary sections would be superimposed, with corresponding loss of resolution between image details.

Reversal of Image Contrast

Occasionally confusing and conflicting image details appear in the topographs. In particular, dislocations usually appear as dark lines in the topograph, since the lattice distortions around the dislocation line lead to enhanced local diffraction intensity. Occasionally, however, dislocations that appear dark in some topographs appear light in others. An example of this reversal of intensity contrast is shown in Figure 10. The reversal of intensity contrast is readily interpretable in terms of slit position. Referring again to Figure 9, note the sketched presence of a dislocation having the necessary geometry to produce diffraction effects. If the traverse section is located (by slit position) to enclose the dislocation, its presence will be reflected in the topograph as a dark line. However, the enhanced diffraction intensity removes energy from the primary beam, so that in the primary beam to the right there is a "shadow" of the dislocation. This shadow will be evident as reduced intensity in the diffracted beam arising from a section such as depicted in Figure 9 and, therefore, the dislocation will appear as a light rather than a dark line. This effect is not rare, but has been observed fairly frequently in large, transparent (to X-rays) samples. Since the light line essentially amounts to a shadow, it is

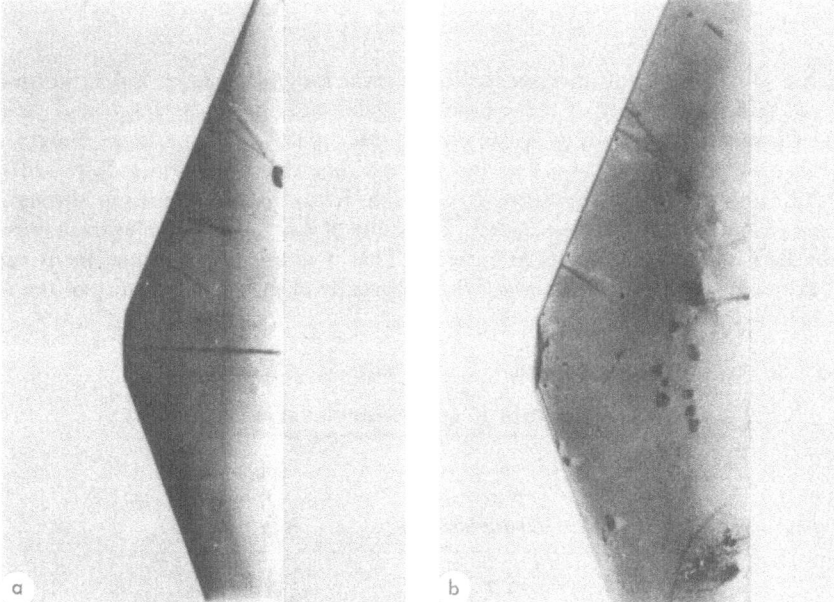

Figure 10. Topographs illustrating intensity contrast reversal, (00.2) reflection from BeO crystal.[10] (a) Traverse section containing an axial screw dislocation (see Figure 9). (b) Traverse section to right of dislocation.

clear that any strongly diffracting defect (not necessarily a dislocation) can give rise to this same diffraction phenomenon. The phenomenon described above is simply a geometrical one, interpretable in terms of straightforward Bragg diffraction. Image contrast reversal also may arise as a dynamical X-ray effect, and from X-ray energy flow through special forms of lattice distortion. See for examples, Figures 4 and 8 of Lang's paper in this volume. Therefore, care should be exercised not to misinterpret the observed topographic details only in terms of kinematic diffraction.

Emulsions. Since relatively penetrating X-rays are commonly used for Lang topography, plates with thick emulsions are advantageous. The thick emulsion allows more effective use (absorption) of the X-rays in producing developed optical density in the image. With Mo K_α and Ag K_α radiation, excellent results have been obtained with 50 μ Ilford Nuclear Research emulsions type L-4 and G-5.* Type L-4 is somewhat slower but has finer grain size than type G-5. The developed grain size in both emulsions is stated by the manufacturer to be less than one micron. The longer wavelength Cu K_α is considerably less penetrating and a 10 μ emulsion thickness is suitable.

The fine-grained nuclear research emulsions are subject to deterioration at ambient room temperature, so the laboratory should be equipped for refrigerated plate storage.

Exposure Time Estimation

An area of considerable trial and error is the selection of a suitable exposure time for making a topograph. This is dependent on the X-ray wavelength, plate emulsion, slit width, sample size, etc. A special problem arises from the variability of diffraction intensity from point to point within a given specimen. A convenient procedure for making an estimate of the proper exposure time involves the use of appropriate parameters and an approximate intensity term derived from the diffraction intensity recorded during sample scan. These parameters and terms are related through

$$P = (HG^N K)/(IS)$$

in which P is the number of traversed passes, H is the sample height, G is the transmission coefficient through absorber, N is the number of absorbers between sample and detector, K is the parameter dependent on X-ray wavelength and choice of emulsion, S is the scale factor on counter-recorder, and I is the average intensity obtained as described next. With the sample and camera fully aligned, the X-ray intensity passing through the emulsion plate is continuously recorded. The value of I is taken as the average recorded intensity (see Figure 7) over the entire scan. This is not a critical value, for it can be varied over a range of two times or so without grossly changing the quality of the topograph.

* Ilford, Ltd., Ilford, Essex, England.

Table I. Parameters K and G for Several Emulsions

		Parameter	
λ	Emulsion (50 μ thickness)	K	G
Mo K_α	L-4	4000	0.3
	G-5	1000	0.3
Ag K_α	L-4	—	0.55
	G-5	1000	0.55

The parameters G and K can be specified for a given system, with values related to X-ray wavelength and photographic emulsion; typical values we have determined are shown in Table I. The values of G were obtained with ordinary microscope glass slides placed between the specimen and the X-ray detector. It should be noted that the values of K are appropriate to a particular counter-recorder instrumentation, and should be empirically evaluated for each individual instrumentation. The values given here should give some idea of magnitude and variations to be expected. Although use of these procedures will not always lead to suitably exposed topographs, final judgment as to proper exposure time (or number of traverses) can be made readily from trial exposures based on these procedures.

PHOTOGRAPHIC PROCESSING

The quality of the finished topograph can be seriously degraded by careless handling and processing of the emulsion coated glass plate from the time it is removed from the original container. Careful handling of the plates to avoid artifact markings that would confuse interpretation or reduce detail resolution is fully as important as the exposure stages, but is more readily achieved.

The emulsions are very sensitive to pressure. Even slight abrasion of paper across the emulsion can result in a family of light lines in the developed topograph. Contact of the emulsion by the topographer's fingers should also be avoided, since the skin oils and perspiration have adverse affects on the developability of the emulsion at the points of contact. While this is appropriate advice for handling any type of photographic materials, it is emphasized here because of the more intimate handling of the plates as compared to most photographic applications, the relative likelihood of mishandling, and the need for subsequent magnification.

The recommended procedures for developing Eastman high-resolution plates are as follows:

1. Develop in 1 : 2 Kodak D-19 for 15 min @ 20°C.
2. Rinse in water for 1 min.
3. Fix 15 min @ 20°C in standard X-ray fixing solution.
4. Wash $\frac{1}{2}$ hr in running water.
5. Swab gently with cotton under running water.
6. Rinse with methyl alcohol from squeeze bottle.
7. Dry in desiccator with emulsion down.

The recommended procedures for developing Ilford Nuclear Emulsion plates are as follows:

1. Soak in water for 10 min.
2. Develop in 1 : 3 D-19 for 30 min @ 0°C.
3. Stop for 10 min in 3% acetic acid solution.
4. Fix for $1\frac{1}{2}$ hr in standard X-ray fixing solution.
5. Wash 90 min in flowing water.
6. Swab gently with cotton under running water.
7. Dip in formalin solution (optional antifungus treatment).
8. Rinse with methyl alcohol from squeeze bottle.
9. Dry in desiccator with emulsion down.

Development at ice-water temperature is for the purpose of suppressing grain growth, which is necessary for maximum resolution.

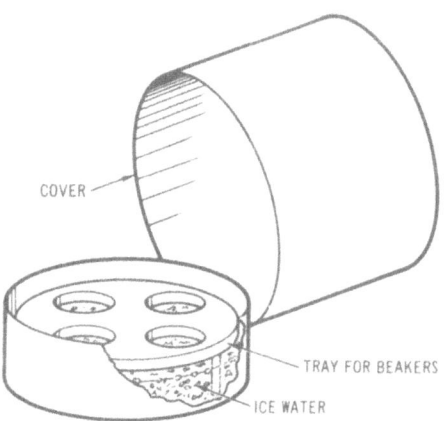

COVER

TRAY FOR BEAKERS

ICE WATER

Figure 11. Sketch of developing tray and cover; 50 ml beakers to hold developing solutions are not depicted.

For consistent results in obtaining good-quality topographs, it is important, throughout the entire development process, to avoid dust or water-borne particles that could settle on, and become embedded in, the emulsion. If necessary, the developing solutions and wash-water should be filtered to remove particulate matter.

The process of developing Lang topographs at ice temperature can be simplified to give economy of both time and processing chemicals by the use of a "kit" illustrated in Figure 11. The principal containers are an 8-in. diameter crystallizing dish (sandblasted and painted black on the exterior) and a closed-bottom carton somewhat larger in diameter (also painted black). With the carton inverted over the dish on a level surface, a sufficiently light tight enclosure is formed. Four 50 ml beakers, containing the process solutions, are placed in the plastic tray so as to partially submerge them in an ice-water bath in the dish.

The processing solutions are stored in capped, plastic bottles in a refrigerator, and are poured into the respective beakers immediately before processing the plates; the solutions are then maintained at ice temperature by placing the beakers in the plastic tray and ice water. Since only small quantities are needed, no effort need be made to save the used solutions for reuse.

To minimize reticulation of the emulsion by rapid temperature changes, the following procedure is used. The plate is placed in a beaker of water initially at room temperature. During the 10- or 15-min soak period, the water and plate are cooled by placing the beaker in the process tray. At the end of the soak period, the plate can be advanced to the other solutions without undergoing significant temperature changes. When the plate reaches the final solution (fix), the plate can be brought back to room temperature by removing the beaker with the fix solution from the ice bath as soon as the plate is inserted. Thus, the plate is slowly warmed during the fixing process.

Sensitivity of the photographic emulsion before and during image development has been already noted. The emulsion of the finished topograph is still tender, however, and should be protected. Protection can be provided with a standard glass microscope slide cover, cemented over the emulsion with suitable mounting fluid. If the topographs are made on short or irregular sections of plate, they can be mounted, also with the mounting medium, on standard size microscope slides. Care must be taken to avoid entrapment of bubbles in the mounting medium when applying either the cover glass or the standard size slide to the topographs. Furthermore, both the slide and the cover glass should be cleaned meticulously before use. Use of a dust-free air jet from an aerosol can is very handy in maintaining cleanliness.

The mounting medium and glass cover over the emulsion has another incidental benefit. In spite of the procedures described above, occasionally reticulation of the emulsion will occur, seriously degrading the quality of the topograph. The mounting medium, sandwiched between the emulsion and the cover glass, however, is very effective in rendering the reticulation invisible.

The final step of preparing photographs for examination or publication is photography of the topograph. This is done most conveniently on an inverted-stage type of microscope arranged with transmitted light and a camera.* With the photographic eyepiece removed from the optical train, and with special macroscope lenses, optical enlargements as low as approximately $10 \times$ can be obtained.

To retain in the photograph as much as possible of the detail and contrast in the topograph, it is recommended that stray and scattered light be eliminated as much as possible. This is facilitated by using an opaque mask over the topograph, with a cutout only as large as the area to be photographed. When this is not used, some of the light in the unmasked beam may be scattered into the lenses and reach the photographic film, thus contributing to some degradation of sharp image detail. This loss of detail is not severe, however, and can be tolerated for most purposes.

ACKNOWLEDGMENTS

The procedures described and recommended in this paper incorporate innovations developed in the authors' laboratories, conclusions from stimulating discussions with colleagues, and advice on procedures developed elsewhere but not previously reported. A few of the procedures have appeared previously in the literature. In particular, the contributions at Atomics International of K. T. Miller (now with Hughes Research Laboratory) and Dr. Michael Hart, Bristol University, were especially valuable. Procedures described here for Berg–Barrett topography were modified and expanded from an earlier publication (Reference 7).

The readers are referred further to instruction manuals accompanying X-ray generator and camera equipment, as well as technical data sheets for photographic emulsions, all of which give detailed information which was therefore not appropriate to report here.

The studies at Atomics International were sponsored by the Fuels and Materials Development Branch, Division of Reactor Development and Technology, United States Atomic Energy Commission under Contract AT(11-1)-GEN-8.

REFERENCES

1. R. W. James, in: *Solid State Physics, Vol. 15,* F. Seitz and D. Turnbull (eds.), Academic Press, New York, 1963, pp. 55–220.
2. B. W. Batterman and H. Cole, "Dynamical Diffraction of X-rays by Perfect Crystals," *Rev. Mod. Phys.* **26**: 681, 1964.
3. W. W. Webb, in: *Direct Observation of Imperfections in Crystals,* J. B. Newkirk and J. H. Wernick (eds.), Interscience Publishers, Inc., New York, 1962, pp. 29–76.
4. U. K. Bonse, M. Hart, and J. B. Newkirk, in: *Encyclopaedic Dictionary of Physics,* to be published.
5. S. Amelinckx, "The Direct Observation of Dislocations" in: *Solid State Physics, Vol. 6,* Academic Press, New York and London, 1964.
6. C. S. Barrett, "A New Microscopy and Its Potentialities," *Trans. AIME* **161**: 15–64, 1945.
7. J. B. Newkirk, "The Observation of Dislocations and Other Imperfections by X-ray Extinction Contrast," *Trans. AIME* **215**: 483, 1959.
8. H. Barth, "Measuring Techniques of Parallel-Beam-Diffraction Micrography," this volume, pp. 80–90.

---

* Such as the Reichart MEF Metallographic Microscope, which the authors have used extensively.

9. A. R. Lang, "Studies of Individual Dislocations in Crystals by X-ray Diffraction Microscopy," *J. Appl. Phys.* **30**: 1748, 1959.
10. S. B. Austerman, J. B. Newkirk, and D. K. Smith, "Study of Defect Structures in BeO Single Crystals by X-Ray Diffraction Topography," *J. Appl. Phys.* **36**: 3815, 1965.

DISCUSSION

D. K. Smith (Lawrence Radiation Lab.): Can you give me some idea of what the useful lower limit of the confining by the slit would be in limiting the volume you are seeing in a topograph? Have you experimentally determined a figure on that?

S. B. Austerman: I have not tried to push that to a limit. Those that I showed would be about 700 μ from tip to tip of the rhombus, but we are using a wide slit. On the Lang camera collimating slit we can go down by a factor of 10. Then we can use a narrower receiving slit, as well, which would go down by, say, a factor of 6. I think we can get down perhaps to section widths of 100 μ without too much trouble.

R. F. Belt (Airtron): Is there another possible cause for the white background on your beryllium oxide crystals? Could it be due to some of the flux precipitating along the screw dislocation or the helix you mentioned?

S. B. Austerman: You mean absorption in the flux material?

R. F. Belt: Yes, absorption of a higher atomic numbered element in your lower atomic numbered beryllium oxide.

S. B. Austerman: This would be another way of removing energy from the beam, to be sure, but the amount of flux material along that line is so very little that we cannot see it by light microscope and I wouldn't image that it would absorb so much X-ray energy that we could see it in the background. It is a remote possibility, though.

J. B. Newkirk: The effect of the increased intensity in the presence of these impurity atoms would probably overwhelm the absorption effect. There would be a dark line.

A. Bhattacharyya (The Foxboro Company): How do you limit the vertical divergence when you use a standard line focus, as in scanning oscillator technique?

S. B. Austerman: Are you referring to the horizontal length of the focus?

A. Bhattacharyya: In the scanning oscillation work mentioned by Dr. Howard (IBM), he uses a line focus. How would you limit the vertical divergence in such a case using a line focus instead of a spot focus? You talked about the importance of vertical divergence.

S. B. Austerman: We have a focal spot that is very small in a vertical direction, so any particular point on the specimen does not see very much divergence from the focal point on the tube.

E. D. Jungbluth (General Telephone and Electronics): I might comment on that question. The method you described, I think, is Schwuttke's. He uses the line focus from a standard X-ray generator and a soller slit system to limit the vertical divergence. However, in the Lang technique, the vertical divergence is limited by the size of the X-ray source which happens to be a spot source from a microfocus generator.

X-RAY DIFFRACTION CONTRAST FROM IMPURITY PRECIPITATES IN CdS SINGLE CRYSTALS

Jun-ichi Chikawa

NHK Broadcasting Science Research Laboratories
361, Kinuta-machi, Setagaya, Tokyo, Japan

ABSTRACT

Impurity-doped crystals $CdS(GaGl_3)$ have been studied by X-ray topography. Some large precipitates are formed close to the crystal surfaces by annealing at 300°C. In the symmetrical Laue case, the precipitates show circular images (30–60 μ in diameter) due to the radial strains around the precipitates which consist of two semicircles separated by a contrast-free plane parallel to the reflecting plane. The observations indicate that the strain field between the crystal surface and precipitate is not responsible for the contrast, and that the images are formed by X-rays which are deviated from the Bragg condition for the perfect region and satisfy the Bragg condition in the strain field on the inside of the precipitate. One of the semicircles is formed by the incident X-rays with larger glancing angles than the Bragg angle and the other with smaller ones. It is concluded that this contrast is due to the strain around a convex lens shaped precipitate.

INTRODUCTION

Diffraction contrast from the strain around vacancy clusters and precipitate particles have been observed by electron microscopy.[1] This contrast has been studied[2,3] and Ashby and Brown[3] have shown that the sign of the strain around precipitates can be determined from the asymmetry of dark-field images of particles close to either surface of the foil (so-called "anomalous image"). Vacancy clusters[4,5] and precipitate particles[6-8] also have been observed by X-ray topography. The present author has described an X-ray topographic method to determine the sign of the Burgers vector of a dislocation.[9] In the present paper, the similar principle will be applied for the determination of the sign of strain around precipitate particles.

EXPERIMENTAL

The crystals were grown in a silica tube with two temperature zones by the sublimation method.[10,11] CdS powder mixed with $GaCl_3$ (10% in weight) was evaporated in the higher temperature zone (1150°C), and the vapor was carried into the lower temperature zone (950°C) by a flow of argon gas, where the wurzite-type crystals grew. Although the impurity content of the crystals was unknown, it was not so high as to change the crystal structure. Some of the crystals had the crystal habits indicated in Figure 1. To observe these crystals by X-ray topography, their lateral surfaces were ground with successively finer grades of carborundum, and finally the crystals were etched. The resulting thicknesses were about 70 μ. Then, the crystal was annealed at 300°C for about 30 min. X-ray topographs were taken by the Lang method[12] (Laue case) using $Mo\,K_{\alpha_1}$ radiation. The thickness is nearly equal to the reciprocal of the linear absorption coefficient.

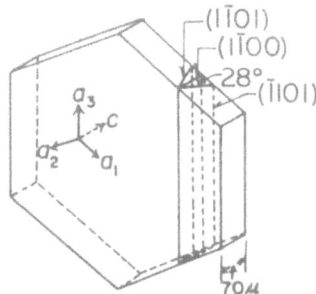

Figure 1. The shape of CdS(GaCl$_3$) crystals.

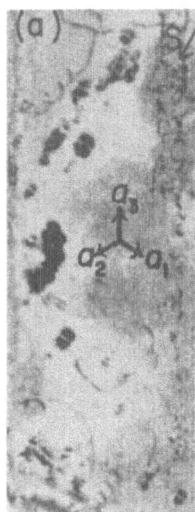
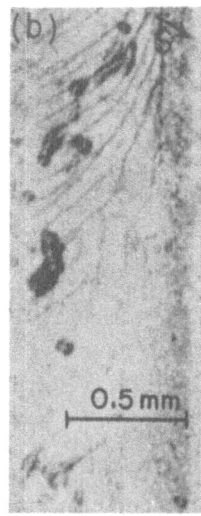

Figure 2. Traverse topographs of the same region in the crystal. (a) 0110 reflection. (b) 2$\bar{1}\bar{1}$0 reflection. The arrows marked "S" indicate the directions of the diffraction vectors. Circular images are visible which consist of two semicircles separated by a contrast-free plane parallel to the reflecting plane used. They are due to radial strains around the precipitates.

OBSERVATION OF THE PRECIPITATES

X-ray topographs of the same area of the crystal are shown in Figure 2. Some circular and irregular dislocation images are seen. The circular images are 30–60 μ in diameter and consist of two semicircles, i.e., they have contrast showing a single contrast-free plane of symmetry. This plane, however, is different in Figures 2a and b, which were taken by the 01$\bar{1}$0 and 2$\bar{1}\bar{1}$0 reflections, respectively. It was established, by taking topographs for the 1$\bar{1}$00 and 2$\bar{1}\bar{1}$0 type reflections, that the contrast-free plane is parallel to the reflecting plane used. This observation shows that the twofold symmetry in the contrast is not due to twofold symmetry of the imperfection itself, but is associated with the orientation of the reflecting plane.

The circular images were never observed before annealing. Some of the images appeared after annealing and aging at room temperature for several days. It is proposed that the imperfections are precipitates of the impurity additive GaCl$_3$.

In order to obtain information about the shape of the precipitates, topographs for the $\bar{1}$101 and the 1$\bar{1}$01 reflections were taken as shown in Figures 3a and b, respectively. In Figure 3 the contrast-free plane also appears parallel to the reflecting plane, but the image is not symmetric with respect to the contrast-free plane, i.e., in Figure 3a the arc on the right side of the contrast-free plane is larger than the one on the left side, and in Figure 3b *vice versa*. The precipitates seem to have a disc or lens shape as illustrated schematically in Figure 4, because any spherical precipitate particles should give symmetrical images.

Figure 3. Traverse topographs of the precipitates in the asymmetrical Laue case. (a) $\bar{1}101$ reflection. (b) $1\bar{1}01$ reflection. The reflecting planes are inclined to the normal to the lateral surface of the crystal by about 28°, as indicated in Figure 1. In each of (a) and (b), two regions of the crystal are shown and the left one is the same region as that in Figure 2. In the images due to the precipitates, the contrast-free planes appear parallel to the reflecting plane, but the images are asymmetric with respect to the contrast-free plane.

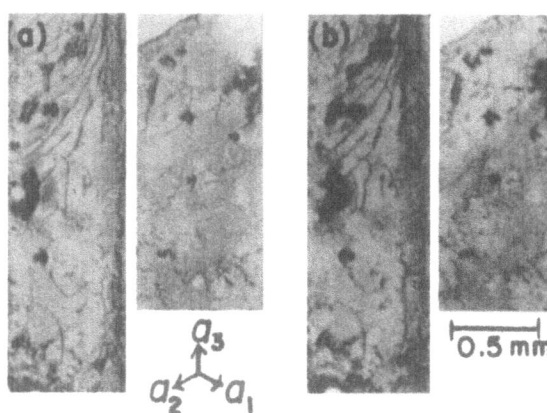

Figure 4. Schematic cross section of the strains around convex-lens and concave-lens shaped precipitates which are close to the crystal surface. The arrows indicate the directions of the displacements of atoms, and the lines illustrate the curved atomic planes.

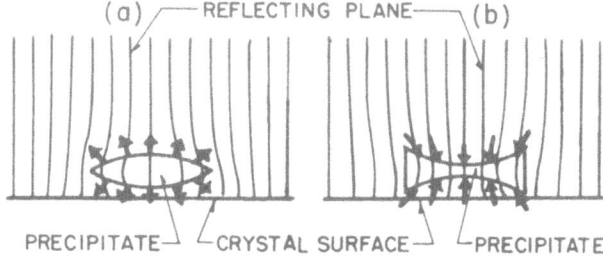

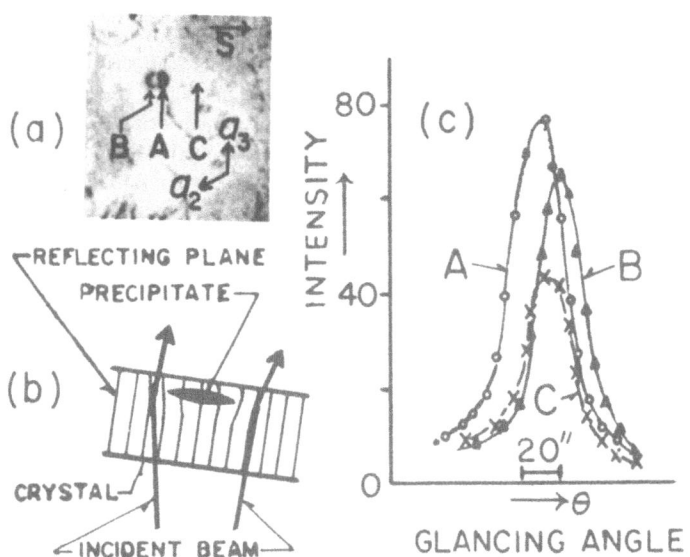

Figure 5. Rocking curves for the selected areas. The $1\bar{1}00$ reflection. Rocking curves, A, B, and C in (c), were obtained for the positions indicated by the arrows, A, B, and C, in the traverse topograph in (a). The cross section of the incident beam was about $30 \times 30\ \mu$. The peak shifts of the rocking curves A and B from that of curve C may be explained by a model of the strain illustrated in (b).

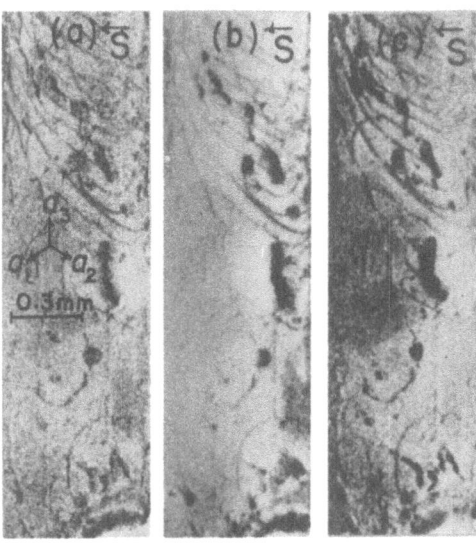

Figure 6. Traverse topographs of the precipitates taken by the slit-collimated beam whose divergency was about 10 sec of arc. The area of the topographs is the same as that in Figure 2, but the crystal was reversed from the orientation in Figure 2, i.e., the precipitates were close to the entrance surface for X-rays. $1\bar{1}00$ reflection. (a) is a topograph for the case where the beam satisfies the Bragg condition for a perfect region. (b) and (c) are those by the beams in a deviation from the Bragg condition, $\delta\theta = \theta_B - \theta \sim +15$ and $\delta\theta \sim -15$ sec of arc, respectively, where θ_B and θ are the Bragg and glancing angles on the reflecting plane. In (a), the circular images show very weak contrast. Only the right semicircles are visible in (b) and the left ones in (c). This observation can be explained similarly to the case of Figure 5.

It was found by stereo observation that the precipitates with a larger diameter are close to the crystal surface (in the case of Figures 2 and 3, close to the exit surface for the X-rays). This was confirmed by the fact that the circular images were extinguished by etching the crystal surface slightly. Smaller precipitate particles seem to be formed in a more interior region of the crystal.

Rocking curves for selected areas were measured by a narrow incident beam with a cross section of about $30 \times 30\ \mu$. Curves A and B in Figure 5c are the rocking curves which were obtained for the cases where the beam passed through the right and left semicircles indicated by the arrows A and B in Figure 5a, respectively. Curve C is that for the perfect region arrowed in C in Figure 5a. The peaks of the curves A and B are shifted from that of curve C in the opposite direction to each other. The peak of curve A is higher than that of curve B. Topographs in Figure 6 were taken by the beam whose divergency was limited to about 10 sec of arc by a slit. Figure 6a is a topograph for the case where the beam satisfied the Bragg condition for a perfect region, and Figures 6b and c are those taken by the beams in a deviation from the Bragg condition $\delta\theta = \theta_B - \theta \sim +15$ and $\delta\theta \sim -15$ sec of arc, respectively, where θ_B and θ are the Bragg and glancing angles on the reflecting plane. Only the right semicricle of each image appears in Figure 6b and the left one in Figure 6c. In Figure 6a contrast of the images is very weak. These observations lead to the conclusion that the images are formed mainly by X-rays which are so deviated from the Bragg condition that the X-rays are hardly diffracted by the perfect region of the crystal.

EXPLANATION OF THE CONTRAST

Eshelby[13] has given expressions for the elastic displacements surrounding inclusions of general character and shown that far from the inclusion the displacements depend only on the volume of the inclusion and the stress-free strain which is expressed as the fractional difference in lattice parameter between matrix and free precipitate if the strain can be treated as a pure dilation. Then, two types of strains around the precipitate particles are produced, as illustrated in Figure 4, depending on the sign of the fractional difference in atomic volume between precipitate and matrix material. Depending on this sign, precipitates have a convex or concave lens shape. Determination of the sign seems to be important in identification of inclusions. The purpose of this section is to distinguish between both types.

The displacements around precipitates are considered to take place in the directions as indicated by the arrows in Figure 4. The directions of the displacements for both the types are similar, although the senses of the directions are opposite. In general, no contrast is produced for the case where the displacement of atoms is parallel to the reflecting plane. When the reflecting plane includes the axis of the lens, the displacement of atoms on the diameter of the lens parallel to the reflecting plane is also parallel to the reflecting plane. Thus, the contrast-free plane appears parallel to the reflecting plane on each circular image as seen in Figure 2.

It is clear that no contrast-free plane should appear for a reflecting plane that is inclined to the axis of the lens, if the contrast is formed by the strains on both the sides of precipitates. Figure 3 shows the contrast-free plane. This fact indicates that the contrast is formed from the strain field on one side of the precipitates. Since the displacement of atoms on the contrast-free plane should be parallel to the reflecting plane, one can conclude from the asymmetric position of the contrast-free plane in Figure 3 that this contrast is formed by the strain field on the inward side of the precipitate rather than by that between the crystal surface and the precipitate. The negligible effect from the latter strain field may be due to the fact that the precipitates are very close to the crystal surface. It should be noticed that the above conclusion is valid for both types of precipitation in Figures 4a and b, because the directions of the displacement of atoms are the same for both types.

Now we shall consider only the strain field on the inward side of the precipitates. The signs of the directions of the displacement for both types are opposite to each other. This means that the atomic planes parallel to the lens axes for both types are inclined in the opposite senses as shown schematically in Figure 4. As has been seen in Figures 5 and 6, the circular images are formed by X-rays which are so deviated from the Bragg condition that the X-rays are hardly diffracted by the perfect region of the crystal. The X-rays in such a deviation from the Bragg condition are considered to pass through the perfect region in the incident direction and to be diffracted in the part of the strain field where the X-rays satisfy the Bragg condition. Then one can determine the sense of the inclination of atomic planes in that part from the sense of the deviation of the X-rays, and the observations in Figures 5 and 6 indicate the type shown in Figure 4a (see Figure 5b).

In Figure 5a, the right semicircle of the image has stronger contrast than the left one. This contrast asymmetry is seen clearly from the rocking curves in Figure 5c and can be explained as follows: In Figure 5, the precipitates are close to the exit surface for X-rays. The left and right semicircles are formed by the X-rays with $\delta\theta < 0$ and $\delta\theta > 0$, respectively. The incident beam with $\delta\theta > 0$ excites a stronger wave on the further branch of the dispersion surface from the origin than that on the other branch, and *vice versa* for $\delta\theta < 0$. Since the absorption of the perfect region for the wave excited on the further branch is smaller than that on the other,[14] the right semicircle has a stronger contrast. This asymmetric contrast corresponds to the so-called "anomalous" images[3] observed by electron microscopy, and is similar to that of dislocations close to the crystal surface.[9,15]

CONCLUSION

It was found by X-ray topography that large precipitates are formed close to the crystal surfaces by annealing at 300°C. In the topographs for the reflections perpendicular to the crystal surface, circular images 30–60 μ in diameter appear which consist of two semicircles separated by a contrast-free plane parallel to the reflecting plane. The observations indicate that the strain field between the crystal surface and precipitate is not responsible for the contrast, and that the images are formed by X-rays which are deviated from the Bragg condition for the perfect region and satisfy the Bragg condition in the strain field on the inward side of the precipitates, i.e., one of the semicircles is formed by the incident X-rays with larger glancing angles than the Bragg angle and the other with smaller ones. From this contrast, the sign of strain around the precipitates can be determined to be the type indicated in Figure 4a.

REFERENCES

1. A. Kelly and R. B. Nicholson, "Precipitation Hardening," in B. Chalmers (ed.), *Progress in Materials Science*, Vol. 10, Pergamon Press, Inc., New York, 1963, p. 148.
2. V. A. Phillips and J. D. Livingston, "Direct Observation of Coherency Strains in a Copper–Cobalt Alloy," *Phil. Mag.* **7**: 969, 1962.
3. M. F. Ashby and L. M. Brown, "Diffraction Contrast from Spherically Symmetrical Coherency Strains," *Phil. Mag.* **8**: 1083, 1649, 1963.
4. F. W. Young, Jr., "The Characterization of Nearly Perfect Copper Crystals," in: H. S. Peiser (ed.), *The Proceedings of the International Conference on Crystal Growth*, Pergamon Press, Inc., New York, 1967, p. 789.
5. A. Authier, C. Malgrange, and J. F. Fetroff, "Etude de Defauts Dans le Fluorure de Lithium par la Methode de Lang," *J. Physique* **24**: 566, 1963.
6. J. M. Fairfield and G. H. Schwuttke, "Precipitation Effects in Diffused Transistor Structures," *J. Appl. Phys.* **37**: 1536, 1966.
7. K. Furusho, "Study on Precipitates in Oxygen-Doped Silicon Single Crystals by X-Ray Diffraction Micrography," *J. Appl. Phys. Japan* **3**: 203, 1964.
8. J. Chikawa, "X-Ray Observation of Clustering of Impurity Atoms in CdS Crystals," *Appl. Phys. Letters* **4**: 25, 1964.
9. J. Chikawa, "X-Ray Topographic Observation of Dislocation Contrast in Thin CdS Crystals", *J. Appl. Phys.* **36**: 3496, 1965.
10. S. Ibuki, "On the Crystal Growth of Cadmium Sulfide," *J. Phys. Soc. Japan* **14**: 1181, 1959.
11. J. Chikawa and T. Nakayam, "Dislocation Structure and Growth Mechanism of Cadmium Sulfide Crystals," *J. Appl. Phys.* **35**: 2493, 1964.
12. A. R. Lang, "Studies of Individual Dislocations in Crystals by X-Ray Diffraction Microradiography," *J. Appl. Phys.* **30**: 1748, 1959.
13. J. D. Eshelby, "The Determination of the Elastic Field of an Ellipsoidal Inclusion, and Related Problems," *Proc. Roy. Soc.* **241**: 376, 1957.
14. See, for example, B. W. Batterman and H. Cole, "Dynamical Diffraction of X-Rays by Perfect Crystals," *Rev. Mod. Phys.* **36**: 681, 1964.
15. A. R. Lang, "X-Ray Topographic Determination of the Sense of Burgers Vectors of Pure Screw Dislocations," *Z. Naturforschg.* **20a**: 636, 1965.

DISCUSSION

J. B. Newkirk: Do I understand that these contrast effects that you have been describing are all kinematic effects?

J. Chikawa: A large part of the contrast may be due to the kinematical image, but some part of the contrast is dynamical. The observed contrast asymmetry (Figure 5) can be explained as a dynamical effect.

LANG X-RAY TOPOGRAPHIC STUDIES OF RUBY GROWN BY DIFFERENT METHODS

AIRTRON, A Division of Litton Industries
Morris Plains, New Jersey

ABSTRACT

Crystals of ruby (Al_2O_3:Cr) are being grown at the present time by several standard procedures, e.g., Verneuil, Czochralski, flux, and hydrothermal. Previous work has indicated wide variations in quality and type of defects. The present study is primarily concerned with ruby grown from PbO–PbF_2 fluxes. All crystals were examined in transmission with Mo $K_{\alpha 1}$ radiation and a Rigaku Denki Lang camera. Samples were either sectioned from larger crystals or obtained as plates with natural growth faces. Results on flux-grown ruby have shown a severely banded substructure due to strain introduced by either the flux or chromium segregation. Crystals with a visible chromium gradient have shown fewer bands in those regions which were depleted of chromium. Annealing studies have been performed on ruby and all banded structure was dispersed into areas of fine particles with a much higher dislocation content. Other features of the substructure are described. These include the observation of Pendellösung fringes in wedge-shaped sections. Areas of nearly 0.5 cm² were found to be dislocation free. The X-ray results confirm etching experiments but the former yield more information on internal details other than dislocations.

Recent data on sapphire (α-Al_2O_3) have confirmed several tentative conclusions. Flux grown crystals have some bands due to flux segregation. However, sapphire crystals show higher perfection when compared to ruby. These results have been confirmed by rocking curves obtained on a double crystal X-ray spectrometer used in the parallel position. Measurements on the $(00 \cdot 12)$ reflection from the natural growth faces were performed with Cu $K_{\alpha 1}$ radiation. Widths at half the maximum intensity were always found to be larger for ruby than for sapphire. Comparisons were made with respect to a perfect silicon crystal in the same geometry. Some preliminary experiments have been performed on Czochralski ruby prepared by the Linde Company and on hydrothermally grown ruby prepared at Airtron. Examples of defects observed in each case are given. They are highly unique to the growth method. The Lang method is shown to reveal more detail than other X-ray data. A brief discussion of ruby quality as a function of growth method is presented.

INTRODUCTION

The growth of large single crystals of ruby is highly desirable for optical and microwave masers. In addition to size, many other factors have increased in relative importance. Among these are optical clarity, homogeneous doping at specified levels, absence of large angle grain boundaries and strain. Most of the preceding have been the subject of individual studies relating to the performance of the crystals as a laser.[1,2] Unfortunately, the existing state of crystal growth limited the source of the crystals to those grown only by the flame-fusion method. It has been well verified that the flame-fusion ruby crystals are poor in crystallographic quality.[3] This arises primarily from the nature of the growth process.

It is interesting that ruby is one of a few crystals that has been grown by alternate techniques. Thus far the flux,[4,5] Czochralski,[6] and hydrothermal[7] methods all have been reported to give single crystals of sufficient size for many applications. Some preliminary work in a few laboratories has indicated that the crystal quality can be very high.[8] Programs have been initiated to obtain large, high quality, uniformly doped crystals by both Czochralski and hydrothermal methods. The purpose of the present work was twofold. First to study and compare crystals grown by the various methods. Second, to examine in more detail flux and hydrothermal rubies for defects incurred during growth. The results can be correlated with experimental data derived from known changes of growth parameters.

EXPERIMENTAL

The single-crystal flux-grown rubies for the present study were obtained from melts composed of PbF_2. The thin crystals were examined with their natural faces intact. These large area faces were of the {0001} type. Flux-grown rubies up to a size of $3 \times 2 \times 1$ cm were also obtained. These crystals were sectioned with a diamond wheel and polished with successively finer diamond grits. The final mechanical polish was performed with a 1μ or $\frac{1}{4} \mu$ diamond abrasive. Hydrothermally grown ruby crystals were obtained from growth on flux ruby seeds. These crystals also were cut and polished with diamond tools and abrasives. The source of the Czochralski ruby was the Linde Company. It was obtained as a $90°$ rod. All crystals were oriented by optical and X-ray back reflection methods. The surface condition of the cut and polished crystals was quite sensitive to technique. All polishing scratches had to be removed by fine diamond pastes prior to chemical polishing. Final chemical polishing was performed with a PbF_2–PbO eutectic mixture at 550–$650°C$ to give a defect-free surface.

All crystals were examined with a Rigaku Denki Lang camera using Mo K_α radiation. A Dunlee commercial X-ray tube served as the source. The focal spot was viewed "end on." The source to crystal distance was increased to 50 cm by means of a brass tube. The horizontal divergence was fixed at 90 sec and the vertical at $1°$ by means of fine slits. The resulting horizontal and vertical resolution were 1μ and 15μ, respectively. Preliminary exposures were made on Kodak No Screen X-ray film. All of the final topographs were recorded on Kodak Type M X-ray films or Kodak Type A Autoradiographic Plates. Total exposure times ranged from 2–10 hr depending on scan length, thickness, and orientation.

The double crystal spectrometer utilized for line widths was assembled by combining a standard Picker X-ray diffractometer with the Rigaku Denki Lang camera. The first crystal was a nearly perfect silicon crystal provided by the Dow Chemical Company. The dislocation content by etch pit count was no more than a few hundred per cm^2. The crystal was cut, lapped, and finally at least 0.1 mm of the surface was removed by chemical means. The major face was (111). The second crystal was always the sample crystal and was mounted on a goniometer head contained on the Lang camera. The latter instrument was capable of measuring directly to 1 sec of arc. The spectrometer was always operated in the parallel arrangement and utilized the (333) reflection from silicon and the (00·12) reflection from ruby and sapphire. The latter was chosen because the growth morphology provided a large area (0001) face on both flux and hydrothermal crystals. This natural face was highly flat and free from any surface damage. The crystals were not polished or etched because surface damage is rather difficult to remove and the measurements are very sensitive to surface defects.

The radiation used was Cu K_α. Soller slits and plain slits were incorporated to limit the horizontal and vertical divergence to 1° and 2°, respectively. The axes of rotation of both crystals were vertical. Two slits were used in front of the second crystal at different times. In one case a 1 mm × 1 mm aperture was employed and in the other a 1 mm × 0.2 mm slit. The distance between crystals was approximately 40 cm. Radiation was detected with a scintillation counter incorporating a pulse height analyzer. The rocking curve was obtained by rotating crystal two through the Bragg angle. The intensity was scaled at each 2 sec of arc after manually turning the crystal. The resulting data were plotted for each crystal and the width at half maximum intensity determined. These widths were readily reproducible to a few tenths of a second.

RESULTS AND DISCUSSION

Flux-Grown Ruby

Some confusion has existed between the designation of planes in the morphological and true structural unit cell of Al_2O_3. Kronberg[9] has presented a valuable description of the proper relationships. In our investigations we have adhered to the use of the hexagonal X-ray structural unit cell. The d spacings and indices have been fully tabulated for this unit cell.[10]

Enough evidence has been published on flame fusion rubies to show that typical dislocation contents are nearly always in the range of 10^5–10^6 per cm². Grain boundaries of 1–2° are present and the crystals are often severely strained. The fruitful application of the Lang method[11] requires a crystal of much lower dislocation density in order to prevent severe overlap of defects. Small rubies of high quality can be prepared by means of the flux growth method. Etching studies have verified that ruby crystals with dislocation contents as low as 10^2 per cm² could be prepared.[12] For this reason, and because flux grown rubies are used as seeds for hydrothermal growth, preliminary topographic work was restricted to flux-grown plates.

The first crystal examined was a flux-grown ruby plate with {0001} faces. The area of the face was about 1.0 × 1.5 cm and the thickness was 0.22 mm. Preliminary microscopic examination showed very few defects on the surface. One tiny particle of ruby adhered to the surface of one face. This was probably caused by sudden nucleation and growth during the cooling cycle. No sign of PbF_2 inclusions was visible. This was confirmed by a later X-ray spectrographic analysis which showed less than 0.25% Pb in the entire plate. The use of crossed polarizers in a view parallel to the optic axis showed no evidence of strain or grain misorientation. The crystal was thoroughly cleaned in HNO_3 before all tests and X-ray topographs were taken with the {0001} natural faces intact. The bounding faces were planes of the type {01$\bar{1}$2}, and {10$\bar{1}$4}. Figure 1 is the X-ray topograph of the crystal taken from the (11$\bar{2}$0) planes which were vertical during the exposure. Some of the tiny defects could be attributed to surface marks but others are apparently dislocations running from the front surface to the back. It is interesting that at the exact point of attachment of the small crystallite, there is only a slight defect on the topograph and it is similar to other surface damage. A possible small angle grain boundary runs vertically from the bottom of the photograph to join a boundary along an a axis. The included angle of the two is very close to 150°. Similar boundaries run along the top of the photograph and upper right-hand corner with no evident relation to the crystal axes. Within each area there is a profuse band structure. The bands in the left grain are parallel to the a-axis while those in the center grain all run perpendicular to the a-axis. The lighter areas of the topograph are from portions of the crystal which were not in diffracting position during the exposure.

Figure 1. X-ray topograph of ruby plate, diffraction from (11$\bar{2}$0).

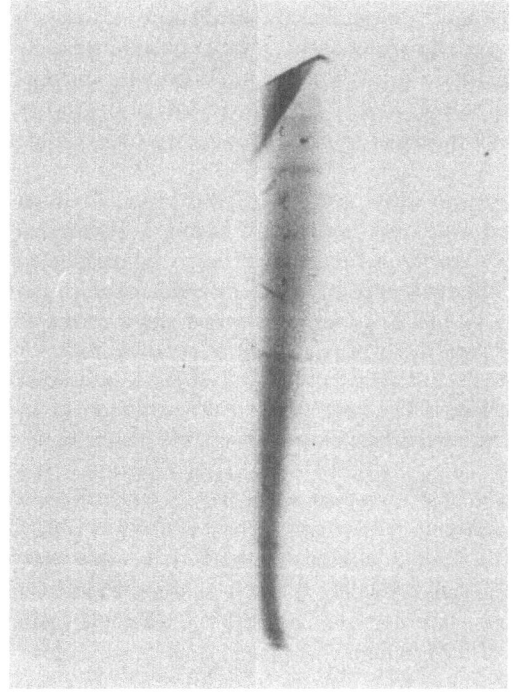

Figure 2. X-ray topograph of ruby plate, diffraction from (03$\bar{3}$0), crystal rotated 30° clockwise from Figure 1.

In order to determine whether a strain was introduced from the mounting procedure, the crystal was removed and remounted in an entirely different manner. At the same time the crystal was rotated 30° in a clockwise direction from Figure 1 and the topograph was taken from the (03$\bar{3}$0) planes. Figure 2 gives the results and now shows the band structure of the central grain of Figure 2 to be horizontal. A portion of the same grain boundary is also in view. The scan length of the crystal was limited to about 2 mm for Figure 2. The remainder of the crystal was out of diffracting position. Further rotations and remounting of the crystals were made at 60°, 90°, and 120° in a clockwise direction to Figure 1. Topographs were recorded from $\{11\bar{2}0\}$ and $\{03\bar{3}0\}$ planes alternately. In all cases the same general features of the banding were obtained. For $\{11\bar{2}0\}$ planes the complete area of the crystal could be recorded on the topographs. For $\{03\bar{3}0\}$ planes only 2 mm of the crystal could be scanned under the same experimental conditions. This may arise from a slight bend about the a-axis, plus the differences in Bragg angles for $\{11\bar{2}0\}$ and $\{03\bar{3}0\}$ planes.

A compilation of the principal dislocation systems in corundum has been given by Scheuplein and Gibbs.[13] Under the conditions of flux growth the temperatures do not exceed 1200–1300°C. The most likely dislocation system is basal slip with (0001) as the slip plane and Burgers vector equal to $\frac{1}{3}$ $(11\bar{2}0)$. The prismatic slip system is apparently only activated at temperatures greater than 1600°C. A special study of flux grown rubies has been made by Janowski, Stofel, and Chase.[14] They gave etching evidence of the effects of entrapped flux and twin boundaries. The latter are 180° rotation type twins about [0001]. The most frequent composition plane is $(11\bar{2}0)$ with $(10\bar{1}0)$ as an alternative. From the preceding data we can arrive at a tentative explanation of the banding observed on Figures 1 and 2. The general absence of birefringence suggests that large strain fields and their association with twin boundaries are unlikely. Back-reflection X-ray photographs also provided no evidence of extensive twin boundaries. The fact that the bands are parallel to the natural growth faces is possibly because of impurity precipitation. The likely impurities are either PbF_2 flux or Cr_2O_3 although both must be on a scale not readily observable at $250\times$ microscopically. Any resultant strains must also be slight and not evident with crossed polarizers. The more powerful X-ray method readily records such defects.

In order to gain more information on the cause of the banding, the crystal plate was annealed in a platinum container for 70 hr at 1400°C. The furnace was left to cool at a rate of about 50°/hr to 500°C. At this point the container was removed from the furnace and allowed to cool to room temperature. A new X-ray topograph was then taken utilizing diffraction from $(11\bar{2}0)$ planes. The orientation of the crystal was exactly the same as Figure 1. Results are given in Figure 3. The striking disappearance of all banding is immediately noticeable. In place of the bands a fine particle structure covers the entire crystal. A faint but noticeable outline of the major grain boundaries has persisted. Many of the particles appear to be dislocation lines with their directions parallel to the $(11\bar{2}0)$ planes. Figure 4 is a topograph of the same crystal rotated 30° clockwise with diffraction from $(03\bar{3}0)$ planes. The same defects are present and again the majority run in a direction parallel to the diffracting planes. In both X-ray topographs there is a similarity to effects noticed microscopically on decorated crystals. Dislocations in Al_2O_3 have been decorated by means of ZrO_2[15] and many straight segments of dislocation lines have been found parallel to $\langle 11\bar{2}0 \rangle$. There is no reason why PbF_2 or PbO could not act in a similar manner. Ultramicroscopic techniques were not used on the annealed crystal to confirm this reasoning.

The next crystal to be examined was a cut from a large flux grown ruby. This crystal had the plate morphology but measured about 5 mm along [0001]. The $\{01\bar{1}2\}$ and $\{10\bar{1}4\}$ planes were highly developed. A cut was made parallel to [0001] and perpendicular to $[21\bar{3}0]$. The large faces were planes of $(11\bar{2}0)$. The thickness of the crystal after mechanical and chemical polishing was 0.4 mm. Figure 5 shows the X-ray topograph of the crystal taken from $(03\bar{3}0)$ planes which are now vertical. One again notes the banding

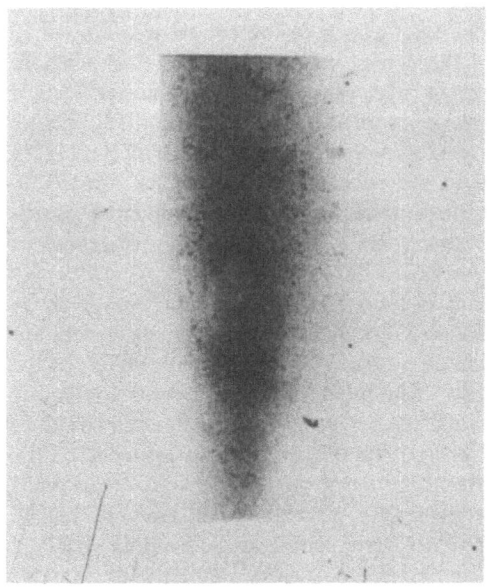

Figure 3. X-ray topograph of annealed ruby plate, diffraction from (11$\bar{2}$0), same orientation as Figure 1.

Figure 4. X-ray topograph of annealed ruby plate, diffraction from (03$\bar{3}$0). Same orientation as Figure 2.

with the lines almost exactly parallel to {01$\bar{1}$2} growth planes. At the bottom of the topograph they appear to intersect the (0001) surface but near the top they meet some dislocations or boundaries which run nearly horizontal across the crystal. The crystal was now rotated in its own plane to get diffraction from (01$\bar{1}$2) planes. The resulting topograph is given in Figure 6, where the (01$\bar{1}$2) planes are oriented vertically. The

Figure 5. X-ray topograph of cut ruby crystal, diffraction from (03$\bar{3}$0).

Figure 6. X-ray topograph of ruby crystal, diffraction from (01$\bar{1}$2).

same banding appears with lines also in a vertical direction. However, a new feature is prominent across the top left area of the topograph. These are a few wide Pendellösung fringes which show a slight bending in the neighborhood of intersecting line defects. The fringes arose from a natural beveling of the crystal in that region during the chemical polishing. The angle of the bevel was approximately $30°$. Figure 7 is a topograph of the same crystal after a net counter-clockwise rotation of about $92°$ to bring the $(10\bar{1}4)$ planes into diffracting position. The latter planes are vertical in the photograph. It may be noticed that the area which contained the banded structure in Figure 6 now shows the same area completely free of any defects. At least six Pendellösung fringes are visible on the right side. All of them bend sharply around a major linear defect. The change of contrast for the banded area in Figure 6 after the rotation, must be caused by a change in orientation of the Burgers vector **b** of the defects with respect to the diffracting planes. Maximum contrast (Figure 6) is usually obtained when **b** is normal to the diffracting planes. Little or no contrast (Figure 7) is seen when **b** is parallel to the diffracting planes. With this criteria in mind we see that Burgers vector of the banded defect may closely parallel the $(10\bar{1}4)$ planes. Such a Burgers vector has apparently not been observed or reported yet for the Al_2O_3 structure.

It is significant from Figure 6 that the crystal contains some defects which do not intersect the basal (0001) planes. This may partially explain the fact that etchants used only on the basal planes consistently show very low (<10) dislocations/cm^2 for flux grown rubies. No detailed study of etching on other planes has been presented. It is clear though that the X-ray method is a more powerful tool for all types of internal defects.

Hydrothermally Grown Ruby

Initially, the growth of hydrothermal ruby was performed on flame-fusion seeds. It was soon apparent that flux-grown ruby would provide a higher quality seed material. While the size of flux ruby was not large, a sufficient quantity could be made without any flux inclusions. Many of the preliminary hydrothermal runs gave crystals with severe cracking, inclusions, bubbles, and other visible defects. It was useless to examine these by the Lang method. Some runs did provide an overgrowth on the seed covering an area of $1–2$ cm^2 with a thickness of $1–3$ mm. The general freedom of visible defects

Figure 7. X-ray topograph of crystal from Figure 5, diffraction from $(10\bar{1}4)$, rotated $92°$ counterclockwise from Figure 6.

in these crystals suggested that they would be worthwhile to examine by X-ray topographic methods. The morphology of the hydrothermal crystals followed those of the flux seeds very closely. Thus the major faces of the crystal were {0001} and the bounding planes were of {01$\bar{1}$2}, {11$\bar{2}$0}, and {10$\bar{1}$4}. All of these did not grow at the same rate and even differences were noted in the growth rates along [0001] and [000$\bar{1}$].

Figure 8 is an X-ray topograph of a ruby crystal which was cut parallel to the c-axis and parallel to an a-axis. In the figure the c-axis is vertical and the a-axis is horizontal. The dimensions were 12 mm along a, and 2 mm along c. The thickness was 0.27 mm. The crystal was oriented to diffract from (11$\bar{2}$0) planes which are also vertical. The topograph shows only diffraction from the hydrothermally grown portion. The seed crystal was entirely out of diffracting position. Obviously, the growth was not strictly epitaxial on a micro scale. A partial rotation of the overgrowth about [0001] must have occurred in the early stages of growth. The angle of rotation was not measured but could be as small as 2–3 min. The presence of many dislocation lines is indicated near the surface of the seed. It also can be noted that the density of these defects becomes less as the crystal grows in thickness. Probably many of the dislocations have grown out of the crystal after a poor initial fit.

Figure 9 is the topograph of the same crystal. A rotation of 30° counterclockwise was made in the plane of the crystal. The diffracting planes were (11$\bar{2}$3) and they run vertically. In this topograph the banding from the flux seed is visible. Again the dislocation content seems to be high at the seed-overgrowth interface. A progressive diminishing of defects occurs through the hydrothermally grown material. A comparison of similar areas of Figure 8 and Figure 9 also shows a change of contrast in many of the dislocation lines. In both topographs the defects in the hydrothermal portion run parallel to [0001]. These may be dislocations of a screw type. The extent of growth along [0001] and [000$\bar{1}$] is seen to be unequal.

Figure 8. X-ray topograph of hydrothermal growth of ruby on a flux seed, diffraction from (11$\bar{2}$0).

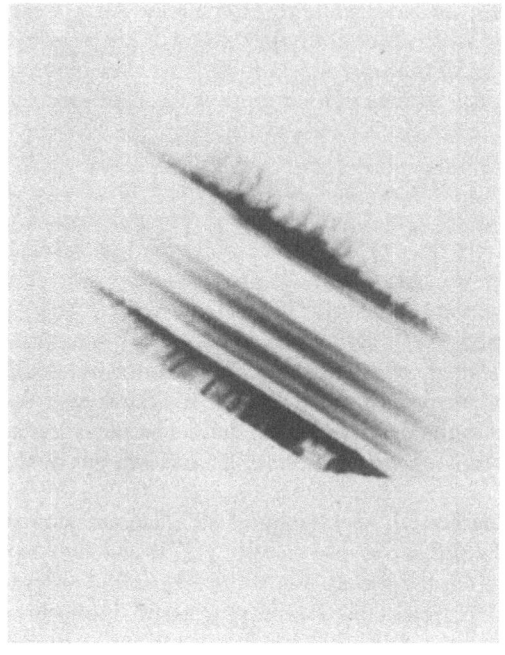

Figure 9. X-ray topograph of same crystal in Figure 8, rotated 30° counterclockwise, diffraction from (11$\bar{2}$3).

Czochralski Grown Ruby

One sample of Czochralski grown ruby was obtained from the Linde Company as a 90° rod. A piece was sectioned in a manner where the large faces were parallel to c and perpendicular to a. Diffraction was obtained from (03$\bar{3}$0) planes. An X-ray topograph of the crystal was taken after mechanical polishing. The thickness of the crystal was 0.6 mm. Diagonal lines which were remnants of polishing scratches were observed. Small precipitates and numerous dislocation lines running horizontally across the crystal also were evident. As the c-axis was vertical in the initial topograph, these defects could lie in the (0001) planes. The latter are base planes upon which slip occurs readily. The c-axis was also the radial growth direction but there appeared to be no great defect gradient from the center of the rod to the edge. The Cr^{+3} concentration is usually larger at the center and edges for these rubies.

Figure 10 is the topograph of the crystal in the same orientation. This topograph was taken after removing nearly 0.4 mm from the thickness by chemical polishing. The same dislocations and precipitates were present in the crystal and all surface damage was completely removed. It is difficult to say what the precipitates might be. They were certainly not visible in the crystal when viewed under a microscope. Two possibilities are metallic iridium or its oxide from the crucible; further impurities may be small chemical species already in the melt. From the X-ray topographs the dislocation content is estimated to be in the range of 100–1000/cm^2. Other rod orientations may contain fewer defects. However, a more thorough examination was not made. The overall crystallographic quality is very good but flux rubies seem to run a factor of ten lower in dislocation content than the Czochralski ruby. However, dislocation content is not the sole criteria for performance quality.

X-Ray Line Widths from a Double Crystal Spectrometer

In addition to X-ray topographic methods for determining crystal perfection,

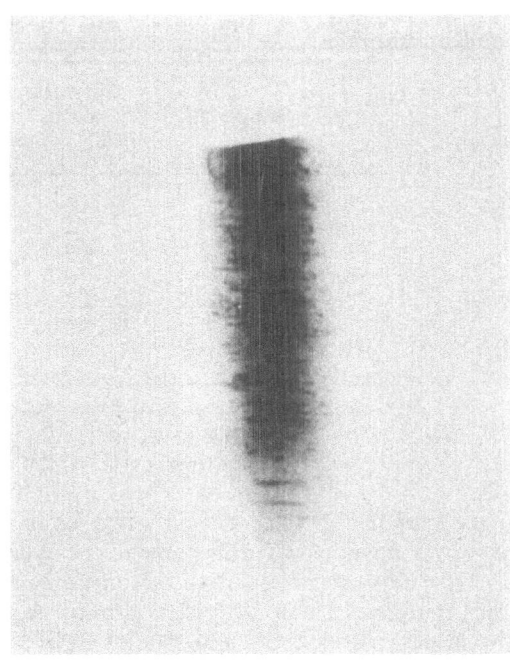

Figure 10. X-ray topograph of Czochralski ruby, diffraction from (03$\bar{3}$0).

X-ray line widths obtained from a double crystal spectrometer are a powerful and reliable procedure. The two techniques are complementary because Lang topography reveals internal imperfections not readily apparent from a linewidth measurement. Likewise, some crystals with imperfections could conceivably give a uniform intensity in transmission and only a line width study could assess the true perfection. The fact that ruby can be grown by several methods has prompted some recent studies on X-ray line widths. One result[16] for Czochralski ruby was mentioned in early 1965. Birks, Hurley, and Sweeney[17] published a more detailed paper on Verneuil, flux, and Czochralski ruby. In recent months, some rather large single crystals of ruby have been grown by the hydrothermal method in the Airtron laboratories. X-ray topographic studies have shown that such crystals are somewhat strained but no effort was made to determine the overall perfection. In order to properly evaluate hydrothermal ruby, double crystal X-ray measurements were made on some of our best samples. At the same time flux grown ruby and sapphire were examined because it was believed that their perfection was higher than indicated in Birks report.[17] The growth of hydrothermal ruby also has utilized flux ruby as seed material. Therefore, the relative perfection is of interest to verify whether defects have propagated from the seed and how certain growth variables may change perfection. To the present time no other reports on hydrothermal ruby have appeared in the literature.

Measurements were performed on various thicknesses of flux grown sapphire and ruby. The hydrothermal ruby was a crystal which was cut perpendicular to [0001]. The (0001) and (000$\bar{1}$) natural faces were examined with no essential differences in perfection. Table I is a summary of results. Under the experimental conditions the flux ruby and sapphire are reflecting as almost perfect crystals with no angular misorientations. Since this was found on many different samples of quite random growth runs, the evidence is certainly conclusive. Birks[17] reported data on one flux ruby and under his conditions the misorientation was estimated to be as high as 10 sec of arc

Table I. Double Crystal Spectrometer Line Widths

Sample	Thickness along [0001] in mm	Width W in sec[a]	Width W in sec[b]
Flux ruby	0.6	6.1	
Flux ruby	1.1	6.5	
Flux ruby	4.9	5.4	9.6
Flux sapphire	0.2	8.6	
Flux sapphire	1.7	5.5	7.0
Flux sapphire	3.0		10.0
Hydrothermal ruby	3.0		16.4

[a] 1 mm × 0.2 mm aperture in front of crystal 2 plus 10 mm × 0.1 mm aperture at crystal 1.

[b] 1 mm × 1 mm aperture in front of crystal 2.

for the local area. No information was given in relation to the source of the crystal and the growth conditions. Turning now to our hydrothermal ruby, it is estimated that after correcting for dispersion due to the unequal spacings of the silicon and ruby crystals, a total misorientation of 8–10 sec is present. A similar result was found for Czochralski ruby in an early report.[16] The data of Birks[17] indicate that his Czochralski ruby was perfect.

Our current work complements and substantiates all of the previous Lang X-ray topographic results. Both flux ruby and sapphire are almost completely free of any dislocations. They still retain some small impurities that segregate along faster growing planes. The (00·12) planes are the slowest growing and therefore may contain the least strain of all. This leads to the almost perfect crystals when these planes are examined. Both the Lang and line width data indicate that hydrothermal ruby must be growing under conditions where small residual strains are readily incorporated into the crystal. At the present time it is doubtful that the strain arises from a thermal origin or a propagation of defects from the seed crystal. The strain can apparently "grow out" as crystal growth proceeds along [0001]. However, the residual misorientation found in crystals may indicate that the diminution of strain is not complete. The results may point to an impurity incorporated in the growing crystal in addition to the Cr^{+3} substituted in the Al_2O_3 lattice.

CONCLUSIONS

Single crystals of ruby prepared by three different growth methods have been examined by Lang X-ray topographic procedures. Flux-grown ruby has by far the lowest dislocation content. However, some strain due to flux entrapment or chromium segregation is still evident in optically perfect crystals. Hydrothermally grown ruby is also subject to strain and cracking at the present stage of development. The encouraging thing is that this probably is not due to chromium concentration gradients. The majority of dislocations are able to grow out as growth proceeds. Czochralski rubies can be grown with very little strain and low dislocation contents. Certain orientations are favorable for dislocations to grow out. The presence of small precipitates, voids, or impurities is a persistent problem.

X-ray line widths obtained from a double crystal spectrometer have verified that flux ruby and sapphire can be grown as nearly perfect crystals. Hydrothermal ruby still retains a misoriented substructure of up to 10 sec of arc.

ACKNOWLEDGMENT

The author wishes to thank the staff members of Airtron for providing the samples of flux-grown ruby and hydrothermally grown ruby. The sample of the Czochralski ruby was kindly provided by Dr. O. E. Nestor of the Linde Company. A critical review of the work and paper was provided by Dr. J. W. Nielsen. The support and permission to publish was graciously received through Litton Industries, Inc. A portion of the experimental work was also performed under Air Force Contract AF33(615)-3160.

REFERENCES

1. R. L. Barns, "Imperfections in Ruby for Maser Applications," in G. E. Brock (ed.), *Proceedings of a Technical Conference on Metallurgy of Advanced Electronic Materials*, Interscience Publishers, New York, 1962, p. 337.
2. G. W. Dueker, C. M. Kellington, M. Katzmann, and J. G. Atwood, "Optical Properties and Laser Thresholds of Thirty-Nine Ruby Laser Crystals," *Appl. Optics* 4: 109, 1965.
3. K. R. Janowski and H. Conrad, "Dislocations in Ruby Laser Crystals," *Trans. Met. Soc. AIME* 230: 717, 1964.
4. J. P. Remeika, "Growth of Single Crystal Rare Earth Orthoferrites and Related Compounds," *J. Am. Chem. Soc.* 78: 4259, 1956.
5. R. C. Linares, "Growth of Refractory Oxide Single Crystals," *J. Appl. Phys.* 78: 4259, 1962.
6. M. N. Plooster, H. M. Dess, and O. H. Nestor, "Czochralski Ruby," Contract No. NONR-4132(00), Linde Division, Union Carbide Corp., July 8, 1964.
7. R. A. Laudise and A. A. Ballman, "Hydrothermal Synthesis of Sapphire," *J. Am. Chem. Soc.* 80: 2655, 1958.
8. D. F. Nelson and J. P. Remeika, "Laser Action in a Flux Grown Ruby," *J. Appl. Phys.* 35: 522, 1964.
9. M. L. Kronberg, "Plastic Deformation of Single Crystals of Sapphire: Basal Slip and Twinning," *Acta Met.* 5: 507, 1957.
10. H. E. Swanson, M. I. Cook, T. Isaacs, and E. H. Evans, "Standard X-Ray Diffraction Patterns," National Bureau of Standards Circular 539, Vol. 9, February 25, 1960, p. 3.
11. A. R. Lang, "The Projection Topograph: A New Method in X-Ray Diffraction Microradiography," *Acta Cryst.* 12: 249, 1959.
12. D. L. Stephens and W. J. Alford, "Dislocation Structures in Single Crystal Al_2O_3," *J. Am. Ceram. Soc.* 47: 81, 1964.
13. R. Scheuplein and P. Gibbs, "Surface Structure in Corundum: I, Etching of Dislocations," *J. Am. Ceram. Soc.* 47: 81, 1960.
14. K. R. Janowski, E. J. Stofel, and A. B. Chase, "Growth Defects in Flux Grown Rubies," *Trans. Met. Soc. AIME* 233: 2087, 1965.
15. H. E. Bond and K. B. Harvey, "Decoration of Dislocations in Aluminum Oxide," *J. Appl. Phys.* 34: 440, 1963.
16. M. N. Plooster, H. M. Dess, and O. H. Nestor, "Czochralski Ruby," Contract No. NONR 4132(00), Linde Division, Union Carbide Corp., January 22, 1965.
17. L. S. Birks, J. W. Hurley, and W. E. Sweeney, "Perfection of Ruby Laser Crystals," *J. Appl. Phys.* 36: 3562, 1965.

DISCUSSION

D. K. Smith (Lawrence Radiation Lab.): In your abstract, you mention that the half breadths are generally found to be wider for ruby than for sapphire, yet on the last slide (Table I), it was just the opposite.

R. F. Belt: At the time the abstract was written, this appeared to be the situation, but since then, many more crystals have been examined and I don't think the conclusion is quite valid. Generally, the pure sapphire, that is the aluminum oxide without the chromium, seems to be a little better quality than when chromium oxide is included.

J. L. Engelke (A. D. Little, Inc.): If I understand you correctly, the annealing treatment you suggested caused a precipitation or segregation of the chromia as fine particles. One would normally expect homogenization instead, if the diffusion kinetics were fast enough. How would you explain that observation?

R. F. Belt: Was I misunderstood on this? I felt that it was not solely the chromia that was doing this. There is also the presence of flux. The lead fluoride flux has a melting point of 850°C. Heating the crystals at 1400°C certainly does something to the flux. The flux has a high vapor pressure at 1400°C, whereas the chromia and alumina have practically none.

J. L. Engelke: What do you suggest the precipitate material is?

R. F. Belt: Primarily, the flux, not the chromia. However, there may be some interaction between the two.

THE ANALYSIS OF BERG–BARRETT SKEW REFLECTIONS AND THEIR APPLICATIONS IN THE OBSERVATION OF PROCESS-INDUCED IMPERFECTIONS IN (111) SILICON WAFERS

E. M. Juleff

Westinghouse Space and Defense Center
Elkridge, Maryland

A. G. Lapierre, III

Computer Control Company, Inc.
Framingham, Massachusetts

and

R. G. Wolfson

P. R. Mallory and Company, Inc.
Burlington, Massachusetts

ABSTRACT

The geometry of Berg–Barrett skew reflections (the normal to the specimen surface and the incident and reflected beam vectors are not coplanar) is analyzed with particular reference to (111) silicon. Angular relationships required for obtaining the 78 most intense such reflections are presented on stereographic projections. Skew reflections are utilized to adapt the Berg–Barrett technique of extinction-contrast micrography to the examination of the (111) wafers generally used in integrated circuit technology. Skew reflections are shown to be more suitable for Berg–Barrett micrography than the zero-layer reflections described by Newkirk; in particular, their versatility in providing a means of varying the angle of incidence of the X-ray beam for a specific reflecting plane is demonstrated. A relatively simple experimental arrangement is described for recording skew reflection images. It permits a high resolution X-ray sensitive plate to be placed parallel to the specimen, and their separation to be increased to as much as 5 mm without excessive loss of resolution; this avoids both image distortion and surface scattering. Furthermore, the specimen area recorded in a single micrograph is 1–3 cm², which is large enough to eliminate the need for scanning. Exposure times are very short, in the order of 10 min. Micrographs of boron-diffused silicon are presented showing device components delineated by solute strain, strain fields induced in epitaxial silicon films by underlying buried-layer diffusions, and diffusion-induced Lomer–Cottrell dislocations. These micrographs demonstrate the resolution and contrast obtainable over large specimen areas. The capability of the Berg–Barrett technique is discussed in the examination of the near-surface regions directly involved in device fabrication and operation.

INTRODUCTION

Extinction-contrast X-ray diffraction micrography has been used extensively in the transmission geometry of the Lang technique[1-3] to examine lattice defects in silicon (see, for example, the work of Schwuttke and his colleagues on dislocations and diffusion effects in silicon wafers[4-6]). Unfortunately, the Lang technique requires elaborate scanning equipment and exposure times in the order of 2–20 hr. Although this method reveals the imperfections distributed throughout the bulk of the specimen, it is often incapable of resolving fault structures within a few microns of the surface i.e., in the region intimately associated with semiconductor device fabrication and operation. A more suitable means for the nondestructive observation of near-surface defects is provided by the Berg–Barrett technique, which utilizes extinction contrast in the Bragg reflection geometry. The advantages of Berg–Barrett micrography for revealing dislocations and diffusion geometries in silicon integrated-circuit wafers have been demonstrated in a previous paper.[7] In particular, it was shown that large surface areas can be photographed without scanning in approximately 10 min.

The Berg–Barrett technique was first applied to the study of undecorated dislocations in silicon by Newkirk,[8] who also analyzed the diffraction conditions for zero-layer reflections. (A zero-layer reflection is one in which the normal to the specimen surface is coplanar with the incident and reflected beam vectors; all others are denoted as skew reflections.) However, silicon devices are generally fabricated on {111} surfaces, which are not capable of yielding useful zero-layer reflections for Berg–Barrett micrography. Although a 13$\bar{1}$ skew-reflection micrograph of growth faults at a (11$\bar{1}$) silicon surface has been published by Lauriente et al.,[9] the geometry of such reflections has not been analyzed.

The purpose of the present paper is to develop a systematic treatment of Berg–Barrett skew reflections from {111} silicon surfaces and to demonstrate their application to the in-process observation of defects in the device fabrication regions.

APPARATUS

The Berg–Barrett apparatus, which is a modification of that described by Newkirk,[8] is shown schematically in Figure 1. A beam of line-focus Cu K_α X-radiation, passed through a 0.7 mil nickel K_β filter and collimated by an adjustable slit system, is diffracted by the specimen onto a photographic plate. The vertical divergence of the beam is controlled to within 0.5°, and the horizontal divergence is limited to 2.2° by a

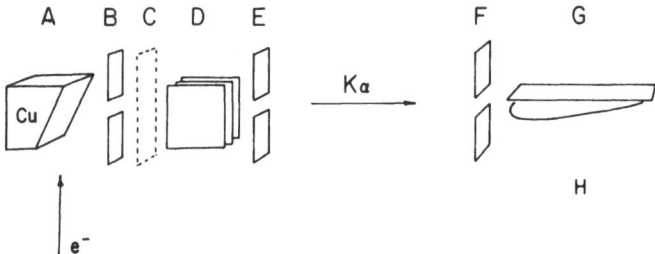

Figure 1. Schematic representation of the Berg–Barrett equipment, where A represents the Cu target, B represents an adjustable slit, C represents a Ni K_β filter, D represents a parallel slit assembly, E represents a scatter reducing slit, F represents a second adjustable slit, G represents an Ilford 25 μ L-4 nuclear plate, and H represents the specimen.

parallel plate assembly; it is not possible to reduce the horizontal divergence much below this value without sacrificing either incident intensity or beam width. The silicon wafer specimen is mounted on a goniometer 15 cm from the X-ray target, with its surface normal tilted a predetermined angle ϕ toward the source. In order to orient the specimen for a specific skew reflection, a Geiger–Muller tube is set to intercept the reflected beam, and the specimen is rotated about its surface normal through the required angle from a ($\bar{1}01$) reference flat. The image is recorded on an Ilford L-4 25 μ nuclear plate placed close to the specimen surface; a modified Meieran and Lemons[10] development procedure is used to enhance contrast in the plate.

ANALYSIS

The analysis of Berg–Barrett reflections is a relatively simple exercise in spherical trigonometry. The relevant angles are defined for zero-layer reflections in Figure 2 and for skew reflections in Figure 3. The angular coordinates for the latter have been derived by the present authors[7]:

$$R = \cos^{-1}\left(\frac{\sin\theta - \sin\phi\cos\alpha}{\cos\phi\sin\alpha}\right) \tag{1}$$

$$e = \cos^{-1}(2\cos\alpha\sin\theta - \sin\phi) \tag{2}$$

$$S = \cos^{-1}\left(\frac{\sin\theta - \cos\alpha\cos e}{\sin\alpha\sin e}\right) \tag{3}$$

where the restriction

$$\alpha + \phi > \theta$$

limits these equations to permissible skew reflections. The significance of this restriction is illustrated by the stereographic projection of Figure 4, which represents the 311 reflection of Cu K_α radiation from a {111} silicon surface. The zero-layer 311 reflection occurs at a negative tilt angle, and the incident X-ray beam is intercepted by the edge of the specimen before reaching the surface. In such a case, however, skew-reflection micrographs can be obtained over a range of positive tilt angles, as shown; values of ϕ of the order of $+3°$ have been found to give the best images.

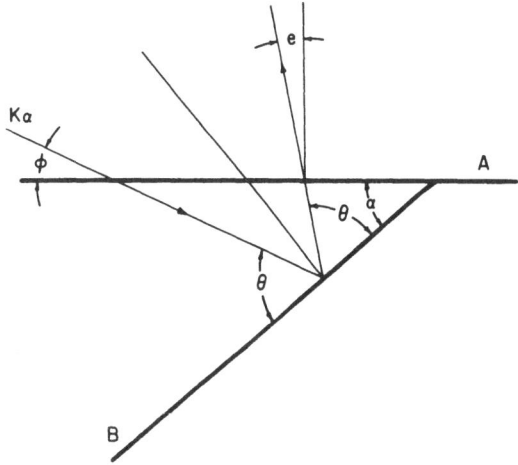

Figure 2. Relationship of the surface and reflecting planes, A and B, to the incident and diffracted beam vectors for a zero-layer reflection, where θ, represents the Bragg angle, α represents the AB interplanar angle, ϕ represents the angle of tilt of the specimen into the beam, and e represents the angle between the reflected beam vector and the normal to the specimen surface A.

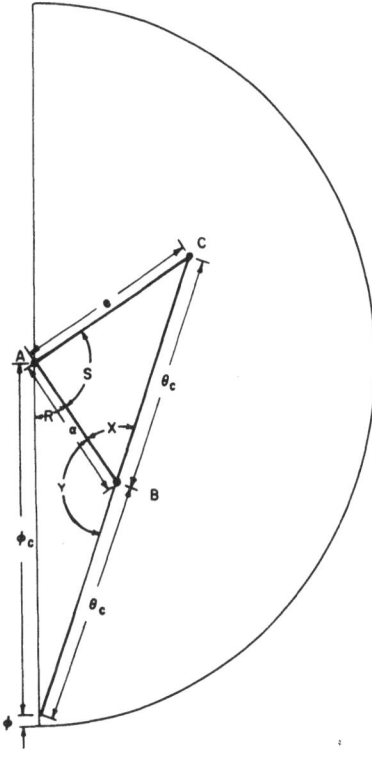

Figure 3. Skew reflection interangular relationships for the general case, where A represents the surface plane pole, B is the reflecting plane pole, and C is the diffracted beam direction. $(R + S)$ represents the angle between the incident and diffracted beam directions as measured on the specimen surface, e represents the angle between the specimen surface normal and the diffracted beam direction, ϕ represents the tilt angle into the beam, and $\phi c = 90° - \phi$, θ represents the Bragg angle, and $\theta c = 90° - \theta$, X represents the interplanar angle between A and B. X represents an angle required for calculation, and $Y = 180° - X$.

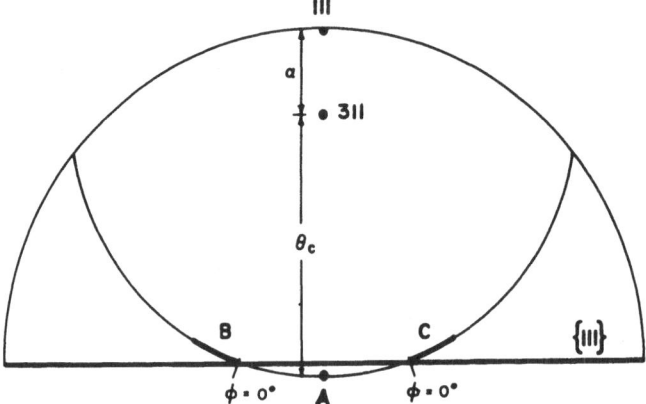

Figure 4. Stereographic projection illustrating low, positive, ϕ angle, skew reflections at B and C for a 311 plane. The zero-layer reflection occurs when the incident beam impinges at A, where ϕ is negative, i.e., below the surface (111). θc represents the Bragg angle complement, i.e., $\theta c = 90° - \theta$, and α represents the (311) (111) interplanar angle.

Since there are no strong zero-layer reflections available from {111} silicon surfaces at small, positive tilt angles, the above equations were used to locate skew reflections capable of yielding relatively undistorted, high-resolution Berg–Barrett micrographs. The Bragg angles were calculated for Cu K_α X-radiation, and the tilt angle ϕ was fixed

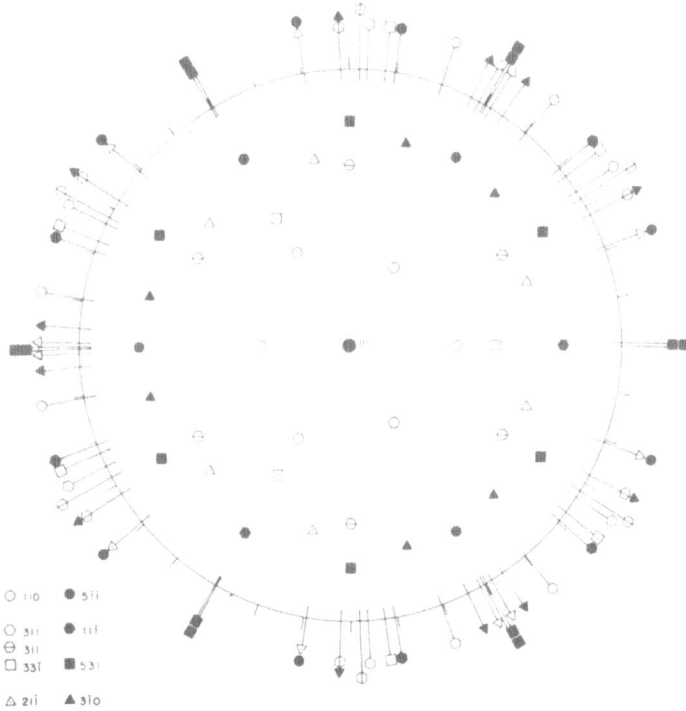

Figure 5a. Stereographic projection showing the incident beam directions for skew reflections from the major low-index reflecting planes of silicon, using Cu K_α radiation and a positive tilt angle, ϕ, of 3°.

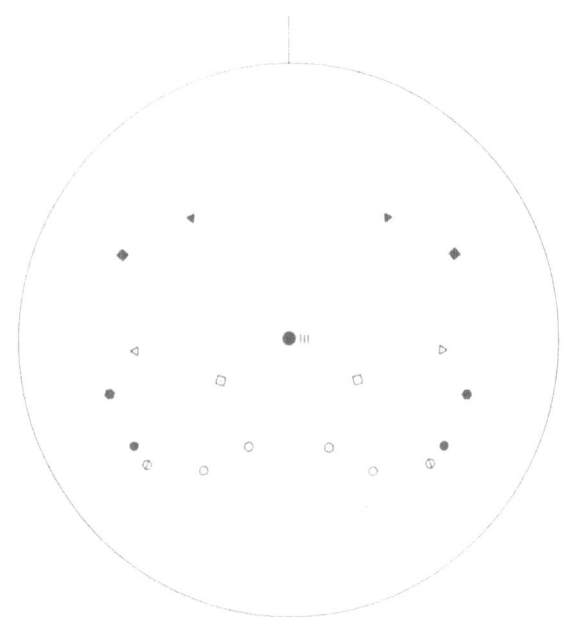

Figure 5b. Stereographic projection showing the coordinates of the corresponding reflected beams from the planes illustrated in Figure 5a. These were derived using equations (1), (2), and (3) in the text.

at $+3°$. The results are presented in Figure 5. The incident beam directions for skew reflections from the major low-index planes are plotted on the stereographic projection of Figure 5a; the coordinates of the corresponding reflected beams are shown in Figure 5b. Except for reflections of the form 220, the angular separation between neighboring directions of incidence is smaller than the horizontal divergence of the X-ray beam. Consequently, multiple images are obtained in all but 220-type micrographs. This can be an advantage rather than a problem for the images do not overlap unless the directions of the reflected beams are similar. For example, distinct $3\bar{1}1$ and $3\bar{1}0$ images are obtained in a single exposure with a common incident beam direction.

APPLICATIONS

In order to demonstrate the usefulness of Berg–Barrett micrography, the technique was applied to the examination of (111) silicon surfaces after various processing treatments. Figure 6a is a 022 skew-reflection micrograph of arrays of diffusion-induced dislocations lying between $1.0\ \mu$ and $1.4\ \mu$ below the surface of a boron-diffused wafer[11]; the reflection geometry is illustrated stereographically in Figure 6b. The three sets of lines, which differ significantly in extinction contrast, have been associated with the presence of Lomer–Cottrell dislocations along those $\langle 110 \rangle$ parallel to the (111) surface.[11] In some cases, the structural disturbance induced by the boron diffusion extends into the surrounding silicon, as shown in Figure 7. The extension of the lattice misfit strain into the undiffused matrix has been corroborated by etching studies.[12]

This specimen also has been used to provide a comparison between the sensitivities of the Berg–Barrett and Lang techniques to near-surface defects. In Figure 8a a 022

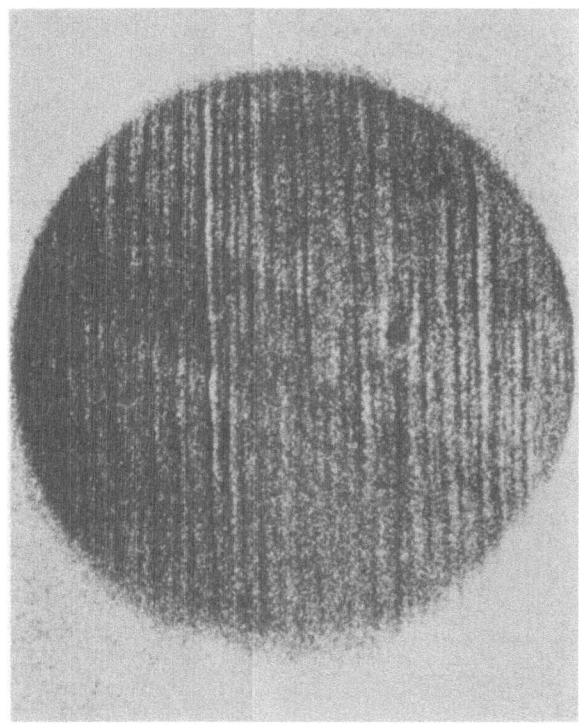

Figure 6a. A 022 Berg–Barrett skew reflection from a chemically mechanically polished, As doped $13–15\ \Omega$-cm silicon wafer, boron diffused at 1250°C through 32 mil-diameter thermal oxide windows. (Magnification: 200×, reduced 47% for reproduction.)

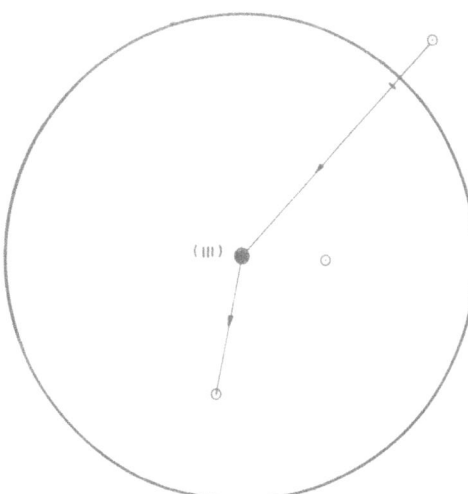

Figure 6b. A stereographic representation of the 022 skew reflection utilized to produce the micrograph seen in Figure 6a.

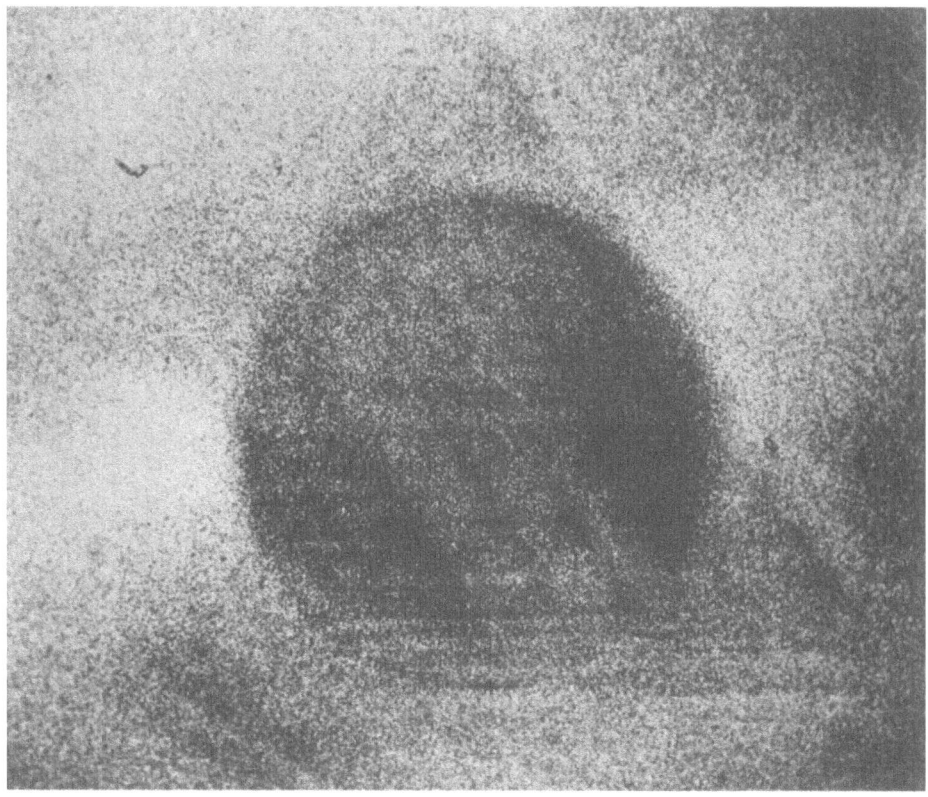

Figure 7. A 022 Berg–Barrett skew reflection from a chemically mechanically polished, As doped 0.003 Ω-cm, silicon substrate, with a 13 μ, As doped 2.0 Ω-cm, epitaxial layer, boron diffused at 1250°C through 20 mil-diameter thermal oxide windows. The specimen thickness was in the order of 7 mils and the back surface had been lapped. (Magnification: 425×, reduced 47% for reproduction.)

Berg–Barrett skew reflection micrograph clearly defines the dislocation arrays within the boron-diffused region, while Figure 8b, a $\overline{2}20$ reflection Lang transmission micrograph shows little difference in defect contrast between the boron-diffused regions and the surrounding silicon. It should be emphasized, however, that this specimen had a lapped back surface, and a low resistivity substrate, both conditions being unfavorable for revealing the subtle changes in extinction contrast caused by the boron diffusion. Nevertheless, 8 min exposure time for the Berg–Barrett recorded image, relative to the 10 hr required for the Lang, demonstrates the advantage of the former technique for observing this type of defect.

The application to device fabrication is demonstrated in Figure 9, where a 311 skew reflection reveals the geometry of boron-diffused isolation regions in a sense amplifier circuit. Here, one strong set of dislocation arrays is visible, again found to have those $\langle 110 \rangle$ parallel to the (111) surface. In particular, it is evident that, in some cases, the dislocation and the diffusion extend beyond the intended isolation boundary; compare geometries in Figure 9. The effect of such observed defects on device electrical properties has yet to be determined, however.

The observation of extended strain is demonstrated in Figure 10, which is a 022 skew reflection micrograph from an integrated circuit wafer processed so as to produce

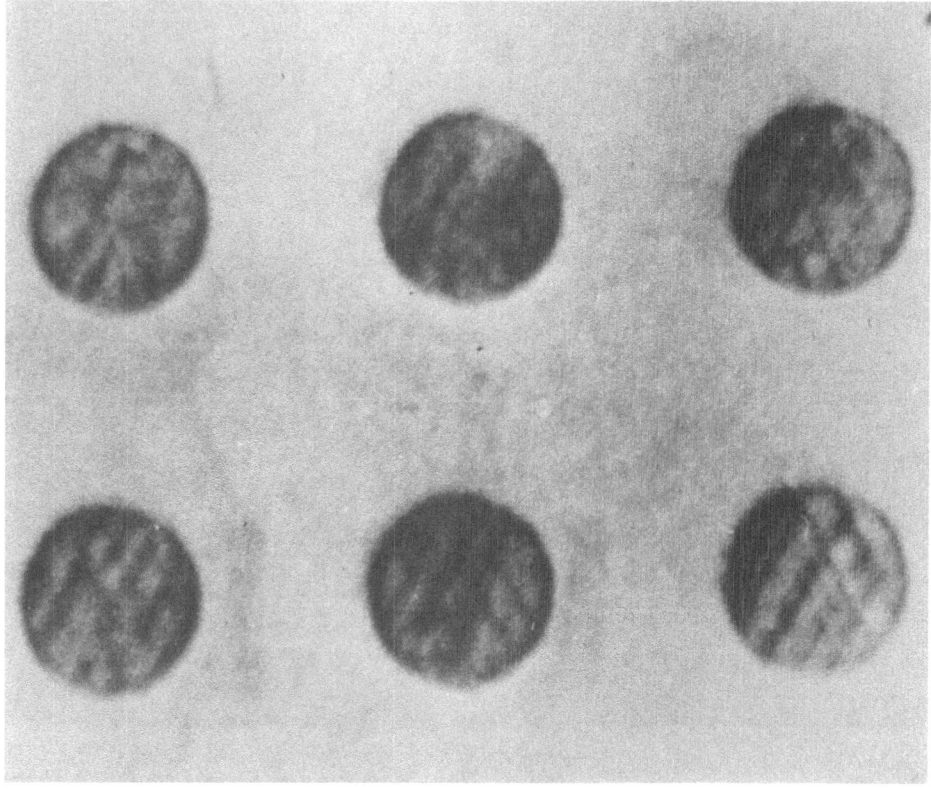

Figure 8a. A 022 Berg–Barrett skew reflection micrograph of the specimen described in Figure 7, but illustrating a larger surface area for comparison with the $\overline{2}20$ Lang transmission micrograph, seen in Figure 8b. (Magnification, 160×, reduced 47% for reproduction.)

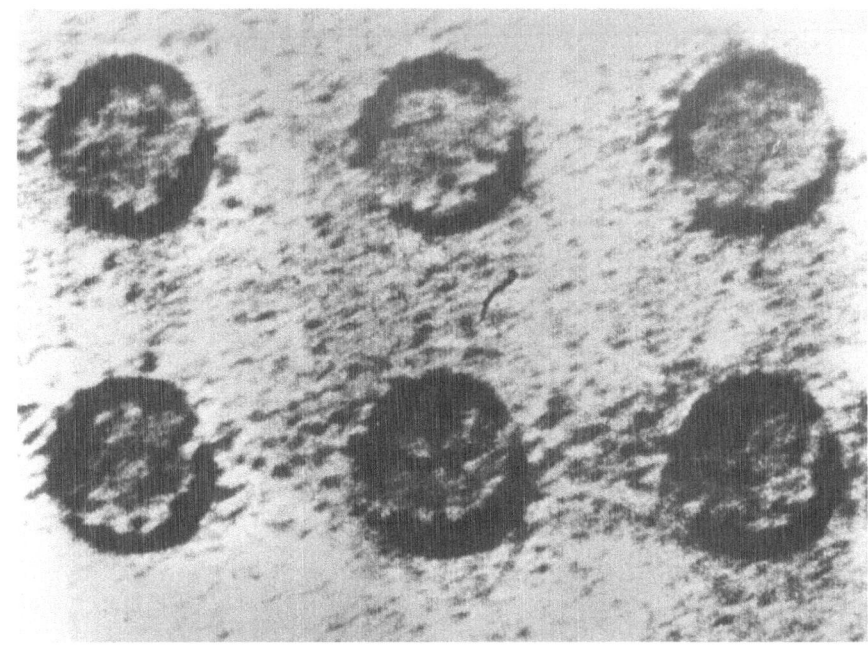

Figure 8b. A $\bar{2}20$ Lang transmission micrograph of the specimen described in Figure 7. (Magnification 160 ×, reduced 52% for reproduction.)

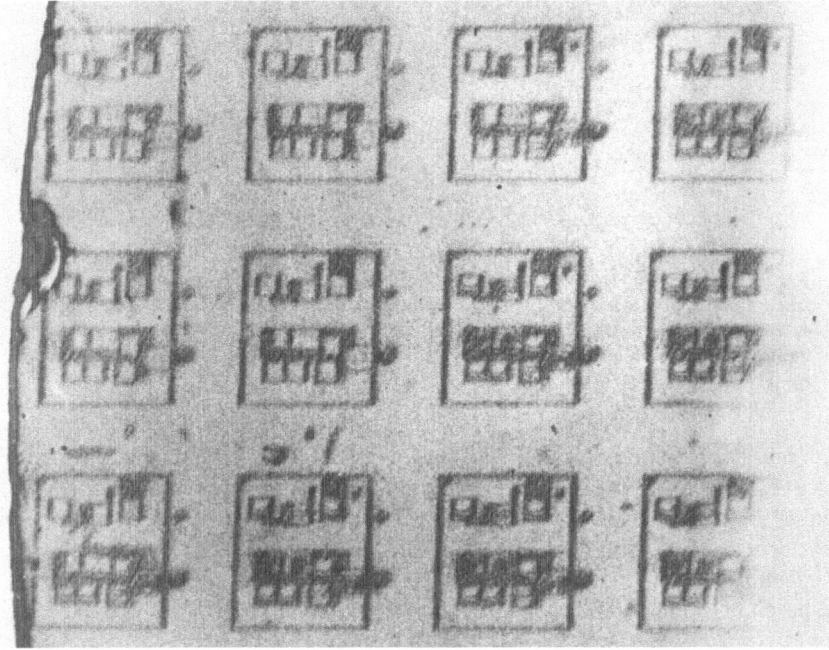

Figure 9. A 311 skew reflection from a sense amplifier circuit wafer processed through all fabrication stages to Al-metallization. This included a boron isolation diffusion at 1250°C through thermal oxide windows in an As-doped, 0.3 Ω-cm, 7 μ epitaxial layer. (Magnification: 42 ×, reduced 52% for reproduction.)

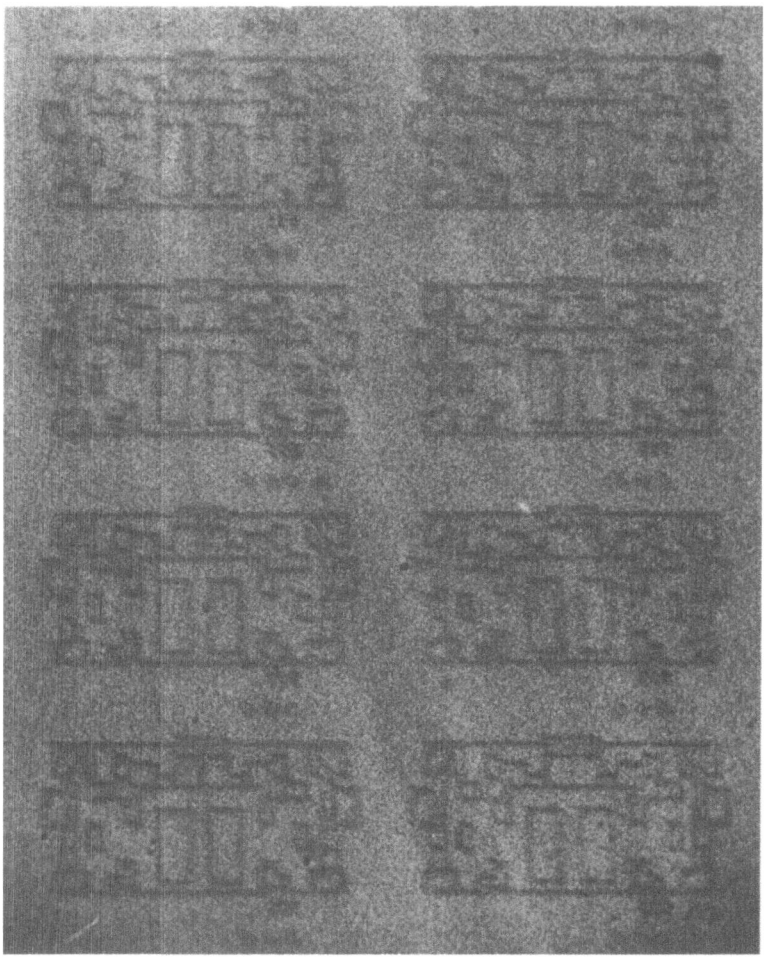

Figure 10. A 022 Berg–Barrett skew reflection from an integrated circuit silicon wafer processed through a buried boron isolation diffusion stage. The substrate was a boron doped, 10 Ω-cm, silicon wafer chemically mechanically polished, boron isolation diffused at 1250°C and subjected to growth of an As-doped, 0.15 Ω-cm, 6 μ epitaxial layer. (Magnification: 50×, reduced 47% for reproduction.)

a buried boron isolation diffusion. The surface relief, corresponding to the isolation regions, and determined by thallium light interferometry to be less than 0.5 μ, was removed by a chemical-mechanical polish. Determination of the depth of penetration of the 3° incident beam of Cu K_α radiation in pure silicon indicates that the primary beam is attenuated 50% at 0.1 μ below the (111) surface.[13] Bevel and strain techniques show that the boron back diffusion during epitaxial growth of the layer is less than 20%. These facts suggest that, since the layer is at least 5 μ thick, the image of the boron isolation regions, seen in Figure 10, is caused by strain extending from the buried boron diffusion through the epitaxial layer.

Finally, the effect of phosphorus diffusion was examined by this technique. A 022 skew reflection micrograph is shown in Figure 11 which clearly illustrates the phosphorus

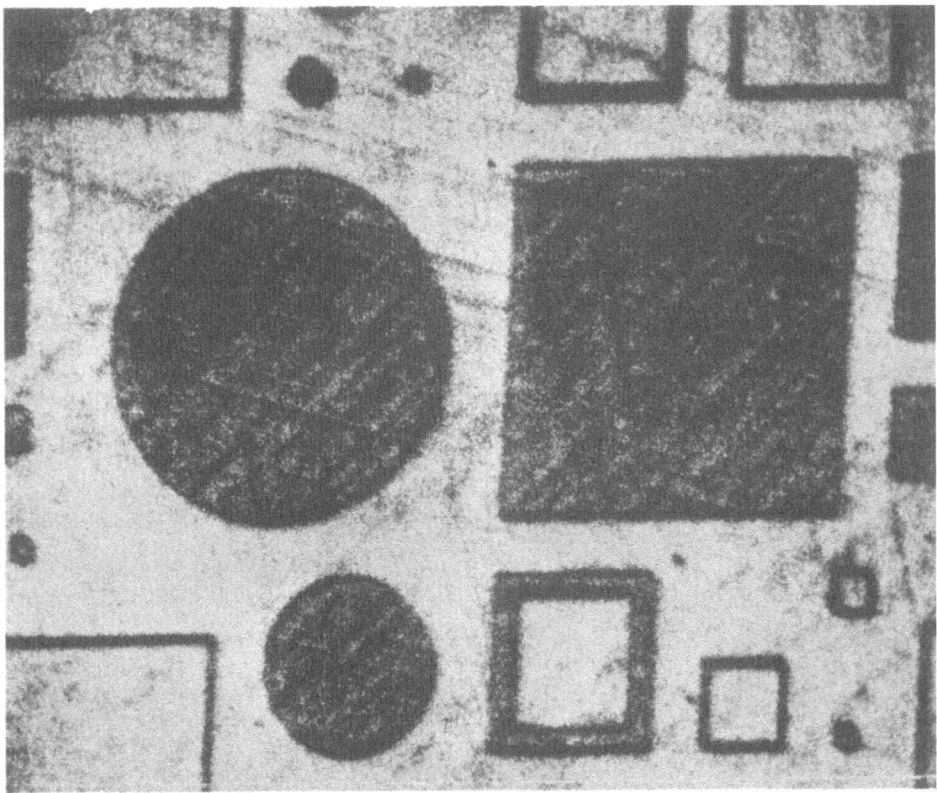

Figure 11. A 022 Berg–Barrett skew reflection from an As-doped 13 Ω-cm silicon substrate chemically mechanically polished and phosphorus diffused at 1250°C through oxide windows of variable geometry. The large square is 32 mils on a side.

diffusion-induced array of dislocation, not only within the diffused regions, but often extending into the surrounding silicon.

In all cases the boron deposition (source BBr_3) was such that the surface concentration was equal to the solid solubility at the deposition temperature (1250°C). The junction depth as determined by sectioning and etching techniques was found to be of the order of 7 μ after 15 min drive at 1250°C.

The phosphorus deposition (source $POCl_3$) was such that the surface concentration was equal to the solid solubility at the deposition temperature (1000°C). The junction depth, in this case, as determined by sectioning and etching techniques, was in the order of 6 μ after 30 min drive at 1250°C.

In conclusion, it may be stated that the Berg–Barrett X-ray diffraction technique described here, and applied to (111) silicon wafers, provides a useful tool for observing, over large areas without scanning, defects produced by large concentrations of boron and phosphorus in the fabrication regions of semiconductor devices.

ACKNOWLEDGMENTS

The authors wish to thank Dr. Pei Wang and R. Berkstresser for their helpful discussions, and Plenum Press for allowing the use of Figures 1, 2, 3, and 4 from a previous paper (see reference 7).

REFERENCES

1. A. R. Lang, "Direct Observation of Individual Dislocations," *J. Appl. Phys.* **29**: 597, 1958.
2. A. R. Lang, "The Project Topograph: A New Method in X-Ray Diffraction Microradiography," *Acta Cryst.* **12**: 249, 1959.
3. A. R. Lang, "Studies of Individual Dislocations in Crystals by X-Ray Diffraction Microradiography," *J. Appl. Phys.* **30**: 1748, 1959.
4. G. H. Schwuttke and H. J. Queisser, "X-Ray Observations of Diffusion-Induced Dislocations in Silicon," *J. Appl. Phys.* **33**: 1540, 1962.
5. G. H. Schwuttke, "X-Ray Observation of Partial Dislocations in Epitaxial Silicon Films," *J. Appl. Phys.* **33**: 1538, 1962.
6. G. H. Schwuttke and V. Sils, "X-Ray Analysis of Stacking Fault Structures in Epitaxially Grown Silicon," *J. Appl. Phys.* **34**: 3127, 1963.
7. E. M. Juleff and A. G. Lapierre, "The Application of Berg-Barrett Skew Reflections for Observing Boron Diffusion-Induced Imperfections in Silicon Wafers," *Intern. J. Electron.* **20**: March, 1966, in press.
8. J. B. Newkirk, "The Observation of Dislocations and Other Imperfections by X-Ray Extinction Contrast," *Trans. AIME* **215**: 483, 1959.
9. M. Lauriente, R. Stickler, and R. W. Armstrong, "X-Ray Diffraction Analysis and Etch Pattern of Faults in Epitaxial Silicon," *J. Appl. Phys.* **35**: 3061, 1964.
10. E. S. Meieran and K. E. Lemons, "The Study of Defects Due to Surface Processing in Silicon by Means of X-Ray Extinction-Contrast Topography," *Advances in X-Ray Analysis*, Vol. 8, Plenum Press, New York, 1964, p. 48.
11. R. G. Wolfson, E. M. Juleff, and A. G. Lapierre, "The Observation of Lomer–Cottrell Dislocations in Boron Diffused (111) Silicon by Berg–Barrett Skew Reflections," *Intern. J. Electron.*, in review.
12. D. P. Miller, J. E. Moore, and C. R. Moore, "Boron Induced Dislocations in Silicon," *J. Appl. Phys.* **33**: 2648, 1962.
13. R. W. James, *The Optical Principles of the Diffraction of X-Rays*, G. Bell and Sons, Ltd., London, 1954, p. 60.

DISCUSSION

N. Spielberg (Philips Laboratories): What wavelength radiation was used in this technique?

E. D. Jungbluth: Cu K_α.

N. Spielberg: Might that account for some of your gain in intensity?

E. D. Jungbluth: It might, although several other factors should be considered such as the short X-ray collimation and elimination of a crystal scanning system. For a more detailed explanation I would refer to the authors.

M. Renninger (Krist. Institut der Universität Marburg): What is the cause of the coarse grain? What kind of photograph material was used? Normal X-ray film?

E. D. Jungbluth: These were Ilford Nuclear plates.

M. Renninger: These plates should have finer grain than the pictures shown by you seem to exhibit.

E. D. Jungbluth: This might be a problem with their development procedures.

J. B. Newkirk: Can you estimate the concentration of the dopant that was diffused into these surfaces?

E. D. Jungbluth: In the case of the phosphorus, the surface concentration is about 10^{21} impurity atoms per cubic centimeter or so. In the case of boron, it is probably 10^{20} per cubic centimeter.

H. W. Pickett (General Electric, X-Ray): I thought I heard you read a remark to the effect that the effective layer was only $\frac{1}{10}\,\mu$ thick. I don't have any tables in my head, but this seems too thin.

E. D. Jungbluth: I was referring to the attenuation of the X-ray beam in silicon. They had calculated that for the case of extinction contrast the primary beam is attenuated by 50% at $\frac{1}{10}\,\mu$ below the surface layer for the case of silicon with copper radiation.

H. W. Pickett: It is conceivable, but I would debate that.

THE EFFECT OF SMALL ADDITIONS OF MAGNESIUM ON THE PREPRECIPITATION BEHAVIOR OF Al-Zn ALLOYS

Robert W. Gould

University of Florida
Gainesville, Florida

ABSTRACT

Magnesium has been shown to have a marked effect upon preprecipitation processes occurring in aluminum–13 wt.% zinc alloys containing 0.025, 0.098, 0.17, 0.19, 0.23, and 0.27 wt.% magnesium. X-ray small-angle scattering and resistance measurements have been used to monitor the rate of growth of Guinier–Preston zones in quenched foils and wires. The following results have been obtained: (a) small additions of magnesium noticeably decrease the rate of growth of G–P zones but the final zone size reached in the magnesium containing alloys is generally larger than in Al–Zn alloys quenched and aged under identical conditions; (b) the general pattern of preprecipitation found in Al–Zn alloys is not seriously changed by these small additions of magnesium; (c) the dependence of the rate of preprecipitation on quenching temperature is shown to be a function of magnesium content and aging temperature; (d) the dependence of the rate of preprecipitation on aging temperature is influenced by the range of aging temperature, magnesium content, and the quenching temperature.

INTRODUCTION

It is now generally accepted that the rapid rate of zone growth in quenched Al–Zn solid-solution alloys is due to a high, nonequilibrium concentration of vacancies frozen-in by quenching.[1,2] The ensuing process of zone growth in these alloys has been well documented by several authors. Using resistometric techniques, Panseri and Federighi[3] and Turnbull, Rosenbaum, and Treaftis[4] obtained activation energies of 0.70 eV for the formation of vacancies and 0.4–0.5 eV for the motion of vacancies in Al–Zn alloys. Reasoning from these results, Panseri and Federighi[3] concluded that the binding energy between vacancies and zinc atoms is of the order of 0.06 eV.

The zone state in Al–Zn and other alloys has been conveniently studied using the technique of small-angle X-ray scattering. Small-angle X-ray studies of the preprecipitation state in Al–Zn alloys were initiated by Guinier,[5] who showed that the zones were spherical. Jan[6] studied the integrated small-angle scattering intensity for Al–Zn and found that it remains constant during aging at a constant temperature. This implies that the volume of zones does not change with time when the aging temperature is constant; a view later expanded by Gerold. Gerold and Schweizer[7] used the small-angle-scattering method on alloys of 15, 20, and 25 wt.% zinc and obtained values of 0.69 eV and 0.43 eV for the formation energy and motion energy of vacancies, respectively, and also showed that the rate of preprecipitation is a sensitive function of Zn concentration. Gerold and Schweizer support Panseri and Federighi's claim that the rate of preprecipitation is governed by the quenched-in vacancy concentration.

Recently, Bonfiglioli and Guinier[25] and Perry[26,27] have published several excellent papers dealing with the effect of Mg on preprecipitation in Al–Zn alloys. Their work appeared too late to be included in the discussion section of this paper.

The Addition of Magnesium to Al–Zn Alloys

It has been known for some time that the small additions of magnesium (less than 1%) do have a pronounced effect on the physical properties of Al–Zn alloys.[8] Herenguel[9] in the early 1940's, has shown that fractional amounts of magnesium may at least double the strength that can be reached by aging Al–Zn alloys at room temperature.

Only recently has some direct experimental evidence been obtained concerning the role played by Mg and a clearer picture is now emerging. Schmalzried and Gerold,[10] and Gerold and Haberkorn[11] have shown that ordering of magnesium and zinc atoms occurs within G–P zones of Al–Zn alloys containing 0.86 at. % magnesium and 7.6% Zn. The type of ordering is dependent upon the Zn:Mg ratio in the alloy. Guinier[12] suggests that there may actually be two types of zones (Zn-rich zones as in binary Al–Zn alloys, and zones rich in both magnesium and zinc) simultaneously present in aged Al–Zn–Mg alloys. During this same period Polmear,[13] using hardness measurements, obtained aging curves for Al–Zn alloys (4, 6, and 8 wt.% zinc) containing 0.08, 0.4, 1.0, 2.0, and 3.0 wt. % magnesium. He concludes that there are at least two stages in the room temperature aging process; namely, the formation and growth of G–P zones followed by an intermediate precipitate. Using his results Polmear[14] has determined a metastable boundary for G–P zone formation in a small portion of the Al–Zn–Mg system. Polmear's results suggest that these small quantities of magnesium greatly prolong the zone-growth process in Al–Zn alloys. A possible explanation, as suggested by Federighi,[15] might reside in the interaction between quenched-in vacancies and magnesium atoms.

Studies of such vacancy–impurity interactions generally require a suitable vacancy controlled process whose rate is strongly affected by small quantities of impurity atoms. Concurrently, a very sensitive and accurate method must be available to monitor this process. The growth of G–P zones from a quenched-supersaturated solid solution has been shown to be a vacancy controlled process, which in the case of certain Al–Zn alloys, occurs at a measurable rate at or near room temperature. A further advantage of the Al–Zn system lies in the apparent low binding energy (weak interaction) that exists between vacancies and zinc atoms.[3] (As pointed out recently by Perry[26] this is not necessarily true for very dilute concentrations of zinc in aluminum.) Thus, the strength of the interaction between a solute impurity element, A, and a vacancy might be measured by noting the effect that small additions of A have upon the zone-growth process in quenched Al–Zn alloys. It is, of course, necessary to assume that the addition of a ternary impurity does not alter the general pattern of preprecipitation found in Al–Zn. There exists ample experimental evidence[9,13,15,16] to suggest that magnesium would serve as a suitable impurity.

Studies by Panseri and Federighi[17] (published during the course of this work), Bartsch,[18] and the recently published results of Ohta and Hashimoto[19] represent the only early stage preprecipitation observations that have been made on Al–Zn alloys containing small ternary (0.1–0.3%) additions of magnesium. However, as stated in the introduction several new papers have appeared or have gone to press since this article was written.[25–27] Panseri and Federighi[17] used resistometric methods on a single alloy of aluminum – 10 wt.% zinc–0.1 wt.% magnesium, and have reached the following conclusions:

1. Magnesium addition of 0.1 wt.% has negligible effects upon the general preprecipitation pattern.
2. A strong binding energy exists between magnesium and vacancies (0.54 eV) so that practically all quenched vacancies are bound to magnesium atoms.
3. Magnesium-vacancy pairs will move easily at room temperature so that diffusion of zinc involves magnesium-vacancy pairs and not single free vacancies. Bartsch,[18] however, is of the opinion that the small quantities of magnesium used by Panseri and Federighi do influence the general pattern of preprecipitation.

Ohta and Hashimoto,[19] using a resistometric method similar to Panseri and Federighi, have recently studied the preprecipitation processes occurring in two aluminum – 10 wt.% zinc alloys containing 0.13 and 0.30 wt.% magnesium. They present the following results:

1. The first stage of annealing is characterized by the clustering of zinc and magnesium into ternary zones (a strong Mg–Zn interaction is postulated).
2. These ternary zones interact with vacancies and thus reduce the free vacancy concentration.
3. After all of the magnesium has been incorporated into ternary zones the remaining zinc atoms now cluster in a manner similar to the binary Al–Zn alloy.

It is evident from the foregoing results that no general agreement exists at present concerning the effects of magnesium upon the early stages of the preprecipitation process in Al–Zn alloys. The present research constitutes a study of the growth of G–P zones in six aluminum – 13 wt.% zinc alloys containing small ternary additions of magnesium (0.025–0.27 wt.%). The rate of growth of the G–P zones has been investigated as a function of magnesium content, quenching temperature and aging temperature. X-ray small-angle scattering and resistometric methods have been used to measure the kinetics of the zone-growth process.

MATERIALS AND PROCEDURES

Alloys were prepared from aluminum, zinc, and magnesium of 99.999% purity by induction melting in a nuclear-grade graphite crucible followed by casting into graphite billet and rod molds. Alloys so produced were then given a long homogenizing heat treatment. Table I gives the nominal and analyzed compositions of the several alloys used in this work.

Billets were reduced by cold rolling (with intermediate annealing at 500°C) until, with some difficulty, the optimum foil thickness[7] of approximately 0.10 mm was reached. It was found that swaging the rod alloys prior to cold rolling considerably reduced the difficulty encountered in reaching the final foil thickness. Wire was prepared by repeated swaging of the $\frac{1}{2}$ in. rods to a diameter of 0.156 in. followed by wire drawing through a set of Gesswein wire dies.

Foil samples were individually mounted in their aluminum holders[20] and placed in the quenching furnace whose temperature was controlled to within ± 1°C. Quenching was done by rapidly dropping the samples into water at room temperature. Prior to every quench the foils were homogenized and stress relieved in an adjacent furnace at 460°C for 30 min.

Table I. Nominal and Analyzed Compositions of Alloys Used in This Research

Alloy designation	Nominal		Analyzed		
	Mg, wt.%	Zn, wt.%	Mg	Zn	Al
Alloy 1	0	13.3	<0.005	13.3	Balance
Alloy 2	0.1	13.3	0.025	13.4	Balance
Alloy 3	0.5	13.3	0.19	13.1	Balance
Rod 1	0	13.3	<0.005	13.3	Balance
Rod 2	0.4	13.3	0.27	13.1	Balance
Rod 3	0.1	13.3	0.098	13.3	Balance
Rod 4	0.3	13.3	0.17	13.1	Balance
Rod 5	0.2	13.3	0.23	12.8	Balance

Much of this work was involved with a measurement of the rate of initial rapid zone growth. This type of study required the proportional counter to remain stationary at $2\theta = 0.5°$ (a position that corresponds to the maximum in the scattering curve when the zones are fully grown) and that the sample be placed in the X-ray small-angle scattering apparatus[20] as soon after the quench as possible. Samples were quenched and under observation within 10–12 sec.

The usual systematic experimental errors were encountered in the X-ray measurements. These are variations which are ordinarily of small magnitude and can be corrected for if necessary. However, the major cause of the variations found in this research arose from the necessity of using a separate foil specimen for each experiment.* It is a common experience that more reproducible results can be obtained if it is possible to utilize the same foils for all experiments. The use of separate foils was necessitated by the tendency of the foils to crack after repeated quenching.

As a result of this procedure it is not always possible to compare the time needed to reach a preselected scattering intensity (for a particular set of variables C_{Mg}, T_q, T_a) from one foil to another. Within one experiment, where the same foil was used, no such problem was encountered.

EXPERIMENTAL RESULTS

Fast Growth Stage

Presentation of data from the initial fast growth stage is conveniently done using the Arrhenius type plot. The log time to reach a preselected value of scattering intensity at $2\theta = 0.5°$ is plotted against the reciprocal absolute quenching or aging temperature. This method of presentation has two advantages; namely, the volume of data is greatly condensed, and the slopes of the lines drawn through the data points have physical significance and are related to energy. The fast growth experiments are of two basic types:

1. The aging temperature is held constant while the quenching temperature is varied; and
2. The quenching temperature is held constant while the aging temperature is varied.

Some of the results of the fast growth experiments may be seen in Figures 1-4. These data are summarized in Table II, where the slopes of the various lines have been calculated.

* Experiment here refers to a set of quenching temperatures or a set of aging temperatures, generally involving 5–6 quenches of the foil.

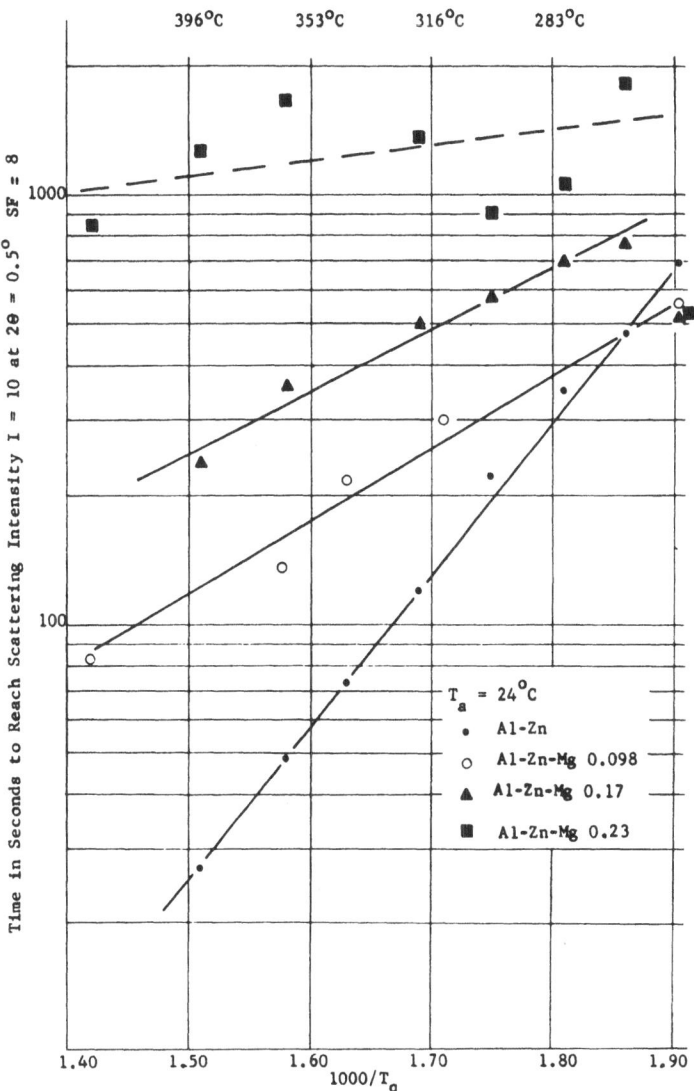

Figure 1. Data on growth of G–P zones as a function of quenching
temperature T_q during the stage of fast growth; $T_a = 35°C$.

As was stated earlier these slopes have been used in the case of Al–Zn[7] to calculate the
activation energies for motion (E_m) and formation (E_f) of vacancies. It will be discussed
later that it is not correct in the case of the ternary alloys to refer to these slopes simply as
E_m and E_f. Therefore, the term S_a is related to the slope of a line on the Arrhenius plot
when the aging temperature is varied. Similarly S_q is related to the slope when the
quenching temperature is varied.

Slow Growth Stage

Figure 5 schematically illustrates that the zone growth process in these alloys takes
place in at least two well-defined stages. The initial stage is characterized by rapid growth

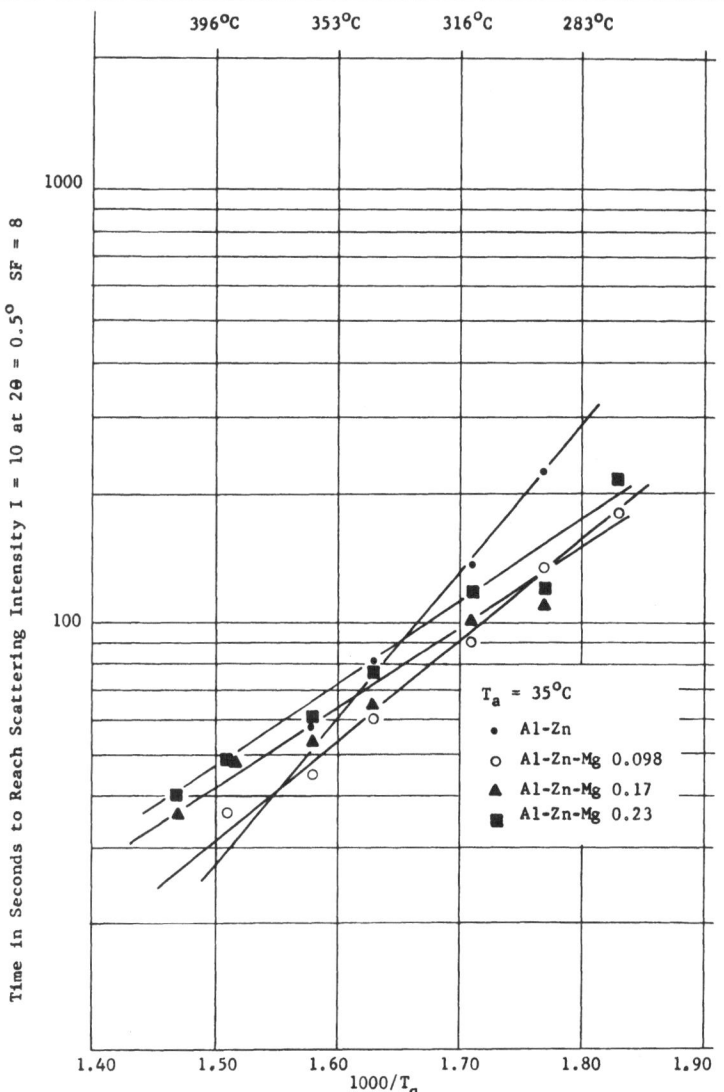

Figure 2. Data on growth of G–P zones as a function of quenching temperature T_q during the stage of fast growth; $T_q = 24°C$.

and short duration while the second, or slow growth stage, may persist for many months in alloys containing small additions of magnesium. Figure 6 and 7 present some of the results of the slow-growth stage for several alloys aged at 35°C after quenching from various temperatures (T_q).

Integrated Intensity

A measure of the volume fraction of zones present may be obtained by calculating a quantity known as the integrated intensity Q_0.[21] Average values of the integrated intensity as a function of T_a and magnesium content are given in Table III. Calculation of the quantity was facilitated by the use of the IBM 709 computer at the University of Florida.[21]

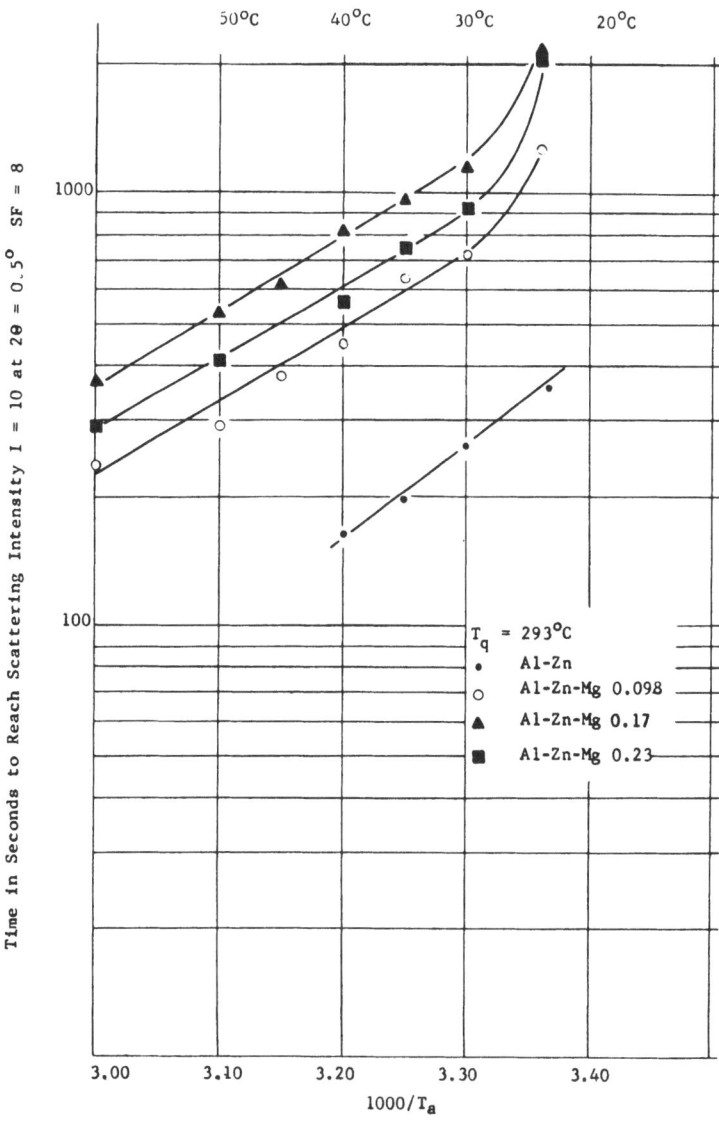

Figure 3. Data on growth of G–P zones as a function of aging temperature T_a during the stage of fast growth; $T_q = 293°C$.

Resistance Measurements

Resistance measurements have been widely used in the study of preprecipitation phenomena[3] due to their relative simplicity. Some results from this research are given in Figure 8 for alloy wires (0.8 mm diameter) quenched from 302°C and aged at room temperature.

INTERPRETATION OF RESULTS

Thomas[22] has outlined the decomposition sequence of the supersaturated solid solution. This schematic form (as shown below) illustrates some of the complexities involved in the process.

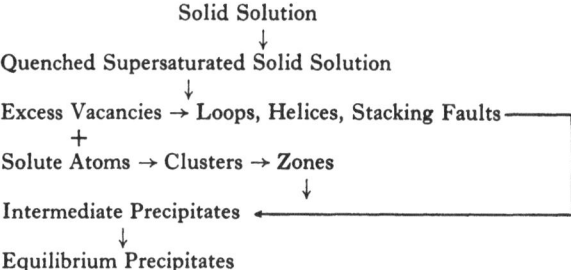

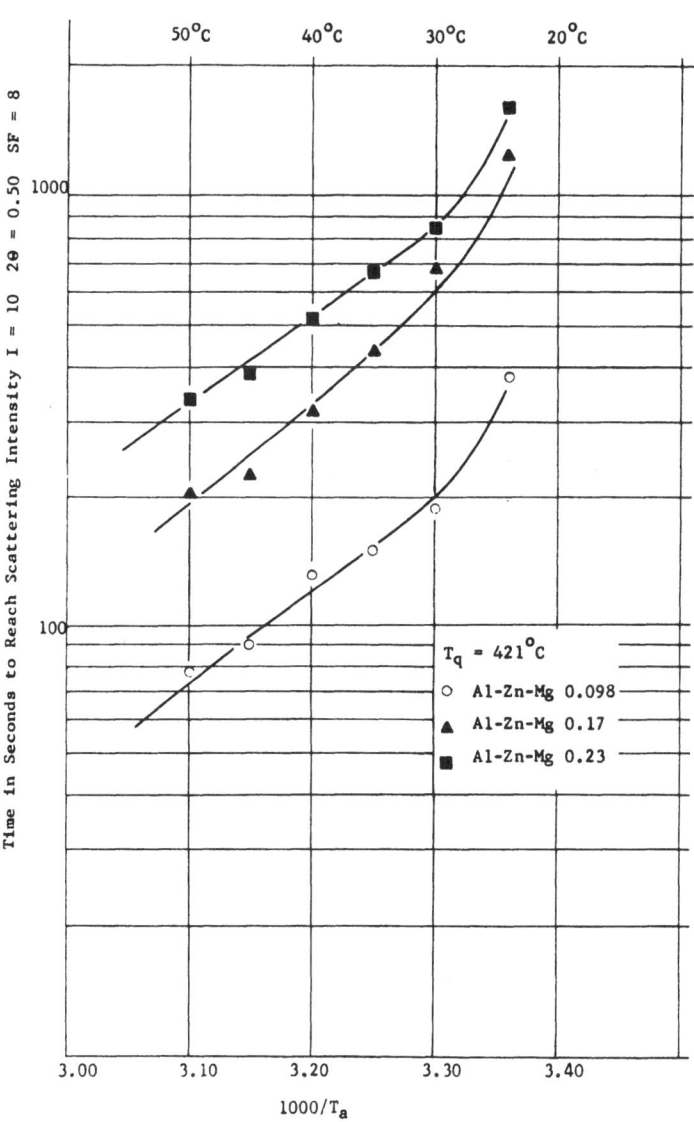

Figure 4. Data on growth of G–P zones as a function of aging temperature T_a during the stage of fast growth; $T_q = 421°C$.

Table II. Calculated Values of S_a and S_q from Slopes of Curves on Arrhenius Plots Where S_a and S_q are Given in Electron Volts

Alloys used	S_q at various values of T_a, eV				S_a at two values of T_q, eV	
	0°C	22°C	35°C	50°C	293°C	421°C
Al–Zn Alloy 1	0.73	0.67	—	0.66	—	—
Al–Zn Rod 1	—	0.72	0.70	—	0.42	—
Al–Zn–Mg 0.025	0.24	0.43	—	0.44	—	—
Al–Zn–Mg 0.098	—	0.34	0.46	0.44	0.33	0.42
Al–Zn–Mg 0.17	—	0.29	0.38	—	0.33	0.44
Al–Zn–Mg 0.19	*	0.20	—	0.33	—	—
Al–Zn–Mg 0.23	—	0.10	0.39	—	0.36	0.41

* Slope zero.

Table III. Average Values of the Integrated Small-Angle Scattering Intensity as a Function of T_a and Magnesium Content of the Alloy

Alloy	Integrated intensity at three values of T_a, $Q_0 \times 10^7$		
	$T_a = 0°C$	$T_a = 22°C$	$T_a = 50°C$
Al–Zn	1.72	1.41	1.10
Al–Zn–Mg 0.025	1.62	1.29	1.02
Al–Zn–Mg	1.65	1.35	1.13

Note: Values of integrated intensity are given as ($Q_0 \times 10^7$).

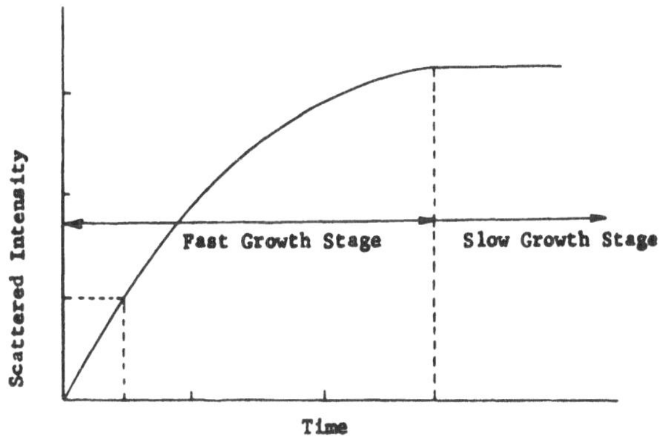

Figure 5. The rate of change of small-angle X-ray scattering intensity at $2\theta = 0.5°$ immediately after quenching from some temperature T_q above the solvus.

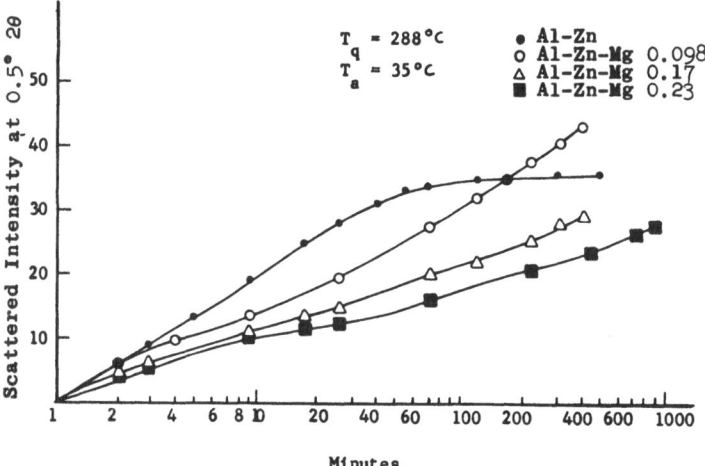

Figure 6. Data on the growth of zones for a series of alloys having various
magnesium contents; $T_q = 288°C$ and $T_a = 35°C$.

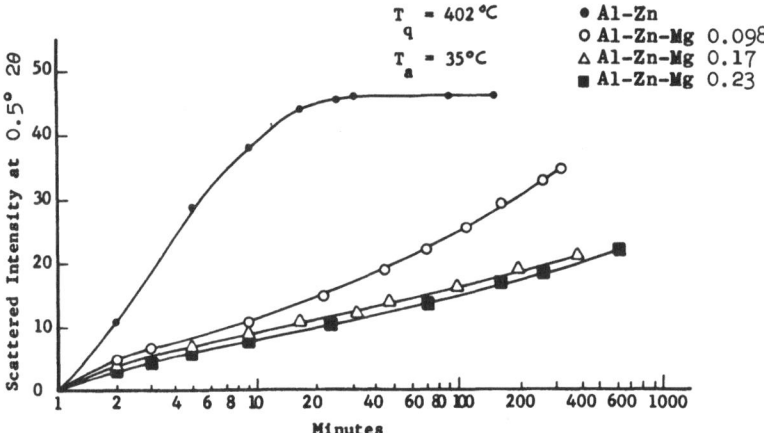

Figure 7. Data on the growth of zones for a series of alloys having various magnesium
contents; $T_q = 402°C$ and $T_a = 35°C$.

The initial purpose of this research was to obtain quantitative information concerning the
effect of minor additions of magnesium upon the decomposition process in quenched
Al–Zn alloys. In part this has been realized, but the processes occurring in the Al–Zn–Mg
alloys have proven far more complex than was anticipated and several details are still not
clear. For instance, it was initially assumed that the rate of zone growth (diffusion of zinc)
in the magnesium-containing alloys would be proportional to the product of the vacancy
concentration and the mobility of the vacancies as is the case in the binary Al–Zn alloys.[7]
In other words, the rate R was assumed to be proportional to the product of $\exp(-E_f/kT_q)$
and $\exp(-E_m/kT_a)$, where E_f and E_m are temperature-independent. This latter statement is a basic assumption which must be made if the simple Arrhenius relationship is to
be used to determine values of E_m and E_f. Unfortunately, the diffusion process is more
complex in the case of the magnesium-containing alloys and the dependence of the
process on aging temperature (T_a) cannot be described by a single activation energy.

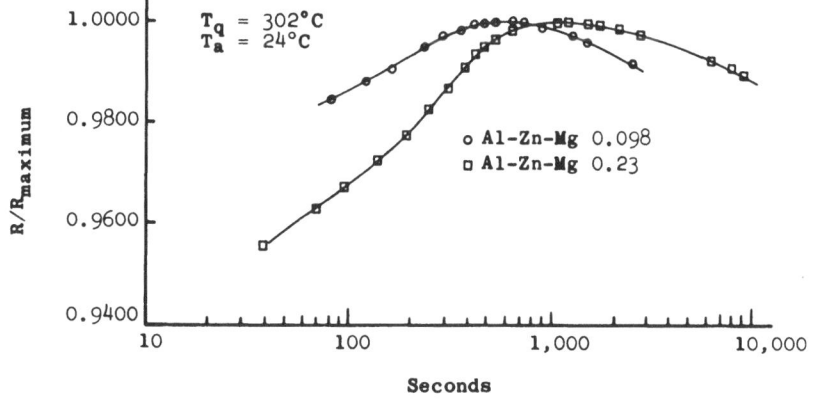

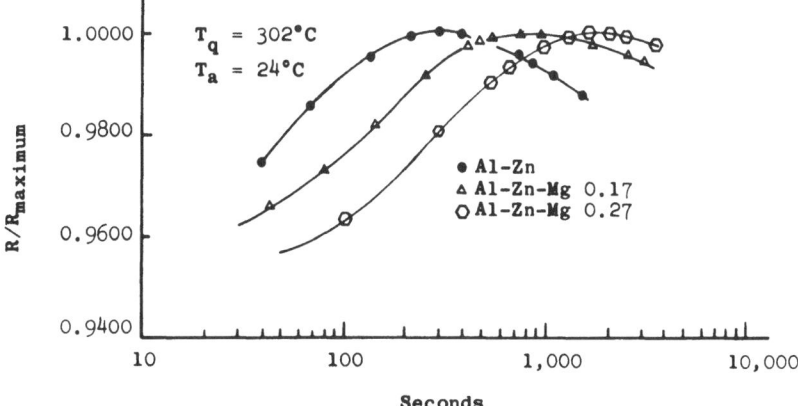

Figure 8. The change in resistance for quenched alloy wires having various magnesium contents; $T_q = 302°C$ and $T_a = 24°C$.

Furthermore, if t_0 represents the time necessary to reach a distinct state (characterized by a distinct value of a property) the slope in a $\ln t_0$ vs. $1/T_a$ plot has been found to be dependent upon quenching temperature as well as the aging temperature. For this reason the term E_m is replaced by S_a (slope at a constant $T_q \cdot k$, the Boltzmann constant). Similarly, it was found that a plot of $\ln t_0$ vs. $1/T_q$ is quite dependent upon aging temperature and therefore E_f has been replaced by the more general quantity S_q.

As pointed out in the preceding paragraph, S_a may be calculated using the simple Arrhenius relationship. This has been done for several Al–Zn and Al–Zn–Mg alloys, and the results as summarized in Figure 9 show clearly that S_a is a function of T_q. It has been shown also that S_a is dependent upon the range of aging temperature. Panseri and Federighi's data[3] for aluminum–10 wt.% zinc are included in Figure 9 and it is seen that they indicate a decrease of S_a with increasing T_q. They explain the decrease in the effective energy of motion at the high T_q as being due to a higher concentration of quenched-in divacancies. For the alloy Al–10 wt.% Zn–0.1 wt.% Mg, Panseri and Federighi indicate no change of S_a and T_q.[17] Furthermore, their value of 0.59 eV for S_a is much higher than the values obtained in this research for similar alloys. Careful examination of

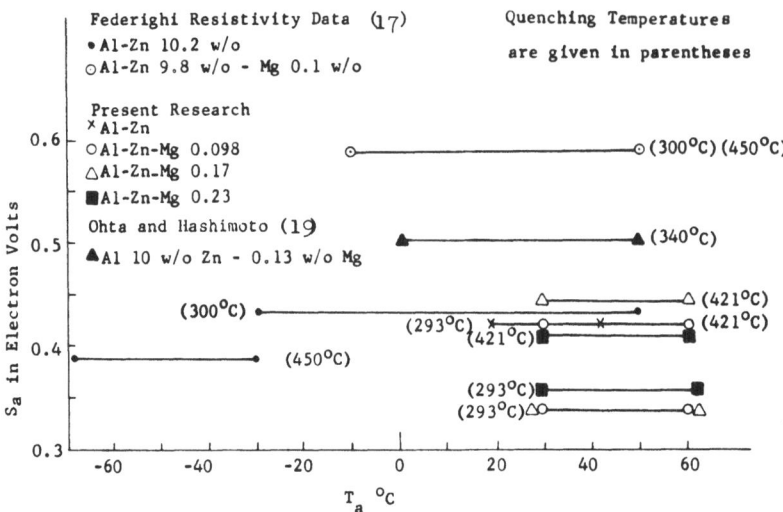

Figure 9. The dependence of S_a on T_q, magnesium content, and range of aging temperature from this research and others.[3,17,19]

Panseri and Federighi's data[17] shows that it is not possible to obtain a single value for the slope S_a which will apply over their entire temperature range ($-10° \leq T_a \leq 50°C$). Furthermore, when Panseri and Federighi's temperature range is restricted to that used in this study, S_a (calculated from their data) will be approximately 0.34 eV for high and low quenching temperatures. Ohta and Hashimoto's[19] recent data for S_a are also presented in Figure 9. Their value of 0.5 eV also will be lower if the temperature range is restricted to the one used in this study.

Using a procedure similar to that described for S_a, the values of S_q for the various alloys have been calculated and the accumulated results of the plot of $\ln t_0$ *vs.* $1/T_q$ are given in Figure 10. Included here also are the results of Panseri and Federighi.[3,17] However, their results are restricted to a single aging temperature of 20° C. Note that S_q values from the present research show a marked dependence upon T_a in the case of the magnesium-containing alloys.

Measurement of Integrated Intensity

Gerold has shown that the integrated intensity[21] is a good measure of the volume fraction of solute atoms that have segregated into zones during preprecipitation. The measured values given in Table III are only relative values of integrated intensity, absolute values being unnecessary for comparison purposes. These measurements indicate that the average zone state is not seriously altered by the addition of small quantities of magnesium. The Q_0 values for the magnesium-containing alloys are similar in behavior (with respect to T_a) to those obtained for the binary Al–Zn alloy. Based upon the theoretical considerations of Gerold, it can be assumed that no appreciable change in density, shape, or structure of the zones has taken place.

General Features of Fast Growth Stage

With the exception of instances where a low quenching temperature has been employed, it is generally the case that the initial rate of growth of G–P zones varies almost inversely with magnesium content. This has been shown clearly in Figure 8 where the

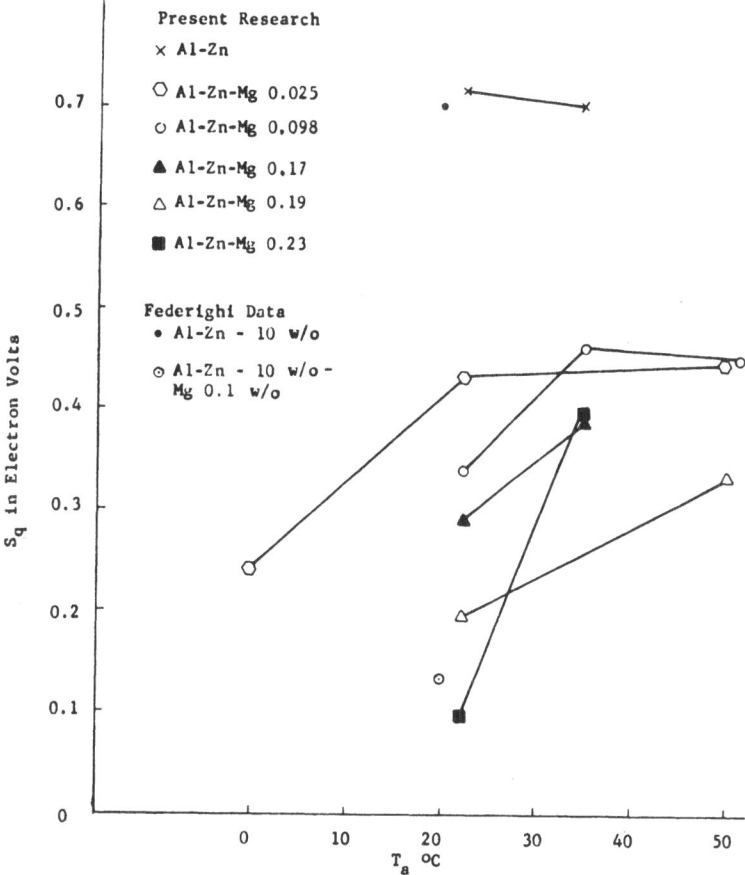

Figure 10. Results from this research and others,[3,17] showing the dependence of S_q on T_a and magnesium content.

time to reach the maximum resistance varies directly but not linearly with magnesium content. Similarly, the time needed to reach a preselected scattering intensity also increases with the magnesium content as can be shown for example in Figure 5.

At low quenching temperatures, notably below 300°C, the higher magnesium-containing alloys may exhibit a sharp initial jump in scattered intensity at $2\theta < 0.5°$. As a result of this initial jump the time needed to reach a preselected scattering intensity may be shortened abnormally (Figure 4). This jump is thought to be due to small precipitates that are stable at the lower quenching temperatures.

The Longtime Zone Growth Behavior

The addition of magnesium in quantities as low as 0.025 wt.% imparts a long duration (>1000 hr) to the zone growth process. This is in contrast to the case observed in the binary Al–Zn alloy where zone growth appears to terminate after approximately 100 min at the lowest quenching temperature used (253°C). Zone growth in some of the magnesium-containing alloys continues for many months at room temperature at a rate which is almost inversely proportional to the magnesium content.

Final zone radius (when measurements were stopped) in the magnesium-containing alloys is generally larger by a factor of nearly two, than zones formed in Al–Zn alloys aged under identical conditions. The only exception is when binary Al–Zn alloys are quenched from very low temperatures. In some instances, the zone radius in the magnesium alloys became so large (approximately 60 Å) that only the steep high-angle side of the scattering curves was visible and no real scattering maximum could be observed.

DISCUSSION

It has been clearly demonstrated that small additions of magnesium have a profound influence upon the preprecipitation behavior of Al–Zn alloys. At the same time, it is evident that this behavior is complex compared to that of binary Al–Zn alloys, a case that has been well documented and for which a satisfactory preprecipitation model exists. The behavior of ternary Al–Zn–Mg alloys reveals a complicated interplay of quenching temperature, aging temperatures, and magnesium content (on the growth of Guinier–Preston zones) that cannot be adequately explained by existing binary preprecipitation models.

Several interesting theories have been proposed to explain preprecipitation behavior in Al–Zn–Mg alloys. One of the earliest of these theories was presented by Hardy[23] and later extended by Polmear.[13] These investigators were of the opinion that magnesium, because of its larger relative size and chemical affinity for zinc, would tend to surround itself with an atmosphere of zinc atoms at temperatures above the solvus. While the model could partially account for many of their observations it could not, for instance, explain the dependence of zone growth rate on the quenching temperature.

It remained for Panseri and Federighi[3] and others[2] to clearly demonstrate the role played by vacancies in preprecipitation processes. In one of their early publications[1] Panseri and Federighi suggested that the addition of a third element (such as magnesium) to Al–Zn alloys could have a marked effect on the formation and growth of G–P zones through an interaction of the large magnesium atom with the vacancies. Subsequent research by these authors indicated that the magnesium-vacancy binding energy was large, i.e., 0.59 eV.[17] Thus, in the single alloy they were studying (Al–Zn 10%–0.1% Mg) it was concluded that most of the quenched-in vacancies would be bound to magnesium atoms and that the diffusion of zinc would then of necessity occur via mobile magnesium-vacancy couples rather than by free vacancies.

In the spring of 1964, as the present research was nearing completion, a new model was presented by Ohta and Hashimoto.[19] Like Hardy and Polmear they assumed that the Mg–Zn interaction is of primary importance but that it takes place not at the solution temperature, but during and after the quenching process. Their model predicted the following sequence of post quenching events: (a) magnesium and zinc cluster into ternary zones until the magnesium supply is depleted; (b) these ternary zones interact with vacancies and lower the quenched free vacancy concentration in the matrix; and (c) the remaining zinc atoms cluster in a manner similar to that observed in the binary alloy.

The results of the present investigation are summarized schematically in Figure 11. It is evident that none of the preceding models is capable of fully explaining these results. Hardy and Polmear's model could explain the first result (Figure 11a) simply on the basis that the "free" zinc concentration in the matrix (after quenching) would be lowered by the presence of magnesium in the alloy. This decrease in "free" zinc concentration would then result in a decreased rate of G–P zone growth. Panseri and Federighi's model was unfortunately based upon a single alloy, and their conclusion that increasing the magnesium content would increase the rate of zone growth has been shown to be in error.[24]

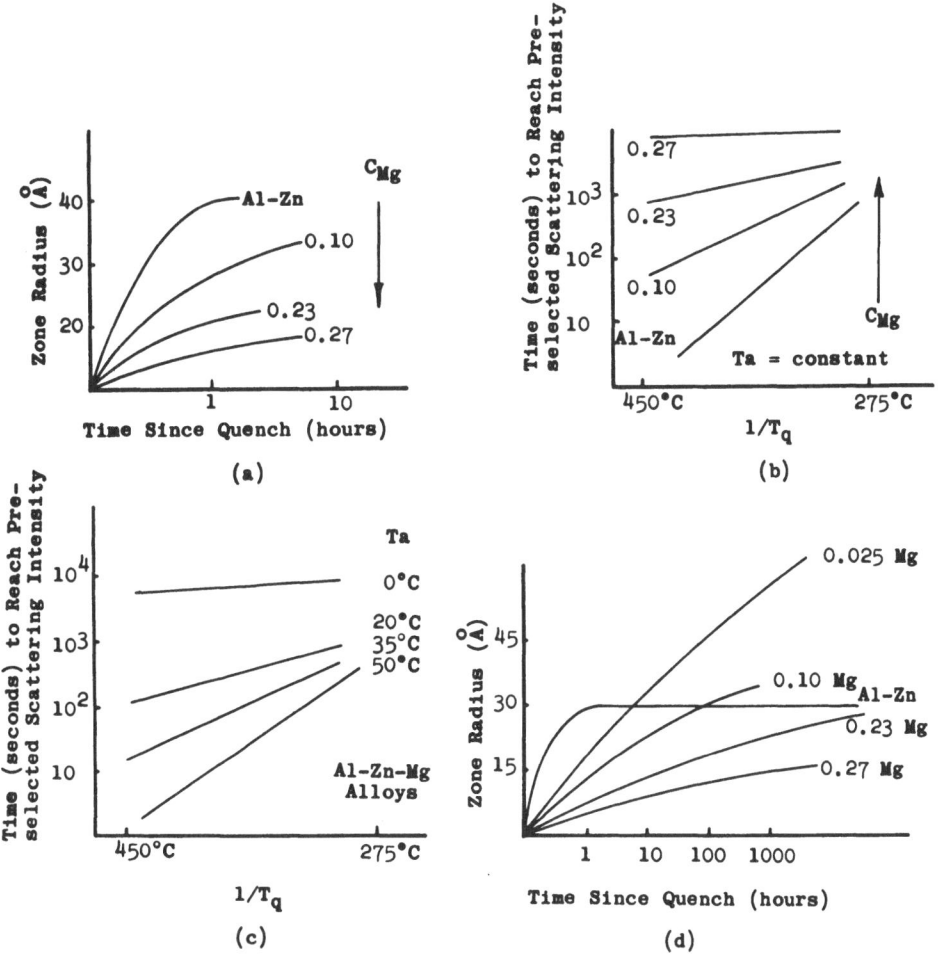

Figure 11. Schematic representation and summary of the results obtained in this research.

A Proposed Explanation

Borrowing first from Hardy and Polmear, the magnesium atoms are pictured as being associated with an atmosphere of zinc atoms at temperatures above the solvus temperature. The number of zinc atoms near a given magnesium atom at any instant should also depend sensitively on the quenching temperature. That is to say, at 300°C one would expect (on the average) more zinc atoms around each magnesium atom than at 450°C. It is also expected, as suggested by Ohta and Hashimoto, that during quenching these embryonic clusters would interact with free vacancies. This interaction should be stronger in the case of clusters quenched from higher quenching temperatures because clusters of this type would not be as stable in terms of lattice strain energy as those clusters in which the magnesium atom is more completely surrounded by smaller zinc atoms. In this system, zinc has the smallest interatomic distance (2.66 Å), followed by aluminum (2.86 Å) and then magnesium (3.20 Å). The shielding effect of zinc is due to the accommodation, by the smaller zinc atom, of the excess volume occupied by the larger magnesium atoms. In the absence of zinc this excess volume would presumably be accommodated by a vacancy. The net result of this effect would be a larger reduction in the expected "free" vacancy concentration (C_V) for high quenching temperatures (T_q) than for low quenching temperatures. The vacancies trapped in this way are not considered to be permanently lost to the preprecipitation process but instead are bound rather loosely to the ternary clusters as proposed by Ohta and Hashimoto. The strength of this binding should depend on the aging temperature and the number of zinc atoms associated with the initial cluster. As this cluster-vacancy binding energy decreases, vacancies will, with increasing frequency, leave the sphere of influence of the cluster and at the same time augment the remaining "free" vacancy concentration in the matrix.

The experimental behavior summarized in Figure 11 can now be explained in light of the proposed model. The effect of magnesium content in reducing the rate of zone growth (Figure 11a) can be largely explained as a lowering of the concentration of free zinc atoms. It is suggested that magnesium acts as a nucleation site for the formation of clusters at temperatures above the solvus. Thus, the total number of clusters quenched into a magnesium-containing alloy will probably be larger than in a binary Al–Zn alloy with the same zinc concentration. Simply allowing each magnesium atom to accommodate 4–5 zinc atoms at a temperature above the solvus will reduce the "free" zinc concentration by about 23% (for the case of the 0.27% magnesium alloy). Reducing the zinc concentration in binary Al–Zn alloys by this amount has been shown[7] to significantly decrease the rate of growth of G–P zones. Superimposed on the lowering of free zinc concentration is the trapping of vacancies by the quenched-in Mg–Zn clusters. This trapping will be very effective at low aging temperatures where the Mg–Zn clusters are believed to be quite stable. It will be assumed throughout this discussion that the further addition of zinc atoms to the initial Mg–Zn clusters will not lower the free energy of the system as is the case with the growth of magnesium free zones. It also is assumed that the path of minimum activation resides in the growth of magnesium-free zones.

The second effect (shown in Figure 11b), the decrease of the slope S_q with increasing magnesium content, also may be explained by the interaction of Mg–Zn clusters with vacancies. At high quenching temperatures, many effective vacancy traps (Mg–Zn clusters) are quenched into the alloy. The number of such traps will be a sensitive function of magnesium content. On the contrary, when T_q is low, the shielding effect mentioned previously will result in a smaller number of trapped vacancies and moreover, the trapping will be relatively insensitive to the magnesium content. Therefore, the net result of increasing the magnesium content will be to pivot the S_q curve about a nearly common value for low quenching temperatures, Figure 11b.

It has been demonstrated experimentally (Figure 11c) that increasing the aging temperature will result in an increase in the value of S_q for those alloys containing magnesium. Stated in another way, the curve for S_q appears to pivot around the value obtained for the lower quenching temperature. This effect may be attributed to the decomposition and concurrent release of vacancies from the initial Mg–Zn clusters. As pointed out previously, this decomposition will be rapid at the higher aging temperatures and almost nonexistent when T_a is very low.

Figure 11c may also be used to illustrate the influence of the range of aging temperatures on the slope S_a. This figure clearly shows that S_a must be dependent on the range of aging temperature and that this dependence will be such that, for $0°C < T_a < 20°C$, a higher value of S_a will result than when $T_a > 20°$. An additional effect which also can be seen with the aid of Figure 11c is the increase in the slope S_a with increasing T_q (i.e., the rate of zone growth has been found experimentally to be a more sensitive function of aging temperature when T_q is high rather than low).

The last effect listed, Figure 11d, refers to the duration of zone growth and to the larger ultimate zone radii that are found in Al–Zn–Mg alloys as compared to Al–Zn alloys quenched and aged under identical conditions. This prolonged growth is undoubtedly associated with the higher initial vacancy concentration in the ternary alloys and subsequent slow release from the Mg–Zn vacancy cluster. At moderate aging temperatures (RT) the quenched-in bound vacancy sources are capable of providing a long continuing supply of free vacancies to the matrix, thus enabling the zone growth process to continue for many months. Contrasted with this, the binary Al–Zn alloys generally terminate the zone growth process within a few hours of quenching.

CONCLUSIONS

Small angle X-ray scattering and resistometric measurements have been applied to the study of the zone forming process in Al–Zn alloys containing small additions of magnesium. A clearer understanding of the process has resulted.

1. The initial rate of growth of G–P zones in the temperature range (0–50°C) decreases with increasing magnesium content. The duration of zone growth is increased by several orders of magnitude with the addition of as little as 0.025 wt.% magnesium.

2. The final zone size reached in the magnesium-containing alloys is generally larger than in Al–Zn alloys quenched and aged under identical conditions.

3. The general pattern of preprecipitation (as revealed by the measurement of integrated intensity) is not greatly altered by these small additions of magnesium.

4. The dependence of the zone-growth rate on reciprocal quenching temperature S_q has been shown to decrease with increasing magnesium content. The value of S_q for alloys containing magnesium increases as the aging temperature is raised through the range 0–50°C.

5. A model is proposed in which the free vacancy assumes the dominant role in the growth of G–P zones. The complicated features of the zone-growth process in the magnesium-containing alloys can be explained by quenched-in magnesium-zinc-vacancy complexes. These complexes are most likely formed during the quench from high temperatures and will be a very stable configuration at low aging temperatures (below 20°C). As the aging temperature is increased, the free vacancy concentration, and therewith, the rate of zone growth, also will increase due to a progressive decomposition of such complexes.

ACKNOWLEDGMENTS

The author wishes to acknowledge the financial assistance given to this work by the Army Research Office in Durham, North Carolina. High purity metals were supplied by Aluminum Company of America and New Jersey Zinc Company. Chemical analyses were kindly supplied by Aluminum Company of America.

REFERENCES

1. T. Federighi, "Quenched-in Vacancies and Rate of Formation of Zones in Aluminum Alloys," *Acta Met.* **6**: 379, 1958.
2. F. Seitz, *L'etat solida*, Brussels, 1952, p. 405.
3. C. Panseri and T. Federighi, "A Resistometric Study of Pre-Precipitation in Al-10% Zn," *Acta Met.* **8**: 217, 1960.
4. D. Turnbull, H. S. Rosenbaum, and H. N. Treaftis, "Kinetics of Clustering in Some Aluminum Alloys," *Acta Met.* **8**: 277, 1960.
5. A. Guinier, *Metaux Corrosion* **18**: 209, 1943.
6. J. P. Jan, "Small Angle X-ray Scattering from Precipitation in Cold-Worked Al–Ag and Al–Zn," *J. Appl. Phys.* **26**: 1291, 1955.

7. V. Gerold and W. Schweizer, "Die kinetik von Entmischungs-vorgängen in übersattigen Al–Zn Mischkristallen," *Zeitschrift Metallkunde* **52**: 76, 1961.
8. *Metals Handbook*, American Society for Metals, Cleveland, Ohio, 1948.
9. J. Herenguel and G. Chaudron, *Metaux Corrosion* **18**: 30, 1943.
10. H. Schmalzried and V. Gerold, "Röntgenographische untersuchungen über die auschartung einer Aluminum–Magnesium–Zink Legierung," *Zeitschrift Metallkunde* **49**: 291, 1958.
11. V. Gerold and H. Haberkorn, "Röntgenographische untersuchungen der Kaltauschartung von Aluminum–Magnesium–Kupfer und Aluminum–Magnesium–Zink Legierungen," *Zeitschrift Metallkunde* **50**: 568, 1959.
12. A. Guinier, "Heterogeneities in Solid Solutions," in: *Advances in Solid State Physics*, Vol. 9, Academic Press Inc., New York, 1960, p. 293.
13. I. J. Polmear, "The Aging Characteristics of Ternary Aluminum–Zinc–Magnesium Alloys," *J. Inst. Metals* **86**: 113, 1957–58.
14. I. J. Polmear, "The Upper Temperature Limit of Stability of G.P. Zones in Ternary Aluminum–Zinc–Magnesium Alloys," *J. Inst. Metals* **87**: 24, 1958–59.
15. C. Panseri, F. Gatto, and T. Federighi, "Interaction Between Solute Mg Atoms and Vacancies in Aluminum," *Acta Met.* **6**: 198, 1958.
16. W. L. Fink and L. A. Willey, "Equilibrium Relationships in Al–Zn–Mg Alloys of High Purity," *Trans. AIME* **124**: 78, 1937.
17. C. Panseri and T. Federighi, "Evidence for the Interaction Between Mg Atoms and Vacancies in Al–Zn 10%–Mg 0.1% Alloy," *Acta Met.* **11**: 575, 1963.
18. G. Bartsch, "Der Einfluss von Mg-Spuren auf die Widerstands-anderungen bei der Aushartung von Al–Zn–Mg Legierungen," *Acta Met.* **12**: 270, 1964.
19. M. Ohta and F. Hashimoto, "Clustering in Al–Zn–Mg Alloys," *J. Phys. Soc. Japan* **19**: 1337, 1964.
20. R. W. Gould and V. K. Gerold, "Adaptation of the Norelco High Angle Diffractometer for Small Angle Scattering Studies of Preprecipitation Phenomenon," *Norelco Reporter* **XII**: 7, January–March, 1965.
21. V. K. Gerold, "The Zone Formation in Aluminum–Zinc Alloys," *Physica Status Solidi* **1**: 37, 1961.
22. G. Thomas and J. Washburn, *Electron Microscopy and Strength of Crystals*, Interscience Publishers, New York, 1963.
23. H. K. Hardy, *J. Inst. Metals* **79**: 321, 1951.
24. R. W. Gould and V. K. Gerold, "The Influence of Mg Atoms on Diffusion Processes in Al–Zn Alloys," *Acta Met.* **12**: 954, 1964.
25. A. F. Bonfiglioli and A. Guinier, "La Structure des Zones G.P. dans les Alliages Aluminum–Zinc au Premier Stade de leur Formation," to appear in *Acta Met.* in 1966.
26. A. J. Perry, "Solute-Vacancy Interaction Energies and the Effect of 0.009 At. % Mg on the Ageing Kinetics of an Al–4.01 At.% Zn Alloy," to appear in *Acta Met.* in 1966.
27. A. J. Perry, "The Effect of Solute-Solute Interaction on the Apparent Vacancy Formation Energy in Dilute Aluminum Alloys," *Acta Met.* **14**: 719, June 1966.

DISCUSSION

B. S. Sanderson (National Lead Co.): Is there not some danger in taking the intensity at a fixed angle, that the maximum will shift with zone growth?

R. W. Gould: You have to be sure that you don't set your detector at a 2θ value where the maximum will go down. As long as your position is one where the intensity is always increasing, you're all right. When we first started the work, we were a little bit too far to one side, in which case the intensity went up and then came down, and this is a problem.

G. A. Walker (The Three M Company): Did you do any microprobe analysis to find out the microdistribution of the magnesium in these zones?

R. W. Gould: The zones are too small. I didn't give any sizes, but the maximum measured radii of the zones were 70 to 80 Å. We would have liked to, but the probe beam, of course, is too large.

M. C. Huffstutler (Bell Telephone Laboratories): Do you know the partial molar volume change of these alloys and how it depends on magnesium composition?

R. W. Gould: No, I do not and this may be significant. If I had to do it again, I would certainly use much more dilute alloys. The comments I have had from people I have talked to indicate that these alloys are much too high in magnesium. Recently, Perry (see Reference 26 and 27) has come out with several very good papers on the effect of magnesium, but at a much, much lower level. His theory for the decomposition in the initial stages seems to be very reasonable and I tend to go along with it. I think what I have measured is something that is intermediate. It is not the initial stage of decomposition, but rather somewhat after the initial stage.

J. B. Newkirk: I might support the interpretation that you give to explain your results by quoting some work that was done by R. W. Hendricks, who is now at the Oak Ridge National Laboratory. He was studying the rate of precipitation of silver chloride in sodium chloride by much the same method, i.e., small angle scattering. He found that the addition of a small amount of a divalent impurity, calcium chloride, increased the rate of precipitation of silver in sodium chloride considerably. He interpreted this effect in terms of extrinsic vacancies that would be included in the matrix because of the addition of this divalent cation impurity in an otherwise monovalent-monovalent system. The extra vacancies were presumed to promote diffusion which apparently limited the rate of precipitation.

R. G. Baggerly (The Boeing Company): Are there any problems introduced by preferred orientation in the foils?

R. W. Gould: Not in this small angle scattering study because the zones are spherical. Also, the small particle size that results when you roll and anneal these alloys generates no problems. If we had platelet-shaped particles and had used single crystals, there would be an orientation problem. It's an ideal system to work with because of the spherical zones and relatively small zone radii.

ANALYSIS OF HIGH ANGLE DIFFUSE SCATTERING FROM SMALL PLATELETS

A. D. Thomas, Jr.

University of Texas
Austin, Texas

and

Gerald L. Liedl

Purdue University
Lafayette, Indiana

ABSTRACT

A detailed analysis of the high angle diffuse scattering from small platelets is given. A large number of statistically centrosymmetric platelets is considered, and it is shown that, in this case, the positive square root of the diffuse intensity from the platelets is proportional to the amplitude of the scattered radiation over particular regions in reciprocal space. The measured amplitude distribution is truncated from the true amplitude distribution by the limits of measurement and the influence of Bragg scattering. A truncation function is introduced to describe this truncated amplitude distribution in terms of the true amplitude distribution. This truncation introduces modulations on the measured electron density distribution. The measured electron density distribution is described in terms of the convolution of the true electron density distribution and the transform of the truncation function. The transform of the truncation function is known analytically, so the true electron density distribution can be found by a relaxation method. The true electron density distribution is given in terms of composition and strain parameters which are independently adjusted during the relaxation procedure to fit the measured values. Examples of the influence of the truncation function are given and the technique is applied to G–P 1 zones in an aluminum – 1.67 at.% copper alloy.

INTRODUCTION

The subject of age-hardening in aluminum–copper alloys has been of interest both commercially and scientifically, since its discovery by Wilm at the turn of the century. Of particular interest is the mechanism of hardening in which three distinct structures have been found in addition to the equilibrium $CuAl_2$, or θ phase. The first two copper-rich structures have been termed "preprecipitates" and are called G–P 1 and G–P 2 zones after Guinier[1] and Preston[2] who, working independently, interpreted the diffuse streaks to be the result of a two-dimensional defect.

Attempts have been made by Toman[3,4] and Gerold[5] to analyze the structure of the G–P 1 zones by postulating models and comparing the calculated intensity distribution from such a model with the measured intensity. These calculations involved assumptions as to the concentration of copper atoms clustering on the (100) planes of the aluminum

Table I. Symbol Definitions

Operation or function	Definition	Transform	Definition
$g(x_i)$	General function	$G(y_i)$	
$g(x - x_0)$	Displaced function	$G(y)e^{2\pi i x_0 y}$	
$\mathrm{rect}\dfrac{x}{a}$	$= \begin{cases} 1 \text{ for } \lvert x\rvert < a/2 \\ 0 \text{ for } \lvert x\rvert > a/2 \\ \frac{1}{2} \text{ for } \lvert x\rvert = a/2 \end{cases}$	$a \text{ sinc } ay$	$a\dfrac{\sin \pi a y}{\pi a y}$
$\sum$	Summation	$\sum$	
$*$	Convolution	$\cdot$	Product

matrix as well as the lattice strain caused by the difference in atomic radii. The first to use the amplitude of scattered radiation to solve directly for the structure without an artificial model was Doi[6] who made use of the fact that the amplitude of scattered radiation is the Fourier transform of the electron density of the scattering material. By including the composition and strain parameters in the mathematical analysis of Toman's data, Doi solved directly for the strain and relative composition of the first five layers of a G–P 1 zone. The only assumption made was that the first layer of the zone contained 100% copper atoms. The results of Gerold,[5] Toman,[3,4] and Doi[6] are shown in Figure 1, and show obvious disagreement.

It is possible, however, by using a different technique of Fourier analysis than used by Doi[6] to determine the copper concentration and strain directly without restrictions of a zone model.

DIFFRACTION ANALYSIS

The basic concept, resulting from considering the interaction of radiation with matter, is that the amplitude of the scattered radiation is the Fourier transform of the electron density. The relationship is expressed as

$$F(y_j) = f_e V_a \int_{-\infty}^{\infty} \int_{-\infty}^{\infty} \int_{-\infty}^{\infty} \rho(x_j) e^{2\pi i x_\alpha y^\alpha} dx_j \qquad (1)$$

where x_j and y_j are the coordinates of matter space and reciprocal space, respectively, F is the amplitude of scattered radiation, f_e is the amplitude of radiation scattered from a classical electron, V_a is the unit cell volume, ρ is the electron density, and $i = \sqrt{-1}$. Equation (1) is referred to as the diffraction image and is in general complex. Only the real part of the image, i.e., the intensity or the square of the modulus, can be measured. Some features of the Fourier transforms, useful in diffraction work, are listed in Table I. The definitions of some of the functions in Table I are not common; therefore, the definitions also have been included.

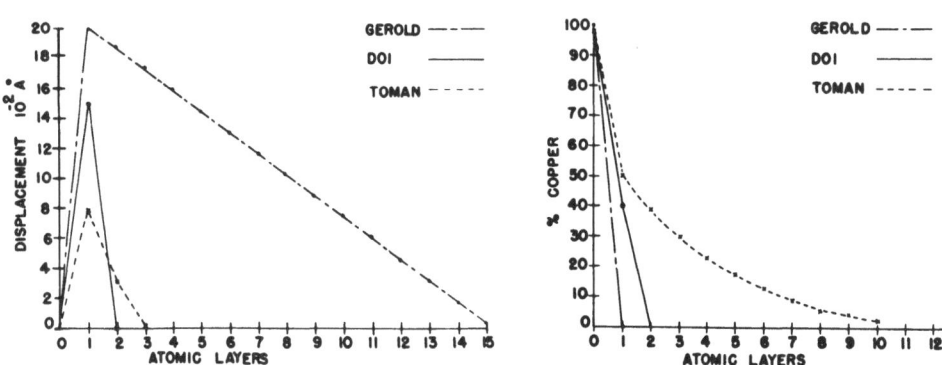

Figure 1. A comparison of the Guinier–Preston zone models resulting from previous investigations.

Some interesting and useful relationships are found by investigating the results obtained when one or more variables in equation (1) are equal to zero. If a plane through the origin of the image, $F(y_1, y_2, 0)$ is to be predicted for a known material, equation (1) becomes

$$F(y_1, y_2, 0) = f_e V_a \int_{-\infty}^{\infty} \int_{-\infty}^{\infty} \int_{-\infty}^{\infty} \rho(x_j) e^{2\pi i (x_1 y_1 + x_2 y_2)} \, dx_1 \, dx_2 \, dx_3 \qquad (2)$$

Only $\rho(x_j)$ is a function of x_3 so that by changing the order of integration and integrating with respect to x_3 equation (2) becomes

$$F(y_1, y_2, 0) = f_e V_a \int_{-\infty}^{\infty} \int_{-\infty}^{\infty} \rho_{\text{proj}}(x_1, x_2) e^{2\pi i (x_1 y_1 + x_2 y_2)} \, dx_1 \, dx_2 \qquad (3)$$

The amplitude distribution of a section passing through the origin of image space is related by equation (3) to the electron density distribution projected onto a plane. By similar reasoning, the amplitude distribution along one axis of image space is the Fourier transform of the electron density distribution projected onto a line.

In many diffraction measurements, the instrument naturally limits the measurement to either a plane or a line. For example, the Bragg–Brentano diffractometer measures along a line in reciprocal space. If the line corresponds to an image axis, the transform will be the projection of the electron density onto one crystal axis. Another example is the precession camera zero-level photograph which records the intensity on a plane in image space. The amplitude of this intensity is the transform of the projected electron density onto a plane as given by equation (3).

Before any quantitative work can be performed, the limitations in measuring must be studied. The size of the image is unlimited; whereas, the technique employed limits the range of measurement. This truncation of the image causes problems in the solution of equation (1).

The truncation problem may be given mathematically by defining a function $T(y_j)$ such that within the boundaries the function is equal to unity and outside the boundaries it is equal to zero. The truncation of the image can then be expressed as

$$F_{obs} = F_{actual} \cdot T(y_j) \tag{4}$$

The transform of equation (4) using Table I is

$$\rho_{obs} = \rho(x_j)_{actual} * t \tag{5}$$

Thus, the observed electron density is a modulation of the actual electron density by the transform of the truncation function. Solving equation (5) is a tedious job which can best be done by an iteration technique on a computer when $\rho(x_j)_{obs}$ and t are known. Since the truncation is inevitable, it is convenient to choose T such that its transform can be easily evaluated.

An additional complication is encountered from instrumental effects. These effects must either be eliminated through an analysis of these effects or be eliminated by experimentally keeping the instrumental effects at a minimum.

The nature of the precipitation hardening process makes it necessary to study two images instead of one in the case of a single crystal of the Al–Cu alloy. The preprecipitate, G–P 1 zone, has an almost two-dimensional lattice coherent with that of the three-dimensional matrix. The diffraction image will be a map of both images although they can be calculated separately as

$$\rho(x_i) = g_z(x_i)[\rho_z * l_z] + g_m(x_i)[\rho_m * l_m] \tag{6}$$

where ρ is the electron density; g is a shape function; l is the lattice; and the subscripts, z and m, refer to the zone and matrix, respectively. The amplitude of the scattered radiation is the transform of equation (6) and is

$$F(y_i) = G_z(y_i) * f_z L_z + G_m * f_m L_m \tag{7}$$

where $f_z L_z$ is a weighted strained lattice, and $f_m L_m$ is a weighted unstrained lattice. If we assume the amplitudes of the zone and the matrix are independent and are concentrated in different regions as a result of the strained and unstrained lattices, then the intensities are additive

$$I_{total} = I_z + I_m \tag{8}$$

The instrumental effects may be represented by a function M which is dependent on the Bragg angle and the geometry of the instrument. The observed intensity is the convolution of I_{total} and M, and is

$$I_{obs} = I_{total} * M = I_z * M + I_m * M \tag{9}$$

If one assumes that the matrix intensity does not change appreciably during the formation of the zone, then the difference in intensity before and after formation of the zone gives

$$\Delta I_{obs} = I_z * M \tag{10}$$

Since the intensity of the zone extends over a wide range and one may experimentally control the range of M to a small region, the influence of the instrument is negligible. Therefore,

$$\Delta I_{obs} = I_z \tag{11}$$

Given a large number of parallel zones randomly positioned along the direction considered, the amplitude of scattering from the zone as given in equation (7) may be written as

$$F_z(y_1, 0, 0) = \sum_{k=1}^{N} f_z(y_1, 0, 0)G_{zk}(y_1 - h_1)e^{2\pi i x_{1k} y^k} \tag{12}$$

where x_{1k} is the position of the kth zone along the x_1 axis and h_1 is the Miller index of the relpoint under consideration.

The intensity is the square of the modulus of equation (12) and for a large number of statistically centrosymmetric zones may be given as

$$I = \sum_{k=1}^{N} ||G_{zk}(y_1 - h_1)f_z(y_1, 0, 0)||^2 \tag{13}$$

The summation in equation (13) is proportional to the average value of the zone amplitude squared. Therefore, the zone amplitude is

$$F_z = \sqrt{I} = f_z G_{\text{ave}} \tag{14}$$

if we consider f_z to be independent of the index k.

Since the G–P 1 zones in the Al–Cu alloy are composed of copper atoms substituting for aluminum atoms on a slightly distorted matrix, ρ_z is the difference between the electron density distribution of the copper atoms and aluminum atoms. The amplitude obtained from equation (14) is the observed amplitude of equation (4) which must now be solved for the actual amplitude of scattering.

PROCEDURE

To determine the structure of a G–P 1 zone in an Al–1.67% Cu alloy after 36 hr of aging at 130°C, both precession camera and diffractometer techniques were employed to measure the intensity of the diffuse streaks between the (200) and (400) matrix Bragg spots. A simple and convenient cut-off function was chosen such that only that portion of the diffuse amplitude between the (200) and (400) relpoints was used. To obtain a real rather than a complex transform of this portion of the curve the symmetrical portion of the amplitude function along the negative y_1-axis also was used so that only a cosine transform was needed. The cut-off function of equation (4) becomes in this case two rect-type functions as shown in Figure 2.

$$T = \left[\text{rect} \frac{y_1}{4} - \text{rect} \frac{y_2}{2}\right] \tag{15}$$

The larger rect function cuts off all the amplitude beyond (400) and ($\overline{4}$00), and the smaller cuts off the amplitude between (200) and ($\overline{2}$00).

The cosine transform was performed on a digital computer and is shown in Figure 3. Since the Fourier transform of a symmetric curve also is symmetric, only the positive half of the truncated electron density curve is shown.

The transform of the truncation function T of equation (15) is

$$t = [4 \sinc 4x_1 - 2 \sinc 2x_1] \tag{16}$$

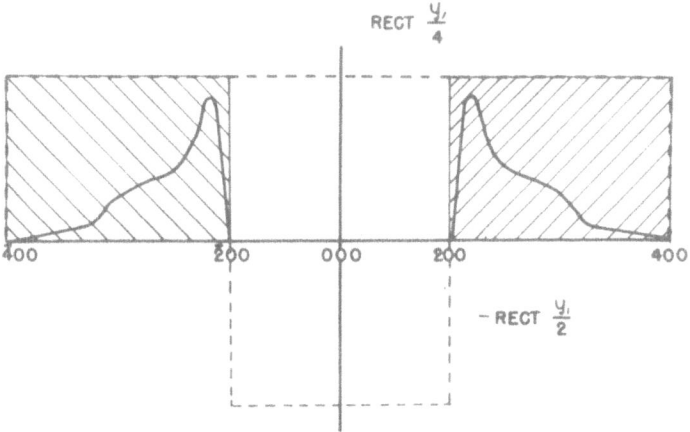

Figure 2. Truncation of the image of the Guinier–Preston zone.

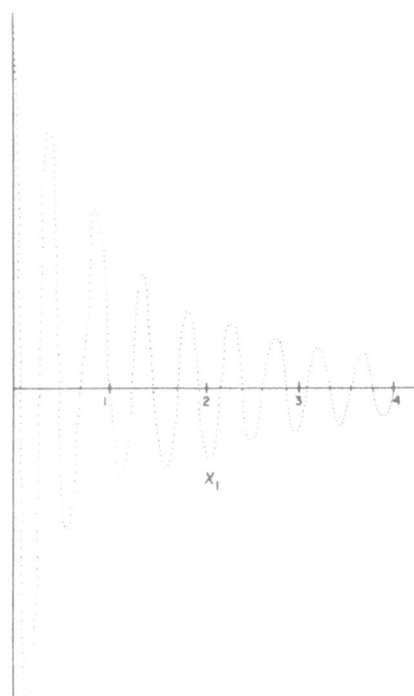

Figure 3. The Fourier transform of the truncated
Guinier–Preston zone.

which is shown in Figure 4. It is easy to see why the observed or truncated electron density function of Figure 3 has negative values while the actual electron density distribution is always positive. Two of the functions of equation (5) are known and the actual projected electron density distribution was solved by trial and error. The shape of the electron density of a single atom was obtained from the transform of the $(f_{Cu} - f_{Al})$ curve, corrected for truncation. The magnitudes of the projected distribution as well as the position of the atom planes was varied systematically until the ρ_{obs} in equation (5) matched the ρ_{obs} in Figure 3.

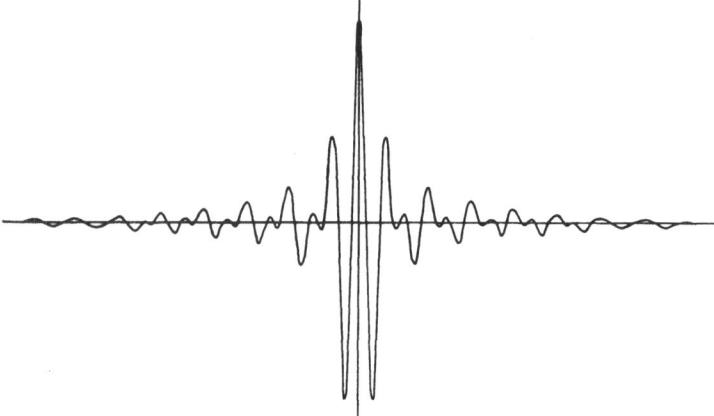

Figure 4. Fourier transform of the truncation function.

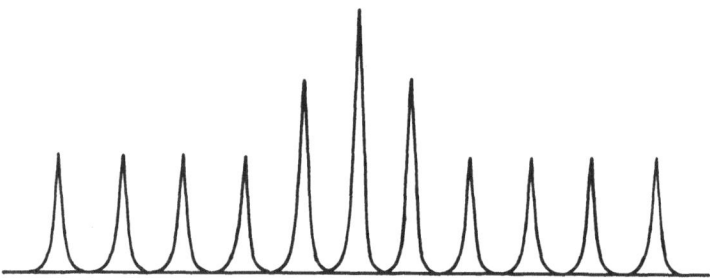

Figure 5. Projected electron density model of the Guinier–Preston zone.

RESULTS AND DISCUSSION

The trial and error solution of equation (5) takes the general form of Figure 5. The peaks represent the actual electron density projected onto the x_1 axis. A comparison of the area associated with each peak is a measure of the relative copper concentration of the zone layer, and the peak positions are a direct representation of lattice strain.

The peak areas eventually reach a constant value at some distance from the zone center, which represents the concentration of copper atoms of the solid solution in equilibrium with the G–P 1 zones. Beton and Rollason[7] found this to be 2 w/o copper, and this value makes it possible to determine by area ratios, the copper concentration in the rest of the zones.

The values of copper concentration in atomic percent formed by aging the alloy 36 hr at 130°C as determined by film and counter techniques are shown in columns 1 and 2 of Table II, respectively. The differences in the results can be attributed to three causes; (a) the greater accuracy of the counter method, (b) the extremely long (60 hr) exposure times of the precession camera technique, and (c) the method of comparing the peak areas to obtain copper concentrations.

The first reason was the justification for a complete analysis of the aging of Al–Cu alloys by James and Liedl.[8] The second reason is subtle in that, although the crystal was aged at 130°C for 36 hr, the room temperature aging while orienting the crystal as well as the aging occurring during the several runs throughout the aging process make a direct comparison of columns 1 and 2 almost impossible. In fact, the data obtained from

Table II. Zone Structure Comparison for G–P 1 Zones After 36-hr Aging at 130°C

Layer	Precession camera		Diffractometer[a]		Diffractometer[b]	
	at.% Cu	$\Delta \times 10^2$ Å[c]	at.% Cu	$\Delta \times 10^2$ Å[c]	at.% Cu	$\Delta \times 10^2$ Å[c]
0	75	—	30	—	51.8	—
1	35	25	2.2	22.5	13.8	22.5
2	1.4	0	1.67	12.5	1.4	12.5
3	1.4	0	1.2	12.5	1.1	12.5
4	1.4	0	1.67	12.5	1.4	12.5
5	1.4	0	1.67	10.0	1.4	10.0
6	1.4	0	1.67	7.5	1.4	7.5
7	1.4	0	1.67	7.5	1.4	7.5
8	1.4	0	1.67	7.5	1.4	7.5

[a] Results from James and Liedl.[8]
[b] Recalculations of the data of James and Liedl.[8]
[c] Displacement of the zone plane from the corresponding matrix plane.

the film technique are considered to be only of sufficient accuracy to provide a qualitative analysis of the zone structure.

Aside from data inaccuracy, the method of computing copper concentrations of the two studies was different. In the film technique, the areas of the peaks were used as a direct ratio of the copper content of the zone layers and the data were corrected for the contribution to background by zones parallel to the (020) and (002) planes. The diffractometer data were analyzed for the copper content by relating the square of the peak areas as given by Cochran[9] for defect structures. Either analysis applied to the other data increases the differences between the two techniques.

Guinier[10] has shown for very dilute solid solutions, the square root dependence used by James and Liedl[8] applies. However, the zones themselves are not dilute solutions and the contribution in the zone should be in direct proportion to the peak area. Using a square root dependence for the background and a direct dependence for the copper-rich layers, the data of James and Liedl[8] were used to obtain the new figures for copper concentration given in column 3 of Table II.

While some question may exist over the absolute value of concentration, this technique can measure changes in concentration to $\pm 5\%$ of the value and strain to ± 0.025 Å as determined from the fit of the solution to equation (5). The analysis of high angle diffuse scattering is then a technique that is applicable to many areas of materials research and is being used to study defect structures after radiation damage.

REFERENCES

1. A. Guinier, "Structure of Age-Hardened Aluminum–Copper Alloy," *Nature* **142**: 569, 1938.
2. G. D. Preston, "Structure of Age-Hardened Aluminum–Copper Alloy," *Nature* **142**: 570, 1938.
3. K. Toman, "The Structure of G–P Zones. I. The Fourier Transform of the Diffuse Intensity Diffracted by a Guinier–Preston Zone," *Acta Cryst.* **8**: 587, 1955.
4. K. Toman, "The Structure of G–P Zones. II. The Room Temperature Aging of the Aluminum–Copper Alloy," *Acta Cryst.* **10**: 187, 1957.
5. V. Gerold, "The Structure of Guinier–Preston Zones in Aluminum-Copper Alloys," *Acta Cryst.* **11**: 230, 1958.
6. K. Doi, "The Structure Analysis of Guinier–Preston Zone by means of a Fourier Method," *Acta Cryst.* **13**: 45, 1960.
7. R. H. Beton and E. C. Rollason, "Hardness Inversion of Dilute Aluminum–Copper and Aluminum–Copper–Magnesium Alloys," *J. Inst. Metals* **86**: 77, 1957.

8. D. R. James and G. L. Liedl, "Variations in the Structure of Guinier-Preston Zones During Aging," *Acta Cryst.* **18**: 678, 1965.
9. W. Cochran, "Scattering of X-Rays by Defect Structures," *Acta Cryst.* **9**: 259, 1956.
10. A. Guinier, *X-Ray Diffraction In Crystals, Imperfect Crystals, and Amorphous Bodies*, W. H. Freeman and Co., San Francisco, 1963, 264.

DISCUSSION

F. W. Lytle (Boeing Scientific Laboratory): What would the presence of vacancies in the zone do to your calculations, say, a concentration of about one vacancy per four copper atoms?

A. D. Thomas: This would materially affect the problem, in fact, one of the things we are looking at now is radiation damage in this very alloy and we would get a decrease in the peaks due to vacancy concentration. This nonuniform vacancy distribution was not taken into account in this particular work.

F. W. Lytle: I reported some work on this same system at the Detroit AIME meeting last fall, using X-ray absorption fine structure in which my results could be explained by the vacancy concentration I just mentioned.

Chairman D. K. Smith: In the use of your precession camera, could you not get a cleaner background, or had you used a monochromator?

A. D. Thomas: I did use a monochromator. The precision was not quite as good in the technique.

X-RAY DIFFRACTION STUDY OF ORDERING IN TWO SIGMA PHASES

R. W. Spor, H. Claus, and Paul A. Beck

University of Illinois
Urbana, Illinois

ABSTRACT

X-ray powder pattern line intensities were measured for the (Cr, Re)σ and (Re, Fe)σ phases by a step-scanning diffractometer, using Cr$K\alpha$ radiation, scintillation counter, and a pulse height analyzer. The measured intensity ratios for all available pairs of adjacent lines were compared by means of a computer with the corresponding calculated intensity ratios based on approximately 1800 different ordering schemes for each alloy. The results showed ordering in both alloys, and indicated that the ordering was based on atomic size. These results are different from those obtained previously by Kasper and Waterstrat (no ordering), and by Ageyev *et al.* [In (Cr, Re)σ the Cr atoms are preferentially in large coordination number positions.]

INTRODUCTION

In the family of phases of variable composition formed by transition elements, the sigma phase is the one that has been most thoroughly investigated. Most binary sigma phases are formed by combining a transition metal X having less than seven electrons outside of its closed shell with a transition metal Y having seven or more electrons outside the closed shell.[1] Among the $(X, Y)\sigma$ phases formed by combining various X and Y metals the composition shifts in such a way as to maintain the electron concentration (average number of electrons outside closed shells for the alloy) approximately 6.5. Thus, sigma may be considered an electron phase.[2-7]

The crystal structure of the sigma phase is well known.[8-10] The 30 atoms in the tetragonal unit cell occupy five crystallographically different types of sites: A, B, C, D, and E. As described by Kasper,[11] the coordination number for sites of type A and D is CN = 12, that for sites of type B is CN = 15, and for sites C and E it is CN = 14. The distribution of atoms X and Y over the various types of sites in the unit cell has been studied by a number of investigators,[11-23] using either X-ray diffraction or neutron diffraction. The results are shown in Table I. As first noted by Kasper,[11] X atoms preferentially occupy sites of CN = 15 (type B) and Y atoms tend to occupy sites of CN = 12 (types A and D). Sites of intermediate coordination number CN = 14 have X and Y atoms in varying proportions, depending mainly on the overall alloy composition. In those sigma alloys, which have a large preponderance of either X or Y, even the CN = 12 and CN = 15 sites may be "invaded" by the "wrong" atoms. This great flexibility in the occupancy of the various crystallographically different sites in the unit cell by the atoms of the two components, which is very typical of sigma and of related phases, allows the sigma phase to accommodate X and Y atoms in widely different stoichiometric ratios according to the requirements of the electron concentration. Table I shows how the occupancy

213

Table I. Atomic Order in Certain $(X, Y)\sigma$ Phases

| Composition | V_X/V_Y | Fractional occupation | | | | | Refs. |
		A 12	B 15	C 14	D 12	E 14	
$Mo_{72}Ir_{28}$	1.1	0.50	0	0	0.81	0.12	19
$V_{70}Ni_{30}$	1.27	0.85	0.025	0.06	0.86	0.01	15
$Mo_{65}Os_{35}$	1.11	0.75	0	0.06	0.94	0.12	19
$Cr_{61}Co_{39}$	1.08	0.65	0.05	0.15	0.62	0.50	20
$Mo_{60}Co_{40}$	1.43	1.00	0	0.12	1.00	0.12	17
$Nb_{60}Ir_{40}$	1.27	1.00	0	0.12	1.00	0.12	19
$Nb_{60}Os_{40}$	1.28	1.00	0	0	1.00	0.25	19
$V_{60}Fe_{40}$	1.18	0.85	0	0.19	0.85	0.25	15
$V_{56.7}Ni_{43.3}$	1.27	0.85	0	0.30	0.87	0.06	15
$Cr_{53}Co_{47}$	1.08	1.00	0	0	1.00	0.50	14
$Mo_{50}Fe_{50}$	1.32	1.00	0.25	0.25	1.00	0.25	16
$Cr_{46}Fe_{54}$	1.02	0.70	0.45	0.55	0.65	0.50	20
$Nb_{45}Re_{55}$	1.22	1.00	0	0.41	1.00	0.41	19
$Mo_{45}Re_{55}$	1.06	1.00	0.25	0.50	0.81	0.37	18
$Mo_{37}Mn_{63}$	1.28	1.00	0	0.62	1.00	0.50	12
$Mo_{33}Re_{67}$	1.06	1.00	0.50	0.50	1.00	0.50	18
$Cr_{25}Mn_{75}$	0.98	1.00	0.75	0.62	1.00	0.56	15
$Cr_{20}Mn_{80}$	0.98	0.90	0.65	0.85	0.95	0.65	20
$V_{19}Mn_{81}$	1.14	1.00	0.55	0.85	1.00	0.66	20

Fractional occupation of sites A to E by Y atoms shifts with alloy composition. V_X/V_Y gives the ratio of the atomic volumes of the component atoms.

of sites A to E shifts in sigma phase alloys with the concentration of Y increasing from 28 to 81 at.%.

Since in most cases the X atom is larger than the Y atom (see ratio of the atomic volumes V_X/V_Y in Table I), it was at first not clear whether the preference of the X atoms for the high coordination number sites is due to their larger size or rather to their lower electron concentration. However, it has been pointed out some years ago[24] that at least in one instance then known, $(Cr, Mn)\sigma$, the X atom is smaller than the Y atom. (The atomic volume ratio V_X/V_Y given in Table I was computed from the atomic volumes of elemental Cr and Mn,[25] but similar volume ratio values may be obtained also from the effective volumes of these metals in the bcc or β-Mn terminal solid solutions they form with one another.) In spite of this inverse size relationship ($V_X/V_Y < 1$), the ordering studies for $(Cr, Mn)\sigma$ both by Kasper and Waterstrat[15] and recently by Algie and Hall[20] show clearly that the (slightly) *larger* Mn atoms preferentially occupy the small coordination number sites (types A and D). It was, therefore, concluded[24] that the electronic structure is, at least in this case, more important than the atomic size in determining atomic ordering. Waterstrat and Kasper[21] later studied ordering in the $(Cr, Re)\sigma$ phase, where the inverse size relationship is much more pronounced than in $(Cr, Mn)\sigma$. They concluded that here the distribution of the X and Y atoms over the various sites is essentially random, Table II. Nevitt suggested[26] that, in this alloy, the conflicting ordering tendencies arising from electron concentration, on the one hand, and from atomic size, on the other, may approximately balance each other. However, Ageev and Shekhtman[22,23] later concluded on the basis of their own X-ray diffraction study of $(Cr, Re)\sigma$ that ordering here is of the same type as that found in other sigma phases: Y atoms prefer A and D type sites and X atoms occupy B type sites (see Table II.) This type of order would suggest

Table II. Atomic Order in (Cr, Re)σ and (Re, Fe)σ

Composition	V_X/V_Y	Fractional occupation					Refs.	RMSD
		A 12	B 15	C 14	D 12	E 14		
(Cr, Re)σ	0.816	0.63	0.63	0.63	0.63	0.63	21	0.56
		1.00	0.0	1.00	1.00	0.09	22, 23	1.03
		0.25	0.81	0.72	0.50	0.66	19	0.87
		0.32	0.92	0.86	0.40	0.54	This work	0.21
(Re, Fe)σ	1.24	1.00	0.0	0.34	1.00	0.34	22, 23	0.48
		0.90	0.10	0.22	0.86	0.57	This work	0.11

Fractional occupation of sites A to E by Y atoms.

that the ordering tendency resulting from electronic structure overrides the conflicting tendency due to atomic size, in spite of the very considerable difference in size between the two components. A third investigation of (Cr, Re)σ was reported by Spooner and Wilson,[19] who found inverse ordering: X atoms preferentially occupy CN = 12 sites [types A and D (see Table II)]. This kind of order would represent a predominance of size effect over electronic effect. Thus, the three investigations so far reported on the type of ordering in (Cr, Re)σ lead to contradictory conclusions. For a reliable assessment of the relative importance of atomic size and of electronic effects in ordering a new study was, therefore, required for (Cr, Re)σ. Accordingly, a careful investigation was carried out and, for comparison, ordering was studied also in (Re, Fe)σ, where the size relationship of the components is normal ($V_X/V_Y > 1$).

EXPERIMENTAL PROCEDURE

Since the weak X-ray diffraction lines are particularly sensitive to ordering in the sigma phase, all these were measured point by point, in increments of 0.02° in 2θ, using a stabilized diffractometer and an automatic step scanning device. Counting was continued at each Bragg angle for 4 min in order to minimize as much as possible the errors due to statistical fluctuations. Cr K_α radiation was used, giving high dispersion, so as to increase the separation between diffraction lines for planes with similar interplanar spacings. Since the temperature factor for the sigma phase was unknown and the measurements were carried out at room temperature, intensity ratios of pairs of diffraction lines with slightly different Bragg angles only were used. All diffraction lines from (200) through (412) were measured, with the exception of the high intensity diffraction lines (410), (330), (202), (212), (420), (411), and (331), which are quite insensitive to ordering. The measured relative intensities of X-ray diffraction lines for (Cr, Re)σ and (Re, Fe)σ are given in Table III, including also lines with vanishing intensity. The intensity values for these undetected lines are given in the table as *less than* the standard deviation of the scatter of the measured intensities over the Bragg angle range where the line should have appeared, multiplied by the average line width at half intensity. The undetected lines were also made use of in forming measured intensity ratios, by taking as their intensity half of the corresponding upper limits listed in Table III.

The 21 intensity ratios J_m/J_n of near neighbor line pairs m, n formed from the data were compared with the ratios of the corresponding intensities I_m/I_n calculated for various

Table III. Bragg Angles and Measured Relative Intensities of X-Ray Diffraction Lines for (Cr, Re)σ and (Re, Fe)σ Cr K_α Radiation

hkl	(Cr, Re)σ		(Re, Fe)σ	
	2θ (degrees)	Intensity	2θ (degrees)	Intensity
110			20.575	<5.6
200	28.536	<16	29.268	35.1
101	30.983	79.4	31.746	206
210	31.990	37.3	32.803	<6.4
111	34.214	<26	35.065	62.7
220	40.797	<18	41.853	<6.4
211	42.622	12	43.710	9
310	45.871	43.8	47.072	19.4
221	49.856	<16	51.162	22.3
301	52.096	190.4	53.470	180
320	52.759	88.5	54.170	120.9
311	54.270	508.4	55.711	474.4
002	56.580	440.1	58.054	465.9
400	59.065	<16	60.680	48.1
222	72.075	583.5	74.098	145.7
421	73.725	<16	75.853	5
312	75.697	144	77.865	124.2
430	76.070	<16	78.306	25.8
510	77.858	<16	80.167	<6.4
322	81.029	<16	83.423	<6.4
431	82.623 ⎫		85.126 ⎫	30.1
501	82.623 ⎬	180	85.126 ⎭	
520	83.151 ⎭		85.692	27.4
511	84.375	23.4	86.958	5.9
402	86.283	<16	88.921	23.1
412	88.026	37	90.747	<6.4

ordering schemes, using the (Cr, Co)σ atomic positions, and the atomic scattering amplitudes as functions of the Bragg angle. The degree of agreement of various ordering models with the data was estimated by means of the root weighted mean square normalized deviations (RMSD) of the calculated ratios from the measured ones

$$\text{RMSD} = \sqrt{\frac{1}{r} \sum_{m,n} w_{m,n} \frac{(I_m/I_n - J_m/J_n)^2}{I_m J_m/I_n J_n}} \tag{1}$$

where $w_{m,n}$ is the relative weight assigned to the normalized squared deviation of the calculated from the measured intensity ratio for the line pair m, n. A digital computer was programmed to give the RMSD values for hundreds of sets of calculated line intensity ratios, resulting from a systematic variation of the assumed fractional occupation of sites of types A to E by the Y atoms. In the region of best fit (near the minimum RMSD values) computation was carried out for fractional increments as small as 0.02 of the Y concentrations at the various types of sites.

RESULTS

Table IV gives the X-ray diffraction line intensity ratios calculated for (Cr, Re)σ according to the ordering scheme giving the lowest RMSD value, in comparison with the measured intensity ratios.

Table IV. X-Ray Diffraction Line Intensity Ratios for $(Cr, Re)\sigma$ with $Cr\,K_\alpha$ Radiation

Diffraction lines		Calculated	Measured
m	n	I_m/I_n	J_m/J_n
(200)	(101)	0.33	0.10 ± 0.13
(210)	(101)	0.18	0.47 ± 0.32
(111)	(101)	0.40	0.16 ± 0.21
(220)	(101)	0.046	0.11 ± 0.15
(220)	(310)	0.12	0.21 ± 0.27
(211)	(310)	0.64	0.27 ± 0.37
(221)	(310)	0.40	0.18 ± 0.24
(221)	(301)	0.16	0.042 ± 0.045
(301)	(320)	1.75	2.15 ± 0.56
(320)	(311)	0.12	0.17 ± 0.04
(311)	(002)	0.67	1.16 ± 0.07
(400)	(002)	0.059	0.018 ± 0.019
(400)	(222)	0.18	0.014 ± 0.014
(421)	(222)	0.088	0.014 ± 0.014
(421)	(312)	0.10	0.056 ± 0.061
(430)	(312)	0.016	0.056 ± 0.061
(510)	(312)	0.014	0.056 ± 0.061
(322)	(312)	0.10	0.056 ± 0.061
(322)	(511)	0.71	0.34 ± 0.55
(402)	(511)	0.36	0.34 ± 0.55
(412)	(511)	0.49	1.6 ± 1.9

Table V. X-Ray Diffraction Line Intensity Ratios for $(Re, Fe)\sigma$ with $Cr\,K_\alpha$ Radiation

Diffraction lines		Calculated	Measured
m	n	I_m/I_n	J_m/J_n
(200)	(101)	0.088	0.17 ± 0.06
(210)	(101)	0.043	0.016 ± 0.016
(111)	(101)	0.54	0.30 ± 0.07
(220)	(101)	0.14	0.016 ± 0.016
(220)	(310)	0.63	0.17 ± 0.22
(211)	(310)	0.65	0.45 ± 0.48
(221)	(310)	0.30	1.15 ± 0.68
(221)	(301)	0.13	0.12 ± 0.04
(301)	(320)	1.38	1.49 ± 0.16
(320)	(311)	0.18	0.26 ± 0.02
(311)	(002)	1.00	1.02 ± 0.05
(400)	(002)	0.14	0.10 ± 0.02
(400)	(222)	0.47	0.33 ± 0.08
(421)	(222)	0.091	0.03 ± 0.03
(421)	(312)	0.11	0.04 ± 0.04
(430)	(312)	0.089	0.21 ± 0.06
(510)	(312)	0.038	0.03 ± 0.03
(322)	(312)	0.020	0.03 ± 0.03
(322)	(511)	0.40	0.5 ± 0.9
(420)	(511)	4.39	3.9 ± 4.7
(412)	(511)	0.90	0.5 ± 0.9

Table V gives the corresponding data for (Re, Fe)σ. The ordering arrangement corresponding to the lowest value of the RMSD is given for both alloys in Table II, in comparison with the ordering arrangements arrived at by previous investigators. For purposes of comparison, the RMSD values are also given for each ordering scheme, calculated on the basis of the alloy composition and the measured line intensities from the present investigation. The table shows clearly that for (Cr, Re)σ the present data can be fitted by far best by the present ordering arrangement. For (Re, Fe)σ the present results are in good agreement with those of Ageev and Shekhtman.[22,23]

Among the previous investigations of ordering in (Cr, Re)σ the present results agree best with Spooner and Wilson's work,[19] which was also based on quantitative intensity measurements by means of a diffractometer. However, the area under the various peaks, as obtained by Spooner and Wilson on a recorder chart by continuous scanning, is a less accurate measure of line intensities than that given by the much more time consuming step-scanning procedure followed in the present work. This is particularly true for the lines of very low intensity. Other differences between the two investigations include the comparison of the experimental line intensity ratios in the present work with a very large number of sets of calculated intensity ratios. The drastic differences between the present results for (Cr, Re)σ and those of Ageev and Shekhtman[22,23] are very likely due to the fact that those authors based their conclusions on the intensities of the strong X-ray diffraction lines from (112) to (331), which depend on atomic ordering only to a relatively small extent. The differences with respect to Waterstrat and Kasper's conclusions[21] are probably largely due to the use by those authors of line intensities estimated by the visual inspection of photographic films.

DISCUSSION

Table II shows that ordering in (Re, Fe)σ corresponds to the usual arrangement: the larger X atoms prefer to occupy sites of CN = 15 (type B) and the concentration of Y atoms is highest at sites of CN = 12 (types A and D). This type of ordering was to be expected here, since the relationship between atomic size ratio and electronic configuration is the same as is encountered in the majority of sigma phases.

Although the present results for (Cr, Re)σ are somewhat different from those obtained by Spooner and Wilson,[19] they support the inverse ordering scheme found by those investigators, corresponding to the inverse size relationship of the components X and Y (that is $V_X/V_Y < 1$). The smaller X atoms preferentially occupy sites of CN = 12 (type A and to a somewhat lesser extent type D) and the larger Y atoms occupy sites of CN = 15 (type B). Thus, in contrast to (Cr, Mn)σ, which orders normally in spite of its inverse size ratio, Table I, ordering in (Cr, Re)σ takes place essentially according to the requirements of atomic size. One may, therefore, conclude that, if the inverse size difference between the two component atoms is large enough, the ordering effect due to size may predominate over that due to electronic structure. In the case of (Cr, Mn)σ, where ordering according to electronic structure predominates, the inverse size difference is rather small.

In view of these findings, it would be of considerable interest to re-examine ordering in the (Cr, Os)σ and (Cr, Ru)σ phases, for which random occupancy was reported by Waterstrat and Kasper[21] as in the case of (Cr, Re)σ, and to study it in (V, Re)σ, for which no ordering data appear to be available. The components of each of these three sigma phases have an inverse size relationship, with the V_X/V_Y ratios falling between the values for (Cr, Mn)σ and for (Cr, Re)σ. Order information for these phases might help in establishing narrower size ratio limits for the changeover from size-controlled to electronic structure-controlled ordering. It may also lead to the discovery of a true cancellation of the two effects in the borderline case, where neither effect predominates over the other, as visualized by Nevitt.[26]

ACKNOWLEDGMENTS

This investigation was supported by grants from the U.S. Army Research Office, Durham, North Carolina, the National Science Foundation, and the U.S. Atomic Energy Commission, COO-1198-385. The authors wish to express their appreciation to Mr. Claus Neuman for his very competent assistance with some of the computer work.

REFERENCES

1. E. O. Hall and S. H. Algie, "The Sigma Phase," *Met. Rev.* **11**: 61, 1966.
2. A. H. Sully and T. J. Heal, "An Electron Compound in Alloys of the Transition Metals," *Research* **1**: 288, 1948.
3. A. H. Sully, "The Sigma Phase in Binary Alloys of the Transition Elements," *J. Inst. Metals* **80**: 173, 1951.
4. S. P. Rideout, W. D. Manly, E. L. Kamen, B. S. Lement, and P. A. Beck, "Intermediate Phases in Ternary Alloy Systems of Transition Elements," *Trans. AIME* **191**: 872–876, 1951.
5. D. A. Bloom and N. J. Grant, "Regarding Sigma Phase Formation," *Trans. AIME* **197**: 88, 1953.
6. P. Greenfield and P. A. Beck, "The Sigma Phase in Binary Alloys," *Trans. AIME* **200**: 253–257, 1954.
7. P. Greenfield and P. A. Beck, "Intermediate Phases in Binary Systems of Certain Transition Elements," *Trans. AIME* **206**: 265, 1956.
8. D. P. Shoemaker and B. G. Bergman, "The Crystal Structure of a Sigma Phase, FeCr," *J. Am. Chem. Soc.* **72**: 5793, 1950.
9. G. J. Dickens, A. M. Douglas, and W. H. Taylor, "Structure of the Sigma-Phase in the Iron–Chromium and Cobalt–Chromium Systems," *Nature* **167**: 192, 1951.
10. J. S. Kasper, B. F. Decker, and J. R. Belanger, "The Crystal Structure of the Sigma-Phase in the Co–Cr System," *J. Appl. Phys.* **22**: 361, 1951.
11. J. D. Kasper, "Atomic and Magnetic Ordering in Transition Metal Structures," in: P. A. Beck (ed.), *Theory of Alloy Phases*, American Society for Metals, Cleveland, Ohio, 1956, p. 264.
12. B. F. Decker, R. M. Waterstrat, and J. S. Kasper, "Formation of Sigma Phase in the Mn–Mo System," *Trans. AIME* **197**: 1476, 1953.
13. G. Bergman and D. P. Shoemaker, "The Determination of the Crystal Structure of the Sigma Phase in the Iron–Chromium and Iron–Molybdenum Systems," *Acta Cryst.* **7**: 857, 1954.
14. G. J. Dickens, A. M. B. Douglas, and W. H. Taylor, "The Crystal Structure of the Co–Cr Sigma Phase," *Acta Cryst.* **9**: 297, 1956.
15. J. D. Kasper and R. M. Waterstrat, "Ordering of Atoms in the Sigma Phase," *Acta Cryst.* **9**: 289, 1956.
16. C. G. Wilson and F. J. Spooner, "Ordering of Atoms in the Sigma Phase FeMo," *Acta Cryst.* **16**: 230, 1963.
17. J. B. Forsyth and L. M. D'Alte Da Veiga, "The Structure of the Sigma Phase Co_2Mo_3," *Acta Cryst.* **16**: 509, 1963.
18. C. G. Wilson, "Order in Binary Sigma Phases," *Acta Cryst.* **16**: 724, 1963.
19. F. J. Spooner and C. G. Wilson, "Ordering in Binary Sigma Phase," *Acta Cryst.* **17**: 1533, 1964.
20. S. H. Algie and E. O. Hall, "Site Ordering in Some Sigma Phase Structures," *Acta Cryst.* **20**: 142, 1966.
21. R. M. Waterstrat and J. S. Kasper, "X-Ray Diffraction Study of the Sigma Phase in the Systems Re–Cr, Ru–Cr, and Os–Cr," *Trans. AIME* **209**: 872, 1957.
22. N. V. Ageev and V. Sh. Shekhtman, "The Crystal Chemistry of the Compounds of Rhenium with Transition Metals," in: B. W. Gonser (ed.), *Rhenium*, Elsevier Publishing Co., New York, 1962, p. 45.
23. N. V. Ageev and V. Sh. Shekhtman, "On the Question of the Nature of the Sigma Phase," *Dokl. Akad. Nauk SSSR* **135**: 309, 1960.
24. P. A. Beck, "Certain Intermediate Phases in Alloys of the Transition Elements," in: *Conference on Basic Research in Metallurgy*, O.O.R., U.S. Army, Durham, North Carolina, March 1958.

25. P. S. Rudman, "The Atomic Volumes of the Metallic Elements," *Trans. AIME* **233**: 864, 1965.
26. M. V. Nevitt, "Alloy Chemistry of Transition Elements," in: P. A. Beck (ed.), *Electronic Structure and Alloy Chemistry of the Transition Elements*, Interscience Publishers, New York, 1963, p. 101.

DISCUSSION

P. R. Morris (Armco Steel Corporation): You dismissed the temperature factor by choosing adjacent reflections or reflections with close Bragg angles. Do you assume then that the temperature factor is isotropic, and do you assume that it is the same for the two different atoms in the material?

P. A. Beck: No, we simply don't have any information on this, so we did the best we could, as everybody else does. To minimize the effect, we took adjacent reflections because the temperature factor depends upon $\sin \theta / \lambda$, so if you have reflections which have similar theta values, you expect the least difference.

P. R. Morris: Not if you don't have equal contributions from the different atoms for the two adjacent reflections, and not if they don't correspond to equivalent directions in the lattice.

P. A. Beck: These refinements could not be taken into account because the temperature factor, let alone its anisotropy, is unknown. This was the best ordering scheme we could find, but the root mean square deviation was still reasonably large. I suspect that most of the remaining deviation was not due to the temperature factor so much as to the uncertainty in the available atomic position parameters. These are also not known for the alloy in question and, in using the powder method for ordering studies, one just has to assume some values. We used the atomic position parameters determined by single crystal methods for the cobalt–chromium sigma phase and small deviations in these parameters for the chromium–rhenium sigma phase probably account for most of the root mean square deviation.

H. K. Herglotz (E. I. du Pont de Nemours and Company): How were these samples prepared? Rhenium melts at about 3000°C and chromium at 1900°C.

P. A. Beck: We struggled with this alloy for a long time. It was prepared by arc melting. This is possible, but one loses a lot of the high vapor pressure component.

H. K. Herglotz: I ask this particular question because I am worried that you might not have a true binary system. There could be additional components from the crucible or from impurities in the components from the crucible or from impurities in the components significant enough to influence the result. Some recent work on rhenium alloys done in Russia has given what appears to be anomalous results and this may be due to contaminants.

P. A. Beck: The arc melting method is better than melting in a refractory crucible. We never have any copper contamination that we can detect spectroscopically. As you say, of course, there are always impurities in any alloy and there is always some possible effect of the impurities. However, the impurity content was reasonably low in these alloys and it seems unlikely that it had an appreciable effect on the type of ordering found. Also, I doubt that the Russian investigators' deviating results were due to impurities. Almost certainly, the discrepancy there was due to the fact that they chose insensitive diffraction lines to base their conclusions on, as set forth in the paper.

A STUDY OF THE UNUSUAL LINE STRUCTURE IN POWDER PATTERNS OF PYROLYTICALLY DEPOSITED BORON COMPOUNDS AND OTHER MATERIALS

Robert L. Prickett

Air Force Materials Laboratory
Wright–Patterson AFB, Ohio

R. L. Hough

Hough Laboratories
Springfield, Ohio

and

Duane Earley

University of Dayton
Dayton, Ohio

ABSTRACT

A prolonged study has been made of the unusual X-ray powder patterns generated by certain pyrolytically formed boron compounds, consisting of two to four moderately broad diffraction lines in the 1–4 Å range. These broad lines are unexpectedly made to disappear by a single sharp blow with a hammer on the specimen in a Plattner diamond mortar, the disappearance being accompanied by the spontaneous generation of the standard diffraction lines expected for normally crystalline materials. The same diffraction pattern changes also may be brought about by regular grinding methods, fracture, or thermal annealing. Transmission electron diffraction yields results identical with those obtained by X-ray. Orientation studies demonstrate that the phenomenon is not orientation dependent. Controlled fracture has been applied in an attempt to develop a hypothesis. It appears from the constancy of the width and position of the diffraction lines among samples that the phenomenon being studied has a definite potential energy, well located somewhere between amorphous and crystalline states. Demonstration of this phenomenon of line broadening has been obtained with boron, boron carbide, and in part with silicon carbide. These materials have been formed by various methods: pyrolytic decomposition of carborane, low pressure electrical discharge in diborane, chemical reaction between boron halide and hydrogen.

INTRODUCTION

The Air Force is concerned with refractory materials and their mechanical strength at elevated temperatures. Recent work includes a study of matrices containing refractory fiber substructures. Some of the components of these fibers are boron carbide, boron, and silicon carbide. These components are being formed as coatings on a heated substrate, such as tungsten wire, by pyrolysis of gaseous mixtures. In the process of an X-ray

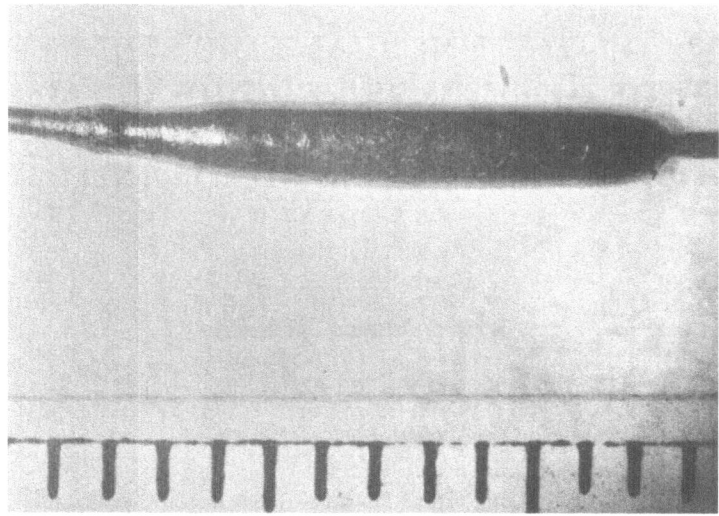

Figure 1. Boron carbide deposited on tungsten wire (each scalar division equals 1 mm).

examination of boron carbide, some very unusual diffraction patterns have been obtained, quite similar to patterns very recently mentioned in the literature for boron. A somewhat similar pattern is likewise derived from silicon carbide fiber coatings. The data derived are given below, along with a deduced hypothesis which purports to explain the data heretofore, regarded as perplexing.

EXPERIMENTAL

Chemical

Carborane-10 crystals are placed in a reaction tube beneath 10 mil tungsten wire heated by electrical current to either 1900° or 2000°F while hydrogen gas is passed through the tube at a moderate rate for varying periods from 10–30 min. The resulting samples, typically shown in Figure 1, are then examined by offset X-ray collimation techniques.[1] Offset collimator techniques assist in the derivation of the diffraction pattern of the coating material without interference by lines from the substrate. Silicon carbide was prepared at 2200°F by pyrolysis of methane, silicon tetrachloride, and hydrogen on hot tungsten wire.

Diffraction

Specimens of boron carbide prepared as described above were examined by offset collimation as prepared, many photographs were taken throughout the length of numerous specimens for statistical validation of the data. Several boron carbide specimens showing typical paracrystallinity* were mounted on goniometer heads. Layer line type photographs were taken, using offset collimation techniques, on these specimens

---

* Paracrystalline is a descriptive term that has been coined from the Greek prefix "para" meaning new, beyond, or aside from, coupled with crystalline to denote the unusual broad line phenomenon, described in this paper, which represents neither crystalline nor amorphous matter in normal, powder pattern crystallography.

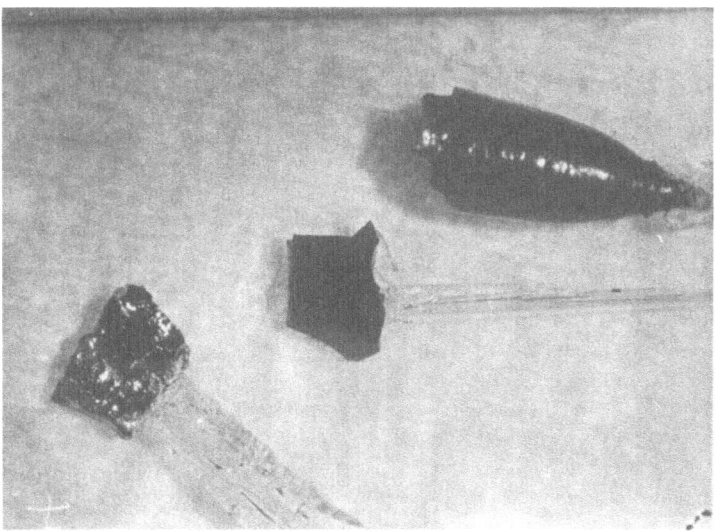

Figure 2. Fragmentary B_4C specimens prepared for X-ray diffraction.

positioned at many different angles to determine the orientation present and the effects thereof. Specimens prepared at two different temperatures were then fragmented by either breaking or grinding. Each grinding operation consisted of a single blow by a hammer on a specimen loaded in a Plattner diamond mortar. Fragments obtained by impact grinding were studied both individually and in groups. Fragments obtained by breaking were examined at the break edge and at varying distances away from the break. Typical fragments prepared for examination by X-ray diffraction are shown in Figure 2. All cases demonstrating paracrystallinity by X-ray diffraction were further examined by electron diffraction for an attempted correlation. A few silicon carbide specimens also received preliminary X-ray examination according to the same procedures used for B_4C as described above.

RESULTS AND CALCULATIONS

A typical paracrystalline X-ray diffraction pattern for boron carbide is shown in Figure 3 and only three to four broad bands are demonstrated as seen in the densitometer scan in Figure 4. Transmission electron diffraction is shown in Figure 5, fully supporting the X-ray data. Numerous repetitious runs on specimens prepared under somewhat differing conditions always showed the same X-ray and electron diffraction pattern for boron carbide. Electron diffraction indicated that there are apparently small fragments of crystalline boron carbide in the paracrystalline specimens, shown by infrequent spot patterns derived as the filament is scanned. Boron carbide specimens tilted at 0°, 15°, and 30° on either or both X and Y goniometer arcs and then remounted at 90° for a repetition of angular studies, showed no difference in the paracrystalline pattern for any angular position, indicating that there was no orientation measurable.

All paracrystalline type patterns derived from boron carbide showed identical d spacings and line width regardless of preparative methods or temperature of preparation, and were quite similar to patterns derived by others for boron filaments,[2-6] a typical

Figure 3. X-ray diffraction pattern taken by offset collimation on specimen shown in Figure 1.

boron pattern derived by Lockheed is shown in Figure 6. Table I shows a comparison between boron and boron carbide with respect to paracrystalline d spacings and corresponding body-centered cubic indices. The slide rule technique,[7] used for indexing both boron and boron carbide lines, showed a lattice parameter of 5.88 Å for B_4C, and an average lattice parameter for boron of 6.25 Å.

Boron and boron carbide line widths are approximately equal. The calculation of apparent particle size from X-ray data for B_4C is given by

$$t = \frac{K\lambda}{\beta \cos \theta}$$

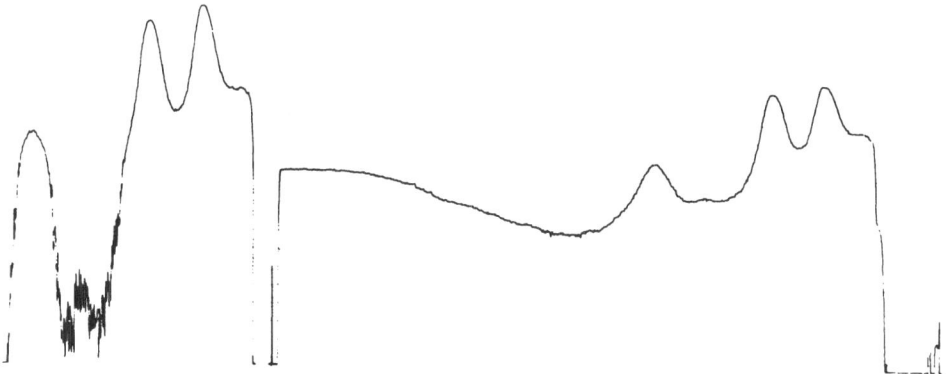

Figure 4. Densitometer scan of X-ray diffraction pattern shown in Figure 3, with increased sensitivity introduced to demonstrate a fourth peak in the partial second scan shown on the left.

where t is the particle thickness in angstroms, K is the shape factor taken equal to 0.9, β is the line broadening in radians, θ is the Bragg angle, and λ is the wavelength:

$$t = \frac{(0.9)\,(1.542)}{(0.0437)\,(0.9467)} = 34\,\text{Å} \qquad \text{for B}_4\text{C}$$

For comparison, boron shows $t = 45\text{Å}$ and 20Å for X-ray and electron diffraction calculation, respectively.[2] In summary, particle size calculations, based on line widths, would indicate that boron and boron carbide crystallites average 35–40 Å in diameter. Insofar as both boron and boron carbide can be indexed as body centered cubic in the paracrystalline condition, the broadness of the paracrystalline lines for both boron and boron carbide are essentially equivalent, and the average lattice parameters for paracrystalline boron and boron carbide are, respectively, 6.25 and 5.88 Å, it is postulated that the two paracrystalline materials have similar structures—the difference in lattice parameters is considered to be probably the result of the smaller atomic radius for four bonded carbon in boron carbide as compared to three bonded boron in both boron and boron carbide, while both the unbonded boron and carbon atoms are said to have equal atomic radii of 0.77 Å.[8-9]

Table II shows the results of average strain and mean stress calculations[10] from boron carbide data for the diffraction pattern shown in Figure 3. Average strain $e = (\beta/4)\cot\theta$ and mean stress $\bar{\sigma} = Ee$ where E (Young's modulus) is 66×10^6 for B_4C,[11] β is the line broadening expressed in radians and θ is Bragg's angle. The exceptionally high mean stress figures and the variation in strain and stress with θ should be noted.

To test the effects of modest changes in paracrystalline specimen character such as those resulting from singular breaks, boron carbide specimens, such as shown in Figure 1, were broken by hand. Those specimens formed at 1900°F show no change at the break edge after breaking, still yielding the four broad bands; in contrast, those specimens formed at 2000°F show normal crystallinity after breaking, identifiable as hexagonal B_4C, at the break edge, but still retaining the abnormal band structure a very short distance away from the break.

Boron carbide specimens, placed in a Plattner diamond mortar and given one blow with a hammer, develop a regular hexagonal B_4C crystalline structure. A statistical study showed that there are still a considerable number of pieces which exhibit the body

Figure 5. Transmission electron diffraction pattern of sample shown in Figure 1.

Table I. Comparison of Boron Carbide and Boron d Spacings and bcc Indices

Boron carbide		Boron					
		Lockheed		Otte & Lipsitt		G. E.	
d	hkl	d	hkl	d	hkl	d	hkl
4.16	110	4.43	110	4.35	110	4.30	110
2.42	211	2.55	211	2.56	211	2.50	211
1.70	222	1.74	222	1.73	222	1.75	222
1.38	411	1.42	411	1.42	411	1.40	411

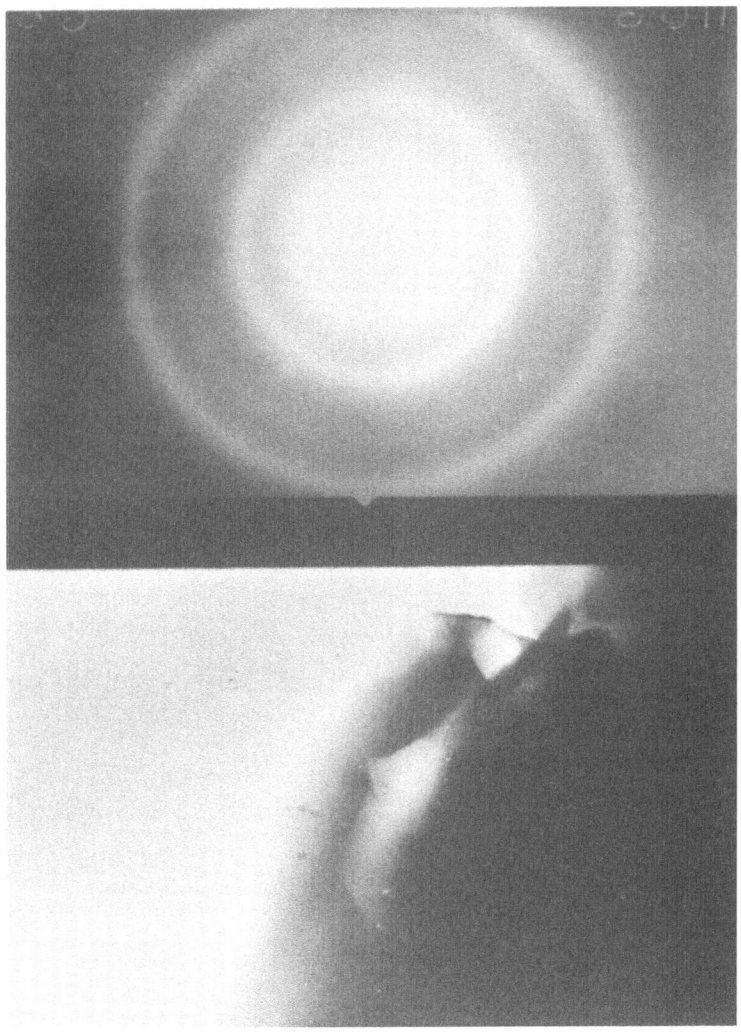

Figure 6. Electron diffraction pattern of chemically thinned boron filament, filament image shown at bottom.

Table II. Stress and Strain Calculation Data for Boron Carbide

2θ	Degrees line breadth, 2θ	θ	Corrected for normal line breadth $(0.15°\theta)$	Radians	Average strain, e	Mean stress, $\bar{\sigma}$
21.4	5.4	2.7	2.55	0.0445	0.0585	3.6×10^6
36.9	7.0	3.5	3.35	0.0585	0.0438	2.7×10^6
68.2	6.3	3.15	3.00	0.0524	0.0192	1.2×10^6

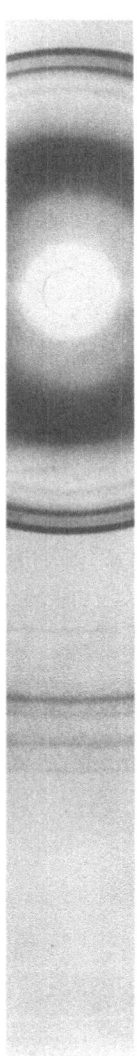

Figure 7. X-ray diffraction pattern of B_4C crushed by single impact in Plattner diamond mortar.

centered cubic broad band structure. The overall volume average, though, is predominantly hexagonal boron carbide, as shown in Figure 7.

Silicon carbide (Figure 8) exhibits six bands considerably less broad than those found for boron or boron carbide discussed above. These six bands are all shown to be face centered cubic β-silicon carbide. Earlier work with silicon carbide indicates that a higher temperature of formation can generate two-dimensionality for certain lattice planes.[1]

EXPLANATION AND THEORY

Although there appears to be no explanation based on existing X-ray diffraction theory for the broad bands described in this work, the most prominent characteristics

Figure 8. β-Silicon carbide filament formed at 2200°F, exhibiting intermediate broad band structure.

connecting boron and boron carbide paracrystalline diffraction patterns is the constancy of the band width and position. The constancy and reproducibility strongly suggest a distinct thermodynamic well in potential energy similar to that found for regularly crystalline materials. Various explanations and hypotheses will be discussed below.

Particle Size

An attempt to explain the broadness of the bands by particle size broadening furnishes a particle size of approximately 35 Å. This explanation might prove satisfactory if one were not faced with two pieces of data which conflict with the concept of particle size being the explanation.

1. The specimen developed a normal diffraction pattern upon single impact grinding, indicating that rapid crystal growth had occurred if small particle size accounted for the line broadening. The sharpness of the lines developed indicates a particle size of 1000 Å or larger. This represents instantaneous crystal growth, to the extent of a 2500% increase in size in less than a second from 35 Å to over 1000 Å, if crystal growth is being observed. A change of this extreme degree might best be considered doubtful unless further proven.

2. Heating boron which exhibits paracrystallinity to 1000°C for 50 hr causes no change in the paracrystalline pattern.[5] For comparison, other small particle materials, with apparently equivalent, or even greater, particle size, change by recrystallization to larger particles at much lower temperature, even if their absolute melting point is higher than that of boron (M.P. 2040°C). For example, 50 Å ZrO_2 (M.P. 2700°C) changes to grossly crystalline from X-ray amorphous (average 50 Å by electron diffraction) after heating 150 hr at only 305°C or 24 hours at 400°C.[12] Also amorphous Al_2O_3 (M.P. ~ 2000°C) with a particle size of approximately 30 Å turned crystalline at 100°C,[13] while pure finely divided iron sinters (with accompanying crystal growth) at temperatures slightly over 300°C.[14] The above examples along with numerous others in the literature suggest that a particle size of approximately 35 Å would not be likely to withstand 1000°C for 50 hr without becoming noticeably more crystalline.

Since crystal growth and recrystallization are normally directly temperature or energy dependent, it would seem that the level of energy introduced by fracture of a brittle material would be noticeably less than that available at 1000°C, thereby indicating that the development of crystallinity through fracture, as described herein, is not a true growth of crystals or recrystallization in the commonly understood sense.

Strain

Strain calculations show a variation by a factor of three, with change in θ on the diffraction pattern, indicating that some other variable is more likely to be the determining factor. The stress found to be apparently present from these calculations exceeds one million which appears to be too large by a factor of ten to one hundred.

Dislocations

If the broad bands were caused by a high density of dislocations, partial or complete annealing out of the dislocations, particularly if the dislocation density were sufficiently high to cause the extreme broadening found in paracrystalline specimens, would be expected to occur if the samples were heated at 1000°C for 50 hr, particularly since 1000°C equals 55% of the absolute M.P. Such annealing would result in a sharpening of the diffraction lines, but the pattern does not change upon heating to 1000°C, in the case of boron. Boron carbide has not yet been similarly tested. Therefore, it is improbable that dislocations are responsible for the broadness of the bands found in paracrystallinity.

With regard to grinding and dislocations, it is generally understood that standard grinding procedures would increase the number of dislocations, thereby broadening the already broad lines. Therefore, it would appear illogical to relate the extreme line sharpening resulting from single impact grinding to a lessening of dislocations. On the contrary, the fact that sharpening, rather than broadening, occurs during grinding, or breaking, serves as evidence that the broadness of the lines is not caused by dislocations.

Peak Overlap

An attempt has been made to integrate the standard diffraction pattern for either α or β crystalline boron into the abnormal broad band paracrystalline diffraction pattern.[3]

The most noticeable discrepancy between this theoretically integrated crystalline pattern and the measured broad bands discussed herein is the shape of the peaks. The integrated peaks are required to tail, similar to a two-dimensional structure, while the measured peaks are perfectly symmetrical. See Figure 4 for boron carbide. A second discrepancy is present—there is no proper correspondence between the broad peak and crystalline peak locations throughout the pattern. It is believed by the authors that peak integration is an unsatisfactory explanation.

Curved Crystal

If a set of bent lattice planes in polycrystalline material, randomly oriented, is considered to develop this broad band structure and, further, it is postulated that the constraints which make the bent planes possible are released upon breaking or grinding, the data seems to be satisfied. The randomly oriented bent lattice planes may be the result of the pyrolytic decomposition preparative methods.

This curved crystal concept is best illustrated in the case of the bent mica-crystal monochromators used in many X-ray laboratories. These bent crystals allow a diffraction image to be expanded to any desired broadness through proper control of crystal curvature by mechanical means. To differentiate this concept from the strain concept discussed above, it is necessary to note that bending of these mica crystals does not require abnormal strain introduction.

As an interesting laboratory experiment, to develop the bent mica-crystal concept along the line of randomization, thin, ground mica flakes, spread out on a surface could be sprayed with a material which would shrink on drying, have good adhesion to mica, but would yield a negligible diffraction pattern. Then these bent mica flakes could be randomly mounted on a greased spline to see if this synthesized system would not approximate the broad band structure shown by boron carbide, boron and silicon carbide.

CONCLUSION

Boron carbide, similar to boron, has been found to yield an unusually broad lined diffraction pattern which is difficult to explain. It is indicated by data and calculation that stress, particle size, and dislocation theory are not very satisfactory explanations, and that randomly oriented bent lattice planes may be another explanation.

ACKNOWLEDGMENTS

The authors wish to express their sincere gratitude to: Lockheed Missile and Space Company, Palo Alto, California; General Electric Company, Valley Forge Space Technology Center, Philadelphia, Pennsylvania; Martin Company, Materials Research Laboratory, Orlando, Florida; Dr. W. F. Stuhrke and Mr. R. M. Neff of the Air Force Materials Laboratory, Wright–Patterson AFB; Dr. Harry A. Lipsitt of the Metallurgy and Ceramics Research Laboratory, Wright–Patterson AFB; and to the Technical Photographic Division of Wright–Patterson AFB.

REFERENCES

1. R. L. Prickett and R. L. Hough, "Generation of Two Dimensional Silicon Carbide Lattice," in *Advances in X-Ray Analysis*, Vol. 9, Plenum Press, New York, 1966, p. 144.
2. Lockheed Missile and Space Company, "Research to Study the Structure of Non-Crystalline Boron," Contract AF33(615)-3140, third quarter progress report, April 1966, p. 13, 37.
3. Lockheed Missile and Space Company, "Research to Study the Structure of Non-Crystalline Boron," Contract AF33(615)-3140, second quarter progress report, January 1966, p. 13.
4. H. A. Lipsitt, "A Study of Boron Filaments," Metallurgy and Ceramics Research Laboratory, Wright–Patterson AFB, 1965.

5. H. M. Otte and H. A. Lipsitt, "On the Interpretation of Electron Diffraction Patterns from 'Amorphous' Boron," *Phy. Stat. Solidi* **13**: 439, 1966.
6. General Electric Company, "Research in Improved High Modulus, High Strength Filament and Composites Thereof," *Tech. Rep.* AFML-TR-65-319, September 1965, pp. 129 and 159.
7. A. Guinier, *X-Ray Crystallographic Technology*, Hilger and Watts, Ltd., London, 1952, p. 107.
8. M. C. Neuburger, "Gitterkonstanten fur das Jahr, 1936," *Z. Kristallographie Mineralogie* **93**: 1, 1936.
9. A. F. Wells, *Structural Inorganic Chemistry*, Oxford, London, 1962, pp. 55 and 849.
10. A. Taylor, *X-Ray Metallography*, John Wiley and Sons, Inc., New York, 1961, p. 786.
11. General Technologies Corporation, Air Force Materials Laboratory, *Tech. Rep.* 65–265.
12. K. S. Mazdiyasni, C. T. Lynch, and J. S. Smith, "Meta-Stable Transitions of Zirconium Oxide Obtained from Decomposition of Alkoxides," *J. Am. Ceram. Soc.* **49**: 286, 1966.
13. J. W. Newsome, H. W. Heiser, A. S. Russell, and H. C. Stumpf, *Alumina Properties*, Aluminum Company of America, Pittsburgh, 1960, p. 70.
14. W. G. Frankenburg, V. I. Komarewsky, and E. K. Rideal (eds.), *Advances in Catalysis*, Vol. 2, Academic Press, New York, 1950, p. 100.

DISCUSSION

P. A. Beck (University of Illinois): I think you mentioned that in order to explain this very large line broadening by strains, you would have to assume unreasonably large strains. But, the line broadening effect of the "curved crystals" you propose is also due to "strain" and to "particle size," is it not?

R. L. Prickett: The curved crystals should not have excessively large strains inherent just because they are curved. The strains calculated shown in this paper are too large to exist according to normal crystallographic theory.

G. A. Walker (The 3-M Company): In your calculation of crystallite size and strain you assume the total width was caused by either effect. Couldn't it be a coexistence of small crystallite size, strain and stacking faults? My question would be, did you do a total line analysis?

R. L. Prickett: Let us assume that stacking faults were present. Let's further assume that something happens to the stacking fault at the time the hammer impacts. By definition, work must be introduced to do this. A brittle material is almost unable to receive work in a single impact, so I would say that the stacking fault number before and after should be essentially equivalent. Therefore, if I am able to generate a diffraction pattern which is crystalline from a noncrystalline or paracrystalline material by a hammer blow, stacking fault effects must be subtracted from the picture. If it is not subtracted, then how can it be said to have had work done on it when it violates physical principles? Does that answer your question?

G. A. Walker: I just don't see why they couldn't all coexist.

R. L. Prickett: I will agree they can all exist, but if it is small particle size, then how do you grow big ones out of little ones when you hit it with a hammer? If it is defect structure, how do you decrease the defect structure with a hammer? These are neither one ameliorated or made more crystalline by a hammer. Therefore, they must drop out of the picture.

H. W. Pickett (General Electric, X-Ray): I think we are agreed that these are a fascinating collection of observations. I should like to suggest a couple of continuing experiments which might relate to verifying your present—let's call it tentative—explanation. You could very well deposit the material on a fine, flat wire rather than one of cylindrical section and then see what you can find in relation to that part of the material that is deposited on the flat side as against around the edge.

R. L. Prickett: This is helpful.

H. W. Pickett: The other suggestion is that the shape of the samples as they are would easily lend itself to some small angle scattering, and you could evaluate the shape of these curves before and after the shock treatment. Another approach would be to induce the equivalent of the mechanical shock by sudden thermal shock. You could dump a few of these in liquid nitrogen and see if you can shock them enough to develop the sharper lines without actually shocking them off the wire. Or you could attempt to freeze the wire fast enough so it shrinks away from the inside, and so on. I think later we will have a chance to correlate some of this with the effects observed by Frank Karioris in his extreme thermal gradients in powders from "exploded" wires.

H. J. Garrett (Aerospace Research Labs.): I'd like to add a few observations to this and also what would appear to be some contradictions. In working with 10 mil diameter boron filaments, we found radial cracks down the entire length, i.e., the whole filament was split. The cracks can be explained in terms of the peripheral strains or stresses in growth. When we grind this material, we do not see crystalline patterns in X-ray diffraction photograms. When we examine some of the same material by electron microscopy, we find that some of the particles are thin enough to give beautiful electron interference spot patterns in hexagonal arrays. The problem is that we can't index the spots. The crystal structure seems to be some kind of hybrid hexagonal and body-centered. If we say it is body-centered, then we have missing reflections, if we say it is hexagonal, we have too many. Lipsitt and Henry Otte were able to demonstrate by tilting the crystals that very large relrods are present, as one might expect in a light element of this kind. It would appear that we have some very definite stacking fault array here. This is our explanation for not being able to get an X-ray diffraction pattern from ground, crushed boron filaments.

H. K. Herglotz (E. I. du Pont de Nemours and Company): I just would like to add something from the very old literature. In the 1930's, several publications about explosive antimony were written. Explosive antimony does not even need a hammer blow to be converted from the amorphous to the crystalline state. Scratching suffices to initiate the conversion, which propagates with explosive speed. There are also several publications about crystallization of oxides under emission of light due to the heat of crystallization (J. Böhm, *Z. Anorg. Anal. Chem.* **149**, 217, 1925). This seems to be something very similar, again requiring less than a hammer blow.

Then, I would also like to add something to another subject, namely, to the argument about the curved crystals and the tremendous stresses. If you deal with disorder on an atomic level, it is unrealistic to describe the forces required to displace atoms from their regular lattice sites by a stress in pounds per square inch. Stress figures become astronomical if one transforms distances and areas of an atomic scale into square inches.

H. W. Pickett: In response to remarks made by H. K. Herglotz (from E. I. du Pont de Nemours and Company) regarding explosive antimony, with which the authors were unfamiliar, a check of the literature shows that explosive antimony does have exceedingly interesting properties but also does not appear to be too similar to the materials under discussion in this paper. The pictures shown by Cohen and Ringer* for explosive antimony are similar to those for boron carbide in Figure 1, but the diffraction pattern derived by Böhm† shows only amorphousness.

It is noted that explosive antimony becomes crystalline with the evolution of heat and only with an appreciable lapse of time, similar to pyrophoric materials. In contrast, single impact grinding of boron carbide is essentially instantaneous and apparently nearly adiabatic.

---

* E. Cohen and W. Ringer, "Physikalisch-chemische Studien am sogenannten explosiven Antimon," *Z. Physikal Chem.* **47**: 1, 1904.

† J. Böhm, *Z. Anorg. Chem.* **149**: 217, 1925.

THE EXPANSION UPON COOLING OF THICK Cu$_2$O FILMS GROWN ON COPPER SUBSTRATES

T. F. Swank and K. R. Lawless

University of Virginia
Charlottesville, Virginia

ABSTRACT

During the course of catalytic experiments on bulk copper single crystals, several crystals were intentionally oxidized to form thick (6000 Å) Cu$_2$O films on the copper substrates. These oxidized crystals were investigated by means of a high temperature chamber installed on a General Electric XRD-5 X-ray diffractometer. It was discovered subsequently that the lattice spacings of the Cu$_2$O decreased upon heating and increased upon cooling. Bulk single crystals and polycrystals of copper were oxidized at 3 Torr of air for several hours. All of the oxidized crystals were examined with copper and chromium radiation and both showed similar results. Typical of the results was an oxidized (110) copper disc which showed a net contraction upon heating of 1% for the (110) Cu$_2$O planes. This slightly oriented Cu$_2$O film was distinguished because it contracted on heating to 440°C from room temperature, then expanded from 440°–540°C, and then expanded again when cooled from 540°C to room temperature. CuO also was detected in the diffraction pattern and the CuO and copper spacings were behaving normally with the temperature changes. A polycrystal of Cu$_2$O was examined and that, of course, also acted normally as its temperature was varied. Borie and co-workers have reported and explained very nicely similar anomalous behavior for thinner (500 Å) (110) oriented Cu$_2$O films grown on (110) copper substrates. They showed that the epitaxial forces would cause an oxide film grown at high temperatures to contract parallel to the metal interface and expand normal to the interface as the copper cools and contracts. The oxide would expand normal to the surface in order to keep its unit cell volume constant.

It is felt that epitaxial forces are not causing the anomalous behavior in the present work mainly because the 6000 Å of Cu$_2$O is too thick for epitaxy to exert a meaningful force. The oxide film on the (110) copper was slightly (110) oriented but all of the Cu$_2$O reflections behaved similarly. An additional reason to discount epitaxy is that this Cu$_2$O film expanded upon heating from 440°–540°C. For these thick oxide films epitaxial forces do not seem to be the controlling factor; therefore, a point defect mechanism must be the cause. Changing oxidation and diffusion rates with temperature would produce various vacancy concentrations in the oxide layer and cause the spacings to vary.

INTRODUCTION

During a research program designed to study the surface species produced on bulk copper single crystals during catalytic reactions it was discovered that the lattice parameter of thick Cu$_2$O layers grown on copper substrates did not behave normally as the temperature changed. This observation warranted more work and it subsequently was verified that the Cu$_2$O structure contracted when heated and expanded when cooled. This paper will describe these experiments and compare the conclusions reached with these thick films with those reached by other authors concerning thin Cu$_2$O films.

Borie et al.[1,2] have described their observations of the lattice parameters of thin (100–500 Å) (110) oriented Cu_2O films grown on (110) faces of copper single crystals. They concluded that the (110) spacings of the films are greater than that of the bulk oxide and that the unit cell of the Cu_2O is distorted because of the epitaxial force. These epitaxial forces would make a Cu_2O film grown on copper at high temperatures contract parallel to the metal interface as the copper cools and contracts. Therefore, the oxide would expand normal to the metal surface in order to keep its unit cell volume constant.

While Borie's mechanism might well be the controlling factor for thin films, the present research will attempt to show that epitaxial forces cannot account for the abnormal thermal behavior of thick Cu_2O films grown on copper.

EXPERIMENTAL PROCEDURES

Single crystal discs were obtained by machining $\frac{3}{16}$-in.-thick and $\frac{3}{4}$-in.-diameter slices from a (110) oriented copper (99.999% Cu) single crystal rod. After etching, the discs were mechanically polished and then electrolytically polished in phosphoric acid by procedures previously described.[3] A polycrystalline disc was prepared similarly from an OFHC crystal.

The oxidations were performed in a special high temperature chamber mounted on a GE XRD-5 diffractometer. The basic design of this chamber is similar to that of Intrater.[4] Its most important features were: (a) external alignment controls, (b) an 180° beryllium window, (c) a temperature range to 600°C and (d) O-ring seals so the chamber could be evacuated and the atmosphere about the crystal controlled.

The copper crystals were oxidized in 2 Torr of air at the various temperatures ranging from 350°C–550°C for at least 1 hr prior to the initial X-ray observations. The oxidations were necessarily continued while the diffraction patterns were being recorded at the higher temperatures. During these oxidations the crystal temperature varied ±4°C in very slow cycles. Copper and chromium radiations were used.

RESULTS

The anomaly of expansion during cooling of the Cu_2O layer was first noted accidently during a routine experiment with an oxidized crystal. This observation led to more careful experiments designed to observe more closely the heating-cooling behavior.

The polycrystalline disc was oxidized at 390°C, cooled to room temperature, reheated to 450°C, and then recooled; diffraction patterns were taken with copper and chromium radiations at all four temperature steps. Table I gives the data obtained for

Table I. d Spacings for Oxidized Polycrystal Chromium Radiation

390°C	Room temperature	450°C	Room temperature	Bulk room temperature values	Cause
2.526	2.523	2.526	2.522	2.530, 2.523	002 and $\bar{1}11$ CuO
2.467	2.475	2.465	2.477	2.465	111 Cu_2O
2.332	2.328	2.336	2.325	2.323, 2.312	111 and 200 CuO
2.142	2.151	2.139	2.154	2.135	200 Cu_2O
2.096	2.085	2.101	2.085	2.088	111 Cu
1.817	1.807	1.821	1.807	1.808	200 Cu
1.512	1.516	1.510	1.517	1.510	220 Cu_2O

Table II. *d* Spacings for Oxidized (110) Single Crystal Copper Radiation

Room temperature	440°C	540°C	Room temperature	Bulk room temperature values	Cause
3.025	3.015	3.020	3.052	3.020	110 Cu_2O
2.520	2.526	2.540	2.526	2.530, 2.523	002 and $\bar{1}11$ CuO
2.463	2.466	2.469	2.493	2.465	111 Cu_2O
2.322	2.336	2.336	2.325	2.323, 2.312	111 and 200 CuO
2.144	2.137	2.140	2.164	2.135	200 Cu_2O
—	1.863	1.865	1.868	1.866	$\bar{2}02$ CuO
1.513	1.510	1.511	1.528	1.510	220 Cu_2O

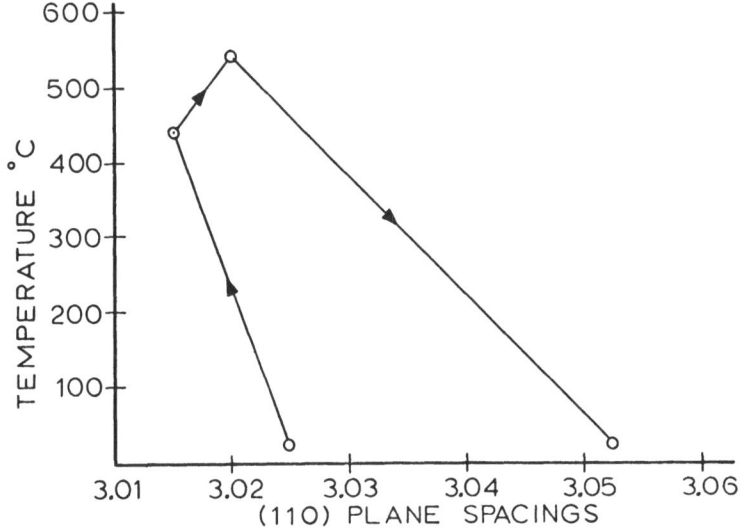

Figure 1. Variation of Cu_2O (110) plane spacings with temperature.

Cu_2O using chromium radiation. The *d* spacings for copper and CuO are also included to show that they expanded and contracted normally with temperature.

The oxidation at 390° had proceeded for 20 hr before the diffraction pattern was taken. The total oxidation time for the 450° pattern was about 24 hr (broken, of course, by the cooling step). Neither the oxides nor the metal showed any preferred orientation.

The peak shapes of the copper and CuO reflections remained sharp throughout the experiment. However, the Cu_2O peaks broadened slightly upon cooling and then sharpened again with heating. The (111) Cu_2O reflections broadened the least.

In another experiment the (110) oriented copper disc was oxidized under the same vacuum as above, at 440°C and then 540°C. It was examined by copper and chromium radiation initially at room temperature, at 440°C, 540°C and again at room temperature after cooling. It was oxidized at 440°C for three hours and for an additional 2 hr at 540°C before recording the respective patterns. The copper radiation results are shown in Table II. The chromium results are essentially identical. All of the original room temperature Cu_2O lines were a trifle broad, but they sharpened on heating to 440°C and stayed sharp throughout heating and cooling. As before, the CuO line profiles stayed sharp, and its structure expanded and contracted normally. It should be noted in

Table II that change in the d spacings of the major planes of Cu_2O varied from 0.024 Å–0.032 Å and that both radiations indicated that the d spacings of Cu_2O increased on heating from 440°C–540°C. Figure 1 shows the variation of (110) plane spacings with temperature. The large increase in plane spacings for this crystal on cooling from 540°C would indicate an expansion coefficient of 5×10^{-5} for the (111) and (200) planes and 6×10^{-5} for the (110) planes of Cu_2O.

At the beginning of the oxidation (R.T. and 440°C), the Cu_2O was somewhat (110) oriented but as the oxidation progressed this slight orientation essentially disappeared.

Another experiment involved the use of a completely oxidized single crystal of copper and this was used to check the behavior of bulk Cu_2O. This disc was principally a polycrystal of Cu_2O with a small amount of CuO on its surface. Upon heating this crystal in an oxidizing atmosphere at 550°C, it was found that the Cu_2O and CuO expanded and contracted normally. X-ray examination at room temperature proved that both oxides had almost exactly the standard lattice parameters. The expansion coefficient for Cu_2O was calculated and the coefficient at 550°C was 4.5×10^{-6} Å/°C. This figure was obtained by using the 311 Cu_2O reflection of chromium radiation which occurred at about 125.5° (2θ). It was calculated that the expansion coefficients of monoclinic CuO are $\alpha = 6.6 \times 10^{-6}$, $\beta = 3.5 \times 10^{-5}$, and $\gamma = 2.5 \times 10^{-5}$ Å/°C.

X-ray diffraction measurements of a crystal in a vertical position, such as the above, determine the spacings of planes parallel to the crystal surface. In an attempt to determine the lattice parameters for planes not parallel to the metal surface, the partially oxidized single crystal was tilted 40° and examined with X-rays. This tilting defocuses the diffractometer because sections of the crystal are inside and outside the focusing circle of the diffractometer. Such a misalignment resulted in inconclusive data because the peak position varied greatly with the exact translation alignment.

DISCUSSION

It is estimated that the total thickness of the Cu_2O on the oxidized crystals was at least 6000 Å. This figure was obtained from observations that the films had grown through the interference fringe series and had assumed the color of bulk Cu_2O. The CuO was probably about 1000 Å thick. This would make the films at least 6000–7000 Å thick. The relative oxide-metal peak intensities for both oxidations indicate the films are greater than 1 μ thick.

The results of this work indicate that epitaxial forces are probably not responsible for the peculiarities of the oxide layer for three reasons: (a) the thickness of the layer, (b) the polycrystalline nature of the layer, and (c) expansion of the oxide between 440°C and 540°C.

The first reason is obvious because epitaxial forces would not be expected to operate through 6000 Å of oxide. Even if there were a perfectly oriented film, there is no indication that the metal in contracting or expanding could influence oxide planes 6000 Å away. It is hard to estimate exactly how far the metal force could be extended into the oxide before it became negligible, but it is generally accepted that 6000 Å would be too great a distance over which to exert epitaxial forces.

The fact that polycrystalline layers were grown on the two oxidized crystals was proven, and all of the oxide orientations (110, 111, and 200) exhibited roughly the same percentage of expansion and contraction for any one temperature change. For all the oxide orientations to be affected by epitaxial forces, they would have to be present in individual grains which extended through several thousand angstroms toward the metal interface; and this is considered unlikely.

The observation that the spacings of all the oxide planes expanded on heating from 440°C–540°C also indicates that epitaxial forces are not controlling. If they were, then a contraction of the plane spacings normal to the surface should have been observed upon further heating of the crystal to 540°C, if the Borie thin film mechanism were in effect.

If we proceed to a macro scale and consider the case of the bimetallic strip as applying to our situation, then we could consider how this would bring about the expansion and contraction. The expansion coefficient of copper is almost 20 times greater than that of cuprous oxide.[5] Therefore, it can easily be understood how, if the copper expanded with heating, it would stretch the oxide parallel to the metal surface and thereby tend to shrink it normal to the copper surface. However, if this were happening with our thick oxide layer, the cupric oxide should behave the same as the cuprous oxide and this is contrary to our results.

Since the above reasons indicate strongly that epitaxy is not the cause, these changes may be brought on by variations in the point-defect structure of the oxide. Changing oxidation and diffusion rates with temperature would produce various vacancy concentrations in the oxide layer and cause the plane spacings to vary. The experiment with the bulk Cu_2O crystal proved that the copper substrate played an important role as a source of copper atoms. The exact mechanism by which the vacancy changes occur is not understood at this time.

There is evidence that large amounts of defects can form at lower temperatures during oxidation of copper. Wieder and Czanderna[6] and O'Keefe and Stone[7] observed that low temperature ($< 200°C$) and low pressure oxidations produced large supersaturated concentrations of vacancies which did not agglomerate because of the low temperatures involved.

The fact that, for any one temperature change, roughly the same percent deviation occurs for the major Cu_2O orientations signifies that the cause of the expansions and contractions is isotropic. This lends substance to the vacancy mechanism, since such a mechanism would be isotropic in effect. A lattice dilatation of 1% or less has been observed in these studies and this is in agreement with studies indicating that an expansion of 1% would be expected with the presence of vacancies in semiconductors.[8]

Following this viewpoint further, it also would be expected that the exact stoichiometry of Cu_2O would change as the oxidation progresses and the defect concentration changed. This would be important for any other chemical reactions such as in catalysis because the reactivity as derived from the Fermi energy level of Cu_2O would change as the exact stoichiometry changed.

It is unfortunate that additional experiments were not performed to elaborate on these results and to definitely establish whether the defect mechanism or the epitaxy mechanism was the controlling force. Several experiments have been suggested that would establish the controlling influence on this anomalous expansion and contraction. These would include:

1. Fast temperature changes and rapid X-ray measurements which would reveal the need for a certain time for the point defects to reach their equilibrium concentration. If epitaxy were the controlling factor, an equilibrium time should not be necessary.

2. After an oxidized crystal is cooled to room temperature and the Cu_2O spacings are shown to be greater than bulk material, the oxide should be scraped off the copper substrate and a Debye–Scherrer pattern obtained from it. This would remove any epitaxial forces from the oxide, and prove conclusively whether or not the expansion and contraction is defect controlled or epitaxy controlled.

SUMMARY

It should be made clear that the authors place great faith in the work performed by Borie *et al.* on thin Cu_2O films. However, we have presented significant arguments to show that for thick layers of Cu_2O the epitaxy theory does not appear to hold and, therefore, a defect mechanism should be considered as an explanation for the anomalous behavior.

ACKNOWLEDGMENTS

The authors wish to thank Professor B. E. Warren and Dr. C. J. Sparks for reading the manuscript and offering their interesting comments. This investigation was supported by the Office of Naval Research and a General Motors' Fellowship.

REFERENCES

1. B. Borie, "A Diffraction Measurement of the Structure of Cu_2O Films Grown on Copper," *Acta Cryst.* **13**: 542, 1960.
2. B. Borie, C. J. Sparks, and J. V. Cathcart, "Epitaxially Induced Strains in Cu_2O Films on Copper Single Crystals," *Acta Met.* **10**: 691, 1962.
3. J. B. Wagner and A. T. Gwathmey, "The Formation of Powder and its Dependence on Crystal Face During Catalytic Reactions of Hydrogen and Oxygen on a Single Crystal of Copper," *J. Am. Chem. Soc.* **76**: 390, 1954.
4. J. Intrater, "High Temperature, High Vacuum Diffractometer Attachment," *Rev. Sci. Instr.* **32**: 982, 1961.
5. R. F. Tylecote, "The Adherence of Oxide Scales on Copper," *J. Inst. Metals* **78**: 301, 1950.
6. H. Wieder and A. W. Czanderna, "The Oxidation of Copper Films to $CuO_{0.67}$," *J. Phys. Chem.* **65**: 816, 1962.
7. M. O'Keefe and F. S. Stone, "The Magnetochemistry and Stoichiometry of the Copper–Oxygen System," *Proc. Roy. Soc.* **267**: 501, 1962.
8. R. A. Swalin, *Thermodynamics of Solids*, John Wiley and Sons, Inc., New York, 1962.

NEW RESULTS ON THE IRON–NICKEL EQUILIBRIUM DIAGRAM—THE GAMMA/GAMMA-PLUS-ALPHA BOUNDARY

N. I. Ananthanarayanan

University of Kansas
Lawrence, Kansas

and

R. J. Peavler

State Teachers' College
Kirksville, Missouri

ABSTRACT

Solid state equilibria in iron–nickel alloys containing 0–65 at.% nickel were studied by X-ray diffraction techniques, and the gamma/gamma + alpha boundary has been relocated.

Alloys were prepared by thermal decomposition and hydrogen reduction of mixed-crystal iron–nickel formates and by distilling off mercury from mixed iron–nickel "amalgams," both methods of preparation yielding equilibrium alloys in relatively short time at temperatures below 500°C. Alloys were studied at various temperatures by use of a high-temperature diffraction camera, or at room temperature as quenched from the temperature of preparation.

The study showed that the gamma/gamma + alpha boundary lies at generally higher temperatures than in previous diagrams and is concave downwards instead of upwards. The boundary shows three discontinuities which have been located to lie at approximately (a) 3 at.% nickel and 760°C, (b) 35–43 at.% nickel and 525°C, and (c) 50 at.% nickel and 330°C. These discontinuities indicate the possible existence of three isobaric invariant three-phase equilibria, heretofore, undetected in the iron–nickel system. Further studies covering the entire range of iron–nickel alloy compositions are in progress.

INTRODUCTION

Examination of the literature on the iron–nickel system[1,2] indicates that available phase diagrams for the solid state region, particularly below 500°C, are inadequate. This is because, for the most part, transformations in this system are often extremely sluggish, with the result, that attempts to establish equilibrium by classical anneal methods using specimens from massive alloys are laborious and generally fall short of yielding true equilibrium.

The generally accepted iron–nickel phase diagram is that published by Hansen[2] (Figure 1). It is based primarily on X-ray diffraction data by Owen and Liu,[3] who studied

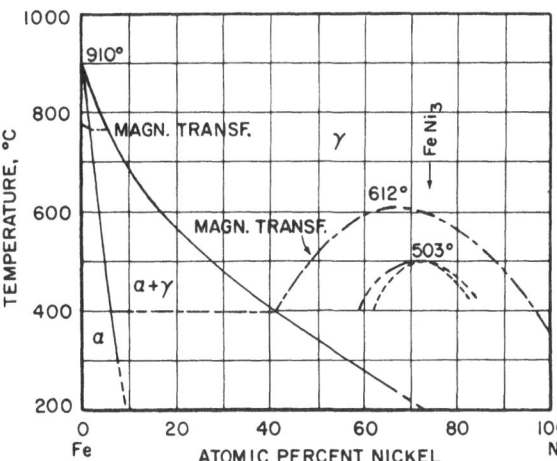

Figure 1. Solid state region of the iron–nickel phase diagram (after Hansen[2]).

alloys in the composition range of 0–65 at.% nickel and on the dilatometric data of Josso,[4,5] who used alloys near the Ni_3Fe composition. Owen and Liu,[3] recognizing the slow approach to equilibrium in this system, annealed their samples for an unprecedentedly long time (more than a year at temperature in some cases) to bring about near-equilibrium states. Their data, used for locating the $\alpha/\gamma + \alpha$ and the $\gamma/\gamma + \alpha$ boundaries, are thus, by far, the best available to date in the composition range studied by them. The significance of the dilatometric data of Josso[4,5] was emphasized upon by Geisler,[6] who pointed out that these data should be considered together with the proposal by Rhines and Newkirk,[7] namely, that the order-disorder transformation should be regarded as a classical phase change and should therefore be represented as such in the phase diagram. Hansen's diagram (Figure 1), while incorporating the most reliable experimental data available, still appears to be far from satisfactory.

Recent development of the mixed-crystal formate reduction as well as amalgamation processes[8–11] for preparing ultra-fine alloy powders, in which equilibria are readily achieved at temperatures even as low as 100°C, offers a new approach to the study of solid state equilibria in the iron–nickel and similar alloy systems with sluggish transformations.

Iron–nickel alloy powders prepared by these new techniques were used for a reexamination of solid state equilibria in the iron–nickel system by X-ray diffraction methods. The present paper deals with only that part of the above study* pertaining to a relocation of the $\gamma/\gamma + \alpha$ boundary.

EXPERIMENTS

Equilibrated iron–nickel alloy powders containing up to ~65% nickel were prepared directly by using the mixed-crystal formate-reduction[9] and amalgam-distillation[11] techniques and were used for X-ray diffraction studies to identify the phases present. Powders by formate-reduction were prepared at 200°C–700°C and over a composition range of up to 59 at.% nickel. Powders by amalgam-distillation were obtained over the

---

* Summary reports made by the authors in the past are available in the Program and Abstracts of (a) The Pittsburgh X-ray Diffraction Conference, Mellon Institute, Pittsburgh, Pennsylvania, 1955, and (b) The Annual Meeting of the American Crystallographic Association, University of Colorado, Boulder, Colorado, 1961.

temperature range of 118°C–450°C and over the composition range 49–62 at.% nickel.

Quenching the alloys to room temperature in benzene served the twofold purpose of eliminating the pyrophoricity of the freshly prepared powders of all compositions, and retaining at room temperature, the structures corresponding to the temperature of preparation for compositions containing more than ~31 at.% nickel. Alloys containing less than 31 at.% nickel were generally susceptible to partial or total martensitic transformation when quenched from temperatures lying in or close to the γ-field. Hence, from a study of these quenched alloys, one may, at best, only infer the true equilibrium structures prevailing at the temperatures of preparation. Room-temperature X-ray diffraction studies of these quenched alloys were, therefore, supplemented by high temperature X-ray diffraction studies as well.

Room-temperature X-ray diffraction patterns were obtained using a Philips Camera of 114.7 mm diameter. High-temperature X-ray diffraction studies were made using an Unicam high-temperature X-ray diffraction camera which had been temperature-calibrated against the thermal expansion of gold. The X-ray source was a high-intensity rotating-anode X-ray generator, the use of which permitted exposure times as short as $\frac{1}{2}$ hr for photographing the diffraction patterns. Quenched alloy samples were reheated to the desired temperatures and allowed to equilibrate at temperature for periods ranging from $\frac{1}{2}$–4 hr, the longest time being used at the lowest holding temperature. It has been found that the fine-particle iron–nickel alloy powders transform readily upon heating. In so far as possible, separate samples were used for each temperature selected for study.

Since alloys containing more than 31 at.% nickel did not transform on quenching, room temperature X-ray diffraction studies were sufficient to obtain the equilibrium information. However, alloys, in the composition range 35–46 at.% nickel, also were studied by high temperature X-ray diffraction.

All of the phase-identification data were assembled and the $\gamma/\gamma + \alpha$ boundary was located by the disappearing phase method.*

RESULTS

Room-temperature X-ray diffraction patterns obtained with a low-nickel iron–nickel alloy (13 at.% nickel) as quenched to room-temperature from preparation temperatures of 695°C and 474°C are shown in Figures 2a and b, respectively. Figures 3a–e are high-temperature X-ray diffraction patterns obtained with the above alloy as quenched from the preparation temperature of 695°C, and as reheated to and held at 487°C, 657°C, 710°C, and 737°C, respectively.

Phase identification data relevant to a relocation of the $\gamma/\gamma + \alpha$ boundary are plotted in Figure 4, in which the relocated $\gamma/\gamma + \alpha$ boundary is also drawn (curve drawn as solid line) together with Hansen's boundary[2] (drawn with dashes and dots) for comparison. The magnetic transformations as well as the order-disorder transformation as given by Hansen[2] (Figure 1) also are shown (drawn as dashed lines). The boundary compositions obtaining at various temperatures are given in Table I together with corresponding data of Owen and Liu[3] used by Hansen (Figure 1).

DISCUSSION

Iron–nickel alloy with 13 at.% nickel serves as a typical example to illustrate the behavior of all alloys in the composition range up to ~31 at.% nickel. When this alloy

---

* Extensive lattice parameter measurements could not be made, but measurements were made whenever feasible. Additional lattice parameter data have been obtained recently. The results are to be considered in a later publication.

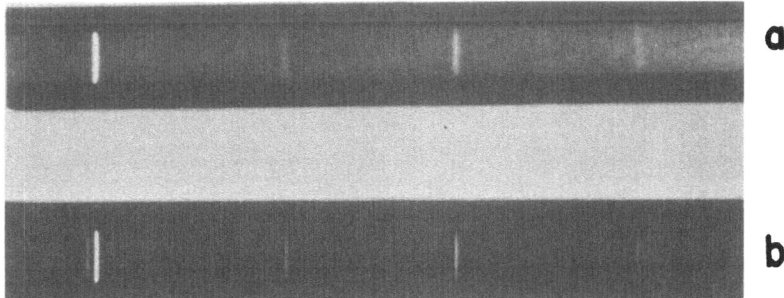

Figure 2. X-ray diffraction patterns (cobalt radiation) of quenched iron–nickel alloy powders (13 at.% nickel) prepared at 695°C and 474°C, showing the martensitic transformation of low-nickel γ-phase and the resistance to transformation of high-nickel γ-phase.

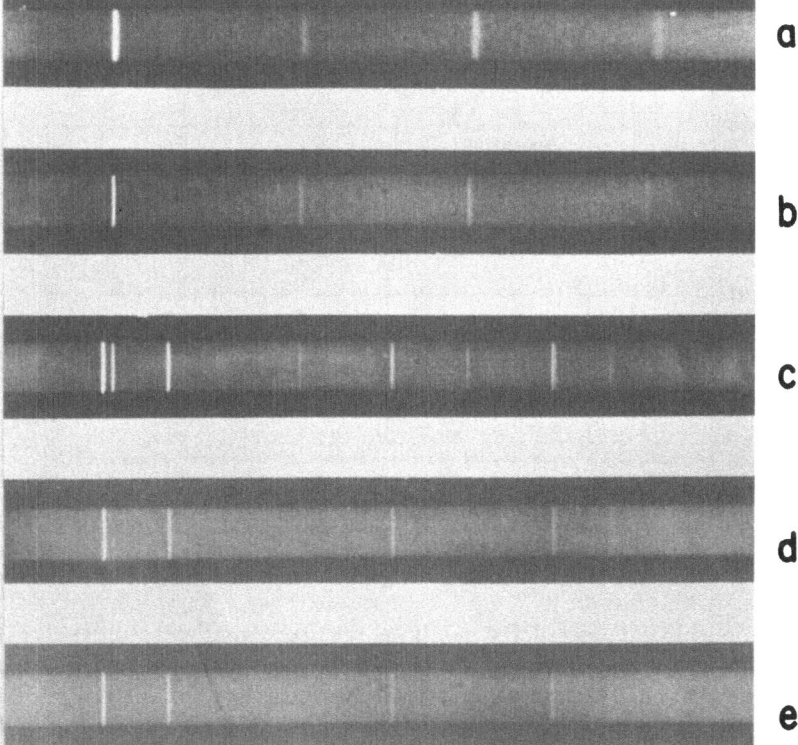

Figure 3. High-temperature X-ray diffraction patterns (cobalt radiation) of an iron–nickel alloy powder (13 at.% nickel) showing the gradual disappearance of the α-phase as temperature is raised.

is prepared by formate-reduction at 695°C and quenched to room-temperature, its X-ray diffraction pattern (Figure 2a) shows the structure to be all martensitic. This pattern does not permit one to determine, unambiguously, the equilibria that prevailed in the alloy at the preparation-temperature because the broad diffraction lines, characteristic of

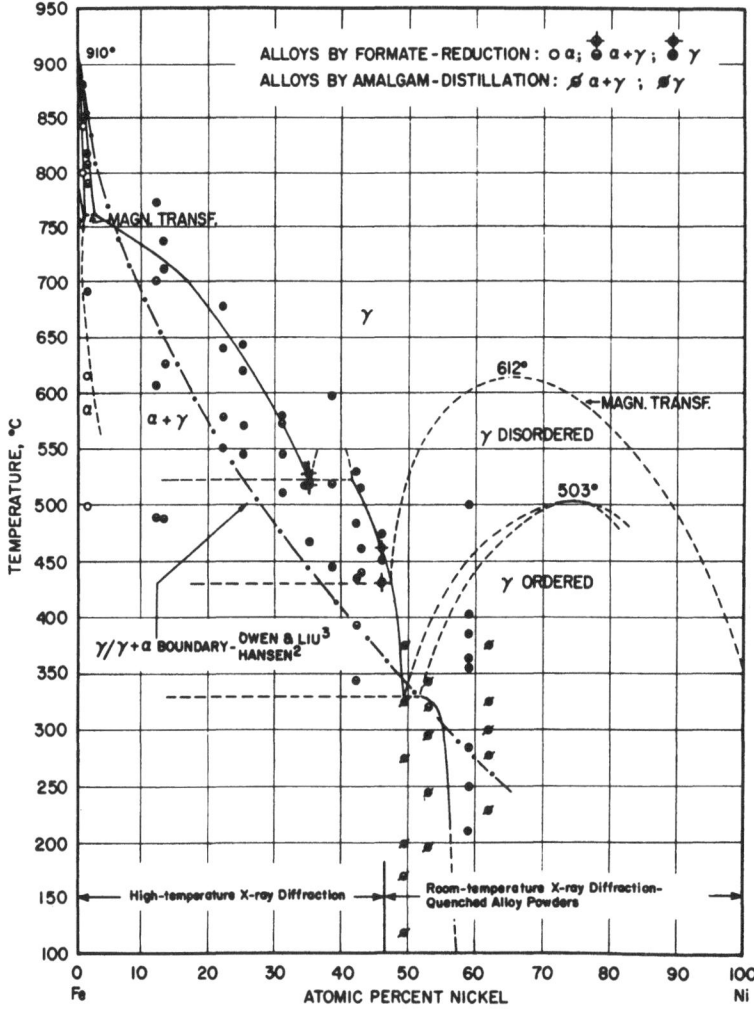

Figure 4. The $\gamma/\gamma + \alpha$ boundary in the iron–nickel system as relocated from the results of the present investigation.

the martensitic structure, mask diffraction lines due to traces of any α-phase that might have been present in equilibrium with the γ-phase at this temperature.

When the alloy of the same composition as quenched to room temperature from the much lower temperature of preparation namely, 474°C, is examined by X-ray diffraction, no martensite is found; but the structure is found to be a mixture of α- and γ-phases (Figure 2b). The result shows that the equilibrium structure at 474°C is $\gamma + \alpha$ for this composition and that the γ-phase containing appreciably more than 13 at.% nickel (estimated as $\sim$45 at.% nickel) does not undergo the martensitic transformation when quenched to room temperature. Thus, while room-temperature X-ray diffraction of these quenched alloys is useful for studying the equilibria within the $\gamma + \alpha$ field, it is inadequate for an exact location of the $\gamma/\gamma + \alpha$ boundary.

The tendency to form martensite is exhibited by alloys containing up to $\sim$31 at.% nickel. The amount of martensite obtained by quenching to room temperature from

Table I. The $\gamma/\gamma + \alpha$ Boundary Compositions at Various Temperatures

Temperature, °C	Composition–atom per cent nickel	
	Present investigation	Owen and Liu[3]–Hansen[2]
850	1.2	1.6
800	2.0	3.8
750	5.5	6.0
700	16.5	9.4
650	23.2	13.0
600	28.5	17.3
550	33.0	22.2
500	43.0	27.5
450	46.8	34.3
400	48.2	41.5
350	49.0	48.5
300	55.0	56.0
250	56.5	65.0*
200	56.8	73.0*
150	57.0	—
120	57.4	—

* Extrapolated data.

sufficiently elevated temperatures of preparation (where these alloys may be presumed to in the γ-field) was found to decrease with increasing nickel content of the alloys. While alloys with 13 at.% nickel contained only martensite, alloys with 22 at.% nickel contained about half martensite and balance untransformed γ, and alloys with 31 at.% nickel were practically all γ with only a barely perceptible amount of martensite.

For locating the $\gamma/\gamma + \alpha$ boundary in alloys with nickel content less than ~ 31 at.%, the high-temperature X-ray diffraction techniques become indispensable. Figures 3a–e are presented to show the usefulness of the high-temperature X-ray diffraction techniques for this purpose.

When the all-martensitic structure of the alloy of 13 at.% nickel obtained by quenching to room temperature from the preparation temperature of 695°C (Figures 2a and 3a) is heated and held at 487°C, the high temperature X-ray diffraction pattern (Figure 3b) shows that the martensite has decomposed to form an equilibrium mixture of α and γ phases, with only a trace of γ. Further heating to and holding at 657°C results in an increase in the amount of γ in the $\gamma + \alpha$ mixture as shown by the X-ray diffraction pattern in Figure 3c. The X-ray diffraction pattern in Figure 3d, corresponding to a holding temperature of 710°C, indicates that most of the α-phase has disappeared and that the structure is made up practically of all γ-phase. The X-ray diffraction pattern obtained for 737°C (Figure 3e) shows the alloy to be in the single-phase γ-field. The $\gamma/\gamma + \alpha$ boundary point for this composition thus must lie between 710°C and 737°C.

In Figure 4, are plotted the experimental data relevant to locating the $\gamma/\gamma + \alpha$ boundary which has been drawn to conform to the results of the present study. The data used for the plots up to 46 at.% nickel are derived from high-temperature X-ray diffraction results. Above this composition, room-temperature X-ray diffraction results from quenched alloys are used. For the compositions with ~ 35 and ~ 46 at.% nickel, the room-temperature X-ray diffraction data for the quenched alloys are also shown and these points may be distinguished from the corresponding high-temperature X-ray diffraction plots by the crosses drawn over the plotted points. It is to be especially noted that there is close agreement between the two sets of data.

The $\gamma/\gamma + \alpha$ boundary data obtained in the present study are believed to represent equilibrium data. The striking result is that the boundary *does not agree* with that in Hansen's diagram (Figure 1). Appreciable differences, not only in the shape and location but also in the course of the boundary, are found.

In the first place, the boundary is concave downwards instead of being concave upwards as in Hansen's diagram (Figure 1). Secondly, between 0–6 at.% nickel and above 50 at.% nickel, the boundary lies at lower temperatures whereas, in the composition range of 6–50 at.% nickel, the boundary lies at much higher temperatures (e.g., ~100°C higher at 20 at.% nickel) than in Hansen's diagram (Figure 1). Thirdly, beyond 42 at.% nickel, the boundary falls abruptly and continues to do so even to the lowest temperature of 118°C used in this study. The composition coordinate at 250°C is ~57 at.% nickel whereas, Hansen's diagram (Figure 1) yields an extrapolated value of ~65 at.% nickel at the same temperature. The boundary compositions for various temperatures are summarized in Table I which also includes the boundary data of Owen and Liu[3] used by Hansen.[2] It may be noted that boundary data below 300°C are being reported for the first time.

The magnetic transformation boundary of the γ-phase alloys meets the boundary at 47.5 at.% nickel and 430°C. The magnetic transformation horizontal of the γ-phase of alloys in the $\gamma + \alpha$ field has been, therefore, shown at 430°C (dashed isothermal line) rather than at 400°C as drawn by Hansen.[2]

The course of the boundary, instead of running smooth as in Hansen's diagram (Figure 1), is marked by three discontinuities which may readily be distinguished as occurring at: (a) ~760°C and ~3 at.% nickel, (b) ~525°C and ~35–43 at.% nickel, and (c) ~330°C and ~50 at.% nickel. These discontinuities have been interpreted as indicating the existence of three concealed isobaric-invariant equilibria in the iron–nickel system.

The first discontinuity occurring at ~760°C and ~3 at.% nickel is difficult to explain by any construction of the phase diagram other than one which includes a transformation in the α-phase. Temperature and concentration at the discontinuity strongly indicate that the magnetic transformation of the α-phase is the one concerned.

One interpretation of the second discontinuity at ~525°C and 35–43 at.% nickel is that it may arise from a concealed, monotectoid type of reaction and its associated two-phase field involving two γ-phases. A diagram including such a reaction would appear similar to but not identical with that given by Kaufman and Ringwood[12] for the iron–nickel system at high pressures. Although in an alloy with 42.9 at.% nickel, a broadening of the γ-lines was observed in the X-ray diffraction pattern obtained at 526°C, sufficient data are not available over the composition and temperature ranges concerned, to conclude that such a reaction does exist. Further, there are alternative ways of interpreting the single observation. For example, recently Kachi, Bando, and Higuchi[13] have, from specific heat and magnetic intensity measurements on fine-particle iron–nickel alloy powders similar to those used in this work, presented evidence for the existence of a superstructure based on Fe_3Ni, with a disordering temperature of 800°C for the stoichiometric composition. It is entirely possible to construct the $\gamma/\gamma + \alpha$ boundary in such a manner as to include the two-phase fields that would exist[7] on either side of this composition by the use of suitable isobaric-isothermal reactions. The indications are that, in the composition range of ~22–45 at.% nickel, the iron–nickel phase diagram may actually be even more complex than what can only be inferred from the results of the present study.

Indeed, examination of several alloys, in the composition range of 22–45 at.% nickel and more closely spaced in composition than those used in this study, at temperatures between 500°C and up to at least 800°C appears to be needed. It also would appear that, while high temperature X-ray diffraction may be useful, more discriminating methods like magnetic susceptibility measurements and neutron diffraction may have to be used. Such studies may reveal structural complexities that perhaps remain, as yet, undiscovered and hence, at present, unknown in the iron–nickel system.

The third discontinuity at $\sim330°C$ and ~50 at.% nickel may be due to interaction between the order-disorder transformation based on $FeNi_3$ and the γ/α transformation, leading to a eutectoid type of reaction. Evidence for such a reaction is difficult to obtain by the use of X-ray diffraction techniques because of the insufficient scattering-factor differential between iron and nickel. Again, the need for other methods of study are indicated. Recently, Marchand and Chamberod,[14] through electrical and magnetic measurements on iron–nickel alloys in the composition range of 35–65 at.% nickel, report that order-disorder reactions based on FeNi and $FeNi_3$ may tend to occur simultaneously. In addition, the present authors[15] have reported the evidence for another transformation in the iron–nickel system in the composition range of 30–45 at.% nickel. The indications are that this part of the iron–nickel system also may be quite complex.

Although some of the solid state reactions that may occur at the discontinuities are shown in Figure 4 (by dashed curves and isotherms), conclusive evidence remains to be obtained for determining whether or not they actually occur. However, the need and importance for a closer reexamination of the iron–nickel system are strongly indicated by the results of the present study.

SUMMARY

1. The $\gamma/\gamma + \alpha$ boundary in the iron–nickel system has been relocated to as low as 118°C by room-temperature and high-temperature X-ray diffraction studies of equilibrated fine-particle iron–nickel powders, prepared by the formate-reduction[8,9] and amalgam-distillation[10,11] techniques.

2. Boundary data below 300°C are being reported for the first time for this system.

3. The boundary data *do not* agree with the presently accepted data of Owen and Liu used by Hansen in drawing the iron–nickel phase diagram.

4. In the region ~6–50 at.% nickel the boundary lies more than 100°C above the temperatures than shown by Hansen,[2] even by as much as over 100°C at certain compositions.

5. The magnetic transformation boundary of the γ-phase alloys meets the $\gamma/\gamma + \alpha$ boundary at 47.5 at.% nickel and 430°C..The magnetic transformation horizontal for the γ-phase in the $(\gamma + \alpha)$ field thus is shown at 430°C instead of at 400°C as drawn by Hansen.

6. Below ~6 at.% nickel and above ~50 at.% nickel, the boundary lies below Hansen's boundary.

7. The course of the boundary is not smooth as shown by Hansen, but is marked by three discontinuities in addition to a precipitous drop starting at ~43 at.% nickel.

8. The discontinuities occur at: (a) ~3 at.% nickel and 760°C, (b) ~35–43 at.% nickel and 525°C, and (c) ~50 at.% nickel and 330°C.

9. The present boundary data are believed to represent equilibrium data for the iron–nickel system.

10. The discontinuities are considered to be indicative of concealed isobaric-isothermal reactions heretofore unsuspected to exist in the iron–nickel system. Some possible reactions are discussed.

11. Of special interest is the possibility that interaction between the magnetic transformation of the α-phase and the γ/α transformation may give rise to an isobaric-isothermal reaction.

12. When the results of the present study are considered together with some of the recent literature[12-16] on the iron–nickel system, the indications are that the iron–nickel phase diagram may be exceedingly complex in contrast with the strikingly simple diagram given by Hansen.[2]

13. A closer examination of the iron–nickel system, using closely spaced compositions and temperatures for study is desirable and is in progress.

ACKNOWLEDGMENTS

Most of the research presented in this paper was conducted during 1954–1959 while the authors were associated with the Westinghouse Electric Corporation, East Pittsburgh, Pennsylvania. Westinghouse support of the work and Westinghouse permission for publication are equally appreciated. The authors take this opportunity to thank the University of Kansas, Lawrence, Kansas, and the Iowa State University, Ames, Iowa, for providing many facilities for the preparation of this paper in its present form.

REFERENCES

1. J. L. Haughton, *Constitutional Diagrams of Alloys: A Bibliography*, The Institute of Metals, London, 1956.
2. M. Hansen, *Constitution of Binary Alloys*, McGraw-Hill Book Company, New York, 1958.
3. E. A. Owen and Y. H. Liu, *J. Iron Steel Inst.* **163**: 132, 1949.
4. E. Josso, *Compt. Rend.* **229**: 594, 1949.
5. *Ibid.*, **230**: 1467, 1950.
6. A. H. Geisler, *Trans. Am. Soc. Metals* **45**: 1051, 1953.
7. F. N. Rhines and J. B. Newkirk, *ibid.*, p. 1029.
8. F. Lihl, *Metall.* **5**: 183, 1951.
9. N. I. Ananthanarayanan and R. J. Peavler, *Metals Eng. Quart.* **2**: 43, 1962.
10. F. Lihl, *Z. Metallk.* **44**: 160, 1953.
11. R. J. Peavler and N. I. Ananthanarayanan, *Advances in X-Ray Analysis*, Vol. 7, Plenum Publishing Corp., 1963, p. 117.
12. L. Kaufman and A. E. Ringwood, *Acta Met.* **9**: 941, 1961.
13. S. Kachi, Y. Bando, and S. Higuchi, *Japan. J. Appl. Phys.* **1**: 307, 1962.
14. A. Marchand and A. Chamberod, *Compt. Rend.* **261**: 3113, 1965.
15. N. I. Ananthanarayanan and R. J. Peavler, *Nature* **192**: 962, 1961.
16. J. I. Goldstein and R. E. Ogilvie, *Trans. Met. Soc. AIME* **233**: 2083, 1965.

DISCUSSION

J. I. Goldstein (Goddard Space Flight Center): I would like to come to the defense of the "Classical Method" for determining the phase diagram. In 1965, we published a reevaluation of the iron–nickel phase diagram using bulk samples, but in this case using an electron probe to measure the compositions of the phases that were grown. We found in these solid alloys down to temperatures of 500°C that the gamma/gamma + alpha-phase boundary was almost identical to that given by Owen and Liu. In fact, it fell within the error limits that you can find in their paper for those alloys which they had available. We found, as you suggested, that the alpha/alpha + gamma boundary was greatly expanded at the high temperatures. However, we find that in the bulk alloys which we studied, these were not powders, these were bulk, homogeneous alloys which were quenched and annealed for very long times and on which we very easily could make measurements with an electron probe.

N. I. Ananthanarayanan: In massive alloys, our general feeling is that equilibria are seldom reached in this system. It is likely that you have observed the early stages of the reactions and their partial progress, but, whether you have observed equilibria is doubtful. The micro-probe examination is equivalent to examination of small regions in the massive alloys. What we have done is to prepare the alloys directly as particles in sizes equivalent to and generally much smaller than those examined by you.

J. I. Goldstein: Using the electron-probe technique, we can indeed just find the beginning of equilibrium. We only need particles on the order of 5 μ size to take a measurement with an electron probe. This we did. We annealed these alloys for long times where we got 100 μ particles, the composition did not change. These were indeed the equilibrium values. X-ray diffraction showed that we had alpha and gamma phases.

N. I. Ananthanarayanan: The effect of particle size becomes appreciable when you closely approach micron size or go below this size. Particles on the order of 5 μ size would be expected to behave like massive alloys and, therefore, Owen and Liu's results would be duplicated. With our powders, prepared directly in ultra-fine sizes, the reaction kinetics have been so accelerated as to give equilibrium easily. Your samples apparently had not attained equilibrium.

J. I. Goldstein: Yes, I would suggest that these are not equilibrium—these are metastable and equilibrium will come if enough annealing time is allowed.

N. I. Ananthanarayanan: Yes, I agree with you. I should say that we have reached equilibrium states for all practical purposes in our experiments.

J. B. Newkirk: These are intriguing and imaginative ways of making alloys at low temperatures. However, I don't understand the basis for your assumption that that phase which first forms need be the equilibrium phase. It is possible that the first metallic phase might very well be metastable and may in time relax to the true equilibrium phase that you're after. Can you tell me what is the basis for the assumption that you're observing the equilibrium phase?

N. I. Ananthanarayanan: What I have stated is that when the powder is prepared at any given temperature in the manner described by us, equilibrium is reached readily. Perhaps metastable states occur in the beginning. In a sluggish system, such states could occur and may even be observable. In a system which is not sluggish, one may not readily observe these states even if they occur. We have not made any attempt to look at the initial stages. We find that even in times at temperature just sufficient to complete the reduction of the formates (or the distilling off of the mercury from the amalgams), equilibrium structures are already achieved. No metastable structures have been found.

Lastly, I would like to point out—although not as an answer to any specific question—that the data given in this paper may be of considerable use in the understanding and interpretation of the behavior of thin alloy-films (e.g., Permalloy films) and of the structures of certain meteorites.

LATTICE CONSTANT AND CRYSTALLITE SIZE OF CONDENSED GOLD VAPOR

Frank G. Karioris

Marquette University
Milwaukee, Wisconsin

Jerome J. Woyci

Allen Bradley Company
Milwaukee, Wisconsin

and

Richard R. Buckrey*

Wayne State University
Detroit, Michigan

ABSTRACT

Gold wires were vaporized by the exploding-wire phenomenon using a 20 μF capacitor bank charged to voltages up to 14 kV. The resulting condensate, an aerosol or metallic smoke, was collected on membrane filters and subjected to X-ray analysis to determine lattice constant, crystallite size, and behavior with isothermal annealing. Wire explosions were conducted in an ambient atmosphere of air or nitrogen at barometric pressure. It is estimated that the quench rate for this material is of the order of 10^6 deg/sec from the melting point although no substrate is involved and it is expected that any effects of epitaxial origin on the structure would be minimized.

Before annealing, diffractograms showed broad peaks apparently shifted to the high-angle side. Line breadth may be attributed primarily to particle size broadening, since it correlates well with size determined by electron microscopy, $(\beta \cos \theta)$ is linear with θ, and $[(\beta \cos \theta)/\lambda]$ is approximately constant for three radiations used. Crystallite size is of the order of 400 Å and is observed to decrease roughly with increasing voltage used for vaporization. The observed lattice decrement, approximately 0.2%, generally increases with voltage used for vaporization, and apparently correlates rather well with the inverse of size as has been reported in some work on thin gold films. However, studies of colloidal gold particles do not show significant lattice shifts, although the particle size is less than 100 Å so that the decrements observed may be due to factors other than size alone. For this black, particulate material, some lattice decrement apparently persists even after protracted isothermal annealing below the melting point. Crystallite size increases with annealing but remains below about 1000 Å. Results suggest that the lattice decrements observed in condensed gold vapor are due to surface tension effects and the presence of vacancy aggregates.

* Work done for partial fulfillment of the requirements for the degree of Master of Science at Marquette University.

INTRODUCTION

In the exploding wire phenomenon,[1,2] a large capacitor is charged to high voltage and then rapidly discharged through a fine wire to produce a dramatic explosive vaporization of the material. In the explosion, temperatures of the order 10,000°K may be achieved.[3] With very rapid heating by the surge current, superheated liquid explodes to form a metallic vapor of high density and the current is interrupted.[4] The metallic vapor then condenses to form a dense aerosol or smoke which can be collected and studied.[5] Previous studies have shown that, although many exploded metals react with oxygen in air[6] or with a nitrogen atmosphere,[7] the noble metals condense out as spherical crystallites of the pure metal.

The gold aerosol produced by this technique makes a matte-black deposit when collected on a membrane filter and is of interest because (a) the basic particle size is small so that any effect of surface on bulk properties would be enhanced, (b) it has condensed by self-nucleation without any possible interferring substrate which would introduce severe strain of epitaxial origin, regardless of its temperature coefficient, and (c) it represents material which has been quenched very rapidly from the melting point after solidification of liquid droplets.

It generally is assumed that rapid quenching preserves high-temperature states, and it has been shown that very rapid quenching from a melt may reveal the existence of high-temperature phases in certain alloys.[8,9] No structures other than the normal fcc were observed in gold in this study. However, the rapid quench from the melting point may be significant in retaining[10] or producing defects in the crystallites. The quench rate from the melting point may be estimated by assuming radiation cooling of an isolated particle or of the "fire-ball" according to the Stephan–Boltzmann law

$$E = \sigma(T^4 - T_0{}^4) = \frac{mc}{A}\left(\frac{dT}{dt}\right) \tag{1}$$

where m is the mass of the particle, c is the specific thermal capacity, σ is the Stephan–Boltzmann constant, and A is the area.

For an isolated particle of 300 Å diameter, this turns out to be about 13×10^6 deg/sec. As in the case of conventional quenching of small wires, the cooling rate is inversely proportional to specimen diameter[11] and the particles may be considered practically isothermal during the quench.[12] It has been shown that for very rapid quenching of very small metal pieces, no thermal strain should be involved.[11] If one estimates the cooling rate from equation (1) taking the observed 12 cm diameter of the "fire-ball" to calculate the area and 0.010 gm for the mass of material, the initial quench rate from the melting point of gold is of the order 5×10^6 deg/sec. In either case, the quench rate is of the order 10^6 deg/sec, considerably higher than the 10^3–10^5 deg/sec obtainable by conventional techniques.[13,14]

The small size of the aerosol particles from the condensed gold vapor is of interest since it represents an opportunity to study possible surface effects on bulk properties. For a 300 Å particle, approximately 20% of the bulk is included in the three outermost atomic layers so that any significant surface distortion would be detectable by X-ray analysis. Numerous authors have attempted experimentally to determine a relation between lattice parameter and particle size by X-ray diffraction and electron diffraction, but discrepancies exist[15] and such a relation is difficult to establish clearly. Much of the work on "particle size" has been done actually on thin films.

EXPERIMENTAL PROCEDURES

In the exploding wire aerosol generator previously described,[16] a 20 μF capacitor bank, charged to voltages up to 14 kV, was discharged through 2.5 cm lengths of 0.007-in. diameter wires in an atmosphere of air or nitrogen at barometric pressure. The authors believe that the material which remains suspended as aerosol in the explosion chamber represents the fraction of the wire which vaporized and condensed.[5] Recently, it has been pointed out that some spectral information may be retained in far-infra red reflectance studies of well-compacted fine particles.[17] Figure 1 is an electron micrograph of a specimen of gold aerosol produced by the exploding wire technique. The basic particle appears spherical indicating that the material probably persisted as liquid droplets before solidification. The particle size distribution for exploded wire aerosols is approximately log-normal with a mass median corresponding rather closely to that determined by X-ray line broadening.[18] The aerosols formed by exploding wires are similar to those formed in a DC arc[19] with the exceptions that, for the former, particles tend to be smaller and polyhedral particles are not observed. Because of the high concentration of particles in

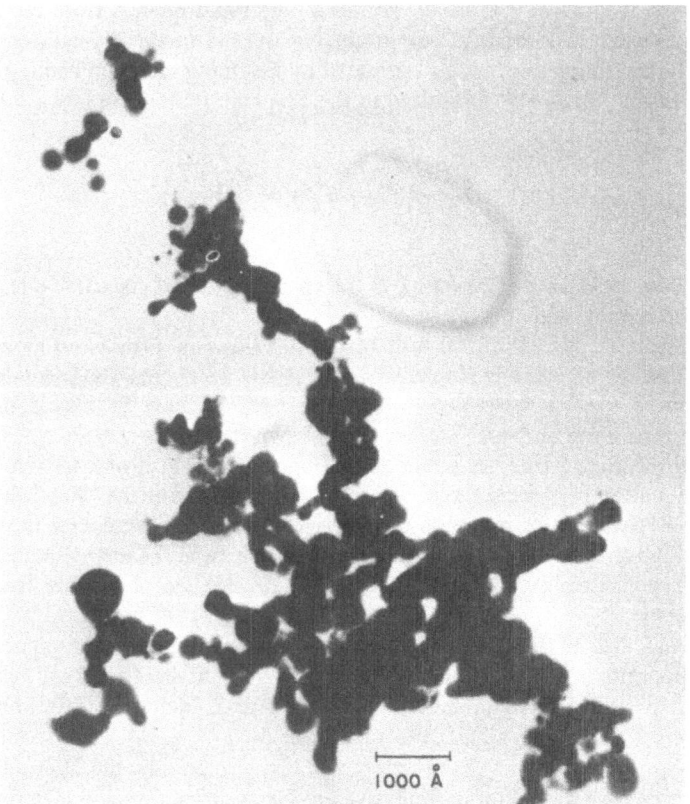

Figure 1. Electron micrograph of gold aerosol particles produced by exploding wires.

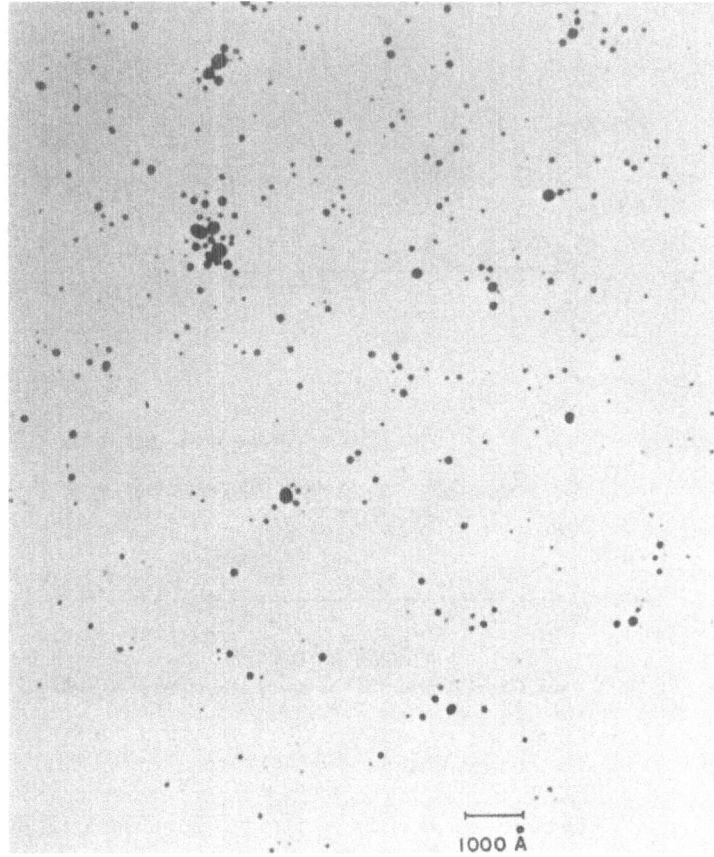

Figure 2. Electron micrograph of colloidal gold particles. Mass median diameter from micrograph, 90 Å. Particle size from line broadening, 80 Å.

the explosion chamber, the particles tend to form large, feathery aggregates in the aerosol phase.[16]

Samples of the gold aerosol were prepared for diffractometry by three techniques: (a) sections of membrane filter on which aerosol had been collected were mounted directly on glass microscope slides, (b) a thin layer of diluted rubber cement was painted onto a glass slide, allowed to dry, and then pressed against the deposit to lift off a thin sample, and (c) a section of the membrane filter was dissolved in ethyl acetate, the residue washed repeatedly with ETA and acetone, and then deposited on a small area of a quartz slide. Results were essentially the same for all three techniques, but the small-area deposit gave higher diffracted intensities in the back-reflection region.

Colloidal gold prepared by a chemical method was obtained from the Abbott Laboratories. This material is supplied as a deep cherry-red liquid which includes varying amounts of ascorbic acid and gelatin to stabilize the gold colloid. Figure 2 is an electron micrograph of a specimen with particle size about 80 Å which had been in our possession for 12 years. Another sample was prepared within a month of this study. Samples of this material were prepared for diffractometry by repeatedly evaporating liquid on a glass slide until a flat, uniform layer of residue was obtained.

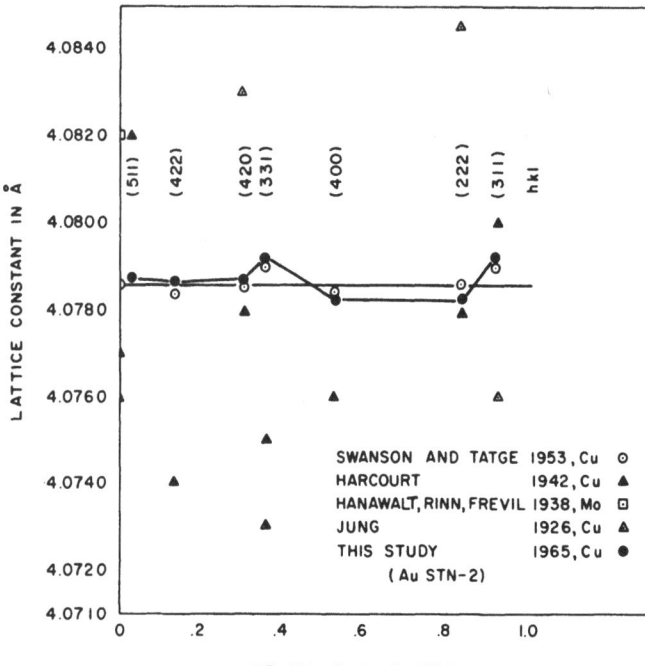

Figure 3. Data for calibration of the spectrogoniometer to read "true 2θ" and comparison with some previous results.

X-ray diffraction spectra were obtained on a General Electric XRD-5 recording diffractometer using 50 KVP, 10 mA, nickel-filtered copper radiation and the xenon proportional counter. The target was viewed at $4°$ with a $3°$ beam slit, a $0.2°$ detector slit, and medium resolution Soller slits. When diffracted intensity was sufficient, automatic scanning was done at $0.2°$ (2θ)/min, chart speed 12-in./hr and time constant 1 sec or less. Numerous peaks were step-scanned at approximately $0.25°$ intervals taking 10^4 counts, and in some cases 10^5 counts to attain reasonable precision.

Standards were prepared of gold and LiF powders on quartz slides to simulate as closely as possible the configuration of the test specimens. The gold standards were annealed at 450°C for several hours and then allowed to cool very slowly, the LiF was annealed at 200°C. The gold standards showed but little deviation from accepted values as illustrated in Figure 3, which plots α_0 determined from various d_{hkl} against the Nelson–Riley function and compares our results with those of others.[20] Thin, transparent specimens of highly absorbing materials are generally considered suitable for precision determination of lattice parameters.[21] Using the accepted values for the various d_{hkl} of the gold and LiF, "true 2θ" positions were calculated and a calibration curve drawn to correct the "observed 2θ" for the test specimens. The maximum correction was of the order $0.03°$ (2θ) for $2\theta > 90°$. These standards were used also to determine instrumental line broadening. The temperature was monitored and corrections amounting to about one part in 10^4 were applied to reduce the data to 25°C.

Emission and microprobe analyses of the gold aerosol samples showed that there was no observable contamination from the material of the electrodes used for the explosion of the wire. For one of the specimens, diffraction spectra were obtained using silver and iron as well as copper radiation and the function ($\beta \cos \theta / \lambda$) was calculated for each line. This function was found to be strongly wave-length dependent indicating that line broadening is due primarily to small particle size. Differential thermal analysis of the aerosol specimens showed a small exoergic peak at about 225°C, but the sample was too small to give significant results. Specimens of gold aerosol on quartz were annealed isothermally at 200°, 300°, and 450°C for times ranging up to 2000 min. Diffractograms were run periodically to determine particle size and lattice parameter at various stages of the anneal.

DATA AND TREATMENT OF DATA

All of the materials studied showed the normal fcc structure of gold with lines considerably broadened and, apparently, somewhat shifted toward the high angles. For some of the preliminary work, peak position was determined by extending the sides of a line tangentially upward[22] or taking the center of the chord at half maximum and using the weighted average wavelength of the doublet, but for broad peaks it is felt that precision of this method is inherently poor.[23] Also, to determine line breadth by this technique, it was necessary to correct for the instrumental broadening and the doublet spacing according to the procedure outlined by Klug and Alexander.[24]

Although the Stokes method of determining the line produced by the singlet α_1 can be applied, it has been suggested[25] that a simpler method proposed by Rachinger[26] would be adequate. This procedure assumes that the composite peak is comprised of

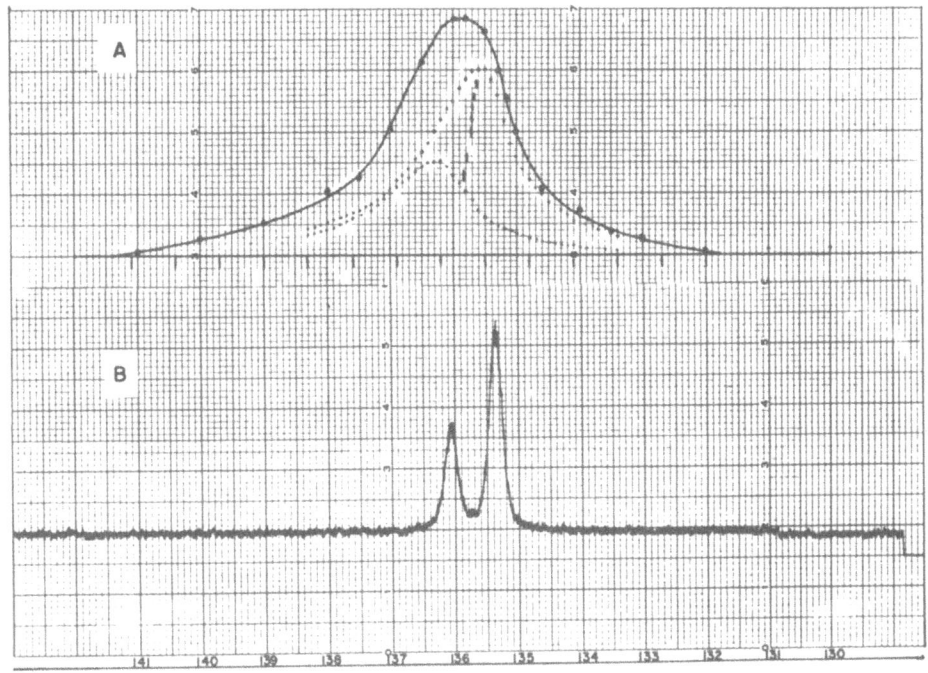

Figure 4. The (422) line of a gold sample from explosion of wire by 13 kV. Graphical resolution of the $\alpha_1\alpha_2$ doublet and the trace of the same line from the standard are shown. Ordinates are not to scale.

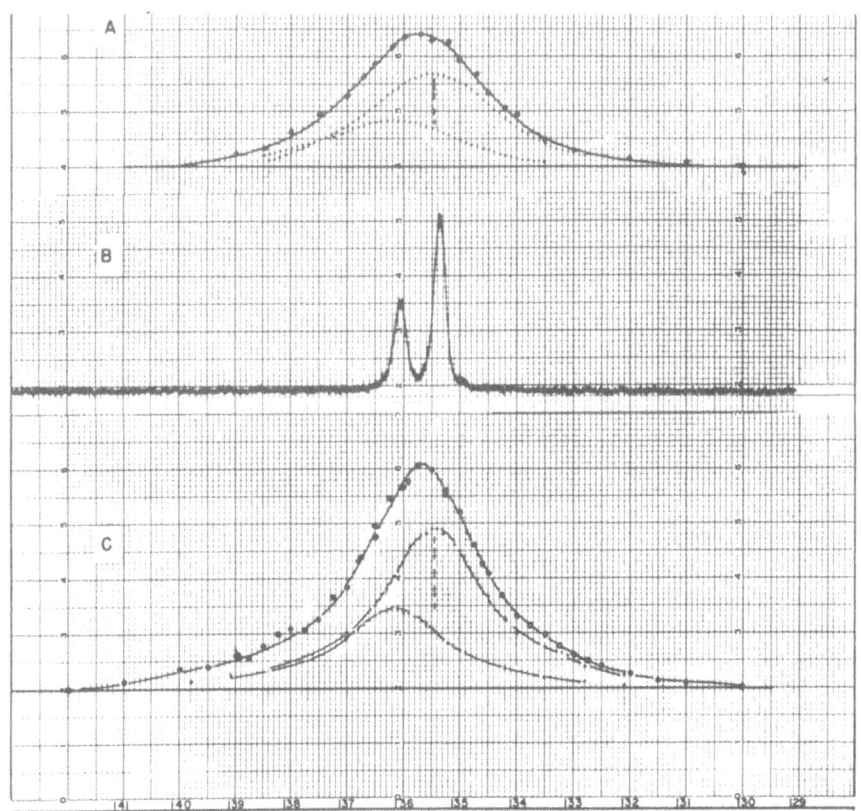

Figure 5. The (422) line for (A) colloidal gold aged 12 years. (B) gold standard and (C) fresh colloidal gold. Ordinates are not to scale.

two lines of the same shape, separated by the known $\alpha_1\alpha_2$ separation and with relative intensities 2 : 1. The graphical procedure was used in X-ray scattering earlier by DuMond and Kirkpatrick[27] and developed further by DuMond.[28] The method tends to reduce errors, or at least not introduce serious new errors and the most probable source of difficulty is estimation of the background.[29] Figures 4 and 5 show this graphical technique applied to diffraction lines of condensed gold vapor and to colloidal gold to determine the position and breadth of the α_1 line.

Line breadths β were determined by measuring the chord at midpeak B of the resolved α_1 line and correcting for the instrumental broadening by using the breadth b of the well-resolved standard line according to the relation[30]

$$\beta^2 = B^2 - b^2 \qquad (2)$$

The diameter of the particle was calculated from the Sherrer equation modified for spherical particles

$$D = \frac{1.107\lambda}{\beta \cos \theta} \qquad (3)$$

From the Sherrer equation, it can be seen that for line broadening due to particle size alone, the function ($\beta \cos \theta$) should be constant for all values of θ. It can be shown

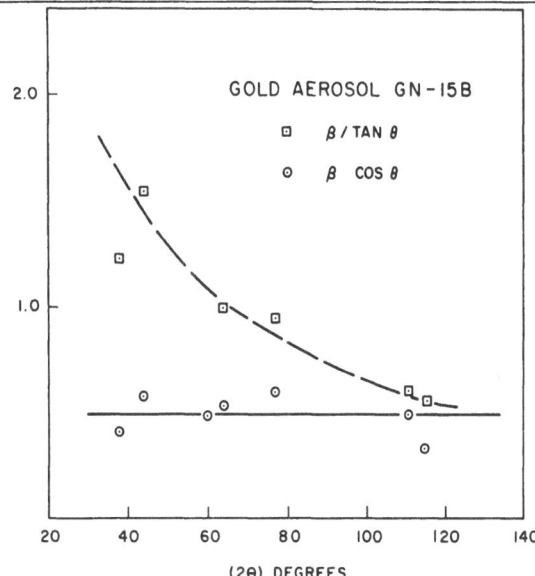

Figure 6. Plot of ($\beta \cos \theta$) and ($\beta/\tan \theta$) for a condensed gold vapor sample from a wire explosion at 9 kV.

also that for isotropically strained crystallites the broadening is proportional to $\tan \theta$ so that for pure strain, the function ($\beta/\tan \theta$) is constant. The degree of linearity of either function is an indication of the relative contribution to the broadening of size or strain.[31] Figure 6 shows ($\beta \cos \theta$) and ($\beta/\tan \theta$) plotted versus 2θ for a gold aerosol sample. From this, it may be deduced that the line broadening is due primarily to particle size. When stacking faults are present, the apparent particle sizes which result from the Sherrer equation are not isotropic but differ by a factor of two or more for different reflecting planes.[32,33] Since this effect was not observed, the existence of a significant percentage of stacking faults was discounted.

After peak positions were determined and corrected to the "true 2θ" by the calibration curve, d spacings were obtained by interpolation from the tables. In general, eight of the first nine peaks of the gold spectrum were used. Values of a_0 were then calculated from each d_{hkl} and plotted against the Nelson–Riley function. Extrapolation of these data to the ordinate was used to determine the lattice constant for each of the several specimens which had been exploded at various voltages. Although this technique may not necessarily be correct for the systematic errors characteristic of the diffractometer,[34] Wilson[35] shows that, for moderate precision, the errors due to sample displacement, absorption and axial divergence of the beam are most significant and do vanish at $\theta = 90°$. The most significant source of error is probably the displacement of the sample surface from the diffractometer axis.[36] The Nelson–Riley function,[37] besides being highly linear with respect to errors in film data, has been used also in high-precision diffractometry.[38]

RESULTS

Twenty-four samples of condensed gold vapor from wire explosion were studied. No significant differences were observed between those exploded in nitrogen, argon, or air. Since the energy available in a charged capacitor depends on the square of the voltage, some effect on the properties of the aerosol produced by explosions of identical wires by different voltages is to be expected. There is, however, only a general tendency for the X-ray determined particle size to decrease with increasing voltage used for explosion. The results here are unaccountably erratic since great care was used to reproduce experimental conditions of the explosions in each case. The trend suggests that much smaller particles might be produced by higher voltages, or perhaps by modifying the apparatus to increase the speed of the discharge by reducing the inductance, and this work is in progress. Also, there is an apparent trend for the lattice decrement to increase with voltage and this suggests again that interesting results might be obtained by going to higher voltages. The lattice decrements observed are generally less than 0.3%, but they are readily observable at high angles. For example, a decrease in lattice constant of 0.1% will shift the (422) line of gold about $0.28°(2\theta)$ to the high angle side of $135.39°$ (2θ) and the (511) line will be shifted by about $0.58°$ to the high angle side of $157.81°$ (2θ). Line shifts of this order can be observed, for example, in Figure 7 in spite of considerable line breadth.

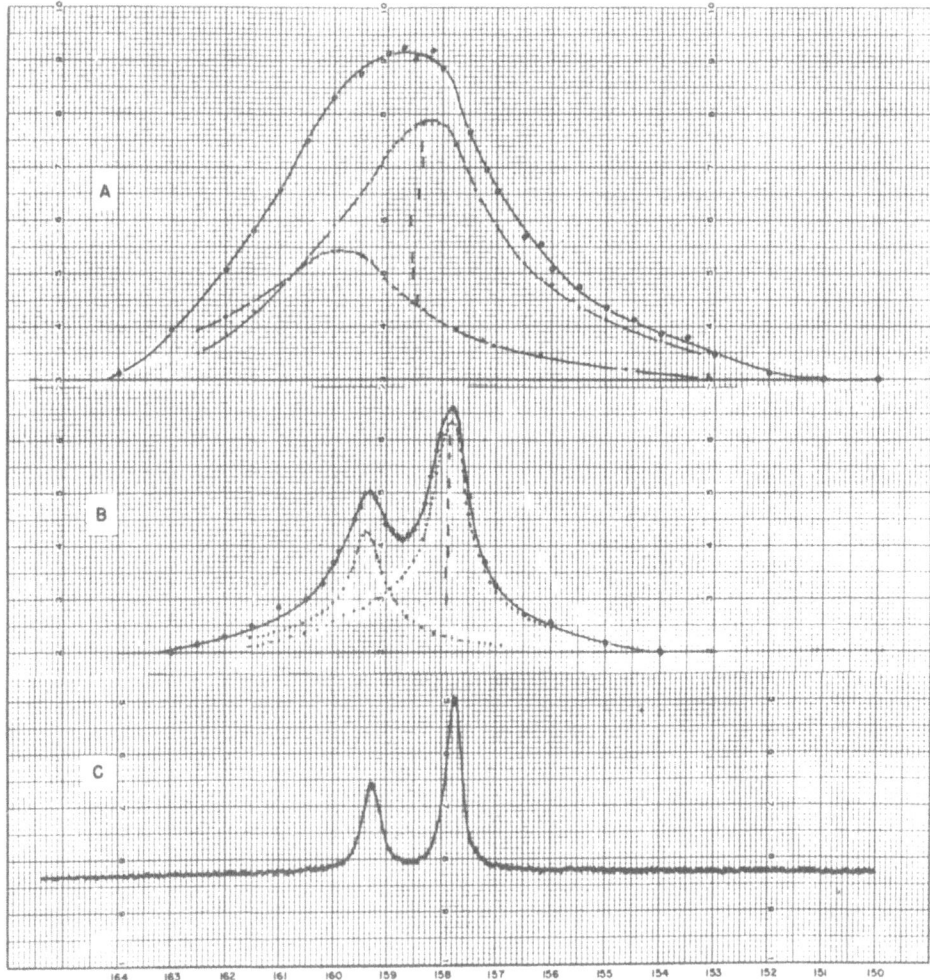

Figure 7. The (511) line of a gold aerosol sample formed by explosion at 13 kV compared with that of the gold standard. (A) gold aerosol before anneal, (B) same sample after annealing at 600°C for 3 hr and (C) gold standard. Ordinates not to scale.

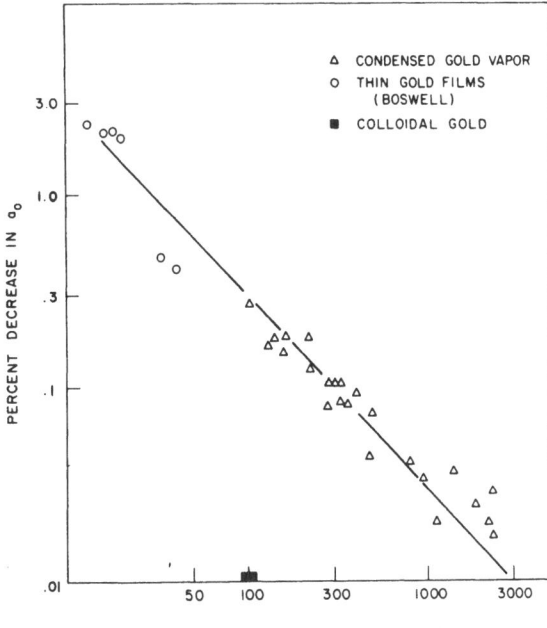

Figure 8. Percent decrease in lattice parameter of gold aerosols shown as a function of the X-ray determined particle diameter. The thin, gold film data are from Reference 15.

Some uniformity is observed in the data when the lattice decrement is plotted as a function of particle diameter as shown in Figure 8. The straight line showing the trend of these data has the form

$$\frac{da}{a} = 0.29 \times 10^{-8} \left(\frac{1}{D}\right) \tag{4}$$

where D is in cm. The mean deviation of the proportionality constant is ± 0.077 so that the data fit to within about $\pm 30\%$. The electron diffraction data of Boswell[15] on thin gold films show at least qualitative agreement with these results although it has been suggested that the results of Boswell may be due to visual error in determining line positions.[39] Electron diffraction applied to vacuum-deposited gold film often show extra lines characteristic of hcp structure but X-rays do not yield extra lines and do show Bragg line shifts on a variety of substrates with differing thermal expansion coefficients.[40] Nicolson[41] has derived an expression relating the fractional change in lattice spacing to the compressibility and surface tension of a solid. This equation states that for spheres

$$\frac{da}{a} = \frac{4}{3}\beta\gamma \left(\frac{1}{D}\right) \tag{5}$$

where β is compressibility and γ is surface tension in cgs units. The expression turns out to be identical to the calculation for a fluid droplet in that it assumes a uniform strain throughout the volume of the particle. For a solid, distinctions exist between the surface free energy, surface stress and surface tension.[42] For crystals, the polar plot of surface tension as a function of orientation of the surface normal is usually anisotropic, but small crystals and cavities in crystals often change their shape in the direction of lower $\int \gamma \, dA$ under heat treatment[43] indicating a sort of average surface tension effect.

If the value of the bulk compressibility of gold from the International Critical Tables is taken as 5.95×10^{-13} cm^2/dyne and used to calculate the apparent surface tension of gold from equation (5)

$$\text{apparent } \gamma_{Au} = 3280 \text{ dyne/cm} \tag{6}$$

Shaler[44] considers a value of 1350 dyne/cm fairly reliable for the surface tension of gold near the melting point in the presence of its vapor. In a review of the measurement of the surface tension of gold from the data of Tamman and Boehme, using the creep method on gold foil, Fisher and Dunn[45] calculate a mean value of 1780 ± 10 dyne/cm. The work based on the creep of wires in vacuum has been summarized[45] to give a mean value of 1450 ± 80 dyne/cm in the temperature range 920–1020°C. With electron diffraction Rymer and Butler[46] have reported decrease of lattice parameter with decreasing film thickness that could be accounted for by assuming a value of 500 dyne/cm.

Some caution is necessary in interpreting the dependence of da/a with $1/D$ as being due solely to surface tension. The work of Nicolson[41] on MgO prepared in air did not show this effect and for MgO prepared in vacuum the effect was only 28–60% of that expected. Harrison, *et al.*[47] report that there is no significant dependence of lattice parameter on particle size for NaCl and Libowitz and Bauer[39] report that the lattice parameter of NiO checked to within 0.001 Å for crystallites in the 80–100 Å range. It has been pointed out that the quench rate has a $1/D$ dependence. This may be a factor in retaining vacancies which may be able to decrease the lattice constant by an amount of the order of the defect concentration.[48] The Bragg line shift in thin film, which does not anneal out at temperatures below the melting point, has been attributed to vacancy complexes.[40] Large concentrations of vacancies may be retained by rapid quenching from the melting point which at room temperature can be observed as black spot defects ($\sim$50 Å vacancy clusters) or tetrahedra.[49] Another factor affecting the interpretation is the possibly inhomogeneous stress. Since the fraction of the bulk in any arbitrary thickness of surface of particle also varies as $1/D$, radial inhomogeneity would tend to show such a dependence.

In our data, the colloidal gold studied does not show significant line shift in spite of its small size. This may be explained by pointing out that samples were prepared for diffractometry simply by evaporating the colloidal suspension on a glass slide so that a considerable amount of foreign material, such as gelatin, etc., was included. Undoubtedly, much of this foreign material is adsorbed onto the surface of these particles and it is generally recognized that adsorption of a component at the surface is always accompanied by a decrease of the surface energy or surface tension.[50] Another study of the surface tension of gold by a fugacity method suggests that the apparent surface free energy of colloidal gold particles depends on the method of preparation and history.[51]

The most evident effect on the diffractograms, due to annealing of these samples, was a narrowing of the lines and a line shift toward the normal position as illustrated in Figure 7. The line shift at low-temperature anneals observed in these samples is different than in the case of thin gold films where it has been observed that the lines do not shift to normal positions until the melting point is reached.[40]

The general effect of annealing isothermally on the X-ray determined particle size and lattice constant is shown in Figure 9. With annealing, the particle size seems to approach an equilibrium value roughly two or three times its original size. All samples studied showed a lattice constant somewhat lower than the bulk value for annealing below 600°C, but all showed shift toward the bulk value with heat treatment.

In Table I are the results of all heat treatments of the condensed gold vapor samples including a calculation of the apparent surface tension from equation (5) for each. Attributing the entire lattice shift of the condensed gold vapor (or perhaps thin films)

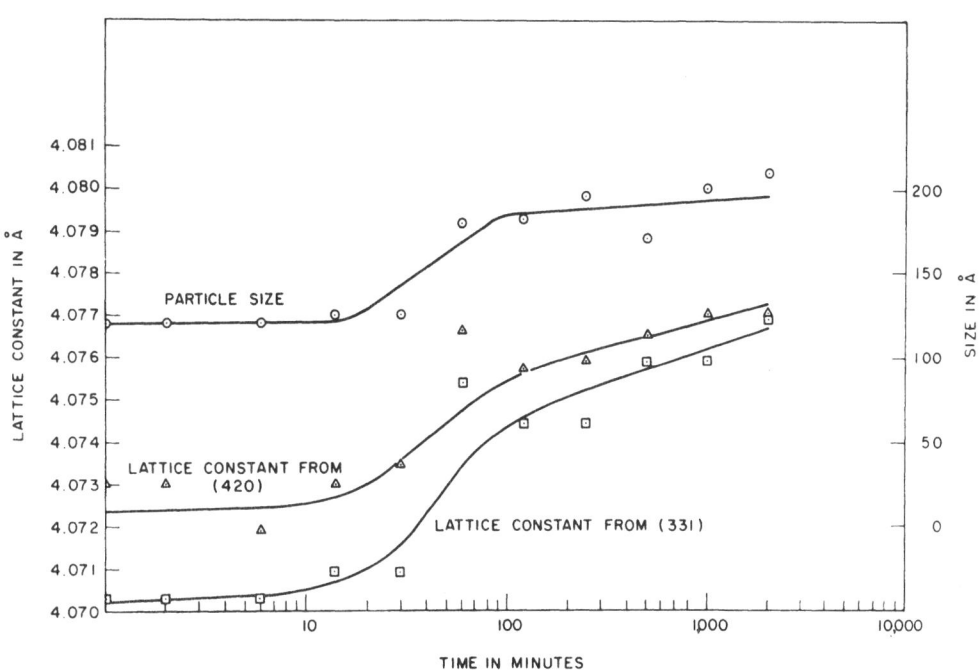

Figure 9. Changes of particle size and of lattice constant with isothermal annealing at 200°C. Data are for a sample from wire exploded at 9 kV.

Table I. Effects of Annealing on Particles of Condensed Gold Vapor [Apparent Surface Tension is Calculated from Equation (5)]

	Pre-anneal				Post-anneal		
Voltage used for wire, kV	Particle diameter by X-ray, Å	Percent decrease in a_0, %	Apparent surface tension, dyne/cm	Treatment	Particle diameter by X-ray, Å	Percent decrease in a_0, %	Apparent surface tension, dyne/cm
9	119	0.21	3100	200°C 2000 min	209	0.04	1050
9	119	0.21	3100	300°C 1000 min	380	0.03	1420
10	160	0.08	1520	200, 300, 400°C 1 hr ea	770	0.03	2890
12	140	0.18	3150	400°C 1 hr	500	0.02	1250
12	160	0.18	3600	300°C 1 hr	270	0.05	1720
12	270	0.05	1720	400°C 1 hr	365	0.03	1360
13	140	0.05	870	450°C 40 min	280	0.02	690
13	144	0.22	4000	450°C 40 min	260	0.01	320
13	126	0.26	4100	600°C 1 hr	570	0.026	2000
14	105	0.27	3550	375°C 1 hr	475	0.09	5200

solely to surface tension effects may be erroneous, since the values thus obtained are considerably higher than those reported in the literature. After annealing, the apparent surface tension agrees fairly well with reported values for particle sizes below 450 Å where reasonably accurate values can be obtained. The larger lattice decrements before anneal may be due to vacancy clusters acting in addition to surface tension. At temperatures near the melting point, nearly all of the equilibrium vacancies are monovacancies[52] and rapid quenching will tend to preserve the total concentration.[10] The aggregation of vacancies may be minimized when the quench is very rapid.[53,54] It is known that at room temperature and for several hundred degrees above, mono- and divacancies diffuse rapidly through a lattice and eventually form tetravacancies and other vacancy clusters.[55] The lattice relaxation around a tetravacancy is smaller than for other vacancies, especially when an atom is centered in the void.[55] Tetravacancies are extremely stable, annealing out at above 600°C,[52,53] and the removal of vacancies does not affect line broadening since point defects do not influence this property.[52]

ACKNOWLEDGMENTS

We wish to thank Mr. Birney R. Fish, Health Physics Division, Oak Ridge National Laboratory for his very kind permission to use the exploding wire facility. Thanks are due also to Dr. Jack B. Greene, Physics Department, Marquette University, for many helpful discussions, and to Dr. Warren J. Deshotels, Chairman, Physics Department, Marquette University, for his encouragement of this work. This project was supported in part by a grant-in-aid from the Marquette University Committee on Research.

REFERENCES

1. W. G. Chace and E. M. Watson, A Bibliography of the Electrically Exploded Conductor Phenomenon, Air Force Cambridge Research Laboratories, AFCRL-62-1053, L. G. Hanscom Field, Mass., 1962.
2. W. G. Chace and E. M. Watson, A Bibliography of the Electrically Exploded Conductor Phenomenon, Supplement No. 1, Air Force Cambridge Research Laboratories, AFCRL-65-384, L. G. Hanscom Field, Mass., 1965.
3. W. G. Chace and E. H. Cullington, Instrumentation for Studies of the Exploding Wire Phenomenon, Air Force Cambridge Research Center, Technical Report 57–235, ASTIA Document No. AD-133842.
4. W. G. Chace and H. K. Moore (eds.), *Proc. Conf. Exploding Wire Phenomenon*, Vol. 1, Plenum Press, New York, 1959, p. 9.
5. F. G. Karioris, B. R. Fish, and G. W. Royster, Jr., "Aerosols from Exploding Wires," in: W. G. Chace and H. K. Moore (eds.), *Proc. Conf. Exploding Wire Phenomenon*, Vol. 2, Plenum Press, New York, 1962, p. 299.
6. A. G. Barkow, F. G. Karioris, and J. J. Stoffels, "X-Ray Diffraction Analysis of Aerosols from Exploding Wires," in: *Advances in X-Ray Analysis*, Vol. 6, Plenum Press, New York, 1963, p. 210.
7. F. G. Karioris and J. J. Woyci, "X-Ray Investigation of Aerosols from Wires Exploded in Nitrogen," in: *Advances in X-Ray Analysis*, Vol. 7, Plenum Press, New York, 1964, p. 240.
8. P. Duwez, R. H. Willens, and W. Klement, Jr., "Continuous Series of Metastable Solid Solutions in Silver-Copper Alloys," *J. Appl. Phys.* 31: 1136, 1960.
9. P. Duwez, R. H. Willens, and W. Klement, Jr., "Metastable Electron Compound in Ag–Ge Alloys," *J. Appl. Phys.* 31: 1137, 1960.
10. R. O. Simmons and R. W. Balluffi, "Measurement of Equilibrium Concentrations of Lattice Vacancies in Gold," *Phys. Rev.* 125: 862, 1962.
11. J. Takamura, "Specimen Size and Quenched-In Lattice Defects," in: R. M. J. Cotterill, M. Doyama, J. J. Jackson, and M. Meshii (eds.), *Lattice Defects in Quenched Metals*, Academic Press, New York, 1965, p. 521.
12. J. J. Jackson, "Strains in Quenched Metals," in: *ibid.*, p. 479.

13. A. Seeger and D. Schumacher, "Quenching Effects in the Noble Metals, and in Nickel and Ni–Co Alloys," in: *ibid.*, p. 15.

14. J. E. Bauerle and J. S. Koehler, "Quenched-In Lattice Defects in Gold," *Phys. Rev.* **107**: 1943, 1957.

15. F. W. C. Boswell, "Precise Determination of Lattice Constants by Electron Diffraction and Variations in the Lattice Constants of Very Small Crystallites," *Proc. Phys. Soc.* **A64**: 465, 1951.

16. F. G. Karioris and B. R. Fish, "An Exploding Wire Aerosol Generator," *J. Colloid Sci.* **17**: 155, 1962.

17. J. R. Asonson, A. G. Emslie, and H. C. McLinden, "Infrared Spectra from Fine Particulate Surfaces," *Science* **152**: 345, April 15, 1966.

18. F. G. Karioris and A. J. Moll, "Production of Submicron Aerosols," Health Physics Div. Ann. Prog. Rept. ORNL-3347, 1962, p. 145.

19. J. Amick and J. Turkevich, "Electron Microscopic Examination of Aerosols Formed in a Direct Current Arc," in: W. E. Kuhn, H. Lamprey, and C. Sheer (eds.), *Ultrafine Particles*, John Wiley & Sons, Inc., New York, 1961, p. 146.

20. National Bureau of Standards Circular 539–1, Table 16, p. 33.

21. R. Bergin, "Quantitative Diffractometric Analysis of X-Ray Transparent Specimens," *J. Sci. Instr.* **41**: 588, 1964.

22. R. H. Geiss, "Determination of Accurate Lattice Parameters Using A Diffractometer," in: *Advances in X-Ray Analysis*, Vol. 5, Plenum Press, New York, 1961, p. 71.

23. R. Asimow, "Precision Lattice Parameter Measurements of Cold-Worked Metals, *J. Appl. Phys.* **31**: 410, 1960.

24. H. P. Klug and L. E. Alexander, *X-Ray Diffraction Procedures*, John Wiley & Sons, Inc., New York, 1954, p. 504.

25. A. Guinier, *X-Ray Diffraction*, W. H. Freeman & Co., San Francisco, 1963, p. 147.

26. W. A. Rachinger, "A Correction for the $\alpha_1\alpha_2$ Doublet in the Measurement of Widths of X-Ray Diffraction Lines," *J. Sci. Instr.* **25**: 254, 1948.

27. J. W. M. DuMond and H. A. Kirkpatrick, "Experimental Evidence for Electron Velocities as the Cause of Compton Line Breadth with the Multicrystal Spectrograph," *Phys. Rev.* **37**: 136, 1931.

28. J. W. M. DuMond, "X-Ray Scattering and Momenta of Electrons," *Rev. Mod. Phys.* **5**: 31, 1933.

29. R. S. Pease, "The Resolution of X-Ray Diffraction Lines into α_1, α_2 Components," *J. Sci. Instr.* **25**: 353, 1948.

30. H. P. Klug and L. E. Alexander, *op. cit.*, Reference 24, p. 500.

31. F. Schossberger, "Amorphous Solids, Small Particles and Thin Surface Films," in: *Advances in X-Ray Analysis*, Vol. 1, Plenum Press, New York, 1957, p. 73.

32. B. E. Warren, "X-Ray Studies of Deformed Metals," *Progress in Metal Physics* **8**: 147, 1959.

33. E. N. Aqua and C. N. J. Wagner, "X-Ray Diffraction of Deformation by Filing in bcc Refractory Metals," *Phil. Mag.* **9**: 565, 1964.

34. L. F. Vassamillett and H. W. King, "Precision X-Ray Diffractometry Using Powder Specimens," in: *Advances in X-Ray Analysis*, Vol. 6, Plenum Press, New York, 1962, p. 142.

35. A. J. C. Wilson, *X-Ray Diffraction by Polycrystalline Materials*, Chapman Hall, London, 1955, p. 392.

36. B. D. Cullity, *Elements of X-Ray Diffraction*, Addison–Wesley, Reading, Mass., 1956, p. 334.

37. H. P. Klug and L. E. Alexander, *op. cit.*, Reference 24, p. 464.

38. H. M. Otte, "Lattice Parameter Determinations with an X-Ray Spectrogoniometer by the Debye–Sherrer Method and the Effect of Specimen Condition," *J. Appl. Phys.* **32**: 1536, 1961.

39. G. G. Libowitz and S. H. Bauer, "Electron Diffraction Determination of Lattice Parameters of Polycrystalline Specimens Giving Broad Diffraction Peaks. 1. Discussion of Technique and Analysis of Errors," *J. Phys. Chem.* **59**: 209, 1955.

40. J. E. Davey and R. H. Deiter, "Structure in Textured Gold Films," *J. Appl. Phys.* **36**: 284, 1965.

41. M. M. Nicolson, "Surface Tension in Ionic Crystals," *Proc. Phys. Soc.* (*London*) **A228**: 490, 1955.

42. R. Shuttleworth, "The Surface Tension of Solids," *Proc. Phys. Soc.* (*London*) **63A**: 444, 1950.

43. C. Herring, "The Use of Classical Macroscopic Concepts in Surface-Energy Problems," in: H. Gomer and C. S. Smith (eds.), *Structure and Properties of Solid Surfaces*, The University of Chicago Press, Chicago, 1953, p. 5.

44. A. J. Shaler, "The Mechanical Properties of Crystalline Metal Surfaces," in: R. Gomer and C. S. Smith (eds.), *Structure and Properties of Solid Surfaces*, The University of Chicago Press, Chicago, 1953, p. 120.

45. J. C. Fisher and C. G. Dunn, "Surface and Interfacial Tensions of Single-Phase Solids," in: W. Shockley, J. H. Halloman, R. Mauer, and F. Seitz (eds.), *Imperfections in Nearly Perfect Crystals*, John Wiley & Sons, Inc., New York, 1952, p. 317.

46. T. B. Rymer and C. C. Butler, "Determination of Structure of Gold Leaf by Electron Diffraction," *Proc. Phys. Soc. (London)* **59**: 541, 1947.

47. L. G. Harrison, J. A. Morrison, and G. S. Rose, "An Investigation of Chloride Ion Diffusion in Subsurface Layers of Sodium Chloride by an Isotopic Exchange Technique," *J. Phys. Chem.* **61**: 1314, 1957.

48. P. H. Miller, Jr., and B. R. Russell, "Effect of Internal Strains on Linear Expansion, X-Ray Lattice Constant, and Density of Crystals," *J. Appl. Phys.* **23**: 1163, 1952.

49. G. Thomas and J. Washburn, "Precipitation of Vacancies in Metals," *Rev. Mod. Phys.* **35**: 992, 1963.

50. A. S. Michaels, "Fundamentals of Surface Chemistry and Surface Physics," in: Symposium on Properties of Surfaces, ASTM Materials Science Series-4, ASTM Special Technical Publication No. 340, American Society for Testing and Materials, Philadelphia, Pa.

51. J. D. Haygood, unpublished dissertation, in: *Dissertation Abstracts* **25**(5): 2779, 1964 (Eng.).

52. R. W. Balluffi, J. S. Koehler, and R. O. Simmons, "Present Knowledge About Point Defects in Deformed Face-Centered-Cubic-Metals," in: L. Himmel (ed.), *Recovery and Recrystallization of Metals*, Interscience Publishers, New York, 1963, p. 38.

53. J. S. Koehler, F. Seitz, and J. E. Bauerle, "Interpretation of the Quenching Experiments on Gold," *Phys. Rev.* **107**: 1499, 1957.

54. L. M. Clarenbrough, R. L. Segall, M. H. Loretto, and M. E. Hargreaves, "The Annealing of Vacancies and Vacancy Aggregates in Quenched Gold, Silver and Copper," *Phil. Mag.* **9**: 377, 1964.

55. M. Doyama and R. M. J. Cotterill, "Stable and Metastable Tetravacancies in an fcc Metal," *Phys. Rev.* **137**: A994, 1965.

LINE SHAPE ANALYSIS OF DEFORMED Cu–Ge ALLOYS

M. Ahlers and L. F. Vassamillet

Mellon Institute
Pittsburgh, Pennsylvania

ABSTRACT

The asymmetry of diffraction peaks of deformed α-Cu–Ge alloy filings is determined by the center of gravity and Fourier coefficient method with computer calculations. It is shown that it is not possible to describe the asymmetry by a simple value that is a characteristic of the extent of the deformation and the Miller indices of the peak. Instead the asymmetry is highly dependent upon the initial arbitrary conditions for evaluation, which throws some doubt on the reliability of published twin fault and stacking fault probabilities.

INTRODUCTION

The use of the X-ray diffraction pattern from deformed metal samples to determine the density of stacking faults in face-centered cubic metals and alloys has been well established.[1-7] However, a number of problems in measurement have become apparent in the course of investigations using this method.[3,8,9] So that there will be no misunderstanding, let us begin by defining some of the terms that we would like to use. By X-ray diffraction profile, we mean the shape of the X-ray diffraction pattern associated with a particular *hkl* reflection. The term "peak" will be used to designate this part of the diffraction pattern. The peak has a maximum value and when the peak maximum is meant the word maximum will be used. The individual peak also may have other measurables associated with it, such as, for example, its center of symmetry, if it has one, or its center of gravity or centroid. Because of the difficulties associated with the measurement of peak maxima and, in particular, the measurement of the shift in peak maxima on heavy deformation of highly alloyed α-phase fcc materials where the line breadth is very large and because it seemed as though the use of the entire profile would be a simpler and more direct method of getting a reproducible measurable, the centroid was originally chosen[7] as the means of determining the relative positions of the peaks before and after deformation. This procedure has been criticized on the ground that the presence of any asymmetry in the peak would markedly affect the position of the centroid. In fact, the difference in position of the maximum and the centroid has been used as a method for measuring the twin faulting parameter under the assumption that peak asymmetry is due to the presence of twin faults.[8] That there is peak asymmetry is obvious, but the source of such asymmetry is still somewhat open to debate. It was noted[10] in some studies on the α-phase Cu–Ge alloys that the shifts in the individual peak positions were not consistent with those predicted from the Paterson theory,[1] with nearly all of the shift taking place on the (200) peak of the (111)-(200) measured pair. This effect is probably due to a change in lattice parameters on deformation.[11,12]

The method of analyzing the data obtained from the measurement of the peak to determine the characteristic parameter to be associated with the peak shift has been reinvestigated in our laboratory. To summarize our conclusions briefly before launching into a discussion of the particular nature of these results, we find that the procedure of measurement of the centroids was oversimplified, and that some of the early results obtained using this method may have been a consequence of the accidental choice of the parameters used in the procedure of determining the centroids. This is not to say that the peak shifts as reported previously are necessarily erroneous but does indicate that the conclusions derived from them may be open to some criticism.

In carrying out the determination of the centroid of a diffraction peak, one must first determine the starting and ending points for the analysis. Thus, the limits have to be assigned at the beginning of the calculation. One usually takes "symmetrical limits"; that is, one picks a center of the line as nearly as can be judged by eye and then takes limits equally distant from this center; we have found that the measured centroid of such a peak is dependent not only upon the limits chosen or to put it another way, the number of half-widths over which the calculations are carried out, but also on the original choice of the center. The technique originally used for centroid determination is that developed by Pike and Wilson[13] and is based on the statements made by them that determination of a peak extending over a region of five half-widths should be sufficient for accurate determination of the centroid and for which it is essentially insensitive to the limits. However, Pike and Wilson's experiments were performed on well-annealed sharp diffraction profiles, whereas, our concern here is in almost entirely with the very broad peaks that one obtains in heavily faulted and deformed materials. Although one can include in each of the (111)–(200) peak pair nearly all of the diffracted energy in four half-widths without having the regions of analysis between the two lines overlap (i.e., be included in the analysis of both peaks). Nevertheless, the peaks clearly do overlap in the regions between them and a method for correcting this overlap is obviously necessary.

As an alternative procedure, it is possible to analyze each peak independently in terms of some expansion, such as the Fourier series expansion that we have used. Using the method of Stokes,[14] a correction of the profile for instrumental line broadening can be performed using the profile obtained from a carefully annealed specimen of the same material. From the corrected Fourier transforms of the peak, the broadening profile can be constructed and the pertinent parameters such as the centroid and the maximum determined. We will consider this technique and the results that may be expected from its use.

UNCERTAINTY IN THE CENTROID DETERMINATION

The difficulties that are encountered in determining the center of gravity of a diffraction line are shown in Figure 1. The peak being analyzed here is the (200) reflection from a copper–9 at.% germanium alloy. We have plotted here the center of gravity vs. the number of half-widths used in the integration range for a series of different starting points or centers. The center values are labeled N_0. Note that the calculated values of the center of gravity tend to converge toward a common value for numbers of half-widths in the vicinity of four to five. However, there is still a large uncertainty as to the center of gravity depending on the number of half-widths and on the point, N_0. The various N_0 values differ from each other by one step in the step scanning process. In the second half of this figure, we have plotted the calculated center of gravity vs. the value of N_0 for two specified half-width ranges. One is for a four half-widths range and the other corresponds to the *value obtained by extrapolation to zero* half-widths. It can be seen that the dependence of the center of gravity on the point chosen for center is not very large for the case

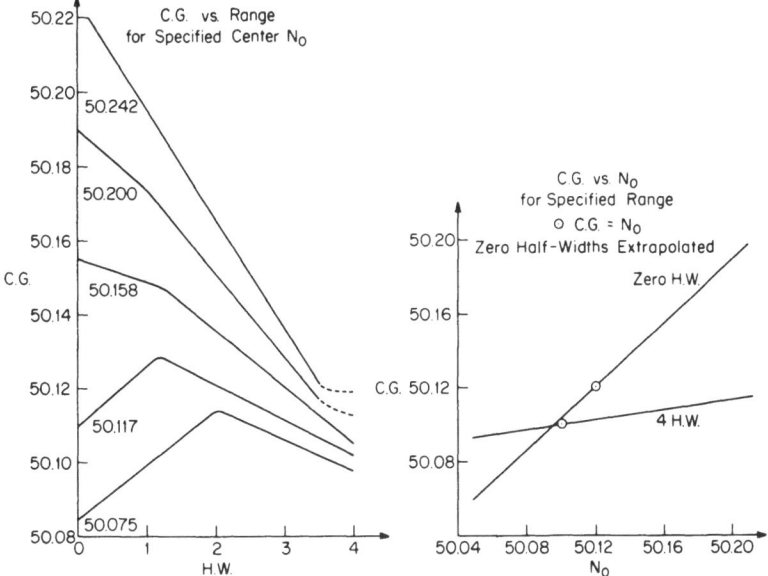

Figure 1. For Cu–9 at.% Ge, (200) peak: (a) Peak center of gravity (C.G.) *vs.* integration range in half-widths (H.W.) for specified peak center N_0. (b) Peak center of gravity *vs.* peak center N_0 for 4 half-width range and an extrapolation to zero half-width range.

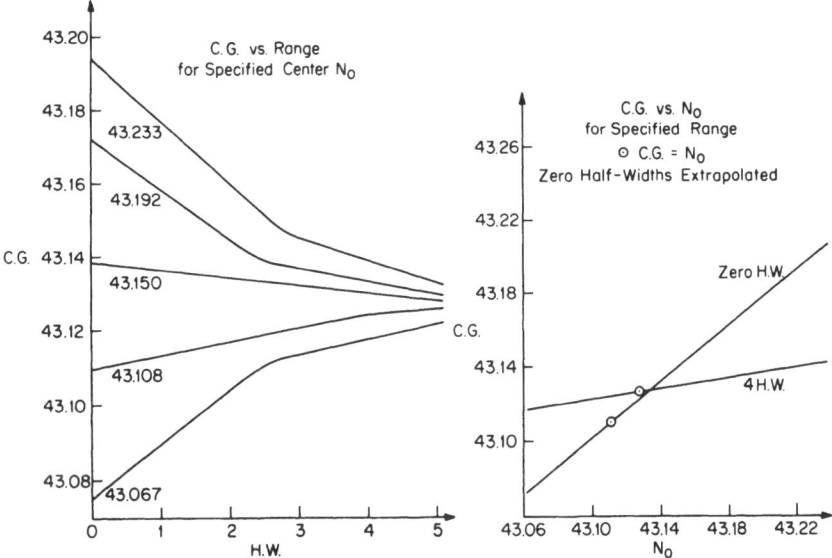

Figure 2. For Cu–9 at.% Ge, (111) peaks: (a) Same as 1a. (b) Same as 1b.

of the four half-width range. It does, nevertheless, exceed in uncertainty the values that have been assigned for the determination of the position of the centroid of the peak. The uncertainty is shown to be at least $0.02°$ 2θ. The dependence of the center of gravity on N_0 in the case of extrapolation to zero half-width is so large that this procedure is essentially meaningless. Figure 2 shows the results obtained for the (111) reflection of the same alloy. Here the dependence of the centroid on the number of half-widths is much smaller probably due to the greater intensity in the (111) peak, a smaller half-width and, consequently, less influenced by the adjacent (200) peak. May I draw your attention here to the fact that the form of the centroid *vs.* number of half-widths curve is different.

The essential result of this analysis and the uncertainty that evolves from it is demonstrated in Figure 3 where the variation in peak separation $\Delta 2\theta$ is plotted *vs.* the alloy composition for deformation in the Cu–Ge alloy series. The various curves represent the results obtained in using different means of identifying the position of the peak. Note that there is a spread in possible choices of the amount of peak shift obtainable depending on the way in which the peak position has been identified, and the amount of peak shift is strongly dependent on the number of half-widths chosen as the range of numerical integration.

Fourier Analysis

It was anticipated that in turning to the Fourier analysis in order to identify the pertinent parameters of the diffraction peaks, that the extrapolation processes would be simplified by the elimination of the instrumental broadening asymmetry components. In performing the Fourier analysis, the measured intensities have been corrected for Lorenz polarization as well as the θ dependence that comes from dealing with a powder.[3] As in the case of the previous part of the study, a trapezoidal background scattering has been assumed. Since the deconvolution process involves a correction of the broadened peak using the sharp profile from the well-annealed sample, it becomes necessary to establish the accuracy of the two 2θ scales on which both have been measured. There is, of course,

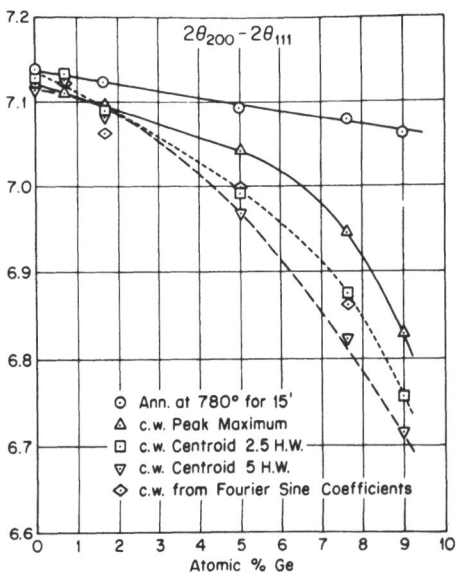

Figure 3. Variation of peak separation $2\theta_{200} - 2\theta_{111}$ *vs.* composition. Circles—annealed 15 min at 780°C; triangles—cold-worked, peak maxima; squares—cold-worked, centroid, 2.5 half-widths (H.W.); inverted triangles—cold-worked, centroid, 5 half-widths; diamonds—cold-worked, centroid from Fourier sine coefficients.

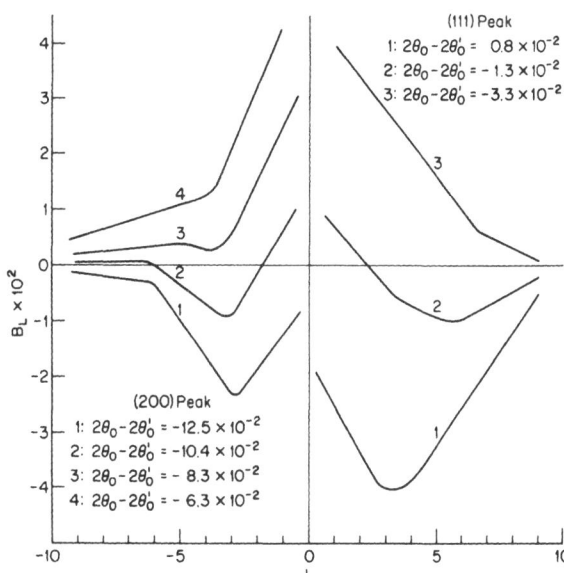

Figure 4. Variation of sine coefficients B_L vs. L for various separation of the centers of expansions $2\theta_0$-$2\theta'_0$, of the cold-worked and annealed peaks.

an unknown difference ϵ in the relative positions of the two 2θ scales. However, it can be assumed that the two sets of profiles in the peak pair (111)–(200) will have the same displacement in the two theta scales. For each profile, the coefficients have been determined in terms of an expansion around an origin, $2\theta_0$, for the deformed peak and $2\theta'_0$ for the annealed peak. Symmetrical limits have been chosen in the 2θ scale. If the only asymmetry in the broadened peak were due to the asymmetry from instrumental causes, and if the two origins were properly matched, then the resultant pure diffraction profile would be symmetrical and there would be no sine terms in the expansion. If there were no asymmetry in the deformed peak, and the two expansion centers were not properly chosen, then there would be sine terms in the Fourier expansion. Figure 4 shows the sine coefficients $B(L)$ for both the (111) and (200) peaks calculated for a series of varying expansion centers. The general similarity of the Fourier coefficients for the two peaks should be noted, one set being essentially the negative of the other. Also note the manner in which the magnitude of the coefficients shift with changes in the value of $\theta_0 - \theta'_0$ (that is, the displacement of the two origins). We have found that the values of the coefficients, the magnitudes, are dependent not on the position of the origins, but on their relative displacements.

It is easily shown that the center of gravity of the true diffraction profile with respect to the centers of expansion θ_0 and θ'_0 is given by

$$\langle x \rangle_{\text{c.g.}} = -\frac{(dB/dL)_{L \to 0}}{2\pi i H(0)}$$

Thus, in principle, it should be easy to obtain from the low order values of the expansion terms, the value for the position of the center of gravity. You will note that an extrapolation of the B's to the origin at $L = 0$ is not meaningful because of the very rapid change in the magnitude of the B's of low order. In addition, the value of $B(0)$ is indeterminate. Thus, the hope that the procedures of extrapolation would be superior in the Fourier series analysis of the diffraction profile has not been borne out. We have, however, estimated the values of the slopes using the additional assumption that all the asymmetry is due to twin faulting and is of the same order for both peaks; i.e., the value of the slope is the same for (200) and (111) peak for equal $B(L)$'s. The resulting value for the peak shift so obtained has also been plotted on Figure 3 and is indicated by the diamonds. Note that the magnitude of the peak shift is equal to that obtained by the previous method, but it should be clear that the uncertainty in the values obtained in this way are easily as great or greater than those obtained by a straight integration of the diffracted intensity.

It is well known that one of the limiting conditions in the use of a Fourier series expansion is the experimental difficulty of establishing the true background. We are faced with the same problem here, irrespective of the method used for the analysis of the peak. The difficulties in establishing a background and the effect upon the choice of the

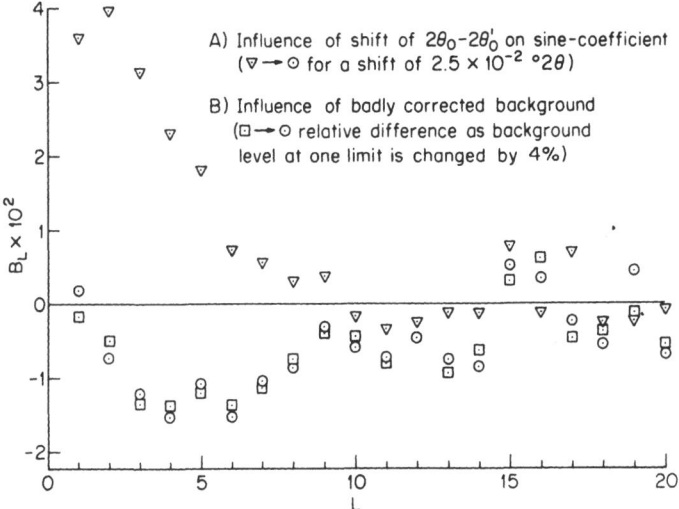

Figure 5. Dependence of the sine coefficients B_L for different analytical conditions for the (111) peak of Cu–5 at.% Ge. Triangles—$2\theta_0$ displacement of the 2θ scales; circles—2θ scales shifted by $0.025°$ 2θ; squares—background level at one limit of integration range is changed by 4%.

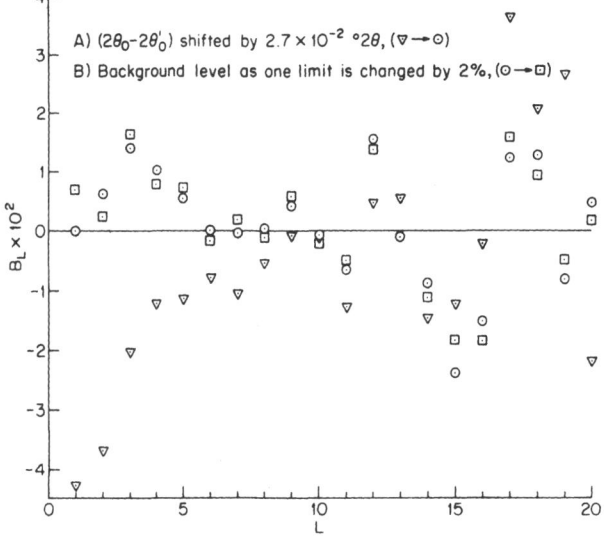

Figure 6. Same as Figure 5 analysis of (200) peak. Background level at one limit of integration range changed by 2%.

background levels are shown in Figures 5 and 6 where the sine coefficients have been plotted over a considerably larger range than those which we have seen in the previous figures. In Figure 5, the shift in the values of the B's in accordance with a change in the relative displacement of the origin $\theta_0 - \theta'_0$ is shown. In addition, the shift in the values of the B's when a slightly different slope in the background level also is shown. The effect upon the lower orders of the B's, B_1 and B_2 is quite large making even more uncertain the value of dB/dL extrapolated to $L = 0$. Figure 6 contains the same data for the (200) peak. The effect of a slight change in the background is even more dramatic in this case.

CONCLUSIONS

Thus, we must conclude that with standard diffraction procedures, the measurables of a broadened peak are such that they cannot be determined to better than $0.02°$ of 2θ, and that the centroid of a peak can be determined with an accuracy not significantly greater than the peak maximum can be determined. Particular difficulty is encountered when using the centroid method since the asymmetry of the lines and the overlap of the peaks between them introducing a further effect on the asymmetry may give misleading results as to the extent of the individual peak shifts.

ACKNOWLEDGMENT

This research has been supported in part by the United States Air Force Office of Scientific Research, Washington, D.C.

REFERENCES

1. M. S. Paterson, "X-Ray Diffraction by Face-Centered Cubic Crystals with Deformation Faults," *J. Appl. Phys.* **23**: 805, 1952.
2. B. E. Warren and E. P. Warekois, "Stacking Faults in Cold Worked Alpha-Brass," *Acta Met.* **3**: 473, 1955.
3. B. E. Warren, "X-Ray Measurement of Twin Faulting in Face-Centered Cubic Crystals," *Australian J. Phys.* **13**: 384, 1960.
4. C. N. J. Wagner, "Stacking Faults by Low Temperature Cold Work in Copper and Alpha-Brass," *Acta Met.* **5**: 427, 1957.
5. H. M. Otte, "Lattice Parameter Studies of Annealed, of Aged, and of Cold-Worked Alpha Brass," *J. Appl. Phys.* **33**: 1436, 1962.
6. R. G. Davies and R. W. Cahn, "Stacking Fault Densities in Filmings of Some Copper- and Silver-Base Solid Solutions," *Acta Met.* **10**: 621, 1962.
7. L. F. Vassamillet, "Stacking Fault Probability of Noble Metal–Zinc Alloys," *J. Appl. Phys.* **32**: 778, 1961.
8. J. B. Cohen and C. N. J. Wagner, "Determination of Twin Fault Probabilities from the Diffraction Patterns of fcc Metals and Alloys," *J. Appl. Phys.* **33**: 2073, 1962.
9. D. E. Mikkola and J. B. Cohen, "Effects of Thermal-Mechanical Treatments on Faulting in Some fcc Alloys," *J. Appl. Phys.* **33**: 892, 1962.
10. L. F. Vassamillet and T. B. Massalski, "X-Ray Measurements of Faulting in α-Cu-Ge Alloys," *J. Appl. Phys.* **35**: 2629, 1964.
11. H. M. Otte, "Lattice Parameter Determinations with an X-Ray Spectrogoniometer by the Debye–Scherrer Method and the Effect of Specimen Condition," *J. Appl. Phys.* **32**: 1536, 1961.
12. C. N. J. Wagner and J. C. Helion, "X-Ray Measurements of Stacking Faults and Internal Strains in α-Cu–Zn and α-Cu–Sn," *J. Appl. Phys.* **36**: 2830, 1965.
13. E. R. Pike and A. J. C. Wilson, "Counter Diffractometer—The Theory of the Use of Centroids of Diffraction Profiles for High Accuracy in the Measurement of Diffraction Angles," *Brit. J. Appl. Phys.* **10**: 57, 1959.
14. A. R. Stokes, "A Numerical Fourier Analysis Method for the Correction of Widths and Shapes of Lines and X-Ray Powder Photographs," *Proc. Phys. Soc. London* **61**: 382, 1948.

DISCUSSION

J. J. Wert (Vanderbilt University): On the previous paper and also on this one, too, it might be well to eliminate the K_{α_2} component by using K_β radiation rather than the K_{α_1} and K_{α_2}.

L. F. Vassamillet: This is a fine idea, but for that I would want a really high-powered source, because of the low-intensity level of the broadened lines. The line broadening gets to be so great with respect to the intensity available for reasonable data collecting times that rather strenuous requirements are placed on the generators, possible sources of noise and, of course, on the number of such analyses that can be made. This is a horribly time-consuming process. I would hate to go to the K_β for that purpose.

ANOMALOUS RESIDUAL STRESSES

R. E. Ricklefs and W. P. Evans

Caterpillar Tractor Company
Peoria, Illinois

ABSTRACT

Residual stresses were measured in hardened and tempered specimens after unidirectional plastic extension. X-ray and strain gage-layer removal methods were compared. Anomalous residual stresses were found in extended samples at hardnesses of Rc 32–35. The X-ray method indicated compressive residual stresses of nearly constant magnitude through $\frac{1}{3}$ the thickness of flat samples, while the strain gage-layer removal method indicated that no macrostress existed. A constant anomalous residual stress was also seen by X-ray through $\frac{2}{3}$ the thickness of a cylindrical specimen deformed uniformly in tension. Little or no anomalous stress was found in an extended specimen at Rc 55 or in a specimen at Rc 44 after uniform bending.

INTRODUCTION

When a metal is unidirectionally and uniformly deformed, the X-ray diffraction method may indicate a residual stress to exist while mechanical methods do not. For example, Donachie and Norton[1] reported an "anomalous" X-ray stress in annealed aluminum and ingot iron strips after plastic extension. Taira and Yoshioka[2] reported similar anomalies in soft plain carbon steel. Hyler and Jackson[3] reported an anomalous component of stress induced by bending in annealed rail steel. Macherauch,[4,5] on the other hand, found a real residual macrostress in the surface layers of small, cylindrical specimens of aluminum, nickel, and copper after uniaxial extension.

The anomalous stress evidently requires unidirectional plastic flow. It also may depend upon specimen geometry or size as this affects restraint during extension. Uniform bending should be included because the surfaces are deformed predominantly in one direction; however, in this case a macrostress component also would be present due to the nonuniform straining. When residual stresses are induced by the distortions accompanying hardening, shot peening, or machining, good agreement is obtained between measurements of these stresses by X-ray and mechanical methods.[6,7]

Because much of the reported work had been on soft metals, it was of interest to investigate the effect in steel at higher hardness. Flat specimens of SAE 1045 steel were used, water quenched and tempered to cover a hardness range of Rc 32–55. Some specimens were unidirectionally, plastically extended a few percent, and one was deformed by uniform bending. Residual stresses were measured both by X-ray diffraction and by a strain gage-layer removal method. To investigate the possible effect of geometry, a cylindrical tensile specimen of SAE 10B35 steel at Rc 32 hardness also was deformed uniaxially in tension. Residual stress was measured by X-ray after layer removal. Possible reasons for the anomaly, when present, are discussed.

METHODS

Specimen Preparation

Compositions of the steels used are given in Table I. Flat tensile specimens were machined as shown in Figure 1 from ⅜-in. SAE 1045 plate stock. The longitudinal axis of the specimens corresponded to the rolling direction of the plate. The specimens were austenitized at 1500°F in a slightly oxidizing atmosphere and water quenched. Various tempering temperatures were used to give the hardnesses shown in Table II. No further surface preparation was given these specimens. The cylindrical tensile specimen was machined from forged SAE 10B35 steel (plain carbon plus boron) as shown in Figure 1. It was austenitized at 1560°F in a neutral atmosphere and water quenched. Approximately 0.010 in. was ground from the radius of the reduced section after heat treatment, and threads were machined. To remove grinding stresses, 0.005 in. was removed from the radius of this specimen by electropolishing in a bath of concentrated 60% phosphoric plus 40% sulfuric acids.

Specimen Deformation

Flat specimens Nos. 1, 2, and 3 and the cylindrical specimen No. 5 were strained 4–6% plastically in uniaxial tension in a tensile machine. The strain was approximately uniform along the gage length except in specimen No. 3, which "necked" to some extent. Specimen No. 4 was uniformly strained 6% plastically by bending over a steel mandrel of 8-in. diameter.

Table I. Chemical Analyses of Steels (wt. %)

	C	Mn	S	Si	Ni	Cr	Mo	Cu	B	Ti	Al
SAE 1045	0.44	0.76	0.02	0.22	0.06	0.04	<0.01	0.07		0.02	0.04
10B35	0.35	0.70	0.03	0.24	0.22	0.09	0.04	0.13	0.0005	0.03	0.03

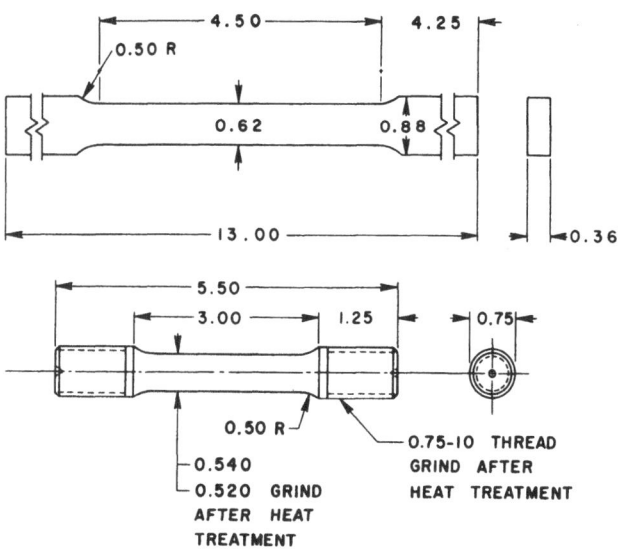

NOTE: ALL DIMENSIONS IN INCHES.

Figure 1. Dimensions of flat and cylindrical specimens.

Table II. Specimens Used

Specimen number	Tempering temperature, °F	Hardness, Rc	Approximate plastic strain, %	Strain gage type
1 and 2, flat	800	32–35	4, extension	ABFX-11 rosette
3, flat	325	55	6, nonuniform extension	AB-7, longitudinal in necked area
4, flat	600	44	6, bending	AB-7, longitudinal and transverse
5, cylindrical	800	32	$4\frac{1}{2}$, extension	

Strain Gage Technique

Bonded wire resistance strain gages of the types recorded in Table II were applied to one face of uniaxially strained flat specimens, and to both the concave and convex faces of the bent specimen near opposite ends of the gage length. Gages were oriented in both the longitudinal and transverse directions except on specimen No. 3, in which only longitudinal strain was measured. Strain was measured with a Baldwin SR-4 strain indicator after removing layers from the face opposite the gage by electropolishing. The gages were protected with stop-off lacquer.

The stress state existing before layer removal was calculated from strain gage readings using the method of Leeser and Daane.[8] Using modified terminology, the initial strain in the longitudinal direction at any point h is given by

$$\epsilon_L = 2\epsilon_{gL} + \frac{1}{2}h\frac{d\epsilon_{gL}}{dh} - 3h\int_h^H \frac{\epsilon_{gL}}{h^2}\,dh \tag{1}$$

where H is the original thickness, h is the thickness remaining after layer removal, ϵ_g is the change in strain at the gage surface, v is Poisson's ratio, 0.29, and E is Young's modulus, 30×10^6 psi. Using subscripts L and T to refer to the longitudinal and transverse directions, respectively, the corresponding stress for the uniaxial case is given by

$$\sigma_L = E\epsilon_L \tag{2}$$

and for the biaxial case by

$$\sigma_L = \frac{E}{1 - v^2}(\epsilon_L + v\epsilon_T) \tag{3}$$

Strain and stress in the transverse direction are obtained from the same equations with the subscripts L and T interchanged.

X-Ray Technique

Longitudinal residual stresses in all flat specimens and transverse stresses in specimens Nos. 1 and 2 were calculated using the two-exposure tilt method described in an SAE Technical Report [7] and the equation

$$\sigma = \frac{E}{1 + v} \cdot \frac{1}{\sin^2 \psi} \cdot \cot \theta(\theta_\perp - \theta_\psi) \tag{4}$$

where $\theta_\perp$ and θ_ψ are the angles of diffraction of crystal lattice planes, respectively parallel to the surface, and at an angle ψ from the surface about an axis normal to both the surface and the direction of σ along the surface. A General Electric XRD-5 X-ray diffractometer was used with chromium $K\alpha$ radiation, the 211 diffraction peak of iron, and a ψ angle of 45°. Bulk values of 30×10^6 psi for E and 0.29 for ν were again used. These have been shown to yield results in agreement with stresses determined by strain gage methods when chromium radiation and the 211 iron peak are used.[7] Surface stresses were determined, and stresses in depth were obtained in the electropolished surfaces. Stresses in the longitudinal and tangential directions were measured in the cylindrical specimen at the surface and at depths until the center of the bar was approached. A few measurements were made using cobalt radiation and the 310 and 220 iron peaks, and were compared to the results using chromium and the 211 peak.

The stress states existing before layer removal, assuming that stresses measured were macrostresses balanced by other layers through the specimen, were calculated for specimens No. 3 and 4 by the method also described in the SAE Report.[7] Using slightly modified terminology, the initial stress existing before layer removal at a point h in the longitudinal direction is given by

$$\sigma_L = \sigma_{mL} + 2 \int_h^H \frac{\sigma_{mL}}{h}\, dh - 6h \int_h^H \frac{\sigma_{mL}}{h^2}\, dh \tag{5}$$

where σ_m is the measured stress at position h.

The sums of the principal stresses in specimens Nos. 1 and 2 were calculated from shifts in diffraction angle of X-ray peaks from 211 planes normal to the surface by

$$\epsilon_\perp = -\cot\theta\, \Delta\theta \tag{6}$$

obtained by differentiating Bragg's law, and

$$\epsilon_\perp = -\frac{\nu}{E}(\sigma_L + \sigma_T) \tag{7}$$

for a biaxial stress state from elasticity theory, where $\epsilon_\perp$ is the strain normal to the surface. Combining these equations, we obtain

$$\sigma_L + \sigma_T = \frac{E}{\nu}\cot\theta\, \Delta\theta \tag{8}$$

A reference specimen, heat treated in the same manner as the deformed specimens, was electropolished to remove the as-heat-treated surface and used to obtain the stress-free diffraction angle. A correction was made for small stresses in this specimen determined by the two-exposure tilt method previously mentioned.

RESULTS

The results for flat specimens Nos. 1, 2 and 3 strained in uniaxial tension are given in Figures 2–8. Specimens Nos. 1 and 2 were plastically strained approximately uniformly over at least a 1-in. gage length and exhibited anomalous residual stresses as shown by Figures 3 and 6. Residual stresses determined by X-ray were 30,000–35,000 psi compression in the longitudinal direction, which was the direction of straining. This stress was approximately constant through more than $\frac{1}{3}$ the specimen thickness. As "stressed" layers

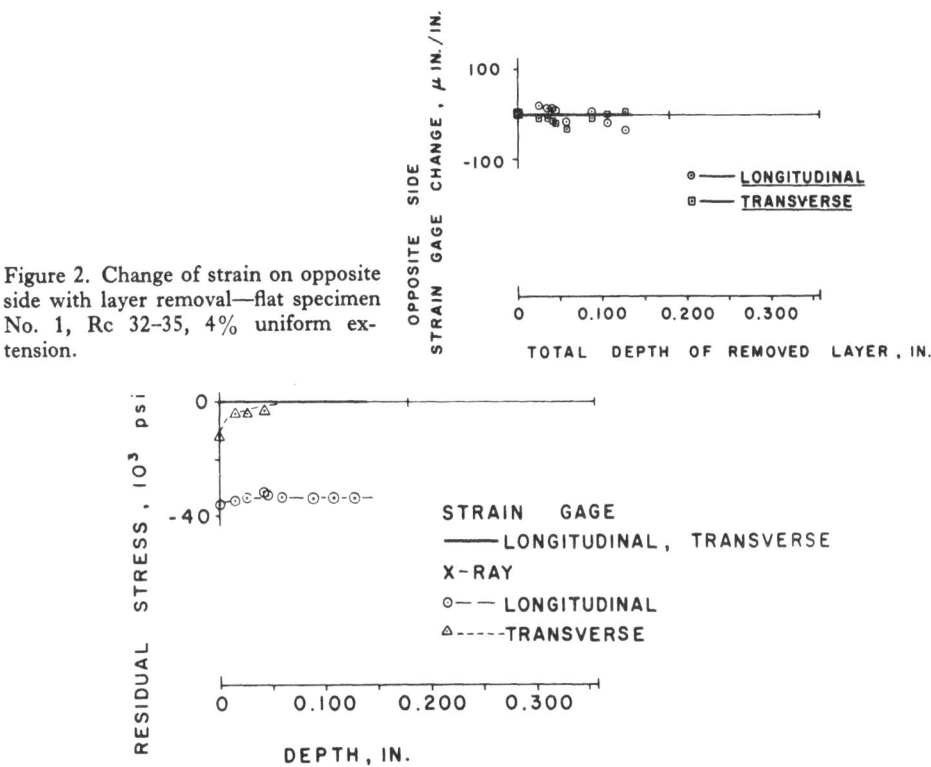

Figure 2. Change of strain on opposite side with layer removal—flat specimen No. 1, Rc 32-35, 4% uniform extension.

Figure 3. Residual stress by strain gage and X-ray—specimen No. 1.

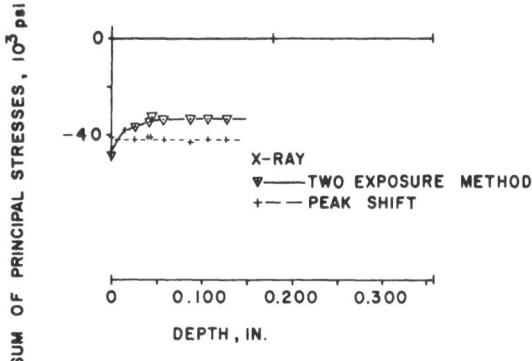

Figure 4. Residual stress by two X-ray methods—specimen No. 1.

were removed by electropolishing, no bending of the bar occurred. The strain gages mounted on the opposite side of the specimen indicated approximately no change in either the longitudinal or transverse directions as shown in Figures 2 and 5. There was, therefore, no macrostress in the usual sense and consequently no justification for correcting the X-ray results for "stressed" layers removed. In addition, the X-ray method indicated a small surface stress in the transverse direction of 12,000 psi compression.

X-ray peak shift results agreed well with results from the standard two-exposure method, as shown in Figure 4. This means that the peak shift was due entirely to residual strain (of whatever origin), and was not a result of faulting.

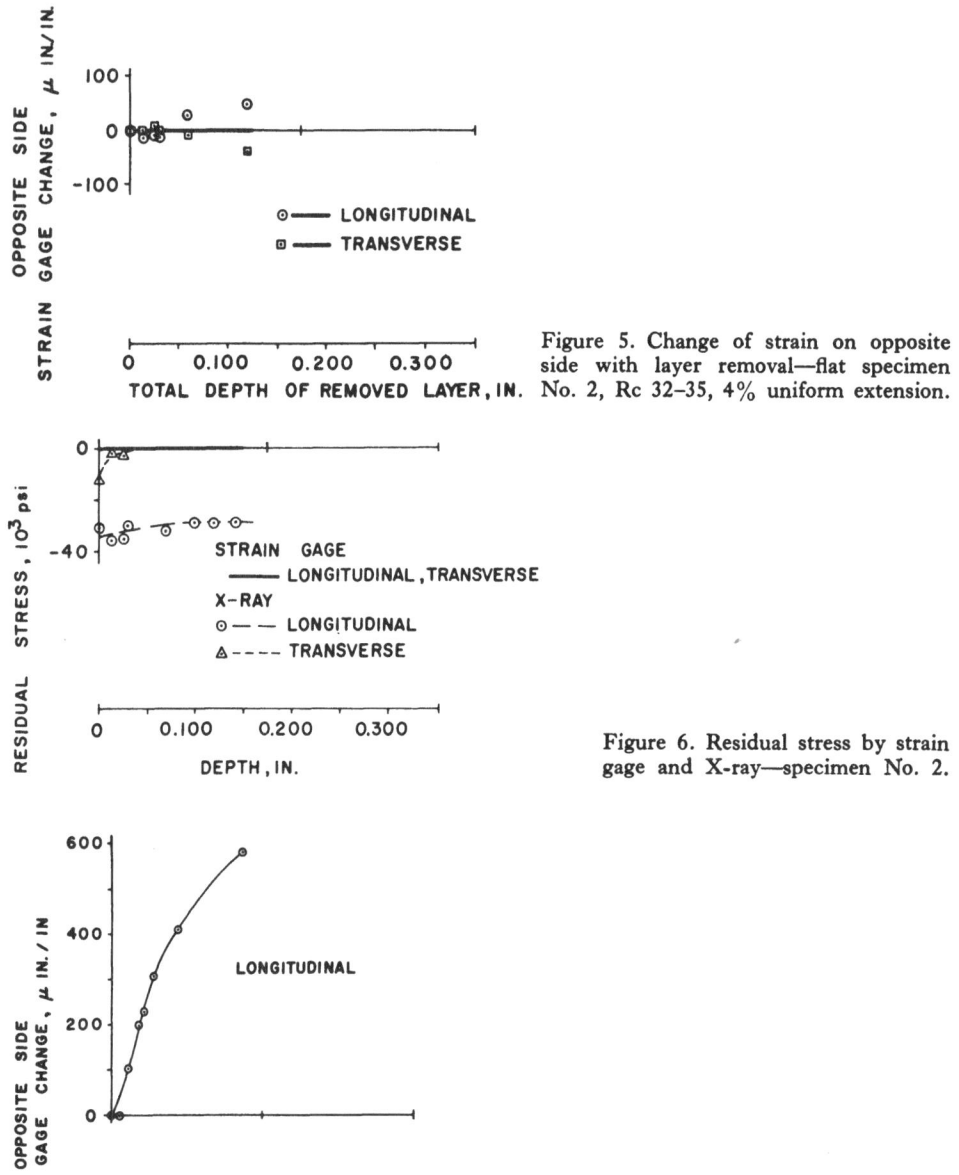

Figure 5. Change of strain on opposite side with layer removal—flat specimen No. 2, Rc 32–35, 4% uniform extension.

Figure 6. Residual stress by strain gage and X-ray—specimen No. 2.

Figure 7. Change of strain on opposite side with layer removal—flat specimen No. 3, Rc 55, 6% nonuniform extension.

The plastic straining of specimen No. 3 was nonuniform, and the anomaly, if it existed, was not as definite. The results for the necked region of this specimen are shown in Figures 7 and 8. Strain gage results of Figure 7 indicated the existence of a macrostress. Although the reason for this is not definitely known, a biaxial stress may have been developed by nonuniform deformation to cause the effect. Residual stress was calculated from the strain gage data assuming a uniaxial stress state because only a longitudinal gage had been mounted.

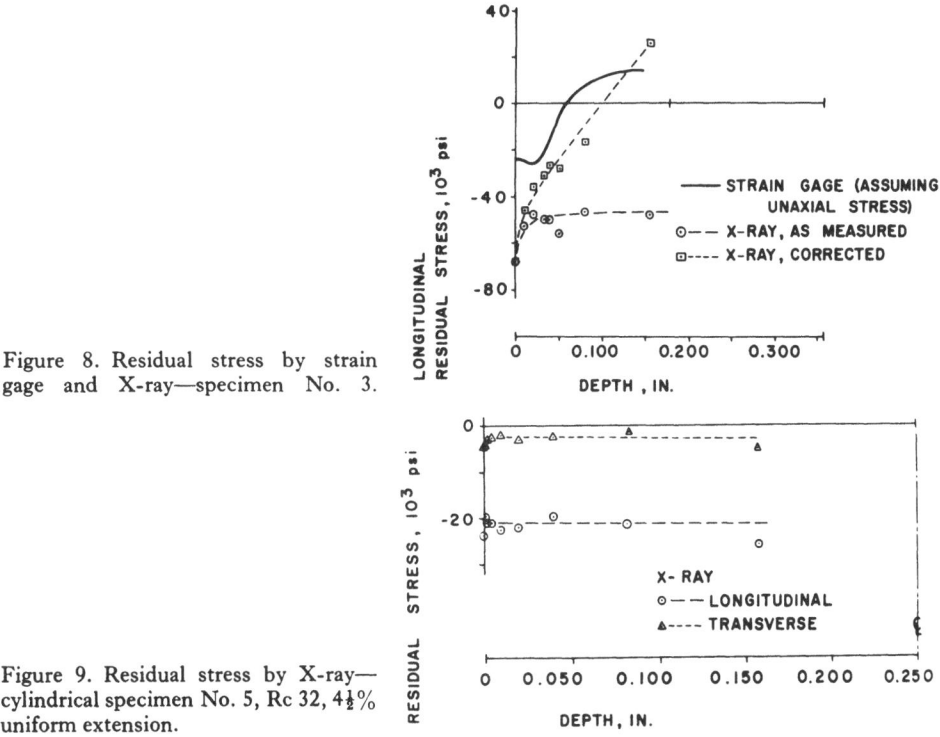

Figure 8. Residual stress by strain gage and X-ray—specimen No. 3.

Figure 9. Residual stress by X-ray— cylindrical specimen No. 5, Rc 32, 4½% uniform extension.

Surface stress as measured by X-ray was 68,000 psi compression. This value decreased to approximately 50,000 psi compression below the surface and again remained nearly constant over ⅓ the specimen thickness, as shown in Figure 8. Except for magnitude, the stress-depth profile was similar to that obtained from specimens Nos. 1 and 2. However, because the strain gage indicated a macrostress which was apparently balanced by other layers of the specimen, the X-ray measurements were corrected for stressed layers removed. Strain gage results follow only approximately the trend of the corrected X-ray curve, leaving the possibility that an anomalous component of stress was present near the surface.

Residual stress in cylindrical specimen No. 5 was approximately zero before deformation. After deformation, a residual stress of about 20,000 psi compression was indicated which was uniform over the depth investigated, as shown in Figure 9. A compressive tangential stress was indicated which was close to zero. Since the longitudinal "stress" did not approach zero as the center of the specimen was approached, this was also considered an anomalous stress rather than a macrostress.

Residual stresses calculated using several different crystallographic planes, at the surface and at a position below the surface, are shown in Table III. The 310 and 220 results were corrected by a factor to obtain agreement with 211 results for specimens in which no anomalous component was thought to be present. Longitudinal stress calculated using 310 diffraction peaks was three times as high in compression as that indicated by the 211 peaks, both at the surface and in depth. The 220 peaks indicated a slight tensile stress. Tangential stresses close to zero were indicated in all three crystallographic directions.

Figures 10 and 11 show the results for specimen No. 4, which was strained by bending. Layers were removed from positions opposite strain gages on both faces near

Table III. Residual Stress in Different Crystallographic Planes—Cylindrical Specimen No. 5, Rc 32, 4½% Uniform Extension

| Depth, in. | hkl | Radiation | Residual stress, 10^3 psi | |
			Longitudinal	Tangential
Surface	211	Cr	−22	
	220	Co	+ 9	
	310	Co	−65	
0.039	211	Cr	−20	− 2
	220	Co		−13
	310	Co	−67	+ 1

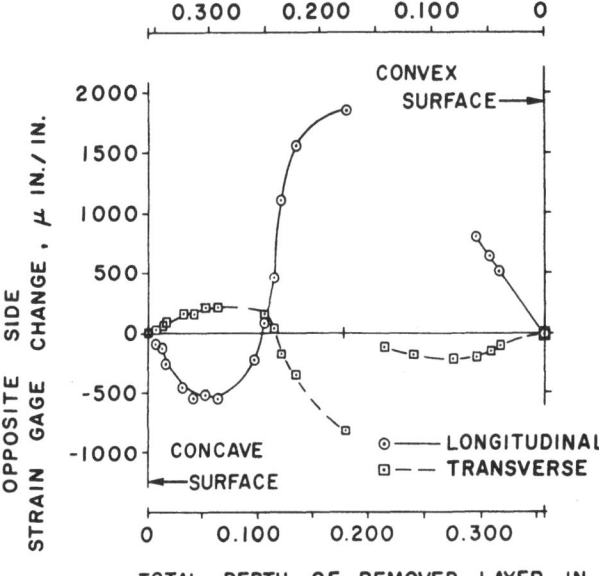

Figure 10. Change of strain on opposite side with layer removal—flat specimen No. 4, Rc 44, 6% strain by uniform bending.

opposite ends of the gage length. Strains in the transverse direction were opposite in sign and smaller in magnitude than those in the longitudinal direction, as shown in Figure 10. They differed by a factor of approximately Poisson's ratio. This indicates that the stress was approximately uniaxial. The ratio of width to thickness may have been too small for much restraint during bending, which would have produced a biaxial stress.

Stresses from the strain gage results are shown in Figure 11 along with stresses determined by the X-ray method. Because a macrostress existed, the X-ray results were again corrected for the effect of stressed layers removed. On the concave side of the specimen, strain gage and corrected X-ray results were in fairly good agreement. Good agreement also was seen on the convex side, although strain gage results near the surface may not be accurate because the strain *vs.* depth curve was not well established in this region. The results indicate that no anomalous stress component was produced by bending; only a macrostress, which normally results from plastic bending.

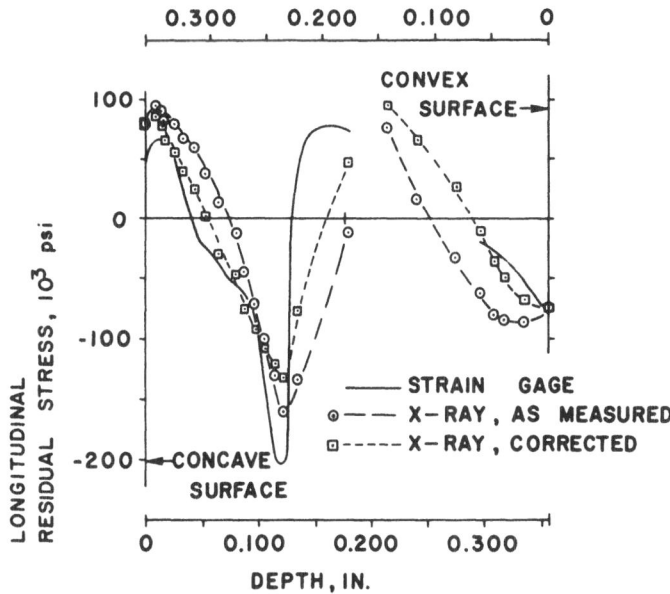

Figure 11. Residual stress by strain gage and X-ray—specimen No. 4.

DISCUSSION

The results for the specimens deformed uniformly in tension indicate the presence of some form of anomalous stress distribution only and no macrostress, in agreement with the results of Donachie and Norton.[1] They found that the anomalous stress increased approximately with the square root of the plastic strain. Taira and Yoshioka[2] saw an anomalous component, with macrostress superimposed near the surface. The anomalous component increased with increasing carbon content in annealed carbon steel. They reported a value of about 8000 psi compression in a 0.39% C specimen after $3\frac{1}{2}$% plastic extension using chromium radiation and the 211 iron peak. Comparing this value with the compressive stress of 20,000–35,000 psi observed in this work, it is seen that the anomalous stress is evidently greater in a hardened and tempered structure. The effects of carbon and hardness indicate a contribution of phase differences to the anomalous stress. One of the arguments advanced to explain the anomalous stress, as enumerated by Macherauch,[5] is that flow occurs in the weak ferrite and not in the stronger iron carbide during deformation. After release of the load, the ferrite goes into compression and the carbide into balancing tension. However, since Taira and Yoshioka saw the anomaly in low carbon steel which contained almost no carbide, this explanation cannot account for the anomalous stress in all cases.

It also has been suggested that the anomalous stress is due to the presence of weak and strong regions in material of one phase. After uniaxial deformation, the stress in the deformation direction in the weak regions is balanced by that in the strong regions. It is thought that the weak regions comprise most of the material, with the result that the X-ray method measures mostly the stress in those regions. Cullity[9,10] suggests that the weak regions correspond to subgrain interiors and the strong regions to dislocation tangles making up the subgrain boundaries. Since stresses then would be balanced over microregions, removal of layers would not produce bending in a specimen because the layers would contain their own balancing stresses. The X-ray method measures stress in the subgrain interiors only, which comprises most of the material. The stress is also fairly uniform in these regions. The higher stresses in the boundaries are not uniform and do not produce a sharp diffraction peak. A stress distribution in which most of the material contains stresses of one sign accounts for certain magnetic measurements as described by Cullity,[9] and Cullity and Puri.[11]

Macherauch[4,5] further observed the occurrence of macrostress only and no anomalous component in annealed face-centered cubic materials in the form of cylindrical specimens 0.2-in. in diameter. He attributes the macrostress to preferential yielding at the surface, and to inhomogeneous work hardening in which the flow stress at the surface never reaches that at the interior. After release, a macrostress state results which is compressive at the surface. These results, however, contradict those of Donachie and Norton[1] using the same material. One difference was that Donachie and Norton used flat specimens. However, the present investigation showed that the specimen shape alone is not the reason for the variation in results.

These arguments do not explain the variation of results with crystallographic direction, however, which also has been observed by others.[5,12,13] The direction, but not the magnitude, of the difference between 310 and 211 measurements is that which would be expected from the lower Young's modulus in the $\langle 310 \rangle$ crystallographic direction. However, the results in Table III have already been corrected for this effect. The modulus in the $\langle 110 \rangle$ direction is the same as in the $\langle 211 \rangle$ direction, so this would not explain the difference between 220 and 211 results.

A possible explanation of the crystallographic directionality of residual stress is the fact that $\{110\}$ and $\{211\}$ planes are slip planes, while $\{310\}$ planes are not. In the case of uniaxial deformation, flow evidently occurs preferentially in macrodirections as well as $\langle 111 \rangle$ crystallographic directions. In a more random deformation process, such as shot peening, there is no resultant macrodirectionality of flow and evidently no resultant crystallographic directionality of stress.

It is not known why no anomalous stress component was seen after bending. An increase in the amount of deformation would possibly enhance the effect, if present.

SUMMARY AND CONCLUSIONS

An anomalous stress determined by the X-ray method was seen accompanied by no macrostress as determined by mechanical methods in quenched and tempered steel specimens uniformly deformed by uniaxial extension. The effect evidently increases with hardness. In a nonuniformly extended specimen and a bent specimen, a macrostress was seen with no definite anomalous component.

It is concluded that the anomalous stress is real and represents a situation of local balance in microregions. It is called "anomalous" because of lack of agreement between X-ray and mechanical methods of measurement. It is caused by a particular mode of plastic deformation characterized by unidirectional yielding. Most of the practical methods of residual stress inducement do not cause anomalous stress.

ACKNOWLEDGMENTS

The authors wish to thank Prof. B. D. Cullity of Notre Dame University for discussions and encouragement of this work. They also wish to acknowledge the assistance of Mr. J. C. Hickman in performing some of the experiments.

REFERENCES

1. M. J. Donachie, Jr., and J. T. Norton, "Lattice Strains and X-Ray Stress Measurement," *Trans. Met. Soc. AIME* **221**: 962, 1961.
2. S. Taira and Y. Yoshioka, "X-Ray Investigation on the Residual Stress of Metallic Materials (On the Residual Stress of Stretched Carbon Steel)," in: T. Nishihara *et al.*, *Proceedings of the Seventh Japan Congress on Testing Materials*, The Society of Materials Science, Japan, 1964, p. 31.
3. W. S. Hyler and L. R. Jackson, "Precautions to be Used in the Measurement and Interpretation of Residual Stresses by X-Ray Technique," in: W. R. Osgood (ed.), *Residual Stresses in Metals and Metal Construction*, Reinhold Publishing Corp., New York, 1954, p. 297.
4. E. Macherauch, private communication, to be published.
5. E. Macherauch, "X-Ray Stress Analysis," *Exp. Mech.* **6**: 140, 1966.
6. D. E. Martin, "Evaluation of Methods for Measurement of Residual Stress," SAE Technical Report 147, SAE, Inc., New York, 1957.
7. A. L. Christenson (ed.), D. P. Koistinen, R. E. Marburger, M. Semchyshen, and W. P. Evans, "The Measurement of Stress by X-Ray," SAE Technical Report 182, SAE, Inc., New York, 1960.
8. D. O. Leeser and R. A. Daane, "Residual Stresses in a Strip in Terms of Strain Gages During Electropolishing," *Proc. SESA* **12**: 203, 1954.
9. B. D. Cullity, "Residual Stresses After Plastic Elongation and Magnetic Losses in Silicon Steel," *Trans. Met. Soc. AIME* **227**: 356, 1963.
10. B. D. Cullity, "Sources of Error in X-Ray Measurements of Residual Stress," *J. Appl. Phys.* **35**: 1915, 1964.
11. B. D. Cullity and O. P. Puri, "Magnetostriction and Residual Stress in Nickel After Plastic Elongation," *Trans. Met. Soc. AIME* **227**: 359, 1963.
12. R. I. Garrod and G. A. Hawkes, "X-Ray Stress Analysis on Plastically Deformed Metals," *Brit. J. Appl. Phys.* **14**: 422, 1963.
13. S. Taira and Y. Yoshioka, "X-Ray Investigation on the Residual Stress of Metallic Materials (On The Residual Stresses of the Quenched and the Plastically Deformed Carbon Steels)," in: T. Nishihara *et al.*, *Proceedings of the Eighth Japan Congress on Testing Materials*, The Society of Materials Science, Japan, 1965, p. 1.

DISCUSSION

H. W. Pickett (General Electric, X-Ray): Have the data involving the use of different radiations been corrected for depth of penetration as they relate to interior distribution of the strain?

R. E. Ricklefs: Yes, but this effect is evidently constant through a considerable layer of material, so one would expect no difference except that which may show up in any other measurement because of the difference in penetration from the diffractometer axis. But the 220 and the 310 results were both obtained using cobalt radiation. Of course, the diffraction angles are different, so there would be a difference of penetration in the two cases. However, I don't think the effect would be any greater than would be observed in other specimens in which there was no "anomalous" stress, and that has been accounted for already by the correction factor.

EXPERIMENTAL FACTORS CONCERNING X-RAY RESIDUAL STRESS MEASUREMENTS IN HIGH-STRENGTH ALUMINUM ALLOYS

Michael E. Hilley

James J. Wert

and

Robert S. Goodrich

Vanderbilt University
Nashville, Tennessee

ABSTRACT

X-ray diffraction as a means of determining stresses has found increasing application in the last few years. This is primarily because it is the only technique by which stresses can be determined without making measurements on the specimen or structure in the unstressed condition and, consequently, it is the only truly nondestructive technique for determining residual stresses. The principles of determining macrostresses on surfaces with commercially available equipment is quite well known and employs either the X-ray diffractometer or back-reflection camera techniques. The diffractometer technique was selected for this investigation because of its accuracy and because it allows both macrostresses and microstrain to be analyzed from the change in position and shape of the diffraction peaks. The X-ray analysis actually consisted of two separate phases. The first dealt with the X-ray determination of the elastic constants (Young's modulus and Poisson's ratio) for several aluminum alloys, including 5083. These values were compared with the theoretical or published values as determined by standard tensile tests and used later in stress calculations. For these tests, a unique stress stage was used which allowed the specimen to be stressed while positioned in the diffractometer, and also have angular rotation about the diffractometer axis that is independent of the rotation of the counter and receiving slit system. The second phase consisted of analyzing different groups of 5083—aluminum alloy specimens which had been subjected to various degrees of cold working by rolling. This analysis consisted not only of the computation of macrostresses, but also of microstrain and change in particle size as a function of percentage reduction in thickness. The final portion of this phase dealt with electropolishing successive layers from the surface of each sample and relating the measured relaxation to the thickness of the layers removed. In this way, stress distribution in depth was obtainable as a function of cold working.

INTRODUCTION

The modern, space age design of hardware requires a clear understanding of the origins and effects of residual stresses on the mechanical behavior of materials in order to control these stresses and improve factors such as the fatigue performance. To accomplish such an understanding requires accurate and dependable techniques for residual stress

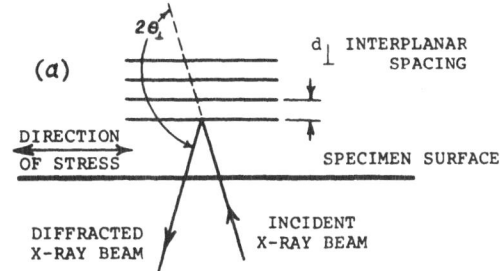

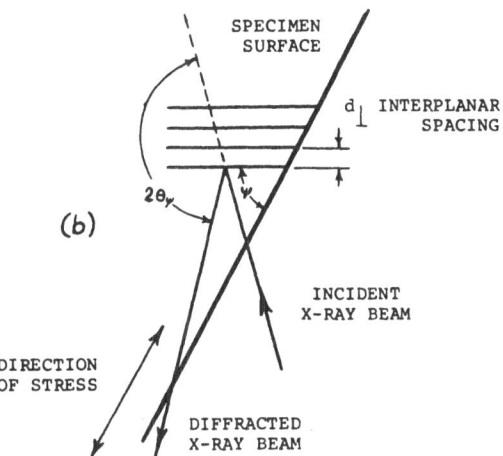

Figure 1. Schematic showing orientation of measured lattice planes with respect to specimen surface: (a) specimen normal to beam, (b) specimen rotated ψ degrees.

studies. A review of pertinent literature reveals that most X-ray diffraction advances in the study and analysis of residual stresses have been concerned with hardened steels, and very little application of techniques has been made on aluminum alloys and other nonferrous materials. It was, therefore, deemed important that the discussion presented in this paper, along with the results of the current research program, be made available for review.

The paper will be confined, for the most part, to a discussion of the X-ray diffraction technique as used in our laboratory for analyzing residual stresses in high strength aluminum alloys. This includes a discussion of possible errors in data acquisition and stress calculations, the method of analysis, and interpretation of data. Experimental data will be presented to describe the magnitude and distribution of residual stresses that are induced by rolling 5083-H323 sheet aluminum. This is of particular importance when a given stress level must be maintained after forming "as received" tempers of stock.

EXPERIMENTAL ERRORS IN DATA ACQUISITION

The fundamental principles and equations upon which macrostress determinations are based can be found in textbooks[1-4] on X-ray diffraction and related publications.[5-7] In the diffractometer "two-exposure" technique, the desired component of stress in, and parallel to, the specimen surface is determined from two measurements, one with the diffractometer aligned in its normal position and the other with the specimen rotated at an angle ψ from its normal position (see Figure 1). The component of stress to be found can be related to the angular position 2θ of the diffracted beam by

$$\sigma_\phi = \frac{E \cot \theta}{2(1 + \nu) \sin^2\psi}(2\theta_\perp - 2\theta_\psi) \tag{1}$$

where E is Young's modulus, ν is Poisson's ratio, $2\theta_\perp$ is the observed value of the diffraction angle in the "normal" measurement [$\psi = 0$ (Figure 1a)], and $2\theta_\psi$ is the observed value of the diffraction angle in the inclined position [$\psi = \psi$ (Figure 1b)].

Most investigators assume the cot θ term to be constant and rewrite equation (1) in the simplified form

$$\sigma_\phi = K(2\theta_\perp - 2\theta_\psi) \tag{2}$$

K is usually referred to as a proportionality constant or stress factor, and is assumed to be constant for a given material.

It should be noted that assuming the cot θ term to be constant will induce a finite error into each stress calculation where the original peak position has shifted due to deformation or straining of the sample. For example, if the (511)(333) reflection of the 5083 aluminum alloy (using Cu K_α radiation) is caused to shift from 159.7° 2θ to 161.2° 2θ, this 0.5° shift will induce an error of 2% when using a stress factor with a constant cot θ term. If a larger shift and higher stress level are involved, this can be quite significant. When studying aluminum alloys, the cot θ term is approximately three times as sensitive as for steels with a reflection occurring at about the same 2θ angle (assuming the same ψ rotation). For this reason, it should not be included as a part of the stress factor calculation and equation (2) would then be written more accurately as

$$\sigma_\phi = K' \cot \theta(2\theta_\perp - 2\theta_\psi) \tag{3}$$

where

$$K' = \frac{E}{2(1 + \nu) \sin^2\psi}$$

One can deduce from equation (3) that in establishing optimum experimental techniques it is also of utmost importance to determine the 2θ shift and the new stress constant K' as accurately as possible. Since the Koistinen and Marburger[8] method for calculating peak positions has proven to be quite reliable, the discussion here will be restricted to possible causes of error when calculating the stress as given by equation (3). Contributing factors will be: (a) selection of maximum usable 2θ diffraction peak, (b) selection of maximum usable ψ rotation, and (c) calibration of elastic constants for the particular set of crystallographic planes being studied.

One way of minimizing the contribution of the first two factors is by essentially minimizing K', thus making the change in peak location as large as possible for a given stress level. This can be accomplished to a large extent by making the rotation angle ψ as large as possible, therefore reducing the term $(1/\sin^2 \psi)$, and by selecting 2θ large so as to reduce the cot θ term. The maximum usable value of ψ is physically fixed at θ, but normally 60° is the maximum ψ value used. Also, the value of θ is obviously a function of the material and the radiation wavelength. Table I contains a tabulation of reflections for aluminum when using copper, cobalt, or chromium radiation. When considering 130° 2θ as the minimum usable diffraction angle, the possible reflections that may be used have been

Table I. Diffraction Angles for Aluminum with Copper, Cobalt, and Chromium Radiation

d	(h, k, l)	Diffraction angle, deg 2θ					
		Cu K_α	Cu K_β	Co K_α	Co K_β	Cr K_α	Cr K_β
2.238	(111)	38.5	34.58	45.04	40.56	58.66	52.96
2.024	(200)	44.7	40.24	52.50	47.22	68.9	62.00
1.431	(220)	65.1	58.22	77.44	69.00	106.26	93.48
1.221	(311)	78.2	69.50	94.34	83.20	139.46	117.30
1.169	(222)	82.4	73.08	99.93	87.80	156.9	126.20
1.0124	(400)	99.1	86.88	124.28	106.40		
0.9289	(331)	112.0	97.06	148.9	121.44		
0.9055	(420)	116.6	100.48	162.6	127.20		
0.8266	(422)	137.5	114.74				
0.7793	(333) (511)	162.5	126.56				
0.7158	(440)		153.04				

Table II. Aluminum Diffraction Angles for Stress Studies

X-ray target	Diffraction peak, 2θ	Specimen angle, ψ	Stress factor K, psi/0.01 deg 2θ	Radial focus,† in.
Cr*	(311) 139.5°	30	1027.31	2.75
		45	513.66	2.64
Cr	(222) 156.9°	30	568.56	4.52
		45	284.28	3.78
		60	189.52	2.73
Co	(331) 148.9°	30	774.16	4.14
		45	387.08	3.24
		60	258.06	2.00
Co*	(420) 162.6°	30	425.73	4.80
		45	212.86	5.53
		60	141.91	3.33
Cu	(422) 137.5°	30	1081.93	3.63
		45	540.96	5.73
Cu(K_β)*	(440) 153.0°	30	667.08	4.33
		45	333.54	3.51
		60	222.36	2.36
Cu*	(511) (333) 162.5°	30	428.20	4.79
		45	214.10	4.20
		60	142.73	3.32

* Recommended peaks.

† $L = 5.73 \dfrac{\cos(90 - \theta + \psi)}{\cos(\theta + \psi - 90)}$.

tabulated in Table II. Those reflections found best for stress-analysis measurements in aluminum have been noted by an asterisk. Of the recommended radiations, Cu K_β for the 440 line has been included for deformation processes where preferred orientation will not prohibit an accurate measurement of $2\theta_\psi$. When the Cu K_β radiation is used, the problem of diffused K_{α_1} and K_{α_2} peaks that is common to deformation is eliminated since the K_β reflection is monochromatic.

The final contributing factor that might cause error in the evaluation of a stress constant is the selection of values for the elastic constants. X-ray measurements are dependent on the determination of the change in interplanar spacings of a particular set of crystallographic planes at particular orientations to the direction of stress. Therefore, it is important that a calibration be performed to determine the proper elasticity constants for the specific crystallographic planes being studied. In some steels, for example, the value of Young's modulus for particular crystallographic directions may vary by more than a factor

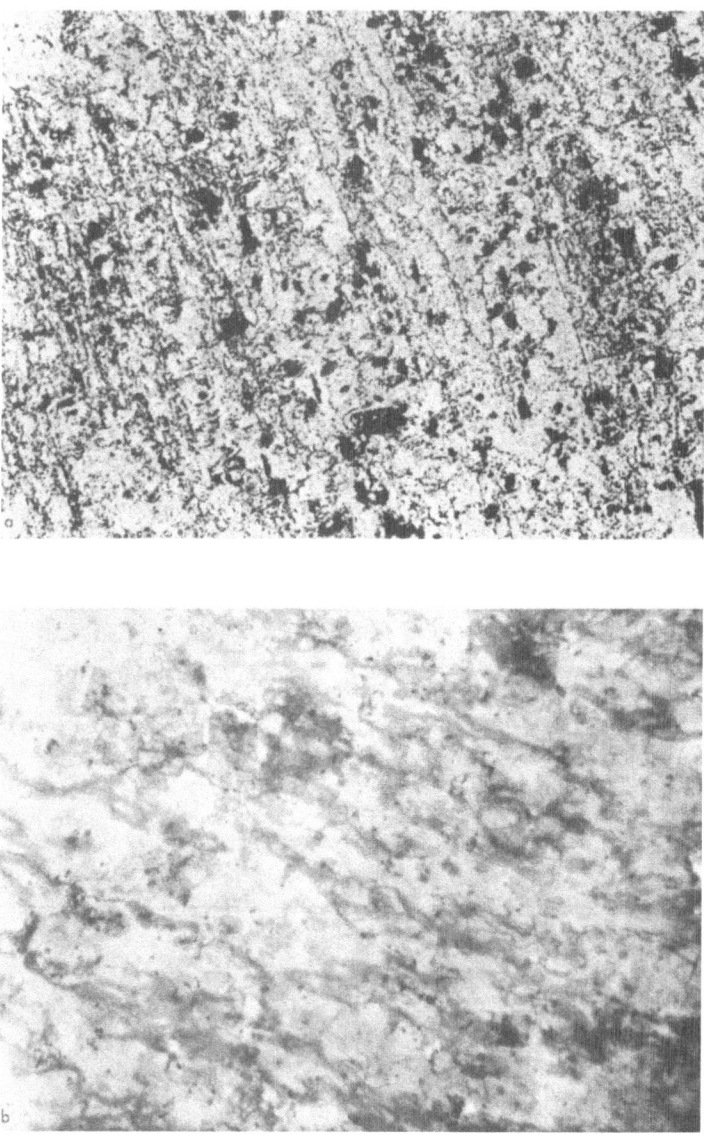

Figure 2. Micrographs of 5083-H323 aluminum alloy; (a) light-optical micrograph 200 ×, (b) transmission electron micrograph 10,000 ×.

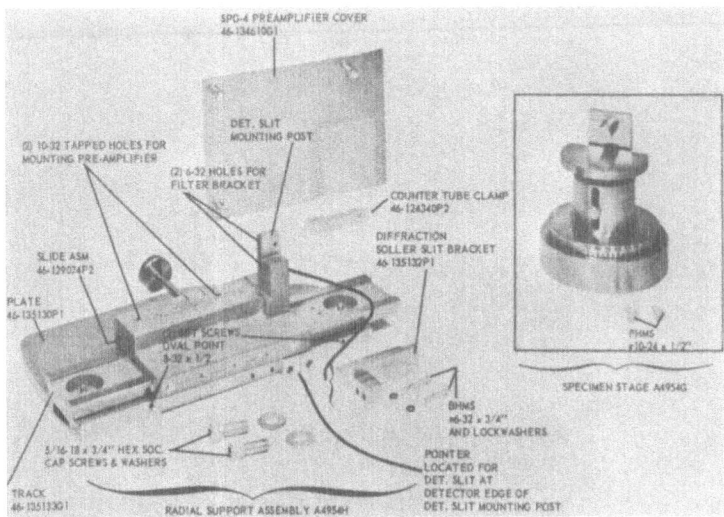

Figure 3. General Electric residual-stress specimen stage and adjustable radial support for the detector.

of two. While Schmidt and Boas[9] indicate that aluminum is quite isotropic in this respect, the calibration used in our laboratory will be reviewed in the discussion of experimental procedures.

EXPERIMENTAL PROCEDURES

The material used for this study was a 5083 aluminum alloy obtained from ALCOA, and contained 4.58-wt.% magnesium, 0.70-wt.% manganese, and slight traces of iron and silicon. The alloy was received in the H323 temper as 0.094-in. thick sheet. Subsequent deformation was introduced by cold rolling additional amounts of 10, 20, and 30% reduction in thickness, both by single pass and multiple passes of 10% (or total thickness) increments. All rolling was carried out in the same direction as in the original temper treatment.

Light optical and transmission electron micrographs of the H323 temper alloy appear in Figure 2. An elongated grain structure, together with inclusions and a background of etched substructure, are evident in the optical micrograph. No difference could be detected optically between the as-received material and that given an additional 30% rolling reduction. The transmission electron micrograph of the as-received material shows the fine dislocation structure arranged in elongated cells that one would expect in a heavily worked material. There was some evidence of a coarsely disbursed precipitate phase throughout the matrix of the as-received material.

A General Electric XRD-5 diffractometer, equipped with a CA-8L high-intensity copper tube and proportional counter, was used for this investigation. All measurements were made using the (511) (333) peak and a ψ rotation of 45°. A 3° beam slit, coupled with a 0.5° detector slit at $\psi = 0$, and a 0.2° detector slit at $\psi = 45°$ were employed. The specimen positioning, ψ rotation, and focusing of the detector were accomplished by employing a General Electric residual stress specimen stage and radial support for the detector as shown in Figure 3. A modification to the preamplifier housing, as can be seen in Figure 4, enabled the maximum diffraction angle obtainable to be increased by several degrees. For example, with this modification, it was possible to go to 164° 2θ at $\psi = 45°$ and a target

Figure 4. General Electric XRD-5 unit with residual-stress stage, straining jig, and detector radial support installed. The preamplifier housing modification can also be seen.

angle of 4°, thus enabling us to use the (511) (333) reflection for our residual stress determinations. The maximum value of 2θ is physically limited by the detector position and the target angle. The radial focus distance, shown in Figure 5 as L, has been tabulated in Table II for the diffraction angles listed and the ψ angles that are commonly used.

The peak positions for $2\theta_\perp$ and $2\theta_\psi$ were calculated using the Koistinen and Marburger[8] three-point parabola fitting technique. Reproducibility of values obtained was found to be best if the three readings were taken in the 85% intensity range above background and if the difference in intensity readings corrected for Lorentz polarization and absorption on either side of the center data point did not differ by more than a factor of two.

In our laboratory, calibrations for the elastic constants were made according to procedures outlined in the SAE Technical Report 182,[6] using the straining jig shown in Figure 4. Straining the calibration specimen in uniaxial tension, instead of bending, enables more of the specimen surface to be tangent to the focusing circle of the goniometer.

For determining the distribution of the stress magnitude with depth below the specimen surface, successive layers were removed by electropolishing in a 25 vol.% solution of perchloric acid in absolute alcohol. Following each electropolishing treatment stresses were determined in both the direction of rolling and the transverse direction. These measurements were carried out on 1 in. × 1½ in. coupons of which a ⅝ in. × ⅝ in. window in the center was electropolished to the desired depth. The irradiated area corresponding to about 0.3 in. × 0.3 in. in the center of the window.

EXPERIMENTAL RESULTS AND DISCUSSION

The results described in this section concern the initial stages of a research program currently in progress to study the effects of residual stresses on the fatigue behavior of

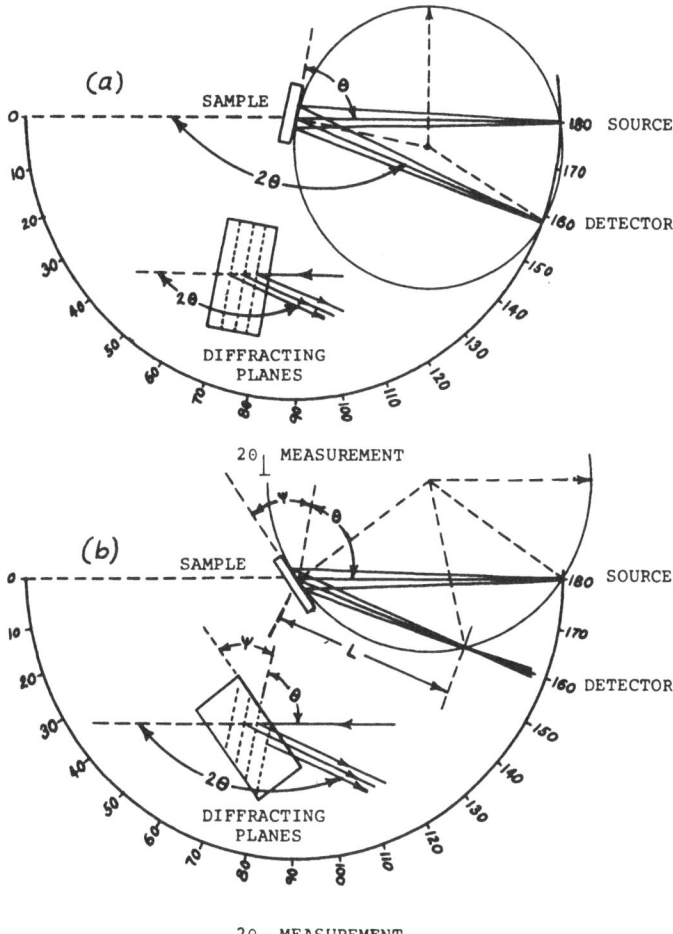

Figure 5. Schematic showing sample and detector positions for:
(a) $2\theta_\perp$ measurements, (b) $2\theta_\psi$ measurements.

high strength aluminum alloys. The first part of the program, to develop reliable and accurate techniques for acquiring and analyzing residual stress data, has been described in the previous sections. Here we shall discuss the results of initial experimentation to determine the state of surface-residual stresses in the H-323 temper of the 5083 aluminum alloys, the effect of further cold reduction by rolling on the surface stresses, and the distribution of stress in depth as a function of cold reduction.

Figure 6 shows the longitudinal and transverse (with respect to the original rolling direction) residual-surface stresses as a function of single-pass rolling reduction. The H-323 temper in the as-received condition exhibited a longitudinal stress of 2600 psi tension and a transverse stress of 4800 psi tension. Standard deviation calculations on several points indicated an error of ± 800 psi for these readings; however, not all points were measured at a constant total count. With increasing amounts of deformation, the surface stresses increase rapidly to a maximum of approximately 10,000 psi tension at 20% reduction in the longitudinal direction, and 8800 psi near 14% reduction in the transverse direction. While appearing to remain saturated near the maximum stress in the longitudinal

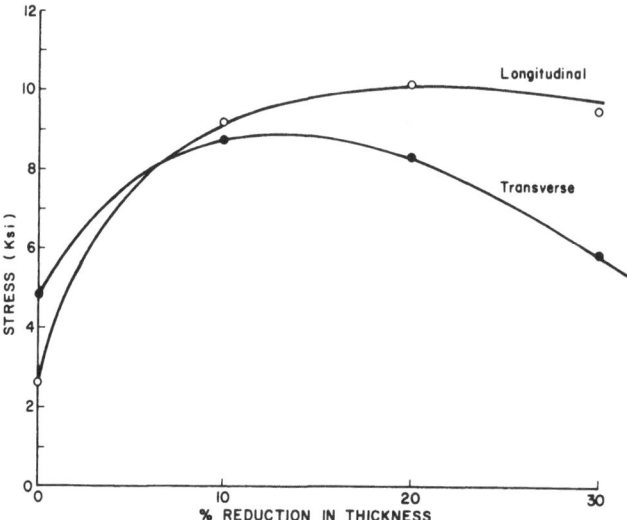

Figure 6. Variation of longitudinal and transverse-residual surface stresses as with room-temperature rolling reduction.

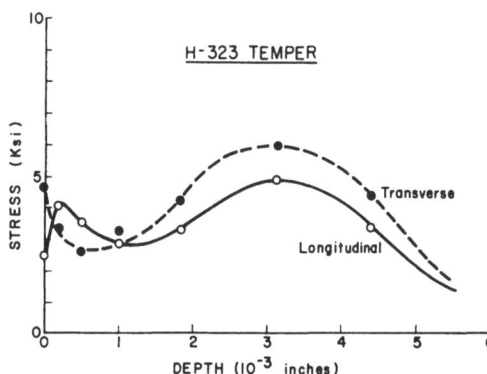

Figure 7. Distribution of longitudinal and transverse residual stresses with depth for H323 temper.

direction, the transverse stress drops rather rapidly with increasing deformation. The reason for this deformation-induced recovery is not yet known, but is at least partly due to the appearance of edge cracks at higher rolling deformations. Measurements of surface-residual stresses in samples deformed the same amounts by multiple-pass rolling of 10% (of total reduction) increments showed similar curves, but were shifted 2500–3000 psi higher in stress value.

Residual stresses as a function of depth below the rolling surface for the H-323 temper alloy are shown in Figure 7. As we were primarily concerned only with the relative distribution of stresses at this stage of the study, no attempt was made to correct for stress relaxation due to layer removal. It is noted, however, that a second maximum in the stress level occurs approximately 3 mils below the rolling surface. We are unable at this time to explain the appearance of a minimum in the stress level in the range $\frac{1}{2}$-mil to 1-mil below the surface. However, the same type of stress distribution with depth is retained after additional rolling of the as-received material. Figures 8 and 9 show this for the 10%-reduction specimens as compared to the H-323 temper in the longitudinal and transverse directions, respectively. The stress levels in the first 5 mils below the surface are considerably higher and the maximum and minimum positions are shifted to somewhat greater depths after 10% reduction. Samples of 20% and 30% reduction by single-pass rolling were also studied in this manner and showed similar curves, but with higher stress levels and more shift to greater depths.

Work is currently in progress to extend the residual-stress measurements to depths much closer to the center of the sample, where compressive stresses are expected to be found. The high degree of tensile residual stresses found on and below the surface of this alloy in its current state should certainly prove detrimental to its fatigue behavior. Deformation programs that will induce compressive residual stresses in the surface of this alloy in different temper conditions and in similar high strength aluminum alloys are now being investigated.

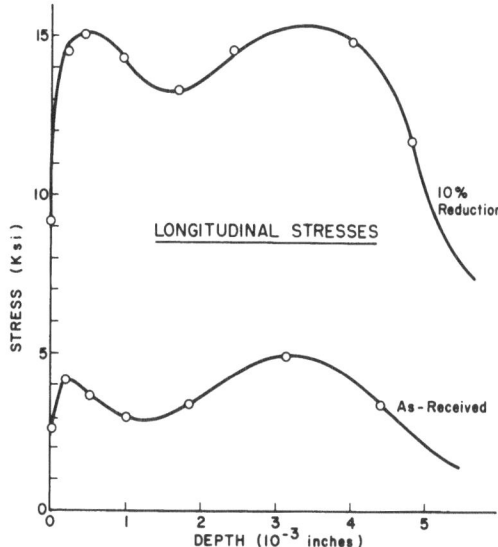

Figure 8. Comparison of longitudinal-stress distribution with depth for as-received and 10% cold-rolled material.

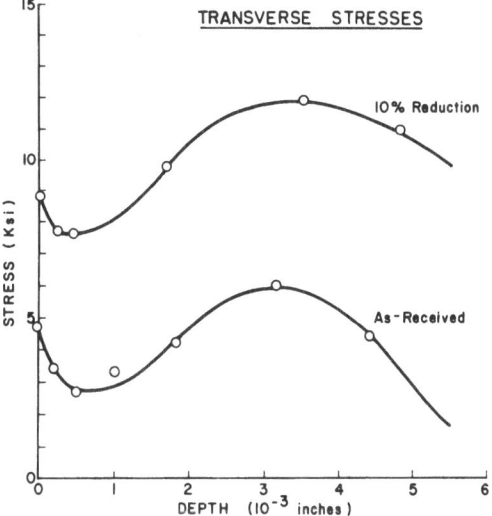

Figure 9. Comparison of transverse stress distribution with depth for as-received and 10% cold-rolled material.

ACKNOWLEDGMENT

Acknowledgment is made to the National Science Foundation for its support of this project—Research Grant #GK-955.

REFERENCES

1. B. D. Cullity, *Elements of X-Ray Diffraction*, Addison-Wesley Publishing Co., 1956, p. 431.
2. C. S. Barrett, *Structure of Metals*, McGraw-Hill Book Co., New York, 1952, p. 316.
3. H. P. Klug and L. E. Alexander, *X-Ray Diffraction Procedures*, John Wiley and Sons, Inc., New York, 1957.
4. A. Taylor, *X-Ray Metallography*, John Wiley and Sons, Inc., New York, 1961, p. 724.
5. R. E. Ogilvie, *Stress Measurement with X-Ray Spectrometer*, M.S. Thesis, MIT, 1952.
6. A. L. Christenson (ed.), D. P. Koistinen, R. E. Marburger, M. Semchyshen, and W. P. Evans, *Measurement of Stress by X-Ray*, SAE TR-182, Society of Automotive Engineers, 1960.
7. E. Macherauch, "X-Ray Stress Analysis," *Experimental Mechanics*, March, 1966, p. 140.
8. D. P. Koistinen and R. E. Marburger, "A Simplified Procedure for Calculating Peak Position in X-Ray Residual Stress Measurements of Hardened Steel," *ASM Trans.* 51: 537, 1959.
9. Schmidt and Boas, *Plasticity of Crystals* (English translation), F. A. Hughes and Co., London, 1950.

DISCUSSION

F. H. Totman (Rockwell–Standard Corporation): In the early part of your talk, I was glad to hear you call attention to the fact that saying stress is proportional to $\Delta\theta$, is an approximation which neglects a cotangent θ term. This approximation is often made, I am sorry to say, but is acceptable only when the stress level is quite low, as in the case of your work here. In our laboratory, however, it is not uncommon to measure stresses well in excess of 200 ksi in steels, and in such cases this approximation is not negligible. At higher stresses, the errors due to this approximation may well exceed over 10% and the error is very simply eliminated by converting the peak 2θ values to d and using the correct expression that stress is proportional to Δd.

M. E. Hilley: Considering this assumption at low stress levels with aluminum as an example and using the (511) (333) peak with copper radiation, a peak shift of 0.5° could induce an error of approximately 2%. At the higher stress levels which you are working with, it could cause a very appreciable error and should certainly be taken into consideration.

D. N. Braski (NASA Langley): What was the thickness of your specimens in these last results in which you showed stress *vs.* depth?

M. E. Hilley: As indicated previously, the specimens were prepared from 0.094 in.-thick sheet of 5083-H323 temper. Different groups were subsequently deformed 10, 20, and 30% in single and multiple passes. For example, in the group that was deformed 20%, there were two classes—one deformed by 20% in a single pass and a class composed of specimens deformed 20% by two 10% passes. Because the original starting thickness was the same for all specimens, the deformed specimens were, therefore, slightly thinner. The results presented were for an electropolished depth of approximately 5 mils.

D. N. Braski: Were the stresses in these figures tensile or compressive?

M. E. Hilley: They were tensile. It is quite possible to have tensile residual stresses at the surface when the specimen has undergone a large reduction in thickness such as that of the H323 temper. Dr. Macherauch discusses the signs obtained from various types of deformation in the paper* that he presented at this conference last summer.

---

* Eckard Macherauch, "Lattice Strain Measurements on Deformed fcc Metals " in: G. R. Mallett, M. J. Fay, and W. M. Mueller (eds.), *Advances in X-Ray Analysis*, Vol. 9, Plenum Press, New York, 1966, pp. 103–114.

X-RAY MEASUREMENT OF RESIDUAL STRESSES IN TITANIUM ALLOY SHEET

David N. Braski and Dick M. Royster

NASA Langley Research Center
Hampton, Virginia

ABSTRACT

An X-ray diffraction technique was used to measure residual stresses in Ti-6Al-4V and Ti-8Al-1Mo-1V sheet created by glass-bead peening, sand and aluminum oxide blasting, and a vibratory tumbling treatment. For peening and blasting, the use of larger particle sizes produced greater compressive stresses. In the case of the vibratory treatment, an increase in vibratory frequency or treatment time increased the compressive stress. Glass-bead peening caused a 10% reduction in yield strength while the other treatments had little effect on the tensile properties. Significant stress relaxation occurred in all the treated Ti-6Al-4V coupons exposed at 600° and 800°F.

INTRODUCTION

The concept of inducing residual compressive stresses in a metal alloy for the prevention of stress corrosion is generally well known.[1] Recent studies at the NASA Langley Research Center which have utilized this concept indicate that certain surface treatments will[2,3] protect titanium alloys against salt-stress corrosion at temperatures below 600°F. Other research at Langley has been aimed at alleviating the stress corrosion of Ti-6Al-4V in nitrogen tetroxide by glass-bead peening the exposed surfaces. However, the use of surface treatments on titanium alloys has raised questions about the magnitude and distribution of residual compressive stresses that are created.

In this investigation, residual surface stresses produced in Ti-6Al-4V and Ti-8Al-1Mo-1V sheet by glass-bead peening, sand and aluminum oxide blasting, and a vibratory tumbling process have been determined by X-ray diffraction. The analysis was facilitated by use of a high-speed digital computer. The distribution of residual compressive stresses was determined by chemically milling off thin layers of material and taking X-ray measurements. Treatment parameters such as particle size and treatment time were varied to study their effect on the magnitude and distribution of residual stresses. The effect of exposure in the temperature range from room temperature to 800°F on the residual stresses and the effect of surface treatment on tensile properties also were investigated. Light and electron microscopy, as well as surface roughness measurements, were used to study surface characteristics produced by the different treatments.

PROCEDURE

Materials

The materials used in this investigation were 0.046-in.-thick Ti-6Al-4V and 0.050-in.-thick Ti-8Al-1Mo-1V sheet in the annealed and duplex annealed heat-treated

Table I. Treatment Parameters

Alloy	Treatment	Commercial designation	Particle composition	Average particle size	Pressure, psi	Nozzle size, in.	Treatment time
Ti-6Al-4V (annealed)	Bead peening	MS-P	Glass	425 μ	60	$\frac{3}{16}$	2, 5, and 10 sec
		MS-L	Glass	60 μ	60	$\frac{3}{16}$	2, 5, and 10 sec
		MS-XL	Glass	30 μ	60	$\frac{3}{16}$	2, 5, and 10 sec
	Blasting	80 grit	Al_2O_3	200 μ	50	$\frac{9}{32}$	5, 15, and 30 sec
		120 grit	Al_2O_3	110 μ	50	$\frac{9}{32}$	5, 15, and 30 sec
		220 grit	Al_2O_3	70 μ	50	$\frac{9}{32}$	5, 15, and 30 sec
		160 grit	SiO_2	75 μ	50	$\frac{9}{32}$	5, 15, and 30 sec
	Vibratory–1375 cpm		Al_2O_3	$\frac{1}{8}$-inch triangles, $\frac{1}{4}$-in. thick	—	—	13, 26, and 52 hr
	Vibratory–1750 cpm		Al_2O_3	$\frac{1}{8}$-inch triangles, $\frac{1}{4}$-in. thick	—	—	13, 26, and 52 hr
Ti-8Al-1Mo-1V	Bead peening	MS-XL	Glass	30 μ	60	$\frac{3}{16}$	2, 5, and 10 sec
	Vibratory–1375 cpm		Al_2O_3	$\frac{1}{8}$-inch triangles, $\frac{1}{4}$-in. thick	—	—	13, 26, and 52 hr

conditions, respectively. The chemical composition of each alloy in percent by weight, as supplied by the producer, is as follows:

Alloy	C	Fe	N	Al	V	Mo	H	Ti
Ti-6Al-4V	0.023	0.1	0.01	5.9	4.1	—	0.007	Balance
Ti-8Al-1Mo-1V	0.026	0.08	0.012	7.8	1.0	1.0	0.006	Balance

Specimen Preparation

The specimens were fabricated by first shearing the sheet into $1\frac{1}{2}$-in. × $1\frac{1}{2}$-in. coupons. The coupons were then glass-bead peened, blasted, or given a vibratory treatment by using the treatment parameters listed in Table I. Table I also lists the commercial designation, particle composition, and particle size in microns for each treatment.

Glass-bead peening and blasting were accomplished manually on one side of each coupon by using a nozzle-to-specimen distance of approximately 6 in. The vibratory-tumbling treatment is a commercial cleaning and deburring process in which the specimens were vibrated in a bath containing aluminum oxide triangles ($\frac{5}{8}$ in. on each side by $\frac{1}{4}$-in. thick) and a detergent. Both faces of the coupons were, therefore, affected by the vibratory treatment. Several specimens were treated on one side by spotwelding two coupons together at the corners. After treatment, the coupons were separated and used for stress-distribution determinations. Vibrational frequencies of 1375 and 1750 cpm were used to vary the degree of tumbling action.

Specimens representative of each treatment were selected for measurement of residual stresses through the treated surface layers and were lacquered on the untreated face. X-ray measurements were then taken on the treated surface and also below the surface in 0.0001-in. increments until the residual stress was zero. The layers of material were removed by chemical milling with a solution consisting of 65 cc H_2O, 25 cc HNO_3 (conc), and 8 cc HF (45% conc). Milling times varied from 3–30 sec, depending on the strength of the solution. The amount of material removed was monitored by taking thickness measurements near each corner and at the center of the coupons with a micrometer.

Surface Examination

To gain insight on how the various treatments affect the surface, light and electron photomicrographs were made on both untreated and treated Ti-6Al-4V coupons. The electron micrographs were made by using two-stage plastic-carbon replicas shadowed at 45° with chromium. Surface-roughness measurements were obtained with a commercial profilometer which employed a 0.0005-in.-radius diamond-tipped exploring stylus. Root-mean-square (RMS) values of surface roughness were obtained directly in micro-inches.

Tensile Tests

To determine the effect of the surface treatments on the tensile properties of Ti-6Al-4V and Ti-8Al-1Mo-1V, a series of ASTM standard tensile specimens were fabricated and given the surface treatments listed in Table II. The tensile specimens were treated on both flat surfaces, as well as on the edges of the reduced section. The tensile tests were performed in a 100-kip hydraulic testing machine.

During the tests the load was recorded autographically against strain. Strain was

Table II. Tensile-Test Results

Surface treatment	Sheet no.	Young's modulus, psi	Yield strength, psi	Ultimate strength, psi	Elongation, percent in gage length	
					1 in.	2 in.
(a) Ti-6Al-4V						
As-received material	1	16.37×10^6	149.2×10^3	163.3×10^3	8	10.5
		16.22	149.8	162.7	8	10
			152.1	162.8	6	10
			153.0	163.2	8	11.5
Glass-bead peened	1		151.1	167.8	7	9.5
30 μ 15 sec			150.3	167.0	6	9
Glass-bead peened	1		156.8	168.0	8	9
30 μ 30 sec			151.1	167.2	5	9
Glass-bead peened	1		135.7	166.0	6	9
425 μ 15 sec			134.5	165.7	8	8.5
Glass-bead peened	1		135.7	166.0	5	8.5
425 μ 30 sec			134.1	165.9	4	8
Al$_2$O$_3$ blasting	1		149.2	162.2	4.5	9.5
200 μ 15 sec			149.0	159.5	6	10.5
Al$_2$O$_3$ blasting	1		151.3	162.0	5	10
70 μ 15 sec			153.2	164.5	5	9
Sand blasting	1		151.4	165.5	4	9
75 μ 15 sec			151.0	164.0	5	10
Vibratory treatment	1		153.0	163.0	5.5	10
26 hr 1375 cpm			154.5	164.5	6	10.5
Vibratory treatment	1		155.0	167.1	7	10.5
26 hr 1750 cpm			155.5	169.0	5	6.5
(b) Ti-8Al-1V						
As-received material	1	18.20×10^6	134.3×10^3	149.1×10^3	8	12.5
		18.12	134.7	149.9	12	13.5
			135.7	149.9	12	13
			133.9	149.7	12	13
Glass-bead peened	1		134.7	150.9	8	12.5
30 μ 15 sec			133.7	151.0	10	12
Glass-bead peened	1		134.7	152.1	12	12.5
30 μ 30 sec			134.2	150.3	8	12
Sand blasting	1		132.0	150.7	8	12
75 μ 15 sec			131.8	150.5	8	11
Vibratory treatment	1		134.2	149.8	8	12
26 hr 1375 cpm			134.7	150.2	10	13
Vibratory treatment	1		134.4	150.4	10	13
52 hr 1375 cpm			134.8	150.9	10	13
As-received material	2	17.25	130.0	142.5	8	11
			130.0	143.5	9	11.5
Glass-bead peened	2		114.2	145.0	8	12
425 μ 30 sec			114.8	145.0	8	10

measured over a 1-in. gage length by means of two, linear variable differential transformers attached to the faces of the specimens.

Stress Relaxation Tests

In order to determine the effects of elevated-temperature exposure on the residual stress produced by three representative treatments, single coupons of each treatment were exposed at room temperature, 200°, 400°, 600°, and 800°F for 1000 hr. The 425 μ glass-bead-peening, 75 μ sandblasting, and 1750-cpm-vibratory treatments were selected as being typical and also because they produced substantial residual compressive stresses. The residual surface stress was determined for each specimen at room temperature before exposure, and also at selected times throughout the exposure period.

X-Ray Stress Measurements

In the two-exposure X-ray diffraction technique, stresses are obtained by measuring elastic strains associated with a selected set of crystallographic planes. This technique[4] utilizes the following expression

$$\sigma = K(2\theta_0 - 2\theta_\psi) \tag{1}$$

where σ is the residual surface stress, K is the stress factor (determined experimentally), $2\theta_0$ is the position of selected peak with specimen in normal position, and $2\theta_\psi$ is the position of selected peak with specimen at angle ψ relative to normal position; $\psi = 45°$ for this study.

The three-point parabola method for determining peak position[4] was used on the reflection from favorably oriented (213) planes. A General Electric XRD-5 diffractometer using nickel-filtered copper K_α radiation at 45 kV and 16 mA was used with a 3° medium-resolution soller slit and a 0.2° detector slit. The (213) reflection was first scanned at 2°/min to locate the approximate position of the peak and to enable the selection of three suitable points at which to count intensity. The three points were selected above 75% of the maximum intensity to minimize errors due to peak shape. The interval selected between points varied from 0.1–0.8° depending on the amount of line broadening. Pulse height discrimination was utilized to obtain peak-to-background ratios of at least 3 : 1. Intensity was measured as reciprocal intensity by recording the time for 40,000 counts. A preset count of 100,000 was used for some subsurface measurements of stress at a time when the pulse height discriminator was malfunctioning. Results of these latter measurements were comparable to those made by using pulse height discrimination because the peaks obtained in subsurface measurements were relatively sharp.

In order to achieve a reasonable degree of accuracy in the X-ray measurement of stress, it was necessary to take into account certain correction factors. The Lorentz–polarization and absorption factors[4] were applied to the values of reciprocal intensity. Corrections for X-ray beam penetration and for stress relaxation created by the removal of layers were also applied to stress measurements made beneath the surface.[4]

Stress-Factor Calibration

In this investigation the stress factor K [see equation (1)] was obtained by measuring the peak shift $(2\theta_0 - 2\theta_{45})$ with the X-ray diffractometer on 6 in. $\times$ $\frac{5}{8}$ in. strips of each alloy bent around dies of different radii. The tensile strain in the outer fibers of each strip was measured with a strain gage. The X-ray measurement was taken directly adjacent to the strain gage. Young's modulus was used to convert the strain-gage values of strain to stress. Young's modulus was determined for the Ti-6Al-4V and Ti-8Al-1Mo-1V sheet by using standard ASTM tensile specimens of each alloy and Tuckerman optical strain gages.

Computer Program

A computer program used to convert the X-ray data into residual stress values is shown schematically in Figure 1. First, the values of reciprocal intensity counted with $\psi = 0°$ and $\psi = 45°$ were fed into the program and corrected by the Lorentz polarization

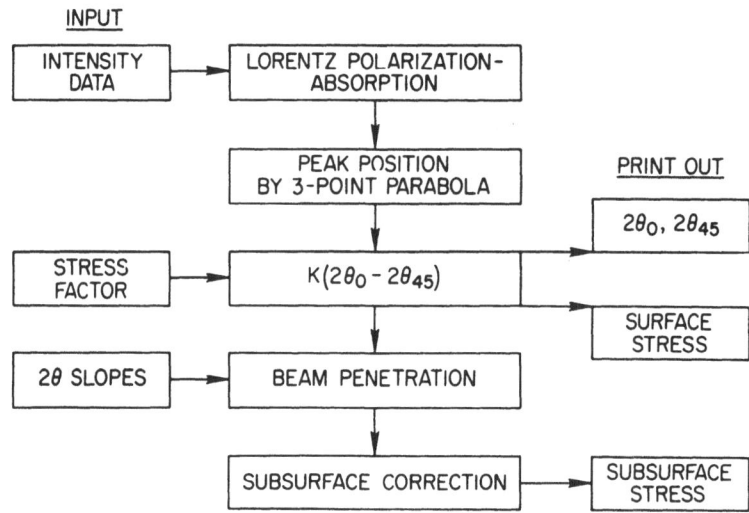

Figure 1. Schematic of computer program for analysis of X-ray data for stress determination.

and absorption factors. Then parabolas were fitted to the corrected intensity points to determine the peak positions $2\theta_0$ and $2\theta_{45}$. The peak shift was multiplied by the stress factor to obtain the residual stress. Both $2\theta_0$ and $2\theta_{45}$, along with the surface stress, were printed out by the computer. When surface layers were removed from the coupons, two additional corrections were applied: the beam-penetration correction and the subsurface correction. Because of the variation in the shape of stress profiles, it was believed that more reliable results could be obtained by using the computer to calculate only part of the beam-penetration correction. The portion of this correction which called for slopes on curves of $2\theta_0$ and $2\theta_{45}$ plotted against depth[4] was carried out manually and then fed into the computer. Finally, the stress, after all corrections were made, was printed out by the computer.

RESULTS AND DISCUSSION

Stress-Factor Calibration

The results of the calibration to obtain stress factors for Ti-6Al-4V and Ti-8Al-1Mo-1V are shown in Figure 2. The stress computed from the strain-gage readings is plotted against the peak shift or $(2\theta_0 - 2\theta_{45})$ measured by X-ray diffraction. Straight lines were fitted to the data points by the method of least squares and displayed slopes or stress factors of 80 and 86 ksi per degree of peak shift for Ti-6Al-4V and Ti-8Al-1Mo-1V, respectively.

Effect of Treatment Parameters on Residual Surface Stress

The results of the stress measurements on Ti-6Al-4V and Ti-8Al-1Mo-1V sheet indicated that the variation in surface stresses produced by the treatments was small enough to permit a meaningful analysis by X-ray diffraction. This is demonstrated in Figure 3 for glass-bead peening with 425 μ beads after 2- and 10-sec treatment times. Five stress measurements were taken on the first specimen in each group and three measurements were taken on the remaining two specimens. The residual-stress value in

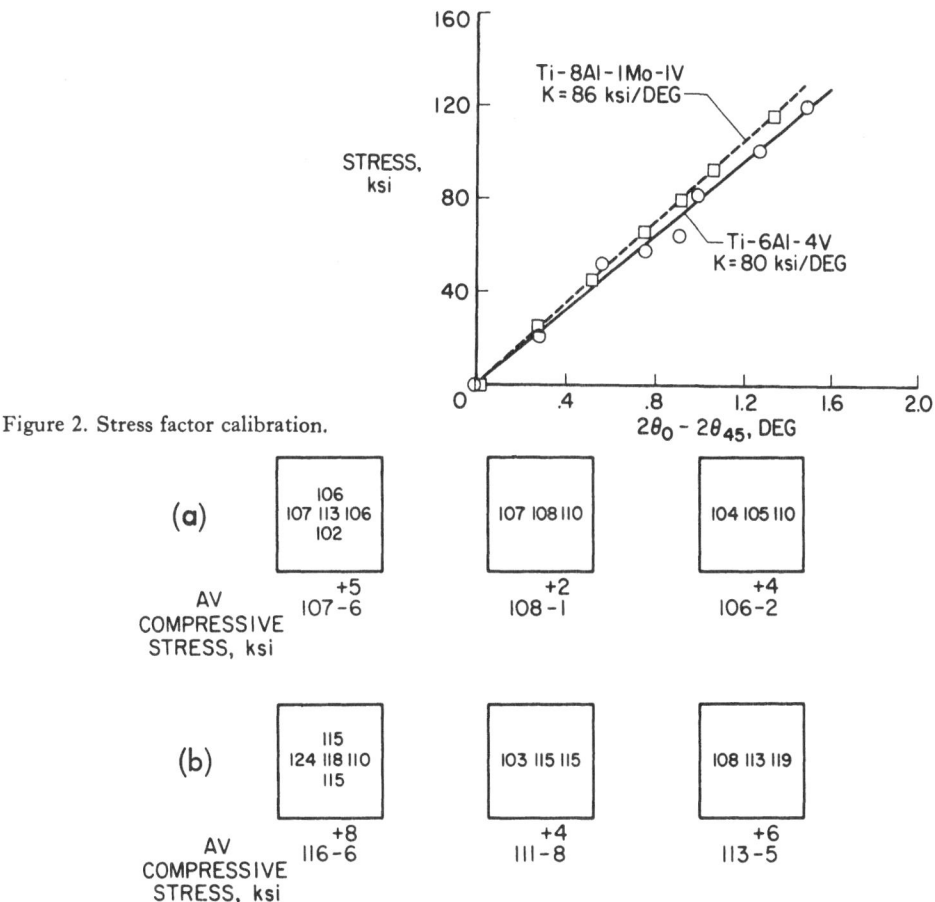

Figure 2. Stress factor calibration.

Figure 3. Variation of residual surface stresses produced by 425 μ glass-bead peening of Ti-6Al-4V. (a) Peening time, 2 sec. (b) Peening time, 10 sec.

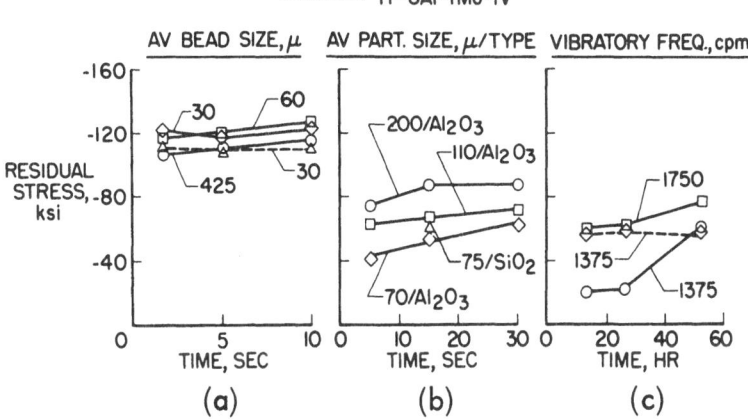

Figure 4. Effect of treatment time on residual surface stress. (a) Glass-bead peening. (b) Sand and aluminum oxide blasting. (c) Vibratory.

the figure appears in the area in which it was measured. It is seen that the range of average residual stresses from specimen to specimen was from 2–5 ksi. The range of variations of single stresses measured across any one specimen was within 1–8 ksi of the average surface stress. It should be pointed out that the surface stress as measured by X-ray diffraction is itself an average of stresses through a thin layer of material. The thickness of the layer is largely dependent on the type of X-ray radiation used.

The effects of varying different treatment parameters on the residual surface stress is shown in Figure 4 where residual stress is plotted against treatment time. Each data point is the average of three stress measurements taken on different areas of the same specimen.

The results for glass-bead peening (Figure 4a) indicate that, for the ranges investigated, treatment time or bead size had little effect on the magnitude of the residual compressive stress at the surface. However, subsurface stress measurements discussed subsequently show that glass-bead size did have an effect on residual stress. Residual-stress measurements on coupons having sand or aluminum oxide treatments (Figure 4b) indicated that increased treatment times caused slight increases in residual compressive stress. Figure 4b also shows that the larger the particle size, the greater the residual compressive stress. For the vibratory treatment on both faces of coupons (Figure 4c), increased treatment times produced increases in residual stress in Ti-6Al-4V while little change was noted for Ti-8Al-1Mo-1V. Frequency had a noticeable effect on residual stress as larger residual stresses were produced at 1750 cpm. Note that for the same frequency and shorter treatment times, higher stresses were produced in Ti-8Al-1Mo-1V than Ti-6Al-4V.

Surface Examination

Light and electron photomicrographs of the surface of Ti-6Al-4V coupons before treatment and after glass-bead peening, blasting, and vibratory treatments are given in Figures 5–8. Figure 5 illustrates the dimpled appearance of the as-received Ti-6Al-4V sheet material at low and high magnification. Several beta platelets are visible in the electron micrograph (Figure 5b). Figure 6 shows the surface after glass-bead peening with 425 μ beads for 5 sec. Relatively large craters are evident as would be expected from the large glass beads. High magnification (Figure 6b) reveals the presence of many slip

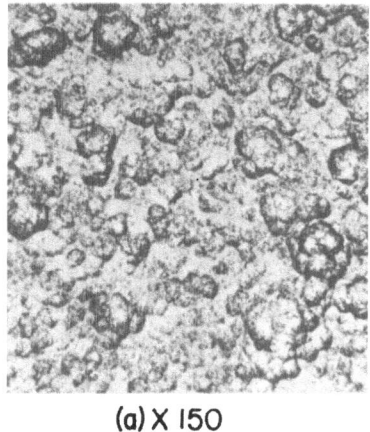

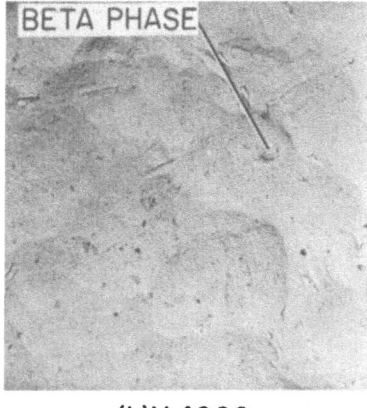

(a) X 150   (b) X 4000

Figure 5. Light and electron photomicrographs of untreated Ti-6Al-4V coupon surface.

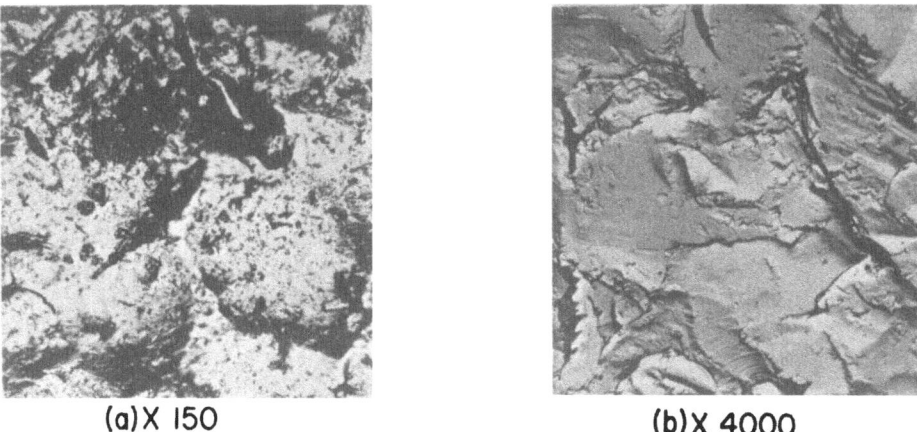

(a)X 150   (b)X 4000

Figure 6. Light and electron photomicrographs of Ti-6Al-4V coupon surface after 425 μ glass-bead peening for 5 sec.

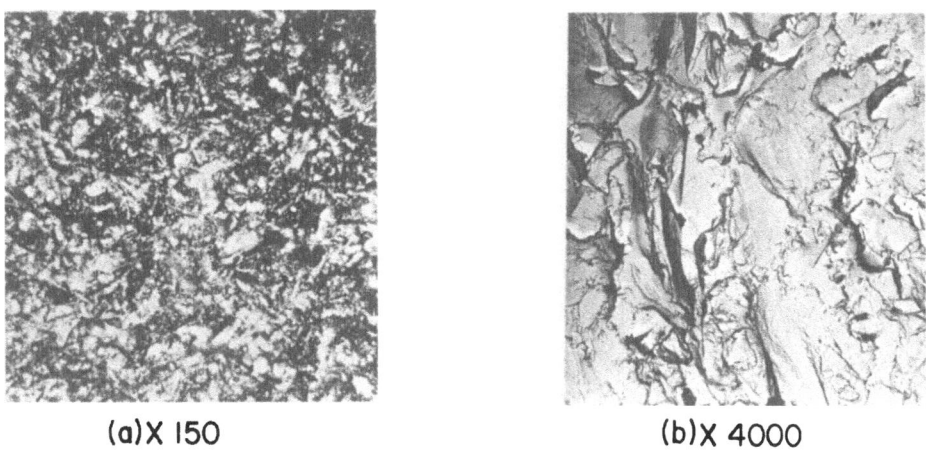

(a)X 150   (b)X 4000

Figure 7. Light and electron photomicrographs of Ti-6Al-4V coupon surface after blasting with 200 μ aluminum oxide for 15 sec.

lines characteristic of rather severe plastic deformation. Figure 7 shows the surface of a coupon which was blasted with 200 μ aluminum oxide for 5 sec. As might be expected from the shape of the particles this treatment produces sharper or more angular irregularities then peening. The effect of a 26-hr vibratory treatment at 1375 cpm on the surface is shown in Figure 8. Both micrographs clearly show the scratches produced by the aluminum oxide triangles in the vibratory treatment.

Figure 9 shows the results of surface-roughness measurements conducted on glass-bead peened and blasted coupons. Surface roughness in root-mean-square (RMS) is plotted against average particle size in microns. Each symbol represents an average of RMS values obtained for the three data points for each treatment. The range of values is given by the bars through each point. The RMS value for the untreated material is plotted on the ordinate. The surface-roughness measurements indicated that the larger the particle size, the rougher the surface finish. On the other hand, treatment time in the

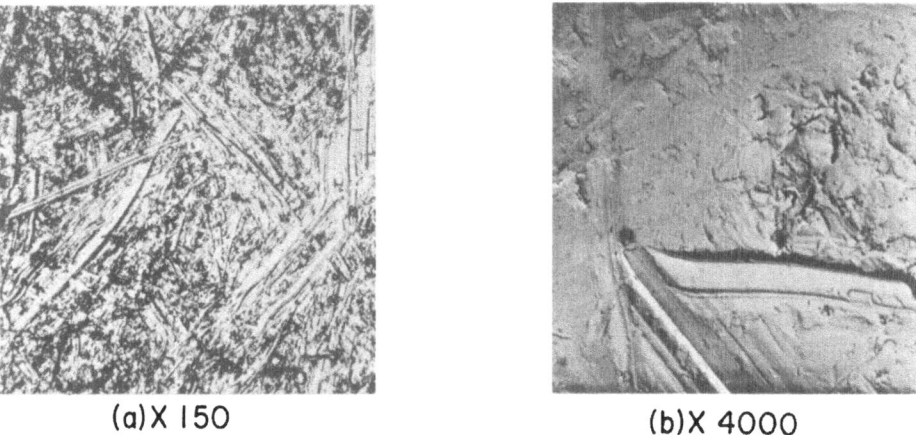

(a)X 150   (b)X 4000

Figure 8. Light and electron photomicrographs of Ti-6Al-4V coupon surface after a 26-hr vibratory treatment at 1375 cpm.

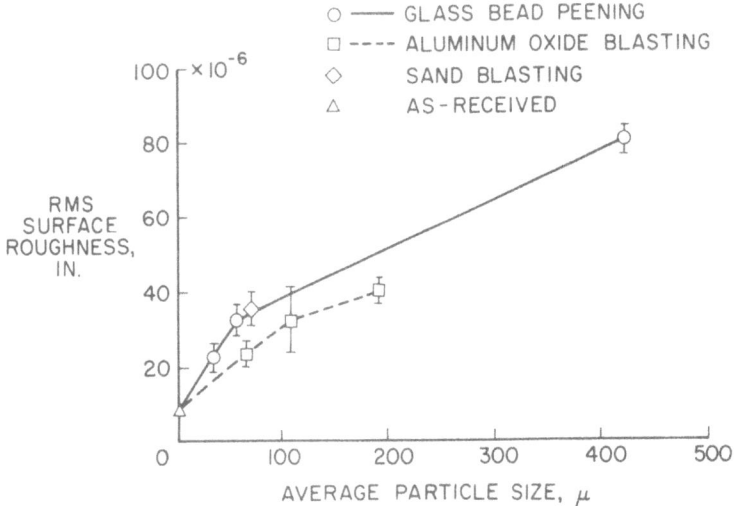

Figure 9. Surface roughness measurements on treated Ti-6Al-4V sheet.

range investigated had little effect on surface roughness. The vibratory treatment produced RMS values of approximately 7 μ in., which was essentially the same as those for untreated material.

Subsurface Stress Measurements

The results of stress measurements conducted beneath the surface of coupons are presented in Figures 10–12. Measurements of residual stress were made in depth increments of 0.0001 in. The resulting stress-distribution curves with depth will, hereinafter, be referred to as "stress profiles."

Figure 10 illustrates the effect of the different correction factors on a stress profile produced by glass-bead peening with 30 μ beads for 5 sec. The stress profile after the Lorentz polarization and absorption corrections is shown by the first curve. (Each symbol represents a single stress measurement.) The beam-penetration correction was then applied to yield the second curve. Note that this correction had a very significant effect on the shape of the stress profile in the layers near the surface. The stress in the first layer is reduced, while the stress in the next two or three layers is substantially increased. As would be expected, the shape of the initial stress profile determines the magnitude and direction of the beam-penetration correction. Finally, the subsurface correction accounts for relaxation effects due to the removal of layers of material and yields the final stress profile as shown by the third curve. This correction tended to reduce the

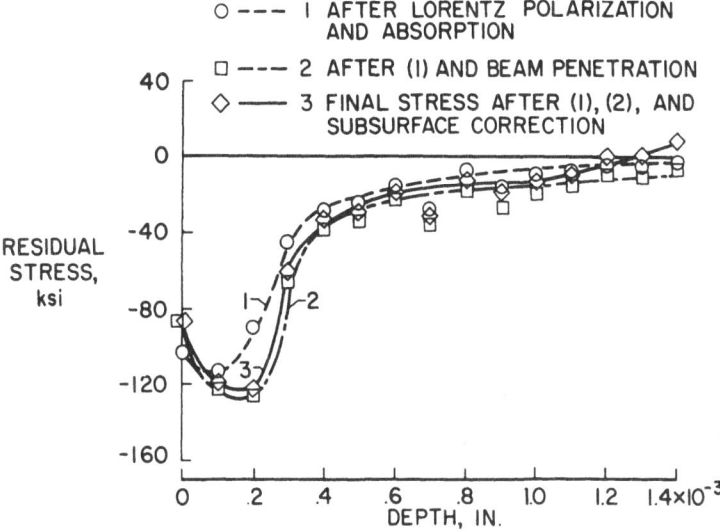

Figure 10. Effect of correction factors on stress profile for Ti-6Al-4V with a 5 sec, 30 μ glass-bead peening treatment.

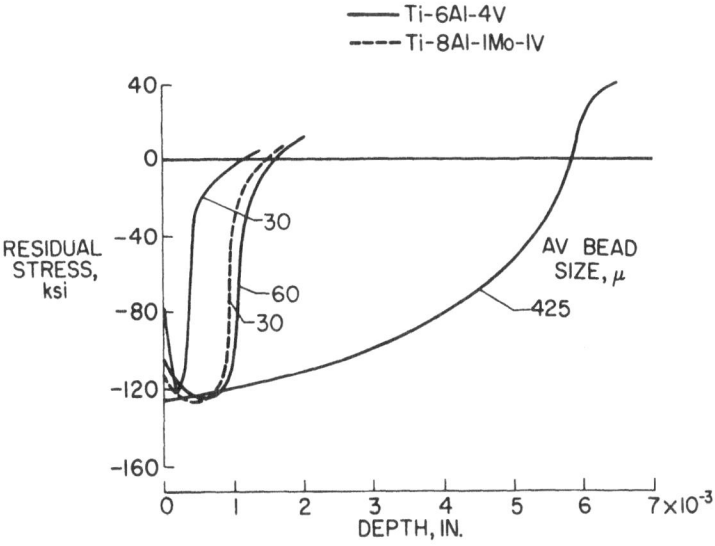

Figure 11. Stress profiles produced by glass-bead peening for 5 sec.

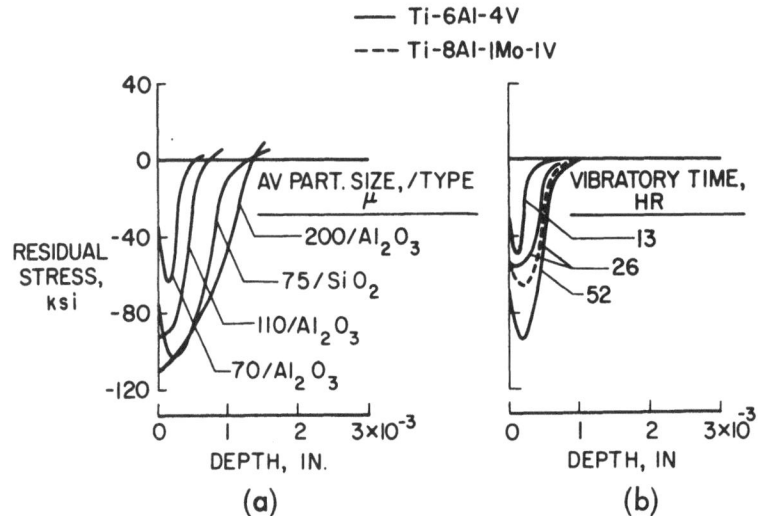

Figure 12. Stress profiles produced by blasting and vibratory treatments. (a) Al_2O_3 and SiO_2 blasting for 15 sec. (b) Vibratory treatment at 1375 cpm.

compressive stress at each point. The effect of the subsurface correction is generally small until rather large penetrations are reached.[5]

The stress profiles obtained after various 5-sec glass-bead-peening treatments are shown in Figure 11. Unlike the results of surface measurements in Figure 4a, particle size is now seen to have a strong influence on the magnitude of the stress profile, with the larger particles producing the higher compressive stresses. The stress profiles for the two 30 μ glass-bead-peening treatments showed that higher compressive stresses were produced in Ti-8Al-1Mo-1V than Ti-6Al-4V. Note also that all the glass-bead-peened coupons except those treated with 425 μ beads exhibited a stress profile in which the maximum stress is slightly below the surface. This shape of stress profile along with the lack of the X-ray beam-penetration correction is the reason the smaller particle-size treatments showed apparently greater surface compressive stresses in Figure 4a.

Stress profiles for coupons with blasting and vibratory treatments are presented in Figure 12. Figure 12a shows that aluminum oxide blasting treatments with larger particles produced greater compressive stresses. For some reason the 75 μ sand-blasting treatment produced a larger depth of compression than treatments with approximately the same size aluminum oxide. Both sand blasting and 70 μ aluminum oxide blasting produced stress profiles in which the maximum stress was slightly beneath the surface. It is interesting to note that this particular shape of stress profile is produced by peening or blasting with particles of 75 μ or less.

The stress profiles for Ti-6Al-4V and Ti-8Al-1Mo-1V coupons after vibratory treatments are shown in Figure 12b. The frequency of the vibratory bath was 1375 cpm. Note that these coupons were treated on only one face to simplify analysis and, therefore, cannot be compared directly with the surface stresses presented in Figure 4c for coupons treated on both sides. Nevertheless, the stress profiles clearly demonstrate the dependence of residual compressive stress on treatment time. Longer treatment times produce a greater magnitude and depth of compressive stress. The maximum compressive stress created by the vibratory treatment was located slightly below the surface. This characteristic was observed in studies of various tumbling processes on steels[6] and was believed to be associated with the scratching action of the sharp points of the abrasive. Again, higher compressive stresses were produced in Ti-8Al-1Mo-1V than in Ti-6Al-4V as demonstrated by the two stress profiles produced by a 26-hr treatment. Although the duplicate tests performed on the two titanium alloys are certainly limited, it does appear that the treatments tended to produce higher compressive stresses in Ti-8Al-1Mo-1V. This could be due, in part, to the higher modulus of elasticity and slightly greater sheet thickness of the Ti-8Al-1Mo-1V.

Effect of Surface Treatments on Tensile Properties

The tensile properties of Ti-6Al-4V and Ti-8Al-1Mo-1V before and after various surface treatments are listed in Table II. The only treatment which caused noticeable change in tensile properties was glass-bead peening with 425 μ beads. This treatment caused a reduction in yield strength of approximately 10% in both alloys. The loss in yield strength was accompanied by a decrease in elongation and an increase in ultimate strength. These results can be justified by considering the stress profile for Ti-6Al-4V having the 425 μ glass-bead-peening treatment (Figure 11). Under equilibrium conditions, a 6-mil layer under an average compressive stress of 80 ksi would produce approximately a 30-ksi tensile stress across the core of the specimen. Therefore, a superimposed tensile load causes the core to yield before the outer layers at a stress below that for the untreated material. Then, after further load is applied, the large plastic strains alleviate the residual-stress effects and the specimen reaches an ultimate stress which is slightly higher than that for the untreated material due to cold working of the surface layers. The other surface treatments affected such a thin layer of material that residual tensile stresses in the core were quite small. Consequently, any reductions in yield strength were also small.

Stress Relaxation of Residual Stresses

The results of the stress-relaxation tests on Ti-6Al-4V are given in Figure 13 where percentage of the residual-surface stress remaining is plotted against temperatures up to 800°F for exposures of 10, 100, and 1000 hr.

Figure 13a shows the results for specimens given a 425 μ glass-bead-peening treatment which produced an initial surface compressive stress of 115 ksi. Noticeable relaxation was observed at temperatures as low as 400°F with almost complete relaxation after 1000 hr at 800°F. Figure 13b presents the results for coupons having a 75 μ sand-blasting treatment and an initial compressive stress of 70 ksi. Measurable stress relaxation occurred in these specimens only at 600°F and 800°F. Similar results were obtained for coupons given a 52-hr vibratory treatment at 1750 cpm (Figure 13c). The initial compressive stress created by this treatment was 80 ksi. It should be pointed out that with a high residual-stress level, the lowering of the proportional limit at elevated temperatures may be sufficient to cause a certain amount of immediate stress relief. This process differs from that described for creep or other stress relaxation phenomena which are time dependent. Although no attempts have been made to separate the two processes in these experiments, the data at 400°F deserve comment. At 400°F the only stress relaxation that was measured occurred in the glass-bead-peened coupons which exhibited a relatively high surface compressive stress of 115 ksi (Figure 13a). Since Ti-6Al-4V shows a reduction of 20% in yield strength[7] at 400°F, it is reasonable to assume that the proportional limit is reduced by approximately the same amount or, in this case, to a value of

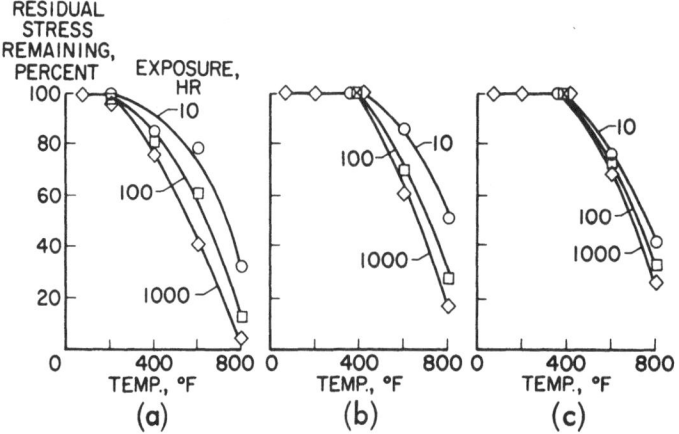

Figure 13. Stress relaxation in Ti-6Al-4V after various surface treatments. (a) Glass-bead peened, 425 μ, 5 sec, −115 ksi. (b) Sand-blasted, 75 μ, 15 sec, −70 ksi. (c) Vibratory-treated, 1750 cpm, 52 hr, −80 ksi.

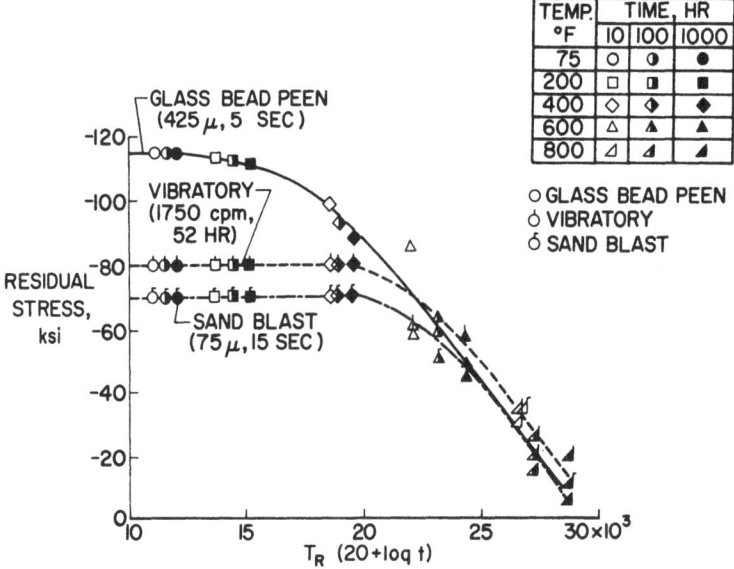

Figure 14. Application of Larson–Miller parameter to stress relaxation data.

about 100 ksi. Therefore, the stress relaxation measured at 400°F for the glass-bead-peened coupons was probably due entirely to the lowering of the proportional limit. The residual surface compressive stresses produced by sand blasting and the vibratory treatment were not relieved because they were below the proportional limit at 400°F.

In order to determine the applicability of rate parameters for correlation of the relaxation data, the data in Figure 13 were plotted using the Larson–Miller time-temperature parameter.[8] The results of this analysis are shown in Figure 14 where residual stress is plotted as a function of $T_R(20 + \log t)$; T_R is temperature in degrees Rankine and t is time in hours. Curves for each of the three treatments are faired through the data. Observe that, as temperatures over 400°F are reached, the curves for all three treatments fall into one band. It was, therefore, concluded that the residual stress induced by glass-bead peening, blasting, or vibratory tumbling underwent the same rate process for relaxation at elevated temperatures. Furthermore, it seems reasonable that any surface treatment which induces compressive stresses in Ti-6Al-4V would also fall in the same band.

CONCLUDING REMARKS

In this investigation residual stresses created in Ti-6Al-4V and Ti-8Al-1Mo-1V sheet by glass-bead peening, sand or aluminum oxide blasting, and a vibratory tumbling treatment have been measured using an X-ray diffraction technique. The residual compressive stress produced by peening or blasting was increased by increasing the particle size. An increase in treatment time or vibrational frequency increased the compressive stress created by the vibratory treatment. The only treatment that noticeably affected the tensile properties of the titanium alloys was glass-bead peening with the largest bead size which caused a 10% reduction in yield strength. Significant stress relaxation occurred in all the treated Ti-6Al-4V coupons exposed at 600°F and 800°F and appeared to be governed by the same rate process.

REFERENCES

1. J. O. Almen and P. H. Black, *Residual Stresses and Fatigue in Metals*, McGraw-Hill Book Co., Inc., 1963.
2. G. J. Heimerl, D. N. Braski, D. M. Royster, and H. B. Dexter, "Salt Stress Corrosion of Ti-8Al-1Mo-1V Alloy Sheet at Elevated Temperatures," NASA paper presented at Pacific Area Meeting of the ASTM, October 31–November 5, 1965.
3. Richard A. Pride and John M. Woodard, "Salt-Stress-Corrosion Cracking of Residually Stressed Ti-8Al-1Mo-1V Brake-Formed Sheet at 550°F (561°K)," NASA TM X-1082, 1965.
4. A. L. Christensen *et al.*, "Measurement of Stress by X-Ray," SAE Information Report, TR-182, 1960.
5. M. G. Moore and W. P. Evans, "Mathematical Correction for Stress in Removed Layers in X-Ray Diffraction Residual Stress Analysis," *SAE Trans.* **66**: 340, 1958.
6. H. R. Letner, "Stress Effects of Abrasive Tumbling," *ASM Trans.* **51**: 1959.
7. V. Weiss and J. G. Sessler (eds.), *Aerospace Structural Metals Handbook, Vol. II–Non-Ferrous Alloys*, ASD-TDR-63-741, Vol. II, U.S. Air Force, March 1963.
8. F. R. Larson and J. Miller, "A Time-Temperature Relationship for Rupture and Creep Stresses," *Trans. ASME* **74**: 771–775, 1952.

DISCUSSION

J. S. Kahn (University of California, Lawrence Radiation Lab.): I am interested in the shape of these particles that you used. Are the glass beads all spherical?

D. N. Braski: Yes.

J. S. Kahn: Are the quartz grains all spherical or angular?

D. N. Braski: All of the glass beads were spherical while the aluminum oxide and the sand particles were angular in shape. The effect of particle shape on the Ti-6Al-4V surfaces can be seen in the optical and electron micrographs in Figures 6 and 7.

G. W. Marks (U.S. Navy Electronics Laboratory): I was wondering about the effect of the forces resulting from the change in velocity of the particles on striking the material. Has this been taken into account?

D. N. Braski: No attempt was made to determine the kinetic energy of the particles striking the surface. However, the results in Figures 11 and 12 indicate that the mass or size of the particle has a significant effect on the depth of residual compressive stress. In our experiments, particle velocity is probably less important since the peening and blasting pressures were held constant.

J. L. Engelke (A. D. Little, Inc.): I was wondering if the subsurface maximum stress might lead to a corrosion effect. Is there any indication in your electron micrographs that there had been a stress relief at the surface by forming microscopic flaws and if so, do these flaws correlate with points of stress corrosion attack?

D. N. Braski: As far as stress corrosion is concerned, these treatments have been quite effective in preventing stress corrosion in both hot sodium chloride and nitrogen tetroxide environments. Apparently the microscopic flaws induced there do not encourage the stress corrosion process. We have attempted to explain the reason for the maximum compressive stress occurring slightly beneath the surface through some type of stress relief mechanism. However, attempts to correlate the depth of pits produced by the various treatments to their respective stress profiles were unsuccessful and no definite conclusions could be reached.

THE APPLICATION OF X-RAY DIFFRACTION
TECHNIQUES TO THE STUDY OF WEAR

T. F. J. Quinn

Southwest Research Institute
San Antonio, Texas

ABSTRACT

The relevance of X-ray diffraction techniques to the investigation of the chemical, physical, and crystallographic changes occurring during sliding wear is discussed in relation to existing theories of wear. By way of illustration, the application of the powder X-ray cylindrical film technique to the unlubricated wear of a low-alloy, medium carbon steel is treated in some detail. The variation of the relative proportions of iron and its oxides in the wear debris with changes in sliding speed is described. It is shown that the results lend partial support to the oxidational hypothesis of the unlubricated wear of steel.

INTRODUCTION

Theories of Wear

When two surfaces slide over one another, they can only, really, come into true contact at only a few points, namely at those points where the asperities on one surface meet the asperities of the other. The average size of this *real* region of contact is about $1 \, \mu^2$. The amount of wear (i.e., the volume V of material removed from either surface) in sliding a distance L is said[1,2] to be proportional to L and to the *real* area of contact A, namely,

$$V = KAL \tag{1}$$

where K is the factor of proportionality. This "K-factor" can take up a wide range of values (from about 10^{-2} down to about 10^{-7}) according to the materials and the conditions of sliding. Since the real area of contact is proportional to the load W (regardless of whether the contact area is formed by elastic or plastic deformation of the asperities), then equation (1) can be written

$$w_L = \frac{V}{L} = K'W \tag{2}$$

where w_L is the wear rate (volume removed in sliding unit distance) and K' is a "coefficient of wear" which is, of course, related to the K-factor of equation (1). Equation (2) bears a strong resemblance to the law of friction which states that F (the frictional force between two sliding surfaces) is proportional to the load W, i.e.,

$$F = fW \tag{3}$$

where f is the factor of proportionality known as the coefficient of friction. This is normally around 0.2 for unlubricated sliding steel surfaces and does not, in fact, vary much from this figure for other sliding surfaces. Although there is still no really satisfactory evaluation of f in terms of basic material constants, (e.g., the hardness of the surfaces, the shear stress of the materials being deformed at the asperities), this is not so important from the practical point of view, since the designer can assume an approximate value for f which will not be very different from 0.2. On the other hand, since K can vary over so many orders of magnitude, the designer can never satisfactorily allow for the wear of the moving parts of his machine. If a method can be found for estimating K from first principles, then this would be of great practical importance.

The Wear of Steels

The present author recently has evolved[3] a model describing the wear of steels in terms of the oxidation of mating asperities at the "hot-spot" temperatures occurring when two surfaces are slid against each other under load. Following Archard and Hirst,[4] the K-factor was interpreted as the probability that an asperity encounter will produce a wear particle, so that, on the average, $1/K$ passes are required (at any particular asperity) for a wear particle to be produced. However, in addition, the present author assumed that these $1/K$ passes are necessary in order to form an oxide film of thickness ξ, and that ξ is the maximum oxide thickness which the metal can support. With these basic assumptions, together with the assumption that oxidational wear follows the normal laws of oxidation, an expression was obtained for K in terms of the oxidational and material constants of the specimens, the temperature of oxidation, and the speed of sliding.

In order to test this model, the author carried out[3] several wear experiments using low-alloy, medium-carbon steel pins sliding, without lubrication, against disks of the same material and found that the variation of wear rate with load and speed could indeed be explained in terms of the oxidational wear model, provided one assumed that a certain parameter (involving the ratio of the square of the maximum oxide film thickness ξ and the distance d along which a wearing contact is made) varied in a well-specified manner with respect to the hot-spot temperature, θ_m. This temperature cannot be readily measured by any known technique—for the purposes of these experiments, it was assumed that Archard's[5] calculations gave a close approximation to the true hot-spot temperature. Since ξ and d are also not readily evaluated (neither experimentally nor theoretically), an indirect (and independent) approach was considered desirable.

The Relevance of X-Ray Analysis to Wear of Steels

Both the wear debris and the worn surfaces should contain different types and proportions of oxides according to the temperatures occurring at the mating asperities. The crystallite size should provide a minimum possible value for ξ. The amount of crystalline strain should provide information about the work hardening (or annealing) of the surfaces or debris. According to the current ideas regarding plastic and/or elastic deformation at asperities coming into contact during sliding, this must affect the *real* area of contact (and, hence, d). Thus, by investigating the crystallography of the worn surfaces and the wear debris, it should be possible to get an independent check on the oxidational wear hypothesis mentioned above.

From metallographic sections, it has been shown that the effects of wear exist as far as $10\ \mu$ below the sliding surface. Hence, a diffraction technique involving the reflection of X-rays from a worn surface should give crystallographic information about the *whole* of the deformed region. It will not, however, give much information about the crystallography of the surface itself. For crystallographic information about the surface and the layers immediately below the surface (i.e., down to about 100 Å), one would normally use a reflection electron diffraction technique. Since the average wear particle diameter will probably be less than the depth of penetration of X-rays into the particle, diffraction patterns obtained by the transmission of X-rays through wear particles supported on a thin, glass fiber should be representative of the bulk of each particle. If information is required about the crystallography of the surface layers of each particle, then the low penetration of electrons through matter makes the transmission electron diffraction technique more suitable to use. In this way, one can see that X-ray and electron-diffraction techniques complement each other—both are necessary for a complete description of the crystallographic changes caused by sliding wear.

The Scope of the Present Paper

This paper describes the manner in which X-ray diffraction techniques have been used in an attempt to obtain additional information about ξ and d, and their dependence on θ_m (the hot-spot temperature). It is concerned with the proportional analysis of the wear debris produced in the speed-variation section of the wear experiments mentioned above, using an X-ray diffraction film technique. The debris was examined in a conventional cylindrical powder camera, and the line positions and profiles obtained using a vernier measuring device and a densitometer, respectively. A special feature of the analysis of the profiles was the use of a variation of a technique first used by Averbach and Cohen[6] (for determining the volume percentage of austenite retained in martensitic steels) in order to determine the proportions of iron and iron oxides present in the debris. Subsequent papers will deal with the application of X-ray diffractometer techniques to the study of the crystallite size and crystallite strain in the worn pin and disk surfaces, as well as the application of the complementary techniques of electron diffraction and electron microscopy to examining both the worn surfaces and the wear debris.

The various correlations between the relative proportions of iron (and its oxides) present in the wear debris and the other wear parameters such as sliding speed, wear rate, friction coefficient, and hot-spot temperature are discussed. Where possible, some tentative conclusions are drawn. The aim of this introductory paper will have been achieved, however, if it has shown that there are many ways in which X-ray (and electron) optical techniques can help in elucidating wear mechanisms.

EXPERIMENTAL

The Wear Experiments

All the details relating to these wear experiments have been reported elsewhere.[3] Essentially, they consisted of measurements of the volume of material removed from the pins, and the frictional force on the pins, as functions of load and speed. In this paper, only the speed-variation experiments are analyzed. Figure 1 shows how the wear rate of this particular steel (designated En26 in British Standards notation) varied with speed, and Figure 2 shows how the coefficient of friction varied with speed. Each point on the wear graph represents the result of taking the slope of the "wear volume versus distance of sliding" curve after the wear had settled down to an equilibrium rate. Each point on the friction graph is the equilibrium friction value attained in each experiment after about 8 hr running. It can be seen that there is a general decrease in wear rate and friction

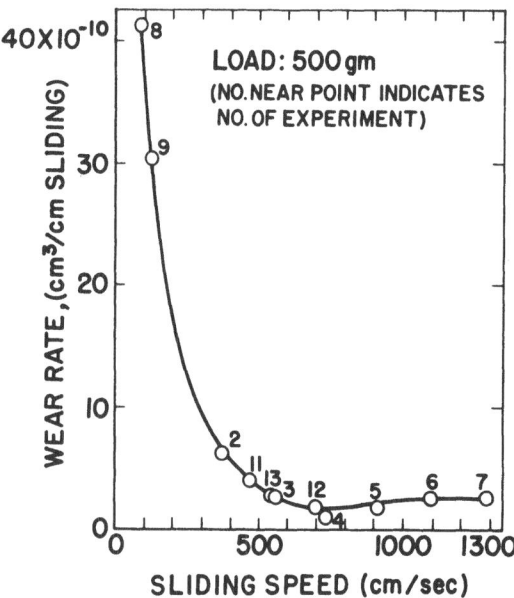

Figure 1. Wear rate (cm³/cm sliding) as a function of speed.

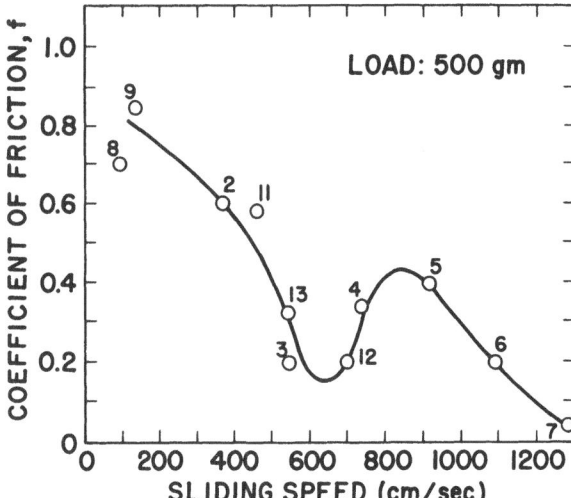

Figure 2. Variation of friction coefficient with speed.

with increasing sliding speed, although there seems to be a significant departure from this general trend for speeds between 600 and 1000 cm/sec.

The author was able[3] to show that the wear rate *vs.* speed variation of Figure 1 (and the wear rate versus load variation *not* being dealt with in the present paper) could be explained in terms of the oxidational wear hypothesis provided one assumes that the parameter ξ^2/d varied with θ_m (the hot-spot temperature) in the uniform manner shown in Figure 3. From this graph, one can see that provided the hot-spot temperature is greater than about 700°C (i.e., close to the softening temperature of steels), then ξ^2/d remains constant at about 10^{-6} cm. Now, according to one of the recent theories of friction between sliding surfaces,[7] the area of contact at an asperity increases under *plastic deformation* until the contact pressure equals the yield stress of the material.

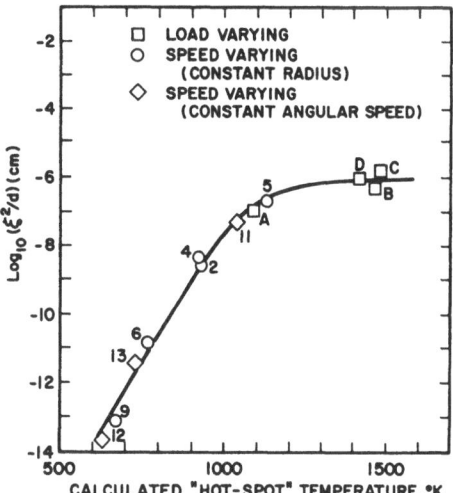

Figure 3. Variation of $\log(\xi^2/d)$ with calculated "hot-spot" temperature (θ_m).

If one assumes that d is of the order of the diameter of an asperity contact area, and that all the load is borne at this one contact, then one obtains

$$A = \pi d^2 = \frac{W}{p_m} \qquad (4)$$

where p_m is the yield pressure (which can be taken as the hardness). Since the hardnesses of the pins and disks (*before* wear) were found to be 300 ± 20 VPN (Vickers Pyramid Number), one can readily see that d would have the value of about 2×10^{-3} cm. This would mean that for $\theta_m > 700°$C, ξ (the maximum oxide film thickness) would have to be about 10^{-4} cm. This is about the size of the typical wear particle. Hence, the proposed model is capable of explaining the wear results for hot-spot temperatures above about 700°C.

When θ_m is less than about 700°C, the value of ξ (the maximum oxide film thickness supportable by the metal) need not be a constant. Furthermore, if plastic flow is *not* occurring at the asperities, then one cannot assume that d is still a constant at about 2×10^{-3} for this range of hot-spot temperatures. All that can be said about this region is that the ratio (ξ^2/d) is strongly dependent on θ_m (see Figure 3). As already mentioned in the introduction, there exists a need for an independent examination of the variations of ξ and d (via crystalline size and crystallite strain) with parameters such as wear rate, velocity, friction, etc. Furthermore, by comparing the components of the wear debris (and their relative proportions) with those to be expected from bulk oxidation experiments, it should be possible to confirm that the hot-spot temperatures were actually occurring at the contacting asperities. The following sections describe how the wear debris patterns have been analyzed with this last aim in view.

The Diffraction Patterns from the Wear Debris

Debris was collected at the end of the first (or second) hour, again at the end of the fifth hour, and, finally, at the end of the seventh (or eighth) hour. It was supported on a glass fiber and irradiated in a conventional cylindrical powder camera for 15 hr with Co K_α radiation using a tube voltage of 35 kV and a tube current of 12 mA. The line positions of the resulting X-ray diffraction powder patterns were then measured using a

Table I. Interplanar Spacings and Relative Integrated Intensities (in the Same Arbitrary Units) as Measured from the X-Ray Powder Diffraction Patterns from the Wear Debris

d(Å)	Relative intensities of possible constituents of the wear debris					Run No. 2, 8 hr	Run No. 3, 8 hr	Run, No. 4, 8 hr	Run No. 5			Run No. 6			Run No. 7			Run No. 8, 8 hr	Run No. 9, 8 hr	Run No. 11, 7 hr	Run No. 12, 7 hr	Run No. 13			d(Å)
	α-Fe	FeO	Fe$_3$O$_4$	γ-Fe$_2$O$_3$	α-Fe$_2$O$_3$	8 hr	8 hr	8 hr	2 hr	5 hr	8 hr	2 hr	5 hr	8 hr	1 hr	5 hr	8 hr	8 hr	8 hr	7 hr	7 hr	1 hr	5 hr	8 hr	
4.82			10	5		—	—	0.031	—	0.032	0.117	0.063	0.080	0.080	0.077	0.079	0.072	—	—	—	—	—	0.103	0.078	4.82
3.68				6	70	0.063	0.067	0.043	—	0.043	0.074	—	0.058	0.047	—	0.208	0.156	0.216	—	0.104	0.035	—	0.192	0.081	3.68
2.96			34	42		0.042	0.075	0.060	0.016	0.107	0.143	0.084	0.192	0.150	0.117	0.191	0.062	0.320	trace	0.127	0.080	—	0.110	0.098	2.96
2.69					100	0.102	0.113	0.126	0.020	trace	0.192	0.032	0.127	0.118	0.104	0.625	0.566	0.298	trace	0.045	0.097	0.045	0.224	0.171	2.69
2.52			100	100	80	0.176	0.234	0.245	0.117	0.271	0.796	0.282	0.638	0.500	0.503	0.468	0.073	—	—	0.137	0.274	0.104	0.477	0.424	2.52
2.46		80				—	0.112	0.110	0.075	trace	trace	0.073	0.077	0.074	0.270	—	—	0.152	trace	0.312	0.466	0.069	—	—	2.46
2.20					70	0.036	0.057	0.050	—	—	trace	—	0.038	0.042	—	—	—	—	—	0.059	—	0.101	—	—	2.20
2.13		100				0.079	0.170	0.113	0.200	0.255	0.656	0.202	0.138	0.170	0.461	0.352	0.254	—	—	0.158	0.084	0.035	0.072	0.084	2.13
2.09			34	33	10	0.035	0.060	0.085	0.032	trace	trace	0.388	0.279	0.178	0.303	0.245	0.213	0.708	0.615	0.208	0.204	0.343	0.154	0.211	2.09
2.02	100					0.126	0.056	0.122	0.330	0.065	0.086	0.525	0.181	0.084	0.449	0.136	0.113	0.160	trace	0.264	0.157	—	—	—	2.02
1.84				2	70	0.045	0.047	0.059	—	—	0.027	—	0.072	0.056	—	0.066	0.041	—	—	0.080	0.248	—	0.181	0.188	1.84
1.79						—	—	—	0.076	—	—	0.123	—	—	0.074	—	—	—	trace	0.038	0.094	0.052	0.075	0.251	1.79
1.70			11	16	80	0.072	0.122	0.076	0.042	0.060	0.106	0.063	0.082	0.125	0.070	0.127	0.126	0.256	—	0.111	0.085	0.046	0.048	0.096	1.70
1.60		60	41	45	40	0.064	0.119	0.092	0.060	0.115	0.175	0.115	0.219	0.189	0.133	0.168	0.263	0.042	—	0.107	0.154	0.115	0.200	0.037	1.60
1.51						0.076	0.144	0.081	0.161	0.169	0.464	0.126	0.115	0.074	0.352	0.226	0.244	0.180	trace	0.120	0.118	—	0.196	0.082	1.51
1.48			62	65	70	0.119	0.248	0.144	0.094	0.178	0.374	0.236	0.354	—	0.306	0.318	0.499	0.256	0.173	0.154	0.104	0.072	0.081	0.094	1.48
1.45					80	0.080	0.135	0.066	—	—	—	—	0.102	—	—	—	—	0.119	0.422	0.066	0.139	—	0.189	0.194	1.45
1.43	19					0.034	—	—	0.099	—	—	0.142	—	—	0.098	—	—	0.314	—	0.042	—	0.099	0.110	0.101	1.43
1.17	30					0.084	—	0.042	0.180	—	—	0.226	—	—	0.142	—	—	—	—	0.103	—	0.126	—	—	1.17

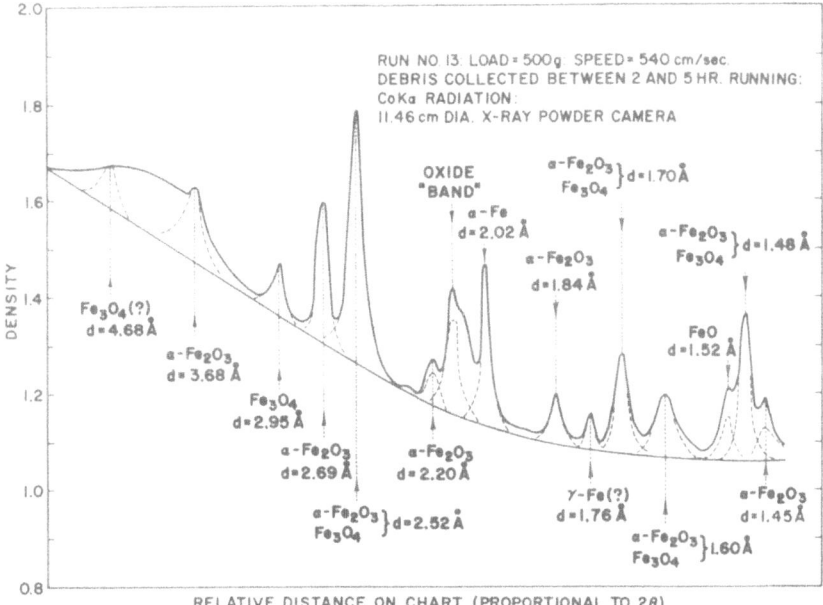

Figure 4. Typical "density *vs.* 2θ" plot.

conventional vernier measuring device. Where necessary, corrections were made for film shrinkage (or elongation) and the interplanar spacings calculated in the usual way. The line profiles also were obtained using a recording densitometer with a calibrated logarithmic aperture. This instrument produced linear density profiles on a 10-in. wide chart, with the zero density being set on the "fog" of the very fast, high-contrast X-ray film used for these experiments. The density of the diffraction peaks rarely reached values in excess of about 1.8. Klug and Alexander[8] consider that by properly choosing the processing conditions, it is possible to preserve the linearity between density of blackening D of the X-ray film and the exposure E to within 3% up to $D \sim 2.5$ for any of the standard X-ray films. Hence, it will be assumed that the "density *vs.* 2θ" curves obtained from the debris are directly proportional to the X-ray intensity. Since the same exposure conditions were used throughout, it is possible to compare the integrated intensities from pattern to pattern, if required. Figure 4 is an example of one of the "density *vs.* 2θ" traces in which the "noise" of the recorder has been smoothed out. In order to obtain the relative integrated intensities, it was assumed that the product of peak height density times the "half width" (in cm) was a good approximation to the relative integrated intensity. This was checked using a planimeter and found to be a valid assumption.

Table I summarizes the interplanar spacings and relative integrated intensities (in the same arbitrary units) obtained from the wear debris diffraction patterns. Runs 5, 6, and 7 were chosen for more detailed analysis because it was calculated that these were the runs with the highest (1190°K), the medium (790°K), and the lowest (390°K) hot-spot temperatures, respectively. Run 13, with a calculated hot-spot temperature of around 770°K, was also examined in more detail as a check on run 6. A "dash" at any point in the table signifies that *no* diffraction line was obtained which could possibly give rise to the relevant interplanar spacing. The word "trace" signifies that it was impossible to measure the relative integrated intensity, although visual examination of the film revealed that a diffraction line was indeed present. All the diffraction lines were successfully identified as being due to either α-Fe, or one of its oxides, with the exception of the line at 1.79 Å. It is possible that this is the 200 line of γ-Fe. However, this identification could not definitely be confirmed—it is definitely not present in the surfaces of the unworn and worn pins and rings. In what follows, it will be assumed that the line at 1.79 Å is due to an insignificantly small volume of material compared to the volumes of α-Fe, or its oxides.

Proportional Analysis of the Wear Debris

The results embodied in Table I reveal strong differences in relative integrated intensities from pattern to pattern. Since the relative integrated intensity is proportional to the volume causing diffraction, it is possible to use these results quantitatively to analyze the debris. As one can see from Figure 4, there is already so much overlap between the lines from the various phases present in the debris, that the addition of yet another pattern (by mixing a known amount of a known element in with the debris) would be completely out of the question. Following a method analogous to that used by Averbach and Cohen,[6] one of the phases already present in the debris was used as an internal standard. The choice of internal standard varied from pattern to pattern, the main criteria for selection being the reliability of a particular profile measurement (as indicated by its lack of serious overlap with adjoining profiles), and the number of reliable profiles obtainable from that phase. This method is best explained in terms of a typical example, namely "run No. 13; Debris collected between 2 and 5 hr running," the "smoothed" densitometer trace of which appears in Figure 4.

From the trace, it is apparent that the 211 and 220 diffraction maxima of α-Fe_2O_3, at 2.69 Å and 1.84 Å, respectively, are reasonably free from overlap. Hence, α-Fe_2O_3 was chosen as the internal standard for this pattern. It is generally accepted that the intensity I'_x of X-rays diffracted by a substance X is given by

$$I'_x \propto \left[\frac{F^2 m(\mathrm{LP}) e^{-2M}}{v_x^2} \right] [V_x A(\theta)] \qquad (5)$$

In this equation, F is the structure factor, m is the multiplicity of the diffracting plane, LP is the Lorenz polarization factor, e^{-2M} is the Debye–Waller temperature factor, v_x is the volume of the unit cell of the substance X, $A(\theta)$ is the sample absorption factor, θ is the Bragg angle, and V_x is the volume of the substance X irradiated. Write equation (5) as

$$I'_x = R[V_x A(\theta)] \qquad (6)$$

where

$$R = \left[\frac{F^2 m(\mathrm{LP}) e^{-2M}}{v_x^2} \right]$$

and where the factor of proportionality in equation (5) has been taken to be unity (since we are only interested in relative values). Now R can be readily calculated for α-Fe_2O_3 from the available literature, V_x is the required unknown, and $A(\theta)$ is an ill-defined absorption factor. By plotting I_x/R (where I_x is the relative integrated intensity given in Table I) against θ, it should be possible to obtain a smooth curve representing $V_x A(\theta)$ as

Table II. Relative $[V_x A(\theta)]$ Values for the X-Ray Powder Pattern Obtained from the Run No. 13 (2–5 hr) Wear Debris

Substance	hkl	$[V_z A(\theta)]$	θ
	110	66.5×10^{-4}	14.2°
α-Fe_2O_3	211	33.4×10^{-4}	19.2°
	220	26.5×10^{-4}	29.1°
	331, $\bar{2}$11	118×10^{-4}	38.5°
α-Fe	110	2.3×10^{-4}	26.1°
FeO	200	23.2×10^{-4}	24.6°
Fe_3O_4	220	72×10^{-4}	17.5°

a function of θ. However, so far we have only obtained *two* good points for α-Fe_2O_3, namely 33.4×10^{-4} for 211 and 26.5×10^{-4} for 220. To obtain the required smooth curve, certain approximations must be made. For instance, the 110 maximum (at 3.68 Å) has been approximated (see the dotted profile) and the resulting profile is 86% of the most intense α-Fe_2O_3 (211) maximum at 2.69 Å. This compares favorably with the 70% value given by the X-Ray powder data file. Accepting this maximum as being approximately correct, a relative value of 66.5×10^{-4} is obtained for $V_x A(\theta)$. To confirm this approximate value, a high angle α-Fe_2O_3 maximum, namely, the $(331, \bar{2}11)$ maximum at 1.45 Å, was also used, and a fairly reliable value of $V_x A(\theta)$ again obtained, namely, 118×10^{-4}.

Figure 5 shows how a smooth curve with a minimum around $\theta = 25°$ can be drawn through all four points. In order to compare proportionate volumes of the various constituents present in this debris, similar calculations of $V_x A(\theta)$ must be carried out for each resolved line present. Due to the similarity between Fe_3O_4 and γ-Fe_2O_3, no attempt is made to distinguish between these two oxides, and, for the calculations, it has been assumed that the Fe_3O_4 spinel structure is the appropriate one to use. The relative $V_x A(\theta)$ values are given in Table II. Now, if all the constituents were present in equal percentage volumes, then all their $V_x A(\theta)$ points would lie along the α-Fe_2O_3 curve. Since they do not all lie on this curve, then the ratio of the ordinates at the particular θ must give the relative percentage volumes. By comparing ordinates, and assuming that only α-Fe_2O_3, α-Fe, FeO, and Fe_3O_4 are present in the debris, one can readily calculate the percentage volumes, namely, $V_{\alpha\text{-}Fe_2O_3} = 26\%$, $V_{FeO} = 30\%$, $V_{\alpha\text{-}Fe} = 3\%$, and $V_{Fe_3O_4} = 41\%$.

As a check on the above percentage volumes, and to show how one can deal with composite profiles, consider the profile at 1.70 Å, due to α-Fe_2O_3 (321) and Fe_3O_4 (422). Assume that the above proportions are correct so that, for this maximum, we have

$$V'_{\alpha\text{-}Fe_2O_3}[1 + (V_{Fe_3O_4}/V_{\alpha\text{-}Fe_2O_3})] = 1$$

where $V'_{\alpha\text{-}Fe_2O_3}$ is the fractional volume of α-Fe_2O_3 contributing to this maximum. Hence,

$$V'_{\alpha\text{-}Fe_2O_3} = 0.422$$

and

$$V'_{Fe_3O_4} = 0.578$$

Since $I_x = (R_{Fe_3O_4}V'_{Fe_3O_4} + R_{\alpha\text{-}Fe_2O_3}V'_{\alpha\text{-}Fe_2O_3})A(\theta)$ for this maximum, one can evaluate $A(\theta)$ and obtain $A(\theta) = 1.92 \times 10^{-2}$ at $\theta = 31.9°$. Thus, one obtains

$$V_{\alpha\text{-}Fe_2O_3}A(\theta) = 26 \times 10^{-2} \times 1.92 \times 10^{-2} = 50 \times 10^{-4}$$

Plotting this value onto Figure 5, one sees that the 321 value lies on the smooth $\alpha\text{-}Fe_2O_3$ curve, thereby, independently, confirming the validity of the $\alpha\text{-}Fe_2O_3[V_xA(\theta)]$ curve and the ratio $(V_{Fe_3O_4}/V_{\alpha\text{-}Fe_2O_3})$.

In a similar way to that described in the last paragraph, it was possible to obtain proportional analyses of all the patterns detailed in Table I. Figures 6, 7, 8, and 9 are graphs of the relative proportions plotted as functions of time during experiments 6, 7,

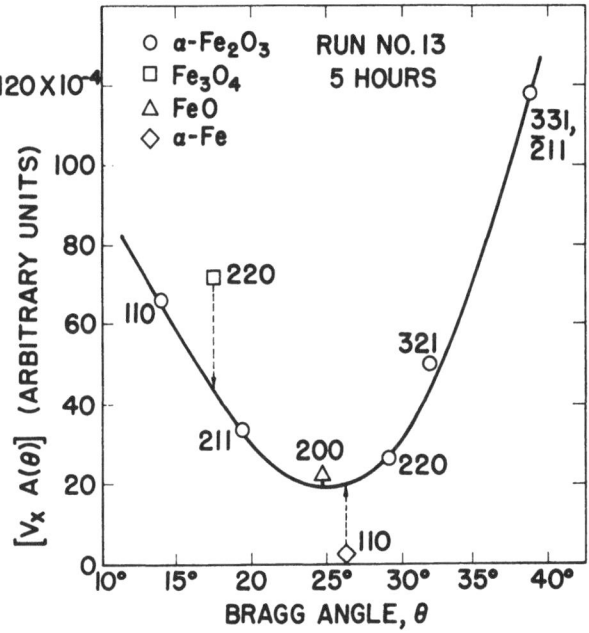

Figure 5. Plot of $[V_xA(\theta)]$ vs. θ.

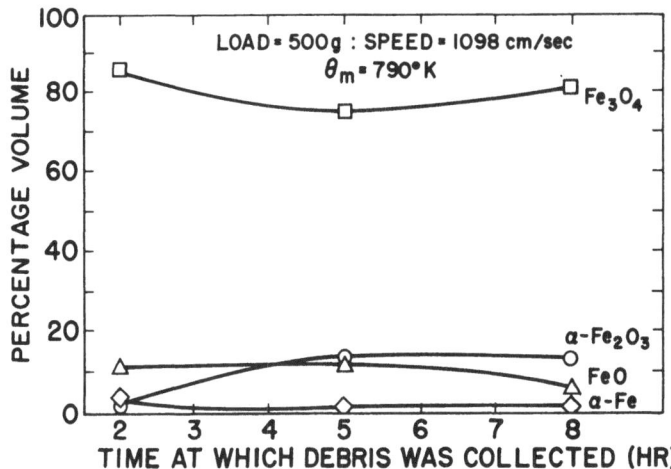

Figure 6. Percentage volumes at various times during run no. 6.

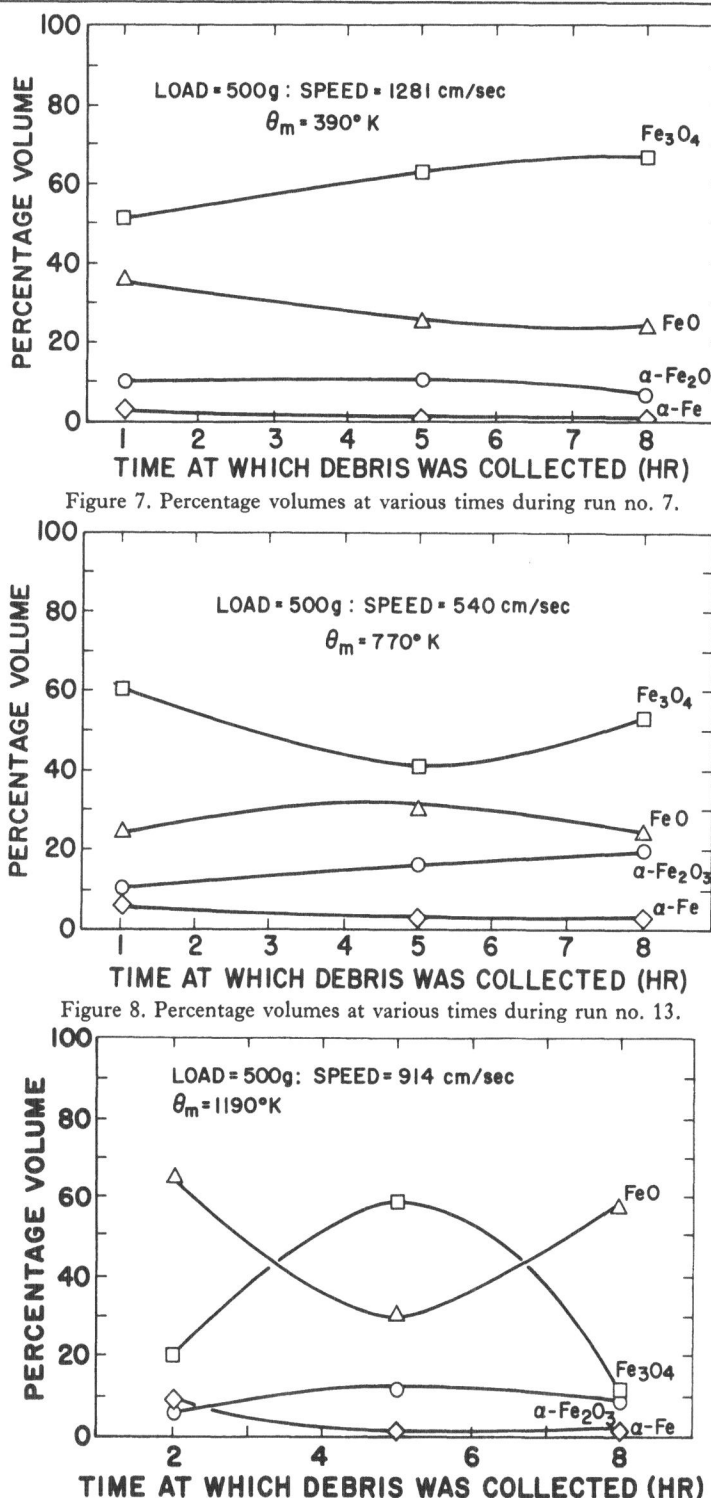

Figure 7. Percentage volumes at various times during run no. 7.

Figure 8. Percentage volumes at various times during run no. 13.

Figure 9. Percentage volumes at various times during run no. 5.

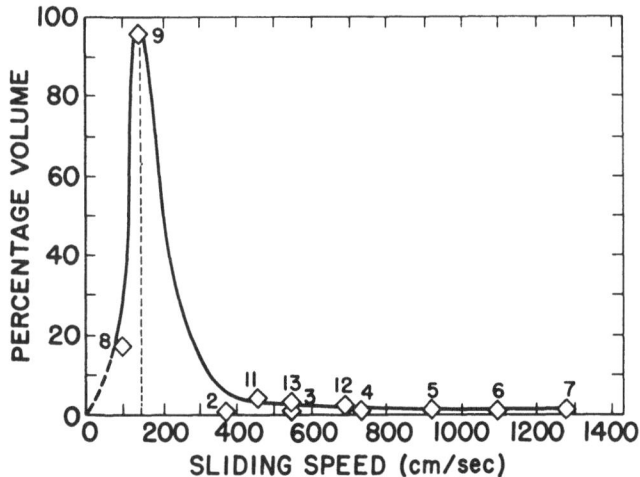

Figure 10a. Percentage volume of α-Fe as a function of the sliding speed.

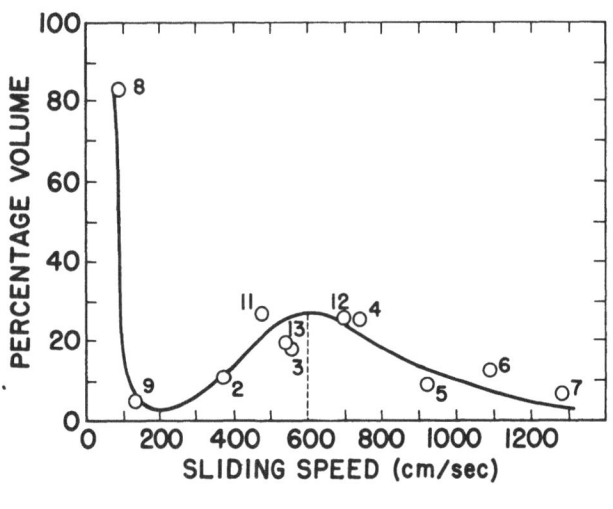

Figure 10b. Percentage volume of α-Fe₂O₃ as a function of the sliding speed.

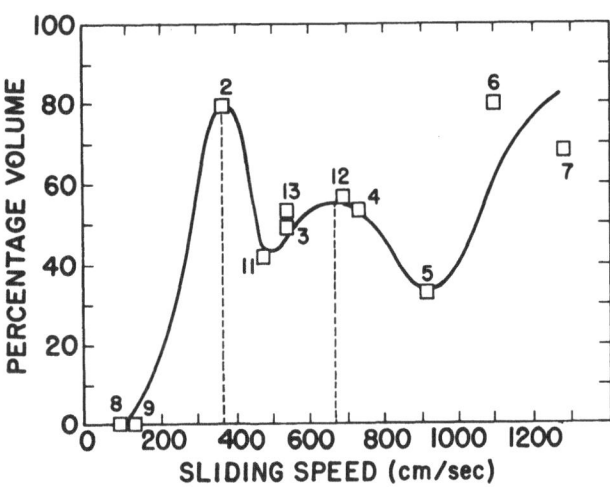

Figure 10c. Percentage volume of Fe₃O₄ as a function of the sliding speed.

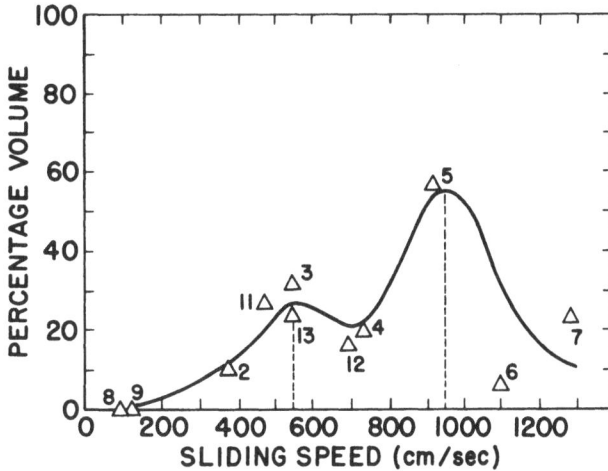

Figure 10d. Percentage volume of FeO as a function of
the sliding speed.

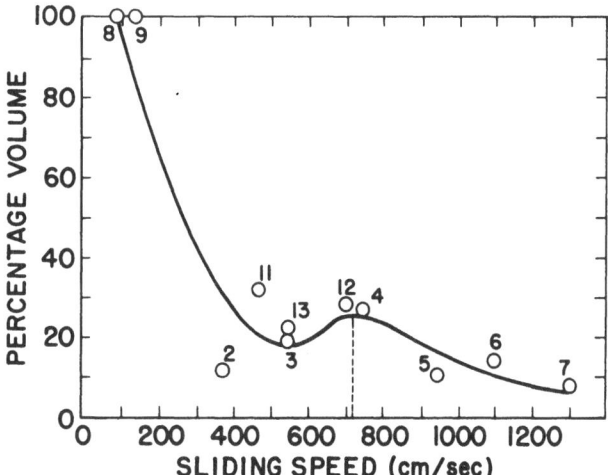

Figure 10e. Combined per-
centage volumes of α-Fe and
α-Fe₂O₃ as a function of
sliding speed.

13, and 15, respectively. With the exception of Figure 9, all the graphs reveal little change
in relative proportions during any one experiment, thereby indicating the validity of
comparing the proportions present in the *final* debris in each case. Figures 10a–d show
the percentage volumes of iron and each of its oxides in the wear debris plotted as a
function of sliding speed. Although the variations revealed in these figures may seem a
little erratic, it is interesting to note that the results for runs 13 and 12 lie very close to
those for runs 3 and 4, respectively. Since runs 11–13 were carried out in a completely
separate set of experiments, using different radii of wear tracks from runs 2–9, and using
a different batch of the same type of steel[3] one can be fairly certain that the apparently
erratic behavior is reproducible.

DISCUSSION

The general trend of Figures 10a–d can be summarized by stating that, at low speeds
of sliding, one gets large amounts of α-Fe and/or α-Fe₂O₃. At high speeds of sliding,
one gets large amounts of Fe₃O₄ and/or FeO. By plotting the combined proportions of

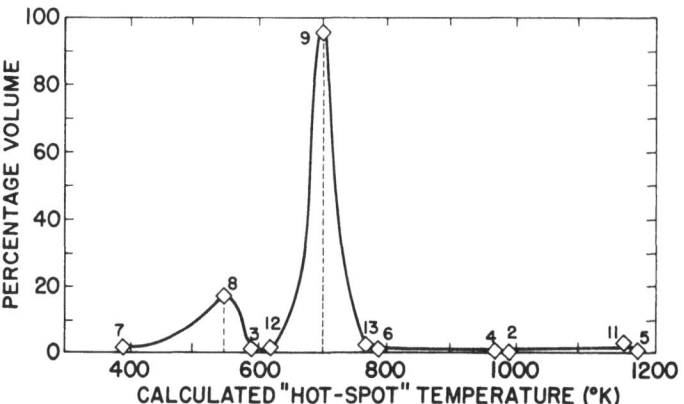

Figure 11a. Percentage volume of α-Fe as a function of "hot-spot" temperature.

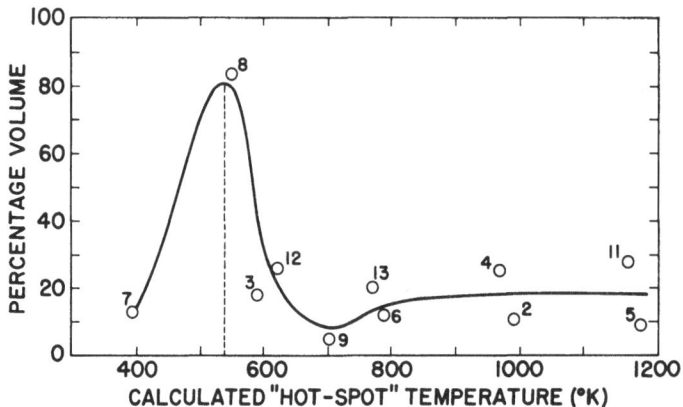

Figure 11b. Percentage volume of α-Fe₂O₃ as a function of "hot-spot" temperature.

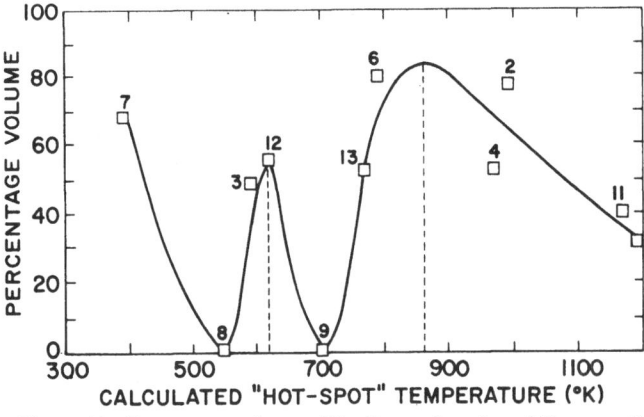

Figure 11c. Percentage volume of Fe₃O₄ as a function of "hot-spot" temperature.

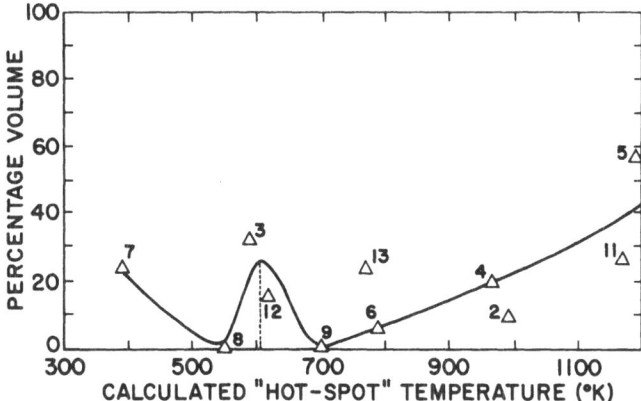

Figure 11d. Percentage volume of FeO as a function of "hot-spot" temperature.

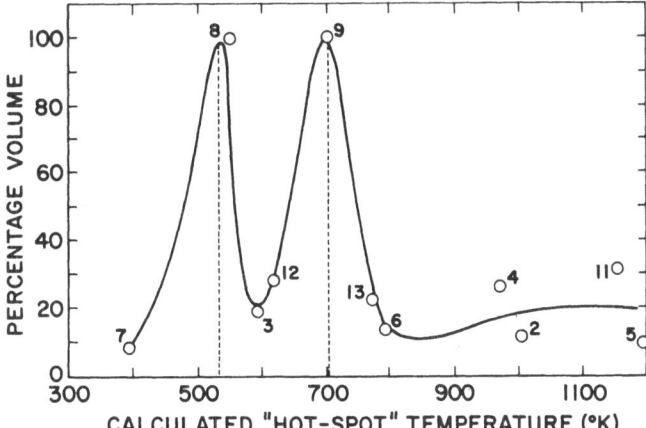

Figure 11e. Combined percentage volumes of a α-Fe and α-Fe$_2$O$_3$ as a function of "hot-spot" temperature.

α-Fe and α-Fe$_2$O$_3$ against sliding speed (see Figure 10e), one obtains a fairly smooth curve which bears a strong resemblance to Figures 1 and 2 (the variations of wear rates and coefficients of friction with speed). This is an important correlation, which implies that high wear rates and high frictions are to be associated with large amounts of α-Fe and/or α-Fe$_2$O$_3$ in the wear debris.

Figures 11a–d are graphs of the percentage volume of each constituent versus the calculated hot-spot temperature. While revealing some disconcerting features, it is possible to see some correlation between the various maxima and minima and the transition temperatures found in bulk oxidation experiments. For instance, we note that the α-Fe$_2$O$_3$ maximum occurs at about 250°C (see Figure 11b) which is approximately the temperature at which iron oxidation experiments[9] reveal that the predominant oxide will be α-Fe$_2$O$_3$. The maximum of the Fe$_3$O$_4$ curve at about 350°C (see Figure 11c) coincides with the temperature at which Fe$_3$O$_4$ begins to be formed beneath the α-Fe$_2$O$_3$ in the oxidation experiments already mentioned. The gradual increase in FeO with θ_m above about 600°C is in accord with the bulk oxidation experiments of Moreau and Bardolle,[10] who found that FeO was formed (as well as Fe$_3$O$_4$ and α-Fe$_2$O$_3$) at about

600°C. The gradual decrease in Fe_3O_4 with θ_m above 600°C could be associated with the increased amount of FeO being formed at these temperatures. On the other hand, the increase of Fe_3O_4 and FeO (the high-temperature phases) with decrease in θ_m below about 300°C is not so readily understood. Equally unpredicted are the peaks in the "α-Fe $vs.$ θ_m" line, and the minima in the FeO and Fe_3O_4 curves at about 300° and 450°C. The position is not resolved (as in the case of the "proportions $vs.$ speed" graphs) by plotting the combined α-Fe and α-Fe_2O_3 proportions (see Figure 11e).

CONCLUSIONS

Obviously, this introductory paper has raised more questions than it has resolved. While there is a need for the X-ray diffraction technique to be checked using an internal standard artificially added to the wear debris after collection, it has been shown to be self-consistent in many ways. For instance, the relative constancy of proportions during an experiment as revealed by Figures 6–9, and the close similarity between Figure 10e and the wear rate and friction variation with speed as shown by Figures 1 and 2.

The maxima and minima in the graphs of relative proportions versus θ_m generally occur at the temperatures to be expected if the wear debris had been formed at the "hot-spot" temperatures. However, there are one or two departures from expected behavior which need further investigation. For instance, why are there large amounts of high temperature oxides (FeO and Fe_3O_4) in the debris from the experiments with the lowest calculated hot-spot temperatures? Again, why does the debris from the high wear and high friction experiments contain only low temperature products (α-Fe and α-Fe_2O_3)? Furthermore, the variation of proportions with θ_m cannot be used *unambiguously* to estimate the hot-spot temperatures occurring at mating asperities.

In spite of these setbacks, the most important result of this paper lies in the fact that an X-ray diffraction technique has been used (with sufficient success to warrant further investigations) in order to analyze the proportions present in wear debris. As far as the author is aware, the only previously reported work of this nature was that carried out by Goldschmidt and Harris[11] in 1941. These authors made visual qualitative estimates of the relative proportions present in wear debris produced under very severe wear conditions using steels of various carbon percentages.

ACKNOWLEDGMENTS

The author would like to thank D. Price, G. Morvay, and G. Morrison for their assistance. He also wishes to record his thanks to Professor C. A. Hogarth, of Brunel College, London, and Mr. P. M. Ku, of Southwest Research Institute, San Antonio, Texas, for their encouragement at various stages during the research.

REFERENCES

1. J. F. Archard, "Contact and Rubbing of Flat Surfaces," *J. Appl. Phys.* **24**: 981, 1953.
2. J. T. Burwell and C. D. Strang, "On the Empirical Law of Adhesive Wear," *J. Appl. Phys.* **23**: 18, 1952.
3. T. F. J. Quinn, "The Effect of Hot-Spot Temperatures on the Unlubricated Wear of Steel," (accepted for publication in *ASLE Transactions*, 1967).
4. J. F. Archard and W. Hirst, "The Wear of Metals Under Unlubricated Conditions," *Proc. Roy. Soc.* **A236**: 397, 1956.
5. J. F. Archard, "The Temperature of Rubbing Surfaces," *Wear* **2**: 438, 1959.
6. B. L. Averbach and M. Cohen, "X-Ray Determination of Retained Austenite by Integrated Intensities," *Metals Technology T.P.* 2342, February 1948.
7. F. P. Bowden and D. Tabor, "The Friction and Lubrication of Solids," Part I, Clarendon Press, Oxford, 1954, pp. 90–121.

8. H. P. Klug and L. E. Alexander, *X-Ray Diffraction Procedures for Polycrystalline and Amorphous Materials*, John Wiley and Sons, Inc., New York, 1954, pp. 373–374.

9. D. E. Davies, U. R. Evans, and J. N. Agar, "The Oxidation of Iron at 175–350°C," *Proc. Roy. Soc.* **A225**: 443, 1954.

10. J. Moreau and J. Bardolle, "An Electron Diffraction Determination of the Constitution of Oxide Films Formed on Iron," *Compt. Rend.* **240**: 524, 1955.

11. H. J. Goldschmidt and G. T. Harris, "An X-Ray Examination of Mechanical Wear Products," *J. Sci. Instr.* **18**: 94, 1941.

X-RAY ANALYSIS OF FATIGUE
DAMAGE IN COPPER

Roy G. Baggerly and Regis M. N. Pelloux

The Boeing Company
Seattle, Washington

and

William F. Flanagan

General Motors Research Laboratories
Warren, Michigan

ABSTRACT

A Warren–Averbach[1-4] X-ray line profile analysis was applied to broadened X-ray diffraction peaks from copper deformed in fatigue. The copper specimens were fatigued by four-point bending at peak-strain amplitudes between 0.00105 and 0.00442 in./in., and measurements were made at various fractions of the total fatigue life. The analysis results in an estimation of (a) an average coherently diffracting domain size normal to the diffracting planes and (b) an rms strain distribution function where the strain normal to the diffracting planes is averaged over a given distance at all points in the diffracting crystals and expressed as a function of averaging distance.

Prior to fatigue cycling, the annealed copper exhibited extinction, which reduced the integrated intensity from the low-angle reflections. After fatigue cycling, the integrated intensity increased with increasing strain amplitude of fatigue. The integrated intensities and the rms strains were established during the first few percent of the fatigue life and were found to increase with fatigue strain amplitude. The measured strains were larger in the $\langle 100 \rangle$ direction than in the $\langle 111 \rangle$ direction, but the absolute values were small. On the basis of transmission electron microscopy of thin foils, these results may be explained by assuming the strains are due to the presence of numerous dislocation dipoles.

INTRODUCTION

During a typical X-ray diffraction experiment a large volume of material is irradiated, which provides quantitative information about the crystal lattice on an atomic scale. Information of this sort aids in the interpretation of transmission electron micrographs from thin foils where the region being examined is limited, and uncertainties are inherent in relating the structure of the thin foils to the properties of bulk specimens. Nearly all previous attempts at using X-ray diffraction in fatigue studies have been concerned with interpreting back reflection X-ray diffraction photographs. Exceptions have been the work by S. Moll,[5] and R. J. Hartmann and E. Macherauch;[6] both applied the Warren–Averbach analysis to the study of fatigue damage in nickel and found that the rms strain reached a limiting value within the first few percent of the fatigue test. Their coherently diffracting domain sizes were too large to be measured with confidence,[7] although they listed values of the order of 1000–4000 Å. Since stacking fault energy is known to influence the endurance limit and also the substructure that forms in fatigue,[8-10] it was believed worthwhile to investigate copper, which has a lower stacking fault energy than nickel, to see whether the rate of hardening and magnitude of rms strains would be affected by this parameter. At the same time, such an investigation would provide more insight into the evaluation of mechanisms of fatigue hardening.

Transmission electron microscopy has shown[11-15] that numerous dislocation dipoles are formed in high stacking fault energy metals fatigued at low strain amplitudes. These dipoles are parallel and form dense clusters on {111} planes that are separated by regions of low dislocation density. Electron diffraction indicates that the dipoles are elongated in a $\langle 112 \rangle$ direction and that the Burgers vectors in the clusters are parallel when the crystal is oriented for single slip; however, when the crystal is oriented for multiple slip, more nearly equiaxed loops are observed. A structure containing dipoles would not be expected to broaden the X-ray diffraction profiles as noticeably as would a structure containing an equal density of dislocations present in the form of tangles, since the strain field decreases as $1/r^2$ for a dipole, compared with $1/r$ for a single dislocation. Also, since dipoles are elongated in a $\langle 112 \rangle$ direction, the strain distribution and coherently diffracting domain size would be expected to be anisotropic in the fatigued grains. Since the substructure is influenced by the strain amplitude, it would be expected that the X-ray diffraction analysis would be similarly influenced.

X-RAY ANALYSIS

X-ray diffraction peaks are represented in the Warren–Averbach analysis[1-4] by a Fourier series, and the coefficients of the cold worked and annealed peaks are used to compute the coefficients for the broadening function by the method of Stokes.[16] All forms of instrumental broadening are eliminated from the broadening function by this technique.

The intensity of an X-ray diffraction peak expressed as a Fourier series is given by

$$I = K(\theta)N \sum_{-\infty n}^{\infty} \{A_n \cos(2\pi n h_3) + B_n \sin(2\pi n h_3)\} \tag{1}$$

where θ is the Bragg angle, N is the number of unit cells in the diffracting crystal, and h_3 is a continuous variable characterizing a position in reciprocal space in a direction perpendicular to the diffracting planes. If A_n and B_n are coefficients derived for the broadening function, they can be defined by the expression

$$A_n = \frac{N_n}{N_3} \langle \cos(2\pi l Z_n) \rangle \tag{2}$$

$$B_n = \frac{-N_n}{N_3} \langle \sin(2\pi l Z_n) \rangle \tag{3}$$

where l is the order of the reflection,* Z_n is the displacement of the nth unit cell from its strain free position relative to the origin atom, N_n is the number of unit cells (having n neighbors) in a column normal to the diffracting planes, and N_3 is the average number

---

* For any (hkl) plane from a cubic crystal it is possible to adopt orthorhombic axes to change the indices of the plane to $(00l)$, thus simplifying the mathematics.

of unit cells in a coherently diffracting column normal to the diffracting planes. Stacking faults lead to an asymmetry in the broadening function,[4] but if stacking faults are absent the broadening is symmetrical about the origin, which allows us to neglect the sin coefficients of the Fourier series. The broadening functions from fatigued copper are symmetrical; hence, the sin coefficients were not used in the analysis. Considering the cos coefficient A_n in more detail, it is generally represented as the product of two coefficients, one relating to coherent domain size and one to rms strain. Thus,

$$A_n = A_n{}^S A_n{}^D \tag{4}$$

where

$$A_n{}^S = \frac{N_n}{N_3} \tag{5}$$

is the domain size coefficient and

$$A_n{}^D = \langle \cos(2\pi l Z_n) \rangle \cong (1 - 2\pi^2 l^2 \langle Z_n{}^2 \rangle) \tag{6}$$

is the distortion coefficient, assuming relatively small values of Z_n. Since $N_n = N_3$ and $Z_n = 0$ for $n = 0$, both coefficients, and thus A_n, are normalized to unity for $n = 0$. N_n is defined as

$$N_n = \int_{i=n}^{\infty} (i - n) p_i \, di \tag{7}$$

where p_i is the fraction of columns of length i cells. Differentiating $A_n{}^S$ with respect to n results in

$$\frac{dA_n{}^S}{d_n} = \frac{-1}{N_3} \int_{i=n}^{\infty} p_i di = \frac{-1}{N_3} \Bigg]_{n=0} \tag{8}$$

which shows that the negative initial slope of a plot of $A_n{}^S$ vs. n gives directly the average coherently diffracting domain size N_3.[4] Writing the distortion coefficient in terms of the planar indices (hkl), where $l_0{}^2 = h^2 + k^2 + l^2$, $\Delta L = a_3 Z_n$, and $a_3/l = d = a_0/l_0$, a_0 being the lattice parameter, the distortion coefficient becomes

$$A_L{}^D = (1 - 2\pi^2 l_0{}^2 \langle \Delta L^2 \rangle / a_0{}^2) \tag{9}$$

The domain size coefficients and distortion coefficients are then separated by plotting $\ln A_L$ vs. $l_0{}^2$ for specific values of L.

EXPERIMENTAL PROCEDURE

Material

OFHC copper sheet was cross-rolled 40% to a thickness of 0.040 in., machined into fatigue specimens, Figure 1, and annealed in a salt bath for 5 min at 750°F. An equiaxed recrystallized grain size of 0.01 mm resulted from this anneal. The specimens were then electropolished in 60% phosphoric acid, 40% N-butyl alcohol solution at a 7 V DC potential.

The degree of preferred orientation in the sheet in the cold-rolled and recrystallized conditions was determined by preparing (111) and (200) pole figures, Figure 2. The texture resulting from cross rolling at room temperature was retained after annealing, although the intensity decreased by a factor of three. It should be noted that the (111)

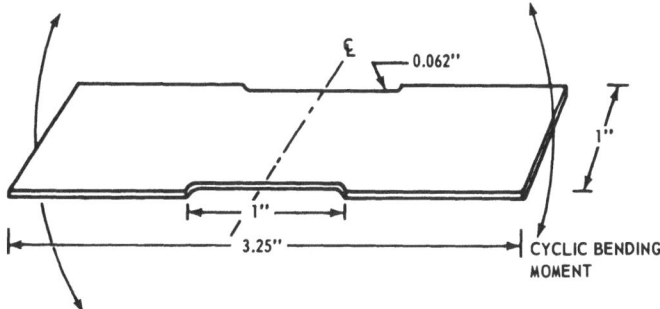

Figure 1. Fatigue specimen configuration.

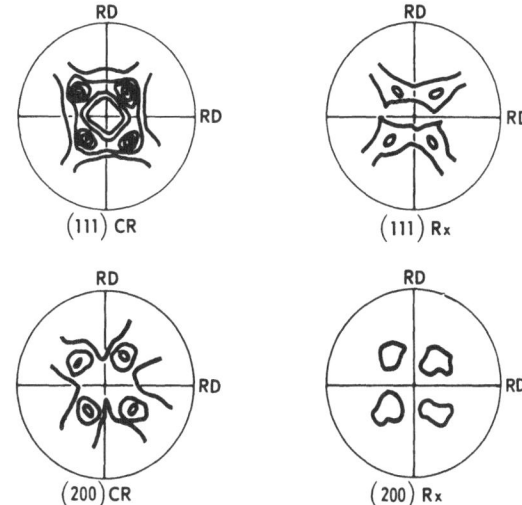

Figure 2. (111) and (200) pole figures.

and (200) pole distributions are similar, which allows for rough comparisons of reflected intensities from these planes.

Testing Procedure

X-ray diffraction profiles were obtained from the fatigue specimens before and after the fatigue test using a Phillips Norelco X-ray diffraction unit and Cu K_α radiation to record the 111, 200, 222, and 400 reflections from the fatigue specimens on strip chart paper. A single-tooth ratchet gear advanced the goniometer one step of 0.01° 2θ for every revolution of the continuously rotating drive motor. Each revolution required approximately 30 sec; hence, a statistical average of the intensity for each step could be obtained.

A commercial BUD fatigue machine was used to fatigue the specimens at a constant peak-strain amplitude and a frequency of 1725 cpm. The specimens were loaded in four-point reverse bending and a zero average strain amplitude was maintained. The maximum peak-strain amplitude could be varied by means of an adjustable cam and the progress of the fatigue test was checked by stopping the fatigue machine and examining the specimen visually. A curve relating the fatigue strain amplitude to the number of fatigue cycles was determined for the copper, and specific levels of strain amplitude were selected for X-ray analysis as indicated in Figure 3.

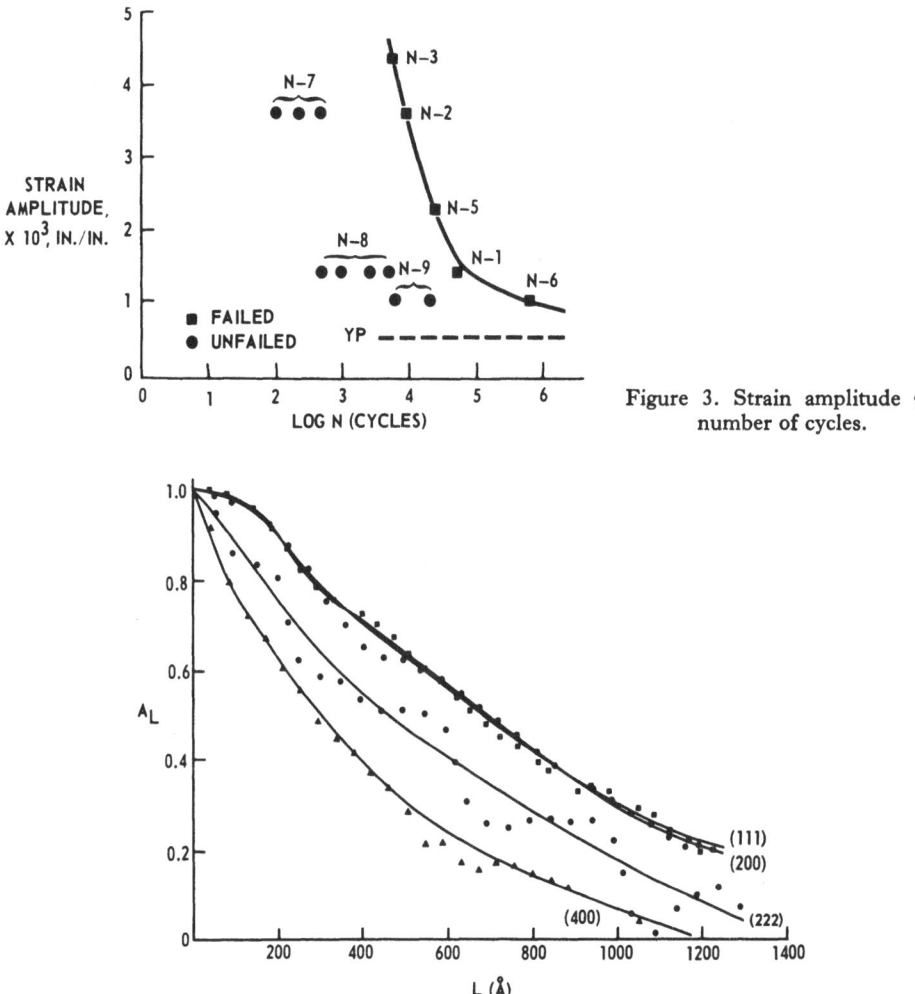

Figure 3. Strain amplitude *vs.* number of cycles.

Figure 4. Cos coefficients *vs.* averaging distance (specimen N-2).

Electron Microscopy

Thin foils were prepared by electropolishing in 30% nitric acid, 70% methanol solution that was cooled with liquid nitrogen. After thinning, the foils were rinsed in methanol and transferred immediately to a JEM electron microscope for examination.

ANALYSIS OF RESULTS

The diffracted X-ray intensity is read directly from the chart recording, in increments of $0.01°\ 2\theta$, with the readings made as far out in the tails of the peak as is necessary to reach the level of background intensity, normally a distance of approximately five to six times the half peak width. The peak maximum is then taken as the origin of the Fourier interval and the data are subsequently analyzed with an IBM 7094 computer. The sin and cos coefficients of the broadening function are thus calculated from the Fourier coefficients of the cold worked and annealed peaks. A typical set of curves

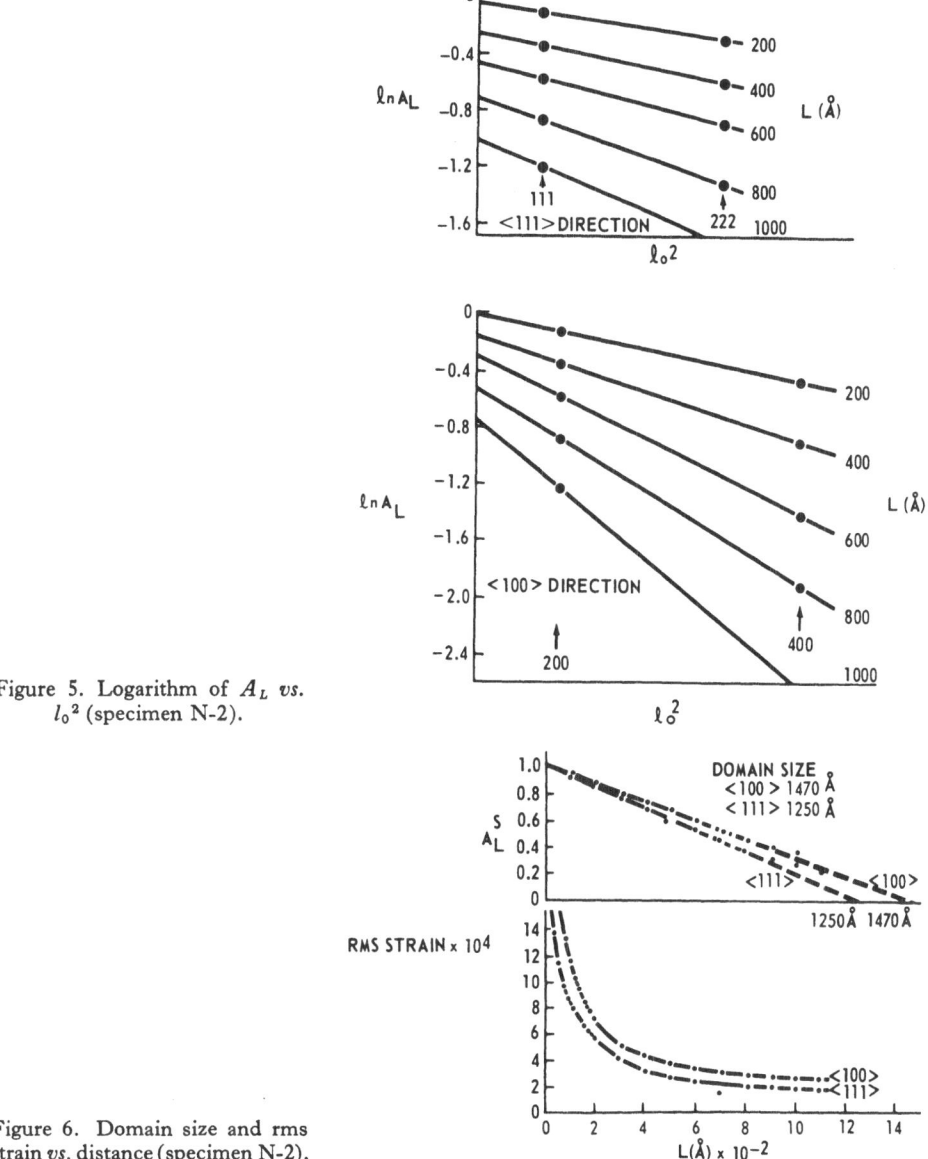

Figure 5. Logarithm of A_L vs. l_0^2 (specimen N-2).

Figure 6. Domain size and rms strain vs. distance (specimen N-2).

showing the cos coefficient A_L as a function of L is presented in Figure 4. Analysis of the broadening function was made difficult by the small degree of line broadening resulting from fatigue damage. Since the fatigued peaks were only 10–15% broader than the annealed peaks, resultant errors were unavoidable in the Stokes analysis, as evidenced by the fluctuations present in Figure 4.[7] An average curve was thus drawn through the coefficients and it was assumed that the smoother curve represented the coefficients of the true broadening function.

When $\ln A_L$ is plotted vs. l_0^2 for a unique L, the resulting points will lie on a straight line if isotropic strains and domain sizes are present; however, where this is not the case

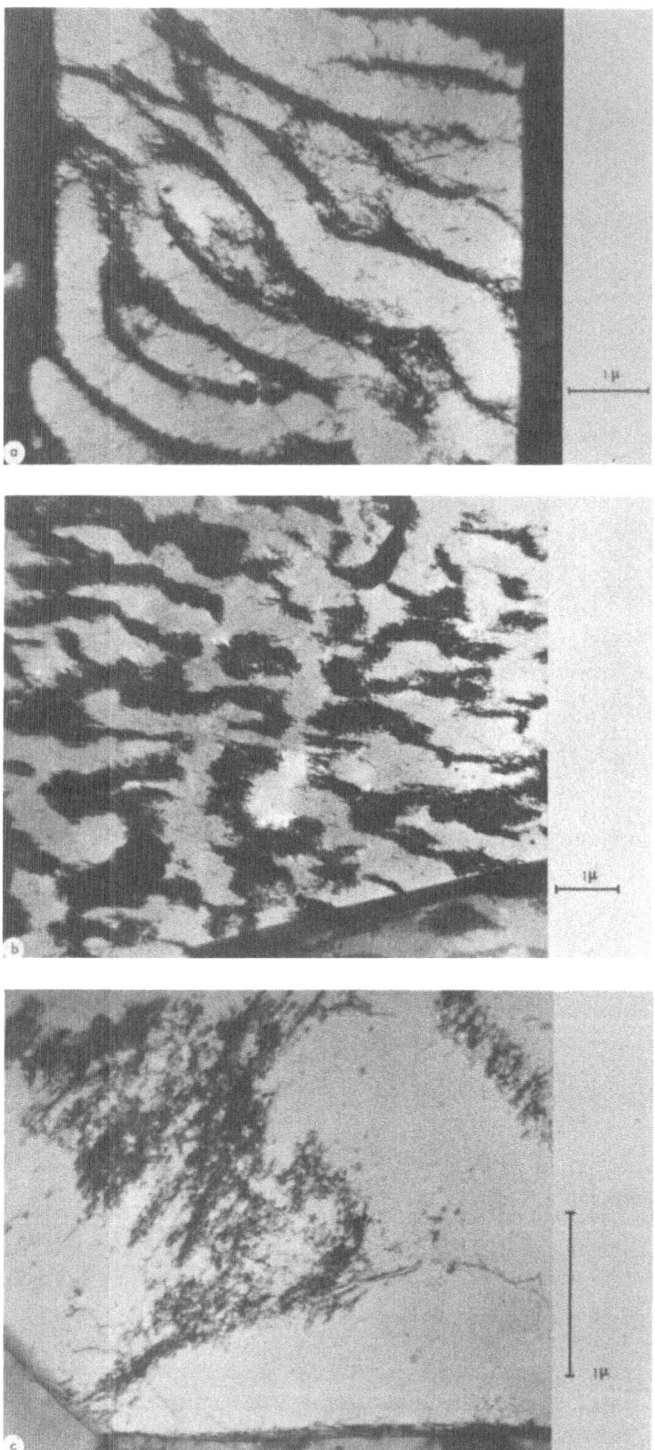

Figure 7. X-ray analysis of fatigue damage in copper: transmission electron micrographs from fatigued copper: (a–c) 10% of fatigue life at a strain amplitude of 0.00145 in./in.; (d–f) 10% of fatigue life at a strain amplitude of 0.0044 in./in.

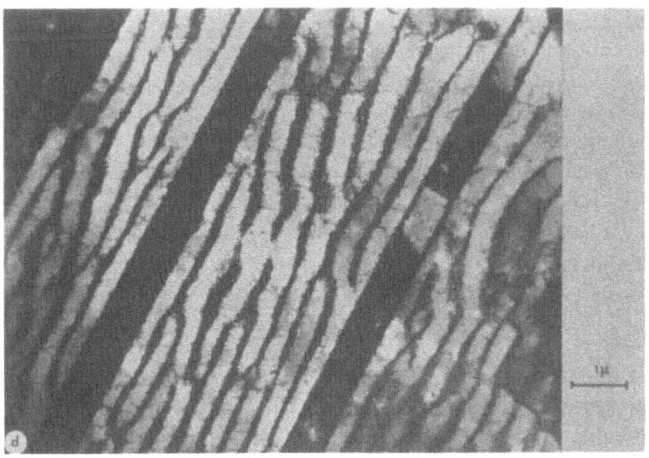

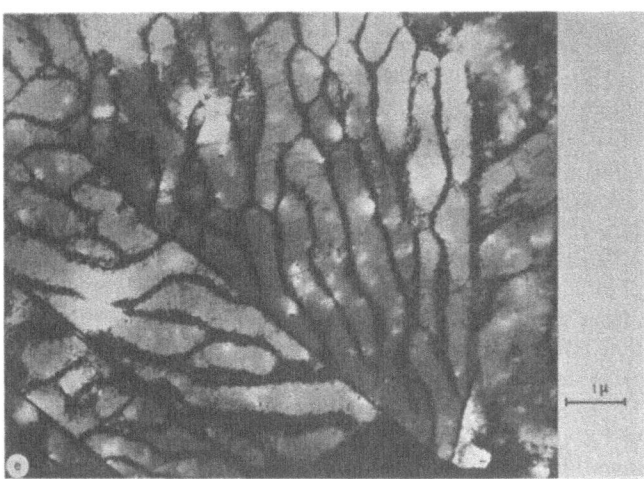

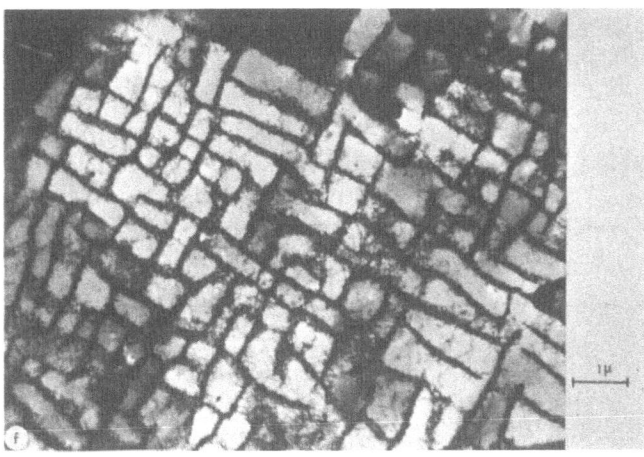

Figure 7 (*continued*).

the coefficients of various orders of the same reflection must be treated independently. Since tests were conducted on bulk sheet specimens, the coefficients for each set of reflections (i.e., 111–222 and 200–400) were treated independently because the reflections originate from different crystals. These different sets of crystals are necessarily affected differently by the experimentally imposed strains, which requires using a certain degree of caution when comparing the indicated rms strains from the $\langle 111 \rangle$ and $\langle 100 \rangle$ directions. The variation of $\ln A_L$ with l_0^2 for the $\langle 111 \rangle$ and $\langle 100 \rangle$ directions is shown in Figure 5, and the domain size coefficients and rms strains are plotted as a function of averaging distance in Figure 6.

DISCUSSION

The state of crystalline perfection was well developed in the individual grains of annealed copper, resulting in a reduction of the diffracted intensity by extinction. Extinction reduces the intensity of the diffraction line but has little influence on a narrow line profile because of the small angular dependence of extinction[17–19]; hence it can be neglected in the Fourier analysis.

Extinction was not entirely eliminated by fatigue damage, but was proportionately reduced by increased strain amplitude of fatigue. This reduction occurred after cycling for as little as 1% of the fatigue life and remained essentially constant throughout the test. The second order peaks remained relatively unaffected, which is characteristic of extinction. This is also evident from the pole figures (Figure 2), where the intensity decreased by a factor of three and the relative pole distribution remained about the same after recrystallization. Thus, extinction and not changes in preferred orientation is effective here. From Table I it is seen that the 200 reflections are more affected than the 111 reflections by increasing strain amplitude of fatigue. The increase in integrated intensity, at small strain amplitudes, represents the situation where fatigue damage has not entirely eliminated the large coherent domains responsible for extinction.

The extinction effect can be understood better by considering the substructure which forms during fatigue. Transmission electron micrographs from thin foils of copper fatigued for 10% of the life at a strain amplitude of 0.00145 in./in. are shown in Figures 7a–c, and for 10% of the life at a strain amplitude of 0.0044 in./in. in Figures 7d–f. Dense clusters of dislocation dipoles are separated by regions 0.5 to 1 μ in size which are essentially free of dislocations.[11–15] Since the extinction distance in copper is of this order of magnitude,[17, 18] it is likely that the substructure is responsible for the variation in integrated intensities. Any line broadening observed in the X-ray diffraction profile would be unaffected by such long range characteristics and would be more related to the nature of the dislocation clusters themselves than to the spacing between them.

Table I. Effect on Integrated Intensities of Increasing Strain Amplitude (Percent)

Strain amplitude, in./in.	111	200
0.00105	30	71
0.00145	28	184
0.00273	51	114
0.00364	107	—
0.00442	129	222

Also, recent work has shown[20] that the average spacing between clusters is inversely proportional to the saturation fatigue stress that is indicated by the increased intensities measured in this work.

Saturation of the rms strain occurs within the first 2% of the fatigue life of copper and the strain is consistently larger in the $\langle 100 \rangle$ direction. The rms strains averaged over distances of 100, 500, and 1000 Å for various fractions of the fatigue life are shown in Figures 8–10. The early saturation of rms strain is consistent with other investigations[21,22] and with a recent theory of fatigue hardening in high stacking fault energy metals.[23] The theory of the rapid hardening stage is based on the premise that the rate of hardening is dependent on the rate of formation of debris obstacles in the active slip planes where debris obstacles include prismatic dislocation loops formed by cross slip of jogged screw dislocations and also point defect aggregates. The rate of hardening is thus dependent on the ease of cross slip or stacking fault energy. It has been shown[23] that the initial rate of hardening in copper single crystals is temperature dependent, the rate increasing with increasing temperature in a way which suggests enhanced cross slip

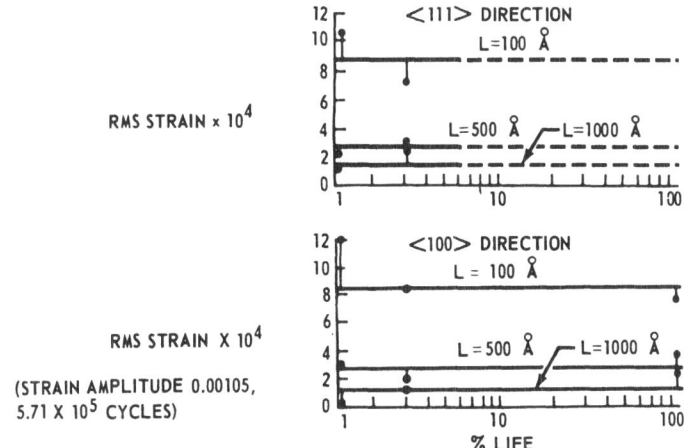

Figure 8. Rms strain *vs.* % fatigue life.

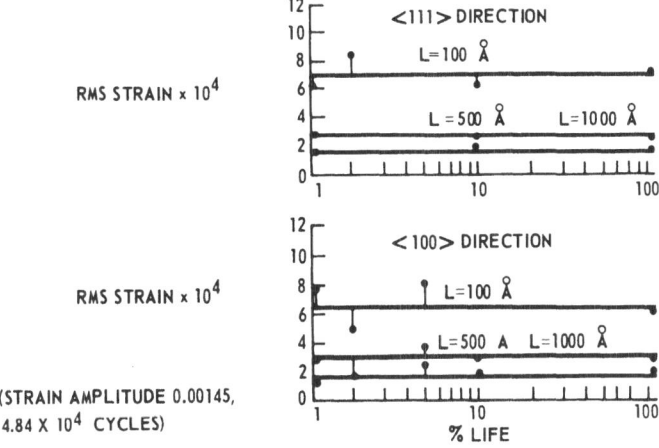

Figure 9. Rms strain *vs.* % fatigue life.

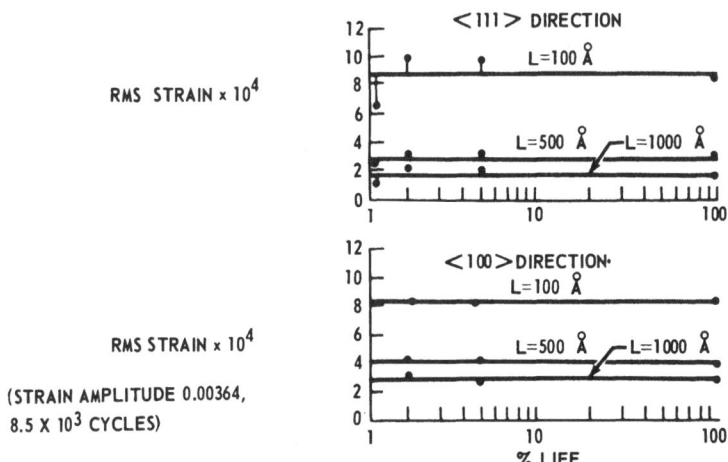

Figure 10. Rms strain *vs.* % fatigue life.

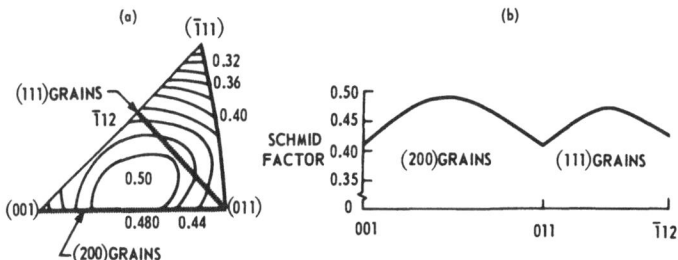

Figure 11. Tension-compression fatigue axes for (200) and (111) grains.

by thermal activation. Saturation of the rms strain, therefore, occurs when the majority of the fatigue strain is accommodated by the flip-flop of the dislocation dipoles from one equilibrium position to another, i.e., from an orientation of 45° to 135°. The density and size of prismatic dislocation loops required to accommodate the strain during a normal fatigue test is consistent with the dislocation arrangements which are observed. The dislocation loop densities in the dense clusters are approximately 10^{10}–10^{11} loop dislocations per square centimeter. This dislocation density, if not in the form of dipoles, would normally result in severely broadened X-ray diffraction lines, but because of the opposite signs of the Burgers vectors in the dipoles their strain fields tend to cancel, producing a minimum degree of broadening and resultant small values of rms strain.

The consistently larger rms strains measured in the $\langle 100 \rangle$ direction can best be analyzed in terms of the crystallite orientations and the deformation process. Deformation would be expected to be more severe in those grains oriented for high resolved shear stress. The range of orientations of tension-compression fatigue axes restricted by the X-ray geometry is indicated on the stereographic triangle in Figure 11a for the (200) and (111) grains.* These are superimposed on equal Schmid factor contours. The approximation that only uniaxial fatigue stresses are present in the crystallites being examined is reasonable since the X-ray data only represent those grains on the surface. From Figure 11, therefore, it is seen that the (200) grains have the largest values of

---

* For the ensuing discussion, the (200) and (111) grains refer to those grains giving rise to 200–400 reflections and 111–222 reflections, respectively.

Schmid factors. Probably more important with respect to fatigue hardening is that these grains are oriented for duplex slip. The (111) grains on the other hand have smaller Schmid factors and are oriented for single slip. It would be expected, therefore, that the (200) grains would result in more complex dislocation interactions and undergo a greater degree of hardening.

The rms strains also can be correlated with the true strains, which are those strains measured in the direction of dislocation Burgers vectors. Referring to the stereographic projection in Figure 12, the (200) grains will deform by slip on ($\bar{1}$11) and (111) planes in the [101] and [$\bar{1}$01] directions, respectively, whereas the 111 grains will deform by slip on the (111) planes in the [$\bar{1}$01] direction. Since strains are measured in directions normal to the diffracting planes, this corresponds to the [$\bar{1}$00] direction for the (200) grains and to the [1$\bar{1}$1] direction for the (111) grains, as defined by the X-ray geometry. Therefore, as a first approximation, the component of strain measured normal to the ($\bar{1}$00) planes is $1/\sqrt{2}\epsilon$ while the component measured normal to the (1$\bar{1}$1) planes is zero. From this overly simplified model one would expect larger strains to be measured in the $\langle 100 \rangle$ direction, which is verified by our data. The fact that a small rms strain is indicated normal to the (1$\bar{1}$1) planes merely shows that the deformation process is more complex than described.

The effect of increasing strain amplitude of fatigue on the rms strains (averaged over distances of 100, 500, and 1000 Å) in the $\langle 100 \rangle$ and $\langle 111 \rangle$ directions is illustrated in Figure 13. The rate of increase of rms strains with increasing strain amplitude of fatigue becomes larger as the averaging distance decreases, and the rate of increase is larger in the $\langle 100 \rangle$ direction than in the $\langle 111 \rangle$ direction. An increased dislocation dipole density or an increased array of dislocation tangles could produce this effect. The substructure developed during high-strain fatigue (see Figure 7d–f), is a very uniform cell structure where the cell boundaries are composed of dense clusters of dislocations. Since the rms strains are quite small, it is believed that the cell walls consist of numerous dislocation dipoles more densely arranged than those observed after low strain fatigue.

This work has not shown any significant differences between copper and nickel[5,6]; the rms strains and coherently diffracting domain sizes after fatigue are of the same order of magnitude for both copper and nickel. The small magnitudes of rms strains are explained by the dense clusters of dislocation dipoles observed in thin foils by transmission electron microscopy. If the rate of fatigue hardening is dependent on the formation of debris obstacles by a cross-slip mechanism, one might expect that copper would require a longer time to achieve saturation of the rms strain than nickel because cross

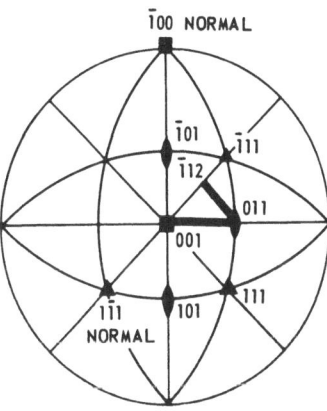

Figure 12. Slip planes and directions for (200) and (111) grains.

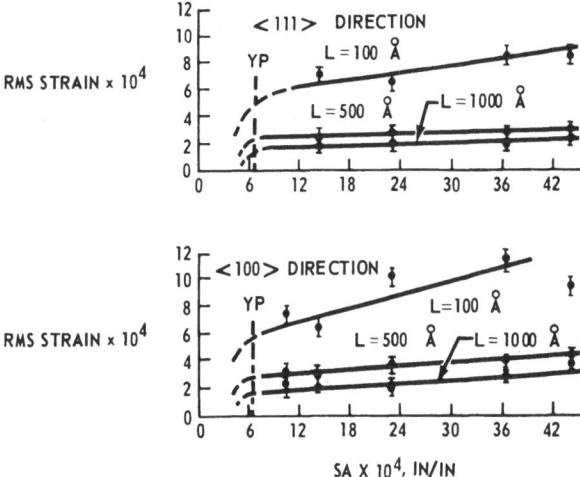

Figure 13. Rms strains *vs.* strain amplitude.

slip is more difficult in copper due to the lower stacking fault energy. In order to critically evaluate this point, examinations would have to be made at a much earlier stage of the fatigue test, i.e., after a very few cycles where the initial hardening is most rapid.

REFERENCES

1. B. E. Warren and B. L. Averbach, "The Effect of Cold-Work Distortion on X-Ray Patterns," *J. Appl. Phys.* **21**: 595, 1950.
2. B. E. Warren and B. L. Averbach, "The Separation of Cold-Work Distortion and Particle Size Broadening in X-Ray Patterns," *J. Appl. Phys.* **23**: 497, 1952.
3. B. E. Warren and B. L. Averbach, "The Separation of Stacking Fault Broadening in Cold-Worked Metals," *J. Appl. Phys.* **23**: 1059, 1952.
4. B. E. Warren, "X-Ray Studies of Deformed Metals," B. Chalmers and R. King (eds.), *Progress in Metal Physics*, Vol. 8, Pergamon Press, New York, 1959, p. 147.
5. S. H. Moll, "X-Ray Studies During Fatigue," Ph.D. Thesis, MIT, 1959.
6. R. J. Hartmann and E. Macherauch, "Die Veranderung von Rontgeninterferenzer, Hysterese und Oberflachenbild bei ein-und wechselsinniger Beanspreechung von Messing, Nichel und Stahl," *Z. Metall.* **54**: 197, 1963.
7. J. B. Cohen, *Diffraction Methods in Materials Science*, The MacMillan Co., New York, 1966, p. 315.
8. N. Thompson and N. J. Wadsworth, "Metal Fatigue," *Adv. Phys.* **7**: 72, 1958.
9. D. H. Avery and W. A. Backofen, "Fatigue Hardening in Alloys of Low Stacking Fault Energy," *Acta Met.* **11**: 653, 1963.
10. J. A. Robertson and J. C. Grosskreutz, "Fatigue of Copper-Zinc Alloys at 100°K," *Acta Met.* **11**: 795, 1963.
11. R. M. N. Pelloux, "The Dislocation Distribution in Face Centered Cubic Metals After Fatigue," Boeing Scientific Research Laboratory Document D1-82-0304, 1963.
12. R. L. Segall, P. G. Partridge, and P. B. Hirsch, "The Dislocation Distribution in Face-Centered Cubic Metals After Fatigue," *Phil. Mag.* **6**: 1493, 1961.
13. J. C. Grosskreutz and G. G. Shaw, "Dislocation Rearrangement in Fatigue-Hardened Aluminum During Preparation for Transmission Electron Microscopy," *Phil. Mag.* **9**: 961, 1964.
14. A. Lawley, J. D. Meakin and K. V. Snowden, "Direct Study of Dislocations in Metals by Transmission Electron Microscopy—Fatigue Deformation," Aerospace Research Laboratories, Document 65-11, 1965.
15. R. L. Segall, "Lattice Defects in Fatigued Metals," G. Thomas and J. Washburn (eds.), *Electron Microscopy and Strength of Crystals*, Interscience Pub., New York, 1963, p. 515.
16. A. R. Stokes, "A Numerical Fourier-Analysis Method for the Correction of Widths and Shapes of Lines on X-Ray Powder Photographs," *Proc. Phys. Soc.* **61**: 382, 1948.

17. A. R. Lang, "Extinction in X-Ray Diffraction Patterns of Powders," *Proc. Phys. Soc. (London)* **B66**: 1003, 1953.
18. G. K. Williamson and R. E. Smallman, "X-Ray Extinction and the Effect of Cold Work on Integrated Intensities," *Proc. Phys. Soc. (London)* **B68**: 577, 1955.
19. S. Chandrasekhar, "Extinction in X-Ray Crystallography," *Adv. Phys.* **9**: 363, 1960.
20. J. E. Pratt, "Dislocation Substructure in Strain Cycled Copper," Department of Theoretical and Applied Mechanics Report No. 652, University of Illinois, Urbana, Illinois, 1965.
21. W. Wood and R. L. Segall, "Annealed Metals Under Alternating Plastic Strain," *Proc. Roy. Soc. (London)* **242A**: 180, 1957.
22. N. J. Wadsworth, "Work Hardening of Copper Crystals Under Cyclic Straining," *Acta Met.* **11**: 663, 1963.
23. C. E. Feltner, "A Debris Mechanism of Cyclic Strain Hardening for fcc Metals," *Phil. Mag.* **12**: 1229, 1965.

DISCUSSION

J. J. Wert (Vanderbilt University): Your electron micrographs showed quite a large change in the substructure. Were you able to observe this at all in the X-ray diffraction profile?

R. G. Baggerly: The X-ray diffraction profiles were broadened very little as evidenced by the resolution of the K_{α_1}–K_{α_2} doublet for the 111 reflection. I believe this small amount of broadening results from dipole formation because the strain around a dipole decreases as $1/r^2$ compared to $1/r$ for a single dislocation.

D. J. Wulpi (International Harvester): On one of the early slides, you showed a strain cycle diagram. Down at the bottom was a line y-s. Was that yield strength?

R. G. Baggerly: That was the approximate yield strength of the material, hence, these specimens were fatigued in the plastic range.

A. Bhattacharyya (The Foxboro Company): In one of your electron micrographs, I saw some directionality in your cell structure formation. How do you explain that?

R. G. Baggerly: The formation of these clusters is one of the interesting aspects of fatigue damage which remains unexplained. It has been observed by several investigators (see References 11–15), that these dipole clusters are elongated and clustered in a directional manner on 111 type planes.

IMPROVEMENT OF ACCURACY IN REPRESENTATION OF CONVENTIONAL POLE FIGURES

K. Aoki, S. Hayami, and M. Matsuo

Tokyo Research Institute
Yawata Iron & Steel Company, Ltd.
Kawasaki, Japan

ABSTRACT

In the conventional pole figure, an accurate representation can be attained by correcting the observed X-ray diffraction intensity for any change in diffraction geometry and by comparing this with the correctly established standard intensity. For the intensity corrections, ASTM has prescribed the method of Decker *et al.* In practice, however, the validity of the correction formula is uncertain, since the prerequisite is difficult to attain for parallelism of an incident beam of sufficient intensity. For the standard intensity, it is desirable to take the intensity obtained with the randomly oriented material of the same composition. However, in most cases an arbitrary unit has been taken because of the difficulty in getting a truly random and uniform sample. Under these circumstances, it is first necessary for an accurate representation of pole figures to make a random sample of uniform thickness for the standard. The authors have obtained satisfactory standard samples by sintering the randomly oriented iron powder made from iron chloride and made use of them to check the method of intensity correction. Satisfactory results are obtained in the randomness tests, such as the comparison of the relative intensity diffracted from the crystal planes parallel to the sample surface and the fluctuation of diffraction intensity during β rotation.

In the reflection case by the Schulz method, the (110) reflection intensity of a random sample, by Co K_α radiation, is independent of the tilting angle up to 50°. The other reflections do not give a constant intensity for the wide range of the tilting angles because of dispersion of the diffracted beam due to the wider separation of K_α doublet in the higher reflection angle and the wider irradiated area at the lower angles. In the transmission case Schulz's correction formula is in good agreement with the observed values for the various diffraction lines and the samples of various μt, while the Decker–Harker formula does not give the absorption change with α-rotation even by an incident beam of $\frac{1}{6}°$ divergence. In both cases, an accurate determination of pole densities is made by comparing the diffraction intensity of the standard sample substituted in place of the test specimen and by correcting the absorption change due to the difference of μt between the standard and test sample, which affords good coincidence in the overlapped region. The pole figure obtained by the above method furnishes an accurate prediction of plastic and elastic anisotropy in sheet metals.

INTRODUCTION

In general, there are two ways to represent the preferred orientation in sheet metal, conventional and inverse pole figures. The information which can be obtained from these pole figures is as follows:

1. The type and degree of preferred orientation of each component of texture.

2. Orientation distribution about a desired direction.
3. Volume percentage associated with each component.

The conventional pole figure can be analyzed to yield information on item (1) but the volume percentage (3) cannot be determined for complex textures such as those of low carbon steel sheets. On the other hand the inverse pole figure gives more direct information on items (2) and (3), but cannot uniquely designate the orientation of each component of the texture. Therefore, both methods of representation should be used to obtain complete information given about the texture.

In the conventional pole figure, accurate representation of contour lines of equal pole density can be obtained by correcting the observed X-ray diffraction intensity for any change in diffraction geometry and by comparing this intensity with the intensity from a correctly established standard. It is desirable to obtain the standard intensity from a randomly oriented material of the same composition, but in most studies to date, an arbitrary unit of intensity has been taken, because of the difficulty in producing a standard sample of satisfactorily random orientation. The method of intensity correction prescribed by ASTM[1] is the method of Decker et al.[2] and the intensity correction factor based on the method is given in the International Tables for X-Ray Crystallography.[3] In practice, however, the validity of this correction formula is in doubt since it is difficult to attain the prerequisite of a parallel incident X-ray beam of sufficient intensity. A few attempts have been made to verify the formula, but, so far, no satisfactory agreement between the formula and the experimental values has been found.

Therefore, to determine an accurate pole figure it is necessary to make a random sample of uniform thickness to be used as a standard. The authors have obtained a satisfactory random standard sample by sintering iron powder and have used it to check the proposed methods of intensity correction and to determine accurate pole figures. From information derived from the pole figure, plastic and elastic anisotropy of aluminum-killed low-carbon steel sheets have been predicted.

MAKING A RANDOMLY ORIENTED STANDARD SAMPLE

Several attempts[2,4,5] have been made to produce a randomly oriented sample, but the degree of randomness has not been reported, and in applying it to check the correction formula, no satisfactory result has been obtained. Grewen et al.[6] integrated the intensity over the pole figure to obtain the random value without using a random sample. However, this procedure is laborious and a random sample is still required in order to apply the absorption correction.

Procedure

A random sample of sufficient quality to be used as a standard was obtained in the following way. Taking into account the importance of the shape of grains in the raw material, pure iron powder was made by a method based on the following chemical reactions:

$$FeCl_3 + 3NaOH = Fe(OH)_3 + 3NaCl \qquad (1)$$

$$2Fe(OH)_3 = Fe_2O_3 + 3H_2O \qquad (2)$$

$$Fe_2O_3 + 3H_2 = 2Fe + 3H_2O \qquad (3)$$

The reduction treatment was done at 550°C for 12 hr, followed by passivation at 880°C for 30 min in a hydrogen atmosphere. Particles 10–30 μ in diameter were obtained

Table I. Randomness Test of the Standard Sample

Sample	A	B
Thickness	125 μ	74 μ
μt(Co $K\alpha$)	4.08	2.46
R^*	70%	70%

Diffraction intensity of standard sample with Mo K_α· radiation. ($\alpha = 90°$)

$h\ k\ l$	$I_{cal}/I_{(110)cal}$	$I_{obs}/I_{(110)obs}$
1 1 0	1.000	1.000
2 0 0	0.165	0.175
2 1 1	0.329	0.334
3 1 0	0.127	0.116
2 2 2	0.029	0.027
3 2 1	0.138	0.110
4 1 1	0.059	0.027
4 2 0	0.031	0.022
3 3 2	0.025	0.018

Variation of diffraction intensity (cps) with β-rotation of standard sample.

$h\ k\ l$	α	I_{obs}	σ_{obs}	Probable error†
1 1 0	0°	1482	38	38.5
	5°	1466	39	38.3
	65°	202	17	14.2
2 0 0	0°	321	18	17.9
	5°	302	17	17.8
2 2 2	0°	202	13	14.2

* Ratio of the μt of the standard sample to the product of the linear absorption coefficient of iron for Co K_α radiation by the thickness of the standard sample.
† Probable error $= \sqrt{I_{obs}}$.

by a screening and suspension method. A disk was formed by compressing this material twice under a pressure of 3.5 t/cm² with an intermediate anneal at 660°C for 1 hr in vacuum. Sintering at 880°C for 2 hr in dry hydrogen was accomplished without grain growth. The disk obtained by the above method was 1.7 mm thick and 40 mm in diameter. The surfaces were polished mechanically, chemically, and electrolytically to the final thickness. Tests of the degree of randomness of the standard sample were carried out in the following ways:

1. By comparison of the relative value of the intensity diffracted from the crystal planes parallel to the samples surface with the theoretical value.
2. By comparison of the fluctuation of diffraction intensity during rotation of the sample about its surface normal with the probable error in counting the X-ray diffraction intensity.

The results of these tests are shown in Table I.

CHECKING INTENSITY CORRECTION METHODS

The conventional pole figure was obtained for the region from the center out to 60° α by the reflection method proposed by Schulz[7] and the outer 30° α by the transmission method.

Reflection Case

The Schulz method gives a constant value of integrated reflected intensity at all angles of tilt for specimens of random orientation. In practice, however, the detector slit has a definite width and does not receive all of the dispersed parts of the diffracted beam which result from defocusing in the parafocusing system, as analyzed by Chernock *et al.*[8] Therefore, accurate geometrical alignment of the system is important for best results. Using our random sample of 500 μ thickness, the variation of reflection intensity with the tilting angle was examined under conditions such that the detector was fixed in the direction of the diffracted beam, the detector slit was set to receive the entire diffracted beam, at a tilting angle of 0°, and the ratio of the diffraction intensity to the background was as large as possible. The result is shown in Figure 1. The intensity of the Co K_α (110) reflection radiation is independent of the tilting angle α between 50° and 90°, but that of the Co K_α (200) and of the Mo K_α (110) decreases with the tilting angle. The decrease can be explained by the wide separation of the K_α doublet in the case of the Co K_α (200) and the wide irradiated area due to the small glancing angle in the case of the Mo K_α (110) reflection.

Although the measured intensity is not constant at all tilting angles under optimum conditions of measurement, such that the ratio of diffraction intensity to the background is large, accurate determination of pole density can be made by comparing the reflected intensity with that of the standard sample at corresponding values of α.

Transmission Case

In the transmission case, the intensity correction is made by taking into account the change in absorption and diffracting volume with the change of the angular position of the sample. We have checked the following two principal methods which have been proposed for this correction.

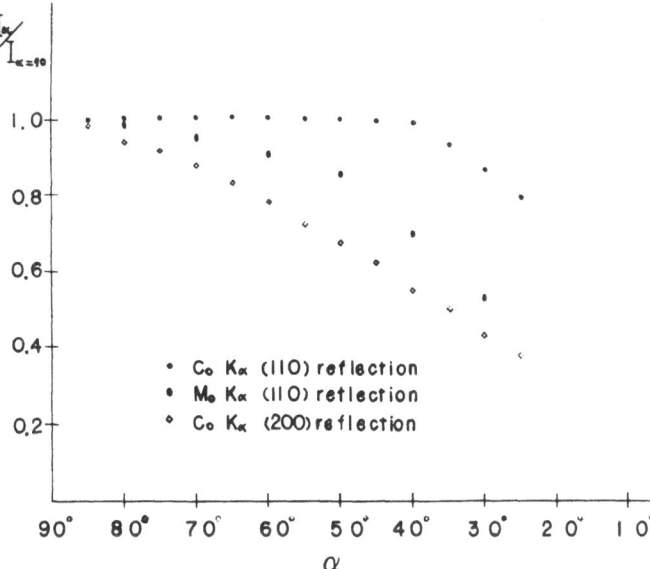

Figure 1. Variation of observed reflection intensity from the random sample with change in tilting angle α for the reflection method of Schulz.

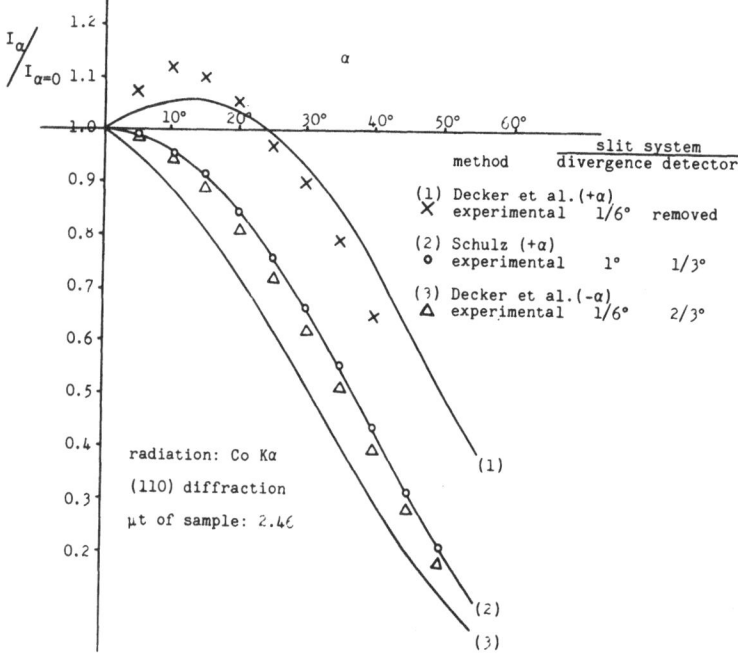

Figure 2. Comparison of observed diffraction intensities with those calculated from Decker *et al.* and Schulz method.

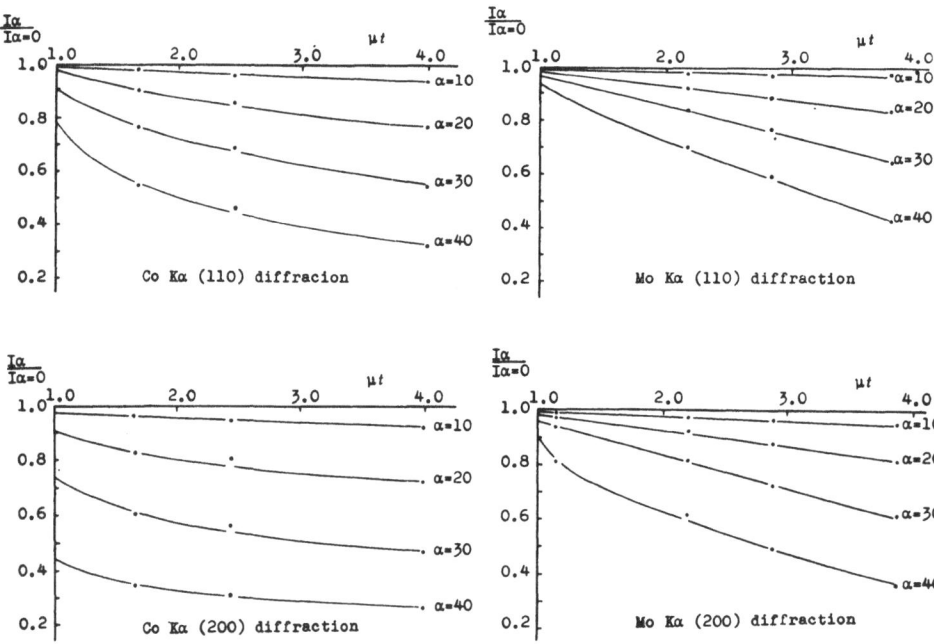

Figure 3. Experimental check of intensity correction factors for various angular positions (α) as a function of absorption factor μt calculated from expression of Schulz for various diffractions.

Decker–Asp–Harker Method[2]. This method assumes that the incident beam is parallel and the measured intensity is that of the whole diffracted beam. Therefore, the correction formula under this assumption is as follows:

$$\frac{I_{\pm\alpha}}{I_{\alpha=0}} = \frac{\cos\theta}{\mu t \exp(-\mu t/\cos\theta)} \frac{\cos(\alpha \mp \theta)}{\cos(\alpha \mp \theta) - \cos(\alpha \pm \theta)}$$
$$\times \left[\exp\left(\frac{-\mu t}{\cos(\alpha \mp \theta)}\right) - \exp\left(\frac{-\mu t}{\cos(\alpha \pm \theta)}\right) \right]$$

where $I_{\alpha=0}$ and $I_{\pm\alpha}$ are the intensities at angles θ and $\theta \pm \alpha$ between the incident beam and the sample normal, μ is the linear absorption coefficient of the material, and t is the thickness of the sample.

Schulz Method[9]. This method assumes an arrangement in which (i) a divergent beam is employed so that the irradiated area on the sample surface is large relative to the sample thickness, and (ii) a narrow slit is used at the detector entrance so that its width is much smaller than that of the diffracted beam at the slit, in order to receive only the portion of the flat maximum which corresponds to the central part of diffracted beam. Under these conditions, the intensity of the beam entering the detector is calculated as follows:

$$\frac{I_{+\alpha}}{I_{\alpha=0}} = \left[\frac{1/\cos\alpha}{\mu t \exp(-\mu t/\cos\theta)} \right] \frac{\cos(\alpha - \theta)\cos(\alpha + \theta)}{\cos(\alpha - \theta) - \cos(\alpha + \theta)}$$
$$\times \left[\exp\left(\frac{-\mu t}{\cos(\alpha - \theta)}\right) - \exp\left(\frac{-\mu t}{\cos(\alpha + \theta)}\right) \right]$$

For the minus α rotation a similar equation is obtained, but in this case the width of the diffracted beam decreases as the rotation increases, therefore the conditions (i) and (ii) become difficult to attain.

Checking the intensity correction methods for the transmission case: The highly parallel incident beam which is required in the method of Decker *et al.* can be obtained by adjusting the divergence limiting slit to the width of the apparent size of the focal spot. But such an incident beam does not give a sufficiently strong diffracted beam. Therefore, an incident beam with a divergence of $\frac{1}{6}°$ was used as an approximation to the parallel beam. In this case, the thickness of specimen and the divergence of the incident beam results in a considerable dispersion of the diffracted beam. In order to receive the total diffracted beam, the detector slit had to be set as wide as possible which resulted in an increase of the background intensity. As the sample is rotated in the plus α direction, the width of the diffracted beam increases, but this width must always be narrower than the detector slit. Measurement of the variation of the $I_{+\alpha}/I_{\alpha=0}$ with a $+\alpha$ rotation was made by removing the detector slit so that the total beam entered over as wide a range of rotation as possible. The results are shown in Figure 2, where the calculated curve (1) and the experimental values ($\times$) are not in good agreement. This is because the detector aperture could not receive the whole diffracted beam, as evidenced from the observation of its profile. To eliminate the disadvantage due to widening of the diffracted beam, measurements were made for the minus rotation which is prescribed by ASTM as a tentative method, the detector slit being kept as wide as possible. The results are shown in Figure 2. In this case, agreement between the observed (Δ) and calculated (3) value is also poor. It is believed that the detector slit might have prevented part of the diffracted beam from entering the detector especially at $\alpha = 0°$. In case of the plus α rotation, however, a discrepancy will arise from the large change in diffracting volume due to the divergence of the incident beam. In the arrangement proposed by Schulz, good agreement was obtained between the calculated curve (2) and the observed values ($\bigcirc$). With this method satisfactory agreement was obtained for various diffraction lines, as shown in Figure 3 in terms of $I_{+\alpha}/I_{\alpha=0}$ vs. μt plots. This arrangement is recommended since the measurement can be made under conditions such that the ratio of the diffraction intensity to the background is kept large.

Overlapped region: In the overlapped region of transmission and reflection, good coincidence of pole density is obtained using the standard intensity of the random sample and the intensity correction described above.

EVALUATION OF PLASTIC ANISOTROPY FROM THE CONVENTIONAL POLE FIGURE

The measurement of the ratio of width to thickness strain (r-value) in a uniaxial tension test has been shown empirically to provide a useful criterion for predicting the ability of metal sheets to undergo a deep drawing operation. Plastic anisotropy, as determined by the r-value, is caused by a preferred crystallographic orientation or texture in metal sheets. Development of methods for predicting the plastic anisotropy of textured metals is an important step toward the improvement of properties by texture control, and therefore, the relation between plastic anisotropy and preferred orientation has received much attention. There are two main difficulties to overcome: (a) the difficulty which arises in relating plastic deformation of single crystals to that of an aggregate of crystals, and (b) the difficulty of description of preferred orientation of an aggregate of crystals with varying orientation. Preferred orientation is best described by a pole figure. Though the use of the information from pole figures, in terms suitable for analysis, to investigate the relation of texture to plastic anisotropy remains an outstanding problem, quantitative pole figures will afford a better understanding of the relation. Elias et al.[10] have explored the possibility of making a graphical analysis directly on the pole figure.

Comparison was made between the D-value* directly measured by the tension test and that calculated from the conventional pole figure after Elias et al. For reference a calculation of average D-value ($\bar{D}$)† from the inverse pole figure (low index crystal axis density figure) was also carried out according to the method of Nagashima et al.[11]

Sheets of aluminum-killed low-carbon steel were cut into ISO tension test specimens whose longitudinal directions made angles of 0°, 45°, or 90° with the rolling direction. D-values were determined after 20% elongation. X-ray pole figures for (222) diffraction of Mo K_α radiation were obtained by point-by-point counting using the transmission method for $\alpha = 0°–34°$ and the reflection method for $\alpha = 32°–56°$ with the β rotation being at intervals of 2°. In the transmission case, measurements (relative to the standard sample) were made employing an incident beam of 1° divergence and the observed intensity was corrected using the Schulz formula. Figure 4 shows the variation of intensity with α rotation for the standard sample used in this experiment for (222) diffraction of Mo K_α radiation. In the reflection case, the arrangement proposed by Schulz was used. Crystal axis densities at eleven points on a stereographic triangle were measured from the reflection intensities of low-index crystal planes lying parallel to the rolling plane, using Mo K_α

---

* D-value is defined as $D = (r - 1)/(r + 1)$, where r is the ratio of width to thickness strain in a uniaxial tension test.

† $\bar{D}$ is defined as $\bar{D} = D_0 + 2D_{45} + D_{90}/4$, where D_0, D_{45}, and D_{90} denote D-values tested in directions making an angle of 0°, 45°, and 90°, respectively, to the rolling direction.

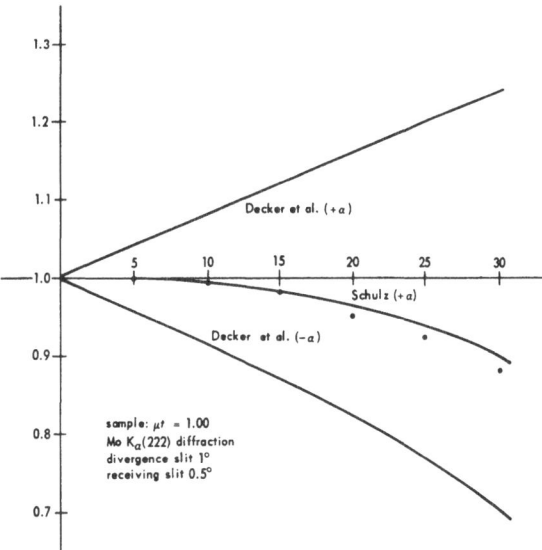

Figure 4. $I_\alpha/I_{\alpha=0}$ vs. α for the (222) diffraction of standard sample used in this experiment. Curves are calculated from expressions of Schulz and Decker et al.

radiation. From these data the average D-value, $\bar{D}$, was calculated according to the inverse pole figure method of Nagashima et al.[11]

Plastic Anisotropy Obtained from a Conventional Pole Figure

A pole figure of one of the specimens is shown in Figure 5. D-values were estimated from the pole figure according to the summation method based on the model proposed by Burns and Heyer.[10] The results are shown in Table II and Figure 6 in which the values obtained by the tension test and the inverse pole figure method are compared.

With regard to the average $\bar{D}$-values, the values predicted from the conventional and inverse pole figures are in agreement with those of the tension test. The conventional pole figure gives additional useful information on planar anisotropy, or D-values in testing directions at 0°, 45°, and 90° to the rolling direction. The best agreement between D-values from the tension test and from the pole figure is for D_{90}, where the greatest difference is 0.04. Poorer agreement for D_0 and D_{45} may be because the pole density in the summation region is lower in the longitudinal and diagonal directions than in the transverse direction. Therefore, improvement of the predictions for D_0, D_{45}, and $\bar{D}$ should be possible by reducing the probable error in the measurement of the diffracted intensity, and by dividing the pole density distribution more finely so as to obtain more detailed contour lines.

As a result of the improvement of accuracy in the representation of conventional pole figures, one of the difficulties previously mentioned above in (b) is eliminated. The quantitative pole figure is proved to be effective in predicting the plastic anisotropy, and the analysis of the present results suggests a key to understanding the relation between plastic deformation of single crystals and an aggregate of crystals. Further detailed discussion on such discrepancies as described above will be discussed elsewhere.[12]

EVALUATION OF YOUNG'S MODULUS

The planar anisotropy of the elastic modulus in sheet metal is thought to be dependent upon its texture and can be predicted from the anisotropy of a single crystal. Therefore, an attempt was made to estimate the gross elastic modulus of a polycrystalline

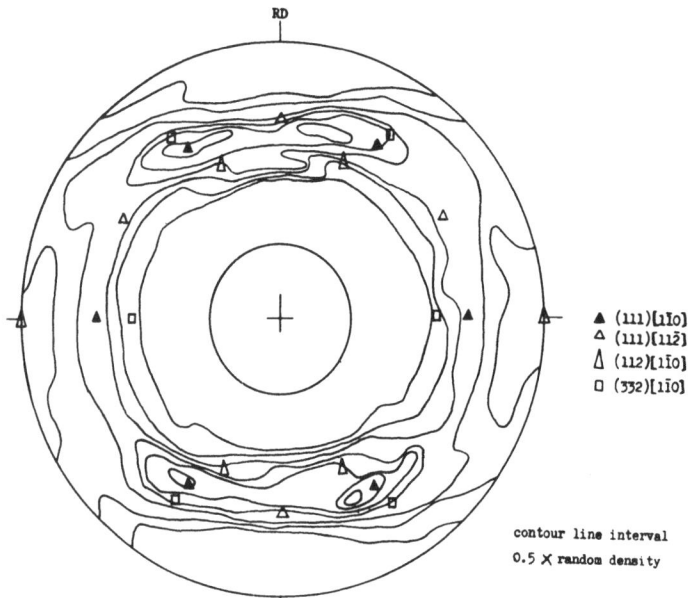

Figure 5. (222) pole figure of specimen 57.

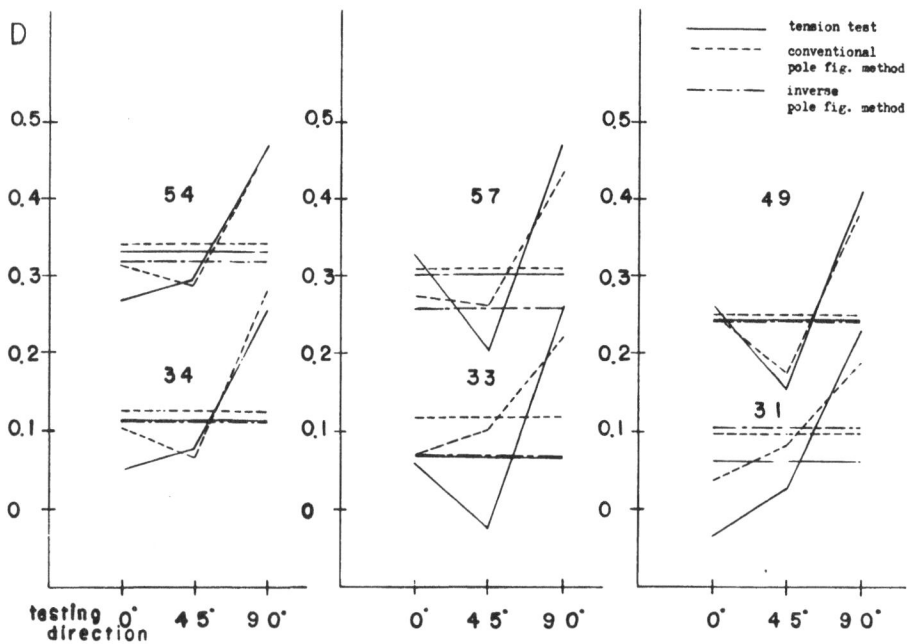

Figure 6. Comparison of D-values obtained by conventional pole figure method and inverse pole figure method with those of tension test.

Table II. Comparison of D-Values Obtained by Various Methods

| Sample | Testing direction | D-values | | |
		Tension test method	Inverse pole figure method	Conventional pole figure method
31	$\bar{D}$	0.059	0.101	0.096
	0°	−0.036		0.036
	45°	0.024		0.081
	90°	0.225		0.187
33	$\bar{D}$	0.067	0.071	0.118
	0°	0.062		0.069
	45°	−0.026		0.093
	90°	0.260		0.218
34	$\bar{D}$	0.113	0.113	0.125
	0°	0.052		0.104
	45°	0.074		0.064
	90°	0.251		0.269
49	$\bar{D}$	0.241	0.237	0.245
	0°	0.259		0.246
	45°	0.149		0.172
	90°	0.405		0.390
54	$\bar{D}$	0.330	0.315	0.337
	0°	0.267		0.310
	45°	0.293		0.284
	90°	0.468		0.468
57	$\bar{D}$	0.301	0.257	0.308
	0°	0.329		0.273
	45°	0.203		0.261
	90°	0.468		0.437

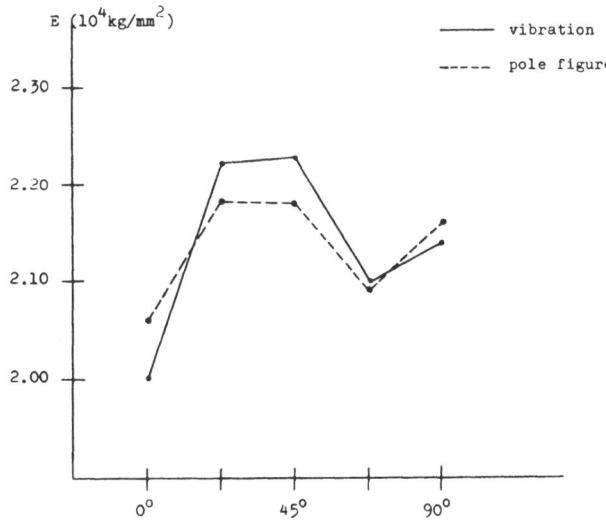

Figure 7. Planar anisotropy of Young's modulus.

aggregate for a low-carbon steel sheet in terms of the moduli of the constituent crystals by use of quantitative pole figures. Since the texture can be represented in terms of principal components, their relative amount was evaluated in part from conventional (222) and (200) pole figures and also from inverse pole figure data. Young's modulus was calculated on the assumption that the reciprocal of Young's modulus of each principal component is additive and is weighted by the relative amount of the component and summing over all of the different orientations of the aggregate. Young's modulus of the sheet was measured directly by a vibration method.

Comparison of the predicted and measured values is shown in Figure 7. The predicted values are in good agreement with the measured values, and further details will be discussed elsewhere.[12]

CONCLUSION

A satisfactory random sample of iron was obtained by sintering randomly oriented powder made of reduced iron from iron chloride. Accurate representation of a conventional pole figure was established using this standard. Application of the quantitative pole figure to predict plastic and elastic anisotropic properties resulted in good agreement with directly measured values on low-carbon steel sheets.

ACKNOWLEDGMENTS

The authors wish to express their sincere thanks to Dr. A. J. Heckler of Armco Steel Corporation for reviewing this paper and for his oral presentation at the Conference. Thanks are due to Professor S. Mizushima and Dr. T. Ikegami for their permission to publish this paper, Professor Z. Nishiyama for his helpful discussions, and Messrs. M. Arakawa, M. Okamoto, and Y. Hayashi for their help in the experiments.

REFERENCES

1. ASTM Committee E-4 on Metallography, "Tentative Method for Preparing Quantitative Pole Figures of Metals," *ASTM Standards* **E81–49T**: 1094–1104, 1949.
2. B. F. Decker, E. T. Asp, and D. Harker, "Preferred Orientation Determination Using a Geiger Counter X-Ray Diffraction Goniometer," *J. Appl. Phys.* **19**: 388–392, 1948.
3. A. E. De Barr, "Determination of the Texture of Polycrystalline Materials," *International Tables for X-Ray Crystallography*, Vol. III, The Kynoch Press, 1962, pp. 302–315.
4. M. Schwartz, "Method for Obtaining Complete Quantitative Pole Figures for Flat Sheets Using One Sample and One Sample Holder," *J. Appl. Phys.* **26**: 1507–1513, 1955.
5. J. B. Newkirk and L. Bruce, "Rapid X-Ray Determination of a Complete Pole Figure," *J. Appl. Phys.* **29**: 151–157, 1958.
6. J. Grewen, A. Segmüller, and G. Wassermann, "Zählrohr-Verfahren mit einem neuentwickelten Probenhalter zur Darstellung von Texturen in quantitativen Polfiguren," *Archiv für das Eisenhüttenwesen* **29**: 115–118, 1958.
7. L. G. Schulz, "A Direct Method of Determining Preferred Orientation of a Flat Reflection Sample Using a Geiger-Counter Spectrometer," *J. Appl. Phys.* **20**: 1030–1033, 1949.
8. W. P. Chernock and P. A. Beck, "Analysis of Certain Errors in the X-Ray Reflection Method for the Quantitative Determination of Preferred Orientations," *J. Appl. Phys.* **23**: 341–345, 1952.
9. L. G. Schulz, "Determination of Preferred Orientation in Flat Transmission Samples Using a Geiger-Counter X-Ray Spectrometer," *J. Appl. Phys.* **20**: 1033–1036, 1949.
10. J. A. Elias, R. H. Heyer, and J. H. Smith, "Plastic Anisotropy of Cold-Rolled Annealed Low-Carbon Steel Related to Crystallographic Orientation," *Trans. Met. Soc. AIME* **224**: 667–686, 1962.
11. S. Nagashima, H. Takechi, and H. Kato, "Plastic Strain Ratios of Textured Sheets of Low-Carbon Steel," *J. Japan Inst. Metals* **29**: 393–398, 1965.
12. K. Aoki, S. Hayami, and M. Matsuo, "Application of Conventional Pole Figure to Prediction of Plastic and Elastic Anisotropy in Low-Carbon Steel Sheets," to be published in *Trans. Japan Inst. Metals*.

DISCUSSION

P. R. Morris (Armco Steel Corporation): I would like to point out that in their evaluation of the Schulz technique, which uses a fine slit and does not receive the entire diffraction peak, that they were using a well-annealed material; and that it is entirely possible that in a commercial sample the shape of the peak may depend on the direction in the piece so that there is a conceivable error that they have not considered.

J. C. Robinson (Lockheed Missiles and Space Co.): I would like to make a few comments on the same thing. Particularly in materials of low atomic number, this distortion occurs and the integrated intensity should be used. Bragg and Packer*† at Lockheed have found that by using the background measured at all the angles in alpha one can estimate the correction factors which, otherwise, must be calculated. If you have a specimen that doesn't have parallel sides or completely fill the sample holder you are in trouble using the calculated correction factors. Estimated from the background the correction takes into account the sample configuration. Secondly, a random sample, which is very difficult to produce, can be calculated from the observed data merely by summing the data and smearing it over a sphere. The final value you get is the intensity for a random sample.

A. J. Heckler: Do you average this intensity over the total pole figure?

J. C. Robinson: That is right. Of course, it says that you have to take a lot of data, but that's not really true since there is a considerable amount of symmetry in almost all pole figures. I know that Don Fraser at Livermore has an automated system and he takes data at about 3400 points on the pole figure in a 10-hr period. Unfortunately, we can't do this, but then we usually estimate it from about 18 or 20 cuts in beta around the pole figure. So you can estimate the random level very well and, as far as that goes, who can tell when they really have a random specimen.

P. R. Morris: We tried to check Bragg and Packer's results and got very poor agreement. We have never been able to correlate the background. I thought this was a godsend when I first saw it.

J. C. Robinson: You haven't been able to check the results with steels?

P. R. Morris: Yes, that's right.

J. C. Robinson: You mean the correction factors?

P. R. Morris: Yes.

J. C. Robinson: You don't have any trouble with geometry or specimen thickness?

J. C. Robinson: We've found noticeable difference with various specimens but we could almost always attribute this to a difference in specimen thickness or some such variable.

Authors' written discussion: (1) As a reply to Mr. Morris' comment, the shape of diffracted beam does not change during β rotation in fixed α angle no matter how specimens are treated—as annealed or as cold-rolled up to 70%. Comparing the measurement through a fine slit with that using the integrated intensity, the difference in pole density is less than 0.1 in the range of the pole density between 0.5 and 7.0 times random. (2) Mr. Robinson wonders whether our random sample can give a true random level. In the pole figures which were constructed by using the random level with our sample and by correcting with Schulz's formula, the average pole density was 1.16 and 0.88 in the rolled and annealed samples, respectively. We think that the positive deviation from 1.0 may be due to lack of the packing of material in the random sample and the negative one to the extinction effect in the annealed sample, but these effects do not make any appreciable errors in representation of the relative pole density. (3) Concerning the correlation between the background intensity and the absorption correction factors according to Bragg and Packer, we could not find any correlation in iron specimens of different thickness.

---

* R. H. Bragg and C. M. Packer, "The Effect of Absorption and Incoherant Scattering on X-Ray Line Profiles," *Rev. Sci. Instr.* **34**: 1202–1207, 1963.
† R. H. Bragg and C. M. Packer, "Quantitative Determination of Preferred Orientation," *J. Appl. Phys.* **35**: 1322–1328, 1964.

PRECISION LATTICE PARAMETER DETERMINATION AT LIQUID HELIUM TEMPERATURES BY DOUBLE-SCANNING DIFFRACTOMETRY

Hubert W. King and Carolyn M. Preece*

Imperial College
London, England

ABSTRACT

The back-reflection double-scanning diffractometer method, by which lattice parameters can be measured with a reproducibility of one part in 150,000 has been applied at liquid helium temperatures. A cryostat attachment is described which enables diffraction profiles to be scanned on both sides of the primary X-ray beam up to 163°, 2θ. Alignment errors may, thus, be eliminated by measuring the included angle 4θ between respective Bragg reflections. The method is illustrated by measuring the lattice parameters of the I.U.Cr. standard specimens of silicon and tungsten at various cryogenic temperatures.

INTRODUCTION

The accuracy of a lattice parameter determination using a modern, commercial X-ray powder diffractometer depends essentially on the ability of the operator to align the instrument with respect to the X-ray source. To facilitate this operation, at normal temperatures and pressures, most manufacturers supply or recommend special alignment devices which are fitted in place of the specimen. Few of these devices can be used at liquid helium temperatures, however, because of the necessity for isolating the specimen from the ambient—usually by enclosing it within an evacuated chamber. Thus, if a cryostat attachment of conventional design is aligned at liquid helium temperatures by using one of these devices in place of the specimen, the cryostat must be returned to room temperature and dismantled (so ruining its alignment) before the powdered specimen can be mounted in position. A cryostat is usually aligned, therefore, at the operating temperature by making adjustments until the specimen surface bisects the beam and the $2\theta : \theta$ setting gives maximum intensity and resolution. The effectiveness of these adjustments, which must be repeated every time a fresh specimen is inserted in the cryostat, depends on the skill and patience of the operator.

The precision of a lattice parameter determination can be improved, of course, if a substance of known lattice parameter is mixed with the powdered specimen. In the present context, however, it is pertinent to ask, "How does one obtain the lattice parameter of the 'calibrated' substance at the selected cryogenic temperature?"

An alternative approach to the problem of eliminating alignment errors in powder diffractometry has been put forward by the authors in a previous publication.[1] It was

* Present address: R.I.A.S., Baltimore, Maryland.

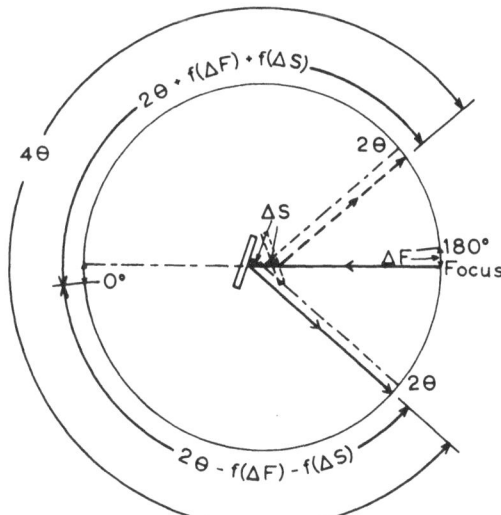

Figure 1. Geometry of the back reflection double-scanning method for eliminating ΔF and ΔS errors by measuring 4θ.

demonstrated that the lattice parameters of the I.U.Cr. standard specimens of tungsten and silicon could be measured to the reproducibility of 1 : 150,000 even though the specimen surface was deliberately displaced as much as 0.012 in. (0.03 cm) from the axis of the goniometer. Further, the values obtained for the lattice parameters agree, to within the same limits, with the diffractometer determinations reported in the I.U.Cr. Project[2] and the subsequent values reported by Delf[3] and Franks[4] for the same specimens. The basis of the authors' back-reflection double-scanning method is that errors caused by a focal spot displacement ΔF or a specimen surface displacement ΔS are eliminated by measuring the included angle 4θ between respective diffraction profiles scanned on both sides of the primary X-ray beam. (The method is in effect the diffractometer equivalent of the Straumanis asymmetrical Debye–Scherrer film method[5] and the powder specimen equivalent of the Bond single-crystal diffractometer method.[6]) The geometry is illustrated in Figure 1 and is achieved by combining a modified Siemens diffractometer with a 3-in. (7.6-cm) Hilger and Watts horizontal microfocus X-ray tube so that Bragg angles can be scanned over the range from $-163°-+163°$, 2θ. The principle of the method has now been incorporated into the design of a liquid helium cryostat attachment for the diffractometer so that the instrument can be aligned at room temperature, taking full advantage of various alignment devices, and any errors introduced on subsequent cooling to liquid helium temperatures are eliminated by the back-reflection double-scanning procedure.

INSTRUMENTAL

Figure 2 shows the Siemens diffractometer aligned against the Hilger and Watts microfocus X-ray tube, with the cryostat mounted on the diffractometer in place of the standard specimen holder. The cryostat is supported on a cylindrical frame with sections cut away to allow the tail-piece containing the X-ray window to be removed at an "O"-ring vacuum joint. This feature permits direct access to the specimen which can be mounted or removed without disturbing the alignment of the cryostat itself. An incidental advantage of a removable tail-piece is that the X-ray window can be rotated through 180^{d} when it is required to scan on the reverse side of the beam, so that the opening

Figure 2. General view of apparatus showing the diffractometer with the cryostat aligned against the short horizontal X-ray tube.

need only extend over 190° of arc. The shape of the cutaway frame is such that at the beam height the cryostat is supported on two legs, so positioned (see Figure 7) that Bragg angles can be scanned over the range from −8°, through zero, to +163°, 2θ, on the normal side of the beam and from −30°−−163°, 2θ, on the reverse side of the beam. The specimen surface can be aligned at all operating temperatures by precision rotation and translation mechanisms incorporated in the base of the support frame.

Details of the apparatus and methods for mounting specimens are given in the Appendix. A multiple specimen holder was used for the present studies, enabling four specimens to be cooled simultaneously and then brought in succession into the X-ray beam. When scanning on the normal side of the beam the specimen holder is held against a reference surface in the cryostat in much the same way as in a conventional room temperature mount. When scanning on the reverse side of the beam, a reference plate replaces the specimen which is then placed against the reference surface from the opposite side (see Figure 7b) using the technique of Vassamillet.

EXPERIMENTAL

The apparatus was aligned in two stages. First, the diffractometer was aligned with respect to the X-ray source using the standard Siemens procedure,[7] which involves making adjustments to the diffractometer until the {111} reflections from a gold specimen, scanned on both sides of the beam, are superimposed when recorded on the same chart paper. This alignment technique is independent of any possible error in specimen displacement ΔS, because the specimen is actually rotated through 180° when scanning on the reverse side of the beam.[1] Secondly, the cryostat was fitted in place of the standard specimen holder and progressively aligned at room temperature by making the following adjustments alternately. The $2\theta : \theta$ setting at 0°, 2θ was adjusted by the horizontal rotation mechanism with the aid of a glass collimator slit of angular aperture 0.08° mounted in place of the specimen. The specimen surface position was aligned with respect to the axis of the goniometer by translating the cryostat until the {111} diffraction profiles from the gold specimen in the normal and reverse positions were superimposed.

Powdered samples of tungsten and silicon were obtained from the same source as the I.U.Cr. precision lattice parameter project.[2] The powders were packed into the

Table I. Wavelengths, Instrumental Settings, and Corrections

$\lambda Cu\,K_{\alpha_1}$ = 1.54051 Å		Scanning speed	$\frac{1}{4}$ deg/min
$\lambda Cu\,K_{\alpha_2}$ = 1.54433 Å		Time constant	4 sec
Takeoff angle	6°	Chart speed	1 cm/min
Divergent slit	2°	Detector	Xe prop. counter
Soller slits	2°	Pulse attenuation	10×
Receiving slit	0.1 mm	P.H.D.: Base 12.0 V; Channel 6.0 V	

Refraction correction[2]: tungsten +0.00016 Å; silicon 0.00004 Å
wavelength correction to Bearden's $Cu\,K_{\alpha_1}$[8]: × 1.000034

specimen holder as described in the Appendix. Temperatures were measured by a Au–0.03% Fe/chromel or a copper/constantan thermocouple embedded in the mount adjacent to the back of the specimen.

Filtered $Cu\,K_\alpha$ radiation was used for all the experiments, the diffracted beam being detected by a Xe proportional counter and recorded graphically. Diffraction profiles in the back reflection region were scanned in a counter-clockwise direction on both sides of the primary beam. Bragg angles were calculated from the included angle 4θ, using the peak positions of the profiles, which were measured to $\pm 0.005°$, 2θ. The wavelengths assumed for $Cu\,K_\alpha$, the various instrumental settings and corrections are listed in Table I.

RESULTS

The I.U.Cr. specimens of tungsten and silicon were cooled simultaneously in the X-ray cryostat and examined at temperatures of 18°, 77°, and 195°K using liquid helium, liquid nitrogen, or acetone-solid CO_2 as refrigerent liquids, and at room temperature. Values of a_0 calculated from the included angle 4θ for the five highest Bragg reflections of the tungsten specimen at 18°K are plotted against $\cos^2 \theta$ in Figure 3. This extrapolation function was selected because, in the absence of alignment errors, the dominant systematic errors[9] are caused by specimen transparency due to absorption $(\alpha \cos^2 \theta)$[10] and axial divergence $(\alpha \cos^2 \theta + \text{const.})$.[11] The linearity of the plot in Figure 3 confirms that the back-reflection double-scanning procedure has been effective in eliminating errors arising from a possible displacement, ΔS of the specimen surface $(\alpha \cos \theta \cot \theta)$[9] caused by uneven contraction of the thin-walled stainless steel support for the helium reservoir. The lack of scatter in the data in Figure 3 enables the lattice parameter of tungsten at 18°K to be established, by extrapolation to $\theta = 90°$, as 3.16233 ± 0.00003 Å. This indicates that the present method can yield absolute lattice parameters accurate to 1 : 100,000 provided, of course, due attention is paid to the value assumed for the wavelength of the X-radiation.[2] The above value for tungsten is based on Lonsdale's[12] wavelength for $Cu\,K_\alpha$ and has not been corrected for refraction.

The lattice parameters of tungsten and silicon at various cryogenic temperatures were derived from similar extrapolation plots corrected for refraction, and related to Bearden's[8] secondary wavelength standard for $Cu\,K_{\alpha_1}$, using the factors listed in Table I. The results are listed in Table II and are plotted against absolute temperature in Figure 4. The values obtained for tungsten follow the usual trend for a metal, there being a steady contraction of the lattice between room temperature and liquid nitrogen, but relatively little change or cooling from liquid nitrogen to liquid helium. The results for silicon, however, fall less steeply with decreasing temperature and pass through a minimum near 100°K, so that the lattice parameter at 18°K is, in fact, greater than that at 195°K. This trend is in agreement with that reported earlier by Batchelder and

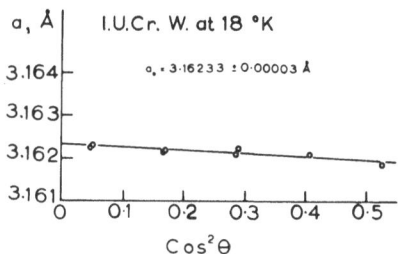

Figure 3. Lattice constants a_0 for I.U.Cr. tungsten at 18°K, calculated from measurements of 4θ.

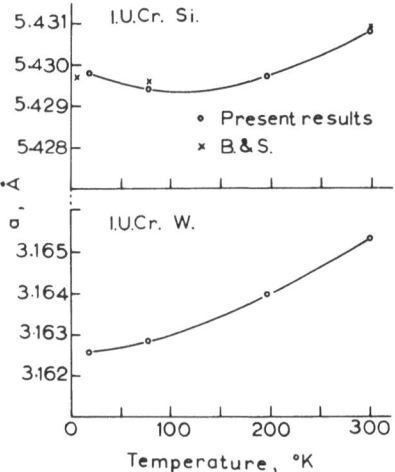

Figure 4. Lattice parameters a of I.U.Cr. tungsten and silicon measured at various temperatures: O present results; × Batchelder and Simmons.[13]

Table II. Lattice Parameters* of I.U.Cr. Tungsten and Silicon

Temperature, °K	Tungsten, a, Å	Silicon, a, Å
298	3.16531	5.43076
195	3.16399	5.42970
77	3.16285	5.42905
18	3.16259	5.42980

* This is corrected for refraction and expressed in terms of Bearden's $\lambda Cu\ K_{\alpha_1}$.[8]

Simmons,[13] who used a single-crystal back-reflection film technique and related the minimum to the temperature variation of the Gruneisen parameter for silicon. It is interesting to note that the differences between the two sets of results in Figure 4 ($\sim$0.00015 Å) fall well within the limits of the different precision determinations of the lattice parameter of silicon at room temperature which have been summarized recently by Isherwood and Wallace.[14]

DISCUSSION

It is fundamental to the back-reflection double-scanning method that the reference surface of the cryostat remains at exactly the same position when the specimen is scanned on both sides of the X-ray beam. Although it may be thought to be more simple in practice to rotate the whole cryostat through 180°, to reverse the specimen surface when scanning on the opposite side of the beam, this operation is quite ineffective in eliminating the effect of ΔS errors. If the specimen surface lies outside the focusing circle on one side of the beam, rotating the cryostat will again cause the surface to lie outside the focusing circle on the reverse side of the beam. Hence, the sign of ΔS will not be altered and the error will not be cancelled by measuring 4θ (see Figure 3 of reference 1 for the geometry of this case). There is, therefore, no easy alternative to returning the cryostat to room temperature and reversing the specimen, using the special reference plate referred to above. The time involved in double-scanning, however, is less than that which would normally be spent in realigning the cryostat for each experiment. The present results confirm that the effort is worthwhile as it does yield absolute lattice parameters to 1 : 100,000.

For a number of applications, the lattice parameter of a crystalline material is not required to be reproducible to better than 1 : 50,000. In this case, the back-reflection double-scanning method need only be applied once—to determine the lattice parameter of a standard specimen at the selected cryogenic temperature. Using the multiple specimen mount, the lattice parameters of up to three unknown specimens may then be measured, at one time, relative to this standard. The selection of a standard specimen presents some difficulty. The I.U.Cr. specimen of tungsten would be suitable because its lattice parameter is insensitive to temperature fluctuations in the region of liquid helium, but there are inadequate supplies of the powder to make a fresh specimen for each experiment. Since the specimens are mounted in a copper disk, the diffraction pattern of copper invariably appears on the chart recording of the unknown specimen. Copper would, thus, appear to be a convenient choice as an internal standard. Since a nondestructible specimen is required, we have used a fine-grained strain-free bulk specimen of 99.999% copper, supplied by Johnson and Matthey. The specimen has been machined to be a snug fit into one of the tapered holes in the specimen mount. The difference in the positions of diffraction peaks on repeated cooling of this specimen to 18°K amounts to $\pm 0.05°$ in the region of 90° and $\pm 0.005°$ in the region of 140°, 2θ. From previous studies of peak shifts caused by inserting shims between the specimen and the reference surface, this degree of reproducibility in 2θ means that the standard specimen is relocated to within 0.003 in. (0.008 cm) in different specimen holders, which is equivalent to an error in an extrapolated lattice parameter of less than 1 : 50,000.[1]

From the geometry of the double-scanning method illustrated in Figure 1, it is evident that the difference between the positions of $\{hkl\}$ diffraction peaks, scanned in the same direction, on both sides of the beam is influenced by both ΔF and ΔS misalignment errors. The two effects can be easily separated, however, because peak shifts $\Delta 2\theta$ due to ΔF are independent of Bragg angle whereas, those caused by ΔS are a function of $\cos^2 \theta$.[9,10] A comparison between the slopes of plots of $\Delta 2\theta_{\text{ccw}}$ $versus$ $\cos^2 \theta$ for the tungsten and silicon results, with those obtained previously when known ΔS errors were introduced by shims,[1] indicates that ΔS displacements of the order of 0.06 in. (0.15 cm) occur on cooling the cryostat from room temperature to 18°K. In order that a temperature gradient be maintained between the boiling liquid helium and the ambient temperature, it is a common practice to suspend the helium reservoir, and hence, also the specimen, at the base of a thin-walled (0.01-in.) stainless steel tube, as described in the Appendix. The ΔS displacements which occur on cooling are the result of uneven contraction of this suspension and may be considered to be inherent in all cryostats of similar design. The magnitude of the ΔS displacement is such that errors as large as 1 : 3,000 will occur in the determination of a lattice parameter from Bragg angles scanned on one side only, unless the specimen surface is realigned at the cryogenic temperature. To a first approximation the specimen surface is displaced by 0.002 in. (0.005 cm) per 10°K change in temperature. Hence, studies of expansion coefficients from relative changes in lattice parameter, again scanning only one side of the beam, will be subject to errors greater than 1 : 50,000 if the specimen surface is not realigned after every 10°K change in temperature.

The displacement of the specimen surface which occurs on cooling from room temperature to 18°K must be accompanied by a tilt of the surface of the specimen. However, because the specimen is suspended at the base of a tube 18-in. (45.7 cm) long, a ΔS displacement of 0.06 in. will result in a surface tilt γ of no more than 0.17°. The effect of a surface tilt is to cause a loss in resolution rather than a shift in peak position,[6,9] the resultant error in $\Delta a/a$ being given by $(1 - \cos \gamma)$. Hence, the error in lattice parameter arising from a specimen tilt of 0.17° is no more than 1 : 250,000. It is, therefore, quite unnecessary to provide an alignment mechanism for tilting the specimen surface in an attempt to correct such errors. To check the effect of a surface tilt on resolution, the {111} reflection from the standardized polycrystalline copper specimen was scanned, at $\frac{1}{8}$°/min. with a fine slit system, after the instrument was carefully aligned at room temperature and again after cooling to 18°K. The resultant profiles, given in Figure 5, indicate no significant loss in the resolution of the K_{α_1}-K_{α_2} doublet. These results also confirm that the contraction causes little, if any, misalignment of the $2\theta : \theta$ setting, since this error will also cause a loss of resolution.[9]

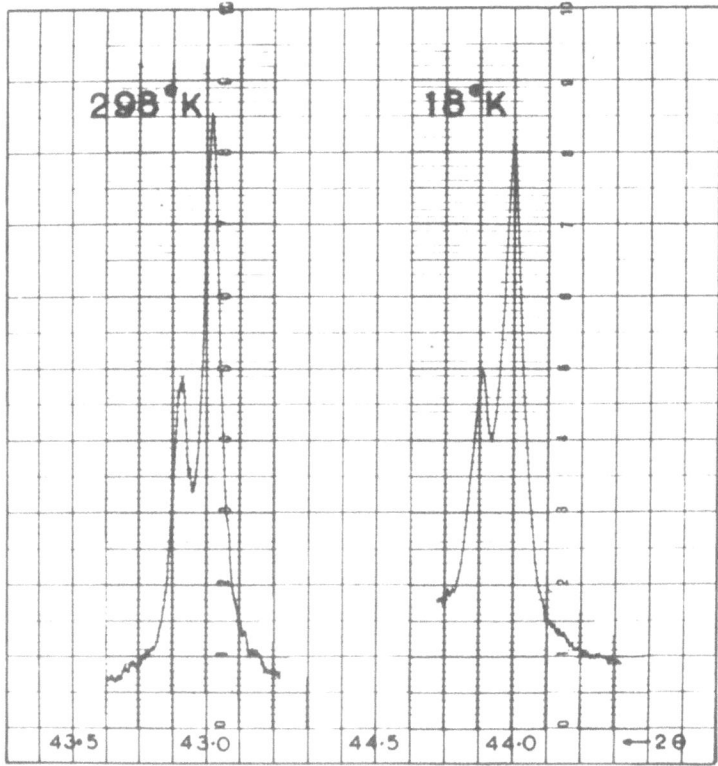

Figure 5. Diffraction profiles at various temperatures of {111} reflections from polycrystalline copper, showing resolution of K_{α_1}-K_{α_2} doublet.

The temperature of the powdered specimens remained constant to $\pm 0.5°$K during the course of the measurements. The lowest temperature achieved, however, was 18°K, i.e., some 14°K above the normal boiling point of liquid helium. This temperature difference is caused by the number of interfaces which lie along the thermal conducting path between the specimen and the liquid helium.* (A similar situation was not encountered at liquid nitrogen temperatures, because the thermal radiation shield was cooled to the same temperature as the specimen by putting liquid nitrogen into both refrigerant vessels.) If the heater is not being used, the number of interfaces can be reduced by attaching the reference surface block directly to the helium reservoir. Even so, the specimen will still be above 4.2°K because of the interface between the specimen mount and the reference surface which is fundamental to the double-scanning method. For most applications, however, the fact that the specimen cannot be cooled to 4.2°K without pumping on the helium reservoir, is a comparatively small price to pay for a precision absolute lattice parameter at an accurately known cryogenic temperature.

APPENDIX

Liquid Helium Cryostat Attachment for X-Ray Diffractometry

The low-temperature attachment used for the present experiments was manufactured by the Oxford Instrument Company, Oxford, England, to a general design[15] in which the cryostat is mounted on a support frame which can be adapted to fit on any commercial horizontal diffractometer without altering the beam height or the X-ray optics. The version for the Siemens diffractometer is shown in the schematic section in Figure 6. The assembly stands 21-in. (53 cm) high with a maximum diameter of $6\frac{1}{2}$ in. (16.5 cm) and a clearance of 4.3 in. (11 cm) at the beam height. It weighs 33 lb (15 kg) a large proportion of which is taken up by the translation and rotation mechanisms built into the base of the support frame. As discussed in the main part of this paper, a tilt mechanism is considered to be unnecessary and so this has not been incorporated in the instrument.

* See the footnote on p. 363.

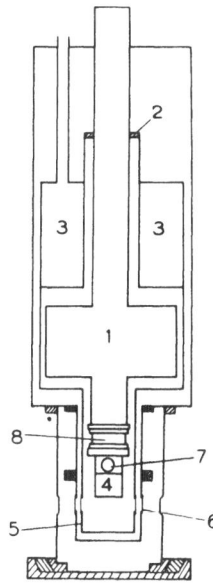

Figure 6. Schematic section of cryostat: (1) Liquid helium; (2) thermal contact; (3) liquid nitrogen; (4) copper reference block; (5) radiation shield; (6) Mylar window; (7) resistance heater; (8) thermal spacer.

The outer case of the cryostat is made of bright nickel plated brass and encloses a common vacuum space within which is suspended a 1.2 liter thin-walled stainless steel liquid helium reservoir labeled 1 in Figure 6. The reservoir is maintained in thermal contact at a joint about half way up its support tube (2), with a 1.4 liter annular section liquid nitrogen vessel (3) made of copper. The joint 2 also gives additional support to the helium reservoir. Specimens for X-ray examination are attached to a copper block (4) which is maintained in thermal contact with the helium reservoir and surrounded by a thermal radiation shield (5) cooled by the liquid nitrogen. The X-ray windows (6) cut in the tail section of the outer case and in the radiation shield are covered with (7) Mylar, aluminized to reduce thermal radiation. The windows are bonded with "Araldite AY111" and withstand a vacuum leak-tested to 10^{-11} cc/sec at N.T.P. helium, yet the four thicknesses of material reduce the intensity of a Cu K_α X-ray beam by less than 1%. A practical advantage of making the windows of this material is that they can easily be replaced by the operator in the event of a breakage.

Under optimum conditions it takes $2\frac{1}{2}$ liter of liquid helium to cool the cryostat and fill the helium reservoir. The cryostat will then remain at helium temperatures for up to 6 hr. Steady temperatures of 77°, 195°K, etc., can be obtained by placing liquid nitrogen or acetone/solid CO_2 mixture etc., in the helium vessel. Intermediate temperatures between the boiling points of the different refrigerents are obtained by using an electrical resistance heating element (7) embedded in the copper block (4). When using this heater, the thermal conduction path between the copper block and the liquid helium reservoir is reduced by inserting a thermal spacer (8) which is of the form of a copper or stainless steel tube. Temperatures of 4.2°K and below are achieved by pumping on the helium bath. Temperatures below 77°K are measured by a Au–0.03% Fe/chromel thermocouple or a germanium resistance thermometer and above 77°K with a copper/constantan thermocouple.

The selection of a specimen mount for use in the cryostat depends on the type of information required, i.e., measurements on many specimens at a fixed temperature or measurements over a range of temperature on a single specimen. The first of these requirements is met by the mount shown in Figure 7. The specimen holder (7) is composed of two high conductivity copper disks which are screwed together. The outer disk is 0.04 in. (0.1-cm) thick and has rectangular holes cut for four powder or bulk specimens $\frac{3}{8}$-in. (1-cm) high × $\frac{1}{2}$-in. (1.25-cm) long. The sides of the holes are tapered so that the specimen cannot fall out once the disks have been screwed together. Powdered specimens are packed into the outer disk, which is placed on a Mylar-covered glass plate, and set with collodion in amyl acetate. When the specimen is dry the Mylar film is stripped off to give a flat specimen with its surface coincident with the mount, which is then screwed against the other part of the holder to give a composite disk containing the four specimens. The specimen holder is pressed firmly against the reference surface (9) in the copper block (4) by a beryllium copper spring anchored in a plate (11) which also serves as an additional thermal contact and radiation shield for the specimen. Each specimen can be rotated in turn into the path of the beam at the operating cryogenic temperature, using the retractable, hollow, stainless steel probe (12) mounted in a sliding "O"-ring vacuum fitting.

The arrangement for scanning on the reverse side of the beam is shown in Figure 7b. A reference plate (13) containing an X-ray window is placed against the reference surface (9). The multiple specimen holder (10) is held against this plate so that it now approaches the reference surface from the opposite side, compared to its position in Figure 7a. Specimens can again be rotated into the beam in turn at the cryogenic temperature, using the lower of the two retractable probes (12). The positions of the

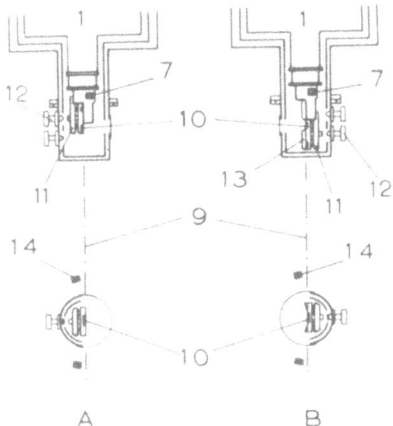

Figure 7. Constant temperature—multiple specimen mount. (A) Normal position. (B) Reverse position: (9) Reference surface; (10) specimen disk; (11) spring anchorage; (12) retractable probe; (13) reference plate; (14) support legs.

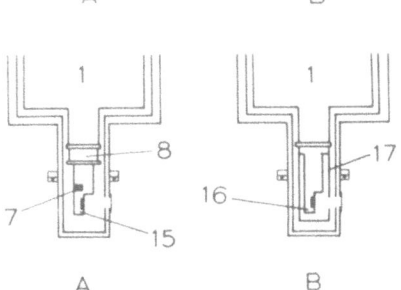

Figure 8. Variable temperature—single specimen mount. (A) Above 4.2°K. (B) 4.2°K and below: (1) Liquid helium; (7) resistance heater; (8) thermal spacer; (15) specimen holder for >4.2°K; (16) specimen holder for ≤4.2°K; (17) liquid helium cooled radiation shield.

legs of the cryostat support at the beam height are labeled 14 in the lower part of the figure and show how the scanning range from $-8°$ to $+165°$, 2θ is achieved in the normal position (a) and from $-30°$ to $-163°$, in the reverse position (b).

When it is required to make relative lattice parameter measurements at a number of different temperatures, or to study phase changes, order-disorder or the fine detail of a diffraction profile as a function of temperature, it is not necessary to scan both sides of the X-ray beam. In this case it is often more convenient to embed the powder (or bulk) specimen directly into a copper block attached to the liquid helium reservoir. If these effects are to be studied at temperatures between 20–77°K, the copper mount (15) is isolated from the helium reservoir by inserting the thermal spacer (8) as in Figure 8a. On the other hand, if it is desired to achieve the lowest attainable temperature by pumping on the helium bath, the specimen is embedded in a copper block (16) attached directly to the liquid helium reservoir and surrounded by a liquid helium cooled copper radiation shield (17) as in Figure 8b.* When the specimen blocks 15 and 16 are fitted in place of the reference block 4 the cryostat must be realigned with respect to the diffractometer. Once this has been accomplished specimens may be exchanged by removing the tail section and radiation shields so that this alignment is not disturbed, i.e., the basic advantage which stems from the method of supporting the cryostat on the special support frame is still effective when using these specimen mounts.

---

* The cooling system is now being modified to enable the temperature of the X-ray specimen to be varied continuously over the range 1.0–300°K. The tail section of the cryostat has also been modified. After removing flange plates from the bottom of the window section and the radiation shield, both of these can now be pushed along sliding fittings up into the main body of the cryostat, to give much easier access to the specimen.

ACKNOWLEDGMENTS

The authors are grateful to Professor J. G. Ball for his helpful advice and encouragement and to the Oxford Instrument Company for many constructive discussions during the manufacture of the cryostat. The liquid helium for these studies was provided by a grant from the Science Research Council, who also provided financial support for one of us (C.M.P.) during the course of these experiments.

REFERENCES

1. Hubert W. King and Carolyn M. Russell, "Double-Scanning Diffractometry in the Back Reflection Region," *Advances in X-Ray Analysis, Vol. 8*, Plenum Press, New York, 1965, p. 1.
2. W. Parrish, "Results of the I.U.Cr. Precision Lattice Parameter Project," *Acta Cryst.* **13**: 838, 1960.
3. B. W. Delf, "The Practical Determination of Lattice Parameters Using the Centroid Method," *Brit. J. Appl. Phys.* **14**: 345, 1963.
4. A. Franks, "The Precision Measurement of Lattice Parameters Using an X-Ray Back-Reflection Focussing Camera," *N.P.L. Report* (to be published).
5. M. E. Straumanis and A. Ieviņš, "The Precision Determination of Lattice Constants by the Asymmetric Method," (translated by K. E. Beu), Goodyear Atomic Corp., Portsmouth, Ohio, 1959, GAT-T-643.
6. W. L. Bond, "Precision Lattice Constant Determination," *Acta Cryst.* **13**: 814, 1960.
7. H. W. King and L. F. Vassamillet, "Precision Lattice Parameter Determination by Double-Scanning Diffractometry," *Advances in X-Ray Analysis, Vol. 5*, Plenum Press, New York, 1962, p. 78.
8. J. A. Bearden, *X-Ray Wavelengths*, U.S. Atomic Energy Commission, Division of Technical Information Extension, Oak Ridge, Tenn., 1964.
9. L. F. Vassamillet and H. W. King, "Precision X-Ray Diffractometry Using Powder Specimens," *Advances in X-Ray Analysis, Vol. 6*, Plenum Press, New York, 1963, p. 142.
10. A. J. C. Wilson, "Geiger-Counter X-Ray Spectrometer—Influence of Size and Absorption Coefficient of Specimen on Position and Shape of Powder Diffraction Maxima," *J. Sci. Instr.* **27**: 321, 1950.
11. E. R. Pike, "Counter Diffractometer—Effect of Vertical Divergence on the Displacement and Breadth of Powder Diffraction Lines," *J. Sci. Instr.* **34**: 355, 1957 and **36**: 52, 1959.
12. K. Lonsdale, *International Tables for X-Ray Crystallography, Vol. III*, Kynoch Press, Birmingham, England, 1962; *Acta Cryst.* **3**: 400, 1950.
13. D. N. Batchelder and R. O. Simmons, "Lattice Constants and Thermal Expansivities of Silicon and of Calcium Fluoride between 6° and 322°K," *J. Chem. Phys.* **41**: 2324, 1964.
14. B. J. Isherwood and C. A. Wallace, *Nature*, in press.
15. H. W. King and C. M. Russell, "Attachments for X-Ray Diffractometers," U.K. Patent Application No. 14138/65, April 1965.

DISCUSSION

K. E. Beu (Goodyear Atomic Corp.): I was interested to see that your room-temperature value for IUCR silicon was lower than D. N. Batchelder and R. O. Simmons (*J. Appl. Phys.* **36**: 2864–68, 1965). We found the same thing on the IUCR silicon. We have about 60 ppm lower than the single crystal values by Bearden, for example. I was wondering if your difference was of the same order of magnitude?

H. W. King: If we correct our results to Bearden's wavelengths we get 5.43076 Å as our room-temperature value for silicon. This is corrected for refraction and also for thermal expansion using Parrish's recommended factors. The value is 24 ppm smaller than that of Batchelder and Simmons.

K. E. Beu: Is this low temperature or room temperature?

H. W. King: This is for room temperature. At liquid nitrogen temperature our result is 19 ppm below that of Batchelder and Simmons. Our liquid helium result is at 18°K, compared to theirs at 6.4°K, and is greater by 1.3 ppm. Incidentally, Isherwood and Wallace[14] have done a survey of the lattice parameters reported for silicon and they also find there is a consistent difference between values obtained with single crystals and those from powdered specimens.

Chairman D. K. Smith: Are you implying in your system that with the same sample you are simultaneously getting both $+2\theta$ and -2θ measurements, or do you have to reset the sample with your second setting?

H. W. King: We have to return the cryostat to room temperature and put a reference surface in place of the specimen and then put the specimen against the reference surface. We have done tests which are described in the paper to show that this does not involve an error which would affect our results; i.e., we were able to reproduce the position of the specimen to within 0.003 in. on subsequent cooling. If I can make one further comment, we are able to short circuit this by having as the fourth specimen a standard which we have previously scanned on both sides of the beam. We can then measure three unknown specimens against the standard so that we need only cool down once and use one side of the beam. This would lower the accuracy to something like 1 in 50,000.

F. G. Karioris (Marquette University): What is the radius of the goniometer?

H. W. King: The standard radius for Siemen's is 17.5 cm. We have had to increase this to 20 cm because, if you know Siemen's geometry, the tube actually overlaps the diffractometer slightly. The Hilger X-ray tube is supported by a vertical pumping line and so we were unable to physically locate our $180° \, 2\theta$ point under the X-ray tube. When the goniometer radius is increased to 20 cm the path of the detector slit lies outside the external dimensions of the goniometer base. The modification is easily accomplished using a special part supplied by Siemens.

Chairman D. K. Smith: It is not uncommon for the tetrahedral compounds of the wurtzite and sphalerite type structures or diamond structures to show this minimum in their thermal expansion. The list contains many which are not semiconductors, but so far only tetrahedral compounds are included.

H. W. King: We are acquainted with this observation, but have not stressed it because our prime aim was to discuss the low-temperature technique.

NUMERICAL CONTROL X-RAY POWDER DIFFRACTOMETRY

R. W. Rex

Chevron Research Company
La Habra, California

ABSTRACT

Recent development of numerical control, N/C, systems for digital X-ray powder diffractometry opens the full potential for rapid and detailed machine processing of digital diffraction data. This capability may prove to be as large an incremental technological advance as the earlier shift from film to counter detector techniques. Our N/C system has the capability of performing essentially all operations possible by manual methods. Data output is on magnetic tape and carries identification and some of the instructions necessary for processing. The data tapes are processed by programs that (a) edit and check, (b) filter through a controlled shape high frequency filter (square, triangular, normal, etc.), (c) remove the minimum background from the diffraction pattern, (d) analyze data points close to background and test for statistical significance deleting those points within limits defined to constitute noise, (e) recognize peaks and record their position and intensity, (f) identify phases, (g) calculate phase concentrations by various methods, including internal standard and mutual standard techniques, and (h) output any desired portion of the data in a variety of digital and analog formats suitable for display, further analysis, and storage in an information retrieval system.

It is possible to record an entire diffraction pattern for a phase on two or three inches of magnetic tape. This record carries detailed peak area and shape information now missing from the ASTM reference system. Furthermore, the new magnetic tape information is directly amenable to computer processing to prepare search oriented record systems of diffraction data that can be inexpensively updated and edited. Currently, an ASTM committee is preparing a magnetic tape format diffraction pattern file. It is suggested that a second generation computer index system be generated based on digital records of actual diffraction patterns. This type of system should prove more versatile than the old system and possibly be the only way to keep abreast of the flood of new diffraction pattern information generated from the vast number of new organic and inorganic compounds being synthesized.

INCENTIVE FOR SYSTEMS DEVELOPMENT

One of the principal barriers to utilization of X-ray powder diffraction analysis for large-scale studies involving thousands of samples is the time and cost of data processing. Conventional powder diffractometry usually involves use of analog rate meter circuitry to integrate the digital X-ray detector output followed by manual redigitization of the analog output. This common process of conversion of digital to analog to digital data loses two to three orders of magnitude of significant data as well as requires extensive and tedious calculations. A solution to this problem is automation of the operation of the scalar and recording output in machine readable language. Automation has been achieved

Figure 1. Numerical control X-ray diffractometer. Present instrument modified from photograph by doubling height of automatic sample changer to hold 140 samples and replacement of balanced Ross-type filter with curved crystal monochromator. IBM modified 026 reader/punch not shown.

by a number of research groups[1,2] as well as at our laboratory. However, full numerical control as well as automation of the system operations is a step beyond automation.

Numerical control X-ray diffractometry opens up the full capabilities of variable operating instructions and control possible with human operators to full time equipment utilization on a 24 hr day and a seven day week. The digital output records are amenable to direct computer data processing utilizing the increasing capabilities of the new high speed computers. The benefits gained by numerical control are both economic and technological. Full time equipment utilization with reduced personnel greatly increases overall output. The relative freedom from human error and elimination of a very tedious task improves analytical quality, and use of machine computation both reduces costs and improves analytical quality by opening up a host of statistical signal-to-noise enhancement techniques.

SYSTEMS DESIGN BOUNDARY CONDITIONS

In order to gain the potential benefits of numerical control, we undertook to design and have fabricated an X-ray powder diffraction system capable of operating unattended for long periods of time with machine-compatible data output. The system described here was a joint development of our laboratory and the Datex Corporation of Monrovia, California (Figure 1).

Design of the system was initiated at the computer interface so that all output would be directly compatible with IBM standard format. In this particular case we designed for an IBM 1401/360-35 computer, although we translate data into IBM 7094/360-65 format for further processing. Accordingly, the system is designed to gap tape and control record length to meet the memory requirements of a 12K memory 1401 computer.

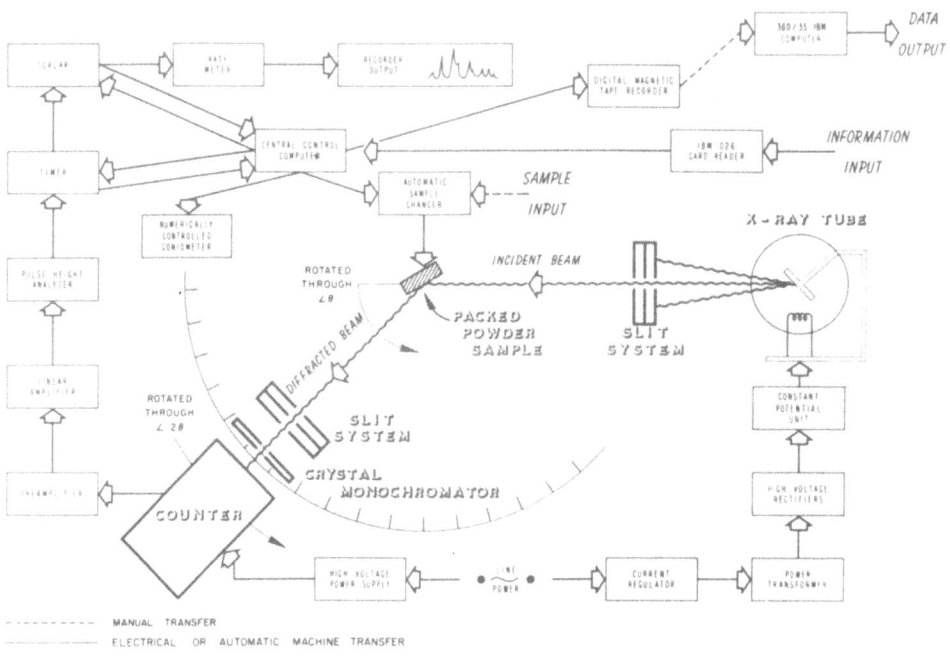

Figure 2. Flow diagram of numerical control X-ray powder diffraction system.

The basic control for the diffractometer is vested in a central program computer built into the system. Numerical command instructions are read from punch cards by a special IBM 026 card reader. The most significant particular modification is that our card reader is designed to emit data one column at a time in response to a "read" signal from the central computer. Therefore, the cards act as effective buffer memory for the system. In addition to command cards, we also utilize transfer cards which are read and the information transferred directly to magnetic tape without further modification. By this means we can transfer sample identification information of any sort in serial sequence ahead of each data record to the data tape.

Samples represent discrete data blocks so we chose punch cards for transfer of sample information onto the data tapes. If a sample is removed or added to the automatic sample changer, then the corresponding suite of transfer and command cards are also removed or added to the card deck in the card reader.

A numerical control diffractometer such as we are describing here could equally well be used on-line with a large time sharing computer to control hardware operation via negative feedback logic loops. However, we find that preprogramming via cards meets the great majority of our current research needs and all of our routine analytical requirements. Nevertheless, the coming availability of large time sharing computers makes their use the logical next growth step in numerical control powder diffractometry.

INSTRUMENTAL SYSTEMS

Our numerical control instrument (Figure 2) consists of a General Electric Corporation XRD-5 power supply, transistorized scalar-rate meter, pulse height analyzer, diffracted beam crystal monochromator, goniometer, and high intensity copper tube

(40 mA, 50 kV); Tempres, Inc. tube mount; Datex Corporation central control computer, Encoderdyne drive motor, 2θ display system; Precision Instrument Co. incremental tape recorder; Leeds and Northrup analog recorder; D. and O. Machine, Inc. automatic sample changer; Tectronix oscilloscope to monitor proportional counter pulse shape; and a modified IBM 026 card reader/punch.

The central control computer commands the various numerical control operations necessary for the operation of the automatic sample changer, the scanning diffractometer, and the timing and integrating operations (Figure 2). All operating variables are numerically controlled by the input information on a command card. Once a deck of cards is loaded into the card reader, the samples inserted into the automatic sample changer, and the system started, no further human intervention is needed until the run has been completed. Uninterrupted runs have been as long as 18 days. Samples may be added to the sample changer and cards added to the deck in the card reader during operation, as desired, to permit runs of indefinite length.

The system was optimized to yield maximum signal/noise possible consistent with 95% availability. Current system availability is actually running at 98% after complete debugging and routine operation for more than a year at near maximum capacity. In order to achieve maximum signal/noise we utilize maximum tube intensity possible without structural damage; a diffracted beam crystal monochromator to remove K_β, tungsten, and secondary fluorescence radiation; pulse height analysis to control circuitry noise; and the maximum time averaging commensurate with the work load.

The card reader function of the IBM 026 can be cut out during the actual X-ray diffraction scan because all necessary operating instructions are retained in memory by the computer and the 026 can then be used as a key punch. In this way transfer and command cards can be prepared on the same 026 used for reading instructions into the control computer. The only requirement is that the appropriate command card be re-inserted by the time that the analytical scan is completed. However, if this is not done, the control computer stops at the end of the scan and awaits further instructions.

The numerical control system can call for scanning speeds of $\frac{1}{4}$–4°/min in a continuous scan mode; integrate over 0.01 to 0.99 degrees 2θ; define starting and stopping

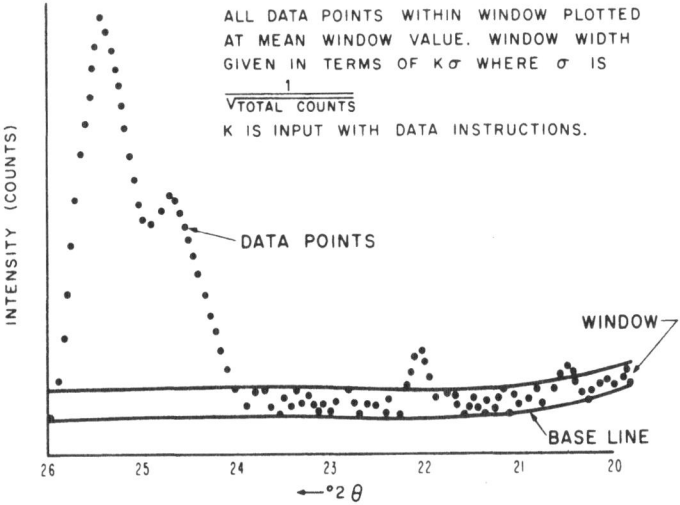

Figure 3. Schematic representation of program which defines the background for diffraction patterns.

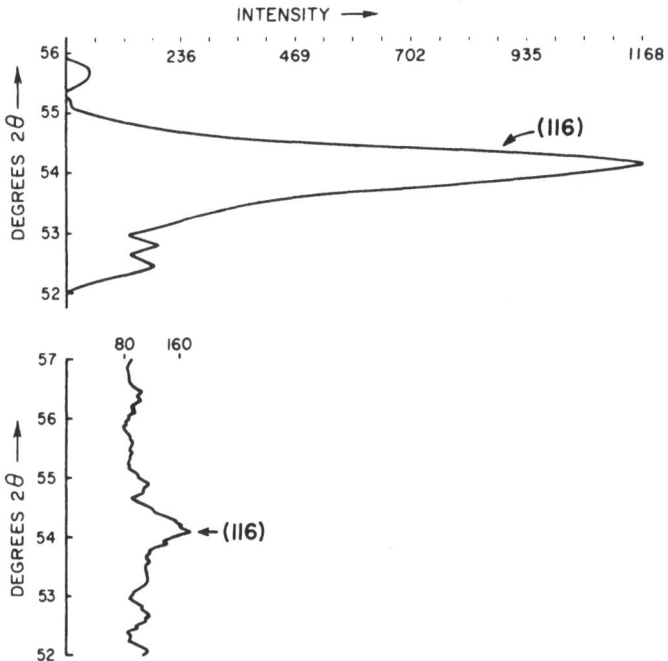

Figure 4. Comparison of the digital and the best analog pattern obtained with Cu K_α radiation of a small amount of an iron mineral.

2θ-0.01°; call for a sample change; or write an end of file mark on tape and shut down the entire operation. In the point to point mode the timing can be preset to any desired number of seconds. The goniometer will then move from point to point and only scale at the defined points. Position is defined to XXX.XX5 $\pm$ 0.005°2θ. Any overshoot in positioning by the Encoderdyne stepping motor is corrected within a few milliseconds. Counting circuit dead time while the scalar output is being written on tape is approximately 1–2 milliseconds. Consequently, no special buffer memory is required between the magnetic tape and the scalar even for continuous scan operation.

Instrument operation is monitored by oscilloscope checking of counter pulse shapes at various stages of amplification; analog strip chart display of the rate meter; and a tape edit/check program which, to a degree, can compensate for some minor hardware errors. Parity or more serious operational errors are detected by the edit/check program and summarized on a computer output. Dumps of the data tapes are then presented for detailed error analysis if necessary. Satisfactory tapes are submitted for processing by a family of programs utilizing our IBM 360/35 and 7094-360/65 computers.

Digital powder diffraction patterns are first smoothed using a modification of the technique of Savitzky and Golay.[3] Our filter function can carry up to 37 points and filter shape is parameterized and, therefore, highly flexible. Computer smoothing is probably one of the most significant improvements of digital diffractometry over older analog technology, in this investigator's opinion.

The smoothed records are then analyzed for peak widths and a running tangent base line drawn (Figure 3). This provides a minimum background. Each data point is then analyzed with respect to this minimum background and then classified by parameterized criteria as signal or noise. If signal, then the points remain as valid data. Those points within bounds defined to be noise are plotted at a running mean position in the noise band directly above the minimum tangency line defined as background. This new running mean within the noise band is next defined as the pattern background and subtracted from the record. The algorithm describing this operation is to be published elsewhere.

The net diffraction pattern is then analyzed by a series of logic loops that recognize peaks, shoulders, and wing edges for peaks. Peak heights and areas are then measured, phases identified, and abundances calculated by internal or mutual standard techniques.

A number of special purpose programs have been and are being written for a variety of purposes. For example, very low crystallinity materials can be rescanned many times and the records stacked to give signal averaging that also eliminates long term drift. The five-digit data word length permits counts up to 99,999 per scaling increment. With record stacking this can be increased to any amount desired. Stacking followed by smoothing and digital filtering to remove unwanted background from the X-ray pattern yields excellent patterns from low crystallinity materials that appear nearly amorphous to conventional analog recorders (Figure 4).

RECOMMENDATIONS

Digital diffractometry adds both the promise and problems of machine information retrieval systems. It is possible to record an entire diffraction pattern for a phase on two or three inches of magnetic tape. This record carries detailed peak area and shape information now missing from the ASTM reference system. Furthermore, the new magnetic tape information is directly amenable to computer processing to prepare search oriented record systems of diffraction data that can be inexpensively updated and edited.

Currently, the Joint Committee on Chemical Analysis by Powder Diffraction Methods (Joint Committee of ASTM, A.C.S., British Institute of Physics, and National Association of Corrosion Engineers) sponsors a project at the Pennsylvania State University for application of computer techniques to the ASTM Powder Diffraction File.[4] The Hanawalt-Davey Index and the Fink Index are being transferred to magnetic tape and search programs are being written in FORTRAN II for identification of phase mixtures; identification of tube contamination lines; and automatic indexing of unknown patterns. The advent of digital diffractometry fed by the prodigious capacities of numerical control diffractometers suggests that it is appropriate to establish a working group to coordinate machine languages and establish analytical standards for digital X-ray patterns. All present X-ray powder diffraction information systems operate on information abstracted and summarized manually from the original patterns. The growing use of X-ray diffraction for studies of organic and polymeric materials as well as the vast number of new inorganic materials being synthesized makes it attractive to shift all standards information to internationally consistent techniques. Direct storage and retrieval of the original patterns accessible to all would, among other things, encourage more research in X-ray diffraction information retrieval systems. It would also provide a basis for using peak areas and shapes, as well as peak heights, in describing reference patterns.

CONCLUSIONS

X-ray powder diffraction technology has evolved from film to counter techniques. Refinement of counter technology has further progressed by component upgrading but the great majority of workers have employed rate meter driven analog charts for X-ray pattern recording. This causes considerable loss of information and requires redigitization for quantitative work. More accurate work is usually done by manual scalar operation. This has been superceded by automation of the scalar operation to perform repetitive analyses. However, research work often requires the greater flexibility of manual operation. We have described here a numerical control system which has the flexibility of nearly all possible permutations of operating conditions possible manually combined with the quantitative capabilities of a production line instrument. We have operated our system for about 6000 hr to date and have obtained a high degree of system reliability. The software developed to process the digital X-ray patterns readily produces quantitative analytical data with much greater sensitivity and accuracy than was possible with analog circuitry and at a much reduced cost. The technological and economic benefits from a numerical control digital powder diffractometer are so great that it seems inevitable that this will be the next growth step in powder diffraction technology for a large number of laboratories. Consequently, it appears appropriate that workers in this field come to some common understanding concerning machine language format in order to maintain interchangeability of digital diffraction pattern data.

REFERENCES

1. G. J. C. Frohnsdorf and P. H. Harris, "Use of Digital Techniques to Aid in the Phase Analysis of Multicomponent Mixtures by X-Ray Diffraction," *Developments in Applied Spectroscopy*, Vol. 3, Plenum Press, New York, 1964, pp. 58–68.
2. S. B. McCaleb, "X-Ray Diffraction Automation and Its Use in Clay Mineralogy," *Clays and Clay Minerals*, Vol. 13, Pergamon Press, London, pp. 123–130.
3. A. Savitzky and M. J. E. Golay, "Smoothing and Differentiation of Data by Simplified Least Squares Procedures," *Anal. Chem.* **36**: 1627–1639, 1964.
4. W. L. Kehl, Announcements: *ACA Newsletter*, June 1966, p.3.

DISCUSSION

B. S. Sanderson (National Lead Co.): I was interested in your method of smoothing your curve. Have you ever tried using the moving average method of Savitzky?

R. W. Rex: Yes. Our particular program is a modification of the technique of Savitzky and Golay. However, we use a more versatile computer program than described in their original paper. We find that the method of Savitzky and Golay works very well for smoothing powder diffraction data.

B. S. Sanderson: I would like to comment on your last statements. I am in the process of trying to get a poor-man's version of this, mainly because I can't raise any more money. We have a very small computer, but thinking in the same terms, we feel that we can come up with a similar concept quite a bit cheaper. Of course, it won't be as versatile.

R. W. Rex: We originally designed a filter that ran on an IBM 1401 by multiple passes. It worked successfully but it wasn't flexible nor fast enough and when we obtained an IBM 360 system, we changed over to a new improved general purpose filter program.

B. S. Sanderson: Yes, I think if you look at Savitzky and Golay's experience, they started with an extremely small computer and have worked up to bigger and bigger ones. The point is that they can be made compatible. What you have to say is quite true, that one ought to get together no matter what size computer system you have. I think most people would look at this system and presume that they will never be able to get anything that expensive. But even if you don't, you still can use the same concepts and be compatible.

R. W. Rex: Yes, definitely. You can trade off machine time for sophistication and you can use multiple passes on small computers to make up for lack of large machines.

K. E. Beu (Goodyear Atomic): When you want to record the entire line profiles, along with *d* values and so forth, how do you propose to handle this in terms of the instrumental factors that affect the profile shape? When you go from one laboratory to another, these can be different.

R. W. Rex: I think this is where we need to work both with the manufacturers and between the laboratories to decide what corrections we want using standard materials to transform our patterns into a standard state. I think what we should do is to develop a family of standard materials with which each analyst can transform or convolve his patterns to the standard state and use the convolutions to transform his particular analytical data to a standard form. We could work out the error function for each of the analytical instruments as operated in different laboratories. I think this is going to be necessary in the future and one of the things we need to do initially is to establish standard reference samples.

Chairman D. K. Smith: This will also affect the computed patterns that will be generated, say, from fundamental cell data and atomic position parameters which should be tied in with the same system.

K. E. Beu: You mentioned that you may have occasional power interruptions from lightning flashes, etc.? Does your program handle it? I mean, does it go back and start over?

R. W. Rex: No. In the interrupted record, assuming that the power isn't off long enough for relays to open up so that the hardware isn't turned off, we usually lose parity. This shows up in the parity checks or we lose synchronization in the timing circuits and get an illegal or misplaced character. The usual effect is an illegal character. Check programs which edit the data, recognize the defective record and print-out the illegal character information and try to analyze it. If we have incorporated the necessary logic for recognizing the error, the program will print out what it is. If the error is something which we don't understand, or is not incorporated into the error recognition routines, then the print-out will only indicate an illegal character or loss of parity. If this happens, we have to repeat the analysis.

THE EFFECTS OF ELECTRONIC STRUCTURE AND INTERATOMIC BONDING ON THE SOFT X-RAY Al K EMISSION SPECTRUM FROM ALUMINUM BINARY SYSTEMS

David W. Fischer and William L. Baun

Air Force Materials Laboratory (MAYA)
Wright–Patterson Air Force Base, Ohio

ABSTRACT

The aluminum K X-ray emission lines and bands from a series of aluminum binary alloys and other binary compounds were investigated using 6 kV electron excitation and a flat crystal vacuum spectrometer. The overall shape of the emission band and its energy position as a function of alloy composition was determined. It appears from the data that the aluminum K band undergoes changes in shape and energy position which are dependent on the electronic configuration of the element with which the aluminum is chemically bonded. These band changes can be interpreted as indicating a change in the bonding character between the metal atoms. In the Al–Ni system, for instance, the Al K band becomes more symmetrical and shifts to lower energy as the nickel content is increased, indicating that perhaps the bonding on the aluminum atoms is becoming less metallic and more covalent in nature as the nickel to aluminum ratio is increased.

The aluminum $K_{\alpha_4}/K_{\alpha_3}$ satellite line intensity ratio also varies in an orderly manner in aluminum binary compounds. In general, these satellite line changes go hand-in-hand with the K band changes. If the K band shifts to lower energy, the $K_{\alpha_4}/K_{\alpha_3}$ intensity ratio will always increase in value.

INTRODUCTION

Although the investigation of soft X-ray emission band spectra of metals and alloys has been the subject of several previous investigations, little is really known of the effect of alloying on the electronic structure of metals. Publications by Appleton[1] and Thompson and Kellen[2] provide a review of much of the work done to date and the problems of interpreting the results. The reader, however, is left with the impression that little change is to be expected in the metal spectrum as a result of alloying. To be sure, this is what has been observed for most of the systems reported to date but we intend to show that there are also alloy systems in which both the band and line spectra undergo rather large changes.

An emission band, such as the aluminum K band, is produced by electron transitions from the valence band to a vacancy in an inner level (K shell). The shape of the band, when properly corrected for a variety of possible distortion effects, essentially reflects the density of states in the outermost occupied shell. Since the valence electrons are the ones most affected by chemical combination, the X-ray emission band should reflect vital information about the chemical bond. It is found that the character of the atomic interaction in compounds and alloys of aluminum has a substantial effect on the shape and position of the K emission band. Since it is the electronic structure of the atoms that determines interatomic bonding and also some of the physical and mechanical properties of the alloys, the X-ray emission band spectrum presents us with a possible method of characterizing this bonding and studying those properties which are dependent upon it.

Previously, we have shown that the aluminum K spectrum, especially the K band and the K_{α_3} and K_{α_4} satellite lines, changes significantly between metal and oxide[3,4] and other aluminum compounds[5,6] and that these changes can be grouped according to bonding type.

EXPERIMENTAL

Instrumentation

The flat crystal vacuum spectrometer used for this work has been described previously.[6,7] Aluminum K emission spectra were produced by electron beam bombardment of the specimen at 6 kV and 1–5 mA, dispersed by an EDDT crystal ($2d = 8.803$ Å) and detected by a flow proportional counter using a formvar window and argon-methane (P-10) flow gas at reduced pressure (100 mm Hg). An anode accommodating four specimens was used so that spectra could be obtained under exactly the same conditions. Usual operating pressure was 1 to 3 $\times$ 10^{-6} torr. All recording electronics were standard Picker items except for a Tennelec low-noise preamplifier. The resultant ratemeter scans and the curves shown in the figures have a mean deviation of $\pm 2\%$.

Specimen Preparation

The alloys were prepared primarily by arc melting in an argon atmosphere. Many of the alloys also were prepared by levitation melting. At compositions where stoichiometric compounds were formed, very fine powders were pressed into pellets and sintered. X-ray diffraction patterns were made for each alloy to check the phases present.

Specimens were usually mounted on the anode in the form of thin pellets. The brittle alloys were also ground into fine powders and spread in a thin layer on the anode surface. No differences in the aluminum K spectrum were observed between these two methods. Four specimens were mounted at one time, one of them always being the pure metal which was used as standard for measuring energy shifts, intensity ratio variations and shape changes in the emission lines and bands.

RESULTS

Aluminum K Band

In previous reports we showed that the aluminum K emission band (also denoted K_β) can change significantly in shape and energy position as a result of a change in the chemical environment of the aluminum atoms, especially when going from pure metal to the oxide.[3-6] For the pure metal the Al K band is very asymmetrical in shape with the intensity maximum at 1557.3 eV. The same band from Al_2O_3 is very symmetrical and the intensity maximum shifts 4.4 eV to lower energy. This change reflects the fact that the nature of the chemical bond is significantly altered going from metallic type in the metal to ionic-covalent in the oxide.

It has been shown that sample self-absorption and electron bombardment energies in large excess of threshold potentials also can affect band shapes,[8] but they appear to have very little or no measureable effect on the results reported here. Nevertheless, all of

Table I. Uncorrected Aluminum K Emission Characteristics in Al–Ni System

Target	Al $K_{\alpha_4}/K_{\alpha_3}$ $\pm$ 0.01	Al K_β position $\pm$ 0.1 eV	Al K_β half width $\pm$ 0.1 eV	Al K_β base width $\pm$ 0.5 eV	Al K_β; A_t $\pm$ 0.2	Al K_β; A_c $\pm$ 0.2	Al K_β edge width $\pm$ 0.1 eV
100Al	0.48	1557.3	6.0	12.3	2.7	3.7	2.1
90Al–10Ni	0.52	1556.9	7.0	12.4	2.2	2.9	3.0
85Al–15Ni	0.54	1556.7	6.9	12.4	1.9	2.3	3.2
75Al–25Ni	0.58	1556.2	6.5	12.4	1.4	1.9	3.7
65Al–35Ni	0.61	1555.8	6.2	12.4	1.4	1.7	3.9
60Al–40Ni	0.63	1555.6	5.7	12.4	1.4	1.6	4.0
50Al–50Ni	0.66	1555.3	6.0	12.5	1.4	1.3	4.2-
40Al–60Ni	0.70	1555.0	6.5	12.4	1.3	1.2	4.2
35Al–65Ni	0.72	1554.7	6.6	12.3	1.3	1.2	4.3
25Al–75Ni	0.76	1554.4	6.8	12.2	1.3	1.2	4.5
20Al–80 Ni	0.77	1554.2	6.8	12.3	1.3	1.2	4.5
4Al–96Ni	0.84	—	—	—	—	—	—
Al_2O_3	0.92	1552.9	6.4	10.8	1.0	1.0	4.0

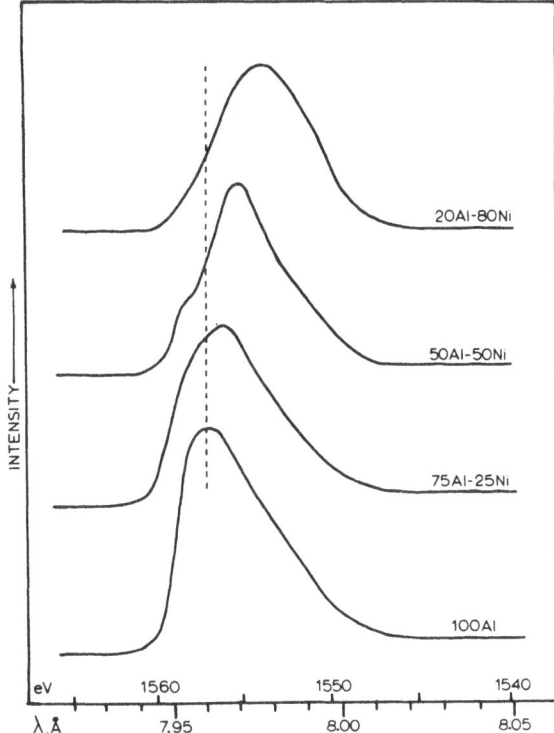

Figure 1. Al K emission band from pure metal and some Al–Ni alloys.

our band measurements were made at the same potential so that any slight bombardment energy effects will be the same for all of them.

All of the band measurements which are shown in this report are in no way corrected for instrumental, temperature, or other broadening effects. They are simply the values measured for the as-recorded spectra and are, therefore, only relative values to be used in comparing band shapes. All of the bands shown are the result of two or more individual runs.

Many aluminum binary compounds and alloys show an Al K band which is somewhere in between those obtained from aluminum metal and Al_2O_3 both in shape and energy position. Some systems even give their own unique band shapes. We will here, however, concentrate mainly on three different aluminum binary alloy systems, Al–Ni, Al–Cu and Al–Mg, which show three entirely different effects on the aluminum K emission spectrum as a result of alloying.

The first type of effect is that observed in the aluminum–nickel binary system. We have examined the Al K band from a large number of Al–Ni alloys, both one- and two-phase structures.[9] Figure 1 gives an indication of the changes which take place in the band as the alloy composition is varied. Notice that, as more and more nickel is added to the composition, the Al K band becomes more symmetrical in shape and shifts to lower energy. This energy shift is, in fact, a linear function of alloy composition as shown in Figure 7 and will be discussed in more detail a little later. The energy positions are listed in Table I.

As the alloy composition varies so does the Al K band half-width and full-width which are listed in Table I. The half-width can be measured fairly accurately but the full band width is subject to more uncertainty because of the slow tailing off of the long wavelength side of the band. It appears, however, that the band width does not change

Table II. Uncorrected Aluminum K Emission Characteristics in Al–Mg System

Target	Al $K_{\alpha_4}/K_{\alpha_3}$ ± 0.01	Al K_β position ± 0.1 eV	Al K_β half width 6.0 ± 0.1 eV	Al K_β base width 12.3 ± 0.5 eV	Al K_β; A_i 2.7 ± 0.2	Al K_β; A_c 3.7 ± 0.2	Al K_β edge width 2.1 ± 0.1 eV
100Al	0.48	1557.3	6.0	12.3	2.7	3.7	2.1
90Al–10Mg	0.49	1557.1	5.8	12.2	2.0	3.7	2.1
82Al–18Mg	0.50	1557.1	5.7	12.0	1.9	3.4	2.3
60Al–40Mg	0.50	1557.0	5.3	11.3	1.7	2.9	2.3
55Al–45Mg	0.50	1557.0	5.2	11.1	1.7	2.9	2.3
50Al–50Mg	0.50	1557.0	4.9	10.9	1.7	2.9	2.3
40Al–60Mg	0.50	1557.0	4.6	10.2	1.7	2.9	2.3
30Al–70Mg	0.50	1557.0	4.6	10.2	1.7	2.9	2.3
20Al–80Mg	0.50	1557.0	4.6	10.3	1.7	2.9	2.3
10Al–90Mg	0.50	1557.0	4.5	10.3	1.7	2.9	2.3

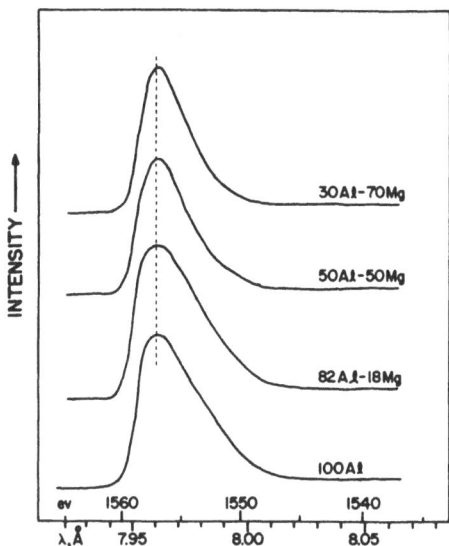

Figure 2. Al K emission band from pure metal and some Al–Mg alloys.

significantly as the alloy composition is varied. In a previous report[9] we had stated that the Al K band appeared to become progressively narrower as the aluminum content decreased but a more careful study of the band indicates that this may not be true. If any narrowing does occur, it is very slight. The band shape change can be measured directly by the broadening of the emission edge and by the asymmetry index (A_i) and the asymmetry coefficient (A_c), the results of which are listed in the last three columns of Table I.[9]

The fact that the Al K band from the Al–Ni alloys approaches the shape and energy position of the band from Al_2O_3 suggests a change in the nature of the bonding on the aluminum atoms. Perhaps as the aluminum content is decreased, the bonding becomes more covalent-like in character as opposed to the predominantly metallic character present in pure aluminum. Similar observations were noted by Nemnonov and Finkel'shtein for aluminum emission spectra[10] and by Das and Azaroff for the absorption spectra.[11]

In contrast to the large changes observed in the Al–Ni system, the Al–Mg system presents an Al K band which remains virtually unchanged throughout the entire composition range. The band from the pure metal and a few of the Al–Mg alloys are shown in Figure 2. About the only change which occurs is in the band width. It appears that, as the aluminum content is decreased, the band becomes progressively narrower. The energy position, edge width and asymmetry constants do not change much as indicated in the compilation in Table II. If the large changes occurring in the Al–Ni system really indicate an increase in the covalent-like nature of the bonding on the aluminum atoms, then the virtually non-changing band in the Al–Mg system probably indicates that the bonding type does not change in this system no matter how much magnesium is added to the aluminum.

A third, completely different type of effect is obtained from the Al–Cu system as illustrated in Figure 3.[12] The most obvious effect of alloying in this system is the splitting of the Al K band into two components. These two components each behave in a different manner. The high energy component remains in approximately the same energy position throughout the entire composition range but the low energy component becomes more

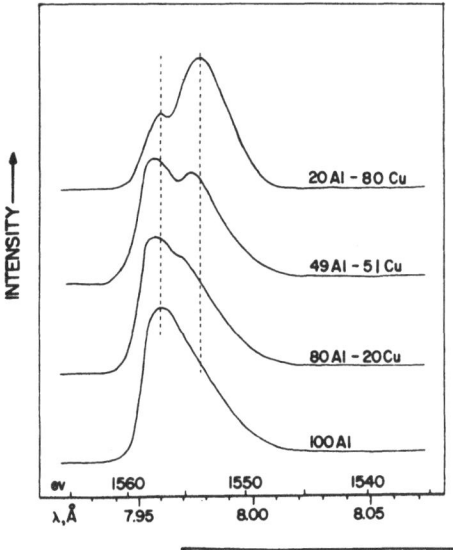

Figure 3. Al K emission band from pure metal and some Al–Cu alloys.

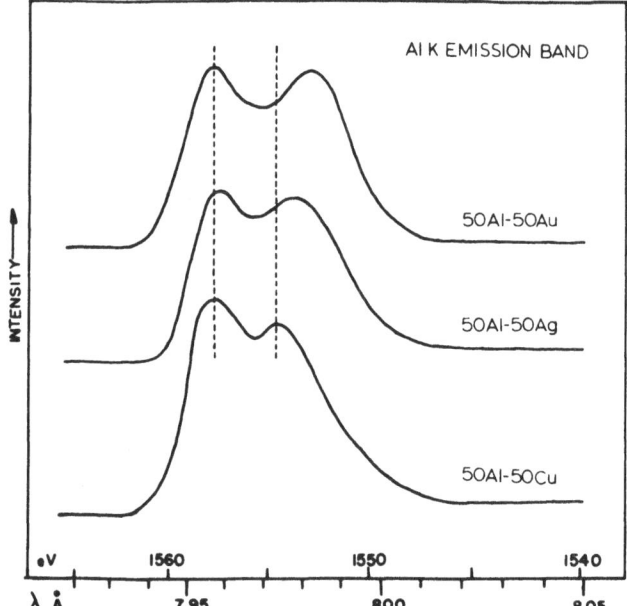

Figure 4. Al K emission band from 1:1 atomic ratios of Al–Cu, Al–Ag, and Al–Au alloys.

prominent and shifts to lower energy as more and more copper is added. The energy separation of these two components is found to be a linear function of alloy composition.[12] In addition, the uncorrected edge width and band width both remain nearly constant over the entire composition range despite the band splitting. These are essentially the same results shown by Yoshida in 1936[13] and Farineau in 1939.[14]

We find that the K emission band from other polyvalent elements also becomes split when they are alloyed with monovalent elements. In addition to Cu–Al, we have investigated alloys of the following systems: Cu–Mg, Cu–Si, Ag–Mg, Ag–Al, Au–Al,

Table III. Uncorrected Aluminum K Emission Characteristics in Al–Cu System

Target	Al $K_{\alpha_4}/K_{\alpha_3}$	Al K_β position (high-energy component)	Al K_β position (low-energy component)	Al K_β edge width
100Al	0.48 ± 0.01	1557.3 ± 0.1 eV	—	2.1 ± 0.1 eV
90Al–10Cu	0.51	1557.4	not resolved	2.2
75Al–25Cu	0.53	1557.6	1555.2 ± 0.1 eV	2.1
70Al–30Cu	0.55	1557.7	1555.1	2.4
67Al–33Cu	0.54	1557.7	1555.2	2.3
49Al–51Cu	0.58	1557.7	1554.6	2.4
40Al–60Cu	0.60	1557.7	1554.4	2.4
33Al–67Cu	0.61	1557.7	1554.4	2.3
30Al–70Cu	0.63	1557.7	1554.1	2.4
25Al–75Cu	0.66	1557.7	1554.0	2.5
20Al–80Cu	0.67	1557.7	1553.8	2.5
10Al–90Cu	0.70	1557.7	1553.7	2.5

and Au–Be. The K band of the polyvalent element was split into two components for each of these systems. Figure 4 shows the similarity of the Al K band from 1 : 1 alloys of Al–Cu, Al–Ag and Al–Au. These group 1B metals are the only ones which we have observed to cause a splitting of the Al K band with formation of a binary alloy. A possible explanation of this phenomenon was proposed by Friedel.[15] The aluminum K emission data for the Al–Cu systems are summarized in Table III.

The fact that the elements of subgroup 1B all have the same effect on the aluminum spectrum when formed as binary alloys with aluminum led to an investigation of the effect of other subgroup systems of the periodic table, with somewhat similar results. Nickel, cobalt, and iron all have many similar properties and all are found to have virtually the same effect on the Al K band for equal concentrations as shown in Figure 5. Metals of subgroup 4B (titanium, zirconium, and hafnium) are all found to have the same effect on the Al K band and so on for every subgroup system investigated. The effect is not restricted to intermetallic compounds as illustrated in Figure 6. Aluminum binary compounds with the subgroup 5A elements, phosphorus, arsenic and antimony, all give the same Al K band shape and energy position. The entire series of aluminum binary systems is discussed in more detail in a recently prepared report.[16] In effect, the results indicate that elements which fall in the same subgroup in any part of the periodic table will have virtually the same effect on the aluminum spectrum but elements of each different subgroup have different effects. This latter point will be discussed more thoroughly a little later in this report.

Earlier it was mentioned that the energy position of the Al K band from the Al–Ni system shifted as a linear function of alloy composition. Similarly, the energy shift is found to be linear in other aluminum binary alloy systems as well and some of them are plotted in Figure 7. There is a significance in the different linear slopes of the different systems as will be explained later but the important point to mention here is that the Al K band shift in any one binary system is a continuous and smooth function of composition no matter how many crystal structure or phase changes are encountered in the system. The bands obtained from two-phase alloys are averages of the band from each phase present. This is one of the reasons why we have not been particularly careful to choose only well characterized single phase structures for our studies.

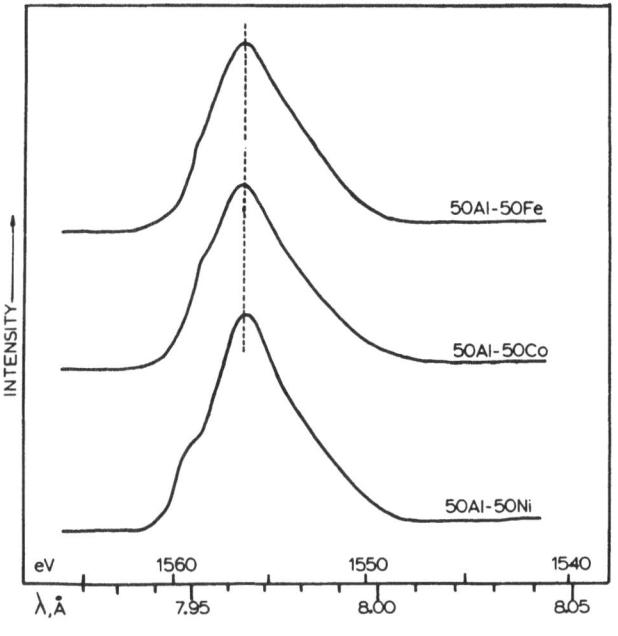

Figure 5. Al K emission band from 1:1 atomic ratios of Al–Ni, Al–Co, and Al–Fe alloys.

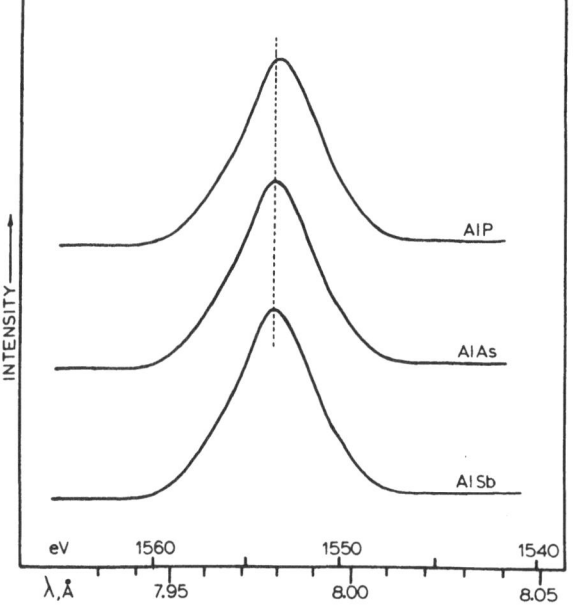

Figure 6. Al K emission band from AlP, AlAs, and AlSb.

Aluminum K Satellite Lines

In studying the Al K spectrum from various aluminum compounds and alloys, our observation has been that any change in the shape and energy position of the Al K band is always accompanied by a corresponding change in the Al $K_{\alpha_4}/K_{\alpha_3}$ satellite line intensity ratio. When going from pure aluminum metal to Al_2O_3, the Al $K_{\alpha_4}/K_{\alpha_3}$ intensity ratio changes from 0.48–0.92 and both lines shift to higher energy.[3-6] Many aluminum binary compounds show a satellite line intensity ratio which is somewhere between those observed for aluminum metal and Al_2O_3. In Figure 8, for example are shown the Al K_{α_3} and Al K_{α_4} satellite lines from aluminum metal and a few Al–Ni alloys. The most notable change in the lines is, that as the nickel concentration is increased the intensity of the K_{α_4} increases with respect to K_{α_3}. The more nickel that is added the more the $K_{\alpha_4}/K_{\alpha_3}$ intensity ratio approaches that observed in the oxide. If one makes a plot of this intensity ratio as a function of alloy composition, not only for Al–Ni but for other aluminum binary systems as well, a linear relationship such as shown in Figure 9 is obtained. For all systems, the intensity ratio increases as the aluminum content is decreased. The intensity

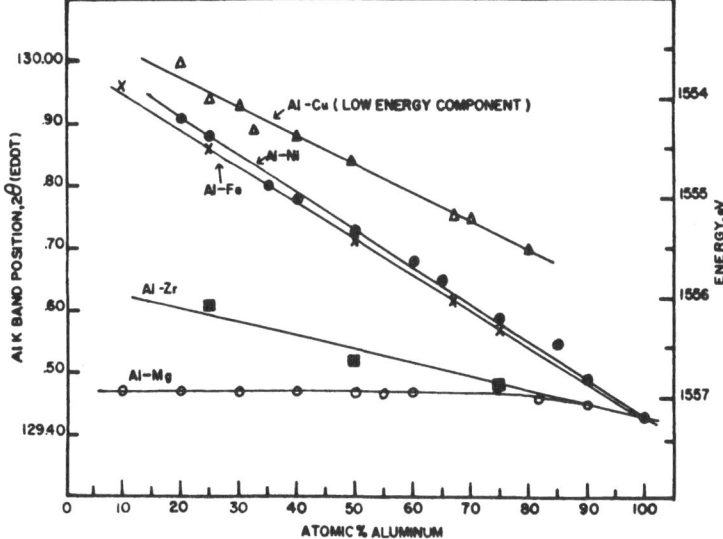

Figure 7. Al K band energy shift as a function of alloy composition for some aluminum binary systems.

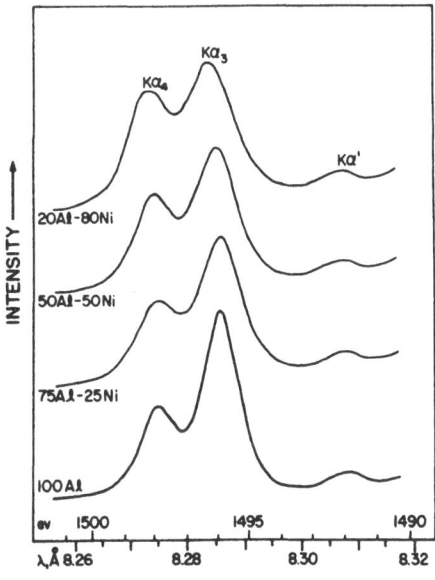

Figure 8. Al K_{α_3} and K_{α_4} satellite lines from pure metal and some Al–Ni alloys.

ratio increase is also accompanied by a shift of the lines to higher energy. Satellite line ratios for the Al–Ni, Al–Mg, and Al–Cu systems are listed in column 2 of Tables I–III.

This satellite line intensity ratio is very reproducible and we have been able to use it as a quantitative tool for determining the aluminum content to within $\pm 2\%$ (atomic) in systems such as Al–Ni. The technique may be compared to a measurement such as a lattice parameter determination in that once curves such as those in Figure 9 have been obtained, it is not necessary to run standards with each sample.

A glance at Figures 7 and 9 will show a striking similarity between the changes in the emission band and satellite lines as the alloy composition is varied, ignoring for the moment the Al–Cu system. We find, in fact, that the Al K band energy position and Al $K_{\alpha_4}/K_{\alpha_3}$ intensity ratio are very closely inter-related in the manner shown in Figure 10. In this figure we have plotted the energy position of the Al K band intensity maximum against the Al $K_{\alpha_4}/K_{\alpha_3}$ intensity ratio. It makes no difference whether the compound is a conductor, semiconductor or insulator. The lowest point on the lower left is from pure aluminum metal; the highest one on the right is from Al_2O_3; other points at the upper right are from other insulator and semiconductor compounds while those at the lower left and at the center are from various binary alloys.

For every 1 eV that the K band shifts, the $K_{\alpha_4}/K_{\alpha_3}$ intensity ratio changes by 0.09. The only aluminum binary compounds which do not follow this relationship appear to be those of the Al–Cu, Al–Ag, and Al–Au systems in which the Al K band becomes split into two components. We have studied approximately 200 different aluminum binary compositions where the Al K band has only one intensity maximum and have not found even one which does not follow the relationship of Figure 10. Why these two spectral characteristics should be so closely interrelated is not obvious to us because supposedly they represent two entirely different phenomena. The Al K band arises from an electronic transition between the valence band and the innermost K shell while the atom is in a state of single K ionization. The $K_{\alpha_4}/K_{\alpha_3}$ intensity ratio is believed to represent two different transition probabilities between the $L_{2,3}$ and K inner levels while the atom is in a state of double KL ionization.

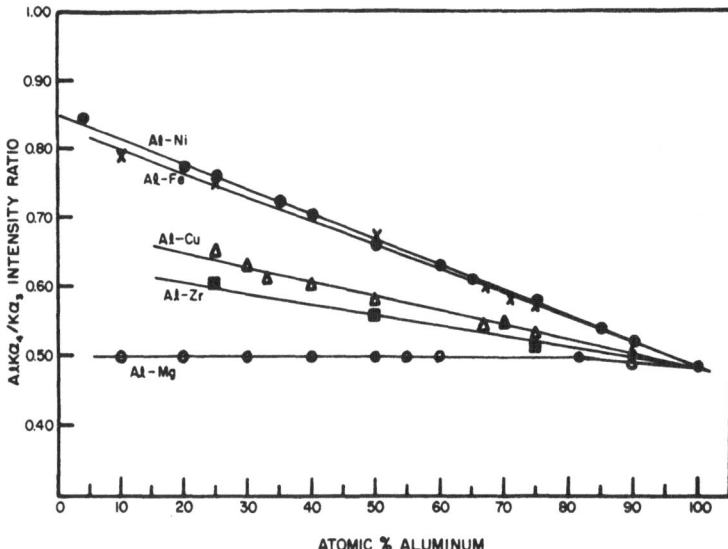

Figure 9. Al $K_{\alpha_4}/K_{\alpha_3}$ intensity ratio as a function of alloy composition for some aluminum binary systems.

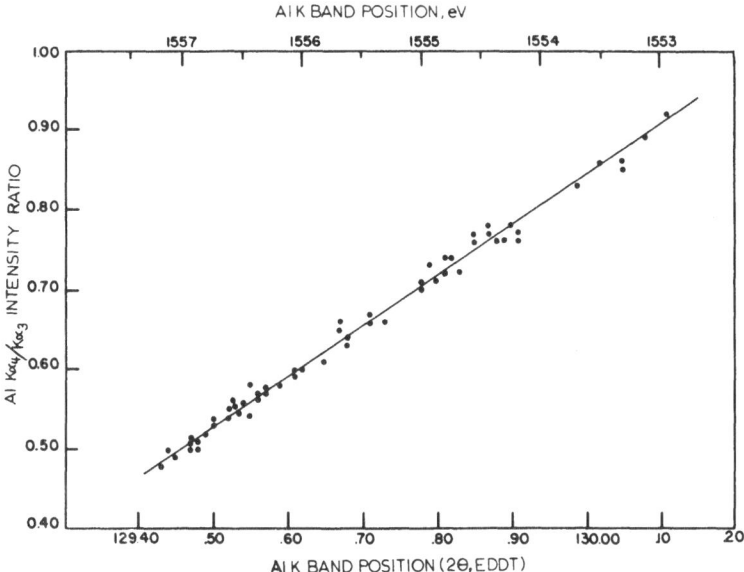

Figure 10. Interdependence of Al $K_{\alpha_4}/K_{\alpha_3}$ intensity ratio and Al K band energy position in aluminum binary compounds.

CONCLUSIONS

There are several conclusions which can be drawn from the results shown here. First of all, it is obvious, from the changes which occur in the Al K emission band, that the electronic structure of aluminum is changed, sometimes in a very pronounced manner, when the chemical environment of the aluminum atoms is altered. It is also evident that the nature of the interatomic interaction changes, not only for each different aluminum binary system, but with varying composition within any one system as well. We have shown in this report, for instance, that the Al K band from Al–Ni alloys (Figure 1) is different from what is observed from the Al–Mg alloys (Figure 2) and that both, in turn, are different from what is obtained from the Al–Cu alloys (Figure 3).

Figures 4–6 give an indication of what we find if we compare the results of binaries in which the second components belong to the same subgroup of the periodic table. The elements within any one subgroup have the same effect on the Al K band for any given aluminum composition. Since elements of the same subgroup have similar outer electronic configurations, it is apparent that the electronic structure of the second component determines the nature of the interatomic bond and, hence, the shape and energy position of the Al K band in aluminum binary compounds. If we form aluminum binary compounds of the same composition for elements in each subgroup, moving from the left to the right of the periodic table, we find that the energy position of the Al K band advances in a regular manner toward lower energy as shown in Figure 11. The farther to the right that a particular element lies, the greater the change it will cause in the Al K band when formed as a binary compound with aluminum. The amount of energy shift observed in the K band corresponds fairly well to the increasing electronegativity of the second component. In general, one can assume that the more electronegative the second component, the lower the energy of the Al K band.

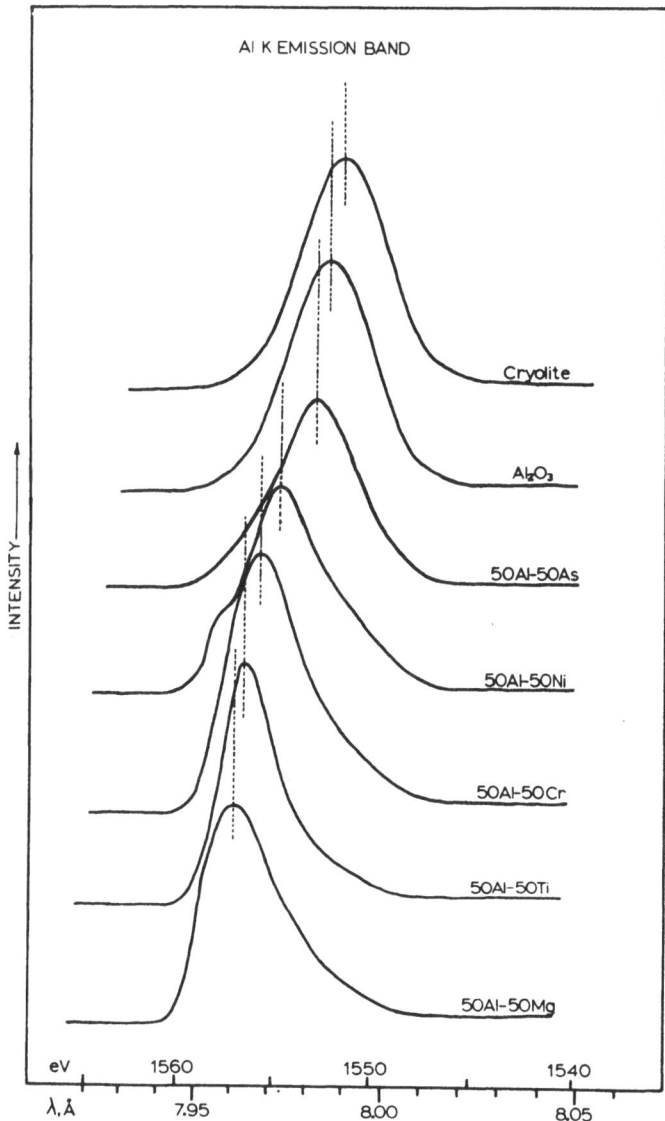

Figure 11. Al K emission band from various binary systems where second component falls progressively from left to right side of periodic table.

We find that these systematic variations are observed not only in the Al K band but in the Al $K_{\alpha_4}/K_{\alpha_3}$ intensity ratio as well, as illustrated in Figure 9.

It appears, then, that we can make two general conclusions concerning the relationship of the Al K emission spectrum from aluminum binary systems with the position of the second component in the periodic table:

1. Elements of the same subgroup each have virtually the same effect on the aluminum spectrum.

2. Elements of each different subgroup have a different effect on the aluminum

spectrum and these effects can be correlated according to the relative positions of each of these subgroups in the periodic table.

These are important points because they illustrate that, at least for the Al K band, the chemical effect on X-ray spectra is quite periodic in nature; just as periodic, in fact, as the Periodic Table of the Elements. For all of the aluminum binary systems which we have reported here, the Al K_α parent and satellite lines shift to higher energy positions when going from pure metal to an alloy or compound. This is just the opposite of the direction of shift of the Al K band which we have always observed to be towards a lower energy position. These different directions of shift point out the different types of information which lines and bands provide. The shift in energy position of inner level transitions (such as the K_α lines of aluminum) are determined by changes in the density of the valence electrons in the inner regions of the atom. The shape and energy position of the emission band, on the other hand, indicates the character of the atomic interaction.

For atoms which behave as electron "donors," the lines due to inner level transitions are shifted to higher energy. For electron "acceptors" these lines shift to lower energy. Since, for all of the aluminum compounds reported here, the Al K_α lines shift to higher energy when going from pure metal to compound, we can assume that aluminum is behaving as a donor of electrons.

Upon alloying aluminum with metals such as iron, cobalt, nickel and many others, the Al K band starts becoming more symmetrical in shape and its intensity maximum shifts to a lower energy. The more of the second metal that is added, the greater the change in the Al K band which now occupies an energy position which is somewhere between that obtained from pure Al and that obtained from Al_2O_3. These changes in shape and position appear to indicate that the predominantly metallic bond present in pure aluminum is acquiring a certain amount of covalent-like nature in the alloys.

These spectral changes which occur are valuable not only from the theoretical viewpoint but can be used as a quantitative analysis tool as well. One has the capability of using the features of the aluminum K X-ray emission spectrum to identify the phase or phases present in an aluminum binary alloy and also determine the exact composition of each phase without requiring the use of an internal standard. The shape and energy position of the Al K band and the intensity ratio of the $K_{\alpha_4}/K_{\alpha_3}$ satellite lines are quite reproducible and once curves such as those shown in Figures 7 and 9 have been obtained for a particular binary system, it is not necessary to run standards with each sample to obtain compositions to within $\pm 2\%$ (atomic).

REFERENCES

1. A. Appleton, "The Soft X-Ray Emission Spectra of Metals and Alloys: I. The Origin and Interpretation of the Soft X-Ray Spectra of Metals," *Contemp. Phys.* 6: 50–67, 1964.
2. B. J. Thompson and P. F. Kellen, "The Soft X-Ray Emission Band Spectra of Metals and Alloys," *Developments in Applied Spectroscopy*, Vol. 4, Plenum Press, New York, 1965.
3. D. W. Fischer and W. L. Baun, "Diagram and Non-Diagram Lines in K Spectra of Aluminum and Oxygen from Metallic and Anodized Aluminum," *J. Appl. Phys.* 36: 534, 1965.
4. W. L. Baun and D. W. Fischer, "High Energy α Satellites in the Aluminum K X-Ray Emission Spectrum," *Phys. Letters* 13: 36, 1964.
5. W. L. Baun and D. W. Fischer, "The Influence of Chemical Combination on Aluminum K Diagram and Non-Diagram Lines," *Nature* 204: 642, 1964.
6. W. L. Baun and D. W. Fischer, "The Effect of Chemical Combination on K X-Ray Emission Spectra From Magnesium, Aluminum and Silicon," AFML-TR-64-350, December 1964.
7. D. W. Fischer and W. L. Baun, "Diagram and Non-Diagram Lines in K Spectra of Magnesium and Oxygen from Metallic and Anodized Magnesium," *Spectrochim. Acta* 21: 443, 1965.
8. R. J. Liefeld, "$L\alpha$ X-Ray Emission Lines of Nickel, Copper, and Zinc," *Bull. Am. Phys. Soc.* 10 (4), 549, 1965.

9. D. W. Fischer and W. L. Baun, "Effect of Alloying on the Aluminum K and Nickel L X-Ray Emission Spectra in the Aluminum–Nickel Binary System," *Phys. Rev.* **145**: 555, 1966.

10. S. A. Nemnonov and L. D. Finkel'shtein, "K_β Emission Band and K Absorption Edge of Aluminum in Some Alloys with Transition Metals," *Bull. Acad. Sci. (U.S.S.R.), Phys. Ser.* **25** (8), 1015, 1961.

11. B. N. Das and L. V. Azaroff, "X-Ray K Absorption Spectra of Ni–Al Alloys," *Acta Metall.* **13**: 827, 1965.

12. W. L. Baun and D. W. Fischer, "Effect of Alloying on the Aluminum K X-Ray Emission Spectrum in the Aluminum–Copper System," to appear in *J. Appl. Phys.*

13. S. Yoshida, *Sci. Papers Inst. Phys. Chem. Research (Tokyo)* **28**: 2431, 1936.

14. J. Farineau, "Contribution à l'Etude Spectrographique de la Structure Electronique des Metaux," *J. Phys. Rad.* **10**: 327, 1939.

15. J. Friedel, "Distribution of Electrons Round Impurities in Monovalent Elements," *Phil. Mag.* **43**: 153, 1952.

16. D. W. Fischer and W. L. Baun, "The Effects of Electronic Structure and Interatomic Bonding on the Soft X-Ray Emission Spectra from Aluminum Binary Systems," *AFML-TR* 66-191, June 1966.

MULTILAYER SOAP FILM STRUCTURES*

R. C. Ehlert and R. A. Mattson

General Electric Company
Milwaukee, Wisconsin

ABSTRACT

Multilayer soap film structures, particularly the lead stearate variety have been used for several years as a dispersing element in soft X-ray spectrometers. These structures have a high scattering power, and if a high order of diffraction is used for the shorter wavelengths they provide good resolution throughout the 10–80 Å range. Structures having a $2d$ spacing smaller than that of lead stearate (100 Å) would provide greater dispersion and, hence, resolution in the first order for radiation in the 10–40 Å range. Details concerning the conditions required to build multilayer structures from the soaps of shorter fatty acids such as lead myristate, lead laurate, lead caprate, etc. are given. The various members of the soap film family are compared regarding their diffracting power both as a function of wavelength and the order of diffraction. Information is given regarding the dependence of the diffracting power, the width of the diffraction peak at half maximum and the peak to background ratio as a function of the number of double layers in a structure. The absorption occurring within a lead stearate and a lead laurate structure has been experimentally measured. Observed spectra can, thus, be corrected for the filtration caused by the soap film structure. The soap film family is evaluated as a dispersing element by comparing the various structures with single crystals such as EDDT and KAP.

INTRODUCTION

Multilayer soap film structures have been shown to be effective dispersing devices for long wavelength X-rays.[1-9] A good deal of effort has been put forth to determine the parameters and techniques needed for building these structures. Most of this effort has been with the stearates, particularly barium and lead stearate. This was natural, since more information was available in the literature concerning the properties of monolayers and multilayers of this fatty acid and its soaps than about any other. Furthermore, these structures are relatively easy to build. Soaps of myristic, palmitic, lignoceric, and melissic acids were studied but less intensively, and little or no information was available regarding their relative quality.

Since the dispersion of a crystal or multilayer structure is proportional to the angle of diffraction, it is useful to have a set of different dispersing elements which would provide high resolution throughout the soft X-ray region. For wavelengths shorter than 25 Å a number of single crystals, such as KAP, mica, and gypsum are available. As yet there are no commercially available single crystals with a $2d$ larger than that of KAP which is 26.6 Å. It seemed useful therefore to determine whether the Blodgett–Langmuir technique could be used to build a soap film structure having a $2d$ of about 50 Å. A structure with a

* This work was supported by the Air Force Materials Laboratory, Research and Technology Division, Air Force Systems Command, Wright–Patterson Air Force Base, Ohio.

Table I. Carboxylic Acids $C_nH_{2n}O_2$

Common name	IUC name	Number of carbon atoms	Length of molecule (Å)	Melting point (°C)	Water solubility (gm/100 ml)
Melissic	Triacontanoic	30	40	91.9–92.1	
	Noneicosanoic	29			
Montanic	Octacosanoic	28	37.5		
	Heptacosanoic	27			
Cerotic	Hexacosanoic	26	35	87.7	
	Pentacosanoic	25			
Lignoceric	Tetracosanoic	24	32.5	81	
	Tricosanoic	23			
Behenic	Docosanoic	22	30	80.7	
	Heneicosanoic	21			
Arachidic	Eicosanoic	20	27.5	76.3	
	Nondecanoic	19			
Stearic	Octadecanoic	18	25	69.4	0.00029
	Heptadecanoic	17			
Palmitic	Hexadecanoic	16	22.5	64	0.00072
	Pentadecanoic	15			
Myristic	Tetradecanoic	14	20	58	0.0020
	Tridecanoic	13			
Lauric	Dodecanoic	12	17.5	44	0.0055
	Undecanoic	11			
Capric	Decanoic	10	15	31.5	0.015
Pelargonic	Nonanoic	9			
Caprylic	Octanoic	8	12.5	16	0.25
Enanthic	Heptanoic	7			
Caproic	Hexanoic	6	10	−1.5–2.0	0.4
Valeric	Pentanoic	5			
Butyric	Butanoic	4		−7.9	5.62
Propionic	Propanoic	3			
Acetic	Ethanoic	2		16.6	∞
Formic	Methanoic	1		8.4	∞

2d of about 150 Å also would be useful. This paper describes the work done to determine the parameters needed to build multilayer structures from fatty acids both shorter and longer than stearic acid and gives information on their relative quality.

FATTY ACIDS

Stearic or hexadecanoic acid is just one member of the aliphatic carboxylic acid family. These acids are listed in Table I. The longer acids are called fatty acids because they are found in animal fats. Both their common and technical names are given along with some of their physical properties. Consider the solubility of a fatty acid in water. It is seen that as the length of the acid increases, the solubility decreases and vice versa. Building multilayer structures requires the formation of an insoluble monolayer of the acid or soap on a water substrate. Working with an acid or soap longer than stearic acid or a stearate soap does not seem to be a problem. On the other hand an acid shorter in length than stearic acid would have less of a tendency to form a stable monolayer because of the increased tendency for the acid to dissolve. The solubility of an organic substance in water can be reduced by using the "salting out" technique. Salting out may be defined as the addition of a highly soluble inorganic salt to an aqueous solution of an organic

Table II. Parameters Used in Building Structures

Soap	Lead acetate conc.	Sodium acetate conc.	Fatty acid solvent	pH	Temper- ature (°C)	τ (dynes/ cm)	Dipping speed (cm/sec)	Miscellaneous
Lead melissate	1×10^{-5} M	—	Benzene	6.5–6.7	31–33	40	0.15	2×10^{-5} M NaOH 1×10^{-2} M CH$_3$CH$_2$OH
Lead myristate	1.5×10^{-2} M	1.5 M	Benzene or hexane	6.7	15	20.4	0.41	1.8×10^{-2} CH$_3$COOH
Lead laurate	1.5×10^{-2} M	1.5 M	Benzene or hexane	6.5	13	16	0.41	Trace Al^{+3}

compound. Since the solubility of an organic substance in a saturated solution of an electrolyte is almost zero, the net result is to prevent the organic substance from dissolving. This technique has been found to be effective and necessary for the building of multilayers of the soaps of myristic and shorter acids. When a fatty acid is spread on a water substrate, a monolayer is formed. Upon the application of pressure to this monolayer, the molecules orient themselves perpendicular to the water surface with the carboxyl end at the water-air interface. The area of the film is minimized; no open areas are present. The monolayer may exist as a one dimensional gas, liquid, or solid depending on the values of the film pressure and temperature, the substrate pH, the cation present, the concentration of ions in solution, and the presence of polyvalent cations. If conditions are chosen to produce a condensed or solid monolayer of a stearate soap, the film will have a certain rigidity. Replacing the stearate soap by a longer soap will increase the rigidity of the film because of the increase in the H—H bonding between neighboring molecules. Film rigidity is the most important difference between monolayers of a stearate and a lignocerate or longer soap.

EXPERIMENTAL CONDITIONS

Multilayer structures are built up using the Blodgett–Langmuir techniques.[1,10] The method of Sher and Chanley[11] is used to apply pressure to the film. Only lead soaps were studied because, in general, structures containing lead will give the highest diffracted intensities.[3] Earlier work had shown that lead lignocerate ($2d = 130$ Å) structures could be built. In order to diffract silicon L radiation ($\lambda = 135.5$ Å), a structure of the lead soap of cerotic, montanic, or melissic acid could be used. These structures have $2d$ values of approximately 140, 150, and 160 Å, respectively, which means that the silicon L diffraction peak occurs at 151, 139.2, and 116° 2θ, respectively. Since most goniometers have an upper 2θ limitation on about 145°, and since it is of interest to observe the shape of a soft X-ray line which necessitates viewing the entire diffraction peak, a melissate structure seems to be the logical choice. Alternatively, one of the acids containing an odd number of carbon atoms could be used. However, these acids are only rarely found in nature and are difficult and expensive to obtain.

Table II lists the conditions used to build lead melissate structures and those soap film structures containing acids shorter than stearic acid. The main problem with lead melissate is its extreme rigidity. To compensate for this property it is necessary to build the structures at a high temperature and a very slow dipping speed. Putting ethanol in the water substrate increases the solubility of melissic acid in the substrate and permits better structures to be built.

One of the problems with lead myristate and shorter soaps are their increased solubility in water. Sodium acetate is used as a buffer to control the pH of the substrate. For these shorter soaps sodium acetate also can be used to salt out the soap. It is found necessary to increase the lead acetate concentration as well as that of the sodium acetate because the high acetate concentration tends to prevent the lead acetate from ionizing. Once the concentrations are high enough, the monolayer properties and their pickup closely resemble that of lead stearate. The table indicates that almost all parameters are different for each soap studied.

Another problem with monolayers of myristate and shorter soaps is their decreased rigidity. This is due to the decrease in H—H bonding between neighboring molecules as the length of the molecule decreases. For a lead caprate monolayer the rigidity is so low that it is difficult to pick up the film. As yet, lead caprate structures have not been successfully built.

EVALUATION OF STRUCTURES

All structures were evaluated by using them to diffract Al K_α radiation. This work was done on a General Electric XRD-6 vacuum spectrometer. Figure 1 shows the dependence of the diffracted intensity as a function of the number of layers present for lead melissate structures. For a perfect structure, the intensity increases as $1 - e^{-aN}$, where N is the number of double layers and a is a function of the absorption coefficient of the structure. Departures from this functional dependence indicate that imperfections are being built into the structures. For lead melissate structures the quality begins to deteriorate after only 15-20 layers. This is due to the high rigidity of the soap film. However, structures containing more than 20-30 layers are not needed for ultra soft X-rays because structures of this thickness appear essentially infinitely thick to the

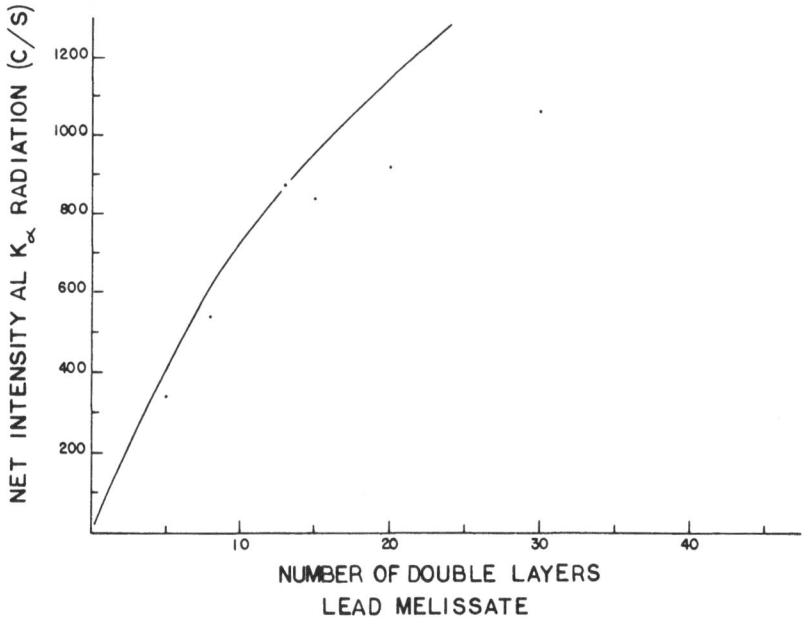

Figure 1. Intensity of Al K_α radiation diffracted by lead melissate structures as a function of the number of double layers present. Background was subtracted.

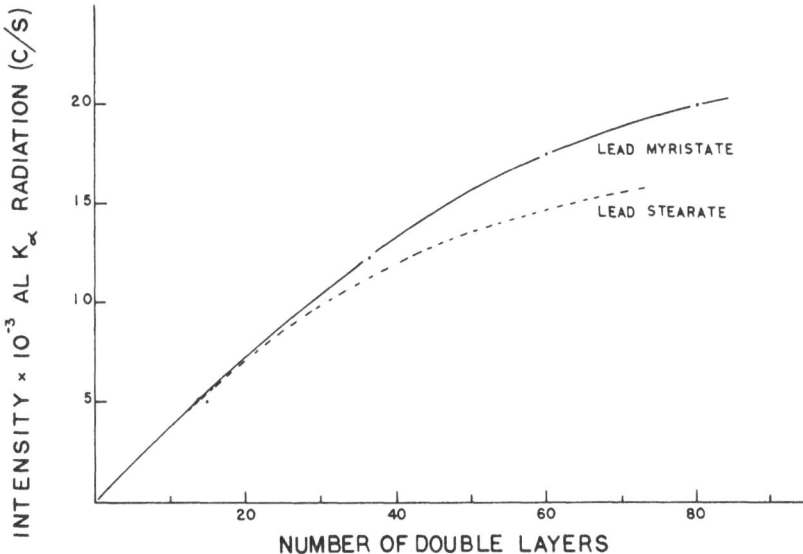

Figure 2. Intensity of Al K_α radiation diffracted by lead myristate structures as a function of the number of double layers present. Similar data for lead stearate structures is given by the dashed line.

radiation. For more energetic radiation lead stearate or myristate structures should be used since good quality structures can be built having a larger number of layers.

Figure 2 shows this data for lead myristate structures with data for lead stearate included for comparison. It is evident that the lead myristate structures have the least imperfections, since they give the highest diffracted intensities and a larger number of layers can be deposited before the maximum diffracted intensity is reached.

Additional comparisons can be made between the various structures. Table III gives the peak intensity as a function of the order of diffraction for aluminum radiation for lead stearate, myristate and laurate structures. The peak intensities were normalized to that obtained in the first order for each particular soap. For each structure the decrease of intensity with increasing order was the same order of magnitude. It is interesting to note that the second and third, fourth and fifth, and sixth and seventh order intensities are the same magnitude.

Table III. Al K_α Intensity vs. Order of Diffraction

Order	Lead stearate	Lead myristate	Lead laurate
1	1	1	1
2	0.35	0.29	0.26
3	0.29	0.22	0.17
4	0.076	0.052	0.042
5	0.075	0.046	0.034
6	0.023	0.013	0.041
7	0.008	0.010	

Table IV. Lead Myristate Widths at Half Maximum and Peak to Background Ratios vs. Number of Double Layers

N	$W^{\frac{1}{2}}$	P/B
6	1.12°	15
20	0.66	49
40	0.5	81
60	0.5	97

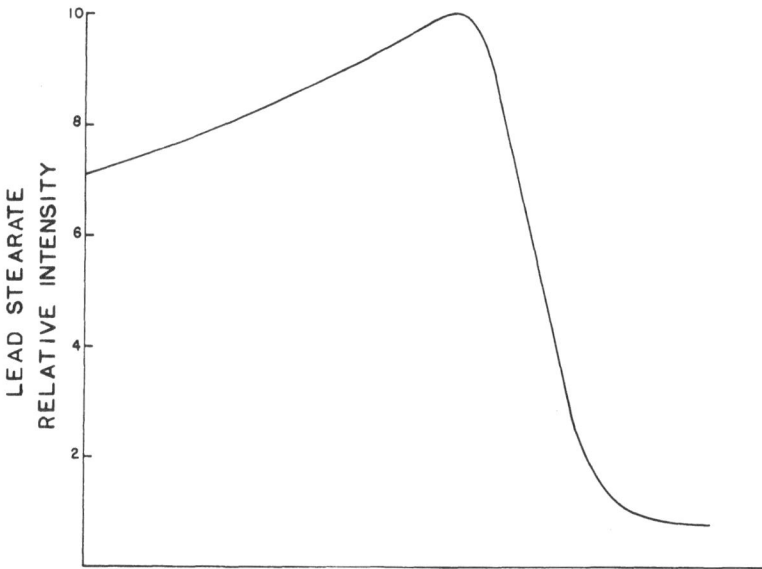

ENERGY ⇨

Figure 3. Relative intensity diffracted by a lead stearate structure as a function of energy in the range 260–290 eV.

Table IV gives data showing how the width of the diffraction peak at half maximum and the peak to background ratio vary with the number of layers in the structure. The data is for a lead myristate soap. Pulse height selection was employed. With a narrower window it is possible to increase the peak to background ratio for a 60-layer structure to 300 : 1. Data has been taken to determine the absorption characteristics of lead stearate and lead myristate soap films in the energy range in which the carbon emission line is found, namely 260–290 eV. Whenever an analyzing crystal contains the element whose characteristic radiation is being observed, the spectra must be corrected for absorption effects taking place within the crystal, since this absorption can appreciably distort the observed spectra. Figure 3 shows the relative absorption for a lead stearate soap film as a function of energy while Figure 4 gives the same type of data for a lead myristate structure. The stearate data is the first order effect. The relative absorption in the second order is given in another paper presented at this conference.[12] The source of continuum for these measurements was a hot tungsten target. It is necessary for the target to be hot to minimize carbon buildup.

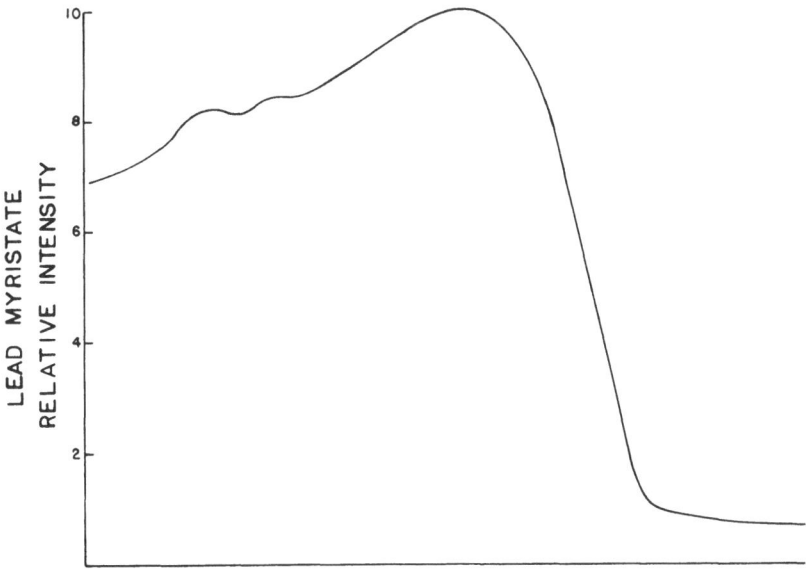

ENERGY ⇨

Figure 4. Relative intensity diffracted by a lead myristate structure as a function of energy in the range 260–290 eV. The structure is due to second order oxygen radiation coming from an oxide contamination of the target.

In practice, a good crystal is one which diffracts a particular radiation with high efficiency or resolution giving a high peak intensity or high peak to background ratio. Although multilayer soap film structures are most useful in diffracting radiation of wavelengths greater than 15–20 Å it is useful to compare a soap film structure with some other crystal using shorter wavelength radiation. A qualitative comparison has been made with Al K_α radiation. The results are listed in Table V. This data was taken using a piece of aluminum filter stock as a sample. Excitation was by a chromium target tube operating at 50 kV and 40 mA. Tight pulse height selection was used: a window 8 V wide was used with a pulse height of 50 V. Although the single crystals were of average quality, the soap film structures were the best that had been made up to the time this data was taken.

Table V. Crystal Comparison Al K_α Radiation

Type	2θ	Peak intensity	P/B	W^1
Gypsum	66.15	680	1350 : 1	0.60°
KAP	36.1	1084	3600 : 1	0.32
EDDT	143.4	4034	1000 : 1	0.47
PET	139.8	5716	420 : 1	0.77
Lead laurate	13.5	3353	350 : 1	0.46
Lead myristate	11.75	3433	350 : 1	0.48
Lead stearate	9.45	3380	350 : 1	0.52

Since multilayer soap film structures are chiefly used to disperse soft X-rays, in normal use they would be used in a vacuum atmosphere. Previous experience with lead stearate structures showed them to have adequate vacuum stability. Soap films built from myristate and shorter soaps have lower melting points and, consequently, would be expected to be less stable in a vacuum. Two lead myristate structures stored in a vacuum and maintained at 40°C for 40 days suffered no deterioration. Two lead laurate structures also were exposed to the same environment. After this time their efficiency in diffracting aluminum K_α radiation was only one-half the original value. This vacuum instability is greater than anticipated since the melting point of lead laurate is only 2.3°C below that of lead myristate. It is possible that the laurate structure had a rather high percentage of lauric acid in it which would evaporate much more readily in a vacuum. On evaporation portions of the structure may have collapsed resulting in the reduced efficiency. No information is available as yet regarding the vacuum stability of the laurates at room temperature. The instability possibly could be reduced by covering the structure with a monolayer of the more stable lead stearate soap. Laurate structures seem to be the practical limit both in using the Blodgett–Langmuir technique to build these pseudocrystals and in using them for the dispersion of soft X-rays in a vacuum spectrometer. Caprate and caprylate salts which have lower melting points would be even more unstable in a vacuum.

The need still exists for a crystal having a $2d$ between that of KAP ($2d = 26.6$ Å) and lead laurate ($2d = 70$ Å). Some work has been done in attempting to build a multilayer structure from a material other than a fatty acid soap. At present 9,10,12,13-tetrabromostearic acid, shown in Figure 5, and pimaric acid, shown in Figure 6, have been studied. Multilayer structures have been built of both materials. The bromine substituted stearic acid was expected to give a structure having a $2d$ of about 50 Å. Instead a poor quality structure with a $2d$ of about 32 Å is obtained. Its quality and $2d$ value are the result of improper orientation of the molecules in both the monolayer and in the multilayer. No diffraction peak was observed with the pimaric acid. Although a structure of the lead salt of pimaric acid was desired, the multilayer probably contained only the acid and not its salt. Lead pimerate structures, if they can be built, would have a $2d$ of about 48 Å.

```
     H
 H C H
 H C H
 H C H
 H C H
 H C H
 H C Br
 H C Br
 H C H
 H C Br
 H C Br
 H C H
 H C H
 H C H
 H C H
 H C H
 H C H
 H C H
   C
O    OH
```

Figure 5. 9,10,12,13-tetrabromostearic acid.

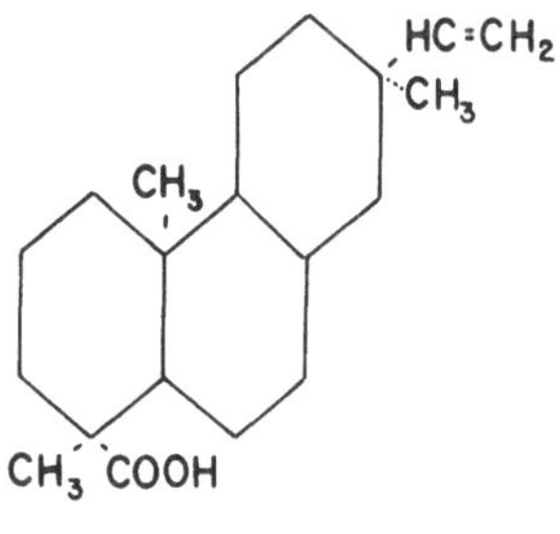

Figure 6. Pimaric acid.

CONCLUSION

The Blodgett–Langmuir technique can be used to build soap film structures from soaps with molecules as long as lead melissate ($2d = 165$ Å) and as short as lead laurate ($2d = 70$ Å). No successful lead caprate structures have been built. Of the various types of soaps, lead myristate gives structures having the highest efficiency in diffracting Al K_α radiation. Lead myristate and longer soaps have adequate vacuum stability for use as an analyzing crystal in a soft X-ray spectrometer. Lead laurate structures can be used provided they are used only for short times and in the vacuum chamber.

ACKNOWLEDGMENT

The authors wish to thank Roger H. Simon and Robert A. Agenten for the technical assistance given them on this project.

REFERENCES

1. K. B. Blodgett, "Films Built by Depositing Successive Monomolecular Layers on a Solid Surface," *J. Am. Chem. Soc.* **57**: 1007, 1935.
2. T. C. Furnas, Jr., and E. W. White, "A Program of Basic Research to Study X-Ray Spectra in the Region 15–50 Å," *WADD Technical Report*, 61–168, 1961.
3. R. C. Ehlert, "The Diffraction of X-Rays by Multilayer Stearate Soap Film Structures," *Advances in X-Ray Analysis, Vol. 8*, Plenum Press, New York, 1965, p. 325.
4. D. W. Fischer and W. L. Baun, "Experimental Techniques for Soft X-Ray Spectroscopy," *WADD Technical Report, RTD-TDR-63-4232*.
5. B. L. Henke, "X-Ray Fluorescence Analysis for Sodium, Fluorine, Oxygen, Nitrogen, Carbon, and Boron," *Advances in X-Ray Analysis, Vol. 7*, Plenum Press, New York, 1964, p. 460.
6. B. L. Henke, "Some Notes on Ultrasoft X-Ray Fluorescence Analysis—10 to 100 Å Region," *Advances in X-Ray Analysis, Vol. 8*, Plenum Press, New York, 1965, p. 269.
7. A. J. Mabis and K. T. Knapp, "Multilayer Soap Films as Analyzing Crystals in X-Ray Spectroscopy," Paper No. 140 presented at the Pittsburgh Conference on Analytical Chemistry and Applied Spectroscopy, March 4, 1964.
8. R. C. Ehlert and R. A. Mattson; "The Characteristic X-Rays from Boron and Beryllium," *Advances in X-Ray Analysis, Vol. 9*, Plenum Press, New York, 1966, p. 456.
9. R. A. Mattson and R. C. Ehlert, "The Application of a Soft X-Ray Spectrometer to Study the Oxygen and Fluorine Emission Lines from Oxides and Fluorides," *Advances in X-Ray Analysis, Vol. 9*, Plenum Press, New York, 1966, p. 471.
10. K. B. Blodgett and I. Langmuir, "Built-Up Films of Barium Stearate and Their Optical Properties," *Phys. Rev.* **51**: 964, 1937.
11. I. H. Sher and J. D. Chanley, "New Technique for Compressing Surface Films," *R.S.I.* **26** (3), 266, 1955.
12. R. A. Mattson and R. C. Ehlert, "Carbon Characteristic X-Rays from Gaseous Compounds," paper presented at Conference on Applications of X-Ray Analysis, Denver, 1966.

DISCUSSION

J. B. Nicholson (Applied Research Labs.): Have you tried adding a short-chain fatty acid to these materials to improve efficiency, as Henke suggested?

R. C. Ehlert: Yes, this doesn't work for us.

J. B. Nicholson: Do you see any differences in the ability to plate mica as opposed to glass?

R. C. Ehlert: We have built lead stearate and lead lignoceric structures on mica.

J. B. Nicholson: How do they compare to glass?

R. C. Ehlert: They are slightly inferior in quality. About 80% as good.

PRODUCTION EFFICIENCIES OF X-RAY EMISSION SPECTRA BY PROTON BOMBARDMENT

A. A. Sterk, C. L. Marks, and W. P. Saylor

American Machine & Foundry Company
Alexandria, Virginia

ABSTRACT

Production efficiencies for the excitation of characteristic soft X-ray emission spectra by proton bombardment were investigated. Results were obtained for K-, L-, and M-shell lines of various elements as a function of proton energies ranging up to 100 keV.

For comparison purposes, helium, nitrogen, and argon ions also were accelerated and used as bombarding particles to yield efficiency data for some of the same spectral lines.

The data demonstrate that:

1. The production efficiencies of spectral lines excited by protons are, in the first approximation, dependent on wavelength only and are independent of the atomic shell with which they are associated.
2. Production efficiencies increase with the atomic weight of the bombarding ions for some of the spectral lines investigated.

X-ray generation by proton bombardment is a powerful method for producing high-intensity monochromatic radiation for wavelengths above approximately 10 Å. The same technique can be used for analytical purposes to determine the concentration of low atomic number elements.

INTRODUCTION

This paper discusses the results of a continuing study of soft X-ray production by proton bombardment. The purpose of the effort has been to develop a more useful soft X-ray (< 10 Å) generator.[1] The fact that characteristic X-ray lines can be produced by proton bombardment with the virtual absence of bremsstrahlung and that the yield increases with proton energy are well established.[1-4] The present data were collected to establish a means of determining the production efficiency or X-ray yield for a given target, wavelength, and proton energy combination. Targets were selected which had K-, L-, and M-shell characteristic lines falling within the range of the detection system (1–20 Å and 44.5 Å).

The data are not intended as absolute cross section measurements for the interactions involved, although, where the data overlap the work done at Lawrence Radiation Laboratory,[2,3] the values compare closely.

APPARATUS AND METHODS

Figure 1 shows the proton accelerator and associated vacuum chamber. An explanation of the proton source used for the present measurements has been discussed elsewhere.[1] There are, however, a number of special features pertaining to the present experimental apparatus and methods which will be discussed briefly.

Referring to Figure 2, the target chamber is the same design as that described by Khan and Potter.[2] The collimator (D) is at 300 V positive potential to prevent backstreaming of secondary electrons from the collimator to the accelerator. The target (F) is held at +90 V relative to the chamber in order to prevent secondary electrons from escaping and, thereby, causing a false indication of beam current.

The X-rays produced by the impinging proton beam are observed at an angle of 90° relative to the beam. After passing through a thin film filter or absorbing foil, the X-rays are detected by a thin window flow proportional counter. A standard gas mixture, 90% argon + 10% methane, is used. The pulses from the counter are then analyzed by a conventional multichannel analyzer.

To provide a good vacuum seal while giving good transmission to the X-ray lines, a 1.1-μ thick, stretched polypropylene film was used as the counter window. The film was supported by an 80%-transmission nickel mesh.

The thin film filters were used to eliminate carbon contamination X-rays (44.5 Å) from the desired data. For the X-ray wavelength range 8–20 Å, a 6-μ aluminum filter was used, and for the X-ray range 1–8 Å, a 6-μ Teflon filter was used. The transmission of these filters is indicated in Figure 3.

The 5.9 keV X-rays from an Fe^{55} source were used periodically to calibrate and check performance of the proportional counter. Figure 4 shows a typical pulse-height spectrum of the counter for copper K- and L-shell X-rays.

A potential source of error in the yield measurement, especially for soft X-rays, arises from the formation of hydrocarbon contamination on the target surface. This contamination comes mainly from the vacuum pumps and O-ring seals. It was found

Figure 1. Proton accelerator and target vacuum chamber.

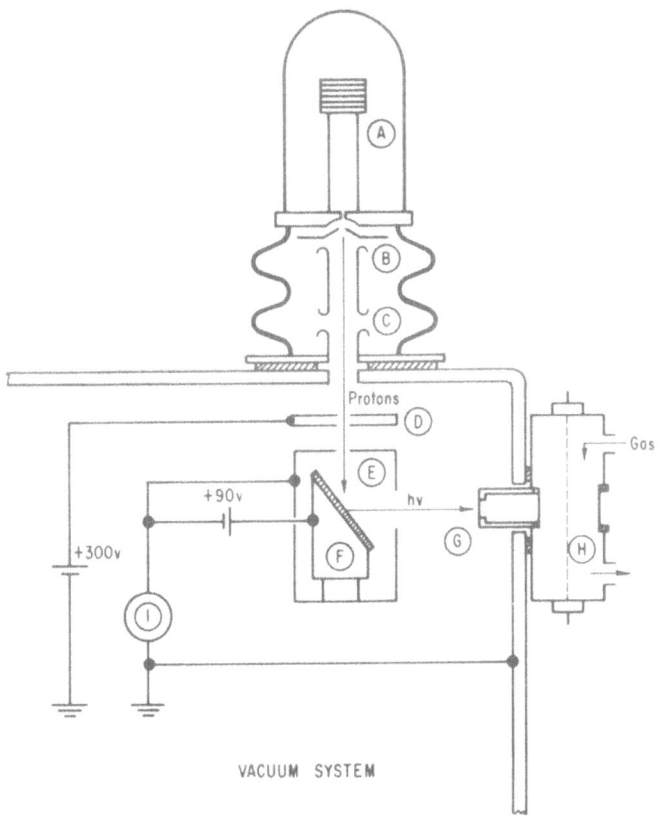

Figure 2. Target chamber: (A) ion plasma bottle; (B) extractor and focusing lens (0–15 kV); (C) main accelerating lens (0–150 kV); (D) collimator and stray electron suppressor; (E) secondary electron shield and collector; (F) target holder; (G) X-ray filter; (H) gas flow proportional counter; (I) target current meter.

that if all measurements were taken within 15 min of initiation of the experiment, the effects of these contaminants on soft X-ray yields was negligible.

DATA

Table I is a list of the target materials and the characteristic spectral lines which were investigated. The data consisted of a pulse height distribution for each target with a simultaneous measurement of proton current at the target. Each line was identified by comparing its pulse height peak with that of the Fe^{55} X-ray line. The count rate under the peak was integrated and divided by the current to give a relative value for X-ray yield. These values were then multiplied by the appropriate geometric and detector efficiency correction factors (as shown in Table I) to give absolute X-ray yield values.

OBSERVATIONS

When the X-ray yield for each target and spectral line is plotted against proton

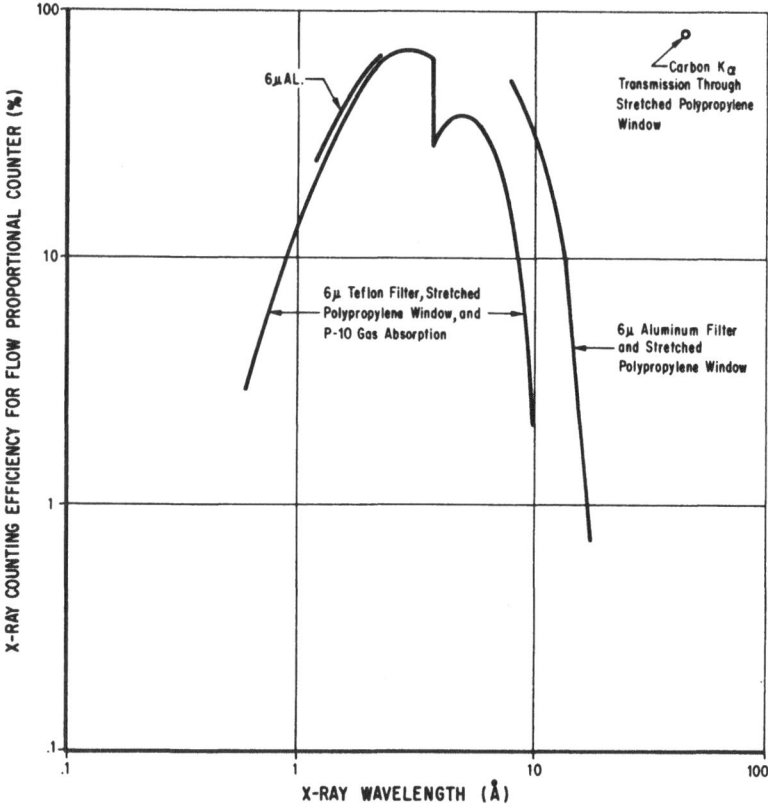

Figure 3. X-ray counting efficiency of flow proportional counter as a function of wavelength. The aluminum filter was used for the copper and iron targets, no filter was used for carbon, and a Teflon filter was used for all other targets.

Table I. Target Elements

Element	Atomic number	Shell	Wavelength (Å)	Correction factor*	X-ray yield† (photons/proton)
Carbon	6	K	44.5	3.35×10^{-8}	1.31×10^{-4}
Aluminum	13	K	8.33	2.85×10^{-9}	2.4×10^{-7}
Vanadium	23	K	2.5	5.55×10^{-10}	9.1×10^{-9}
Iron	26	L	17.6	4.8×10^{-8}	3.3×10^{-5}
		K	1.93	6.45×10^{-10}	3.4×10^{-9}
Copper	29	L	13.3	3.52×10^{-9}	1.97×10^{-5}
		K	1.54	8.8×10^{-10}	1.9×10^{-9}
Molybdenum	42	L	5.4	9.37×10^{-10}	6.1×10^{-8}
		K	0.71	7.4×10^{-9}	5.0×10^{-10}
Tin	50	L	3.6	5.6×10^{-10}	8.2×10^{-9}
Tantalum	73	M	7.2	1.64×10^{-9}	1.35×10^{-7}
		L	1.5	1.13×10^{-9}	1.47×10^{-9}
Lead	82	M	5.2	9.45×10^{-10}	2.1×10^{-8}
		L	1.2	1.83×10^{-9}	1.3×10^{-9}
Uranium	92	M	3.9	1.17×10^{-9}	1.22×10^{-8}
		L	0.91	3.54×10^{-9}	7.9×10^{-10}

* Correction factor $= 4\pi R^2 (1.6 \times 10^{-13} \ \mu a/\text{proton})$ divided by area of counter window; counter window transmission; counter gas absorption.
† This column represents the number of X-ray photons generated per incident proton at 75 keV.

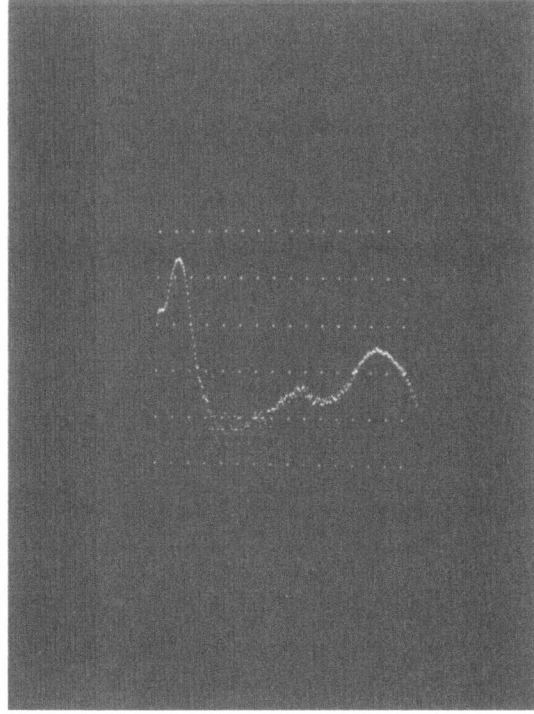

Figure 4. X-ray spectrum observed for thick copper target bombarded with 75 keV protons. From left to right, the first peak corresponds to the Cu L-shell and the third to the Cu K-shell X-rays. Also shown is the escape peak, associated with K-shell X-rays. The vertical scale runs from 1–100,000 counts. The peak to background for the Cu L line is 37000/1.

energy, as shown in Figure 5, all curves indicate a rapidly increasing X-ray yield as the proton accelerating voltage is increased.

When the X-ray yields are plotted against wavelength on log–log paper, with the accelerating voltage held constant, all the data points fall close to a single line (see Figure 6). Thus, the yield is essentially a function of wavelength and approximately proportional to the 3.8 power of the wavelength or

$$N \sim \lambda^{3.8}$$

This would indicate that the X-ray yield is independent of the shell from which the X-ray originates.

When a log–log plot is made of these data as a function of the atomic number Z, it is found that the data fall along three parallel straight lines corresponding to the K-, L-, and M-shell X-ray lines (see Figure 7). The equation for this family of lines is

$$N = k_{K,L,M} Z^{-6.82} \exp(0.062E_p)$$

where N is the X-ray yield in photons/proton and

$$k_K = 0.192 \qquad k_L = 10^2 \qquad k_M = 3.4 \times 10^3$$

The preceding equation represents the proton X-ray yield for proton energies E_p between 25 keV and 250 keV. It can be seen that the yield is strongly dependent on the atomic number of the target material. For any given shell, the yield increases rapidly as the atomic number decreases.

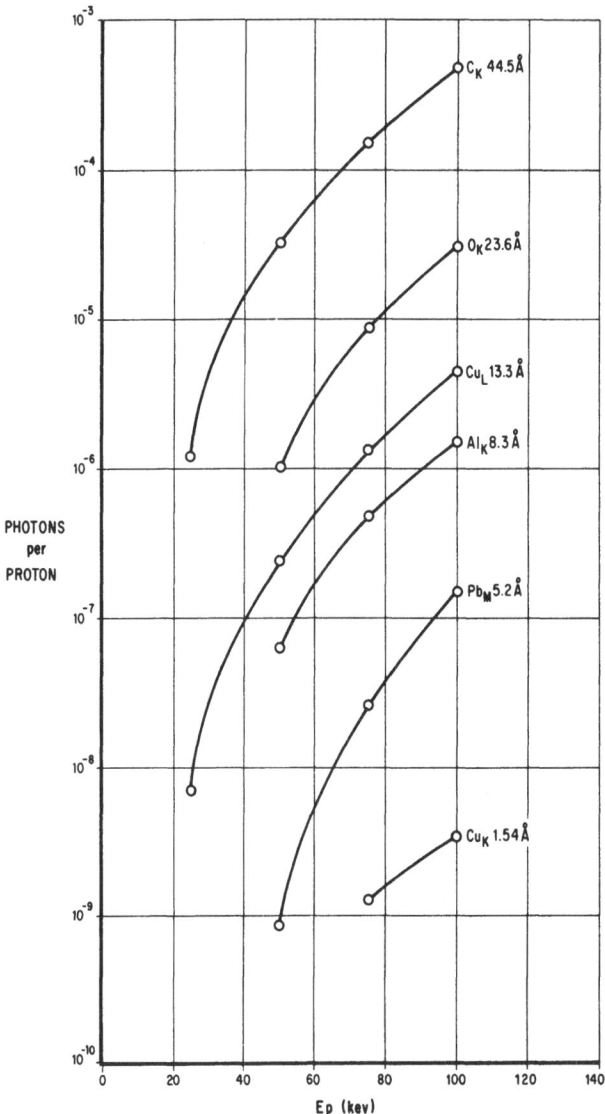

Figure 5. Characteristic K-, L-, and M-shell X-ray yield for various targets as a function of proton energy. (Note that at every proton energy used, the X-ray yield is an increasing function of wavelength.)

X-RAY YIELDS FROM ION BOMBARDMENT

An additional experiment was performed to evaluate the potential of ions other than protons as excitation elements. X-ray yields were measured for helium, nitrogen, and argon ions at the same voltages and for some of the same targets as in the above experiment. The data are shown in Table II.

Two observations can be made from these data. For carbon K-shell X-rays (44.5 Å), yields significantly higher than those for protons were obtained with heaver ions, the

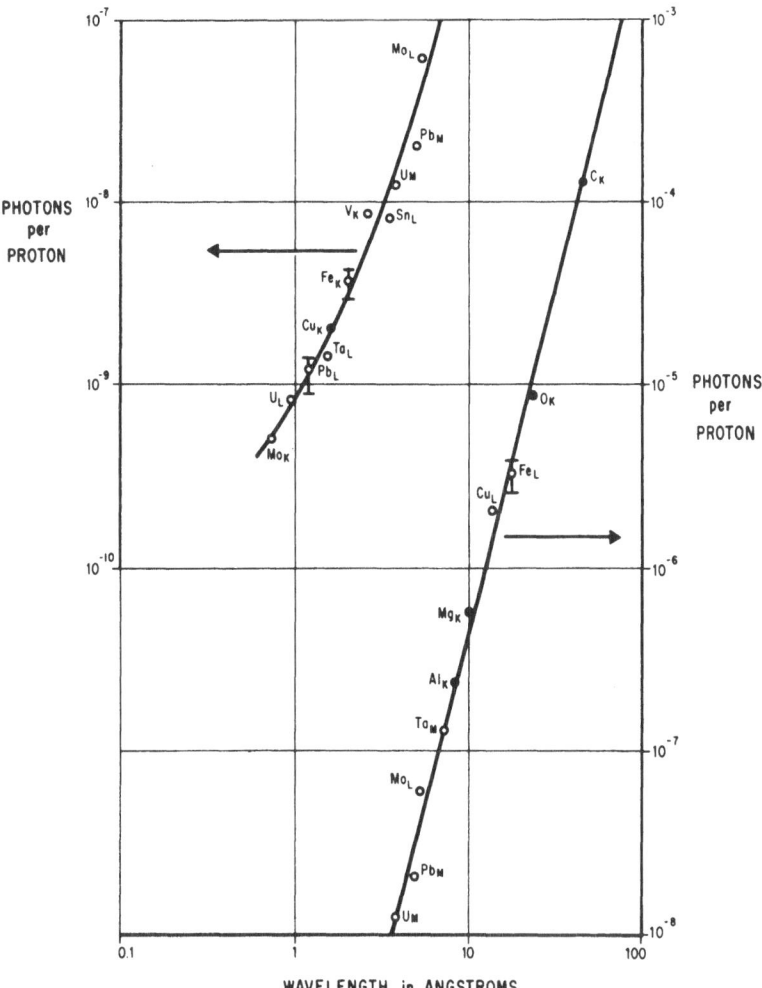

Figure 6. X-ray yield for K-, L-, and M-shell X-rays as a function of wave-length for a proton bombarding energy of 75 keV. (X-ray yields for oxygen and magnesium were obtained from earlier work.[1])

Table II. Characteristic X-Ray Yield for 75-keV Ions

X-ray line	X-ray yield (photons/ion) for indicated particles			
	Protons	Helium ions	Nitrogen ions	Argon ions
Carbon—K	1.5×10^{-4}	1.4×10^{-4}	2.8×10^{-4}	9.8×10^{-4}
Copper—L	1.3×10^{-6}	1.2×10^{-7}	5.5×10^{-8}	7.0×10^{-8}
Copper—K	1.3×10^{-9}	2.1×10^{-9}	3.0×10^{-9}	5.6×10^{-9}
Lead—M	2.5×10^{-8}	—	1.1×10^{-9}	1.4×10^{-9}
Lead—L	9.0×10^{-10}	—	1.3×10^{-9}	2.5×10^{-9}

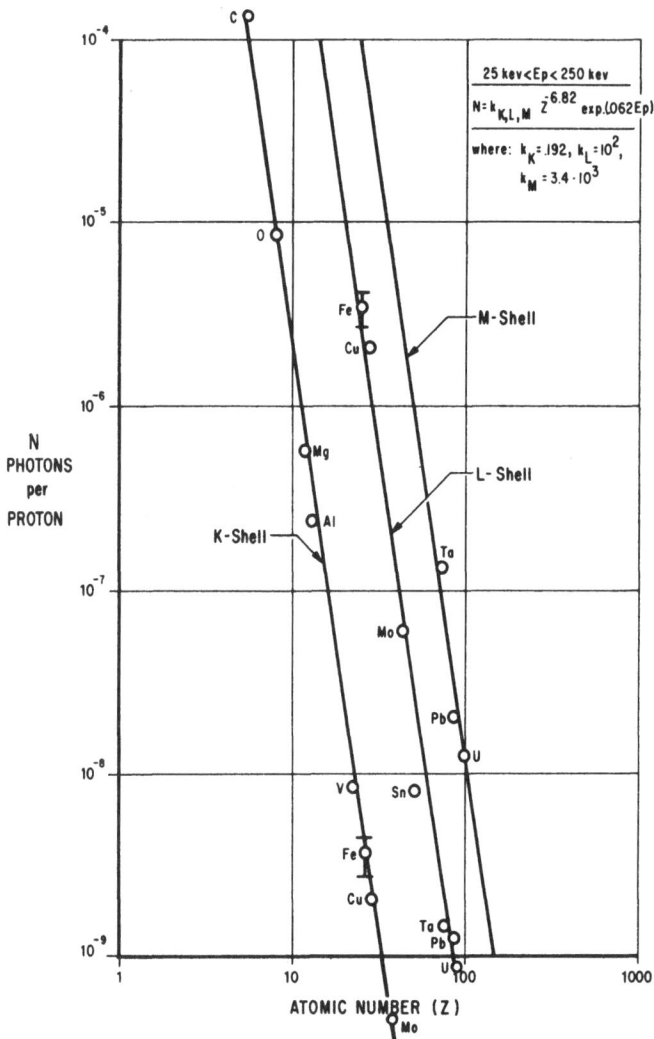

Figure 7. X-ray yield for K-, L-, and M-shell X-rays as a function of atomic number for a proton bombarding energy of 75 keV. The lines connecting the data points are derived by the empirical equation given.

yield increasing with the ion mass. Secondly, the rate of increase of yield with wavelength for a given target is much less than that for proton excited X-rays. For example, the yield for copper L-shell X-rays (13.3 Å) was only a factor of ten greater than that for copper K-shell X-rays (1.54 Å) when argon or nitrogen ions were used compared to a thousand-fold increase for protons.

No attempt was made to analyze the beam to determine the degree of ionization or the amount of nonionized molecules in the beam. However, the results are positive enough to indicate the significant promise in the use of heavy ions for efficient production of soft X-rays. Further evaluation of the technique described is presently being conducted.

CONCLUSION

Probably the most significant feature of the work described in this paper is the demonstration that an approximate production efficiency value for a given spectral line produced by proton bombardment may be obtained from a simple empirical relationship. This relationship is dependent only on the wavelength of the line and the accelerating voltage.

In addition, the results demonstrate that a very high X-ray signal-to-continuum ratio is obtained using the proton bombardment technique. From the copper pulse-height spectrum shown in Figure 4, the ratio of the copper L peak at 13.3 Å to the continuum level at 3.5 Å is 37,000/1 when filter attenuation and counter efficiency are taken into account.

APPLICATIONS

The soft X-ray generator employed in this investigation has found many applications. These may be divided into two general categories.

First as a source of monochromatic X-rays, the generator has been used for the following purposes:

1. The determination of the transmission of ultraviolet filters in the wavelength range 6–90 Å.
2. The development of more efficient X-ray photocathodes for detectors used in spectrometers.
3. The calibration of spectrometers which are designed to analyze the solar X-ray spectrum in the wavelength range 20–100 Å.

Second, the generator is used as an analytical tool where the target itself becomes the object being investigated. The target constituents radiate their characteristic lines which are identified and measured either by dispersive or nondispersive techniques. Presently, corrosion films for nitrogen and oxygen concentrations are being analyzed by this method.

ACKNOWLEDGMENT

The authors wish to thank Mary Spalding for her help in conducting the experiment and in the data reduction.

REFERENCES

1. A. A. Sterk, "X-Ray Generation by Proton Bombardment," *Advances in X-Ray Analysis, Vol. 8*, Plenum Press, New York, 1965, pp. 189–197.
2. J. M. Khan and D. L. Potter, "Characteristic K-Shell X-Ray Production in Magnesium, Aluminum, and Copper by 60- to 500-keV Protons," *Phys. Rev.* **133**: A890, 1964.
3. J. M. Khan, D. L. Potter, and R. D. Worley, "Studies in X-Ray Production by Proton Bombardment of C, Mg, Al, Nd, Sm, Gd, Tg, Tb, Dy, and Ho," *Phys. Rev.* **139**: A1735, 1965.
4. R. C. Jopson, H. Mark, and C. D. Swift, "Production of Characteristic X-Rays by Low-Energy Protons," *Phys. Rev.* **127**: 1612, 1962.

DISCUSSION

J. B. Nicholson (Applied Research Labs.): Do you notice any sputtering of the films when you are bombarding them with the ions? Does this affect precision?

W. P. Saylor: We generally are using thick targets where sputtering doesn't bother us.

J. B. Nicholson: Do you see any problems with precision?

W. P. Saylor: For thin films there probably would be.

J. P. Nicholson: Not for the bulk?

W. P. Saylor: No, not in our measurements. We normally use fairly low currents in the range of 100–200 μA.

J. P. Nicholson: Are your crystals flat or curved?

W. P. Saylor: They are flat.

J. E. Holliday (U.S. Steel): Just a comment that might help you on your carbon contamination. We found that if you thoroughly degas your target before you start exciting it, you will have a lot less trouble with carbon contamination because the gases in the metal, or in the target you are looking at, contain carbon monoxide, which, under action of the beam, will deposit carbon. We found that this will help greatly. Maybe it will help you.

Chairman L. Vassamillet: Are you able to compare the excitation efficiency between protons and electrons?

W. P. Saylor: There probably is some means of doing this. We haven't.

L. Vassamillet: Have you any idea of the heating that goes on in the samples under the bombardment?

W. P. Saylor: We are using a heavy anode holder which is a good heat conductor to minimize the effect. About 20–50 W are dissipated in this region.

L. Vassamillet: Over how much of an area?

W. P. Saylor: Approximately $\frac{1}{2}$-in. beam size diameter. We haven't encountered any problems arising from excessive heat generation. If higher currents were used, a heat dissipation problem might arise.

ELECTRON MICROPROBE ANALYSES AND X-RAY DIFFRACTION STUDY OF SrSi$_2$

G. M. Faulring and E. S. Malizie

Union Carbide Corp.
Niagara Falls, New York

ABSTRACT

Although the phase SrSi$_2$ has been previously recognized, it has been described only briefly in the literature.

An alloy comprised of FeSi$_2$, silicon, and the SrSi$_2$ phase was examined on the electron microprobe. Optical and electron scanning images are correlated with X-ray scanning images for iron, silicon, calcium, and strontium. Quantitative microprobe analyses were made and corrected for absorption and the atomic number effect by several noncomputer methods. The corrections include as variables the X-ray intensities measured at several accelerating voltages. The effects of varying the electron beam size and accelerating potential are included. The results are compared to chemical analyses. The advantages of varying the accelerating voltage when correcting intensity data, increasing the beam size when surface preparation is a factor, and the importance of surface preparation at low accelerating voltages are discussed.

An X-ray diffraction examination showed that the phase SrSi$_2$ has a cubic unit cell with an a_0 of 6.515 Å. There are four molecules per unit cell, and the most probable space group appears to be P2$_1$3. The density was calculated as 3.45 (observed density >3.3).

Metallographic observations with ordinary and polarized light and microhardness measurements are included.

INTRODUCTION

In the course of alloy development studies in the Research and Development Department of Union Carbide (Mining and Metals Division), a strontium silicide (SrSi$_2$) phase was produced, the structure of which could not be correlated with published data.

A survey of the literature showed that the intermetallic phases SrSi and SrSi$_2$ were produced in 1900 by Bradley,[1] and in 1932 by Wöhler and Schuff.[2] Bradley's method of preparation was the carbon reduction of SrO$_2$ and SiO$_2$ in an electric arc furnace. However, sufficient data to characterize strontium silicide phases were not included in these studies because of the experimental techniques available at the time of publication. In 1962, Roctäschel and Weiss[3] reported crystallographic data for SrSi and established the existence of the phase Sr$_2$Si. These investigators also produced SrSi$_2$ by heating a stoichiometric mixture of strontium and silicon under argon at 1180°C, with SrF$_2$ as a flux. They presented chemical analyses and noted the SrSi$_2$ phase was relatively stable in air, but did not include data necessary for its identification. Their primary interests were in the strontium-rich side of the binary system Sr–Si.

The purpose of this study is to establish parameters by which the phase $SrSi_2$ may be identified by X-ray diffraction. Electron microprobe analyses, metallographic and microhardness data are included.

SAMPLE PREPARATION AND DESCRIPTION

The alloy was prepared in our laboratory by a method similar to that of Bradley's [carbon reduction of SrO_2 and SiO_2 in an arc furnace (1600–1700°C) and cooled in air]. Iron and calcium were the major impurities in the alloy. On fractured surfaces, the silicon grains appeared to be blue in contrast with the bright silvery color of the strontium silicide.

CHEMICAL ANALYSES

The amounts of iron, calcium, strontium, acid-soluble and insoluble silicon, as well as total silicon, were determined by chemical analyses. The acid-soluble silicon was considered to be combined with the strontium–calcium and iron, and the acid-insoluble silicon to be free silicon. Since only multiphase samples were available, a direct chemical determination of the silicon combined with the strontium and calcium was not possible. The results of these analyses are shown in Table I.

Table I. Chemical Analyses of the Strontium–Silicon Alloy and Chemical Analyses of the $SrSi_2$ Phase

Chemical Analyses of the Strontium–Silicon Alloy				
Si (total)	Si (acid-soluble)	Sr	Ca	Fe
68.4*	22.0	22.6	1.5	3.5

Chemical Analyses of the $SrSi_2$ Phase (Derived from Chemical Analyses of the Alloy)		
Si	Ca	Sr
39.0	1.5	59.5

* All values wt.%

METALLOGRAPHIC EXAMINATION

The sample was polished with diamond abrasives and examined metallographically immediately after polishing. When the sample was allowed to stand in air for a few hours, a stain developed on the polished surface of the strontium silicide that was difficult to remove. In the photomicrograph, Figure 1, three phases are shown: (a) the darkest gray constituent identified as silicon, was characterized by a somewhat idiomorphic shape and low reflectivity; (b) the intermediate gray phase, identified as $FeSi_2$, occurred as needles in the matrix and was anisotropic; and (c) the lightest gray or matrix phase, identified as $SrSi_2$, was isotropic which made it easily distinguishable from the $FeSi_2$. (The identity of these phases is verified in subsequent sections of this report.) Under a 43-g load, the Vickers pyramid hardness of the silicon phase was 950, the $FeSi_2$ phase 800, and the $SrSi_2$ matrix 1250.

Figure 1. Photomicrograph of the Sr–Si-rich alloy (reflected ordinary light) (500 ×).

Table II. X-Ray Diffraction Powder Data of the Phase SrSi$_2$

d_{obs}	I_{obs}	hkl	d_{calc}
4.60	23	110	4.61
3.75	12	111	3.76
2.91	100	210	2.91
2.66	55	211	2.66
2.30	7	220	2.30
2.17	14	221	2.17
1.96	37	311	1.96
1.879	11	222	1.881
1.802	11	320	1.807
1.741	63	321	1.741
1.580	10	410	1.580
1.536	10	330	1.536
1.496	17	331	1.495
1.421	21	421	1.422
1.390	7	332	1.389
1.278	13	431	1.278
1.256	2	333	1.254
1.210	17	520	1.210
1.189	17	521	1.190
1.154	6	440	1.152
1.087	16	600	1.086
1.071	7	610	1.071
1.057	13	611	1.057
1.030	5	620	1.030

$a_0 = 6.515$ Å.
Density observed, >3.3 g/cc.
Density calculated, 3.45 g/cc.
4 molecules per unit cell.
Possible space group P2$_1$3.

X-RAY DIFFRACTION EXAMINATION

A crushed sample of the alloy was placed in methylene iodide, and the $SrSi_2$ phase was concentrated on the basis of its higher density. Only a small amount of the total sample occurred in the sink fraction (density >3.3 g/cc) since most of the silicon was not freed from the $SrSi_2$ phase by crushing. An optical examination showed that, although the sink fraction consisted primarily of the $SrSi_2$ phase (visually estimated $>95\%$), it also contained a few small particles of attached silicon and $FeSi_2$.

X-ray diffraction powder patterns were obtained of the sink fraction with a diffractometer (Cu K_α) and a Straumanis-type powder camera (114.6 mm diameter—Fe K_α). The X-ray reflections of silicon were used for calibration of the film. The interplanar spacings and the integrated intensities (Cu K_α) of the $SrSi_2$ phase are shown in Table II.

The X-ray diffraction pattern of the $SrSi_2$ phase was indexed as cubic, $a_0 = 6.515$ Å. This is in agreement with the optical observation of isotropy. Assuming there are four molecules per unit cell, the calculated density is 3.45 g/cc. It was previously noted that the density of the $SrSi_2$ phase is greater than 3.3 g/cc (sink in methylene iodide). The space group is believed to be $P2_13$.

The $FeSi_2$ was identified by X-ray diffraction only after its concentration by removing the $SrSi_2$ phase from the sink fraction with acid (HCl).

ELECTRON MICROPROBE DATA AND OBSERVATIONS

Figure 2a is an optical photomicrograph of the area examined in the electron microprobe. The black areas are part of the contamination raster that was not removed

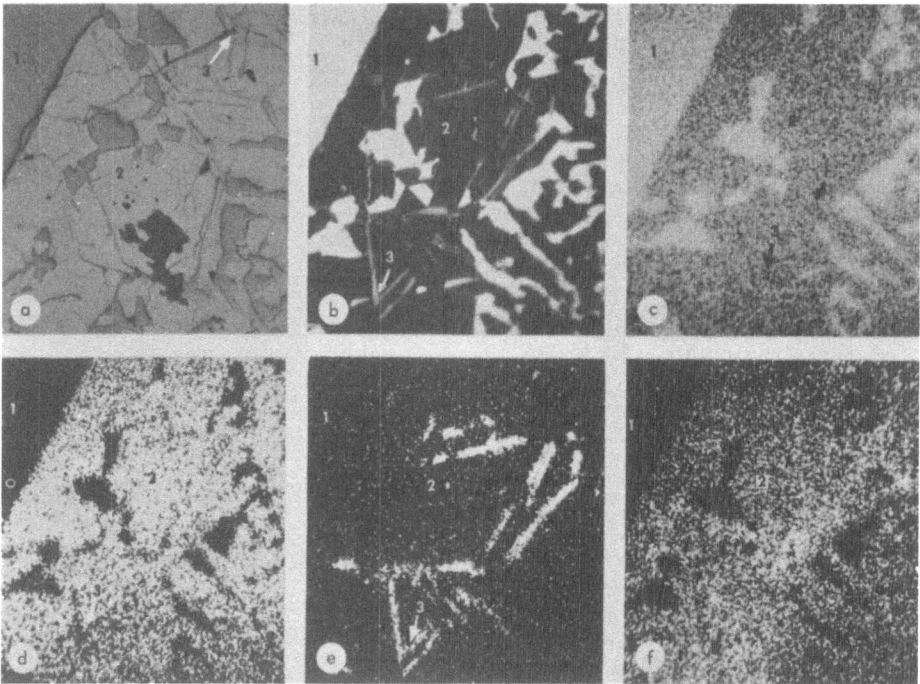

Figure 2. Six views containing the same area in a specimen of Sr–Si-rich alloy. (a) Optical photomicrograph, 200 ×; (b) electron scanning image, 250 ×; (c) silicon distribution, 250 ×; (d) strontium distribution, 250 ×; (e) iron distribution, 250 ×; (f) calcium distribution, 250 ×.

by the light polish after probing. The numbered phases indicated in Figure 2a are (1) silicon; (2) $SrSi_2$; and (3) $FeSi_2$.

The electron scanning image of the specimen surface (Figure 2b) shows contrast by differences in average atomic number and topography. Since the surface of the diamond-polished specimen had little relief, the contrast results almost entirely from differences in average atomic numbers of the phases. It may be noted that the silicon (area 1) appears white (at. no. 14), the $SrSi_2$ (area 2) is black (average at. no. 22), and the $FeSi_2$ (area 3) is gray (average at. no. 18).

The X-ray scanning images in Figures 2c–f show the distribution of silicon, strontium, iron and calcium, respectively, across the area in Figure 2b. The areas most densely populated with white dots indicate enrichment in the selected element in comparison to areas less densely populated on the same micrograph.

The high silicon in the silicon grains (area 1), relative to the amounts in the $SrSi_2$

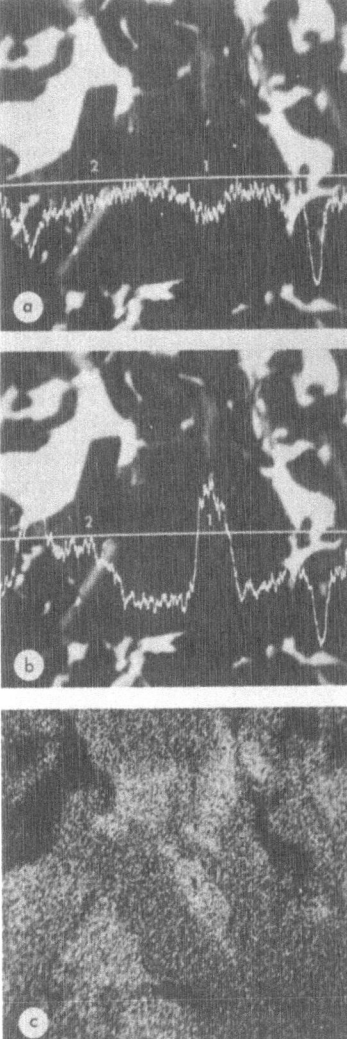

Figure 3. $SrSi_2$ phase with varying calcium and strontium contents (500×). (a) Linear traverse: strontium; (b) linear traverse: calcium; (c) calcium distribution.

phase (area 2) and the $FeSi_2$ (area 3), may be seen in Figure 2c. A comparison of Figure 2d and f shows that strontium and calcium occur chiefly in the matrix or $SrSi_2$ phase and Figure 2e shows that the iron is largely concentrated in the acicular particles ($FeSi_2$).

The measured intensities of calcium and strontium, although different for the two elements, were constant (within experimental variation) across most of the $SrSi_2$ areas examined. X-ray and electron scanning images of one of the few areas found to be an exception may be seen in Figure 3. Superimposed on the electron images are intensity traces characterizing the relative amounts of strontium (Figure 3a) and calcium (Figure 3b) along the linear traverse of the electron beam. The variations in calcium distribution across the total area may be seen in Figure 3c. The pulse counts characteristic of silicon, calcium, and strontium were measured between the numbered points and corrected for background. Absorption corrections were also made. It was found that (a) the silicon content is constant, (b) the calcium-rich area (about 3% Ca) is surrounded by a calcium-poor region (about 1% Ca), and (c) the strontium plus calcium content is constant for all the areas examined. These observations indicate the substitution of calcium for strontium in the $SrSi_2$ phase. No attempt was made, however, to determine the limits of substitution. The nonuniformity of calcium distribution was noted in less than 10% of the specimen examined.

QUANTITATIVE ELECTRON MICROPROBE ANALYSES

Some of the factors that may influence the results in a quantitative determination of the elements in a constituent are: (a) surface irregularities of the specimen, (b) accelerating voltage and electron beam size of the electron microprobe, and (c) mass absorption coefficients and the calculation method used to correct for absorption.

Specimen Descriptions

The intensities of the Sr L_{α_1} and Si K_α X-rays generated in an area of the $SrSi_2$ phase, considered representative of the bulk of the specimen, were compared to X-ray intensities generated in known standards. The calcium content was not determined in the $SrSi_2$ phase with the electron microprobe, but was assigned an average value of 1.5% based on chemical analyses. The grains of silicon in the alloy and a homogeneous sample of chemically analyzed strontianite ($SrCO_3$ containing 54.4% Sr and 5% Ca) were used as standards for the silicon and strontium. Two strontianite specimens were used, one of which contained some minute voids or cracks and its polished surface was slightly irregular. The second was free of voids and no relief effects could be detected on its polished surface. All specimens were polished with a diamond abrasive and then coated with carbon.

Experimental Procedure and Data

The intensities of the Si K_α and Sr L_α X-rays were measured several times for varied intervals in the alloy and standards. The X-rays were generated with a constant beam flux of 0.2 μA, a 1-μ-diameter electron beam, and accelerating voltages of 10, 15, 20, and 25 kV. The peak intensities were corrected for the intensity of the background. The percentages of silicon and strontium calculated from a direct ratio of the measured intensities are presented in Table III and Table IV, respectively.

Since Moll[4] had previously reported that X-ray intensities do not change as the electron beam is defocused, if the focusing conditions are maintained constant and the sample is homogeneous, it was considered that defocusing the electron beam might

Table III. Electron Microprobe Analyses of Silicon in the SrSi$_2$ Phase

Silicon (wt.%)
(Liebhafsky *et al.*—Mass Absorption Coefficients)

kV	(1)*	(2)	(3)	(4)	(5)
10	40.8	41.6	44.9	41.3	44.3
15	38.7	41.1		40.2	44.7
20	36.3	40.5	43.2	38.8	44.6
25	33.7	39.8	42.9	37.4	43.8

(Theisen—Mass Absorption Coefficients)

kV	(6)	(2)	(3)	(5)
10	43.2	41.7	44.5	44.4
15	44.2	41.2		44.8
20	45.4	40.8		45.0
25	45.9	39.2	43.5	44.5

* Calculation Method:
  (1) Ratio of measured intensities (Si K_α ... SrSi$_2$/Si)
  (2) Philibert
  (3) Birks
  (4) Ziebold and Ogilvie
  (5) Duncumb and Shields
  (6) Theisen

minimize the effect of the slightly irregular surface on one of the strontianite specimens. The area selected for examination did not contain any voids that could be noted optically or on an electron scanning image. Several of the SrSi$_2$ phase grains were about $\frac{1}{2}$ mm across. A defocused beam size of 50 μ diameter was arbitrarily selected and the Sr L_{α_1} X-rays from the SrSi$_2$ phase and the strontianite were measured at 10, 15, 20, and 25 kV. For comparison, the Sr L_{α_1} X-rays from the void-free strontianite specimen also were produced with the defocused electron beam and accelerating voltages of 10 and 15 kV. Ratios of the measured intensities (alloy/standard) are presented as the percentage of strontium in Table IV.

The intensities of the Sr L_{α_1} measured with beam diameters of 1 and 50 μ in the SrSi$_2$ phase and the void-free strontianite specimen were nearly constant for each voltage. However, the intensities measured from the strontianite with the irregular surface showed variations (approximately 15% with 10 kV, 10% with 15 kV, and 5% with 20 and 25 kV) indicating that the effect of surface irregularities is most pronounced at lower accelerating potentials. This might have been anticipated since more of the X-rays are generated nearer the surface.

Mass Absorption Coefficients

A nonlinear relationship exists between the measured X-ray intensities and concentration. This is evident in the percentages assigned silicon and strontium based on direct ratios of the measured intensities in Tables III and IV. Nonlinearity is a result of such phenomena as absorption of the emitted X-rays by the specimen itself, excitation of secondary X-rays, and a difference in atomic number of the elements in the specimen. The mass absorption coefficients tabulated by Heinrich,[5] Theisen,[6] and Liebhafsky *et al.*[7] are presented in Table V. These are included, since discrepancies in published mass absorption coefficients may produce uncertainties in analytical results. The largest mass absorption coefficients for Si K_α in silicon and the SrSi$_2$ phase, and Sr L_{α_1} in

Table IV. Electron Microprobe Analyses of Strontium in the $SrSi_2$ Phase
Strontianite Standard—Irregular Surface
(Electron Beam—1 μ Diameter)

| | Strontium (Wt.%) | | | | | | | | |
| | Liebhafsky *et al.*—m.a.c. | | | | | Theisen—m.a.c. | | | |
kV	(1)*	(2)	(3)	(4)	(5)	(6)	(2)	(3)	(5)
10	54.4	50.6	53.4	52.6	53.1	53.0	50.0	53.0	52.5
15	59.4	54.5		57.2	56.9	56.1	53.6		55.7
20	64.5	58.3	62.5	62.3	60.7	58.6	56.9	60.8	58.8
25	68.0	60.5	65.2	66.0	63.2	58.5	58.5	62.7	60.3

(Defocused Electron Beam—50 μ diameter)

kV	(1)*	(2)	(3)	(4)	(5)	(6)	(2)	(3)	(5)
10	62.4	58.2	61.3	60.0	60.9	60.7	57.3	60.8	60.1
15	62.7	57.6		60.6	60.1	59.2	56.6		58.7
20	66.4	60.1	64.3	63.0	62.5	60.3	58.6	62.0	60.5
25	72.5	64.5	69.5	70.0	66.9	62.5	62.4	68.2	64.3

Strontianite Standard—Relief-free Surface
(Electron Beam—1 μ diameter)

| | Liebhafsky *et al.*—m.a.c. | | | | | Theisen—m.a.c. | | | |
kV	(1)*	(2)	(3)	(4)	(5)	(6)	(2)	(3)	(5)
10	60.1	55.9	59.0	58.1	58.6	58.5	55.2	58.5	58.0
15	61.0	56.0		58.7	58.4	57.6	55.0		57.2

(Defocused Electron Beam—50 μ diameter)

kV	(1)*	(2)	(3)	(4)	(5)	(6)	(2)	(3)	(5)
10	61.4	57.1	60.3	59.4	60.0	59.8	56.5	59.8	59.3
15	61.4	56.4		59.2	58.9	58.0	55.4		57.6

* Calculation Method:
(1) Ratio of measured intensities Sr $L\alpha$ . . . $SrSi_2$ phase/strontianite
(2) Philibert
(3) Birks
(4) Ziebold and Ogilvie
(5) Duncumb and Shields
(6) Theisen

the strontianite and the $SrSi_2$ were taken from the tables of Liebhafsky *et al.* and the smallest from Theisen. Since these two sets of data represented the maximum and minimum values, they were used in subsequent calculations. The decrease in the percentages assigned silicon and the increase in the strontium with an increase in voltage (see column 1 in Tables III and IV) correlated with the mass absorption coefficients suggests that absorption is the primary correction to be considered, particularly for the silicon determination.

Absorption Correction Methods

The measured intensities of Si K_α (1 μ beam) and Sr L_{α_1} (1 and 50 μ beam) were corrected by the method of Philibert[8] and the absorption correction methods of Birks,[9] Ziebold and Ogilvie,[10] Duncumb and Shields,[11] and Theisen. The correction method of Philibert includes an atomic number factor which, to a limited extent, considers a difference in the atomic number whereas it is eliminated from the modifications of Philibert suggested by Ziebold and Ogilvie, Duncumb and Shields, and Theisen. (Only the absorption correction factors of the methods suggested by Ziebold and Ogilvie, Birks, and Theisen were used in this study.) Most of the calculations were repeated using the mass absorption coefficients of Theisen as well as those of Liebhafsky *et al.* The results of these calculations are compared in Tables III and IV. The relative amounts of error for each accelerating voltage and calculation method are represented schematically in Figures 4–6. Liebhafsky *et al.* mass absorption coefficients were used in all of the calculation methods represented with the exception of that of Theisen's. The percentage of error is defined as

$$\frac{\text{measured concentration—true concentration}}{\text{true concentration}} \times 100$$

where the true concentration is assigned the percentages determined chemically.

An examination of the silicon percentages (Table III) and the graphical representation of the error in the silicon percentages (Figure 4) shows that the silicon percentages derived from a direct ratio of the measured intensities are most accurate when generated with the lower accelerating voltages, and the percentages calculated using the method of Philibert or the absorption correction method of Ziebold and Ogilvie are most similar to those determined chemically.

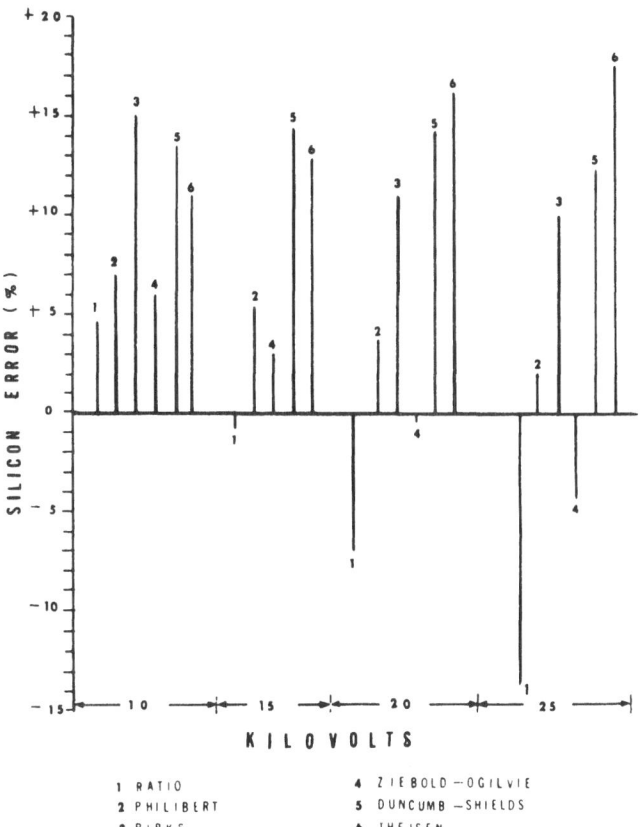

1 RATIO   4 ZIEBOLD—OGILVIE
2 PHILIBERT   5 DUNCUMB—SHIELDS
3 BIRKS   6 THEISEN

Figure 4. Graph of errors in the silicon percentages from intensities generated with several voltages and a microelectron beam.

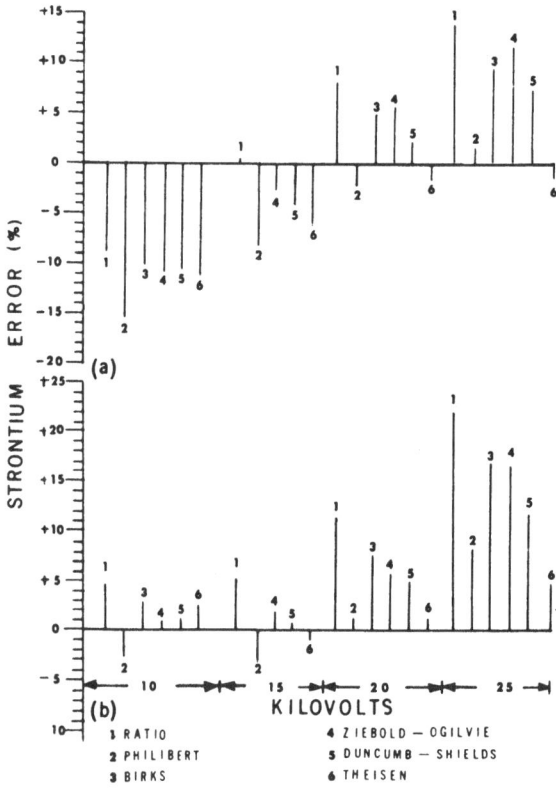

Figure 5. Graph of error in the strontium percentages from the SrSi$_2$ phase—(strontianite standard with an irregular surface). (a) Micro-electron beam, and (b) defocused electron beam.

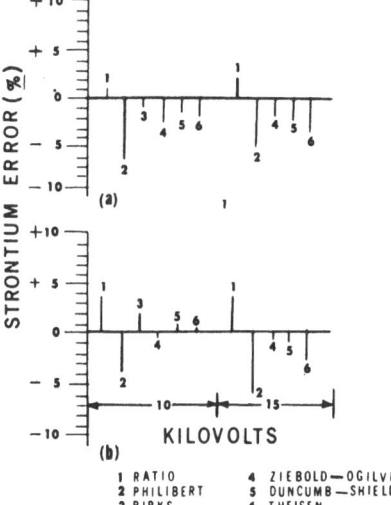

Figure 6. Graph showing error in strontium percentages in the SrSi$_2$ phase—(strontianite standard relief-free surface). (a) Microelectron beam, and (b) defocused electron beam.

An examination of Figures 4–6 and Tables III and IV shows the method of calculation is more important than the mass absorption coefficients considered and at the lowest accelerating voltage, the silicon percentages are larger than determined chemically, whereas the strontium percentages calculated from the micro-electron beam data are less. These percentage differences suggest that an atomic number correction should

Table V. Comparison of Mass Absorption Coefficients for Si K_α and Sr L_{α_1}

	Theisen[6]	Heinrich[5]	Liebhafsky et al.[7]
Si K_α in Si	300	328	398
Si K_α in Sr	945	1092	1072
Si K_α in Ca	1150	1086	1145
Si K_α in SrSi$_2$ (1.5% Ca)	706	744	810
Sr L_{α_1} in Sr	900	990	1178
Sr L_{α_1} in Si	290	296	530
Sr L_{α_1} in Ca	1080	980	1066
Sr L_{α_1} in C	345	321	326
Sr L_{α_1} in O	875	869	886
Sr L_{α_1} in SrSi$_2$ (1.5% Ca)	665	719	934
Sr L_{α_1} in SrCO$_3$ (5% Ca)	808	900	1009

improve their correlation with chemical analysis (Archard and Mulvey[12]). However, the absorption corrected strontium percentages calculated from intensities generated with a defocused electron beam and 10 and 15 kV are slightly greater than chemical analyses indicating that the lack of correlation may be related to surface preparation or absorption.

Surface Irregularities

The strontium percentages (Table IV) or the percentage strontium error graph (Figure 5) shows that the effect of an irregular surface might be minimized by using a defocused beam, and low accelerating voltages (10–15 kV). This may be attributed to an averaging of the effect produced by surface irregularities. This procedure is applicable only when the region being examined is homogeneous and has a volume large enough to contain the penetrating electrons and the generated X-rays. A comparison of the graphs in Figures 5 and 6 shows the importance of obtaining a surface free of relief on microprobe specimens for quantitative analyses. It also may be noted that the intensity ratios of the Sr L_{α_1} X-rays generated with the defocused electron beam and 10 and 15 kV are nearly constant. This suggests a smaller increase in electron penetration into the specimen when the accelerating voltage is increased from 10–15 kV and a defocused electron beam is used instead of a microelectron beam. (The total intensity of the 1 and 50 μ electron beam is equivalent.)

SUMMARY

The phase SrSi$_2$ has been previously recognized but not described in the literature. An alloy comprised of silicon, FeSi$_2$, and the phase SrSi$_2$ was examined metallographically. Optical observations, microhardness measurements, and chemical analyses of the alloy are included.

An X-ray diffraction examination established that the SrSi$_2$ phase is cubic, $a_0 = 6.515$ Å, and the probable space group is P2$_1$3. The calculated density is 3.45 g/cc and the observed, > 3.3 g/cc.

Electron microprobe photographs of two areas of the alloy are presented. One of the areas was selected to show the variable substitution of calcium for strontium in the SrSi$_2$ phase.

The Sr L_{α_1} and Si K_α X-ray intensities generated in the SrSi$_2$ phase and standards were measured using several accelerating potentials. The Sr L_{α_1} X-rays were generated with both a fine and a defocused electron beam. The effect of specimen irregularities on intensity measurements also was considered.

The X-ray intensities were corrected for absorption by several methods with mass absorption coefficients from different sources and converted to percentages. The results of this examination suggest that (a) the effect of an irregular surface on a specimen is most pronounced at lower accelerating voltages, (b) a direct ratio of measured intensities is more accurate when generated with the lower accelerating voltages, (c) if an irregular surface cannot be avoided and the specimen contains sufficiently large grains, a defocused beam at low accelerating voltage should be considered, and (d) in this study, the method of absorption correction was more important than the source of the mass absorption coefficients.

ACKNOWLEDGMENTS

The authors wish to thank Dr. W. D. Forgeng and Mr. J. L. Lamont of the Research and Development Department, Mining and Metals Division, Union Carbide Corporation, for their assistance.

REFERENCES

1. C. S. Bradley, "New Silicon Compounds Obtained by the Use of the Electric Furnace," *Chem. News* **82**: 337, 1900.
2. L. Wöhler and W. Schuff, "The Silicides of Alkaline Earths," *Z. Anorg. Allg. Chem.* **209**: 33, 1932.
3. V. G. Rocktäschel and A. Weiss, "Zur Kenntnis der Strontiumsilicide," *Z. Anorg. Allg. Chem.* **316**: 231, 1962.
4. S. H. Moll, "The Applicability of Theoretically Calculated Intensity Corrections to Practical Metallurgical Problems in the Electron Probe," in Mueller, Mallett and Fay (Eds.) *Advances in X-Ray Analysis, Vol. 7*, Plenum Press, New York, 1963, p. 419.
5. K. F. J. Heinrich, "X-Ray Absorption Uncertainty," *The Electron Microprobe*, John Wiley and Sons, Inc., New York, 1966, p. 296.
6. P. Theisen, *Quantitative Electron Microprobe Analysis*, Springer–Verlag, New York, 1965.
7. H. A. Liebhafsky, H. G. Pfeiffer, E. H. Winslow, and P. D. Zemany, *X-Ray Absorption and Emission in Analytical Chemistry*, John Wiley and Sons, Inc., New York, 1960.
8. J. Philibert, "A Method for Calculating the Absorption Correction in Electron-Probe Microanalysis," *X-Ray Optics and X-Ray Microanalysis*, Academic Press, New York, 1963, p. 379.
9. L. S. Birks, "Calculations of X-Ray Intensities from Electron Probe Specimens," *J. Appl. Phys.* **23**: 387, 1961.
10. T. O. Ziebold and R. E. Ogilvie, "An Empirical Method for Electron Microanalysis," *Anal. Chem.* **36**: 322, 1964.
11. P. Duncumb and P. K. Shields, "Effect of Critical Excitation Potential on the Absorption Correction," *The Electron Microprobe*, John Wiley and Sons, Inc., New York, 1966, p. 284.
12. G. D. Archard and T. Mulvey, "The Effect of Atomic Number in X-Ray Microanalysis," *X-Ray Optics and X-Ray Microanalysis*, Academic Press, New York, 1963, p. 393.

DISCUSSION

H. Yakowitz (National Bureau of Standards): What were the absolute f-chi values at the 10 and 15 kV levels?

G. Faulring: The range of f-chi values calculated by Philibert's method was 0.70–0.80 (10 kV) and 0.66–0.86 (15 kV), by Duncumb–Shields' 0.80–0.93 (10 kV) and 0.67–0.87 (15 kV), and by Theisens' 0.88–0.97 (10 kV) and 0.73–0.89 (15 kV).

H. Yakowitz: If the f-chi values are much below 0.5, Duncumb has stated in print that if f-chi gets below about 0.2, one will be in serious difficulty unless one does Monte Carlo calculations since none of the models work. Heinrich and I showed at the First National Conference on Electron Probe Microanalysis in May 1966 that the models begin to diverge rather badly below f-chi values of about 0.8, and we postulated that one should operate at voltages low enough to achieve such an f-chi in order to reduce the apparent discrepancies between the f-chi curves. As to the Theisen method, it has been well documented that the Theisen method is absolutely useless. This is in print in the *Journal of the Institute of Metals* in a comment by Poole. Furthermore, curves that Heinrich showed in Washington showed that Theisen's method is simply unsatisfactory. Birks' method, as you probably know, is directly based on Castaing's data.

J. I. Goldstein (Goddard Space Flight Center): What was the takeoff angle of your electron probe?

G. M. Faulring: 30°.

THE APPLICATION OF THE ELECTRON MICROPROBE IN THE ANALYSIS OF NUCLEAR FUEL MELTDOWN EXPERIMENTS*

S. J. Stachura and L. Cooper

Atomics International
North American Aviation, Inc.
Canoga Park, California

ABSTRACT

Transient (nuclear) heating experiments were conducted with uranium carbide fuel rods to study the failure characteristics of typical rod designs. Extremely unusual changes in microstructure were observed and the electron microprobe was employed to establish the disposition of materials resulting from the meltdown experiments. The probe results indicated substantial fuel-clad interactions and permitted the resolution of several uncertainties regarding the course of fuel rod failure and material redistribution. The electron microprobe represents a unique capability in the post-test analysis of such meltdown tests.

INTRODUCTION

In order to gain improved safety information on the behavior of advanced nuclear reactors during highly abnormal conditions, experiments were conducted in the Transient Reactor Test Facility† to study the behavior of uranium carbide fuel rods under conditions of severe, transient nuclear heating. This facility provides rapid pulses (periods down to 40 msec) of neutrons which can generate large amounts of fission energy in the encapsulated fuel samples. The objectives of this particular study were to establish the failure limits of stainless steel clad, sodium bonded, hypostoichiometric uranium carbide fuel rods; to determine the mode of failure; and to examine the course of material redistribution resulting from fuel failure. Several tests were conducted at increasing energy levels—the most severe of these resulted in gross melting of fuel into the environment of stainless steel cladding and stagnant sodium coolant. The rather unusual consequences of this test, together with the complex microstructures which were observed in the post-test analysis, made it extremely difficult to determine the disposition of materials in the post-test condition. The electron microprobe proved to be an indispensable tool in this respect, and this paper will describe the probe analysis efforts.

DESCRIPTION OF EXPERIMENT

Figure 1a presents a schematic of the test capsule. It contained a 6-in. long × 0.852-in. diameter uranium carbide fuel slug (4.6% carbon), sodium bonded to 0.952-in.

---

* Based upon studies conducted for the USAEC under Contract AT-(11-1)-GEN-8.
† Operated by Argonne National Laboratory for the AEC.

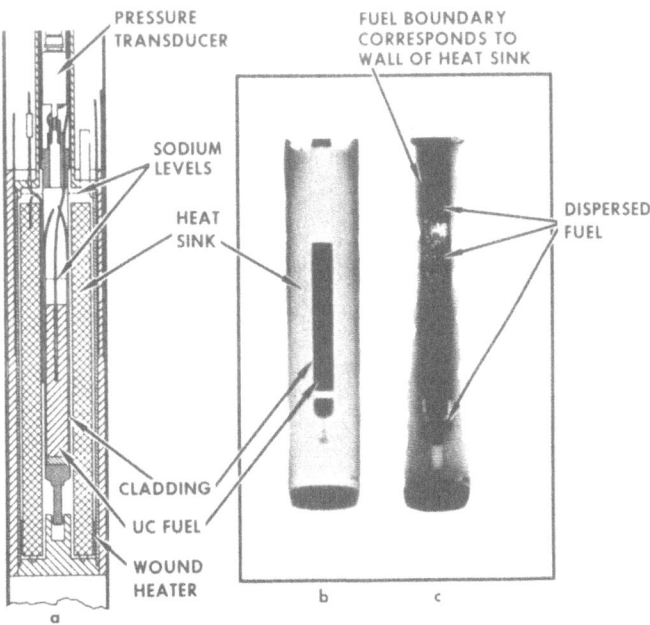

Figure 1. (a) Schematic of test capsule; (b) X-radiograph of capsule before irradiation; (c) X-radiograph of capsule after irradiation.

outside diameter × 0.020-in. wall stainless steel (type 304) cladding. The fuel rod was surrounded by stagnant sodium coolant which was contained in a graphite crucible (which also served as a heat sink). The capsule was highly instrumented, including several fuel and bond thermocouples, coolant and heat sink thermocouples, and a fuel section pressure transducer.

The detailed thermal and pressure data provided considerable information on the sequence of events attendant to fuel failure.[1,2] The deposition within the fuel of approximately 250 Btu/in.[3] (which corresponds to an adiabatic temperature rise to above 7000°F) within 1 sec resulted in the following: boiling of the sodium bond (between fuel and clad); central fuel melting; melting at the surface of the slug and subsequent contact of molten fuel with cladding; reaction or melting of cladding; and finally, flow of molten fuel and clad into the sodium coolant with subsequent gross dispersion of materials due to coolant pressures (resulting from rapid fuel-coolant energy transfer). Figures 1b and 1c show the X-radiographs of the capsule before and after the test—the post-test radiograph shows the dispersion of material, restrained only by the walls (which appear light) of the thick graphite crucible.

Figures 2a and 2b present the photomicrographs of the original uranium carbide structure and a region of one of the meltdown samples. The initial equiaxed structure of UC grains with small amounts of free uranium in the grain boundaries was trans-formed into a wide variety of structures (not all present in this specimen), most of which differed dramatically from the original microstructure. From the metallography alone, it was not possible to characterize the post-test disposition of materials. In addition, the extreme variations in structures which existed over relatively small regions imposed a serious limitation on more straightforward means of chemical analysis, and it appeared that the electron microprobe would be the most appropriate tool for establishing the compositions of the various regions.

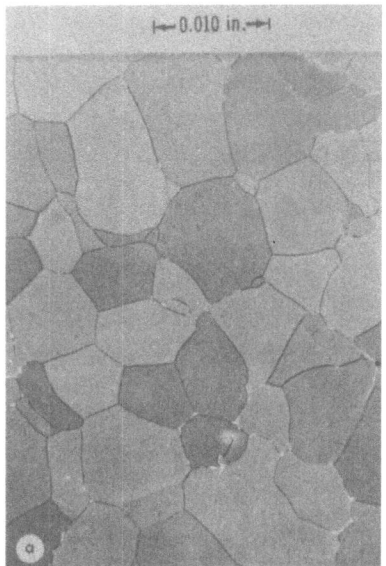

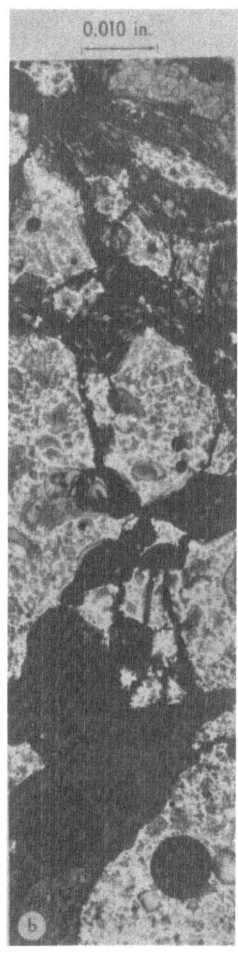

Figure 2. (a) As-cast uranium carbide (4.6% carbon) before irradiation. Water, nitric acid and acetic acid, B. F. Mag. 100 × ; (b) example of post-test microstructure. Water, nitric acid and acetic acid, B.F. Mag. 100 ×.

PROBE ANALYSIS

Scoping

Since the probe had not been previously employed in the analysis of meltdown samples with a broad variety of complex structures, there were considerable uncertainties regarding the most reasonable approach consistent with the acquisition of useful information, and a relatively limited effort. It became clear during the course of analysis that much more accurate methods could be adopted for both the acquisition and reduction of data. However, the semiquantitative results which were obtained proved to be adequate for the purposes of the present study.

The question of radiation levels proved to be insignificant because the transient testing yielded very limited burnup (on the order of $10^{-6}\%$ of the total uranium), and because considerable time elapsed between the tests and the probe analysis.

The system in this study included five major constituents: uranium, carbon, iron, chromium, and nickel. The most serious problem proved to be the analysis of carbon. Initially, in working with uranium, uranium carbide, and uranium dicarbide standards, some success was achieved in reproducible variations in wavelength scans (in the region of uranium N and carbon K lines) for each of these materials. However, this approach did not yield consistent results on the samples and it is believed that the presence of iron, chromium, and nickel and the possible formation of their carbides or mixed carbides was the source of the difficulty. As a result of this problem, little reliance was placed on the carbon analyses. In general, the carbon is expected to exist as primary phase carbides and in some eutectic structures according to most of the reported phase data.[3-5] In view of the difficulties associated with carbon detection in the uranium-carbon system, it is clear that the characterization of ternary and higher systems would require a very extensive effort (which was beyond the scope of this study).

An additional problem which was encountered was the complexity of many of the microstructures, and in order to expedite the analysis, most of the samples were examined in a slightly etched condition. In some cases, apparent edge effects were noted—such regions were avoided in the acquisition of point-count data.

In view of the preceding considerations, the major contribution of the probe in this particular system consisted of the ability to scope the distribution of elements. By use of backscatter, ionization and absorption corrections,* reasonable estimates could be made of uranium and iron concentrations. Because chromium and nickel were usually present in phases with iron, the ratio uranium–total steel components was used in estimating the compositions of specific regions. The next section will discuss the results of area scan, line profile and point-curve data on several meltdown samples and will indicate the interpretation of those data in the analysis of the experiment.

RESULTS

Figures 3 and 4 provide an example of the general procedure for probe data acquisition. The microstructure of Figure 3 appears in the form of an area scan of backscattered electrons at the top of Figure 4. In addition, line profiles taken across a horizontal line about 0.2 divisions above the center horizontal line† indicate the variations in

---

* Correction methods of Duncumb and Shields for backscatter, Nelms for ionization, Philibert with modified Lenard coefficients by Duncumb and Shields for absorption.
† This is the position of the line profiles in all of the area scans presented.

Figure 3. Meltdown sample employed in probe analysis. Water, nitric acid and acetic acid, B.F. Mag. 250×.

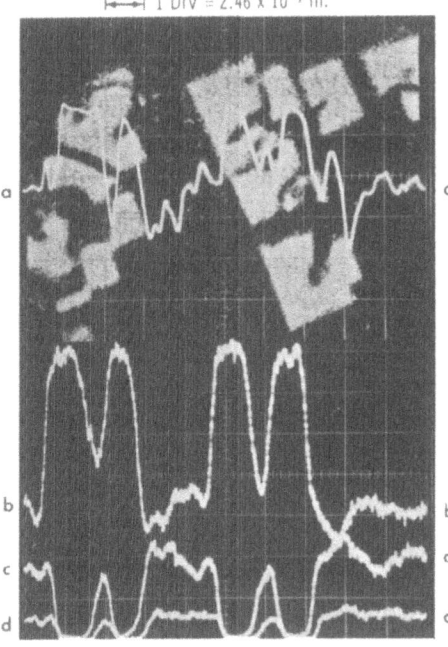

Figure 4. Probe data—upper region shows area scan of backscattered electrons; curve a—line profile of backscattered electrons; curve b—line profile of uranium; curve c—line profile of iron; curve d—line profile of chromium (nickel profile not shown here).

uranium (first order M_β), iron, chromium and nickel (second order K_α). Point counts of the dark grains did not reveal the presence of steel constituents, thus indicating that these grains were uranium carbide. The continuous phase gave uranium to steel component ratios of about one in the level areas (the iron–uranium carbide system has a eutectic of this composition). It is interesting to note the gradual variations in uranium: steel components ratio which occur within the continuous phase of Figure 4. These vary from 0.8–1.6. This analysis was of considerable importance because the structure is representative of hypostoichiometric uranium carbide with very low carbon content, and the probe analysis established the extensive presence of steel constituents in the continuous phase.

The next set of probe analyses were conducted on a region of fuel which had

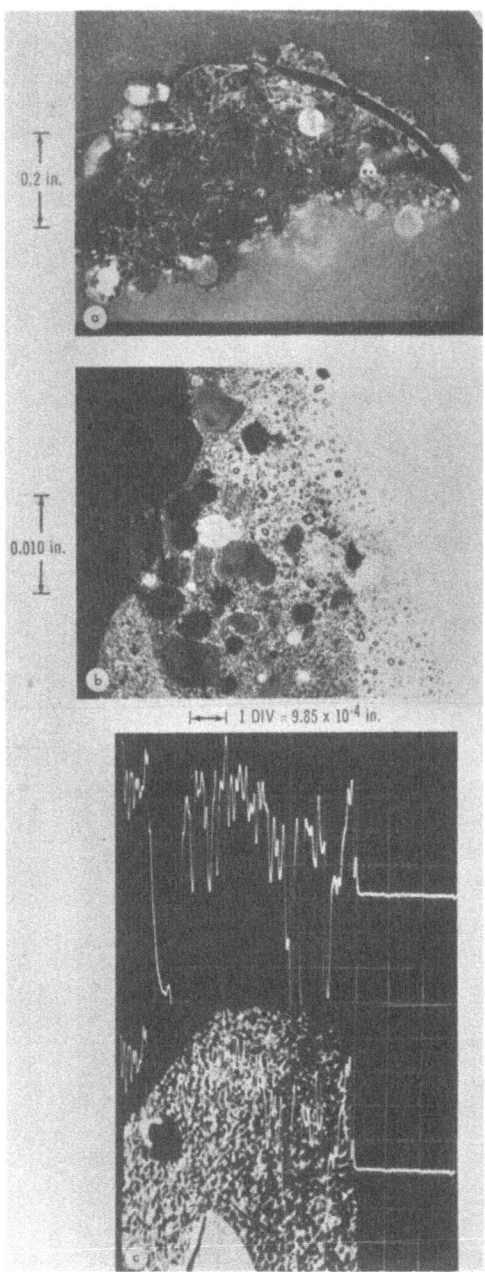

Figure 5. (a) Macrograph of fuel resolidified on cooler portion of cladding. Water, nitric acid and acetic acid, B.F. Mag. 5×. (b) Microstructure of fuel-clad interface. Water, nitric acid and acetic acid, B.F. Mag. 100×. (c) Backscattered electron area scan (and line profile) on fuel-clad interface.

resolidified onto a cooler portion of the clad in the upper region of the capsule. This specimen is relevant to the mode of fuel rod failure because this condition resembled that which existed during the initial failure of the cladding when the molten fuel first slumped to contact the relatively cool stainless steel clad. Figure 5 is a photomacrograph of the region, and Figure 5b shows the microstructure of the fuel-clad interface. An area scan of backscattered electrons appears in Figure 5c; this demonstrates the variation in backscatter across this region. Probe data were then taken on the stainless steel region (which yielded no compositional variations); the region adjacent to, and the region removed from the stainless steel.

Figure 6 (and Figure 7 at higher magnification) presents the backscattered electron area scan and line profiles for the region adjacent to the steel. Figure 8 provides the probe

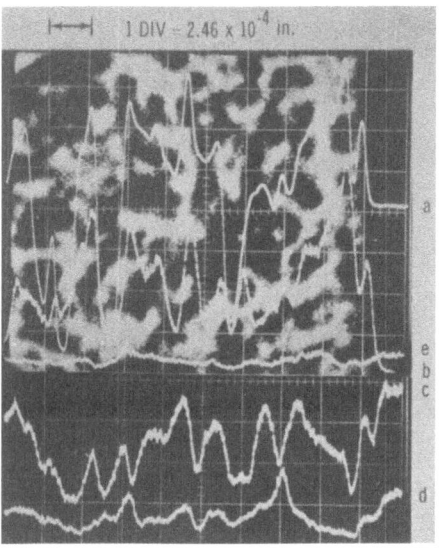

Figure 6. Probe data on region proximate to clad: curve a—line profile of backscattered electrons; curve b—line profile of uranium; curve c—line profile of iron; curve d—line profile of chromium; curve e—line profile of nickel. (Uranium and nickel relative to raised baseline.)

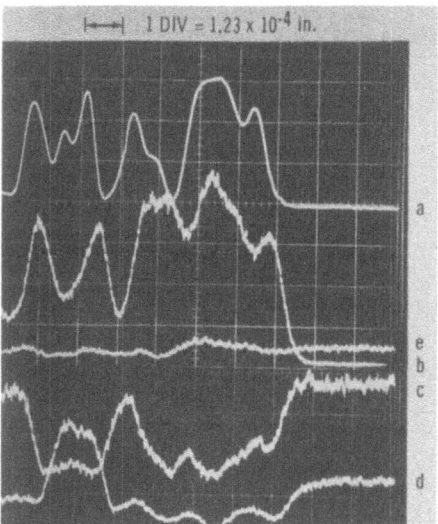

Figure 7. Probe data on region proximate to clad at increased magnification: curve a—line profile of backscattered electrons; curve b—line profile of uranium; curve c—line profile of iron; curve d—line profile of chromium; curve e—line profile of nickel. (Uranium and nickel relative to raised baseline.)

data for the region remote from the steel. The region near the steel yielded the highest steel component–uranium ratios (with the exception of the stainless steel) noted in the meltdown samples (approaching 4:1). The iron and chromium levels in Figure 6 (curves c and d) and Figure 8 (curves b and c) correspond to the same base line, and the increase in these levels as the steel interface is approached can be seen clearly from a comparison of the line profile data. The observed compositional variations are representative of a reaction which occurs via the dissolution of the steel constituents into the molten free uranium at temperatures below the melting point of the stainless steel (and the UC). This reaction occurs until the temperatures decrease sufficiently to precipitate either U or UFe_2 (or Fe or U_6Fe). It might also be noted that there are some regions in these particular samples which show increasing chromium levels with decreasing iron levels— this is a distinct exception to the predominance of probe data and could be associated with the high ratio of stainless steel to uranium carbide in this region and the potential reaction of uranium carbide with iron–chromium solid solutions to form UFe_2 and chromium carbides.[4]

The above observations lead to the conclusion that considerable attack of the steel can be noted under conditions where the steel is not molten. Because of the presence of approximately 4% of the uranium as free uranium in the UC, uranium-clad reactions can occur under conditions where only the uranium is molten. This mechanism has been observed for metallic fuels,[6] and the penetration times are consistent with the delay in fuel failure noted in the transient thermal and pressure data. These considerations have incorporated the results of analog studies of the transient thermal behavior including the effects of nonuniform energy generation in the fuel, effects of bond vaporization and the temperatures which result from contact of molten fuel with relatively cold cladding. While these calculations alone could not establish the question of cladding failure via melting *vs.* reaction, the probe studies established the occurrence of such reactions.

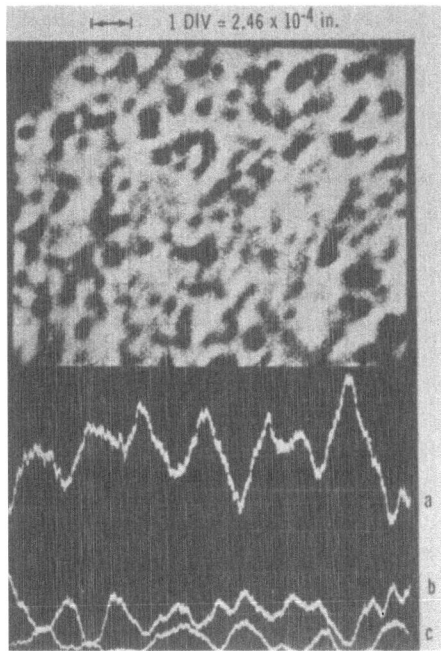

Figure 8. Probe data on region remote from clad: curve a—line profile of uranium; curve b—line profile of iron; curve c—line profile of chromium.

The intimate fuel-clad mixtures provided a general indication that fuel rod failure was a consequence of reaction rather than due to exclusively thermal mechanisms such as high temperature yield or clad melting.

CONCLUSIONS AND RECOMMENDATIONS

This paper represents an initial effort in the application of the electron microprobe to the analysis of nuclear fuel meltdown experiments. Previous studies[6,7] have generally based conclusions entirely on the interpretation of thermal (and sometimes pressure) data together with post-test examination and metallography. The probe provides a means for establishing the disposition of materials which can verify hypotheses determined from other information. This limits the uncertainties in the results and can serve to reduce the required number of experiments.

For the interaction of hypostoichiometric uranium carbide with stainless steel at temperatures above the uranium melting point but below the melting point of the stainless steel, the probe analysis indicates attack of the steel by the molten uranium as a predominant mechanism for clad failure. The intimate mixtures of fuel and clad provide further evidence that fuel rod failure was initiated by other than thermal effects. This coexistence of fuel-clad mixtures and the absence of resolidified regions of stainless steel also provides additional evidence (complementing thermal and pressure data) that coolant (sodium) vapor pressures and not those of the cladding provided the major forces for fuel dispersal.

REFERENCES

1. S. J. Stachura et al., "Uranium Carbide Transient Heating Studies—Phase II," ANS Trans. 8: 305, 1965.
2. S. J. Stachura and M. Silberberg, "The Application of Experimental Data from TREAT Meltdown Studies to Reactor Accident Analysis," Proc. of the International Conference on Safety, Fuels, and Core Design in Large Fast Power Reactors, ANL-7120, 1965, p. 514.
3. J. A. Nichols and J. A. C. Marples, "An Investigation of the U-C-Fe and Pu-C-Fe Ternary Phase Diagrams with Some Observations on the U-Pu-C-Fe Quaternary," in: Proc. of a Symposium held at Harwell, November 1963, Carbides in Nuclear Energy, Vol. I, Macmillan & Co. Ltd., 1964, p. 246.
4. G. Briggs et al., "Phase Relationships in the Systems U-Cr-C, U-Ni-C, and U-Fe-Cr-C," ibid., p. 231.
5. G. Briggs et al., "Systems of UC, UO_2, and UN with Transition Metals," Trans. Brit. Ceram. Soc. 62: 221, 1963.
6. C. M. Walter and L. R. Kelman, "Penetration Rate Studies of Stainless Steel by Molten Uranium and Uranium-Fission Alloy," J. Nucl. Mater. 6: 281, 1962.
7. C. M. Walter and C. E. Dickerman, "Treat Study of the Penetration of Molten Uranium and U/5 wt.% Fs Alloy through Type 304 Stainless Steel," Nucl. Sci. Eng. 18: 518, 1964.

DISCUSSION

R. G. Hurley (Los Alamos Scientific Lab): What was your optimum sample size and what modifications to your probe did you make in order to reduce background and operator exposure caused by the beta-gamma emission?

S. J. Stachura: Because of the low level of burnup produced by the transient testing, which is on the order of 10^{-6} %, and because of the extended time which elapsed between the irradiation and the probe analysis, radiation levels were very close to background and precautions such as shielding, etc., were unnecessary.

PREPARATION OF ELECTRON PROBE MICROANALYZER STANDARDS USING A RAPID QUENCH METHOD

J. I. Goldstein

Goddard Space Flight Center
National Aeronautics and Space Administration
Greenbelt, Maryland

F. J. Majeske

Melpar, Inc.
Goddard Space Flight Center
National Aeronautics and Space Administration
Greenbelt, Maryland

and

H. Yakowitz

National Bureau of Standards
Washington, D.C.

ABSTRACT

Standards for microprobe analysis can be made to serve two purposes: (a) proposed correction models can be tested with them, and (b) analysis can be performed more accurately in the system which includes the standard. Few microprobe standards presently are available because they must be homogeneous on the micron scale and their composition must be known accurately. A modified Duwez splat cooling method is described which enables the investigator to prepare suitable standards in most cases. The apparatus which is relatively simple and inexpensive is described in detail. The systems Au–Si and Al–Mg were chosen as test cases. Suitable standards were prepared at different concentrations in each system. The analytical results for all compositions in Al–Mg are presented and discussed.

INTRODUCTION

More than 15 years ago, Castaing stated that quantitative electron probe micro-analysis could be accurately performed using only elemental standards. Correction procedures for the observed ratios of X-ray flux data from unknown to elemental standard were presented.[1] The correction models and procedures have been the subject of much controversy for 15 years. Uncertainties in the correction models and their input parameters have led many workers in the field of electron probe microanalysis to the belief that quantitative analysis using elementary standards can lead to errors as high as 10% of the amount present.[2]

Therefore, when analysis having a maximum error of 1% relative is necessary, multicomponent standards are required. These standards thus serve a dual purpose: (a) proposed correction models can be tested with their aid, and (b) analysis can be performed more accurately in the system which includes the standard. These standards must be homogeneous on the micron scale and their composition must be known accurately.

Very few well-characterized standards are available because they are difficult and expensive to prepare and to test.[3] Several methods for making standards are available. The first is to melt and chill cast the standards. The difficulty with this procedure is that most castings exhibit segregation, and there are only a few alloy systems which can be homogenized by heat treatment.

Some of these include a few binary systems such as Cu–Zn,[4] Fe–Ni,[5] and Cu–Au.[6]

A second preparation procedure utilizes powder metallurgical techniques. Micron-sized powders are blended and sintered into billets. These are then heavily worked by swaging or rolling and annealed. Extrusion followed by a final anneal at the highest practical temperature is the final step.[7] This method shows promise but is complex, expensive, and time-consuming.

One final approach is to use intermetallic or other inorganic compounds having a narrow range of solubility. However, it has been found that many such compounds either have too large a solubility range to satisfy the stringent homogeneity requirements or are unsatisfactory for other reasons.[2]

The remainder of this paper reports on the development of a modified Duwez method of splat cooling[8] which can be applied directly to the important practical problem of preparing multicomponent electron probe microanalysis standards. The technique allows specimens containing components which are miscible in the liquid phase to be prepared rapidly and with a satisfactory degree of homogeneity.

METHOD AND APPARATUS

In order to adapt the method of Duwez to the preparation of microprobe standards successfully, the following requirements had to be met: (a) the apparatus must be relatively simple and inexpensive, and (b) sufficient material must be produced for complete characterization by both the microprobe and conventional analytical techniques to be carried out. Shards thicker than the penetration of 50 kV electrons are required. It was decided that several cubic centimeters of material would be a satisfactory yield.

These design requirements are not as stringent as those of Duwez. It meant that an explosive charge for accelerating the molten metal was unnecessary. However, for effective cooling of several cubic centimeters of alloy, a much larger and colder hearth, than that used by Duwez, was decided upon. Finally, provision was made to rotate the hearth rapidly during the actual quenching operation. Rapid rotation causes the molten metal to spin out into a number of long relatively flat shards of solid.

The requirement that the apparatus be relatively simple and yet of wide applicability was met by constructing the basic device of glass. (Figure 1.) The central portion is made of glass. The crucible is boron nitride (BN), shaped as a cylindrical bucket with a 20 mil diameter hole in the base. Quartz crucibles were found to be unsatisfactory. The quartz reacted rapidly with many of the molten alloys.

The hearth itself is a hemisphere with a radius of 2.5 in. The outside dimensions are those of a 6-in. diameter by 3.5-in.-high right cylinder. This design is not the only design possible; special modifications may be required for specific alloy systems. Hearth configuration should be questioned when the molten alloy does not spin out into long shards but remains a bead or when the alloy spins completely out of the hearth. The hearth configuration and its rotational velocity are the most critical factors in obtaining the required splat-cooled alloys.

Power for the furnace is supplied by standard induction melting devices. Units rated at both 5 and 7.5 kW, respectively, were tried: the 5-kW unit proved to be adequate.

The plexiglass chamber has inside dimensions of 11 in. × 11 in. × 11 in. This relatively large size permits the separate stand containing the motor driven hearth to be placed in any desired location with respect to the crucible base.

High purity argon is admitted to both the glass column and the plexiglass crucible

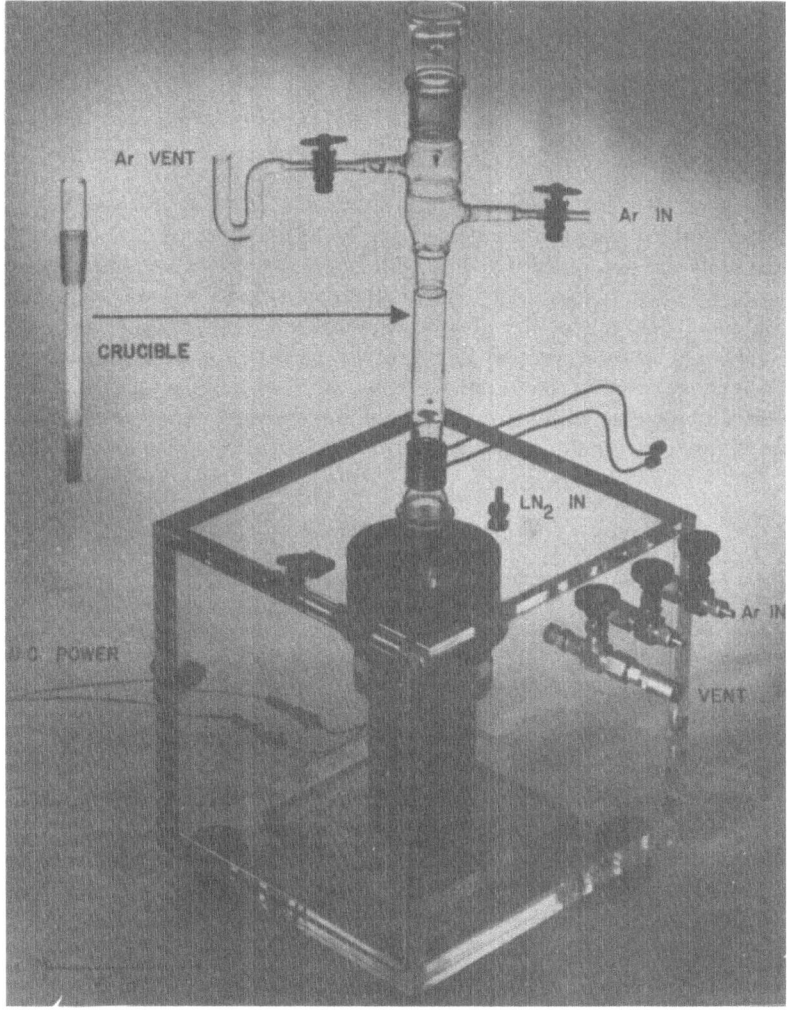

Figure 1. Modified splat cooling apparatus.

chamber at the places indicated. Each of the argon inlets is serviced by a separate gas supply and vent.

The use of the apparatus is straightforward. The weighed components of the desired alloy are placed into the BN crucible. The hearth is filled with liquid nitrogen and the system closed. Argon is allowed to purge the system while the liquid nitrogen in the crucible boils off. When almost all of the liquid nitrogen has boiled off, hearth rotation is begun. Simultaneously, the alloy is melted by induction heating. After melting, about 2 min are allowed for the liquid to homogenize. The hearth is positioned beneath the crucible so that the molten stream will strike the hemispherical surface near the top. Thus, the centrifugal force acting on the molten material is maximized spreading the liquid over a larger area. This leads to a larger overall heat transfer rate than would otherwise be possible.

During the heating, the argon atmosphere in the entire apparatus is at the same pressure. After melting occurs, the surface tension of the liquid seals the small hole in the base of the BN crucible. After the homogenization step, the upper argon vent is sealed by means of the stopcock and the pressure of the incoming argon is increased rapidly. This sudden, large pressure differential forces the molten metal through the hole and into the hearth which is at a temperature of about 77°K. The argon atmosphere in the plexiglass portion is allowed to remain unchanged throughout. The argon pressure in the glass column is reduced by opening the stopcock as soon as freezing of the melt has been completed.

As was previously indicated, the speed of rotation of the hearth was found to be a critical factor in the quality of the results obtained. If the speed of rotation was too slow, the molten alloy did not spread out in a thin layer, and the heat transfer rate was not great enough to obtain homogeneity. If the rotational velocity was too great, the molten alloy spun completely out of the hearth. For example, with Al–Mg alloys containing 50% Al and less, a rotation speed of 1250 rpm was used. For those containing more than 50% Al a rotation speed of 2000 rpm was required. The surface tension of the molten alloys seemed to increase with increasing aluminum content; therefore a higher rotation speed was required to obtain proper spreading of these alloys upon striking the hearth.

RESULTS

Several systems of binary alloys were prepared and examined, including Fe–Ni, Al–Cu, Au–Si, and Al–Mg. The latter two will serve to illustrate the results obtainable as well as the advantages and disadvantages of specimen preparation by splat cooling.

Gold–Silicon

Figure 2 shows the constitution diagram as reproduced from Hansen.[9] The gold–silicon binary system is indicated to have virtually no mutual solid solubility of either element in the other. Preparation of homogeneous intermediate Au–Si standards by chill casting followed by a homogenization anneal is not possible. Such a system represents a stringent test of the splat cooling technique. Alloys containing three and six nominal weight percent of silicon in gold were weighed out using United States mint gold and high purity silicon. The six percent alloy is the eutectic.

On striking the hearth, these alloys spun out into a number of shards of varying length and cross section. Several of the shards containing 6% Si were selected for investigation by X-ray diffraction. Examination of a typical shard in a precession camera using molybdenum radiation yielded the pattern shown in Figure 3. The highly

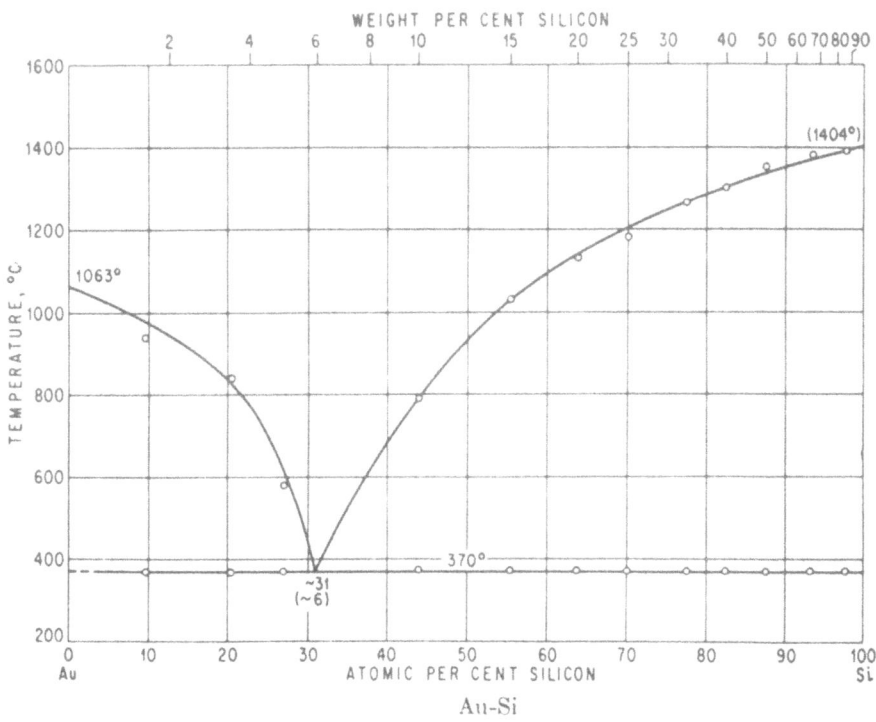

Figure 2. Constitution diagram for Au–Si (after Hansen).

Figure 3. Precession camera diffraction pattern of Au-6 wt.% Si using molybdenum radiation.

broadened lines indicate a very small crystallite size; no attempt to deduce the actual size was made.

Several shards were examined in a Debye–Scherrer camera using monochromated Cu K_α radiation. These showed considerable structural variation as shown in Figure 4. A few showed only a few very broad halos (Figures 4c and d) indicating them to be nearly amorphous while others contained some very broad lines (Figure 4c) and a few contained a great many relatively narrow lines (Figures 4a and b).

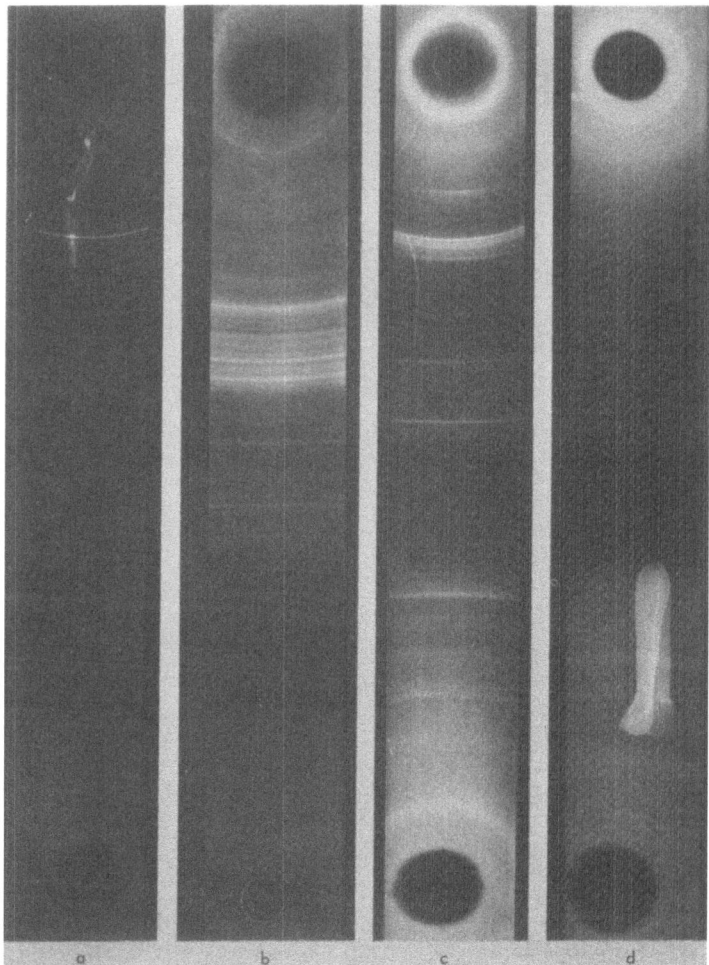

Figure 4. Debye–Scherrer photographs of Au–6 wt.% Si shards: (a) Shard showing probable presence of metastable phases. (b) Shard (2) showing probable presence of metastable phases. (c) Shard (3) showing only a few broad lines. (d) Shard (4) showing nearly amorphous (glass-like) pattern.

Shards for which diffraction patterns were available were chosen for electron probe microanalysis. They were mounted in a cold setting resin which was allowed to harden overnight. Mechanical polishing was carried out on well lubricated papers and wheels in order to minimize sample heating. Final polishing was accomplished using 0.25 μ diamond on selvyt.

The shards were introduced into the microprobe instrument after a bridge of silver based paint was used to assure each a path to ground for the incident electron beam. The lines chosen were Au M_β using an ADP crystal and Si K_α using an EDDT crystal. The probe voltage was 10 kV with a monitor current of one-tenth microampere. The X-ray emergence angle was 52.5°.

Table I. Results for Au–6 wt. % Si Shard (Figure 4d)

Element	Number of points investigated	Average total counts accumulated, N	Maximum counts accumulated in set	Minimum counts accumulated in set	% Coefficient* of variation	$\sqrt{N}$ in %
Au	17	183593	185000	181705	0.45	0.23
Si	17	43307	49191	40011	5.8	0.48

$$* \text{ % Coefficient of variation} = \frac{100}{N} \left[\frac{\sum_{i=1}^{i=n} (X_i - \bar{X})^2}{n-1} \right]^{1/2}$$

Table II. Results for Au–3 wt. % Si Shards

Element	Shard	Number of points investigated	Average total counts accumulated, N	Maximum counts accumulated in set	Minimum counts accumulated in set	% Coefficient of variation	$\sqrt{N}$ in %
Au	1	10	232,304	235,964	228,448	0.89	0.21
Si	1	10	36,476	39,125	35,065	2.91	0.52
Au	2	9	234,481	236,383	232,146	0.56	0.20
Si	2	9	34,734	38,166	32,977	4.73	0.54
Au	3	9	233,391	234,626	232,098	0.43	0.20
Si	3	9	33,468	34,260	33,022	1.20	0.55

Silicon count rates on samples whose structure corresponded to that of Figure 4a varied by as much as 50% on randomly chosen points. When the structure corresponded to that of Figure 4d, the results were somewhat better. Table I presents results for such a shard. While the results do not indicate that complete homogeneity was achieved, the coefficient of variation for both gold and silicon is sufficiently low that the shard in question could be used as a standard. In fact, for silicon at the 6% level, the variability approximates closely that reported by the Washington Electron Probe Users Group for analysis of relatively low concentrations in systems where homogeneity had been achieved.[2]

At the 3% silicon level, Au L_α was the line used for analysis with LiF as the monochromator. The probe voltage was 20 kV; all other conditions were the same. Three shards were examined; the results are shown in Table II. Again, complete homogeneity was not achieved but the coefficient of variation is definitely low enough that any of the three shards would be a useful standard.

It is known that splat-cooled material may undergo room temperature aging; i.e., post-cooling segregation may occur. Splat-cooled alloys have been reported to suffer such aging.[10] Cells of rejected gold and silicon result; this destroys the usefulness of the sample for an electron probe microanalyzer standard. Aging occurred in the alloys described in about a week. Figure 5 shows an electron micrograph of 6w/o Si–Au illustrating the fresh material and the rejection products after one month at room temperature.

Figure 5a. Electron micrograph of fresh Au–6 wt.% Si which is nearly featureless.

The possibility of aging suggests that the following procedure be adopted when using splat-cooled standards:

1. Obtain X-ray diffraction patterns of the shards immediately after splat-cooling.
2. Mount the shards giving the broadest diffraction lines in the cold mount having the fastest curing time and lowest curing temperature.
3. Polish by the technique involving the least heat.
4. Check for homogeneity in the microprobe.
5. If satisfactory, obtain the experimental microprobe intensity data immediately.
6. Then have the sample used in the microprobe analyzed by conventional analytical techniques.

Aluminum–Magnesium

The Al–Mg system is interesting for several reasons. First, there are a great many Al–Mg based commercial alloys. Next, this system permits the most important single

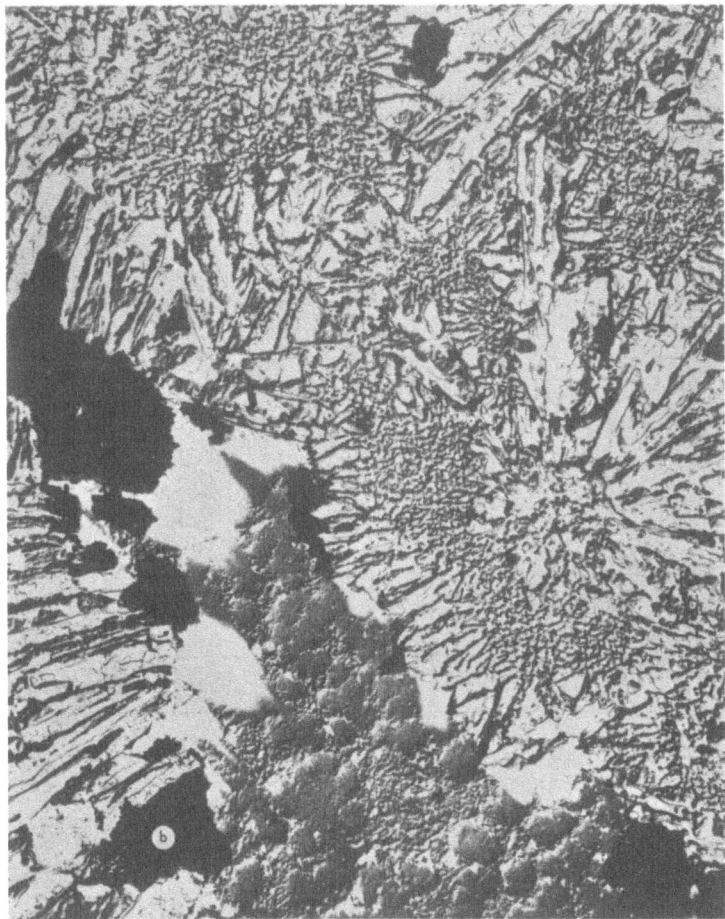

Figure 5b. Electron micrograph of month-old Au–6 wt.% Si showing
dendritic formations amid former featureless structure (×13,500).

microprobe correction procedure, the absorption correction, to be isolated and studied
directly. Finally, the constitution diagram presented by Hansen (Figure 6) shows the
system to be moderately complex with at least one major area of disagreement. Suitable
microprobe standards or models would offer investigators the possibility of a more
accurate redetermination of the range 35–68 wt.% Mg in aluminum.

Seven alloy compositions were prepared ranging from 5–90 wt.% of magnesium.
Diffraction patterns showed substantially the same type of structure for all shards
chosen. Two such patterns are shown in Figure 7. The broad, spotty lines indicate
small crystallites some of which may extend through the very thin shard.

Since the diffraction patterns in this case showed only small differences, shards
were randomly chosen for microprobe analysis. Four shards of each alloy were checked
for homogeneity for both aluminum and magnesium using an ADP crystal at probe
voltages of 10, 15, and 20 kV, respectively. At least five random points were examined
on each shard. Shard lengths were typically 0.5–2 mm while shard widths were 100 μ
or less.

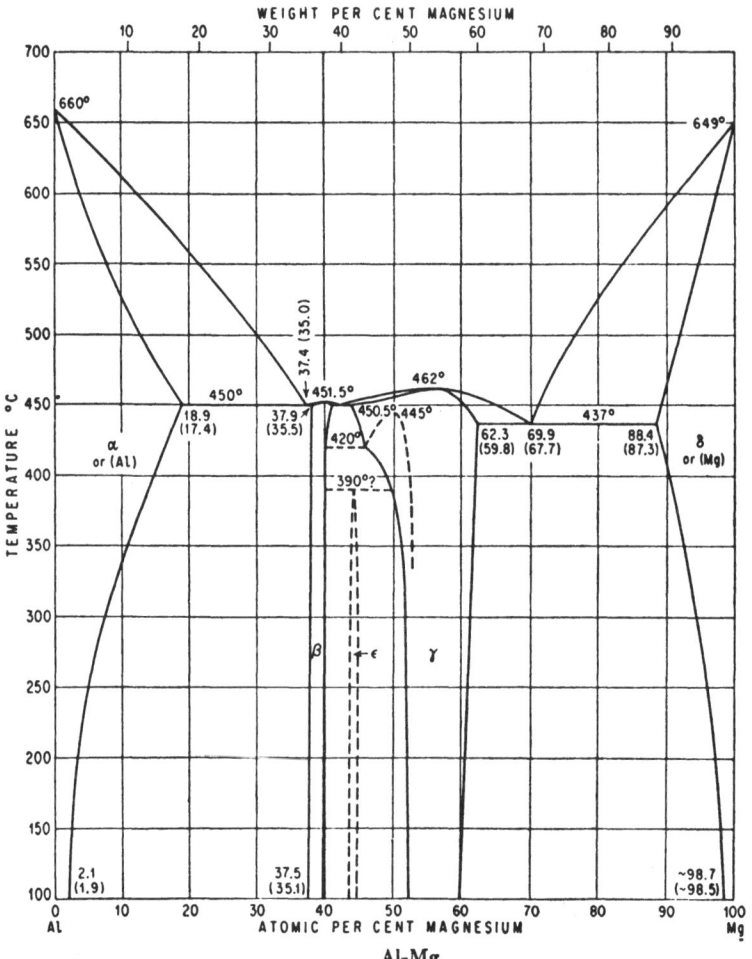

Figure 6. Constitution diagram for Al–Mg (after Hansen).

Virtually all counts for all points in every shard were within $\sigma = \pm 1.2\%$ which is within the statistical counting error for the average count N for the shard. The value of N was set to range from 100,000–200,000 depending on the sample. The agreement of N for different shards of the same nominal composition was well within 1% relative. Thus, each shard was found to be homogeneous to a 1-μ probe and each was found to be of the same composition as other shards from the same melt. Thus, each of the required homogeneity characteristics for standards for electron-probe microanalysis was satisfied.

Some time (approximately one week) after checking for homogeneity, the intensity data from the shards were taken relative to that of pure magnesium and pure aluminum. Background for aluminum was measured on peak with a magnesium sample while background for magnesium was measured on peak with an aluminum sample. Line background ratios for magnesium and aluminum were greater than 250 : 1. Instrumental stability was monitored by counting on the pure elements after each shard was investigated; all such counts were within the statistical counting error of $\sigma = 1.2\%$.

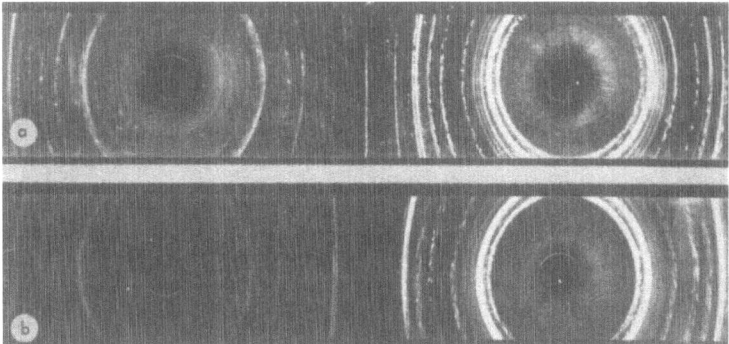

Figure 7. Debye–Scherrer patterns from typical Al–Mg shards (Sample M-32).

Since all shards from the same melt gave the same count rate within the counting error, a single relative intensity value k could be obtained for each composition. In considering the corrections to be applied, atomic number effects were ruled out since aluminum and magnesium are adjacent in the periodic table and since the ratio of atomic number to atomic weight is essentially the same for both.

The Mg K_α lines will be excited by the Al K_α since the Mg K absorption edge lies at a lower energy than the Al K_α energy. A fluorescence correction is required for magnesium in Al–Mg as well as an absorption correction. The magnitude of the correction to be applied is small since the absorption coefficients of both aluminum and magnesium for Mg K_α are nearly equal and since the absolute K fluorescence yield from aluminum is very small.[11] The Philibert–Duncumb relation was used to correct for absorption effects while Castaing's relation was used to correct for fluorescence.[1,12] The total correction calculated at all magnesium compositions is 1% relative or less. Therefore, magnesium in Al–Mg represents very nearly the ideal case in which the measured relative X-ray fluxes correspond to the actual concentrations across the entire range of compositions.

For aluminum, there will be only direct electron X-ray excitation. However, a large absorption correction is to be expected because the mass attenuation coefficient, (μ/ρ) of magnesium for Al K_α is more than an order of magnitude greater than the (μ/ρ) of aluminum for Al K_α.[13] In this case, the large absorption correction for aluminum in Al–Mg can be isolated for experimental investigation. In an evaluation of absorption correction procedures, it was found that the Philibert–Duncumb method gave satisfactory results in many cases.[14] Therefore, this method was adopted for the aluminum in the Al–Mg data.

As a final check, quantitative chemical analysis was performed on four of the seven alloys. An insufficient quantity of material was available in two cases and in the remaining instance, contamination of the specimen submitted caused poor results. The source of this contamination could not be traced. The analyses are claimed to be accurate to within an error of 0.5% relative.

In order to arrive at the true compositions for the three remaining alloys, the Ziebold–Ogilvie method was used. The four analyzed alloys were used to obtain the a value for the system at each operating voltage by means of the relation[15]

$$a = \left(\frac{C}{k}\right)\left(\frac{1 - k}{1 - C}\right) \tag{1}$$

Table III. Results: Al-Mg Alloys

Sample	Nominal pct. Al	k_{10}	k_{15}	k_{20}	C_{10}*	C_{15}*	C_{20}*	Chemical analysis	% Error (rel. to chem.) 10 kV	% Error (rel. to chem.) 15 kV	% Error (rel. to chem.) 20 kV	Average composition by microprobe	Average error: microprobe relative to chemical analysis
								Results: Aluminum in Al-Mg Alloys					
M-38	10	11.1	7.68	5.80	17.1	17.0	17.5	15.3†	+11.8	+11.1	+13.0	17.2	+12.4
M-41	25	18.4	14.0	10.1	27.7	28.0	27.9	25.1†	+10.4	+11.5	+11.1	27.9	+11.1
M-39	50	38.7	31.1	24.5	51.9	51.5	49.8	50.0	+ 3.8	+ 3.0	− 0.40	51.1	+ 2.2
M-40	50	42.0	34.4	27.4	54.8	54.8	53.5	51.9	+ 5.6	+ 5.6	+ 3.1	54.4	+ 4.8
M-32	65	52.5	45.6	38.5	65.5	66.4	64.7	63.5†	+ 3.1	+ 4.6	+ 1.9	65.5	+ 3.1
M-43	90	—	82.2	77.1	—	91.2	90.8	90.4	—	+ 0.88	+ 0.44	91.0	+ 0.66
M-45	95	96.3	95.8	95.9	98.0	98.5	98.3	97.9	+ 0.10	+ 0.61	+ 0.41	98.3	+ 0.41
								Results: Magnesium in Al-Mg Alloys					
M-45	5	1.65	1.64	1.56	1.59	1.61	1.54	2.1§	not computed since range of Chem. Anal. is 1.7 to 2.5% Mg			1.58	×
M-43	10	—	10.1	9.48	—	10.0	9.55	9.6	—	+ 4.2	− 1.1	9.78	+ 1.9
M-32	35	37.3	36.5	36.9	37.2	36.8	37.8	36.5	+ 1.9	+ 0.82	+ 3.6	37.4	+ 2.5
M-40	50	47.6	47.1	47.8	47.6	47.6	48.8	48.1	− 1.0	− 1.0	+ 1.4	48.0	− 0.21
M-39	50	50.6	49.5	49.9	50.6	50.0	50.8	50.0	+ 1.2	0	+ 1.4	50.5	+ 1.0
M-41	75	74.4	71.0	73.6	74.6	71.6	74.5	74.9	− 0.4	− 4.4	− 0.53	73.6	− 1.7
M-38	90	Not measured for Mg						84.7					

* Calculated composition by Philibert–Duncumb equation: $(\mu/\rho)_{Al}^{Al\,K\alpha} = 408$ and $(\mu/\rho)_{Mg}^{Al\,K\alpha} = 4377$.
† Values computed by Ziebold–Ogilvie method (see text).
— Alloy not measured.
§ Mg by difference from aluminum values.

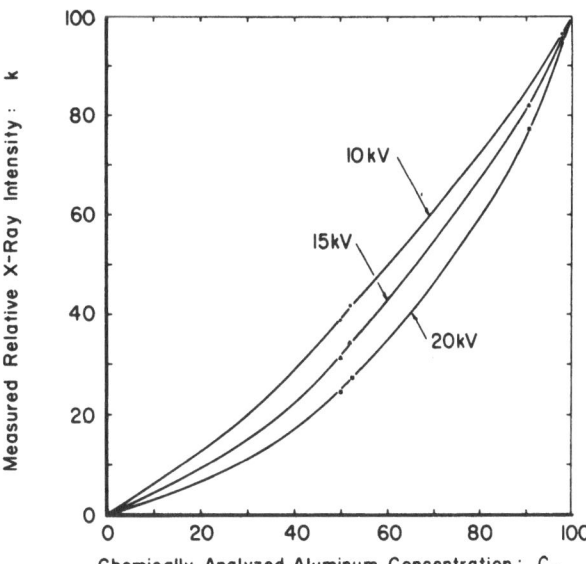

Figure 8. True weight percent *vs.* measured relative X-ray intensities for aluminum in Al–Mg at 10, 15, and 20 kV operating voltages.

From the four values so computed, an average value, $\bar{a}$, was calculated for a given operating voltage. Then using this empirical $\bar{a}$, the compositions were obtained for the remaining alloys by means of equation (2):

$$C = \frac{k\bar{a}}{k(\bar{a} - 1) + 1} \tag{2}$$

The values for C computed at the three operating voltages were nearly equal. As a check on the validity of the average a value, the known compositions were recomputed. Deviations were less than the experimental error for the determination of k. Therefore, the compositions calculated by this means are considered to be valid. The true composition *vs.* k curves are shown plotted in Figure 8.

Final analytical data for both aluminum and magnesium are shown in Table III. Where data are not shown, it was not possible to obtain microprobe data due to specimen instability, e.g., aging prior to measurement of relative intensity data. This again points out that splat-cooled standards should be used as soon after preparation as possible.

Table III shows for magnesium analysis, that the error ranges from 0–4.4% of the amount present and that for fourteen results, six errors are negative, seven positive, while one is zero. The root mean square deviation from the true composition is less than 1.7% over the composition range 10–75% magnesium. This variation is representative of how accurately the relative intensity data k were taken. The fact that no systematic deviation was noted reinforces this viewpoint.

As expected,[14] in the case of aluminum in Al–Mg, systematic errors are clearly indicated by the results (Table III). Nineteen of 20 errors are positive and the remaining error is nearly zero. Furthermore, at lower aluminum concentrations, assuming the Ziebold–Ogilvie calculation to be correct (the aluminum by difference from magnesium tends to confirm this) the deviation from the true value is greater than 10%.

The magnesium results clearly show that the Al–Mg alloys prepared by the splat cooling technique satisfy all of the requirements for a standard suitable for electron probe microanalysis. The aluminum results clearly show the need for standards in the numerous cases where vital correction model input parameters such as (μ/ρ) data or the models themselves are in doubt.

SUMMARY AND CONCLUSIONS

A splat-cooling device is described which is capable of preparing satisfactory standards for electron probe microanalysis. The liquid alloy must be miscible as well. The value of the capability for preparing specifically required standards in two-phase regions is demonstrated.

Some experimentation with hearth design and rotational velocity is usually required. Finally, the apparatus is relatively simple and inexpensive.

The yield is several cubic centimeters of solid which is usually sufficient for both electron probe microanalysis and conventional chemical analysis. The results for Al–Mg alloys show that chemical analysis is definitely required. The actual compositions obtained often differ greatly from the compositions predicted by weighing alone. This is not considered to be a serious drawback since it is only necessary to establish a k vs. true concentration curve. It is imperative that the standards be carefully characterized with regard to homogeneity within shards and from shard to shard by means of the microprobe. The required analysis data should be obtained as soon after preparation of the standards as possible in order to minimize possible difficulties due to aging.

It is concluded that the splat-cooling technique when properly carried out will permit the investigator to prepare standards homogeneous to the extent that the maximum analysis error will be equivalent to the error in measuring relative X-ray intensities. It can be foreseen that with standards available, results from quantitative electron probe analysis will be comparable to those obtained by X-ray fluorescence analysis. Furthermore, it is hoped that such standards will aid in the vital task of evaluation of correction procedures.

ACKNOWLEDGMENTS

The authors wish to thank Mr. C. J. Bechtoldt (NBS) for the X-ray diffraction data, Mr. R. E. Michaelis and the Analytical Chemistry Division (NBS) for the analysis of the Al–Mg samples, Mr. D. B. Ballard (NBS) for the electron microscopy, Mr. D. L. Vieth (NBS) for aid in preparing samples for microprobe examination, and Mr. Paul Soules (Melpar–NASA) for general technical assistance.

Special thanks are due to Dr. H. D. Brody (NASA) for important discussions on the experimental apparatus.

REFERENCES

1. R. Castaing, "Application of Electron Probes to Local Chemical and Crystallographic Analysis," Thesis, University of Paris, 1951.
2. L. S. Birks, J. V. Gilfrich, and H. Yakowitz, "Report of the Washington Electron Probe Users Group," *ASTM Spec. Tech. Publ.*, in press.
3. H. Yakowitz, "Evaluation of Specimen Preparation and the Use of Standards in Electron Probe Microanalysis," *ASTM Spec. Tech. Publ.*, in press.
4. H. Yakowitz, D. L. Vieth, K. F. J. Heinrich, and R. E. Michaelis, "Homogeneity Characterization of NBS Spectrometric Standards II: Cartridge Brass and Low-Alloy Steel," *Advances in X-Ray Analysis, Vol. 9*, Plenum Press, New York, 1966, p. 289.
5. J. I. Goldstein, R. E. Hanneman, and R. E. Ogilvie, "Diffusion in the Fe–Ni System at 1 Atmosphere and 40 k Bar Pressure," *Trans. AIME* **233**: 812, 1965.
6. T. O. Ziebold, "Ternary Diffusion in Cu–Ag–Au Alloys," Ph.D. Thesis (MIT), 1965.
7. H. Yakowitz, D. L. Vieth, and R. E. Michaelis, "Homogeneity Characterization of NBS Spectrometric Standards III: White Cast Iron and Stainless Steel Powder Compact," *NBS Misc. Publ.* 260–12, 13 pp., issued September 19, 1966.
8. P. Duwez, R. H. Willens, and W. Klement, Jr., "Continuous Series of Metastable Solid Solutions in Ag–Cu Alloys," *J. Appl. Phys.* **31**: 1136, 1960.
9. M. Hansen, *Constitution of Binary Alloys* (second ed.), McGraw-Hill Book Co., New York, 1958, p. 105 cf, and p. 232.

10. W. Klement, Jr., "Lattice Parameters of the Metastable Close Packed Structures in Ag–Ge Alloys," *J. Inst. Metals* **90**: 27, 1961.

11. J. Laberrigue–Frolow and P. Radvanyi, "Le Rendement de Fluorescence de la Couche K. Mesures Spectrométriques sur 99 Tc* (6.04 h) et 115, 49 In* (4.5 h)," *J. Phys. Radium* **17**: 944, 43, 1956.

12. P. Duncumb and P. K. Shields, "Effects of Critical Excitation Potential on the Absorption Correction," in: *The Electron Microprobe*, T. D. McKinley, K. F. J. Heinrich, and D. B. Wittry (eds.), John Wiley & Sons, New York, 1966, p. 284.

13. K. F. J. Heinrich, "X-Ray Absorption Uncertainty," in: *The Electron Microprobe*, T. D. McKinley, K. F. J. Heinrich, and D. B. Wittry (eds.), John Wiley & Sons, New York, 1966, p. 296.

14. H. Yakowitz and K. F. J. Heinrich, "Quantitative Electron Probe Microanalysis: Absorption Correction Uncertainty," *Mikrochim. Acta*, in press.

15. T. O. Ziebold and R. E. Ogilvie, "An Empirical Method for Electron Microanalysis," *Anal. Chem.* **36**: 322, 1964.

DISCUSSION

L. Vassamillet (Mellon Institute): Would you be kind enough to give us some of the figures about hearth rotation speed, the maximum diameter of the shards that you were able to use, the cross sectional area—we could see how long they were but didn't really have a feeling for their cross section.

J. I. Goldstein: The shards were approximately $\frac{1}{16}$ in. in cross section, but this will vary from one alloy system to the other. We found that, in the Al–Mg system, we have larger size shards than we did, for instance, in the Au–Si system which you saw in that slide. The hearth rotation speed is of the order of 1500–2000 rpm. This is the order of magnitude that we worked in.

L. Vassamillet: Did you get any feeling for how large a shard could be made before segregation becomes a problem?

J. I. Goldstein: We found that we were getting fairly large specimens, $\frac{1}{8}$ in., typical dimensions when we could see actual dendrites.

L. Vassamillet: You didn't specify the mass absorption coefficient that you used for the case of the aluminum and how much it would have to be changed if you were to reduce that average error?

J. I. Goldstein: We used the mass absorption coefficients according to Heinrich's table.

L. Vassamillet: How much would you have changed them by and what would have been the effect? Could you have got that down to less than a couple of percent?

J. I. Goldstein: We did make the calculation. We didn't put it in the paper because we felt that we were being a little bit too ambitious since we didn't know whether the model itself might have been incorrect. Duncumb says that no model available may be exactly correct and, in the case where $f(\chi)$ is less than 0.5, we had f-χ's on the order of 0.3, so we are dealing with a system where the possibility of something wrong with the model itself exists. Therefore, we are hedging on back calculating. If you wish to back calculate, you may. There are enough data there to do it.

Chairman K. F. J. Heinrich: The absorption correction increases in magnitude with increasing voltage. Would you not then expect that, if the model is proper and the absorption coefficient incorrect, your errors would be worse as your operating voltage increases?

J. I. Goldstein: No, I wouldn't. I would expect that if there were a correct expression in the model itself for f-χ varying as a function of voltage, then this would take care of that and you would expect no change as the function of voltage.

Chairman K. F. J. Heinrich: No, I wouldn't think so. If you have a proper model, and a poor absorption coefficient, then you commit an error. However, if the correction is very small, that error will be small, and if the correction is big, the error will be bigger. Is that not the case?

J. I. Goldstein: Yes, that's right.

Chairman K. F. J. Heinrich: I think really from the considerations all I would say is that one cannot know what is the main source of error in this situation. On the other side, have you tried the curves of Henoc and Castaing? Because this is one case where we do have experimental $f(\chi)$ curves and do not have to rely on Duncumb's model.

J. I. Goldstein: We use Duncumb's correction procedure because we feel it does fit adequately the Castaing correction.

H. Yakowitz (National Bureau of Standards): The Henoc–Castaing measured data is reported at several voltages. If one looks at the curves which Castaing presented in the reprint which you kindly gave to me, it is found that the Philibert–Duncumb values *vs.* those of Castaing–Henoc were off by a couple of percent at 10, 15 and 20 kV. At 10 and 15 kV, Green's data agree with Castaing–Henoc but this is not the case at 20 kV. There will definitely be a difference by using the Castaing–Henoc data. This work is still in a preliminary stage. We would like to splat out some more Al–Mg and start to really study the data correction at a greater number of kilovoltages. What we are trying to show here is the need for standards and that we have successfully prepared them for the Al–Mg system. What you say about the C–H relationships at 20 kV, as Joe I think properly stated, could indeed be a combination of both the model and the μ/ρ value.

QUANTITATIVE MICROPROBE ANALYSIS BY MEANS OF TARGET CURRENT MEASUREMENTS*

J. W. Colby† and W. N. Wise

National Lead Co. of Ohio
Cincinnati, Ohio

and

D. K. Conley

General Electric Co., NMPO
Cincinnati, Ohio

ABSTRACT

In the microprobe analyzer, a portion of the high energy electrons impinging on the surface are backscattered from the sample and re-emitted at high energy levels. Low energy (less than 50 eV) or secondary electrons also are emitted. Both the electron backscatter yield and the secondary electron yield are related to the mean atomic number of the target material and, hence, may be used to provide information about the target composition. Unfortunately, however, the secondary electron yield is very sensitive to the surface condition of the specimen and various instrument parameters. This complicates the otherwise simple linear relationship between sample composition and electron backscatter yield.

It is shown that the effects due to secondary electrons can be minimized by biasing the sample, and that good results can be obtained in the analysis of binary systems. The limitations and utility of the method are discussed, and backscatter yields are determined.

INTRODUCTION

In the electron microprobe analyzer, the accelerated beam of electrons impinging on the surface of the specimen gives rise to a number of signals such as characteristic X-radiation, cathodoluminescence in the visible and near visible range, backscattered and secondary electrons, and sample current. One or all of these signals may be used simultaneously during an analysis, each one providing a unique bit of information. However in some instruments, all of these signals may not be available to the analyst, depending on the construction features of the instrument. In particular, it is quite difficult to separate "true" secondary electrons from backscattered electrons, consequently the microprobe analyst refers to the sum of these signals as the backscattered electron signal, or he may talk about target current measurements as being simply the difference between the beam current and the backscattered electron current. While this may be approximately true at high accelerating voltages in some instruments, it is not

* The work reported herein was performed for the U.S. Atomic Energy Commission under contract AT (30–1)–115.6.

† Present address: Bell Telephone Laboratories, Inc., Allentown, Pennsylvania.

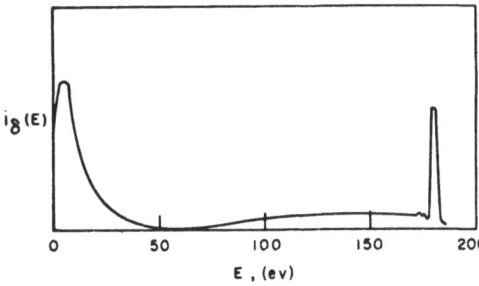

Figure 1. General shape of the energy distribution of secondary and backscattered electrons.

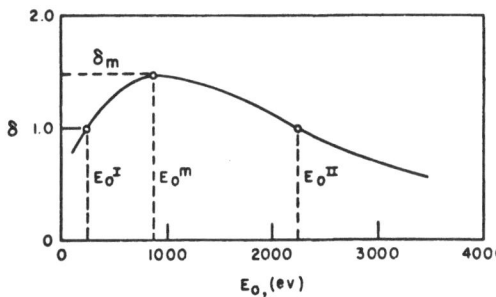

Figure 2. General shape of the yield curve.

exactly true, and it is important to understand the differences and properties of true secondary electrons, and backscattered electrons.

Quantitative analysis by means of target current measurements alone is possible; but is only practicable if the distributions of true secondaries and backscattered electrons have been thoroughly evaluated for the particular instrument being used. Not only are there differences in electron yields between instruments of different manufacture, but there also may be small differences between instruments of similar manufacture, depending on various modifications made to the instrument.

In this paper, the distributions and properties of the true secondary electrons, and backscattered electrons will be discussed, and a method for determining secondary electron yields and of obtaining scanning secondary electron images, independent of the high energy backscattered electrons will be described. It also will be shown that quantitative analysis using target current measurements is possible within certain limits, providing the effects of secondary electrons can be sufficiently reduced.

GENERAL CONSIDERATIONS

If the number of electrons emitted by the target during electron bombardment, having energy E, is plotted against the energy, the result will be a curve[1] similar to that shown in Figure 1. There is a group of electrons having energies equivalent to those of the primary beam, which are elastically reflected primaries. There is a second group, having energies between the energies of the incident primaries and 50 eV, and these are inelastically scattered or rediffused primaries. A third group having energies less than 50 eV, characterized by a sharp maximum in the distribution curve at a few electron volts, and whose total number may exceed the number of incident primaries, are the true secondary electrons. Although there may be true secondaries having energies in excess of 50 eV, and rediffused primaries having energies less than 50 eV, it has become common practice to separate the two groups at 50 eV.

While there is no completely adequate theory covering secondary electron emission, it has been found to be somewhat dependent on the work function of the material.[2]

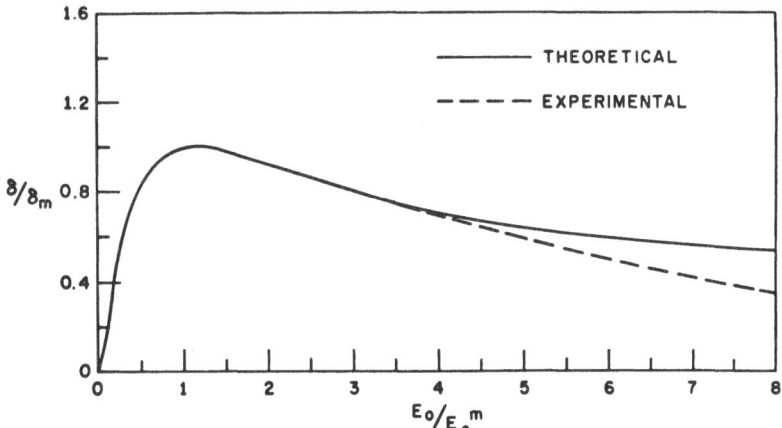

Figure 3. Universal yield curve (experimental and theoretical).

Sternglass[3] also showed that the yield varies as the number of outer shell electrons, hence appears to be a true atomic property, within the periodic system of the elements.

If we define the total secondary electron yield, following convention, as the ratio of the electron current emitted by the target to the impinging primary electron current, a few qualitative remarks may be made concerning the yields of materials. If the yield as defined above, is plotted as a function of the primary incident electron energy, the characteristic yield curve so obtained is shown in Figure 2. The shape of the yield curve is similar for all materials examined, though the various parameters shown may vary considerably from one material to the next. In this figure, δ_m is the maximum yield, $E_0{}^m$ the incident energy at which the maximum yield occurs, and $E_0{}^I$ and $E_0{}^{II}$ the incident energies for which the yield is unity. Maximum yields for metals are between 0.5 and 1.7. They may be broken into two distinct groups: the alkalis, the alkaline earths, and a few other light elements having yields from 0.5–1.0, and the heavier metals whose yields range from 1.1 to 1.7. The elements of the first group also have a relatively low work function.

The dependence of the yield on primary energy was graphically demonstrated by Baroody,[4] who normalized the curves for different metals by plotting δ/δ_m against $E_0/E_0{}^m$, and obtained the universal curve shown in Figure 3. There have been numerous attempts to theoretically derive expressions[4-7] for this universal curve, which have been only partially successful. One of the most successful from the standpoint of fitting experimental data, is the expression due to Lye and Dekker,[8] which gives good agreement up to primary energies of approximately 4 times the "maximum" energy, $E_0{}^m$. Their expression is

$$\delta/\delta_m = \frac{1}{g_n(z_m)} g_n(z_m E_0/E_0{}^m) \tag{1}$$

where

$$g_n(z) = \frac{1 - \exp(-z^{n+1})}{z^n} \tag{2}$$

and z_m represents the value of z for which $g_n(z)$ reaches its maximum value, and $n = 0.35$.

The curve and expression predict that as the primary energy increases beyond $E_0{}^m$, the total yield decreases slowly. However, this expression gives a higher yield at high primary energies than is found experimentally.

To get some idea of the magnitude of secondary electron yields, the curves and expressions may be used with published data[1] for gold to show that at 5 keV, the total yield δ should be approximately 0.70. Since the backscatter yield for gold at 5 keV is approximately 0.49 the true secondary electron yield Δ is approximately 0.21. At 20 keV, however, the true secondary yield is of the order of 0.01.

Thus, it might seem at first that when making analysis at 30 keV, the effects of true secondary electrons are negligible. However, the electrons backscattered from the target impinge on other parts of the instrument, such as the pole piece, causing it to emit secondaries, which may in turn strike the target. Further, while the 30 keV incident primaries, and elastically scattered primaries may produce a negligible number of secondaries the rediffused electrons from below the surface may produce numerous secondaries. Kanter[9] has shown that these inelastically scattered electrons, having lost part of their energy, are approximately five times more effective in producing secondaries. Thus, it may be expected that the backscattered electrons striking the pole pieces, etc., would be even more effective in producing secondaries.

The possible complexity of the currents involved has made quantitative analysis by means of target current measurements alone, appear somewhat less than attractive.

Heinrich[10,11] chose to treat the situation analytically, and investigated the possible electron currents involved. For the massive target normally analyzed, the various electron currents are illustrated in Figure 4. As may be noted, it appears a rather prohibitive task to separate the various electron currents. Heinrich devised a composite target shown in Figure 5, having a small central target, electrically isolated from the massive periphery of the target. Each portion could be independently biased, and the various currents measured. He later modified the composite target, to contain four small central targets, Al, Cu, Ag, and Au.

In the massive target, Heinrich[10,11] showed that the target current i_T is given by

$$i_T = i_B - i_\eta + i_{\eta\eta} + i_{\eta S} - i_S + i_{ST} + i_{SS} \tag{3}$$

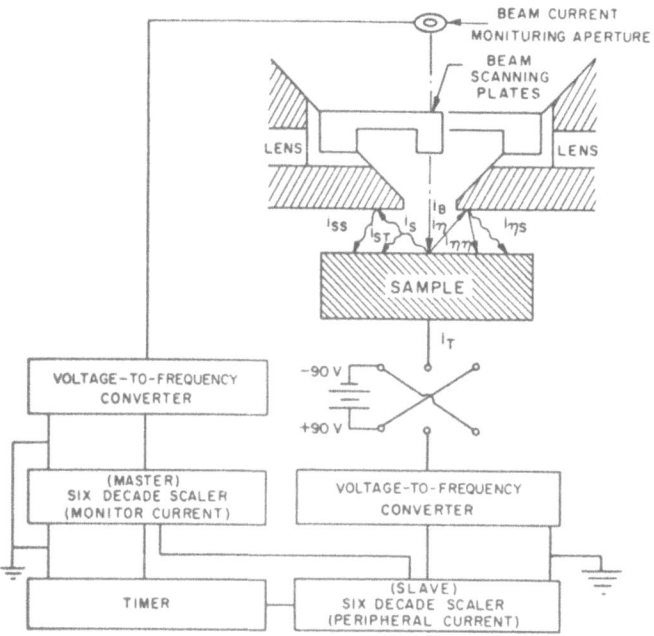

Figure 4. Massive target during analysis showing possible current paths. Parts of the instrument and the measuring circuitry are also shown.

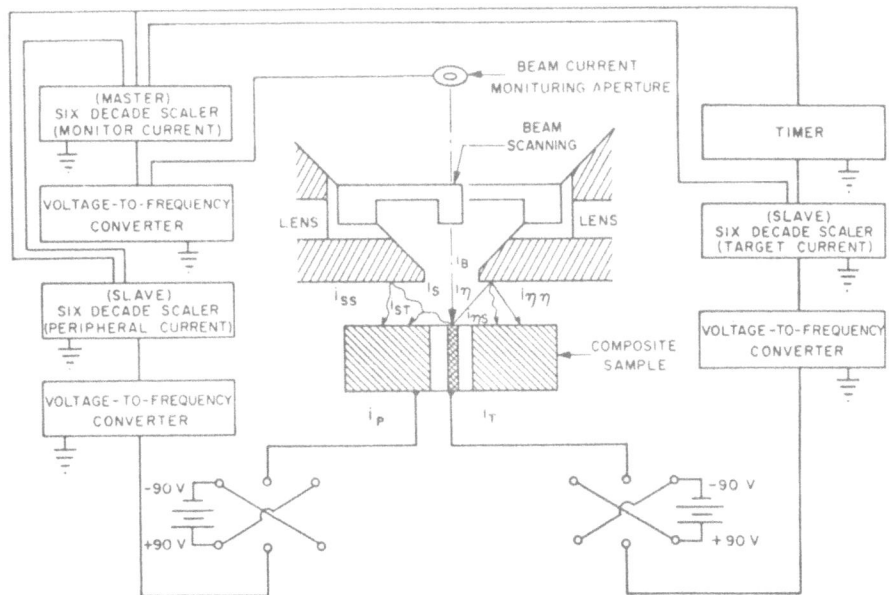

Figure 5. Composite target showing how the possible currents are separated.

Normalizing this by dividing through by the beam current, he obtained

$$\delta = \eta + \Delta - \frac{i_{\eta\eta} + i_{\eta S} + i_{SS} + i_{ST}}{i_B} \qquad (4)$$

where $\delta = 1 - i_T/i_B$, $\Delta = i_S/i_B$ (true secondary electron yield), $\eta = i_\eta/i_B$ (backscattered electron yield), i_B = beam current, i_S = true secondary electron current, i_η = backscattered electron current, $i_{\eta\eta}$ = current due to electrons backscattered first from the target, then from the lens, and finally collected at the target, $i_{\eta S}$ = current due to secondary electrons emitted from the lens due to the backscattered electrons from the target, and collected at the target, i_{ST} = current due to secondary electrons emitted from the target but refocused back onto the target, and i_{SS} = current due to secondary electrons released from the lens due to secondaries from the target.

By various combinations of bias applied to the periphery and the central target, Heinrich[10,11] was able to show that, at 30 keV,

$$i_{\eta\eta} = i_{ST} = i_{SS} \cong 0 \qquad (5)$$

Thus, equation (4) may be considerably simplified for the massive target, to

$$\delta = \eta + \Delta - i_{\eta S}/i_B \qquad (6)$$

and it is easily shown that

$$i_P/i_B = i_{\eta S}/i_B \qquad (7)$$

for the composite target. This peripheral current varies with peripheral bias and with central target composition,[11] and may be quite large. A plot of peripheral current as a function of pheripheral bias for central targets of aluminum, copper, silver and gold is shown in Figure 6. These results[12] were obtained in our laboratory using an instrument

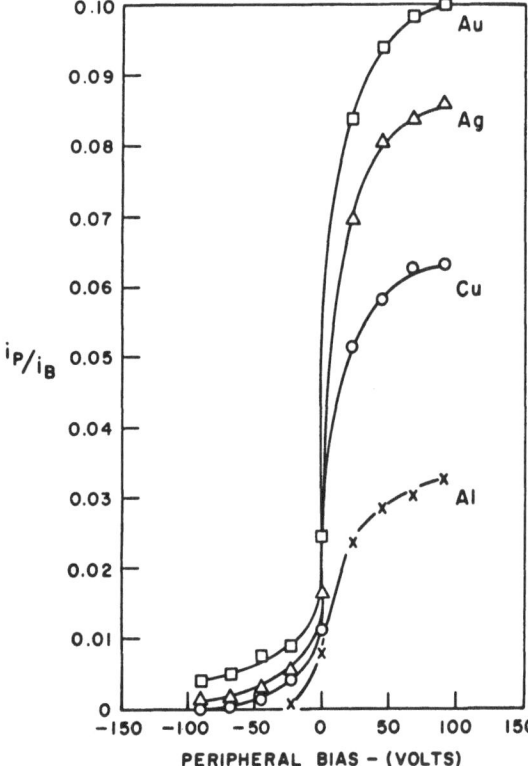

Figure 6. Peripheral current as a function of peripheral bias at 30 keV for composite target.

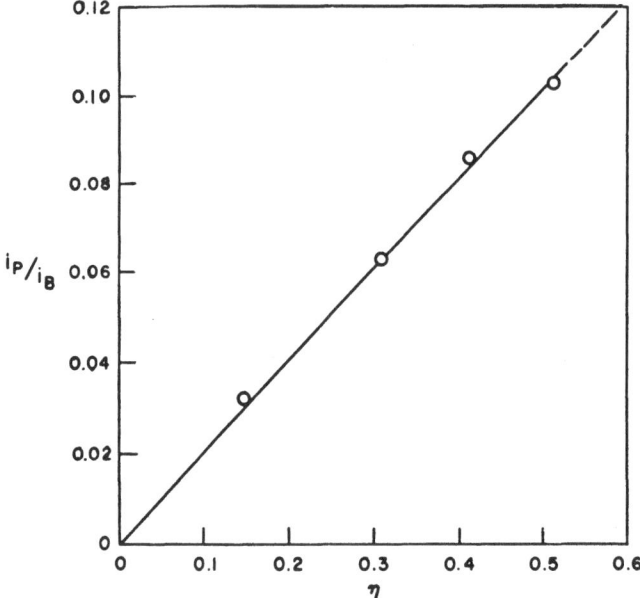

Figure 7. Peripheral current as a function of backscatter yield at 30 keV, for a peripheral bias of +90 V.

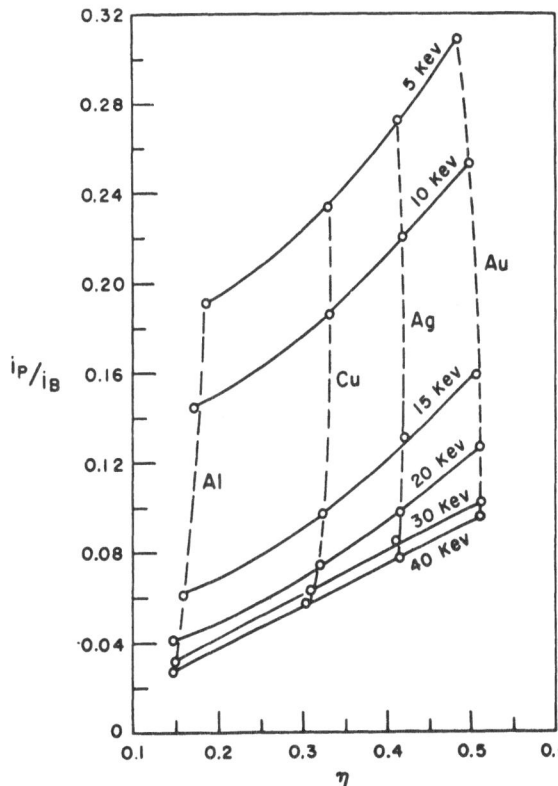

Figure 8. Peripheral current as a function of backscatter yield and accelerating voltage for a peripheral bias of +90 V.

similar to those employed by Heinrich,[10,11] and are in good agreement with his figures. In general we observed a higher peripheral current than he did on his second instrument,[11] but our results for copper agree quite closely with his results from the first instrument.[10]

Neither Heinrich[10,11] nor we[12] found any variation of peripheral current with central bias at 30 keV. However, Wittry[13] using a similar instrument found that peripheral current increased with negative bias to a degree which depended on whether the beam scanning deflection plates were on or off. Neither Heinrich[10,11] nor we found any effect due to the deflection plates. It is not quite clear why this difference exists.

Heinrich also found that at 30 keV the peripheral current was linearly related to the backscatter yield of the central target. This is shown for our instrument[12] at 30 keV in Figure 7, and for the other accelerating voltages in Figure 8. It may be noted that in Figure 8, the peripheral current is linear with backscatter yield at 30 and 40 keV, and may be extrapolated back to zero for zero backscatter. Below 20 keV however, the peripheral current increases rapidly and deviates strongly from the linear relationship with backscatter yield, due to the enhanced secondary electron emission from the lens.

To determine the approximate magnitudes of the secondary electron yields, we employed the composite specimen again, and with zero central bias, we measured the target current first with a negative peripheral bias and then with a positive peripheral bias, at various accelerating voltage from 40 keV down to 5 keV. The results are shown in Figure 9. With a positive peripheral bias, secondaries emitted from the sample are collected on the periphery, consequently the solid curve is probably very nearly equal

Table I. Comparison of Electron Backscatter Yields

	5 keV				10 keV				30 keV			
	Al	Cu	Ag	Au	Al	Cu	Ag	Au	Al	Cu	Ag	Au
Heinrich	—	—	—	—	0.148	0.299	0.428	0.502	0.148	0.306	0.412	0.505
Bishop	0.186	0.352	0.418	0.489	0.177	0.339	0.420	0.501	0.155	0.319	0.420	0.521
Wittry	0.168	0.298	0.361	0.432	—	—	—	—	0.135	0.291	0.388	0.481
Colby	0.185	0.328	0.407	0.482	0.172	0.329	0.416	0.496	0.150	0.311	0.411	0.513

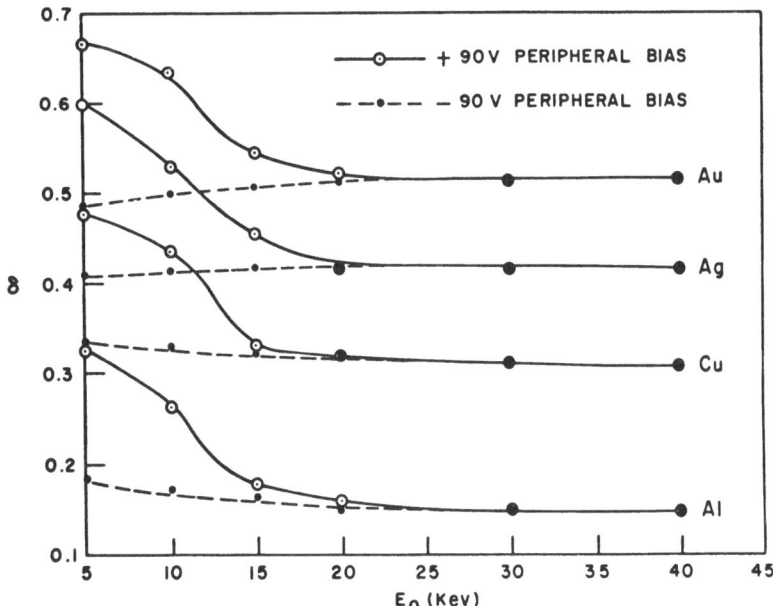

Figure 9. Electron yields as a function of accelerating voltage for the composite specimen.

to the total yield from the target. With a negative peripheral bias, a field is set up which tends to retard the emission of secondaries from the target and the lens, thus, the dashed curve is only due to the backscattered electrons. The difference between the solid and the dashed curves should be the true secondary electron yield. The backscattered electron yields obtained in this way are compared to yields obtained by other investigators in Table I. As may be seen, our measurements are in good agreement with Bishop[14] and Heinrich,[11] but in general are higher than Wittry's[13] values. From Figure 9, we would thus obtain true secondary yields for gold, silver, copper and aluminum, of 0.19, 0.20, 0.15, and 0.14, respectively, at 5 keV. As noted earlier a value of 0.21 was obtained for gold from the universal curve, in good agreement with our experimental results. Similarly from the universal curve, we obtain true secondary electron yields for silver and copper of 0.29 and 0.15 also in good agreement with our results. It was not possible to extrapolate a value for aluminum from the curve but from general considerations it would be expected to be somewhat below the value we obtained, our high value being probably due to the presence of oxide on our sample.

In order to determine the effect of sample elevation on the current distribution, we measured the target current and the peripheral current with a positive 90 V peripheral bias, as the position of the specimen was varied from 1300 μ above the optical focus to 1000 μ below optical focus. In our instrument, the sample was shorted to the lens at approximately 1400 μ above focus. The results are shown in Figure 10. As may be seen δ which consists entirely of the backscatter yield at 30 keV is invariant with sample position, while the peripheral current decreases with increasing height above focus, dropping to about 0.02 with the sample raised as high as practicable.

If we apply these results to Wittry's[13] backscatter determinations the disagreement between his results and Bishop's[14] Heinrich's[11] and ours is understood. Wittry determined backscatter coefficients on massive samples by raising the sample as high as practicable, and applying a positive bias to the sample. Under these conditions, no

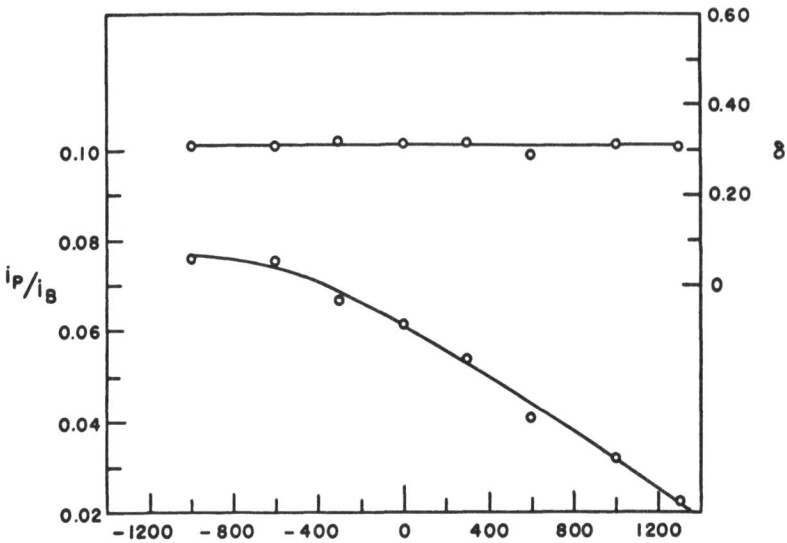

Figure 10. Electron yield, δ, and peripheral current as a function of sample elevation for a peripheral bias of $+90$ V. In this figure, the ordinate is the position of the sample above $(+)$ or below $(-)$ focus.

secondaries would be emitted from the sample, and according to Wittry no secondaries from the pole piece should be collected. However as we see from Figure 10, the secondary current from the pole piece amounts to about 0.02, which is just about the difference between his results and ours.

QUANTITATIVE ANALYSIS

Once the effects of secondary electrons are understood, it becomes possible to use target current measurements for quantitative analysis of binary systems. Poole and Thomas[15] were the first to propose this as an analytical technique in the microprobe, and suggested three different equations for determining composition from backscatter measurements. Their results on binary samples agreed best with an equation proposed by Castaing,[16] although the results weren't conclusive.

Philibert and Weinryb[17] studied the various currents in detail using grids and specimen biasing, and concluded that precise quantitative analysis was not possible using target current measurements.

Heinrich[10] repeated Poole and Thomas' experiments and found as they did that the results agreed most closely with the relation proposed by Castaing, and in most cases, agreement between composition and target current analyses was quite good. Heinrich[11] also showed that, since the peripheral current was linearly related to the backscatter yield of the target, as shown in Figure 7 (i.e., the secondary electron yield of the lens is directly proportional to the number of electrons impinging on it), Castaing's[16] equation

$$\eta_{\text{alloy}} = \sum C_i \eta_i$$

could be reduced to

$$C_A = \frac{i_B - i_{\text{alloy}}}{i_B - i_A}$$

We[18,19] later proved the validity of this expression by analyzing six uranium alloys at various accelerating voltages and found that at 30 keV our results agreed to within 1% of the "true" composition.

Weinryb,[20] however, contends that in the case of two elements of rather different atomic number, the backscatter would originate from two different mechanisms, and it would not be possible to obtain a simple expression relating the backscatter coefficient to the composition, due to the complexity of each mechanism.

However, Brown, using an electron transport equation,[21] has calculated the backscatter yields for copper, aluminum, and two Al–Cu alloys at 29 keV.[22] The calculation in no way assumes a linearity between composition and backscatter yield. The results are shown in Figure 11, and it may be noted that the relationship is quite linear. Further proof of the linear relationship between backscatter yield and composition was obtained by Bishop[14] who measured directly the backscatter yield of several copper–gold alloys, utilizing a specially built apparatus to minimize the effects of secondary emission, and found that backscattered electron yield is linear with composition at 30 keV. At lower accelerating voltages, the curve deviates from linearity and quantitative analysis using target current becomes impractical.

The validity of the linearity between composition and backscattered electron yield having been confirmed, and the utility of this method as an analytical technique having been demonstrated, it is only necessary now to demonstrate the precision of the analyses. For a precision in the sample current measurements of approximately 0.001, and for a difference in backscatter yield of 0.2, such as between copper and gold, concentrations as low as 0.5% can be measured. To illustrate this point, a series of Cu–Au alloys were prepared by quadruple arc melting, inverting between each melt, and chill casting. The resulting alloys were analyzed chemically and by means of target current measurements, using two different microprobes. The results are shown in Table II. Thus, it is seen again that at 30 keV the results agree quite well with the wet chemical analysis and that at the 1% level analysis is possible. At lower accelerating potentials, however, the results are not as good.

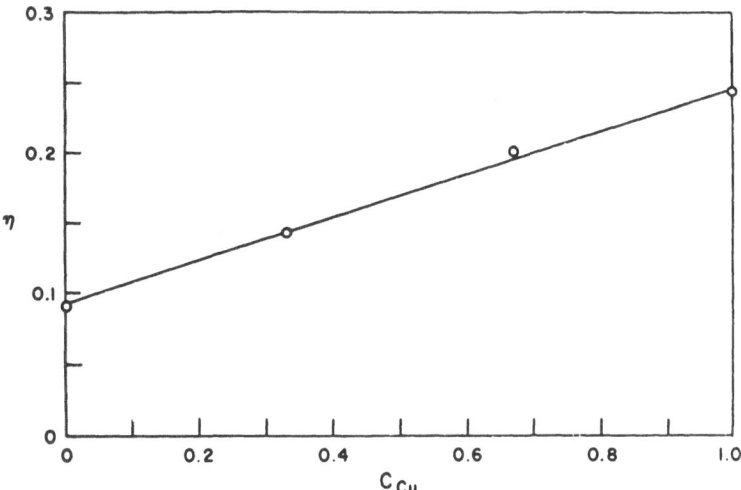

Figure 11. Variation of backscatter yield with composition for the copper-aluminum system at 29 keV, as calculated from the electron transport equation.

Table II. Sample Current Analyses of Gold in Gold–Copper Alloys

Wet chemical analysis, wt.% Au	10 keV	±2σ	Relative error	20 keV	±2σ	Relative error	30 keV	±2σ	Relative error	40 keV	±2σ	Relative error
1.06	1.84	0.18	+74%	1.84	0.25	+74%	0.93	0.13	−12%	0.77	0.11	−27%
2.01	2.49	0.12	+24%	1.81	0.43	−10%	1.87	0.56	−7%	1.66	0.22	−17%
3.07	4.07	0.35	+32%	3.30	0.20	+7%	2.62	0.23	−15%	2.81	0.12	−9%
5.01	6.11	0.15	+22%	5.22	0.12	+4%	5.59	0.14	+11%	5.60	0.06	+12%
9.27	10.20	0.21	+10%	10.56	0.04	+14%	9.79	0.28	+6%	10.15	0.11	+9%
52.41	53.83	0.11	+3%	50.71	0.30	−3%	51.87	0.12	−1%	54.28	0.22	+4%
88.80	95.61	0.88	+8%	92.22	0.91	+4%	90.85	0.59	+2%	92.35	0.46	+2%
95.05	97.53	0.14	+3%	96.91	0.07	+2%	96.25	0.23	+1%	96.68	0.08	+2%
96.83	98.77	0.33	+2%	98.57	0.19	+2%	98.52	0.16	+2%	97.95	0.06	+1%
97.88	99.26	0.19	+1%	97.60	0.16	<1%	98.89	0.33	+1%	98.30	0.11	<1%
98.92	99.39	0.28	<1%	99.45	0.60	<1%	99.12	0.42	<1%	99.30	0.07	<1%

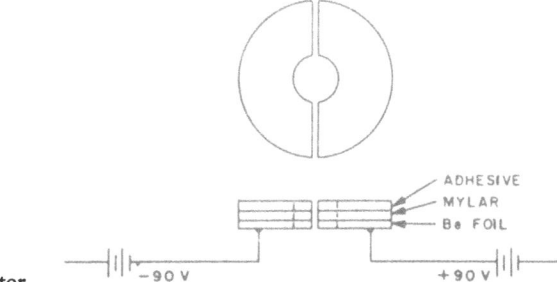

Figure 12. Secondary electron detector.

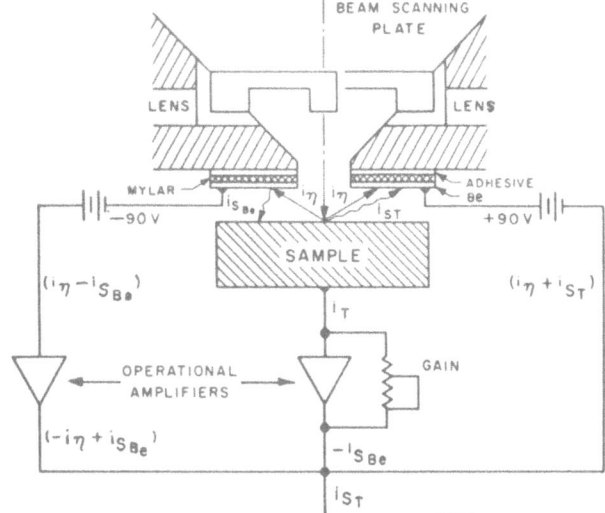

Figure 13. Schematic showing secondary electron detector in position and the associated circuitry.

SECONDARY ELECTRON DETECTION

During the course of this work, it became apparent that a knowledge of the true secondary electron yield would be both desirable and useful. From a knowledge of the various currents in our particular microprobe, it appeared that it might be possible to construct a secondary electron detector[12] which would make it possible to measure the secondary electron yield and to obtain scanning secondary electron images. The detector designed is shown in Figure 12, and its location in the microprobe is shown in Figure 13.

Backscattered and secondary electrons from the target would normally both impinge on the beryllium washer. However the half biased at -90 V would repel secondary electrons from the target, while the half biased at $+90$ V would collect secondary electrons from the target. Both halves would collect high-energy electrons backscattered from the target. In addition, the half biased at -90 V also would reject secondaries arising from bombardment by the backscattered electrons from the target. Beryllium was chosen because it has a low secondary electron yield and a low backscatter yield; hence, it should be efficient as a collector. If the current from one half of the collector is fed through an operational amplifier with unity gain, to reverse the polarity, and added to the current from the other half of the collector, the resulting current will be proportional to the secondary electron yield of the target and the beryllium.

It has been shown that the secondary electron current from the pole piece, and hence the beryllium washer, was proportional to the backscatter yield of the target. Consequently, the target current may also be fed through an operational amplifier with the gain properly adjusted to give a current equal in magnitude but opposite in polarity to the secondary yield of the beryllium collector. If this signal is then added to the two signals from the collector, the resultant signal should be proportional only to the true secondary electron yield of the target material. The secondary electron emission can further be enhanced by biasing the sample negatively, and operating at low accelerating voltages. In this way, true secondary electron scanning images may be obtained, and an estimate of the secondary yield.

CONCLUSIONS

It has been shown both experimentally and theoretically, that although true secondary electron yields from the target are high for low-energy incident primaries, they are negligible for high-energy incident primaries. It has also been shown that electron backscatter yields for a binary system are linearly related to the composition. Through a knowledge of the current distribution in our microprobe, made possible by Heinrich's composite sample, it has been shown that it is possible to do quantitative analysis using target current measurements alone, limited only in precision by the ability to accurately measure target currents. The feasibility of the method has been verified for a series of copper gold alloys.

The composite specimen conceived by Heinrich[10,11] is a much more satisfactory way of studying the various current distributions in the microprobe, than the conventional shielding grids, because the results obtained using grids have to be corrected for secondary emission from the grid wires. This correction is not always simple, and has led to rather erroneous results in many previous examinations, such as those of Philibert and Weinryb.[17] Burkhalter,[23] however, by using shielding grids, was able to achieve good backscatter coefficients, so it is not impossible to use this technique. In most microprobes however the use of the shielding grid is not as practicable as the use of the composite specimens.

ACKNOWLEDGMENTS

The authors would like to thank Dr. K. F. J. Heinrich of the U.S. Bureau of Standards for lending us his composite specimen and for many stimulating discussions; and Drs. D. B. Wittry and D. B. Brown of the University of Southern California for helpful discussions, and for permitting us to use the backscatter calculations on the Cu–Al system.

REFERENCES

1. O. Hachenberg and W. Brauer, "Secondary Electron Emission from Solids," *Advances in Electronics and Electron Physics, Vol. II*, Academic Press, New York, 1959, pp. 413–499.
2. K. G. McKay, "Secondary Electron Emission," *Advances in Electronics and Electron Physics, Vol. I*, Academic Press, New York, 1948, pp. 65–130.
3. E. J. Sternglass, "Secondary Electron Emission and Atomic Shell Structure," *Phys. Rev.* **80**: 925, 1950.
4. E. M. Baroody, "A Theory of Secondary Electron Emission from Metals," *Phys. Rev.* **78**: 780, 1950.
5. H. Salow, "Angular Dependence of the Secondary Electron Emission from Insulators," *Phys. Z.* **41**: 434, 1940.
6. H. Bruining, "Secondary Electron Emission, Parts I, II, and III," *Physica Haag* **5**: 17, 901, 913, 1938.
7. E. J. Sternglass, *Westinghouse Research Lab. Sci. Paper No. 1772*, 1954.
8. R. G. Lye and A. J. Dekker, "Theory of Secondary Emission," *Phys. Rev.* **107**: 977, 1957.
9. H. Kanter, "Contribution of Backscattered Electrons to Secondary Electron Formation," *Phys. Rev.* **121**: 681, 1961.
10. K. F. J. Heinrich, "Interrelationships of Sample Composition, Backscatter Coefficient, and Target Current Measurement," *Advances in X-Ray Analysis, Vol. 7*, Plenum Press, New York, 1964, pp. 325–339.

11. K. F. J. Heinrich, "Electron Probe Microanalysis by Specimen Current Measurement," paper presented at Fourth International Congress of X-Ray Optics and Microanalysis, Paris, France, September 7–10, 1965 (to be published).

12. J. W. Colby and W. N. Wise, "Backscatter and Secondary Electron Measurements in the Microprobe Analyzer," *Summary Technical Report for the Period April 1, 1966–June 30, 1966*, USAEC Report NLCO—980, pp. 1–8.

13. D. B. Wittry, "Secondary Electron Emission in the Electron Probe," paper presented at Fourth International Congress of X-Ray Optics and Microanalysis, Paris, France, September 7–10, 1965 (to be published).

14. H. E. Bishop, "Some Electron Backscattering Measurements for Solid Targets," paper presented at Fourth International Congress of X-Ray Optics and Microanalysis, Paris, France, September 7–10, 1965 (to be published).

15. D. M. Poole and P. M. Thomas, "Quantitative Electron-Probe Microanalysis," *J. Inst. Metals* **90**: 228, 1962.

16. R. Castaing, "Electron Probe Microanalysis," *Advances in Electronics and Electron Physics, Vol. 13*, Academic Press, New York, 1960, pp. 317–386.

17. J. Philibert and E. Weinryb, "The Use of Specimen Current in Electron-Probe Microanalysis," *X-Ray Optics and X-Ray Microanalysis*, Academic Press, New York, 1963, pp. 451–476.

18. J. W. Colby, "Microprobe Analysis of Binary Systems Containing Uranium," *Advances in X-Ray Analysis, Vol. 8*, Plenum Press, New York, 1965, pp. 352–361.

19. J. W. Colby, "The Applicability of Theoretically Calculated Intensity Corrections in Microprobe Analysis," *The Electron Microprobe*, John Wiley and Sons, New York, 1966, pp. 95–188.

20. E. Weinryb, "Etude de la Retrudiffusion entre 5 et 35 keV," *Metaux, Corrosion Industries* **40**: 131–157, 181–201, 1965.

21. D. B. Brown, "Application of an Electron Transport Equation to Electron-Specimen-X-Ray Interactions in the Electron Microanalyzer," Thesis, M.I.T., 1965.

22. D. B. Brown, private communication.

23. P. G. Burkhalter, "Measurements of Backscattered Electrons in an Electron Probe Microanalyzer," *Bureau of Mines Report RI 6681*, 1965.

A COMPARISON OF FOUR SLIT APERTURES FOR SELECTED-AREA ANALYSIS WITH THE X-RAY SECONDARY-EMISSION SPECTROMETER

Eugene P. Bertin

Radio Corporation of America
Harrison, New Jersey

ABSTRACT

Four types of slit apertures—vertical, horizontal, inclined, and single-edge—for selected-area analysis with commercial X-ray secondary-emission (fluorescence) spectrometers are described, evaluated, and compared. The vertical and single-edge slits already have been described by the writer[1] and Rizzo.[2] The other two are described here for the first time. The vertical and horizontal slits are secondary-beam apertures; the other two lie in both the primary and secondary beams. All the slits are mounted on the specimen drawer. No further modification of the spectrometer is required except replacement of the soller collimators with open tunnels or simple slits, and replacement of the flat crystal with a fixed-radius curved crystal to increase sensitivity. The accessories are inexpensive and can be made in even a modest machine shop. They can be installed on or removed from the spectrometer in about 5 min. The accessories are applicable only to linear selected areas where composition varies in the direction normal to but not along the line, for example, linear inclusions and sections of plated surfaces, diffusion couples, and interfaces. The four slits were evaluated and compared for resolution, sensitivity, and spectral-line width. Techniques for use, advantages, limitations, and means for improvement are discussed for each of the slits. The horizontal slit is of little value, but each of the others has features which permit analysis of very narrow selected areas for which pinhole apertures would not be sufficiently sensitive. Slit widths as narrow as 0.00012 in. (0.003 mm) have been used in favorable cases. The inclined slit, used with a pulse-height analyzer, is probably the most useful of the four apertures, combining high sensitivity, high resolution, and narrow spectral-line width.

INTRODUCTION

Commercial flat-crystal X-ray secondary-emission (fluorescence) spectrometers are best applied to analysis of small selected areas by fitting them with fully focusing curved-crystal accessories. Several such accessories have been described,[3-5] and at least one is available commercially.[3,6] However, these accessories are elaborate and expensive, and require an hour or more to install.

Much selected-area work is done on commercial X-ray spectrometers modified only by fitting the specimen drawer or compartment with apertures and replacing the collimators with slits.[7-15] Several-fold increase in sensitivity is realized by replacing the flat crystal with a fixed-radius curved crystal.[7] These accessories are simple and inexpensive, and require only a few minutes to install.

In such applications, slit apertures are superior to pinholes for certain types of specimen. Such specimens include sections of long linear inclusions, diffusion couples, metal-to-metal and metal-to-ceramic interfaces, and other layered structures where composition varies only in the direction normal to the layers. Earlier papers[1,16] describe a vertical secondary-beam slit-probe accessory for the General Electric SPG-2 X-ray spectrometer, techniques for its use, its advantages and limitations, and several applications.

Rizzo[2,17] describes another technique in a study of diffusion of electroplated metal into its substrate. The sectioned specimen is mounted under a single fixed aluminum edge parallel to and just above the specimen surface, and parallel to the plate-substrate interface. The specimen is then translated with respect to this edge.

Two other arrangements are described in this paper: (a) a horizontal slit similar to the vertical slit but with the jaws horizontal and much closer to the specimen plane, and (b) an inclined slit having a pair of slit jaws parallel to and just above the specimen plane.

The horizontal and vertical slits are secondary-beam apertures; the other two lie in both the primary and secondary beams. No evaluation was made of a pure primary-beam slit aperture. The General Electric SPG-2 X-ray goniometer has a horizontal plane and vertical axis, with the specimen plane inclined to the horizontal at a 30° angle. The terms *horizontal*, *vertical*, and *inclined* in this paper refer to the goniometer plane.

This paper compares all four of these selected-area techniques with respect to resolution, sensitivity, and spectral-line width.

EXPERIMENTAL

Equipment

X-Ray Spectrometer. The work was done on a General Electric model XRD-3/SPG-2 X-ray spectrometer. The X-ray tube was a Machlett type AEG-50S having a tungsten target and operated at 50 kV, 45 mA. The source and detector collimators were replaced with open tunnels having the same dimensions but without the soller foils. The analyzer was a G.E. No. A4961 LiF crystal curved to a 14-in. radius; on the G.E. instrument this radius is optimum at 48° 2θ (Ni K_α) and efficient within $\sim 20°$ of this angle. The detector was a G.E. type SPG-2 sealed Kr-filled proportional counter. A G.E. No. 2 preamplifier was used with the original No. 1 ratemeter-scaler, which has no pulse-height analyzer. The capacity of the 14-stage binary scaler was increased 100-fold with a mechanical count register.

Specimen Drawer. The slit accessories were mounted on the G.E. Heinrich miniature-probe specimen drawer,[18] which has a micrometer-driven bidirectional specimen-translation stage. As noted by Rizzo,[2] the micrometer drives do not indicate millimeters directly because of the complex linkages between the micrometers and stage. The relationships between micrometer readings and specimen distances, established by translating a finely scribed millimeter scale, are as follows: lateral-translation micrometer, 1 mm = 0.87 mm; vertical-translation micrometer, 1 mm = 1.16 mm.

Horizontal Slit. Figure 1A shows the horizontal slit in place on the drawer. Figure 2 shows the construction, consisting of two slit blocks A, a base B, various spacers C, and a radiation shield D. The shield prevents X-radiation scattered or emitted by the upper part of the specimen from reaching the crystal by passing over the top slit block. All parts were machined with near gage-block accuracy of steel and given a very thin protective chromium plating. Two 8-32 Allen cap-head machine screws having filed-down heads pass through slots in the blocks and clearance holes in the spacers into tapped

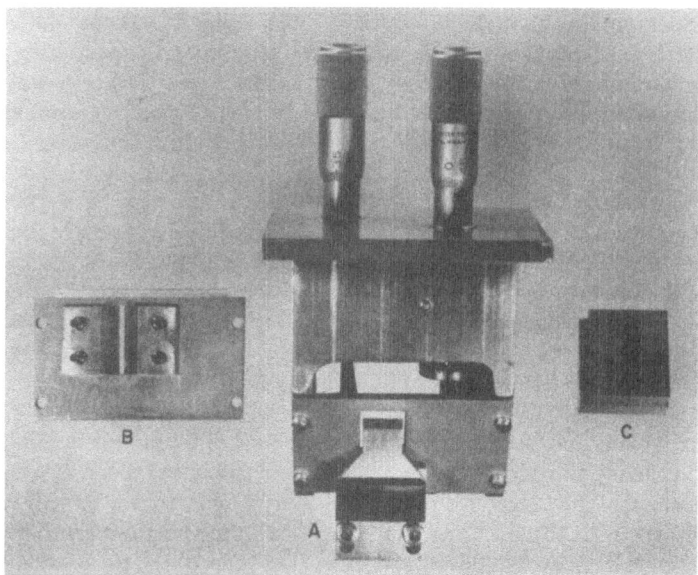

Figure 1. Slit-probe accessories: (A) horizontal-slit assembly in place on General Electric miniature-probe drawer; (B) inclined-slit assembly on specimen-mask plate; (C) vertical-slit assembly.

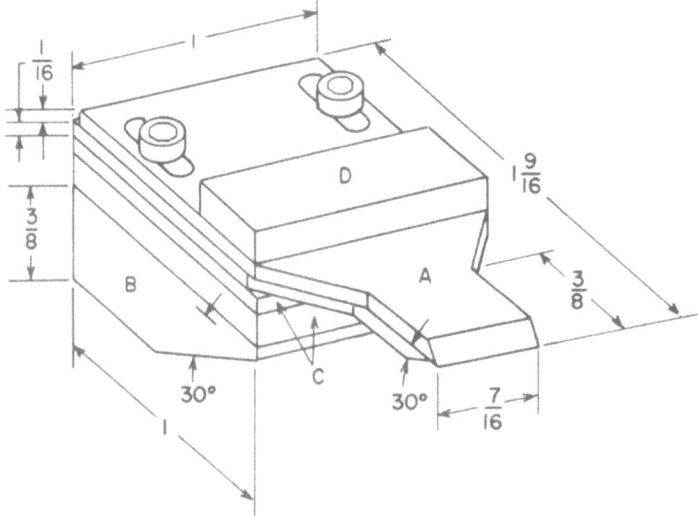

Figure 2. Horizontal-slit assembly: (A) slit blocks; (B) base; (C) spacers;
(D) radiation shield.

holes in the base. The slots allow the slit jaws to be moved with respect to the specimen plane and the top jaw to slightly overhang the lower jaw and shield it from the primary beam. Spacers 1/8-, 1/16-, and 1/32-in. thick allow elevation of the slit to optimum height (see below). The assembly is secured to the mask plate by two 3/16-in. 2-56 flat-head machine screws in the same holes as the original pinhole-mounting block.

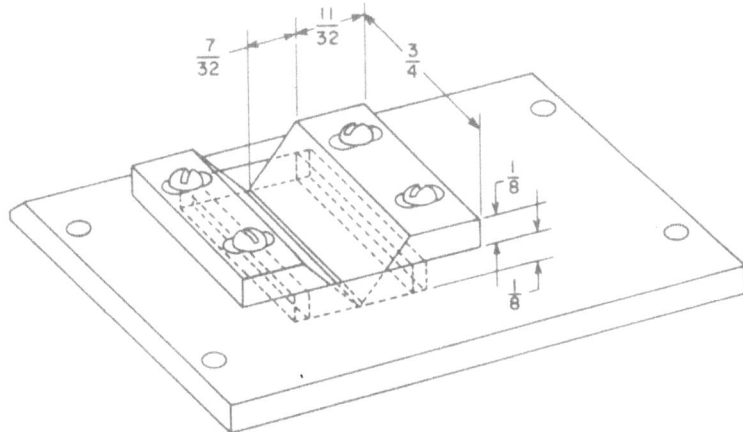

Figure 3. Inclined-slit assembly.

Inclined Slit. Figure 1B shows the inclined slit in place on a specimen mask plate. Figure 3 shows the construction, consisting of two aluminum slit blocks secured to the mask plate with 2-56 machine screws in slots to allow setting the position and width of the slit. The slit jaws are beveled for two reasons: (a) to allow the incremental specimen width under the slit to receive primary radiation from the entire X-ray focal spot, and (b) to allow this incremental width to emit analytical line radiation into the spectrometer over a solid angle. The mating edges are 0.002-in. wide and mate accurately in the plane of the underside of the mask plate. Thus the slit is separated from the specimen surface only by the 0.030-in. thickness of the specimen-translation stage, which lies just under the mask.

Vertical Slit and Single Edge. The vertical slit is described elsewhere[1] and shown in Figure 1C. The single edge consisted simply of the inclined slit with one of the slit blocks withdrawn.

Specimen Translation Motor-Drive. The motor was an Apcor multiratio gear motor, model 2203-1, having six speeds: 1, 2, 4, 12, 30, and 60 rph. The motor was coupled to the specimen-translation micrometers on the drawer by means of a flexible shaft.

Test Specimen

Preparation. Nickel was chosen for the test element primarily because the radius of the only available curved crystal is optimum for Ni K_α, but also because Ni K_α has high instrumental sensitivity and is substantially absent from the primary spectrum. The test specimen consisted of a 1-cm square of 0.0025-in. nickel foil secured between 1-cm copper cubes by an axial machine screw. One side was given a dull matte finish for better visibility of light beams used to align the slit (see below).

Position. Since primary intensity varies over the specimen plane, small specimens and small selected areas on extensive specimens must be placed to receive maximum primary intensity. Optimum positions for the nickel layer of the test specimen in the window of the miniature probe drawer were established as follows. With no aperture accessory on the mask plate, the nickel layer was oriented parallel to the top edge of the window and translated downward, then oriented parallel to a side edge and translated

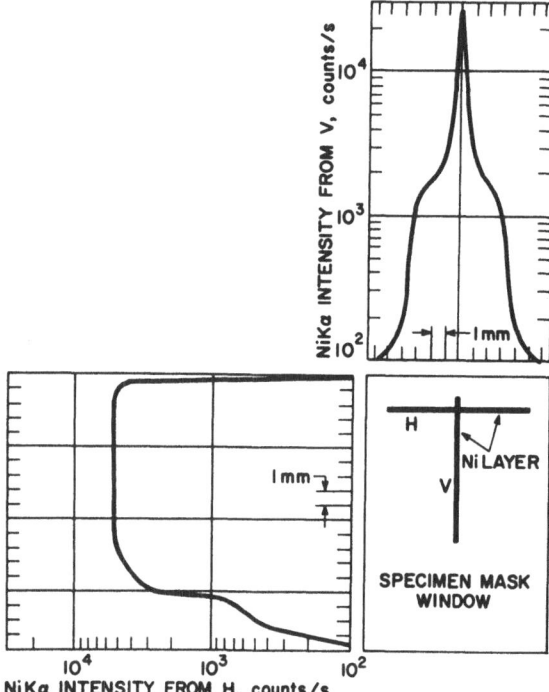

Figure 4. Ni K_α intensity profiles for 1 cm × 0.0025-in. Ni specimen translated horizontally and vertically in the specimen window; optimum specimen positions for horizontal slit (H) and for vertical and inclined slits and single edge (V).

laterally. Ni K_α was recorded during the translations. Figure 4 shows the Ni K_α intensity profiles and optimum specimen positions for horizontally (H) and vertically (V) oriented apertures.

The strong intensity maximum for vertically oriented specimens (Figure 4) does not necessarily lie at the center of the specimen window, but can always be brought there by adjustment of the position of the X-ray tube. Once centered with a given X-ray tube, the intensity maximum remains centered on removal and replacement of the tube provided the X-ray tube collar-adjustment screws and cooling jacket are not disturbed.

Slit Adjustment

All slits were set by use of shims of wrinkle-free Mylar polyester film 0.001-in. thick and 0.5-cm wide. The shims and slit jaws were wiped with chamois just prior to setting to remove dust. After setting, the slits were carefully gaged with 0.001-in. steel shims.

Horizontal Slit. The horizontal slit base (Figure 2B) was mounted on the drawer, and the test specimen was inserted and set with the nickel layer in position H (Figure 4). Spacers were placed on the base as required to elevate the lower slit block to the height of the nickel layer. Two lengths of Mylar shim were laid on the wide rear portion of the block, one at each edge just ahead of the slots. The top slit block was then placed on and advanced until its jaw overhung the lower jaw by ~0.5 mm. The screws were then inserted and the slit blocks advanced, without changing their relative position, as close to the specimen plane as possible without shielding the nickel layer from the primary beam. The screws were then tightened, leaving the shims in place. The drawer was then placed under a low-power microscope and a beam of light directed through the slit onto the specimen plane. As the specimen plane was observed through the microscope, the nickel layer was "nudged" into exact alignment with the light beam, and thereby with the slit.

Vertical Slit. The vertical slit assembly was mounted on the drawer with its slit blocks at maximum spacing ($\sim$3 mm). The test specimen was placed in the drawer and set with the nickel layer in position V (Figure 4). One slit block was set in line with the nickel layer and secured. A Mylar shim was placed between the blocks, the second block advanced to contact and secured, and the shim withdrawn. The nickel layer was nudged into exact alignment in the same way as the horizontal slit.

Inclined Slit and Edge. The specimen mask plate was removed from the drawer and replaced with the inclined-slit assembly shown in Figure 3 with the jaws at maximum spacing ($\sim$4 mm). The test specimen was inserted and set with the nickel layer in position V (Figure 4). One slit block was set in line with the nickel layer and secured, and the nickel layer was nudged into exact alignment. A Mylar shim was placed between the jaws, the second jaw advanced to contact, and the shim withdrawn.

The single edge was formed from the inclined slit by leaving one of the jaws fixed and withdrawing the other as far as its slots would allow, $\sim$2 mm. In Rizzo's work,[2,17] the two edges were separated $\sim$8 mm.

Specimen Resolution

The effective slit width and observed nickel-layer width were evaluated for each slit, prepared as described above, by translating the nickel layer of the test specimen past the slit or edge while recording Ni K_α intensity on the ratemeter-recorder. The specimen was translated downward for the horizontal slit, laterally for the other three apertures. The motor-drive was set at 2 rph, equivalent to 1.0 mm/hr (uncorrected). The effective slit width[8] is taken as the distance on the specimen from the point where the Ni K_α response is just distinguishable above background to the point where the response attains its maximum value. The observed nickel-layer width is the distance on the specimen between the points where Ni K_α response rises above and falls back to background, and is theoretically the sum of the true layer width and effective slit width.

The *measured* effective slit and observed layer widths were derived from the recorder charts using the motor-drive and chart speeds. A scale was prepared from a strip of chart paper for rapid conversion of chart distances to measured specimen distances for the motor-drive and chart speeds used. The *corrected* effective slit and observed layer widths were calculated from the measured widths and the correction factors given above.

Spectral-Line Width and Intensity

Ni K_α spectral-line width and intensity were measured for each slit prepared as already described, but with the nickel layer of the test specimen centered under the slit. The spectrometer was made to scan the Ni K_α peak while response was recorded on the ratemeter-recorder. The peak intensities were then scaled. For the vertical and horizontal slits, the background was measured at a 2θ angle adjacent to the Ni K_α peak without disturbing the specimen. For the inclined slit and single edge, the nickel layer was moved under a slit jaw, leaving a nickel-free area of the test specimen under the aperture. Background was then measured at the 2θ angle for Ni K_α.

Secondary-Beam Projections

The projections of the secondary X-ray beam on the crystal and detector window were photographed for each slit as follows. The horizontal and vertical slits were reset to 0.010-in. width to increase the secondary-beam intensity to realize reasonable photographic exposure time; it was not necessary to alter the inclined slit and edge. The specimen stage was loaded with a nickel strip wide enough to cover the widened apertures. The spectrometer was set at the Ni K_α line (48.64° 2θ) for all exposures. For each slit a film was exposed at the detector window; then the crystal and its two retaining bars were removed, and a film was exposed on the crystal stage. Kodak No-Screen Medical X-ray film was used. The intensity was measured prior to exposing the films, and the exposures at the detector and crystal were based on 7 and 0.5 min, respectively, for each 10,000 counts/sec.

RESULTS AND DISCUSSION

General

Table I summarizes the experimental results. In evaluating the data, it must be remembered that Ni K_α is the optimum wavelength for the fixed-radius curved crystal used. Spectral-line width and diffracted intensity decrease as the analytical line departs from this wavelength. However, the crystal is very satisfactory within $\sim 20°$ of Ni K_α, and one or two other crystals can be provided for adjacent spectral regions.

Figure 5 shows X-ray intensity as a function of distance as the nickel layer is translated past each of the four slits. Figure 6 shows photographs of projections of the secondary X-ray beam on the crystal and X-ray detector for each of the slits. Two detector window outlines are shown: The circular window represents the G.E. type SPG-2 sealed Kr-filled proportional counter used in this work. The rectangular window represents the SPG-6 sealed Xe-filled and the SPG-4 and SPG-7 gas-flow proportional counters.

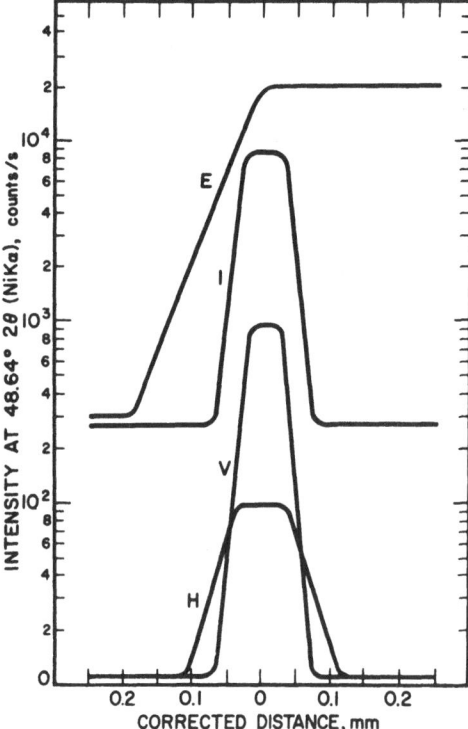

Figure 5. Scans across 0.0025-in. Ni layer with the four slit arrangements: (H) horizontal slit; (V) vertical slit; (I) inclined slit; (E) single edge. Motor-drive speed: 2 rph. Specimen-translation speed (uncorrected): 1 mm/hr. Chart speed: 24 in./hr. Ratemeter time constant: 3 sec. The ratemeter is linear below and logarithmic above $\sim$30 counts/sec.

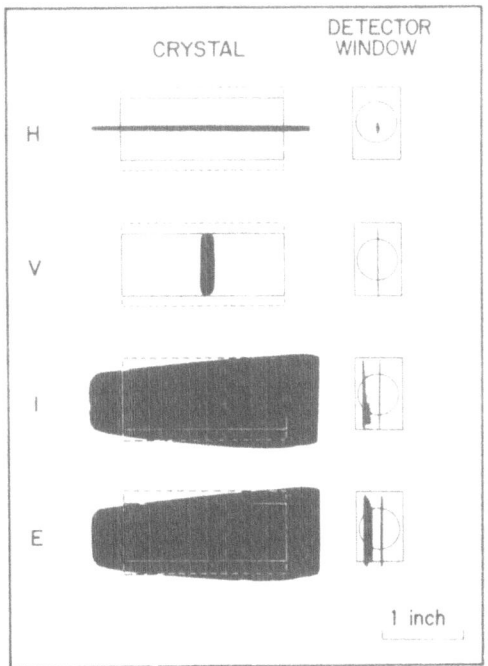

Figure 6. Projections of the secondary X-ray beam at crystal and at detector window with the four slit arrangements. The solid crystal rectangles outline the effective crystal area between the locating edges; the broken lines indicate the top and bottom of the crystal. The rectangular and circular detector windows represent the SPG-7 and SPG-2 detectors, respectively.

Horizontal Slit

The horizontal slit appears to have little value. It does have some advantages: The background is very low, specimen position is not as critical as for the vertically oriented slits (Figure 4), and alignment of the specimen with the slit is relatively easy. However, these features are far outweighed by the disadvantages: The intensity is very low because the selected area, lying in position H rather than V (Figure 4), takes no advantage of the vertically oriented hot spot. The spectral-line width is very great because the unlimited horizontal divergence causes the crystal to intercept a very wide beam which is not fully focused on the detector window. Even for Ni K_α, the optimum wavelength for the crystal, the half width is $\sim 6°\, 2\theta$. The effective slit width is greater than for the vertical and inclined slits because the projection of the horizontal slit on the oblique specimen plane is widened. All these features are evident in Table I and Figures 5H and 6H.

Vertical Slit

Compared with the horizontal slit, the vertical slit gives higher intensity by a factor of ten without increased background and with substantial decrease in effective slit and observed nickel-layer widths, as shown in Table I and Figure 5V. The spectral-line width and spectral purity are excellent, as shown in Table I and Figure 6V. It is easy to align the specimen with the slit, and the specimen may be changed repeatedly without disturbing the slit.

The vertical slit has the disadvantage that the lateral position of the aperture is critical if full advantage is to be taken of the primary intensity hot spot (Figure 4). This feature is common to the other two vertically oriented slits.

The most serious disadvantage of the vertical slit is that, although the selected area receives full excitation of the primary beam, only secondary rays directed straight into

Table I. Summary of Experimental Results

	Horizontal	Vertical	Inclined	Single edge
Ni layer position (Figure 4)	H	V	V	V
Half width of Ni K_α line, deg 2θ	~6	0.2	0.3	0.3
Ni K_α intensity, counts/sec				
Line L	98	920	8,678	20,078
Background[a] B	2	1.5	2,738	3,021
Net $L-B$	96	918	5,940	17,057
Net : background ratio $(L-B)/B$	48	612	2.2	5.6
Effective slit width,[b] mm				
Measured	0.075	0.075	0.071	0.229
Corrected[c]	0.087	0.065	0.062	0.199
Observed Ni layer width,[d] mm				
Measured	0.208	0.183	0.183	—
Corrected[c]	0.242	0.1595	0.1595	—
Theoretical[e]	0.1495	0.128	0.124	—

[a] Measured as described in the paragraph on spectral line width and intensity in the experimental section.

[b] True slit width for horizontal, vertical, and inclined slits: 0.001 in., 0.025 mm.

[c] Correction factors to convert specimen-translation micrometer measurements to corrected specimen distances: Horizontal slit: 1 mm = 1.16 mm; vertical and inclined slits and single edge: 1 mm = 0.87 mm.

[d] True Ni layer width: 0.0025 in., 0.0625 mm.

[e] True layer width plus corrected effective slit width.

the slit aperture enter the spectrometer, Figure 6V shows the narrow beam of rays intercepted by the crystal. This condition is very wasteful of analytical-line intensity.

Inclined Slit

The loss of secondary-beam intensity characteristic of the vertical slit is corrected in the inclined slit. The beveled jaws allow the X-ray tube focal spot to irradiate the selected area almost as effectively as in the vertical slit, and they also allow the crystal to intercept secondary radiation from the selected area over a wide solid angle. The net intensity from the inclined slit is nearly an order higher than from the vertical slit, and with the same spectral-line, effective-slit, and observed nickel-layer widths.

The disadvantages of the inclined slit are as follows. With the horizontal and vertical slits, the X-ray beam intercepted by the crystal contains only the spectral lines of the element(s) under the aperture. With the inclined slit, the beam is highly polychromatic, because of scatter of primary radiation from the slit jaws and possibly secondary excitation of the spectra of the metal of the jaws and its contaminants. Moreover, the beam is highly laterally divergent. The crystal was only curved—not curved and ground—and, thus, incapable of fully focusing the divergent beam and rejecting all extraneous radiation. As a result, the background is very high and the spectral-line width somewhat greater than with the horizontal and vertical slits. Figure 6I shows the divergent beam intercepted by the crystal, and some extraneous radiation at the detector window. However, these limitations are readily remedied. The background from scattered primary radiation is reduced very substantially by making the slit blocks of metal having a higher atomic number, and, more effectively, by use of pulse-height discrimination. An adjustable slit aperture at the detector window excludes the extraneous radiation with little or no loss of analytical line intensity.

If the analytical line is of short wavelength, there may be penetration of the knife edges of the slit jaws. The jaw edges cannot be made thicker without loss in both incident primary and emergent secondary intensity. Here again, a metal of higher atomic number would be beneficial.

A third disadvantage is that the linear specimen cannot be aligned with the slit jaws by the light-beam technique mentioned earlier. In the writer's opinion, the most serious disadvantage is that the specimen cannot be aligned with the slit when set at a narrow width; one of the jaws must be withdrawn to allow observation of the specimen while it is aligned. This presents no problem with an individual specimen. However, when a number of specimens are to be compared, the slit must be opened and reset for each, and it is likely that the analytical results will be impaired by a certain amount of nonreproducibility of slit width.

Single Edge

In Rizzo's single-edge technique, the net intensity is even higher because of less shielding of incident primary and emergent secondary radiation than in the inclined slit. The edge is certainly the simplest of the four types of slit to make. The specimen is easily aligned and may be changed readily.

The single edge has the following disadvantages. At each point of the specimen translation the spectrometer measures the analytical line intensity not from a small incremental width, but from all the specimen area not under the edge. This analytical line intensity is superposed on a high background as in the inclined slit. Thus, point-to-point variation in analytical concentration is derived from small differences in relatively high intensities. All the means cited for reducing background and extraneous radiation with the inclined slit are applicable also to the single edge: use of metal having high atomic number, pulse-height discrimination, and a second slit at the detector window.

The effective slit width (Figure 5E) is substantially wider because the edge-to-specimen spacing allows the spectrometer to receive emission from the nickel layer for some distance after it passes under the edge. This condition could be remedied by reducing the distance between the edge and specimen in either of two ways: The edge could be made to protrude down into the window of the movable specimen stage with the edge lying near the plane of the underside of the stage. This remedy would prevent vertical translation of the specimen stage and restrict its lateral translation. Alternatively,[19] the specimen could be made to protrude up into the window of the stage with the specimen surface lying near the plane of the top of the stage. Then, leaving the edge in its original position in the plane of the underside of the mask, the specimen plane could be brought arbitrarily close to the edge without restricting motion of the stage.

As the specimen layer is translated out from under the edge, the Ni K_α intensity is determined not only by the area of the exposed nickel, but by its position relative to the primary intensity maximum (Figure 4). This effect does not appear in Figure 5E because the nickel layer is sharply defined and very narrow, and was not moved far from the edge. However, if the width of the layer of interest were broad (because of extensive diffusion, for example), the effect would become significant. Finally, the presence of only one edge precludes the study of specimens having three or more layers and, thus, limits the technique to simple binary interfaces.

REFERENCES

1. E. P. Bertin, "Evaluation and Application of an Improved Slit Probe for the X-Ray Secondary-Emission Spectrometer," *Advances in X-Ray Analysis, Vol. 8*, Plenum Press, New York, 1965, pp. 231–246.

2. F. E. Rizzo, "Diffusion of Electroplated Couples," Ph.D. Thesis, University of Cincinnati, 1964.
3. J. A. Dunne, "A New Focusing Vacuum X-Ray Macroprobe Analyzer," *Advances in X-Ray Analysis, Vol. 8*, Plenum Press, New York, 1965, pp. 223–230.
4. M. P. Johnson, P. R. Beeley, and J. Nutting, "Application of X-Ray Fluorescence Probe Analysis to the Study of Segregation in Steels," *ibid.*, pp. 259–268.
5. K. H. Storks and T. C. Loomis, unpublished work.
6. Philips Electronic Instruments, Inc., 750 South Fulton Ave., Mt. Vernon, New York, 10550, commercial literature.
7. E. P. Bertin and R. J. Longobucco, "X-Ray Spectrometric Determination of Composition and Distribution of Sublimates in Receiving-Type Electron Tubes," *Advances in X-Ray Analysis, Vol. 7*, Plenum Press, New York, 1964, pp. 566–583.
8. K. F. J. Heinrich, "X-Ray Probe with Collimation of the Secondary Beam," *ibid., Vol. 5*, 1962, pp. 516–526.
9. J. T. Lynch, "X-Ray Techniques as a Tool to Analyze Intermetallic Coatings," *Plating* 51: 1173–1177, 1964.
10. D. C. Miller, "Norelco Pinhole Attachment," *Advances in X-Ray Analysis, Vol. 4*, Plenum Press, 1961, pp. 513–520.
11. R. D. Sloan, "X-Ray Spectrographic Analysis of Thin Films by the Milliprobe Technique," *ibid., Vol. 5*, 1962, pp. 512–515.
12. K. Togel, "X-Ray Fluorescence Analysis of Small Quantities and Areas," *Siemens-Z.* 36: 497–501, 1962.
13. R. Weyl, "Nondestructive Measurement of Composition and Thickness of Thin Layers by X-Ray Fluorescence," *Z. Angew. Phys.* 13: 283–288, 1961.
14. W. J. Wittig, "Use of an X-Ray Fluorescence Semimicroprobe Attachment in Metallurgy," *Advances in X-Ray Analysis, Vol. 8*, Plenum Press, New York, 1965, pp. 248–258.
15. R. H. Zimmerman, "Industrial Applications of X-Ray Methods for Measuring Plating Thickness," *ibid., Vol. 4*, 1961, pp. 335–350.
16. E. P. Bertin, "X-Ray Probe with Slit Aperture in the Secondary Beam," *Anal. Chem.* 36: 441–443, 1964.
17. F. E. Rizzo, F. Westerman, and R. McDuffie, "Determination of Concentration Profiles by X-Ray Emission," Denver Conf. Applications X-Ray Anal., 1965.
18. General Electric Co., X-Ray Dept., 4855 Electric Ave., Milwaukee, Wisconsin, 53201, *Bulletin 8A-3862*, 1962.
19. F. E. Rizzo, private communication.

DISCUSSION

M. M. Klenck (Atomics International): I think your observation is valid for any spectrometer built around a horizontal goniometer. If you consider any of the spectrometers utilizing a vertical goniometer, you will realize that the "horizontal" slit which you depreciate works very well indeed on them and combines the virtue of getting the slit very close to the specimen with having the slit related to the analyzing crystal in such a way as to gain from good focusing.

E. P. Bertin: This is an absolutely valid comment. I want to emphasize that the terms horizontal, vertical, and inclined are with respect to the type of specimen drawer and goniometer that I described. The General Electric goniometer has a horizontal plane, a vertical axis, and a specimen plane inclined 30° to the horizontal. My terms horizontal, vertical and inclined are with respect to this system.

N. Spielberg (Philips Laboratories): With regard to your inclined slits and the problem of scatter from them, that's probably mostly fluorescence or a substantial part fluorescence of the slit material. If you analyze for materials which are similar in atomic number to the slits, you would still have difficulty even with pulse height analysis.

E. P. Bertin: Absolutely true. This is a disadvantage of the inclined and edge slits. Spectral lines of the slit metal and its contaminants are excited and enter the spectrometer.

F. Bernstein (General Electric Company): What crystal do you use and about what atomic number range can you examine with this system?

E. P. Bertin: I meant to mention that when I enumerated the components of the equipment. I used one crystal, a lithium fluoride crystal curved—just curved, not curved and ground—to a 14-in. radius. This is optimal at 48° 2θ (Ni K_α) for the General Electric instrument. This is one of the reasons I chose nickel as the test specimen. However, I assure you that this crystal works very well on down to indium and on up to chromium, with only slightly poorer performance with respect to spectral line width. With other curvatures of LiF and other types of crystal, all elements from magnesium to uranium can be measured.

THE ANALYSIS OF THE LIGHT ELEMENTS IN FERROTITANIUM ORES AND RESIDUES OF WIDELY VARYING COMPOSITION BY X-RAY SPECTROGRAPHY

Benjamin S. Sanderson and James A. Yeck

National Lead Company
South Amboy, New Jersey

ABSTRACT

Ferrotitanium ores and residues may vary widely in composition. Because of this, interelement effects are very large. An X-ray spectrographic method will be described for the analysis of the light elements in these ores and residues. The method uses a standard helium path Norelco spectrograph with a tungsten target tube. A KAP analyzing crystal is used for all the elements investigated. Appropriate emission lines of the elements magnesium, aluminum, silicon, calcium, titanium, and iron are measured and the results are calculated on a small computer using the empirical corrections for interelement effects of Lucas-Tooth and Pyne. The FORTRAN program for these calculations will be given.

The problems of sample preparation, standardization, interferences and instrumental variables will be discussed. A comparison of X-ray results with chemical analysis is made.

The method is especially adaptable to automated analysis and some of the systems needed will be outlined. The method may also be modified to analyze light elements in other matrices.

INTRODUCTION

The analysis for the light elements in ore samples or residues by chemical means is a long involved procedure requiring several days to complete. X-ray spectrography has many advantages over chemical methods. Speed of analysis, ease of preparation of the sample, nondestruction of sample, and the possibility of automation of the method are a few of these advantages. However, there are many problems which must be overcome before the X-ray method becomes the preferred method. One of the most difficult problems to overcome is the interelement effect. This is caused by the large differences in absorption which occur when the matrix elements differ widely in concentration, along with secondary excitation of some elements by the fluorescent radiation of other elements in the sample. This problem combined with the low counting rates and the high attenuation of long wavelength radiation, makes precise analyses of the light elements very difficult. The methods used in this paper overcome the problems with a minimum expenditure of time and effort.

Interferences

The light elements which are of particular interest in ferrotitanium ores are magnesium, aluminum, silicon, and calcium. The optimum conditions for analyzing each of

Table I. Interferences by Higher-Order Bragg Reflections

Element	$\lambda(\text{Å})$	n	$n\lambda(\text{Å})$
Mg	9.888	1	9.888
Al	8.339	1	8.339
Si	7.126	1	7.126
Ca	3.360	2	6.720
Ti	2.750	3	8.250
Fe	1.937	4	7.748
Fe	1.937	5	9.685

these elements involves three different analyzing crystals and a different setting for the pulse height analyzer for each element. It is not feasible with our present apparatus to change the crystals in the middle of a complete analysis because of the time involved with alignment and flushing with helium. It was decided to use a crystal which could be used for all the elements involved. A potassium acid phthalate (KAP) crystal with a $2d$ value of 26.6 Å was selected. Dunne[1] points out that even though an ammonium dihydrogen phosphate (ADP) crystal has a high reflectivity for magnesium radiation, the fluorescent background caused by the phosphorus gives a poor signal to noise ratio.

While the use of the KAP crystal allows all of the elements to be detected it does have the disadvantage of a low dispersion for wavelength. This makes the problem of interference by high order lines of other elements especially acute. Table I gives a list of the wavelengths of the K_α lines of the elements along with $n\lambda$ which represents the diffracting position of these elements.

It can be seen from this table that the third order titanium line will interfere with the first order aluminum line. Even though pulse height discrimination will eliminate 90–95% of the titanium line, when titanium is a major element the remaining radiation constitutes a substantial interference. A similar situation exists with a fifth order iron line and magnesium; and to a lesser extent with a fourth order iron line and silicon.

Another type of interference also exists. This is the interelement effect. A change in composition of the matrix gives a different absorption for both the entering, exciting radiation, and also for the emerging, excited radiation. Besides this absorption effect there may be an enhancement effect, where the fluorescent radiation from one element excites another element to give off more fluorescent radiation than it would from being excited by the primary radiation alone.

There have been many attempts to minimize these interelement effects by dilution,[2] internal standards,[3] fusion,[4] and other methods designed to get a matrix which is relatively constant.

Standardization

The approach taken in this method is essentially empirical. A second-order equation similar to the one used by Lucas-Tooth and Pyne[5] is used to represent the effect of changes of the matrix on the percentage of the desired element

$$\% \text{ Element} = \alpha_0 + K_0 I_E + K_1 I_E I_{Mg} + K_2 I_E I_{Al} + \cdots + K_6 I_E I_{Fe} \quad (1)$$

where α_0, K_0, and K_1 are coefficients which are determined from standards, and I_E and I_{Mg}, I_{Al}, etc. are the observed X-ray intensities for the desired element and the intensities from magnesium, aluminum, etc. lines, respectively.

Table II. X-Ray Intensities and Chemical Analyses of Standards

Sample	Description	Mg cps	% MgO	Al cps	% Al_2O_3	Si cps	% SiO_2	Ti units	% TiO_2	Fe units	% Fe	Ca units	% CaO
1	T-6129	0.91	0.62	15.34	0.8	31.72	0.67	57.7	59.1	38.5	26.4	4.1	—
2	T-8242	3.41	3.1	10.21	2.05	43.55	2.68	42.9	45.3	59.8	33.7	24.6	1.1
3	T-8255	3.07	3.0	9.99	2.2	43.06	2.49	43.0	45.5	59.9	33.8	26.1	1.1
4	T-8343	2.38	1.4	23.13	4.4	24.17	1.58	9.8	8.4	143.4	61.0	10.0	0.45
5	T-8344	2.96	2.6	4.67	0.82	11.05	0.75	46.7	49.0	60.0	35.4	9.9	0.51
6	T-8349-2	3.49	3.2	8.79	1.9	43.05	3.9	44.3	45.9	60.1	33.3	27.4	1.2
7	T-8426-1	2.83	2.42	4.18	0.94	10.78	0.63	47.5	—	61.7	—	9.7	0.40
8	T-8824	5.57	5.3	4.19	0.96	33.98	2.8	43.4	44.8	62.6	34.8	9.7	0.46
9	T-9083	6.97	7.0	9.36	—	92.05	—	61.1	—	16.9	—	61.9	—
10	T-9084	5.67	5.3	4.63	—	34.55	2.38	43.9	44.7	60.9	28.2	9.4	—
11	T-9093-6	8.52	—	12.65	—	60.63	—	68.7	73.4	14.1	10.2	19.0	—
12	T-9162	3.92	3.02	39.97	7.04	164.35	3.62	17.6	18.6	86.1	40.1	71.3	2.73
13	T-9780	0.41	0.30	8.53	1.84	32.49	4.5	84.8	87.9	4.7	2.62	3.9	1.01
14	T-9851	3.05	2.9	11.81	2.0	54.33	0.51	42.7	—	59.3	—	32.2	1.2
15	T-9951	0.43	0.38	4.55	0.71	12.20	3.59	55.0	54.8	48.1	30.8	3.3	0.61
16	T-10, 167	3.10	1.83	27.33	5.42	55.88	1.3	11.7	10.3	131.7	56.6	22.4	1.18
17	T-10, 640	2.64	2.6	6.06	0.6	16.65	2.80	48.7	—	59.0	—	8.2	0.5
18	D&E306GX8H4	6.02	5.3	4.91	0.96	36.49	—	43.2	44.8	62.3	34.8	9.5	0.46
19	454JC-HCR-10-L4	3.53	3.38	4.51	1.37	22.53	—	80.8	—	9.5	—	14.4	1.22
20	L2A101	5.73	5.34	20.75	4.81	59.77	4.75	64.9	70.2	18.2	11.5	24.3	1.31
21	454AN-H2-8I	0.59	0.4	4.00	1.0	1.27	—	91.8	95.6	3.5	1.7	2.1	0.2
22	454AN-H9-8I	3.71	3.8	1.58	0.1	2.58	—	82.2	84.6	12.9	9.9	2.8	0.1

Notes:
1. cps = counts per second.
2. units = peak height above background in chart paper units.
3. All X-ray results are the average of four analyses corrected to a common base.
4. Iron is present as FeO and Fe_2O_3. The X-ray sees only the Fe, and therefore % Fe figure calculated for all samples.

Since the coefficients for this equation are determined empirically, they take into account the interferences which have been mentioned. In addition, they should correct for instrumental factors such as dead time in counters and electronics, differences of response of the detector for different wavelengths, differences in excitation probabilities of the different elements, and other undefined conditions which are reproducible but do not correspond to the theoretical.

The coefficients for equation (1) were calculated on a small digital computer using a multiple regression program. Details of this calculation and a FORTRAN language version of the program are given in the appendix.

A series of standards was obtained by running accurate chemical analyses on the six elements which are the principal components of these ores and residues of titanium (i.e., magnesium, aluminum, silicon, calcium, titanium, iron). These samples were then analyzed by the standard procedure which will be discussed later. The observed X-ray intensities or their products were then considered to be the independent variables for the multiple regression calculation while the known percent of the desired element was the dependent variable. Twenty-two standards were used. A separate calculation of the coefficients was made for each of the elements. Though the precision in the determination of titanium and iron is not as good for the X-ray method as for the chemical, the percentage of those elements was also determined, since all of the data necessary for this calculation must be obtained to determine the other elements. All of the elements are calculated as their oxides except iron. Table II gives a summary of the samples used. For the determination of each element only the samples which had been analyzed for that element were used. Note that it is not necessary to know the chemical percentage for other than the desired element, since the X-ray intensities are used on the right-hand side of equation (1) rather than the chemical percentages.

Sample Preparation

It is necessary to grind samples well, in order to get reproducible results. Houseknecht and Patterson[6] showed that the addition of a small amount of Boraxo soap helped to grind samples better. It was found for these ores that a 5 min grinding in a Spex 5000 mixer/mill using a lucite-grinding vial with end caps with tungsten carbide inserts and two tungsten carbide grinding balls was adequate. A 2 g sample was used with 0.04 g of Boraxo.

The ground ore was pressed into the recess of a standard plastic sample holder, using a glass plate to form a smoother surface.

Operating Procedures

The X-ray equipment used for this analysis consisted of a Norelco 50-kV X-ray spectrograph, type 52260 with a helium-path attachment; FA-60 tungsten-target X-ray tube; Baird–Atomic single channel pulse height analyzer, model 510 adopted to give a 21 V window; flow proportional counter with stretched polypropylene window; P-10 counter gas; flat KAP crystal; brass holder for standard plastic sample holder and a Norelco electronic circuit panel, type 12049.

The X-ray tube was run at 50 kV and 45 mA. The P-10 gas flowed at 0.08 cubic ft/hr and the helium at 1.0 liter/min. The electronic circuitry was allowed to warm up for a minimum of 45 min with X-rays on. The settings for the counter voltage and pulse height analyzer were checked by means of standards (i.e., magnesium metal for magnesium, etc.). After adjusting the high voltage settings and 2θ position for magnesium the other elements were calculated from the following relations in Table III.

Table III. Instrument Settings
Relative to the Magnesium Settings

Element	2θ	Δ Voltage
Al	− 7.10	− 25V
Si	−12.58	− 50
Ca	—	−120
Ti	—	−115
Fe	—	−155

Table IV. Typical Instrument Settings

Element	2θ	Volts			Scale factor
		High voltage	Base line	Window	
Mg	44.00	1570	6	15	2
Background	40.00	1570	6	15	2
Al	36.90	1545	6	15	2
Si	31.42	1520	6	12	4
Ti	24.20	1455	10	18	8 × 16
Fe	17.10	1415	13	15	8 × 8
Ca	14.86	1450	6	15	8 × 4

Table IV gives the instrumental settings which are typical of an analysis.

Using these settings, corrected to the exact peak position for the magnesium metal standard, five 1-min counts were made at the magnesium, aluminum, and silicon positions and at the background position. The rate meter was used to scan over the titanium, iron and calcium peaks at $\frac{1}{2}°$/min, 2 sec time constant, and 0.8 multiplier. The scans were started approximately 2.0° 2θ before the peak positions. In addition to the unknown samples, a set of standards for each element was run before and after each day's run. Correction factors were used to correct each set of intensities for changes in amplifier gain etc. These factors were the ratios of the calculated intensities for the standards to the average of the two observed intensities. After correction, the intensities were substituted into equation (1) and a percentage of unknown calculated. Normally, the intensities would be punched on paper tape and the calculations done on an SDS-910 computer. When replicate determinations were made the computer program calculated the precisions and typed out all the information for the final report. A FORTRAN II program for this computation is given in the Appendix.

Experimental Data

Table V gives a summary of the results using a straight line equation: % Element = constant + kI_E where I_E is the X-ray intensity for that element corrected for background. A is the chemical value, B is the result using the straight line equation, and C is the result using equation (1). Figures 1–6 show the same data in graphical form.

DISCUSSION

It can be seen from Figures 1–6 that the use of the full equation markedly reduces the scatter of the data. Table VI gives a summary of the statistics of the method.

Table V. Comparison of X-Ray Results with Wet Chemical Results (Column A is for the two parameter straight line equation; B is for the eight parameter equation, and C is for the chemical values)

Sample	% MgO			% Al$_2$O$_3$			% SiO$_2$			% CaO			% TiO$_2$			% Fe		
	A	B	C	A	B	C	A	B	C	A	B	C	A	B	C	A	B	C
1	0.6	0.6	0.8	0.8	2.8	0.9	0.7	2.3	0.8	—	—	—	59.1	60.0	59.4	26.4	21.9	26.3
2	3.1	3.1	3.1	2.1	1.8	1.8	2.7	3.4	3.0	1.1	1.1	1.1	45.3	44.3	44.9	33.7	30.7	33.5
3	3.0	2.7	2.8	2.2	1.9	1.8	2.5	3.1	3.1	1.1	1.2	1.1	45.5	44.4	45.1	33.8	30.7	33.7
4	1.4	2.1	1.3	4.4	4.2	4.3	1.6	1.7	1.8	0.5	0.5	0.5	8.4	9.3	8.1	61.0	65.3	60.7
5	2.6	2.6	2.7	0.8	0.9	0.7	0.7	0.6	0.7	0.5	0.6	0.5	49.0	48.3	47.8	35.4	30.8	34.8
6	3.2	3.1	3.2	1.9	1.6	2.0	3.9	3.1	3.9	1.2	1.2	0.9	45.9	45.8	46.9	33.3	30.8	34.0
7	2.9	2.5	2.6	0.9	0.8	0.7	0.6	0.6	0.7	0.4	0.6	0.5	—	—	—	—	—	—
8	5.3	5.2	5.1	1.0	0.8	0.9	2.8	2.4	2.4	0.5	0.6	0.5	44.8	44.9	44.5	34.8	31.9	33.1
9	7.0	6.6	7.0	—	—	—	—	—	—	—	—	—	—	—	—	—	—	—
10	5.3	5.3	5.2	—	—	—	2.4	2.5	2.3	—	—	—	44.7	45.4	45.0	28.2	31.2	32.4
11	—	—	—	—	—	—	—	—	—	—	—	—	73.4	71.6	73.4	10.2	11.8	9.0
12	3.0	3.6	3.0	7.0	7.3	7.0	—	—	—	2.7	2.7	2.7	18.6	17.6	18.7	40.1	41.6	39.9
13	0.3	0.1	0.3	1.8	1.6	1.7	3.6	2.3	3.4	1.0	0.4	0.8	87.9	88.6	87.6	2.6	7.9	2.9
14	2.9	2.7	2.8	2.0	2.2	2.4	4.5	4.0	3.7	1.2	1.4	1.3	—	—	—	—	—	—
15	0.4	0.1	0.3	0.7	0.9	0.4	0.5	0.7	0.5	0.6	0.4	0.4	54.8	57.1	55.6	30.8	25.9	30.9
16	1.8	2.8	1.9	5.4	5.0	5.5	3.6	4.2	3.6	1.2	1.0	1.1	10.3	11.3	10.7	56.6	60.5	57.1
17	2.6	2.3	2.4	0.6	1.1	1.0	1.3	1.1	1.3	0.5	0.6	0.6	—	—	—	—	—	—
18	5.3	6.2	5.5	1.0	0.9	1.1	2.8	2.6	2.6	0.5	0.6	0.5	44.8	44.6	44.4	34.8	31.7	32.8
19	3.4	3.2	3.5	1.4	0.9	1.4	—	—	—	1.2	0.8	1.3	—	—	—	—	—	—
20	5.3	5.3	5.3	4.8	3.8	4.8	4.7	4.5	5.1	1.3	1.1	1.3	70.2	67.6	70.1	11.5	13.5	12.4
21	0.4	0.3	0.4	1.0	0.8	1.1	—	—	—	0.2	0.3	0.3	95.6	96.0	95.3	1.7	7.4	2.0
22	3.8	3.4	3.7	0.1	0.3	0.4	—	—	—	0.1	0.4	0.2	84.6	85.9	85.2	9.9	11.3	9.5

Table VI. Summary of Precisions and Accuracies of Method

		Standard Deviations				
			Accuracy			
Element	Precision for 66 d.f.	B	d.f.	C	d.f.	Concentration range, %
MgO	0.22	0.36	19	0.15	13	0.3–7.0
Al$_2$O$_3$	0.25	0.61	17	0.26	11	0.7–7.0
SiO$_2$	0.26	0.67	14	0.25	8	0.5–4.7
CaO	0.11	0.24	16	0.16	10	0.1–2.7
TiO$_2$	1.13	1.28	15	0.70	9	8–96
Fe	0.67	3.84	15	1.78	9	2–61

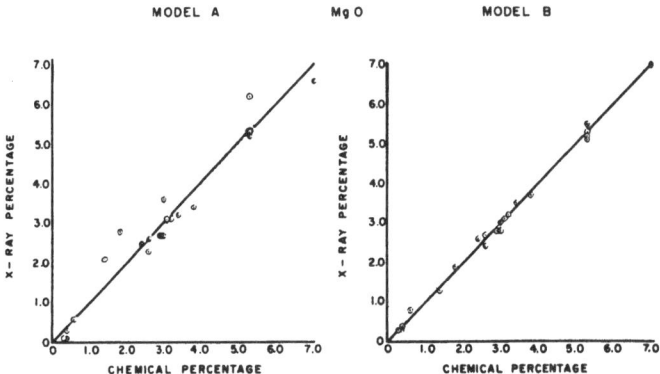

Figure 1. Comparison of results for magnesium using the two parameter straight line relationship of model A against the complete equation of model B. The line represents a perfect agreement with wet chemical results.

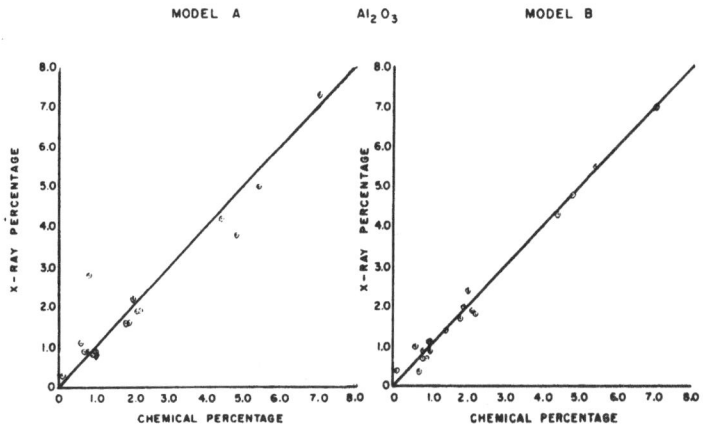

Figure 2. Comparison of the straight line model A with the full equation model B in the analysis of aluminum.

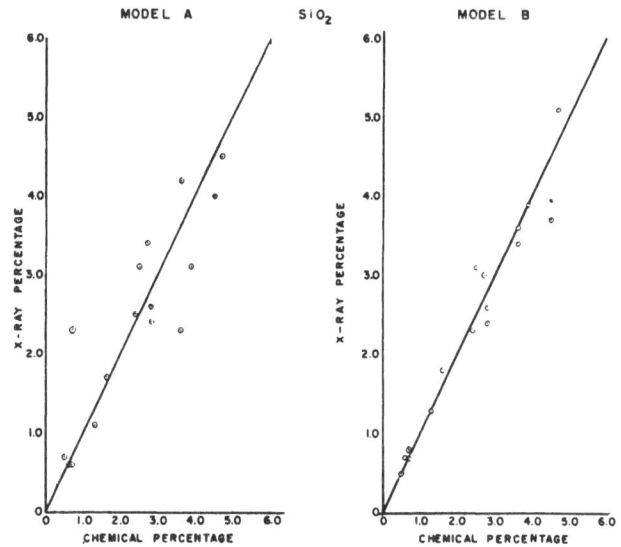

Figure 3. Comparison of the straight line model A with the full equation model B in the analysis of silicon.

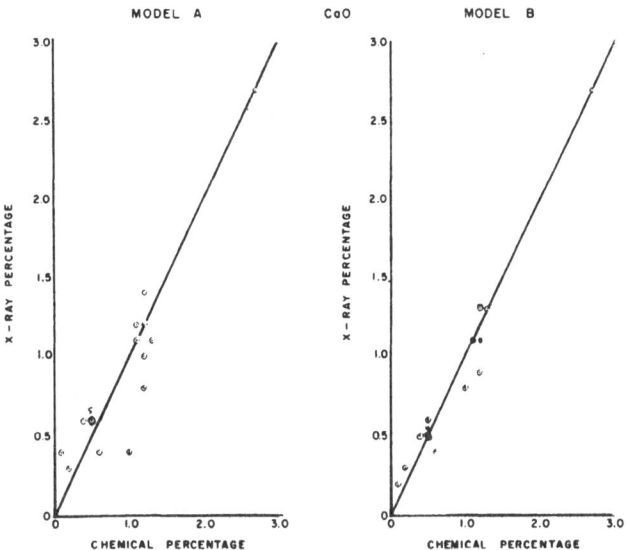

Figure 4. Comparison of the straight line model A with the full equation model B in the analysis of calcium.

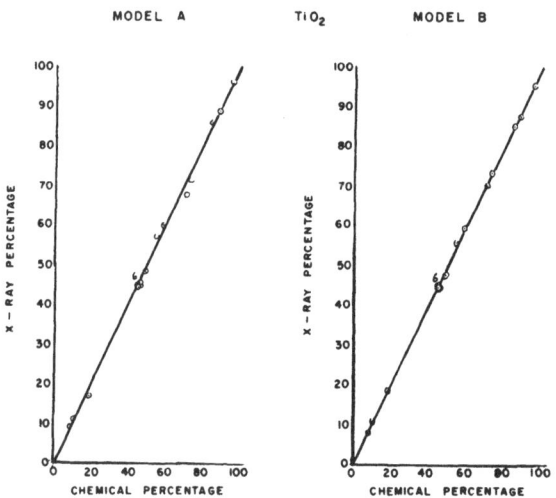

Figure 5. Comparison of the straight line model A with
the full equation model B in the analysis of titanium.

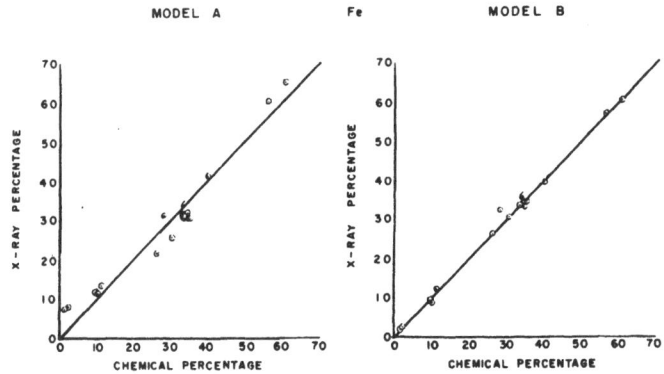

Figure 6. Comparison of the straight line model A with the full equation
model B in the analysis of iron.

The precision is a measure of the ability to reproduce a measurement. For our purpose, the accuracy is defined as a measure of the ability of the X-ray method to reproduce the chemical method. It is obvious that there is a reduction in the standard deviation for accuracy when equation (1) is used. Even though the standard deviations for TiO_2 and iron are high the coefficient of variation is about the same as for all the other elements [i.e., approximately 10% for the straight line calibration and 5% for equation (1)].

There are many possibilities for increasing the accuracy; one would be to expend more time in counting. Another means of improving accuracy would be the use of a vacuum spectrometer and a thin window chromium target X-ray tube. The excitation efficiency of the chromium tube for the light elements is much higher than for the tungsten tube. It also would be advantageous to use two or more analyzing crystals in order to take advantage of the better dispersion at different wavelengths. This is only feasible if the crystal could be changed without breaking the helium flow or the vacuum. The use of different counters in the different wavelength regions should also improve the precision.

Although the method has been described for a particular system, there is no reason why it could not be adapted to many widely different systems. Lucas-Tooth and Pyne[5] have demonstrated the use on alloys. It would appear that many complex mineral systems could be analyzed by these methods.

AUTOMATION

Even though the time to make an analysis by the X-ray method is much shorter than the chemical methods, the counting and recording of all the intensities is tedious and prone to transcription errors. The method lends itself to automation. It is quite feasible to program the goniometer to step to each angle count for the proper length of time, move to the next and repeat. It also is very easy to have the output of the scaler either interface directly with a small computer or punch tape or cards which are compatible with a computer. This has the advantage of freeing personnel from the job of transcribing data. It also makes it possible to run analyses at night unattended thereby gaining additional instrument availability.

REFERENCES

1. J. A. Dunne, "Components for X-Ray Fluorescence Spectroscopy in the 5–65 Å Wavelength Region," *Norelco Reporter* 11: 109–112, 1964.
2. E. L. Gunn, "Fluorescent X-Ray Spectral Analysis of Powdered Solids by Matrix Dilution," *Anal. Chem.* 29: 184–189, 1957.
3. G. J. Lewis, Jr. and E. D. Goldberg, "X-Ray Fluorescence Determination of Barium, Titanium and Zinc in Sediments," *Anal. Chem.* 28: 1282–1285, 1956.
4. F. Claisse, "Sample Preparation Techniques for X-Ray Fluorescence Analysis," *Quebec Dept. Mines Prelim. Report No. 402*, 1960.
5. J. Lucus-Tooth and L. Pyne, "The Accurate Determination of Major Constituents by X-Ray Fluorescence Analysis in the Presence of Large Interelement Effects," *Advances in X-Ray Analysis, Vol. 7*, Plenum Press, New York, 1964, pp. 526–54.
6. T. M. Houseknecht and W. Patterson, "Sample Preparation for X-Ray Analysis," *Spectrographers News Letter* 17: 2, 1964.

APPENDIX

The first program listed here is for the solution of n equations in m unknowns by a least-squares procedure. If a linear set of equations

$$y_j = \sum_{i}^{m} b_i X_i$$

```
  1  *     MULTIREGRESSION PROGRAM
  2  C
  3  C     TAKES X'S LINE BY LINE,N IS NO. OF X'S,MAX OF 20. NN IS NO. OF
  4  C     LINES. THE ANOVA IS FOR SS FOR EACH VARIABLE ADJUSTED FOR ALL OTHERS
  5  C     SENSE SWITCH 1 SET TO TYPE VARIANCE-COVARIANCE MATRIX
  6  C     SENSE SWITCH 2 SET TO CALC. RESIDUALS. DATA TAPE RELOADED.
  7  C
  8        DIMENSION S(20,21),X(20),Z(20),B(20)
  9        COMMON S
 10     41 ACCEPT 1,N,NN
 11        IN=N+1
 12  C-----COMPUTE X'X AND X'Y MATRIX
 13    171 DO 17 K=1,N
 14        DO 17 J=K,IN
 15     17 S(K,J)=0.
 16        SIGMAY2=0.
 17        DO 16 I=1,NN
 18        ACCEPT TAPE 2,(X(J),J=1,IN)
 19        DO 15 K=1,N
 20        DO 15 J=K,IN
 21        S(K,J)=S(K,J)+X(J)*X(K)
 22     15 S(J,K)=S(K,J)
 23     16 SIGMAY2=SIGMAY2+X(IN)*X(IN)
 24  C-----INVERT (X'X)
 25        DO 18 K=1,N
 26     18 Z(K)=S(K,IN)
 27        CALL INVE(N,S)
 28  C-----CALCULATE COEFFICIENTS
 29        DO 31 I=1,N
 30        B(I)=0.
 31        DO 31 J=1,N
 32     32 B(I)=B(I)+S(I,J)*Z(J)
 33     31 TYPE 3,I,B(I),I,Z(I)
 34  C-----COMPUTE THE VARIANCE,SIGMA SQUARED.
 35        NNN=NN-N
 36        XN=NNN
 37        REG=0.
 38        DO 19 I=1,N
 39     19 REG=REG+B(I)*Z(I)
 40        SIGMA2=SIGMAY2-REG
 41  C-----CALCULATE AND TYPE VARIANCE-COVARIANCE MATRIX
 42        DEV=SIGMA2/XN
 43        IF(SENSE SWITCH 1)20,22
 44     20 TYPE 4
 45        DO 52 K=1,N
 46        DO 52 J=K,N
 47        V=DEV*S(K,J)
 48     50 TYPE 5,K,J,V,K,J,S(K,J)
 49     52 CONTINUE
 50  C-----ANALYSIS OF VARIANCE
 51     22 TYPE 6
 52        DO 51 J=1,N
 53        SS=B(J)*B(J)/S(J,J)
 54        F=SS/DEV
 55     51 TYPE 7,J,SS,SS,F
 56        TYPE 8,NNN,SIGMA2,DEV,NN,SIGMAY2
 57        XNN=NN
 58        COR=Z(1)*Z(1)/XNN
 59        R2=(REG-COR)/(SIGMAY2-COR)
 60        TYPE 9,R2
 61  C-----CALCULATION OF RESIDUALS
 62        IF(SENSE SWITCH 2)40,41
 63     40 PAUSE 1
 64        TYPE 11
 65        DO 42 I=1,NN
 66        ACCEPT TAPE 2,(X(J),J=1,IN)
 67        PRED=0.
 68        DO 43 K=1,N
 69     43 PRED=PRED+X(K)*B(K)
 70        RES=X(IN)-PRED
 71        TYPE 10,X(IN),PRED,RES
 72     42 CONTINUE
 73        GO TO 41
 74  C
 75      1 FORMAT(2I3)
 76      2 FORMAT(E15.8)
 77      3 FORMAT($B($,I2,$)=$,E15.8,5X,$X'Y($,I2,$)=$,E15.8)
 78      4 FORMAT(//,3X,$VARIANCE-COVARIANCE MATRIX$,5X,$INVERSE MATRIX$
 79        2 ,/)
 80      5 FORMAT($V($,I2,$,$,I2,$)=$,E15.8,6X,$S($,I2,$,$,I2,$)=$,E15.8)
 81      6 FORMAT(//,30X,$ANOVA$,/,$ SOURCES,12X,$D.F.$,8X,$SSS,16X,$MSS,
 82        2 13X,$F$)
 83      7 FORMAT($ REG ON X($,I2,$)$,6X,$1$,5X,E15.8,2X,E15.8,2X,F8.2)
 84      8 FORMAT($ DEVIATIONS$,6X,I3,5X,E15.8,2X,E15.8,/,$ TOTALS,11X,I3,
 85        2 5X,E15.8)
 86      9 FORMAT(//,$   R2=$,F6.5)
 87     10 FORMAT(3(5X,E15.8))
 88     11 FORMAT(/,9X,$Y OBS.$,14X,$Y CALC.$,13X,$RESIDUALS$,/)
 89  *     END
```

```
COMMON ALLOCATION

  76270 S

PROGRAM ALLOCATION

  00005 X        00055 Z        00125 B        00175 N
  00176 NN       00177 IN       00200 K        00201 J
  00202 I        00203 NNN      00204 SIGMAY2  00206 XN
  00210 REG      00212 SIGMAZ   00214 DEV      00216 V
  00220 SS       00222 F        00224 XNN      00226 COR
  00230 R2       00232 PRED     00234 RES

SUBPROGRAMS REQUIRED

  INV

THE END
```

can be represented by matrices, then the least-square solution is given by the following matrix equations.

$$[b] = [X'X]^{-1}[X'Y]$$

where $[b]$ is a m-dimensional vector and $[X'X]$ is a $m \times m$ symmetrical matrix. $[X'X]^{-1}$ is the inverse of this matrix and is also $m \times m$ and symmetrical. The program not only calculates $[b]$ but gives a complete statistical analysis. The second program is for the calculation of equation (1) and typing out the results with the precisions. An example of the typed report is also given here.

These programs are written in FORTRAN II language for an SDS-910 computer with paper tape input, typewriter input/output, and 4K core memory.

```
 1  *      MATRIX INVERSION ROUTINE
 2  C
 3  C-----INVERSE REPLACES ORIGINAL MATRIX A.
 4  C
 5         SUBROUTINE INV[N,A]
 6  C
 7         DIMENSION A[20,21]
 8         COMMON A
 9         NP=N+1
10         DO 100 I=1,N
11  C-----STORE I-TH COL. OF UNIT MATRIX IN N+1-COL. OF A.
12         DO 5 J=1,N
13    5    A[J,NP]=0.
14         A[I,NP]=1.
15  C-----DIVIDE PIVOT ROW BY PIVOT ELEMENT
16         DIV=A[I,I]
17         DO K J=1,NP
18    K    A[I,J]=A[I,J]/DIV
19  C-----NO CHECK IS MADE FOR ZERO PIVOT
20         DO 99 J=1,N
21         IF[I-J]97,99,97
22   97    FAC=A[J,I]
23         DO 98 K=1,NP
24   98    A[J,K]=A[J,K]-A[I,K]*FAC
25   99    CONTINUE
26  C-----STORE N+1 COL. IN I-TH COL.
27         DO 9 J=1,N
28    9    A[J,I]=A[J,NP]
29  100   CONTINUE
30         RETURN
31  *      END


COMMON ALLOCATION

  DUMMY A

PROGRAM ALLOCATION

  DUMMY A        00012 INV      00013 NP       DUMMY N
  00014 I        00015 J        00016 K        00017 DIV
  00021 FAC

THE END
```

```
COMPILER READY

*    1  *      ORE ANALYSIS BY XRAY FLUORESCENCE
*    2  C
*    3  C         CALCULATES THE PERCENT CONCENTRATIONS OF SIX ELEMENTS AND THE
*    4  C      PRECISIONS FOR REPLICATES.CORRECTS FOR INTER ELEMENT EFFECTS.
*    5  C      THE INPUTS ARE A MATRIX OF COEFFICIENTS,VECTORS OF XRAY
*    6  C      INTENSITIES,
*    7  C      EACH VECTOR PRECEDED BY N THE NO. IN THE GROUP AND THE NO. OF
*    8  C      GROUPS COMES BEFORE THE FIRST VECTOR.
*    9  C
*   10         DIMENSION COEF(8/6,6),ALPHA(6),COUNTS(3/6),SIGMAY(6),
*   11        2 SIGMAY2(6),TSS(6),STDDEV(6),SS(6,25),Y(6,13)
*   12         TSS(J) = 0
*   13         NTDF = 0
*   14         ACCEPT TAPE 1,COEF,ALPHA
*   15       1 FORMAT(6(7F9.6/),6F9.6)
*   16      21 PAUSE 1
*   17         ACCEPT TAPE 2,NGROUPS
*   18         L=NGROUPS
*   19         TYPE 3
*   20       3 FORMAT(21X,$XRAY FLUORESCENCE ORE ANALYSIS$,//,2X,$PERCENTS,4X
*   21        2 $MGO$,6X,$AL2O3$,5X,$SIO2$,6X,$TIO2$,7X,$FE$,8X,$CAO$,5X,$D.F.$,
*   22        3 //,2X, $SAMPLE$,/)
*   23         DO 11 M=1,L
*   24         ACCEPT TAPE 29,N,NSAMPLE
*   25         DO 8  K=1,N
*   26         ACCEPT TAPE 4,( COUNTS(J),J=0,6)
*   27         DO 6 J=1,6
*   28         SUM=0
*   29         DO 5 I=0,6
*   30       5 SUM=SUM + COUNTS(I)*COEF(I,J)
*   31         Y(J,K) = ALPHA(J) + SUM*COUNTS(J)
*   32       6 CONTINUE
*   33      23 TYPE 7,NSAMPLE,(Y(J,K),J=1,6)
*   34         IF(N-1)11,11,12
*   35      12 NDF=N-1
*   36         DO 18 J=1,6
*   37         SIGMAY(J)=0.
*   38         SIGMAY2(J)=0.
*   39         DO 9 K1=1,N
*   40         SIGMAY(J)=SIGMAY(J)+Y(J,K1)
*   41       9 SIGMAY2(J)=SIGMAY2(J)+Y(J,K1)*Y(J,K1)
*   42      18 SS(J,M)=SIGMAY2(J)-SIGMAY(J)*SIGMAY(J)/N
*   43       8 CONTINUE
*   44      14 TYPE 18,(SS(J,M),J=1,6),NDF
*   45      18 FORMAT(/,2X,$S.S.$,4X,6(F7.4,3X),I3,//)
*   46         NTDF=NTDF+NDF
*   47      11 CONTINUE
*   48         DO 19 J=1,6
*   49         L=NGROUPS
*   50         TSS(J)=0.
*   51      13 DO 38 M=1,L
*   52      38 TSS(J)=TSS(J)+SS(J,M)

*   53      19 STDDEV(J)=SQRT(TSS(J)/NTDF)
*   54         TYPE 17,STDDEV,NTDF
*   55      17 FORMAT(/,1X,$STDDEV$,3X,6(F7.4,3X),I3)
*   56         GO TO 21
*   57       2 FORMAT(I3)
*   58      29 FORMAT(I3,I4)
*   59       7 FORMAT(2X,I4,4X,6(F7.2,3X))
*   60       4 FORMAT(7F7.2)
*   61  *      END

PROGRAM ALLOCATION

00005 COEF      00131 ALPHA    00145 COUNTS    00163 SIGMAY
00177 SIGMAY2   00213 TSS      00227 STDDEV    00243 SS
00717 Y         01107 J        01110 NTDF      01111 NGROUPS
01112 L         01113 M        01114 N         01115 NSAMPLE
01116 K         01117 I        01120 NDF       01121 K1
01122 SUM

SUBPROGRAMS REQUIRED

  SQRT

THE END
```

DISCUSSION

L. S. Gray (Armour Industrial Chemical Co.): In using the Lucas-Tooth method, on a wide range of ores, do you have to account for everything in every one of the ores? For example, if you get an ore from Africa and an ore from someplace else, one might include, say, a lot of arsenic and the other might not.

B. S. Sanderson: We have found that the six elements that we are interested in are, as far as we have determined, the important ones. To be exact, the iron and the titanium, being mostly in large quantities, are the ones that we are principally correcting for. I have no doubt that there are probably other systems in which you would have to change the elements that you consider. We have not found any others that are particularly affected.

L. S. Gray: Did you try leaving out any of the major elements, just to see what would happen?

B. S. Sanderson: Yes. You will find that a number of these coefficients are quite small. If you are going to do the calculations on a computer, you will find that the best unbiased estimate of these coefficients is the one using all of them. Some of them may be statistically nonsignificant. That is, they are so small that the effect of random errors is larger than the effect of using those coefficients. But you gain very little by leaving them out, except ease of computation. Now, if you are going to do this on a desk calculator, this may be a significant reason to omit them. However, I suspect that if you have some equations with four or five coefficients and some with eight, you will get more confused doing it by hand than if you left all eight of them in even though it didn't have much effect on the final value. We examine the equation statistically to see the significance of each of the coefficients. However, we leave them there whether they are significant or not.

N. Spielberg (Philips Laboratories): This comment is not in the way of a criticism because, as you point out, you had a number of severe experimental constraints. But I thought it would be useful to show the sensitivity curves for different counter tube gases. In the figure below, Curve 1

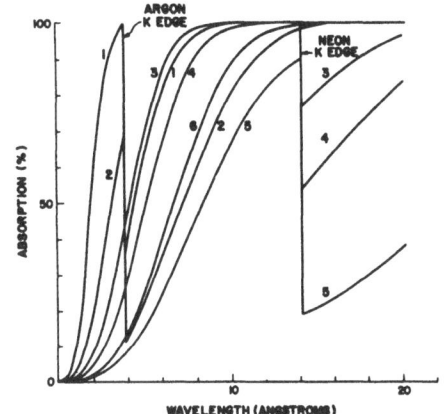

Absorption curves for 22.2-mm-diameter counter tube for various gas fillings: (1) 90% argon–10% methane; pressure, 760 mm Hg. (2) 90% argon–10% methane; 200 mm. (3) 90% neon–10% methane; 760 mm. (4) 50% neon–49% helium–1% butane; 760 mm. (5) 90% neon–10% methane; 200 mm. (6) 46% nitrogen–46% helium–8% methane; 760 mm.

shows the sensitivity for the P-10 gas mixture at normal atmospheric pressure, in a counter of 22.2-mm absorbing path length. Curve 2 shows it at 200 mm pressure. The titanium will appear still on the high side of the absorption edge, so you may not get too much gain, for example, discriminating against titanium or iron which were among the problems which you mentioned earlier in your talk. If we look at Curve 3 and Curve 4, Curve 3 is a 90% neon–10% methane mixture at atmospheric pressure. Curve 4 is a 50% neon–49% helium–1% butane mixture at atmospheric pressure that I reported in my paper.* You will see that they discriminate fairly well against the harder radiation. Now the only problem with using those, as I mentioned also, is the question of stability. But assuming that that could be worked out, and assuming also that your counter tube window is fairly impervious to, say, helium gas, so that the composition doesn't change, there may be some hope to still maintain the discrimination against the harder element. You would probably still need your calculations for a number of other things, but such mixtures might be

---

* N. Spielberg, "Characteristics of Flow Proportional Counters for X-Rays," this volume, pp. 534–545.

helpful sometime in the future. (Curve 5 is 90% neon–10% methane at 200 mm pressure and Curve 6 is 46% nitrogen–46% helium–8% methane at 760 mm pressure.)

B. S. Sanderson: Yes, any of these improvements would help. We don't worry too much about the complications of a long equation. As long as the computer is going to take care of it for us, this isn't a big problem. We feel that probably the way to get around the Al–Ti interference is to use an analyzing crystal which will give you a good wavelength dispersion which, of course, the KAP crystal does not do. We can just barely separate the two, but the tails of the titanium curve, of course, do interfere with the aluminum one. With better dispersion, one can actually separate them by the optics of the system.

E. M. Proctor: I was just going to make the comment that in studying the absorption effects of different instrumental parameters, we found that the most prohibitive absorption factor was the detector window and the difference between using a vacuum and a helium path was negligible for elements down to sodium.

B. S. Sanderson: I agree. Probably our reason for wanting to go to a vacuum instrument is not so much that we would be able to use the vacuum, but that we would have a very nice sample changer. Also, the fact that you can interchange two different detectors, the flow detector being in place all the time. The scintillation counter is more effective for the titanium and the iron radiation than the flow counter.

A GLASS FUSION METHOD FOR X-RAY FLUORESCENCE ANALYSIS

J. O. Larson, R. A. Winkler, and J. C. Guffy

Chevron Research Company
Richmond, California

ABSTRACT

A thermally tough, low-melting, low-viscosity glass composition is described that has been used as a fusion mixture in sample preparations for X-ray fluorescence analysis.

The fusion mixture is cast into a glass disk which, after annealing, is used directly for the X-ray measurements. Problems arising from disk surface imperfections have, in the past, been minimized by either prolonged polishing of the glass surface, or grinding the disk and pressing the powder into a pellet. This practice has been eliminated through the incorporation of appropriate internal standard elements into the glass fusion mixture. This approach also minimizes interelement effects and errors due to instrumental drift.

A variety of corrosion products have been successfully analyzed by this method. Results will be given.

INTRODUCTION

Depending upon the problem at hand, many methods of sample preparation may be used in the X-ray fluorescence analysis of solids. Simple solution, grinding (with or without pelleting), and powder dilution methods may be adequate for a wide variety of analyses. However, for many materials, particularly for those that will not dissolve easily, matrix and particle size effects can become very troublesome for accurate work. Claisse[1] showed how a Borax glass fusion technique eliminated particle size effects, minimized matrix effects, and allowed many analyses to be performed without an internal standard. Townsend[2] extended this concept and applied it to the analysis of silica–alumina based catalysts and to corrosion products. Rose *et al.*[3] used the method at the U.S. Geological Survey Laboratory to determine light elements in rocks; and Luke[4] reported on determining refractory metals in ferrous alloys and high alloy steels.

The experimental sections of most of the work cited above contain statements concerning the fragile nature of the glasses used. These statements include: "The cast disk must not be removed at this stage as shattering will occur because of thermal shock," and "If the hot disk comes in contact with a cold metallic object, it will crack; if chilled too rapidly, it may explode when handled." If the glasses are to be reground and pelleted prior to examination, this factor is not necessarily too troublesome. However, for our purposes, we wanted to use the glass disks directly in the instrument and therefore a tougher material was desired. The tendency of some of the Borax glasses to stick to the metal casting plates was also noted in this early work.

This paper describes a glass which overcomes the shattering problem, a simplified method for casting the disks, and how the use of an internal standard removes the necessity for very smooth pellets.

EXPERIMENTAL

Glass Flux

A glass made of low atomic number elements, which was thermally tough, and had a low-melt viscosity was desired. Combinations of B_2O_3, Na_2CO_3, Li_2CO_3, Al_2O_3, SiO_2, and MgO were tried. A number of promising combinations were encountered, but the boron–sodium–lithium melts seemed most useful, particularly from a low-melt viscosity point of view. The final composition selected—six parts B_2O_3, two parts Na_2CO_3, and one part $Li_2B_4O_7$—was one which showed a minimum tendency to fracture during cooling. It also exhibited good solvent properties as a fluxing material and did not lose its advantageous thermal properties when various internal standard materials were added. Melt viscosity increases were noted when silica-based samples were added in large amount, but the 9 : 1 dilution most commonly used in this work gave no significant problems. While this combination was developed in a reasonably logical fashion, the final recommendation should by no means be regarded as a truly optimum solution because the work was stopped as soon as a satisfactory useful flux composition was established.

For a period of time after this glass was developed, small batches were prepared for use as needed in the laboratory. The technique rapidly became useful on a broad application front, and this small batch practice became too time consuming. Arrangements were made with the Corning Glass Works, Corning, New York, to prepare larger batches of the glass. At this time they prepared the material according to the specifications and ground to a −100 mesh at a cost of about $200 for a 10-lb batch. This is enough material for approximately 600 samples when used in the manner to be described later.

Preparation of Glass Disk

The exact amount of sample and flux to be used will, of course, depend upon the particular analytical problem. For many applications we have found that 7.2 g of glass and 0.8 g of finely ground sample is a good combination. These materials are mixed intimately, placed in a 30-ml platinum crucible, and fused over a Meeker burner. Slow heating at first is desirable to make sure that gases evolved from the sample do not cause sputtering. This is followed by a 10–15 min fusion at full burner heat and occasional swirling of the crucible to ensure thorough mixing. A muffle furnace set to operate between 1050 and 1100°C also is equally acceptable. The molten glass is poured into a graphite mold and pressed with a graphite block while the melt is fluid and flows easily. The mold is made by drilling a 1-in. diameter hole $\frac{1}{8}$ in. deep in a $\frac{1}{4}$-in. thick graphite block. The disk is transferred while still hot to an annealing oven consisting of a Pyrex Petri dish placed on an electric hotplate set to about 200°C. The disk may be allowed to cool by turning off the hotplate or by placing it after a few minutes in an asbestos cloth sandwich.

After the disk has cooled to approximately room temperature, the overflow glass is trimmed with a pair of forceps and is now ready for use in the X-ray spectrometer. This relatively simple cooling procedure is all that is needed to prevent shattering of the pellet. Figure 1 shows the equipment used in the preparation.

Preparation of Standards

Standards for the calibration curves were prepared by doping 5 g of an appropriately selected base material with standard solutions of the metals of interest. Mixtures were

Figure 1. Equipment for preparing glass pellets.

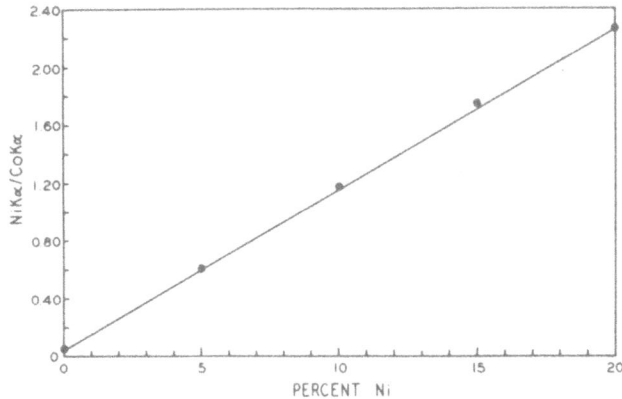

Figure 2. Analytical curve for nickel.

dried in an oven at 105°C and mixed thoroughly in a Wig-L-Bug Mixer. The standard pellets were prepared by adding 0.8 g of the synthetic standards to 7.2 g of the glass flux and proceeding with the fusion described above.

RESULTS

Figure 2 shows a typical calibration curve for nickel in which the ratio of the Ni K_α to Co K_α quantity is plotted against percent nickel. It is to be noted that the plot remains linear over a wide concentration range.

Table I shows comparative X-ray and chemical analyses of a series of corrosion deposits. The results in nearly all instances are well within the expected accuracy of the two methods.

Table II indicates the reproducibility of the pellet method for both nickel and chromium. In this instance, cobalt was used as an internal standard for both metals.

After a little experience with making the glass pellets, the upper surface comes out almost perfectly smooth. However, once in a while, pellets are produced which show a surface roughness which can cause a variation in fluorescent intensity. Figure 3 shows three pellets representing a normally smooth product, one from a poor casting, and a third from a mold which was made very rough on purpose. Table III shows how the use of an internal standard eliminates the variations due to surface imperfections. These results are based on pellets made with the disfigured mold.

Table I. Comparison of X-Ray and Chemical Analyses

Sample no.	% Ni	
	X-ray	Chemical*
1	0.60	0.45
2	0.75	0.85
3	3.20	3.15
4	4.90	4.70
5	5.40	5.50
6	6.40	6.45
7	8.00	7.60
8	13.1	13.0, 13.2
9	15.6	15.6, 15.7

* Dimethylglyoxime and polarographic methods.

Table II. Determination of Precision

Sample no.	% Ni	% Cr
1	5.50	8.60
2	5.80	8.50
3	5.80	8.60
4	5.65	8.40
5	5.70	8.60
6	5.65	8.60
Average	5.68	8.55
Standard deviation	0.121	0.092

Table III. Effect of Pellet Surface Conditions on X-Ray Measurements

	Counts/sec		$\dfrac{\text{Ni } K_\alpha}{\text{Co } K_\alpha}$	% Ni found
	Ni K_α	Co K_α		
Smooth Pellet No. 1	5185	8150	0.636	5.35
Rough Pellet No. 1	4725	7215	0.653	5.50
% Change	8.9		2.7	
Smooth Pellet No. 2	3430	6730	0.510	4.30
Rough Pellet No. 2	3350	6680	0.503	4.20
% Change	2.3		1.4	
Smooth Pellet No. 3	3345	7860	0.426	3.45
Rough Pellet No. 3	3100	7460	0.415	3.35
% Change	7.3		2.6	
Smooth Pellet No. 4	2230	6680	0.334	2.60
Rough Pellet No. 4	2128	6400	0.333	2.60
% Change	4.5		0.3	

Figure 3. Glass disks showing surface roughness.

CONCLUSIONS

In our experience the glass described, by being thermally tough, simplifies the pellet technique. Only moderate precautions are necessary to prevent shattering, and a fairly simple technique of casting the pellets can be used. Normally, the disks are smooth enough so that no surface roughness effect can be noted. However, for the most accurate work an internal standard is recommended to remove the effect of any small physical variations and to minimize interelement effects not taken care of by the 10 : 1 sample dilution.

REFERENCES

1. F. Claisse, "Accurate X-Ray Fluorescence Analysis Without Internal Standard," *Norelco Reporter* **4**: 3, 1957.
2. J. E. Townsend, "X-Ray Spectrographic Analysis of Silica and Alumina Base Catalyst by a Fusion-Cast Disc Technique," *Appl. Spectr.* **17**: 37, 1963.
3. H. J. Rose, Jr., I. Adler, and F. J. Flanagan, "X-Ray Fluorescence Analysis of the Light Elements in Rocks and Minerals," *Appl. Spectr.* **17**: 81, 1963.
4. C. L. Luke, "Determination of Refractory Metals in Ferrous Alloys and High-Alloy Steel by the Borox Disk X-Ray Spectrochemical Method," *Anal. Chem.* **35**: 56, 1963.

DISCUSSION

F. Bernstein (General Electric Co.): Approximately, what is the melting point of the mixture you recommend?

J. C. Guffy: It is difficult to give a melting point for a glass. I would guess around 550°. It's a fairly low melting glass.

R. W. Gould (University of Florida): Will this glass dissolve phosphate?

J. C. Guffy: Yes, it works very well.

R. W. Gould: Phosphate rock?

J. C. Guffy: Yes. In fact, you can use phosphate (phosphorus) as an internal standard for certain jobs if you like.

G. Edwards (Colorado School of Mines Research Foundation): Have you tried this on any of the rare earths, particularly with zircon and cerium present?

J. C. Guffy: Yes. Lanthanum, cerium, hafnium, zirconium were all right. Some of these get a little stubborn—you have to make sure it's dissolved.

G. Edwards: Were you going to an increased temperature then in fusion?

J. C. Guffy: No.

QUANTITATIVE X-RAY EMISSION ANALYSIS OF MAGNESIUM THROUGH FLUORINE WITH X-RAY AND ELECTRON EXCITATION

F. Bernstein and R. A. Mattson

General Electric Company
Milwaukee, Wisconsin

ABSTRACT

The analysis of dry powder samples for magnesium, sodium, and fluorine by X-ray and electron excitation has been studied. As in the case of heavier elements, the form of chemical combination influences the elemental sensitivity; sensitivity changes due to self-absorption can be adequately predicted using published absorption coefficients. Where both absorption and enhancement effects are possible, selection of X-ray target or excitation potential can eliminate the enhancement problem. Matrix effects were found to be extremely variable and unpredictable. Finally, X-ray and electron excitation results are compared for the three elements in a series of geological samples. Efficiency of excitation was far better for electron excitation, but limits of detectability were lower for X-ray excitation due to significantly lower backgrounds.

INTRODUCTION

In the past three years, there have been a number of technological advances in X-ray instrumentation which have made it possible to apply X-ray methods to quantitative analysis of elements below sodium. These include soft X-ray sources, with and without windows,[1,2] good artificial crystals built up of successive monolayers of long chain fatty acids,[3] and extremely thin organic windows to transmit soft X-rays to the detector. There have been numerous papers published recently related to fine structure observed in spectral lines of elements through boron,[4] and some work also has been reported on quantitative analysis of light elements.[5] Instrumentation used has been quite varied in nature; commercial vacuum spectrometers have been applied down through fluorine, while electron probes and units with soft X-ray sources have extended the range to beryllium, at. No. 4.

This paper is concerned with quantitative analysis of magnesium, sodium, and fluorine, which are all measurable when excited by sealed X-ray tubes, soft X-ray sources, and electrons. The primary objective of this work is to explore and evaluate the problems related to quantitative analysis of the aforementioned elements in powder samples with the three different types of excitation available. The importance of form of chemical combination, matrix, and interelement effects will also be investigated.

Particle size effects for elements zinc through oxygen were reported by Madlem,[6] who concluded that these effects were intensified for low atomic number elements. Matrix effects also were studied with mixtures of sodium and magnesium carbonates. It was reported that the intensities measured from these mixtures were not readily

494

explainable. Earlier work[7] has shown that for heavy elements (at. No. 12 and above) elemental intensities as well as the degree of interaction between components in a mixture were related to particle size. When the particles in a powder are large compared to the depth to penetration of the X-rays, it was found that interactions, such as absorption and enhancement, were minimized. This is the so-called heterogeneity effect in powders.

The form of chemical combination in which an element occurs has been shown to have a significant effect on measured intensities in powder samples. This is also described as the mineralogical effect and is caused by localized matrix effects. The relative intensity of an element in different forms of chemical combination can be calculated simply and directly when only absorption effects are present in the compound. By studying intensities obtained from several compounds of magnesium, sodium, and fluorine, using different energy X-ray and electron excitations, the following factors will be evaluated:

1. The relationship between form of chemical combination and X-ray intensity for these elements.
2. The degree of correlation obtainable between observed and calculated intensities, the latter being based upon published absorption coefficients[8,9] which are somewhat inexact at the present time.
3. The magnitude of enhancement effects, which can be determined by choosing compounds containing elements which can excite the elements being measured.

There is a direct method of measuring enhancement which was used in this investigation. It involves the successive selection of X-ray targets such that their characteristic lines are above and below the excitation energy of the enhancing element, but above the excitation energy of the element being measured in both cases. This was readily accomplished in the unit used in this work which permitted easy interchange of target materials.

EQUIPMENT

A conventional General Electric vacuum spectrograph was used for X-ray excitation with a sealed chromium target X-ray tube. Electron and X-ray excitation measurements were made in an ultrasoft X-ray spectrometer in which there is no window between the sample and anode during excitation. This system has been described by Mattson.[2] Figure 1 depicts the essential features of the spectrometer, which is enclosed in a vacuum bell jar.

When operating in the X-ray or secondary excitation mode, the sample and cathode are held at ground potential, while the anode is made positive. Backscattered electrons are, thereby, caused to return to the anode and create no undesirable background. Three solid anodes were employed in these experiments: molybdenum, copper, and aluminum. The electron current was held constant by an electronic current regulator and a line voltage stabilizer maintained the DC high-voltage supply at a constant value.

Electron excitation is accomplished by impressing a positive voltage on the sample, drawing the electron beam to it. This is the mode depicted in Figure 1. The same voltage stabilizer and current regulator are used. A small anode bias allows the beam to be positioned to the optimum location. The cathode can be rotated in its mount to bombard the sample directly.

Other essential features of the system are an enclosure in which the anode and cathode are housed and the stainless steel mask to which the sample makes contact. A variable slit on top of the enclosure was set at 0.020 in. for these measurements. A thin

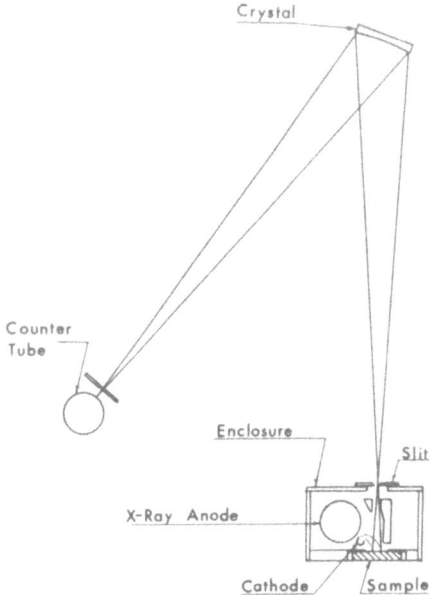

Figure 1. Schematic of ultrasoft X-ray spectrometer.

polypropylene window counter tube and a curved crystal on a fully focusing spectrometer make up the remainder of the system.

PROCEDURES AND RESULTS

Chemical Combination and Interelement Effects

For the studies on the effects of chemical combination with X-ray excitation, compounds of pure salts which were finely ground (in the 1–5 μ range) were used. These were compacted into rings at a pressure of about 10,000 psi. For electron excitation, similar samples were used with the exception that they were blended with 20% by weight of graphite in order to obtain conductivity. For the study of the magnesium, five different compounds were used, two of which contained elements capable of enhancing magnesium. There were the fluosilicate and the sulfate; the samples were excited with Cr K_α from a sealed, thin window chromium target tube at 60 kV, with electrons at 3.5 kV and with Mo L_α radiation from a target operated at 10 kV. The Cr K_α and the electron excitation are capable of exciting both silicon and sulfur, while the Mo L line will excite only silicon. In view of the fact that the molybdenum target was operated at 10 kV, some excitation of sulfur by the continuum is possible, but this is felt to be small enough to be neglected.

Table I shows the relative intensities which were obtained on each of the five compounds. It can be noted that there are variations in the figures for the different modes of excitation. The changes in X-ray values are attributable to three factors, namely, differences in effective excitation energies, takeoff angles, and presence or absence of enhancement effects. The takeoff angle for chromium data was 30°, while for all other targets it was 90°. The relationship of calculated intensities with those obtained by chromium excitation is shown in Figure 2; the results with molybdenum are not shown but were essentially the same. The calculated values do not take enhancement effects into account and the effective exciting wavelength is assumed to be that of the characteristic

Table I. Relative Intensities of Magnesium, Sodium, and Fluorine Compounds as a Function of Excitation

Compound	Chromium	Molybdenum	Aluminum	Copper	Electron
$Mg(OH)_2$	0.59	0.60	—	—	0.63
MgO	1.0	1.0	—	—	1.0
$MgCO_3$	0.33	0.49	—	—	0.40
$MgSiF_6$	0.20	0.21	—	—	0.24
$MgSO_4 \cdot 7H_2O$	0.16	0.19	—	—	0.15
$NaC_8H_{11}N_2O_3$	0.23	0.33	—	—	0.31
NaF	1.0	1.0	—	—	1.0
CH_3COONa	0.57	0.67	—	—	0.77
$Na_2B_4O_7$	0.38	0.45	—	—	0.35
$NaBiO_3$	0.09	0.09	—	—	0.20
$NaCl$	1.07	1.06	—	—	0.62
LiF	1.0	1.0	1.0	1.0	1.0
NaF	0.45	0.47	0.48	0.72	0.76
CaF_2	0.16	0.19	0.24	0.28	0.69
PbF_2	0.06	0.04	0.04	0.06	0.29

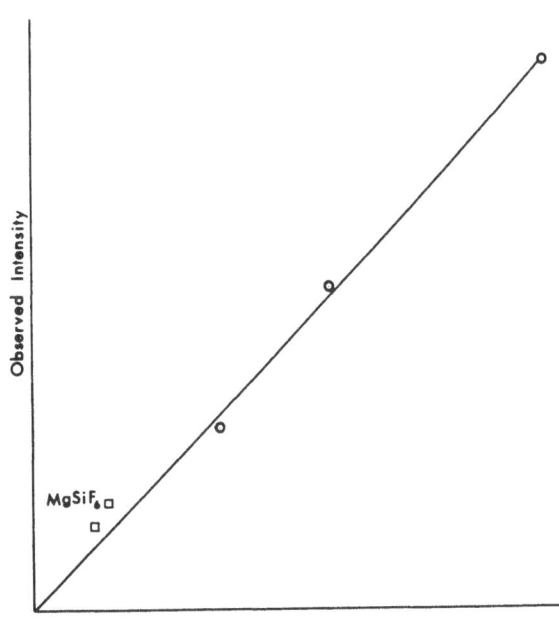

Figure 2. Relationship between calculated and observed intensities for magnesium compounds with chromium excitation.

line of the target being used. The magnesium oxide was arbitrarily assigned a value of one. Data points where enhancement can occur are plotted as squares and other points as circles. Since there was little difference in the fit of the magnesium sulfate data with and without sulfur enhancement, the latter being the case with Mo L excitation, it would appear that sulfur enhancement is small or of the same order of magnitude as the uncertainties in the absorption coefficients. The magnesium fluosilicate intensity was above the line for both sets of data, indicating probable enhancement by silicon. The

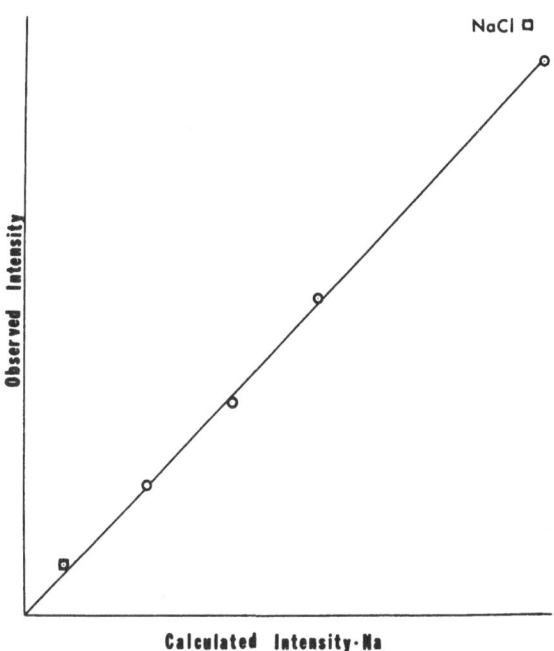

Figure 3. Relationship between calculated and observed intensities for sodium compounds with chromium excitation.

relative intensities with electron excitation were qualitatively the same as the X-ray results, but no attempt was made to interpret the values quantitatively. It should also be noted that the spot size with electron excitation was about $\frac{1}{2}$ in. × $\frac{1}{4}$ in. A similar test was conducted on sodium using six different sodium compounds, two of which contained elements capable of enhancing sodium. These were sodium metabismuthate, in which both the L and M lines of bismuth can excite sodium, and sodium chloride, in which the chlorine can effectively excite sodium. The relative intensities measured on the six compounds are shown in Table I. Included here are excitation with chromium at 60 kV, a windowless molybdenum X-ray source at 10 kV, and electron excitation at 3.5 kV. Chromium characteristic radiation can excite Cl K and Bi M lines, which can enhance sodium; Mo L radiation will not excite chlorine or bismuth so no enhancement should occur.

As was the case with the magnesium salts, there are significant differences amongst the relative intensities (Table I) which were obtained as a function of the type of excitation. Figure 3 shows a plot of the predicted intensity for the different compounds based upon absorption effects against the observed intensities. The correlation is very good, with only the chloride showing a significant deviation from the curve. This can be expected because of enhancement of sodium by chlorine radiation within the sample. A similar plot for the sodium compounds with Mo L excitation is shown in Figure 4. Agreement between calculated and observed intensities is good for all the compounds. With no enhancement, the sodium chloride point is now essentially on the line. In addition, it would appear that the Bi M radiation had no measureable effect on the sodium intensity. It is interesting to note that the relative intensity of the sodium chloride with electron excitation was considerably lower than with X-ray excitation, although the kV applied was sufficient to excite chlorine.

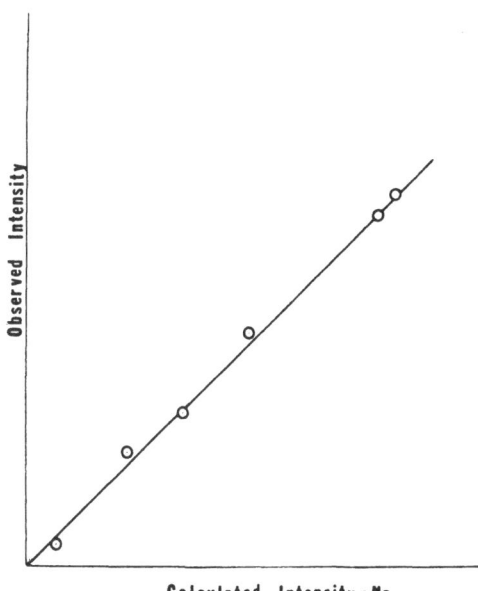

Figure 4. Relationship between calculated and observed intensities for sodium compounds with molybdenum excitation.

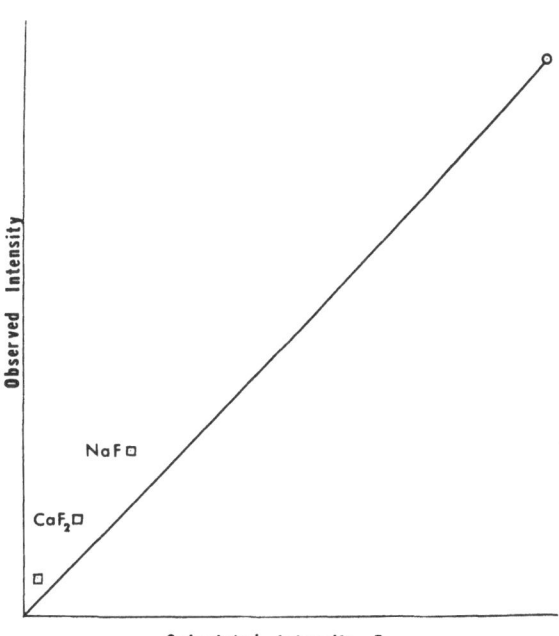

Figure 5. Relationship between calculated and observed intensities for fluorine compounds with chromium excitation.

The investigation of fluorine was carried out with four different fluorine compounds. These were lithium fluoride, sodium fluoride, calcium fluoride, and lead fluoride. Excitations used were Cr K_α from a sealed tube at 60 kV, molybdenum, copper and aluminum windowless X-ray targets at 10 kV, and finally electron excitation. There are some interesting interactions when using the above forms of excitation for fluorine. With chromium excitation, all the compounds except lithium fluoride produced both absorption and enhancement effects. With molybdenum and aluminum excitation, only the sodium fluoride will show both absorption and enhancement, the Ca K and Pb M lines having higher excitation energies than Al K_α or Mo L_α. Finally with the Cu L_α excitation none of the compounds should show enhancement effects. The electron excitation, which was run at 2.5 kV, is capable of exciting only the sodium in the sodium fluoride, insofar as interelement effects are concerned. Although the excitation potential for the Pb M line is 2.3 kV, it is not felt that there is significant excitation using 2.5 kV. Figure 5 shows the relation between predicted intensities for these salts based upon absorption calculations and the measured intensities using chromium excitation at 60 kV. The lithium fluoride was arbitrarily assigned the relative intensity of unity. The fluorides of sodium, calcium, and lead all show rather significant deviations from the 45° line which would be expected if only absorption effects were present. These deviations can be attributed to enhancement by the lead, calcium, and sodium in the salts.

This is further borne out by Figure 6 which shows the relationship between the predicted and the observed intensities using aluminum radiation. Only the sodium fluoride is significantly above the line and this can be attributed to the enhancement of fluorine by the sodium. A similar set of data was obtained using Mo L_α excitation. This would be expected because molybdenum and aluminum characteristic lines will excite sodium but not the other elements causing interelement effects. The final run with X-ray excitation was performed using a copper target. Since Cu L radiation is incapable of exciting any of the elements which were enhancing fluorine, it would be expected that all the intensities could be adjusted to the line by simply making absorption corrections. This is confirmed by Figure 7.

The relative intensities observed for the fluorine compounds as a function of excitation are listed in Table I. There are some marked differences between X-ray values and those for electron excitation at 2.5 kV. No explanation is offered for these differences since

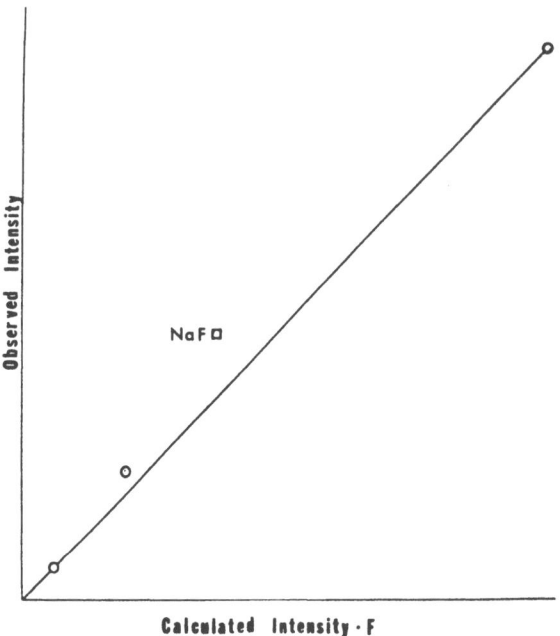

Figure 6. Relationship between calculated and observed intensities for fluorine compounds with aluminum excitation.

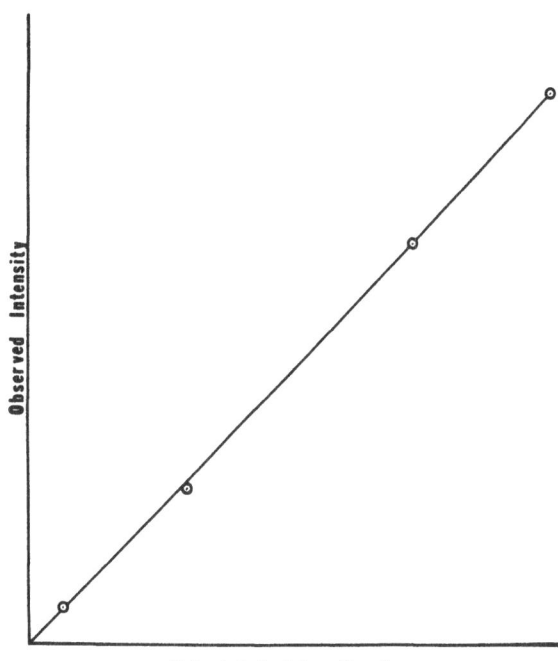

Figure 7. Relationship between calculated and observed intensities for fluorine compounds with copper excitation.

a comprehensive study and calculation of electron excitation effects is beyond the scope of this paper.

Matrix Effects

In order to study matrix effects for the measurement of fluorine, mixtures consisting of 50% by weight of lithium fluoride powder and seven different compounds were prepared. Two of the matrices, namely, lithium carbonate and sugar, produced only absorption effects. The remaining five, which included carbonates and iron oxide, were capable of causing both absorption and enhancement effects with sufficiently energetic excitation. All of the materials used in making these mixtures had a particle size in the order of 1–5 μ since it was felt that this would be sufficiently small to produce the absorption interactions in these powders for fluorine radiation. This was based upon the effective depth of penetration for fluorine in pure lithium fluoride, which is of the order of 12 μ with X-ray excitation. Fluorine intensities were measured from the mixtures using Cr K, Mo L, and electron excitation at 3 kV. The relative intensities and the calculated mass absorption coefficients are shown in Table II. Comparison of these coefficients shows that the values for the diluents are ten to twenty times that of lithium fluoride. It is obvious that the relative intensities shown in Table II are substantially higher than they should be, based upon the absorption coefficients of the mixtures. This is true in every case except possibly that of the magnesium carbonate mixture in which magnesium can enhance fluorine. This sample shows values which are reasonably close to that which would be expected from absorption effects alone. It is difficult to explain the other results unless the 1–5 μ particle size which was used for these powders is relatively large compared to the overall depth of X-ray penetration. If this were so, then the interactions between components would be reduced to the point where the results are completely scattered and could not be interpreted.

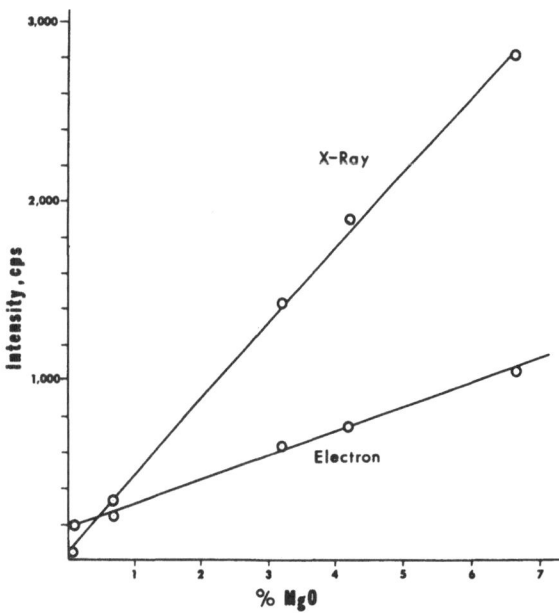

Figure 8. Working curves for magnesium in geological samples with X-ray and electron excitation.

Table II. Fluorine Intensities from 50% LiF in Various Matrices

Matrix	Mass absorption coefficient (F K_α) of matrix	Excitation		
		Chromium	Molybdenum	Electron
Li_2CO_3	8870	0.47	0.46	0.46
$C_{12}H_{22}O_{11}$	8600	0.29	0.35	0.34
Fe_2O_3	9200	0.24	0.26	0.22
$MgCO_3$	14500	0.11	0.17	0.18
Na_2CO_3	6700	0.39	0.46	0.48
Al_2O_3	7100	0.18	0.28	0.22
SiO_2	8400	0.41	0.48	0.52
Pure LiF	714	1.0	1.0	1.0

Table III. Comparison of X-Ray and Electron Excitation Results

Magnesium	cps/% MgO	(cps/% MgO)/watt	Limit of detectability
X-ray (Al anode), 8 kV, 30 mA	430	1.8	0.0031%
Electron, 4 kV, 31 μA	137	1100	0.028%

Sodium	cps/% Na₂O	(cps/% Na₂O)/watt	Limit of detectability
X-ray (Al anode), 9 kV, 40 mA	253	0.7	0.005%
Electron, 4 kV, 31 μA	88	710	0.027%

Fluorine	cps/% F	(cps/% F)/watt	Limit of detectability
X-ray (Cu anode), 7 kV, 40 mA	31	0.11	0.027%
Electron, 40 kV, 50 μA	67	335	0.045%

X-Ray and Electron Excitation

In order to compare results obtainable with X-ray and electron excitation, a set of five geological samples were chosen. These were powders which contained different concentrations of magnesium, sodium, and fluorine. The powders were prepared by grinding to a particle size of approximately 1–5 μ and for the electron excitation they were diluted with 15% by weight of graphite. The results are shown in Figure 8 for magnesium with both types of excitation.

X-ray excitation was accomplished by means of an aluminum anode operated at 8 kV and 30 mA. The electron excitation was run at 4 kV and 31 μA. The conditions under which the samples were run were not necessarily optimized, but serve as a basis for comparison of the two approaches. The results are summarized in Table III. It is interesting to note that a higher background was observed with electron excitation under conditions which produced approximately half the count rate obtained with aluminum X-ray excitation. Limit of detectability with X-ray excitation is about one-tenth that of electron excitation (based upon a net count equal to three times the standard deviation of the background for 100 sec counting time). However, the efficiency of X-ray production with electrons is far better since counts per second per watt per percent MgO is about 600 times the value for X-ray excitation.

There is no particular significance to the fact that the curve for X-ray excitation has a larger slope than the electron excitation, since the power for the latter could easily be increased to the point where the slope was equivalent or higher, but with proportionately higher background.

Figures 9 and 10 show similar sets of curves for sodium and fluorine using aluminum and copper anodes, respectively, for X-ray excitation. Correlation is good for all curves. As in the previous case for magnesium, limits of detectability are lower with X-ray excitation, but efficiency of characteristic photon production per unit of power input is far better for electron excitation. These results are summarized in Table III.

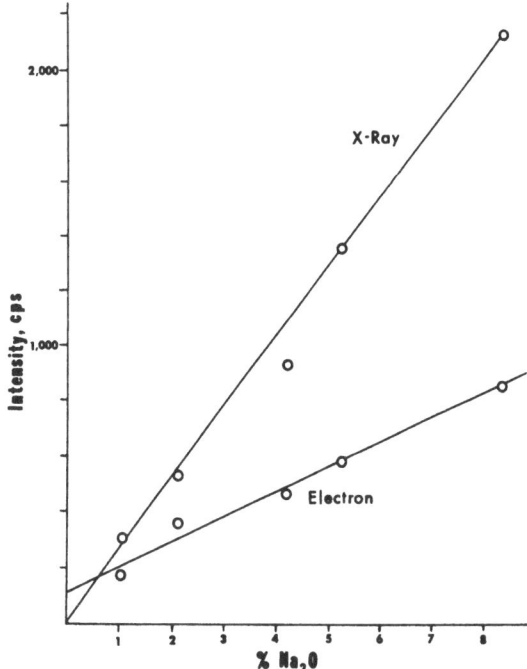

Figure 9. Working curves for sodium in geological samples with X-ray and electron excitation.

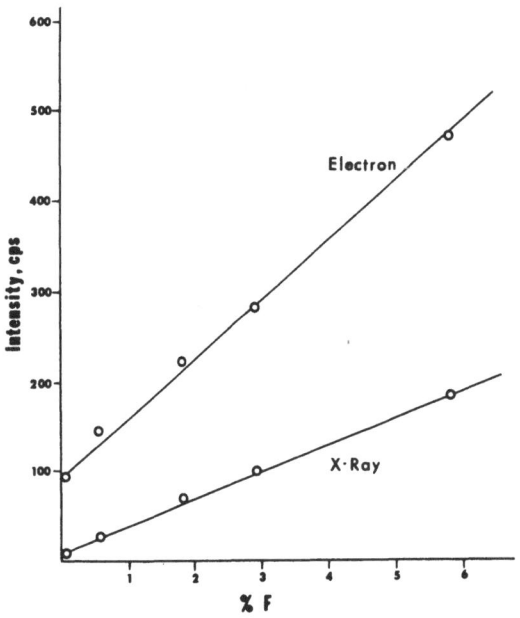

Figure 10. Working curves for fluorine in geological samples with X-ray and electron excitation.

DISCUSSION

The foregoing results are indicative of the fact that the X-ray absorption coefficients used to calculate relative intensities must be reasonably close to the correct values. In any case, they are certainly usable for determining order of magnitude of predicted intensities for compounds for the light elements of sodium, fluorine, and magnesium. It also appears that interelement effects in powders are predictable when the enhancing element is combined chemically with magnesium, sodium, and fluorine. These effects can be handled either by calculation methods or by selection of excitation energies so as to eliminate them. The results are not as well defined regarding matrix and interelement effects in mixtures. In spite of the fact that the powders used in the fluorine studies were quite fine by normal standards, the resulting intensities were relatively unrelated to absorption and enhancement considerations.

The five geological samples tested gave reasonably good working curves for magnesium, sodium, and fluorine with both electron and X-ray excitation. However, in view of the findings above, it may be optimistic at the present time to expect good correlation on powder samples in general, unless more elaborate methods of sample preparation are used. With the improved instrumentation now available, it is feasible to consider fusion techniques for eliminating sample heterogeneity without undue concern over losses in intensity. Further work is required along these general lines. In addition a more complete study of sample preparation and excitation parameters for electron excitation is needed.

CONCLUSIONS

1. The form of chemical combination affects X-ray sensitivities for magnesium, sodium, and fluorine. These changes can be satisfactorily predicted using available absorption data.

2. Enhancement effects are readily eliminated by selection of primary excitation energy so that it is below the excitation energy of the enhancing element.

3. Matrix and interelement effects for fluorine in 1–5 μ powders are erratic and unpredictable.

4. Electron excitation of powder samples is feasible but higher background and poorer limits of detectability were observed as compared to X-ray excitation.

ACKNOWLEDGMENT

The writers wish to thank Messrs. W. W. Kochheiser and R. A. Agenten of the X-Ray Department for their assistance in gathering the data, and Mr. H. J. Rose, Jr., of the U.S. Geological Survey for furnishing samples used in this investigation.

REFERENCES

1. B. L. Henke, "Some Notes on Ultrasoft X-Ray Fluorescence Analysis—10–100 Å Region," *Advances in X-Ray Analysis, Vol. 8*, Plenum Press, New York, 1965, p. 269.
2. R. A. Mattson, "Some Measurements of Carbon *K* Excitation in a New Ultrasoft X-Ray Spectrometer," *Advances in X-Ray Analysis, Vol. 8*, Plenum Press, New York, 1965, p. 333.
3. R. C. Ehlert, "The Diffraction of X-Rays by Multilayer Stearate Soap Films," *Advances in X-Ray Analysis, Vol. 8*, Plenum Press, New York, 1965, p. 325.
4. D. W. Fischer and W. L. Baun, "Effect of Chemical Combination on the Soft X-Ray *K* Emission Spectrum of Boron," *J. Appl. Phys.* **37**: No. 2, 1966.
5. C. J. Toussaint and G. Vos, "Quantitative X-Ray Spectrographic Determination of Traces of Elements Using Direct Electron Excitation," *Anal. Chem.* **38**: 711, 1966.
6. K. W. Madlem, "Matrix and Particle Size Effects in Analyses of Light Elements, Zinc Through Oxygen, By Soft X-Ray Spectrometry," *Advances in X-Ray Analysis, Vol. 9*, Plenum Press, New York, 1966, p. 441.
7. F. Bernstein, "Particle Size and Mineralogical Effects in Mining Applications," *Advances in X-Ray Analysis, Vol. 6*, Plenum Press, New York, 1964, p. 436.
8. L. S. Birks, *Electron Probe Microanalysis*, Interscience Publishers, New York, 1963.
9. R. Thiesen, *Quantitative Electron Microprobe Analysis*, Springer–Verlag, New York, 1965.

DISCUSSION

N. Spielberg (Philips Laboratories): I thought I saw a slide there which indicated that the choice of chromium, molybdenum, and aluminum targets made no difference in the intensity of the excitation of fluorine radiation. Did I misread that?

F. Bernstein: The numbers we showed were merely relative intensities. There certainly are differences in excitation efficiencies.

N. Spielberg: I got the impression that these three targets all excited with the same effectiveness.

F. Bernstein: No, they were run on different units with different power inputs. The closer you get in your exciting energy to the excitation potential, the more efficient you are.

N. Spielberg: In your comparison of the limits of detectability by electron and X-ray excitation, I have a feeling that the slope of the curve is also important. In fact, I think that the limit probably goes linearly with the slope and inversely with the square root of the residual background. Yet, I get the impression from you that you feel that the residual background is the most important thing.

F. Bernstein: What you say is true, but if you increase the excitation energy you are going to increase the background and possibly you increase the slope at a faster rate than the background. We haven't looked into this, so there may be some merit in what you say. This is not put forth as a final paper but is a preliminary report on some exploratory work we did. We did not optimize or investigate that particular aspect. There can be a great deal more work done on electron excitation, particularly on relative importance of diluent and how much to use. We haven't looked into this in any detail.

A METHOD OF LIQUID ANALYSES PROVIDING INCREASED SENSITIVITY FOR LIGHT ELEMENTS

D. W. Beard and E. M. Proctor

Picker X-Ray Corporation
Cleveland, Ohio

ABSTRACT

A method for analyzing solutions using a sample surface directly exposed to the primary X-ray beam is discussed. This method eliminates the need for the conventional Mylar covered liquid cells. The advantages of this method are the elimination of the scattering of the longer wavelength X-rays and the absorption effects due to the Mylar covering, thereby giving significant improvement in peak-to-background ratios and peak intensities for the light elements. This increased sensitivity can be used to improve the limits of detectability for light elements in solutions, broaden the range of practical elemental determinations, and reduce the counting time for any light element analysis in liquids.

A new liquid cell, developed for this technique, provides easily repeatable setting of target-to-sample distance and simplified preparation and handling of samples. A comparison between results obtained with conventional method and this uncovered sample surface method is made for typical solution applications.

INTRODUCTION

Today, liquid analysis by X-ray fluorescence is an accepted procedure. The petroleum industry, in particular, employs X-ray instrumentation to control the amount of additives and to check for trace impurities in its products. Besides the many applications in which the sample is a liquid in original form, solution techniques have broadened the scope of X-ray fluorescence. When particular types of bulk solids or powders are not adaptable to a repeatable sample preparation method or when, in their original forms, these samples exhibit effects, other than compositional changes, which cause variations in X-ray intensities, a solution technique with the choice of a proper solvent may allow routine, practical analyses by X-ray fluorescence. Because of the leveling nature of the solvent, the solution technique also helps to eliminate matrix effects and the resultant need for mathematical or graphical corrections to the X-ray data. The lower absorption of the liquid sample allows dilution factors which would be prohibitive in powder samples because of intensity losses. Also, liquid samples are readily adaptable to the addition of internal standards. Several excellent papers have been written on the subject of solution analysis by X-ray methods and a comprehensive review of specific techniques and papers written on solution applications has been made by E. P. Bertin.[1]

This paper will show the advantages of an uncovered liquid surface in the analysis of low atomic number elements and the increased sensitivity available with the elimination of the commonly employed Mylar covering. A discussion will also be given on the instrumental criteria necessary for trace analyses and instrumental conditions necessary for an uncovered liquid surface.

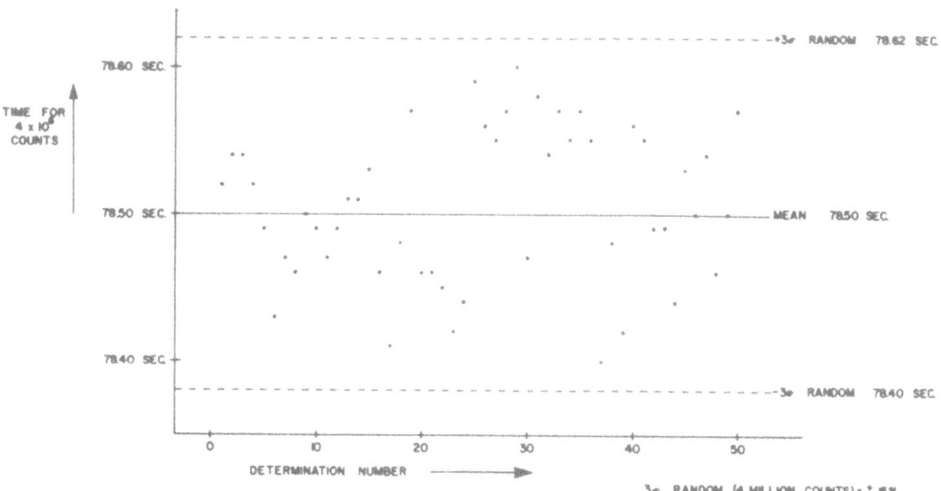

Figure 1. Repeatability of individual measurements.

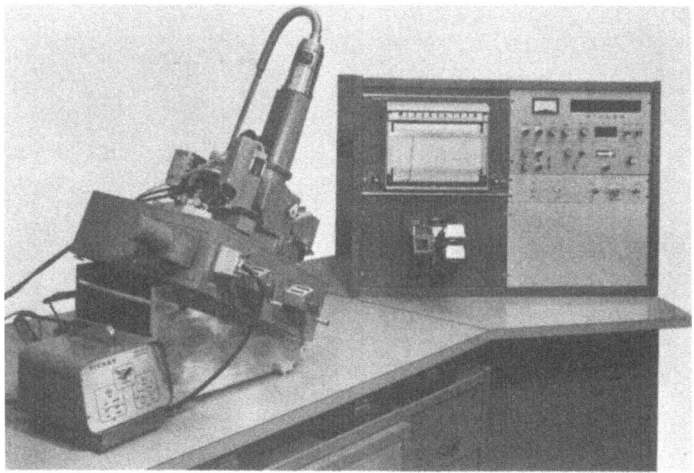

Figure 2. Picker spectrodiffractometer adapted for uncovered liquid analyses.

PRELIMINARY CONSIDERATIONS

For valid trace analyses when small changes in intensity reflect the concentration changes of interest, certain criteria of instrumentation must be met to insure that the recorded changes in X-ray intensities are truly indicative of a difference in elemental concentration and not the result of a variation in an instrumental parameter. For valid application of statistics to the recorded X-ray data, both the systematic and random errors arising from instrumentation must be accounted for. The exacting nature of low concentration work demands an electronically stable and mechanically precise system. The system must provide a constant X-ray output during the period of analysis and the

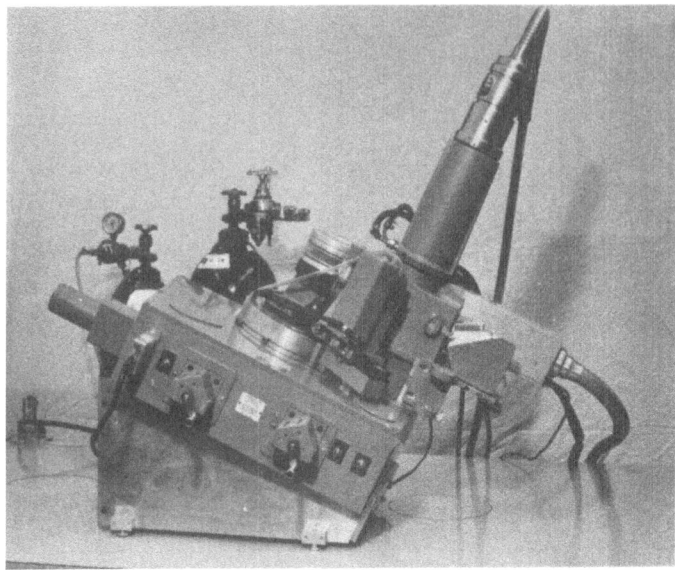

Figure 3. Adaptor base for Picker spectrodiffractometer.

Figure 4. Liquid cell assembly; micarta mask and crosshair assembly.

detecting system must be stable and unaffected by sources of noise. The method determining X-ray beam and system stability has previously been covered in "The Measurement of X-Ray Source Stability."[2]

The second system requirement is mechanical precision. This includes the reproducibility of 2θ settings, the precision of crystal bisecting, and the repeatability of sample positioning. When a helium or vacuum path is used, there is the added need of controlled and repeatable path for constant absorption effects. In the case of the very light elements,

Figure 5. Sample holder with crosshair reference in place.

the absorption in the X-ray path from sample to detector is the most significant parameter affecting the X-ray intensity. With a helium path, the effect of temperature and pressure changes on X-ray intensity[2] can be eliminated with a constant gauge pressure of helium and, for practical purposes, the constant temperature bath provided by a large tank of helium.

Prior to collecting the data reported in this paper, a test had been made to determine the overall mechanical precision, X-ray beam stability, detector stability, and helium flushing efficiency. A sample of high calcium content was removed and replaced for 50 measurements. The time necessary to collect 4,000,000 counts at the Ca K_α peak was recorded for each trial. These results are shown in Figure 1. With random statistics, 3σ limits are $\pm 0.15\%$ for the 4,000,000 counts collected. The observed 3σ limits for these 50 measurements were $\pm 0.20\%$ and indicated that the system provides the precision and stability needed for accurate trace analyses.

For an uncovered liquid surface, the sample surface must, of course, be level. Because of the geometry of the system in which the end window tube and primary collimation are at right angles to each other, the sample surface in normal operation is 25° off the horizontal. To provide the level sample surface while still retaining the design geometry for maximum intensities, the spectrodiffractometer was adapted as shown in Figure 2. At the bottom of the adaptor base, shown in Figure 3, are four leveling screws for exact adjustment of the instrument.

Another requirement for any system used in liquid analysis is efficient cooling of the X-ray tube to prevent compositional change caused by evaporation. In this system, the OEG 60 chromium target X-ray tube gives off a minimum of heat and further cooling of the sample chamber keeps the chamber below room temperature and allows for the analysis of low melting-point materials, polymers and liquids, without vaporization problems. This cooling and the constant flushing of the system with helium protects crystals which exhibit hygroscopic characteristics and eliminates any errors arising from temperature caused shifts in reflection angles as reported by Lee and Campbell.[3]

SAMPLE HOLDER AND SAMPLE PREPARATION

A special sample holder was designed to allow height adjustment of the liquid sample level and to provide a reference surface for maintaining a constant target-to-sample distance between samples. The reusable cells chosen for use in the holder are shown in Figure 4. They consist of an inner plexiglass core, flexible polyethylene bottom, and outer plexiglass ring. For the inner cylinder, plexiglass tubing of $1\frac{1}{8}$-in. inner diameter and $1\frac{1}{4}$-in. outer diameter was cut to $\frac{7}{8}$-in. lengths. The polyethylene sheets that served as the bottom of the cells were stretched tight and held securely by means of $\frac{1}{4}$-in. rings cut from plexiglass tubing of $1\frac{1}{4}$-in. inner diameter and $1\frac{1}{2}$-in. outer diameter. To validly compare the two methods of analyses, these same cells were used to collect data for the Mylar covered method. The same $\frac{1}{4}$-in. rings used to hold the polyethylene bottoms in place were used to secure the Mylar covering. Also shown in Figure 4 are the micarta mask and

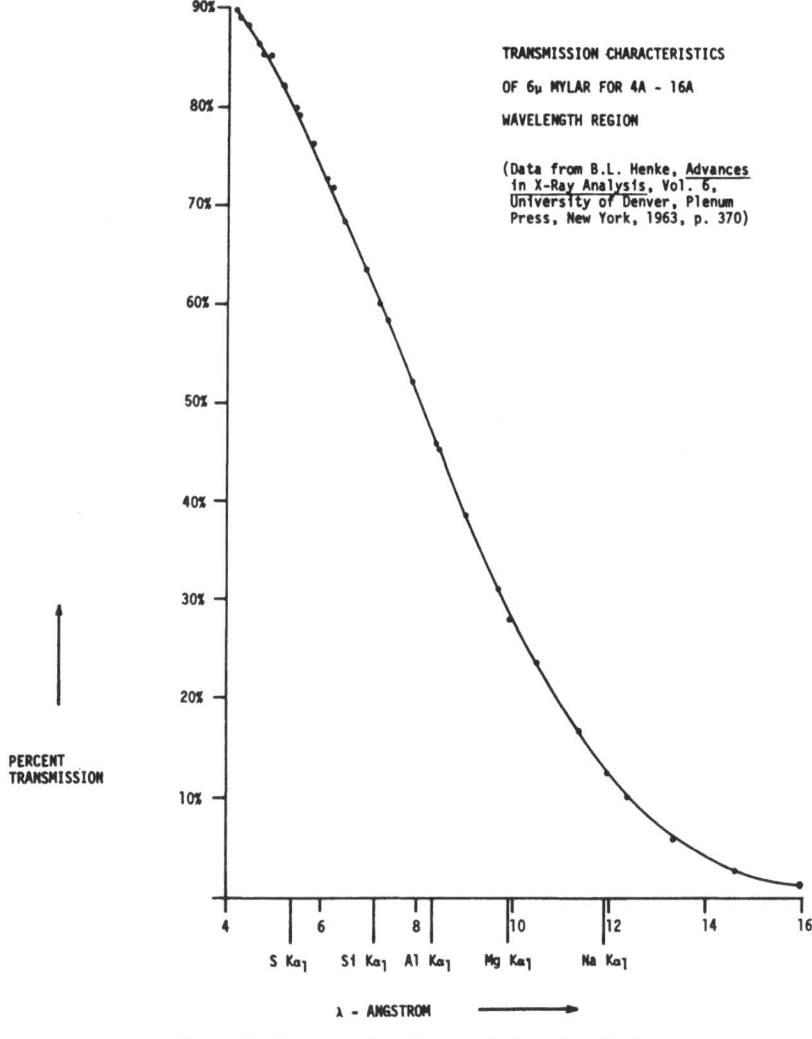

Figure 6. Transmission characteristics of 6μ Mylar.

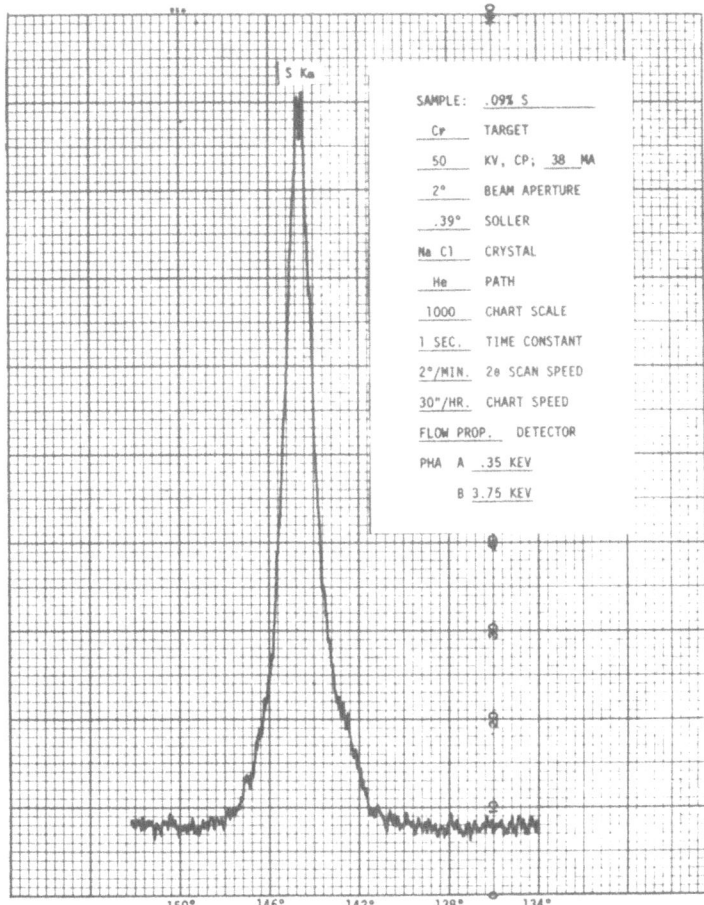

SAMPLE: .09% S
Cp   TARGET
50   KV, CP; 38 MA
2°   BEAM APERTURE
.39°   SOLLER
Na Cl   CRYSTAL
He   PATH
1000   CHART SCALE
1 SEC.   TIME CONSTANT
2°/MIN.   2θ SCAN SPEED
30"/HR.   CHART SPEED
FLOW PROP. DETECTOR
PHA A .35 KEV
     B 3.75 KEV

Figure 7. Spectrometer scan: S K_α peak.

removable crosshair assembly used to establish liquid level for repeatability between samples. The micarta mask exposes a $1\frac{1}{16}$ in. sample area and has two parallel channels for positioning the nylon crosshairs.

Figure 5 shows the sample holder with cell and crosshair assembly in place. A movable aluminum plate served as a piston for changing the level of the liquid by deforming the polyethylene bottom of the cell. An O-ring at the bottom of the cell holder compensated for any variations in cell height and ensured tight contact with the micarta mask.

For the uncovered method, a cell was filled to a manageable height, put into the holder, and the micarta mask tightened over the cell. The crosshair assembly was then placed over the holder and the liquid level adjusted by the pressure of the aluminum plate against the cell bottom until the liquid sample just wet the crosshairs. The crosshairs were removed for the analysis and cleaned for the next sample. For use as a covered cell, the cell assembly was placed on flat surface, filled with the sample, and Mylar placed over the surface of the sample. The ring stretched the Mylar and sealed the cell. The repeatability of the height adjustmcnt for the uncovered method was checked by repeated trials and the intensity recorded on the strip chart recorder was within $\pm\frac{1}{4}$ division at 80% of full scale.

EXPERIMENTAL PROCEDURE AND INSTRUMENTAL PARAMETERS

Known amounts of aluminum nitrate, sodium silicate and ammonium sulfate crystals were dissolved in distilled water. These served as base solutions for further dilution with distilled water to prepare ten samples covering the following concentration ranges: (a) 0.650% Al–0.0032% Al; (b) 0.732% Si–0.0037% Si; and (c) 0.12% S–0.00045% S. These samples were analyzed both by the directly exposed surface and Mylar covered methods.

A thin window chromium target X-ray tube operating at 50 kV, DC and 38 mA, a flow proportional detector, 2° beam aperture, 0.39° soller and helium path were used for the three elements. Data was recorded on the basis of 100 sec present time.

An upper level discriminator was used in the three analyses to eliminate the multi-ordered higher energy scatter, thereby substantially improving the peak-to-background ratios. For the silicon and aluminum analyses, the upper level was 3.00 keV. For sulfur, a 3.75 keV setting was used. In all three cases, a 0.35 keV lower level setting in the pulse height analyzer was used for noise discrimination.

On the basis of its reflectivity, a PET crystal was used for silicon and aluminum. Also because of its higher reflectivity, a NaCl crystal was used for sulfur. The choice of an organic crystal, such as EDDT, would have given better peak-to-background ratios

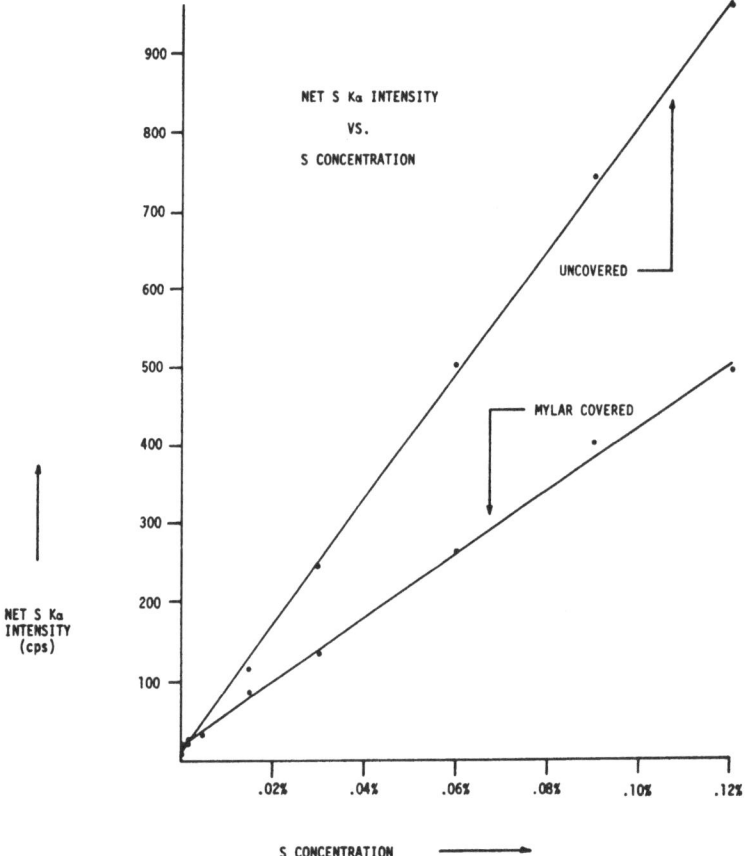

Figure 8. Comparison of calibration curves: sulfur analyses.

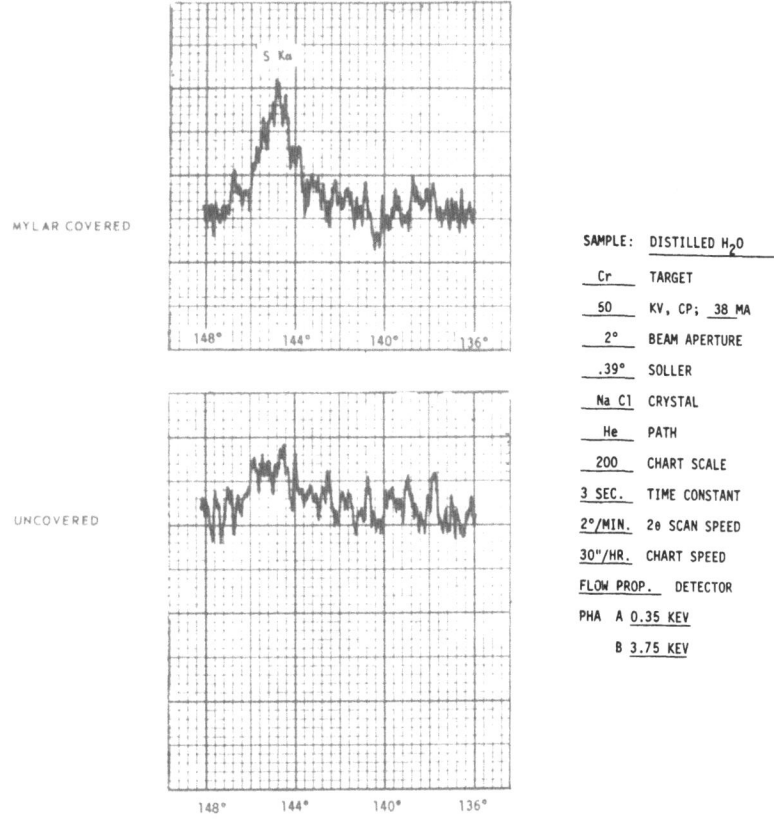

SAMPLE: DISTILLED H_2O

_Cr_____   TARGET

_50____   KV, CP; _38 MA

_2°____   BEAM APERTURE

_.39°___   SOLLER

_Na Cl__   CRYSTAL

_He____   PATH

_200____   CHART SCALE

_3 SEC.__   TIME CONSTANT

_2°/MIN.__   2θ SCAN SPEED

_30"/HR.__   CHART SPEED

_FLOW PROP.__   DETECTOR

PHA   A _0.35 KEV_

      B _3.75 KEV_

Figure 9. Spectrometer scan showing the sulfur in the Mylar.

for sulfur, since the high-energy scatter from the samples caused the chlorine in the crystal to fluoresce and resulted in a relatively high background level of about 80 cps. Pulse height discrimination could not be used to reduce this background level because of the proximity of the S K_α radiation at 2.31 keV and the Cl K_α radiation at 2.62 keV. These cannot be effectively resolved with a proportional detector having 18% resolution for Fe[55]. The improvement in peak-to-background using an organic crystal did not compensate for the accompanying loss in intensity.

Although a Formvar counter window would have given a substantial increase in the silicon and aluminum intensities, the standard $\frac{1}{10}$ mil Mylar window was used on the proportional detector and did not provide prohibitive absorption even in the concentration ranges studied. P-10 gas was used for the detector. Another parameter that could have been changed to give increased counting rates was the 0.39° soller collimator. To change from this $3\frac{3}{4}$ in. × 0.021 in. soller to a coarser one would have been advantageous for these specific samples, since they contained only one or two elements that are detectable. The presence of other elements with L or M series lines in the neighborhood of the elemental line of interest, e.g. analysis of sulfur in the presence of molybdenum, would require at least the resolution provided by the 0.39° soller.

RESULTS

The transmission characteristics for the Mylar covering used in the comparison are shown in Figure 6.[4] For the three elements analyzed, the theoretical transmissions through 0.00025 in. of Mylar are as follows: (a) sulfur—0.797, (b) silicon—0.597, and (c) aluminum —0.459. For practical purposes, however, the absorption of these wavelengths by the Mylar is much higher. The angles formed by the incident beam and the emergent beam with the normal to the sample surface must be taken into account as the path length through the Mylar is increased as a function of the cosecants of these angles. The actual absorption can be seen in the slope change in the calibration curves.

Sulfur

In Figure 7, a representative scan using the directly exposed sample surface method is shown for the sample containing 0.09% sulfur. The instrumental parameters are listed. In Figure 8, the results for the ten samples compared are shown in the net sulfur K_α intensity *vs.* sulfur concentration curves. The noticeable difference in the zero intercepts between the uncovered and Mylar covered calibration curves is caused by the presence of sulfur in

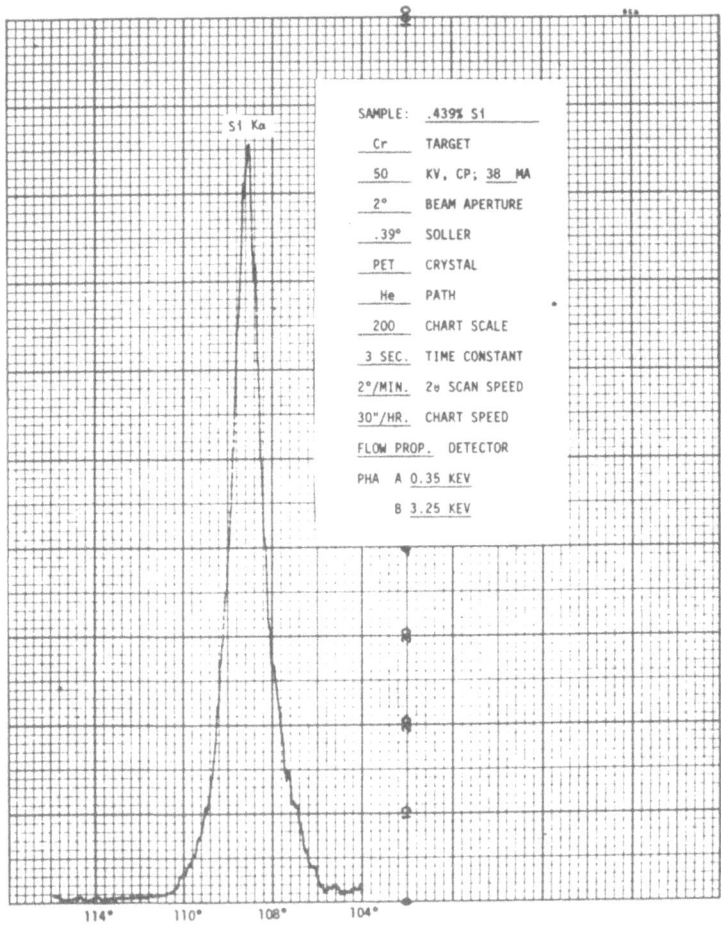

Figure 10. Spectrometer scan: Si K_α peak.

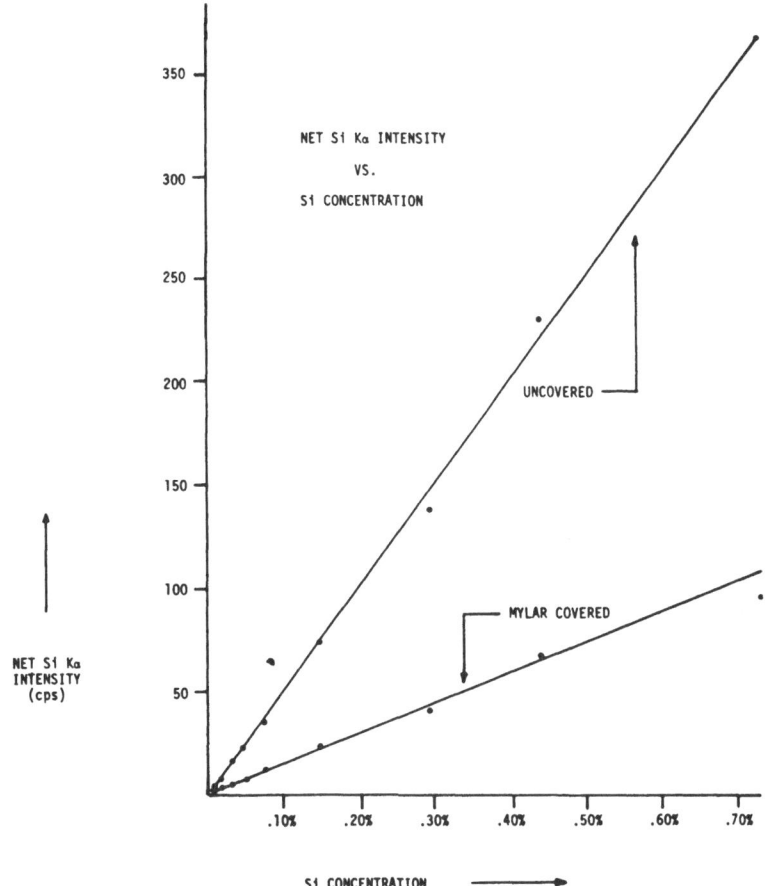

Figure 11. Comparison of calibration curves: silicon analyses.

the Mylar. Figure 9 shows the sulfur intensity from the Mylar covering over a distilled water sample. Different sheets of Mylar caused the same increase and it was concluded that rather than the presence of an impurity or contamination on the surface of the Mylar, the sulfur was present as a trace constituent in the Mylar.

Silicon

Figure 10 shows a spectrometer scan and the instrumental parameters used for a sample containing 0.439% silicon. The results for the comparison are shown in Figure 11, where the net silicon K_α intensity is plotted against silicon concentration.

Aluminum

Figure 12 shows the Al K_α peak for the sample containing 0.260% aluminum. Figure 13 gives the resulting calibration curves for the two methods being compared.

CONCLUSIONS

The difference in slopes is apparent in the three calibration curves shown and the advantage of the greater net change in intensity per given change in concentration available with the uncovered method is clearly indicated. A figure of merit can be assigned to each curve based on the method of N. Spielburg and M. Bradenstein.[5] To calculate a

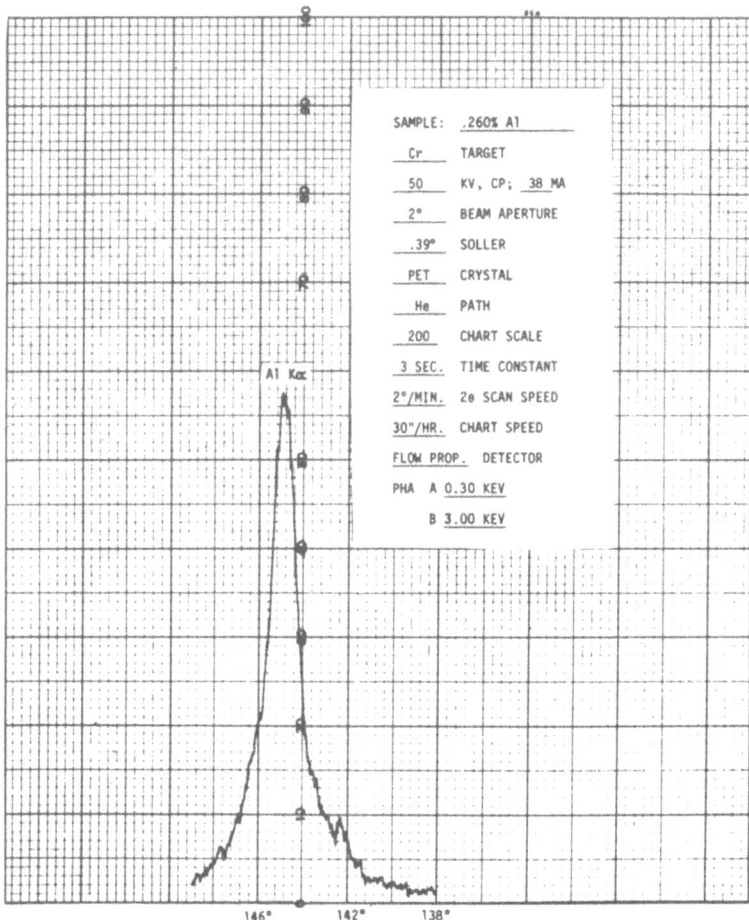

Figure 12. Spectrometer scan: Al K_α peak.

Table I. Comparison of Methods: Figures of Merit

Element	Method of analysis	Slope of calibration curve	Intercept at zero concentration (cps)	Figure of merit
Sulfur	Uncovered	7970	7.9	2840
Sulfur	Mylar covered	3970	22.1	840
Silicon	Uncovered	502	1.0	502
Silicon	Mylar covered	150	2.0	106
Aluminum	Uncovered	420	7.8	151
Aluminum	Mylar covered	70	7.0	26.5

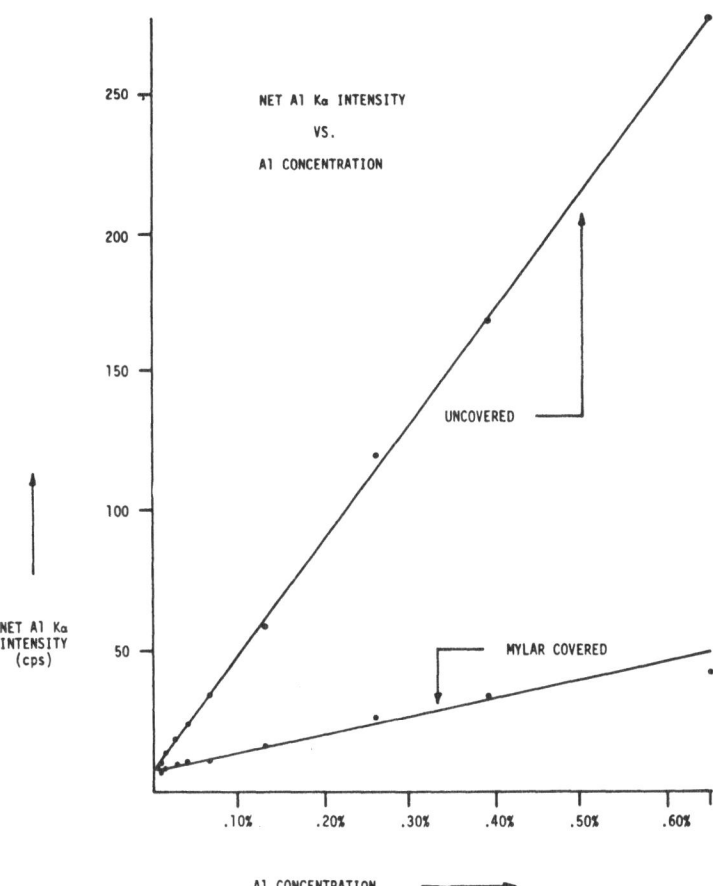

NET Al Kα INTENSITY
VS.
Al CONCENTRATION

UNCOVERED

MYLAR COVERED

NET Al Kα
INTENSITY
(cps)

Al CONCENTRATION

Figure 13. Comparison of calibration curves: aluminum analyses.

Table II. Comparison of Methods: Limits of Detectability in Water Solution

Element	Concen- tration (ppm)	Method	Peak intensity (cps)	Background (cps)	Net peak intensity (cps)	P/B	Limit of detectability* (ppm)
Sulfur	150	Uncovered	201.4	84.0	117	2.4	3.5
Sulfur	150	Mylar covered	176.4	89.4	87	2.0	4.9
Silicon	146	Uncovered	9.5	1.5	8.0	6.3	6.6
Silicon	146	Mylar covered	6.0	1.7	4.2	3.5	13.6
Aluminum	130	Uncovered	16.7	2.6	14.1	6.4	4.4
Aluminum	130	Mylar covered	11.3	2.8	8.5	4.0	7.8

* Based on 100-sec counting time.

given figure of merit, the slope of the calibration curve and the intercept at zero concentration are calculated as follows

$$m = \frac{\Delta I}{\Delta C}$$

$$I_0 = I_i - mC_i$$

where m = slope of calibration curve, ΔI = intensity change for corresponding ΔC, ΔC = concentration change, I_i = intensity coordinate for given C_i, C_i = concentration coordinate, and I_0 = intercept at zero concentration.

These values can be used to determine the respective figure of merit by use of the equation[5]

$$Z = A(B)^{-\frac{1}{2}}$$

where Z = figure of merit, $A = m$ = slope of calibration curve, and B = intercept at zero concentration.

These results are tabulated in Table I. The advantages offered by the uncovered method, namely, higher net intensities with the elimination of the absorption caused by the Mylar and the greater slopes of the working curves, can reduce the analytical time necessary for a desired degree of accuracy and extend the practical working range of concentrations. The figures of merit for the two methods compare as follows for the three elements used:

For sulfur

$$\frac{Z \text{ uncovered}}{Z \text{ Mylar}} = 3.4$$

For silicon

$$\frac{Z \text{ uncovered}}{Z \text{ Mylar}} = 4.7$$

For aluminum

$$\frac{Z \text{ uncovered}}{Z \text{ Mylar}} = 5.7$$

To show the increased sensitivity available, the limit of detectability of sulfur, aluminum, and silicon in H_2O was determined according to the method outlined by L. S. Birks.[6] The equation used was

$$MDL = \frac{3W\%(B)^{\frac{1}{2}}}{P - B}$$

where MDL = minimum detectability limit of the element in weight percent of solution, $W\%$ = weight percent of element, B = counts collected at background position, and P = counts collected at peak position.

These results are shown in Table II. Again, the advantages of the uncovered or directly exposed surface method are shown; and as the degree of advantage offered by this method increases with increasing wavelength, the elimination of the Mylar would give a 10–20-fold increase in sensitivity for magnesium and sodium with the same instrumental parameters.

ACKNOWLEDGMENTS

The authors wish to thank Mr. W. P. Segulin for the design of the liquid sample holder and Messrs. I. L. Gatautis, T. J. Pszenny, K. J. Page, D. H. Kerrigan, J. F. Harrington, and Miss P. A. Walters for their assistance and cooperation.

REFERENCES

1. E. P. Bertin, "Solution Techniques in X-Ray Spectrometric Analysis," *Norelco Reporter* **12**: 15–26, 1965.
2. W. D. Ashby and E. M. Proctor, "The Measurement of X-Ray Source Stability," *Picker Analyzer* **4**: 1965. Paper delivered at ACA meeting, Bozeman, Montana, 1964.
3. F. S. Lee and W. J. Campbell, "Variation of X-Ray Spectral Line Position with Ambient Temperature Change: A Source of Error in X-Ray Spectrography," *Advances in X-Ray Analysis, Vol. 8*, Plenum Press, New York, 1965, pp. 431–442.
4. B. L. Henke, "Sodium and Magnesium Fluorescence Analysis—Part I: Method," *Advances in X-Ray Analysis, Vol. 6*, Plenum Press, New York, 1963, p. 370.
5. N. Spielburg and Mechtilde Bradinstein, "Instrumental Factors and Figure of Merit in the Detection of Low Concentrations by X-Ray Spectrochemical Analyses," *Appl. Spectr.* **17**: 6, 1963.
6. L. S. Birks, *Electron Probe Microanalysis*, Interscience Publishers, New York, 1963, pp. 136–138.

DISCUSSION

D. C. Bishop (Medusa Portland Cement Co.): In what commercial products have you identified light elements? Have you identified them in the rock products? What kind of rock products? I notice that you have done most of your work on sulfur, silicon and aluminum.

E. M. Proctor: Yes, I have done light element analyses in commercial samples. An example of an application using the uncovered solution technique was the checking of petroleum samples for aluminum, picked up as contamination, and also for the sulfur content. The purpose of presenting the data on the three elements were to compare methods and to illustrate the merits of the uncovered technique.

D. C. Bishop: Have you ever done this in higher concentrations?

E. M. Proctor: My purpose was to show the increased sensitivity available with an uncovered surface. One way was to show the improvement in detectability limits. Therefore, low concentration ranges were chosen. The same factors of increased sensitivity are available for analyses of higher concentrations.

THE DEMOUNTABLE TUBE IN LIGHT ELEMENT FLUORESCENCE ANALYSIS

M. A. Short and M. J. H. Ruscoe

Associated Electrical Industries, Limited
Urmston, Manchester, England

ABSTRACT

In view of the importance of obtaining optimum conditions for the X-ray fluorescence analysis of light elements, an investigation has been made of the effects of varying the X-ray tube target, the target take-off angle, the tube voltage, and the tube window thickness. The effects of these parameters have been observed by measurement of the intensity of fluorescence of two light elements using the Raymax 60 demountable tube and vacuum path spectrometer. The results obtained are compared with those given by theoretical calculations based on consideration of the relevant parameters; good qualitative agreement has been obtained. It is shown that a high primary X-ray intensity is obtained with a high target take-off angle, a low angle of incidence of the electron beam on the target, and an optimum setting of tube voltage. It is further shown that the most suitable target to use for the fluorescence analysis of light elements is markedly dependent on the thickness of the X-ray tube window.

INTRODUCTION

There are two types of X-ray tubes available for use in fluorescence analysis; these are commonly known as sealed (or sealed-off) tubes and demountable (or continuously evacuated) tubes. Most manufacturers of X-ray equipment supply sealed tubes with their X-ray fluorescence analysis units. Demountable tubes have the advantage that they permit the ready replacement of targets, filaments, and windows at a minimal cost. The targets may be cleaned as required, thus preventing the production of spectrally contaminated radiation; it is also possible to use custom made targets and very thin windows both of which can be of particular importance in the analysis of the lighter elements.[1,2]

In the design and use of an X-ray tube for fluorescence analysis there are a number of parameters to be considered: the electron beam current, the electron beam accelerating voltage, the target composition, the angle of incidence of the electron beam on the target, the target take-off angle, and the tube window composition and thickness. For a sealed tube, the target angles, the tube window, and to some extent the target composition are fixed by the manufacturer; on the Raymax demountable tube, however, these parameters can be readily varied with the exception of the angle between the electron beam and the midray of the useful X-ray beam which must be maintained at 90° under normal instrumental geometry.

For relatively short wavelength exciting radiations such as Cr K_α, theoretical considerations of target take-off angle do not suggest that this will have any profound influence on the intensity of the primary radiation. In addition, beryllium tube windows as thick as 0.01 in. (250 μ) will still transmit over 80% of such radiation. For the analysis of lighter elements, specifically fluorine through chlorine, with their relatively low fluorescence yields, it is important to use exciting radiations of a wavelength as close as possible to the absorption edge of the light element. It has, in fact, been said that for small concentrations of light elements the fluorescence conversion efficiency varies with the cube of the wavelength of the primary radiation.[3] Table I lists the absorption edges of fluorine through chlorine together with the wavelengths and excitation potentials of a number of selected characteristic radiations. The suitability of a longer wavelength characteristic radiation (longer than Cr K_α) for light element analysis is largely determined by the transmission of the X-ray tube window; the transmission of the selected radiations through various thicknesses of beryllium is also given in Table I. Theoretical considerations for these longer wavelength radiations suggest that the target take-off angle will have a marked effect on the intensity of the primary radiation.

In view of the importance of obtaining optimum conditions for the fluorescence analysis of the lighter elements, an investigation has been made of the effects of varying the target take-off angle, the electron beam accelerating voltage, and the tube window thickness. To determine the dependence of the intensity of the useful beam of longer wavelength characteristic radiation on target take-off angle and tube voltage, a series of measurements have been made on Al K_α and Ag L_α radiations; these measurements have been compared with relative intensities derived theoretically. To determine the dependence of the fluorescence intensity of light elements on the target composition and the tube window thickness measurements have been made of the intensity of fluorescence of two light elements, specifically magnesium and chlorine, using a number of targets and tube windows over a wide range of tube voltages.

INSTRUMENTATION

The experiments detailed below were carried out using a standard A.E.I. Raymax 60 Fluorescence Analysis Unit; besides a continuously evacuated tube this incorporates a vacuum path spectrometer and a constant potential generator. A cross section of the demountable tube is shown in Figure 1. An NaCl crystal was used for dispersing Cl K_α radiation (from rocksalt—sodium chloride) and an ADP crystal for dispersing Mg K_α radiation (from pure magnesium); 0.36° Soller slits were used. A demountable flow proportional counter with a 6-μ Melinex window (for Cl K_α) or a 2-μ Makrofol window held in place by an 80%-transmission nylon mesh (for Mg K_α) was used; standard P-10 gas was employed. A wide pulse height analyzer window was used so that all of the characteristic radiation was counted. Background counts were averaged from measurements made at 4° 2θ on each side of the characteristic radiation reflection; counting loss corrections were made to intensity measurements where appropriate.[7]

EFFECT OF VARYING THE TARGET TAKE-OFF ANGLE

The intensity of the useful X-ray beam was, for longer wavelength characteristic radiations, thought to be particularly dependent on the target take-off angle and the tube voltage. To examine the magnitude of this dependence measurements were made on two relatively long wavelength radiations: Al K_α (8.340 Å) and Ag L_α (4.154 Å). Aluminum and silver targets suitable for the demountable tube were made giving take-off angles of 20°, 30°, 40°, 50°, and 60°. The relative intensities of the useful X-ray emission from these targets were obtained by using the Al K_α radiation to excite Mg K_α from a sample of magnesium, and by using the Ag L_α radiation to excite Cl K_α from a sample of rocksalt

Table I. Absorption Edges, Characteristic Wavelengths, Excitation Potentials, and Window Transmissions

Element	K absorption edge, Å [a]	Target	Characteristic line	Excitation potential, keV [b]	Wavelength of line, Å [a]	Transmission through beryllium [c]			
						0.001 in.	0.002 in.	0.004 in.	0.010 in.
		W	$K_{\alpha_{12}}$	69.5	0.211	1.000	1.000	1.000	1.000
		Ag	$K_{\alpha_{12}}$	25.5	0.561	1.000	0.999	0.999	0.997
		Mo	$K_{\alpha_{12}}$	20.0	0.711	0.999	0.999	0.998	0.995
		W	L_{β_1}	11.5	1.282	0.997	0.993	0.987	0.967
		W	L_{α_1}	10.2	1.476	0.995	0.990	0.980	0.951
		Cu	$K_{\alpha_{12}}$	9.0	1.542	0.994	0.989	0.978	0.945
		Cr	$K_{\alpha_{12}}$	6.0	2.291	0.983	0.966	0.932	0.839
		Ag	L_{β_1}	3.5	3.935	0.921	0.848	0.720	0.440
		Ag	L_{α_1}	3.4	4.154	0.908	0.825	0.681	0.383
Cl	4.397	Mo	L_{β_1}	2.6	5.177	0.835	0.697	0.486	0.165
S	5.018	Mo	L_{α_1}	2.5	5.407	0.815	0.665	0.442	0.130
P	5.784								
Si	6.738	W	M_{α_1}	2.1	6.983	0.654	0.428	0.183	0.014
Al	7.948	Al	$K_{\alpha_{12}}$	1.6	8.340	0.494	0.244	0.060	0.001
Mg	9.512	Mg	$K_{\alpha_{12}}$	1.3	9.890	0.318	0.101	0.010	0.000
Na	11.569								
F	18.02								

[a] Values for absorption edges and wavelengths are taken from ref. 4.
[b] Excitation potentials are taken from ref. 5.
[c] Transmissions calculated from data given in ref. 6.

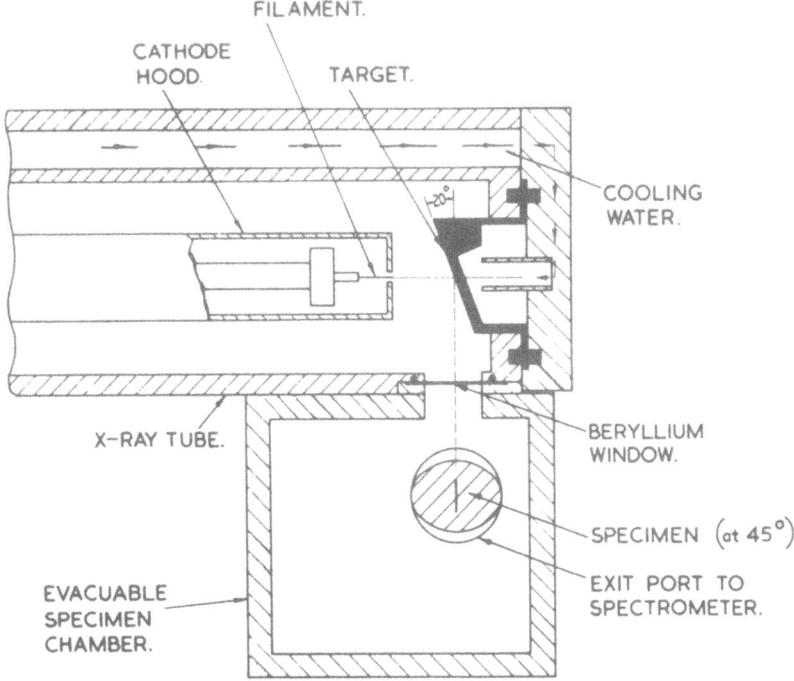

Figure 1. Cross section of the Raymax 60 demountable tube and sample chamber.

(NaCl); the intensities of the Mg K_α and Cl K_α fluorescence were then measured in the usual way. For both targets the intensities were measured as a function of the tube voltage: from 10–55 kV for the aluminum targets and from 10–60 kV for the silver targets. An X-ray tube window of 0.001-in. beryllium was used for the Al K_α radiation and of 0.0025-in. beryllium for the Ag L_α radiation.

The excitation efficiencies, in counts per second per watt, of the different targets were then evaluated from the measured intensities, thus permitting the dependence of the excitation efficiency on take-off angle and tube voltage to be ascertained. This excitation efficiency, used also by Thatcher and Campbell,[2] gives the optimum tube voltage for each target take-off angle for unit power input to the X-ray tube. This is of particular importance in the design of a matching X-ray generator as there is usually no difficulty in matching the power output of the generator to the maximum permissible input power to the target at high-tube voltages. At low-tube voltages, however, where a high-tube current is required this may not be possible unless the generator has been especially designed with this in mind. This measure of excitation efficiency is also useful because it permits a direct comparison to be made with theoretically calculated intensities.

The experimental results for the Al K_α and Ag L_α radiations are shown in Figures 2a and 2b, respectively. It can be seen that the optimum tube voltage and maximum excitation efficiency increase from 18 kV and 13.5 cps/W at 20° take-off angle to 25 kV and 19.1 cps/W at 60° take-off angle for Al K_α, and from 23 kV and 53.5 cps/W at 20° to 33 kV and 64.5 cps/W at 60° for Ag L_α. Thus, both the excitation efficiency and the optimum tube voltage increase with increasing target take-off angle. It is worth noting that at the optimum setting of the tube voltage there will be a minimum in the rate of change of intensity due to fluctuations in the voltage.

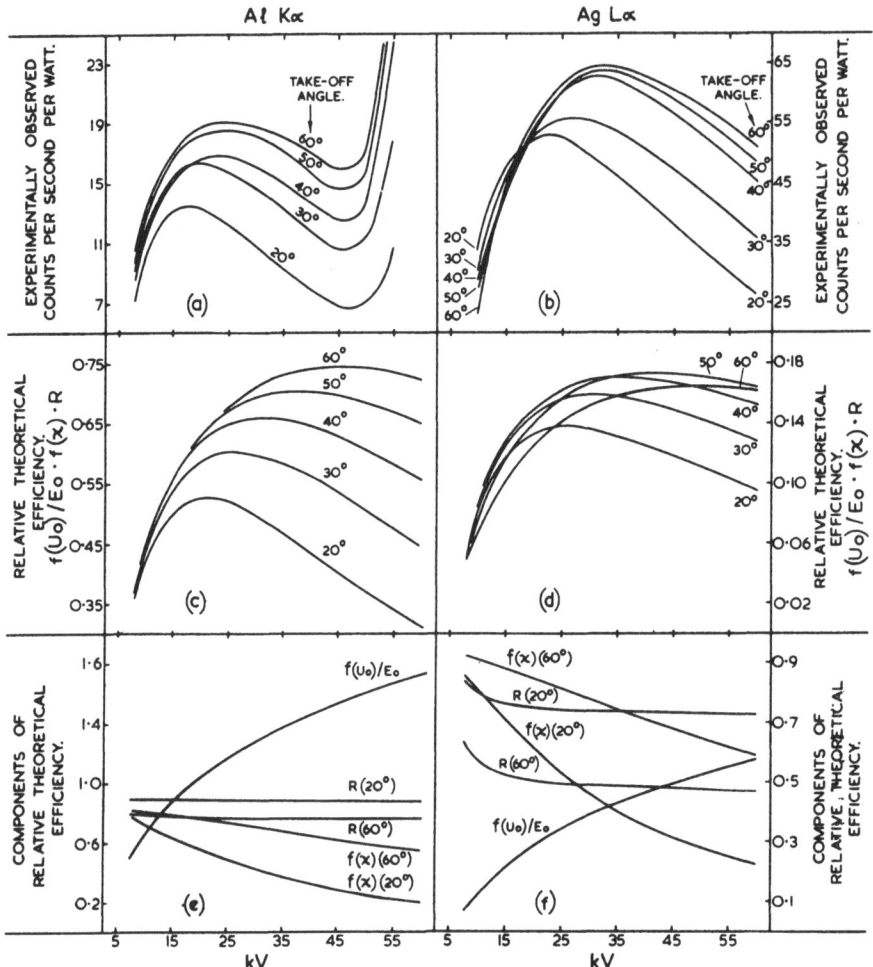

Figure 2. Experimental and theoretical excitation efficiencies for Al K_α and Ag L_α radiations as functions of tube voltage and target take-off angle.

The sudden increase in excitation efficiency of the Al K_α around 45 kV is due to the passage of backscattered electrons through the 0.001-in. beryllium window giving rise to direct electron excitation of the magnesium sample. The increased excitation efficiency was accomplished by a marked rise in the intensity of the background.

The experimental intensity curves for Al K_α and Ag L_α shown in Figures 2a and 2b may be compared with the curves shown in Figures 2c and 2d which have been derived from theoretical considerations. The theoretical calculations have been made on the assumption that the intensity/mA of X-rays emitted by the target (as distinct from the tube) is dependent on three parameters: the overvoltage U_0, the transmission of generated X-rays to the surface of the target $f(\chi)$, and the effective electron current factor R.

The intensity of excitation varies directly with $f(U_0)$, a function of the overvoltage,[8] where

$$f(U_0) = U_0 \ln U_0 - U_0 + 1$$

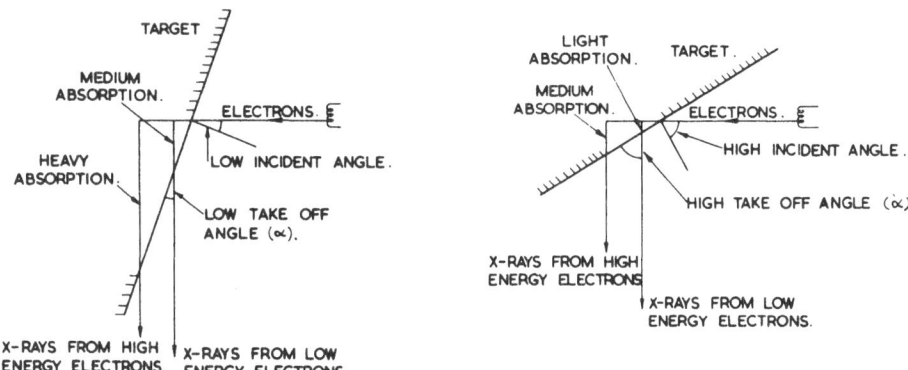

Figure 3. Diagrammatic representation of the effects of high and low take-off angles and high and low energy electrons on X-ray absorption in the target.

and $U_0 = E_0/E_c$ where E_0 is the accelerating potential and E_c is the excitation potential of the characteristic line of the target.

The transmission of the X-rays from the point of generation to the surface of the target is dependent on the target take-off angle and the penetration depth (i.e., energy) of the incident electrons. It should be noted that in the Raymax tube geometry the penetration depth of the electrons below the target surface is itself affected by the target take-off angle. Figure 3 illustrates the ways in which take-off angle and electron energy influence self-absorption in the target of the Raymax tube. The transmission of the generated X-rays through the target, $f(\chi)$, may be calculated using the expression due to Philibert[9] which takes into account absorption due both to electron penetration[10] and target take-off angle

$$\frac{1}{f(\chi)} = \left(1 + \frac{\chi}{\sigma}\right)\left[1 + h\left(1 + \frac{\chi}{\sigma}\right)\right]$$

where $h = 1.2A/Z^2$, A = atomic weight, Z = atomic number, and

$$\sigma = \frac{2.39 \times 10^5}{E_0^{1.5} - E_c^{1.5}}$$

$$\chi = (\mu/\rho) \cot \alpha$$

the term μ/ρ is the mass absorption coefficient for the characteristic radiation in the target, and α is the target take-off angle.

The beam current meter in the Raymax generator control circuit registers the filament emission current as distinct from the current actually passing into the target; because of this it is necessary to make a correction for the backscattering of electrons from the target. In making this correction we have used the theoretical values of $(1 - R)$ evaluated by Bishop[11] for copper as a function of U_0 for a number of different values of the electron angle of incidence. These are then converted to values of $(1 - R)$ for aluminum and silver using the expressions

$$(1 - R)_{Al} = (1 - R)_{Cu} \times \frac{\eta_{Al}}{\eta_{Cu}} \quad \text{and} \quad (1 - R)_{Ag} = (1 - R)_{Cu} \times \frac{\eta_{Ag}}{\eta_{Cu}}$$

where values of η, the backscattering coefficient, have been taken from the graph of Burkhalter.[12] No additional correction has been made to take into account those backscattered electrons which, due to the force field, return to the target.

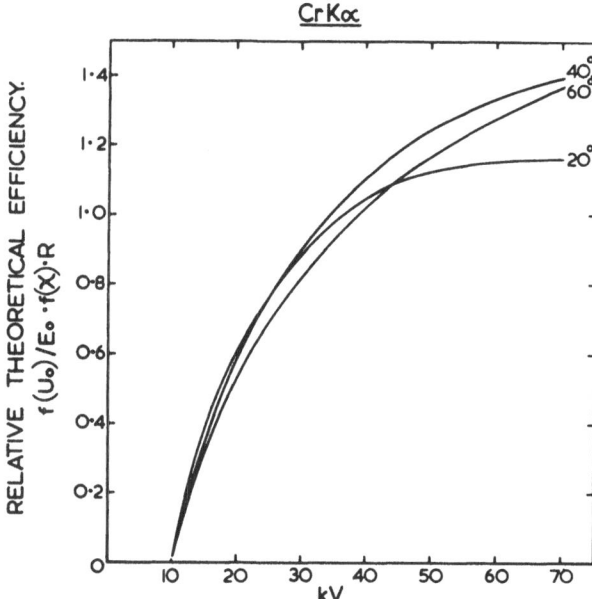

Figure 4. Theoretical excitation efficiencies for Cr K_α radiation as a function of tube voltage and target take-off angle.

The experimental intensities have been evaluated in counts per second per watt; the comparative theoretical intensities are consequently given by the expression

$$\frac{f(U_0) \cdot f(\chi) \cdot R}{E_0}$$

The relative excitation efficiencies obtained by means of this expression have been used to plot the theoretical curves for Al K_α and Ag L_α in Figures 2c and 2d, respectively.

It can be seen that there is good qualitative agreement between the experimental and theoretical curves; in particular, the telescoping of the curves for Ag L_α at high take-off angles predicted theoretically is confirmed experimentally. This telescoping can be explained to some extent by reference to Figures 2e and 2f which show separately the components $f(U_0)/E_0$, $f(\chi)$, and R for aluminum and silver $20°$ and $60°$ targets; the increase in the backscatter at high angles of electron incidence is responsible for a considerable drop in the intensity of the Ag L_α radiation.

Theoretical excitation efficiency curves have also been calculated for Cr K_α radiation for take-off angles of $20°$, $40°$, and $60°$; these are shown in Figure 4. It would appear that the target self-absorption changes comparatively little over a wide range of take-off angles for shorter wavelength radiation.

EFFECT OF VARYING THE TUBE WINDOW THICKNESS

In selecting a target for fluorescence analysis we have to take into account the intensity of the primary radiation, the transmission of the tube window, and the closeness of the primary wavelength to the absorption edge of the element in which we are interested. For example, if magnesium is the element to be analyzed there is a choice of the following primary characteristic radiations: Al K_α, W M_α, Mo L_α, Ag L_α, Cr K_α; if chlorine were to be analyzed the corresponding choice would be between: Ag L_α, Cr K_α, Cu K_α, W L_α; for other light elements the useful radiations can be readily selected using the data given in Table I. In both the examples given, the radiations have been listed in order of increasing distance from the absorption edge of the element to be analyzed and, hence, in order of decreasing effectiveness. This generalization is, however, complicated in two ways. Firstly, K radiation is usually more intense than L radiation which is, in turn, more intense than M radiation. Secondly, the more effective, longer wavelengths tend to be more heavily absorbed in the X-ray tube window. Thus, for a given tube window, it would be most difficult to predict which target would be the most suitable for a particular analysis.

If a sealed tube is used for light element analysis it is generally necessary to use a chromium target because the longer wavelengths listed above are too heavily absorbed in the thick beryllium tube windows that are used. A continuously evacuated tube, however, is not limited to the use of such windows which are generally of the order of 0.010 in. as the pumping system will remove any gas which may leak through a thin window. Raymax demountable tubes are currently fitted with 0.004-in. beryllium windows as standard, but in recent months 0.002-in. and 0.001-in. beryllium windows have been used quite successfully. These thin windows may, of course, easily be replaced should the need arise.

To demonstrate the dependence of target selection on tube window thickness and to illustrate the improvement that can be achieved using long-wavelength characteristic radiations and thin tube windows for the analysis of light elements, two experiments have been carried out. One is on the analysis of chlorine, the first light element for which there is a possibility that Cr K_α is not the most suitable radiation. The second is on the analysis of magnesium, which for many years was the lightest element normally analyzed by X-ray fluorescence methods. The X-ray fluorescence intensity of these two elements has been measured for different targets, different beryllium windows, and at different voltages as listed below:

Sample	Cl in NaCl	Mg
Targets	Ag, Cr, Cu, W	Al, W, Mo, Ag, Cr
Be windows	0.0025 in., 0.0048 in.	0.001 in., 0.002 in.
	0.0103 in.	0.0043 in., 0.0101 in.
Tube current	10 mA	10 mA
kV	10 to 60 kV in	10 to 55 kV in
	steps of 5 kV	steps of 5 kV

To keep the number of experiments within reasonable bounds, targets with a take-off angle of 20° only were used.

The excitation efficiencies measured using chlorine in NaCl are shown in Figure 5 and using magnesium in Figure 6. It can be seen that the efficiency is markedly dependent on the tube voltage and, for longer wavelength radiations, on the tube window thickness. It is particularly noteworthy that when a 0.001-in. beryllium window was used, Ag L_α was only marginally more efficient than Cr K_α for the analysis of chlorine whose absorption edge is relatively close to Cr K_α; in the case of magnesium, however, whose absorption edge is far from Cr K_α, Al K_α is much more efficient than Cr K_α. On the other hand, when a 0.010-in. beryllium window was used, Cr K_α radiation was more efficient than Ag L_α for both chlorine and magnesium. As might be expected, therefore, the advantage of using a longer wavelength characteristic radiation increases with decreasing atomic number—provided that the tube window thickness can be decreased at the same time.

To highlight the importance of the transparency of the tube window we have shown in Figure 7 the excitation efficiency of the different targets at their optimum tube voltages as a function of tube window thickness. It may be seen that the most efficient radiation

CHLORINE IN ROCKSALT; 20° TAKE-OFF ANGLE.

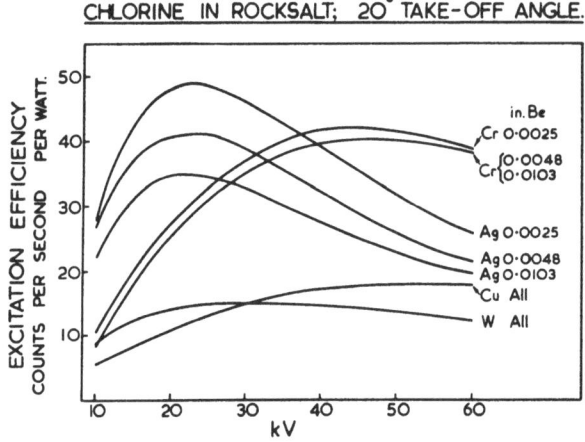

Figure 5. Experimental excitation efficiencies for silver, chromium, copper, and tungsten targets obtained using a chlorine (NaCl) sample as a function of tube window thickness and tube voltage.

MAGNESIUM SPECIMEN, 20° TAKE-OFF ANGLE.

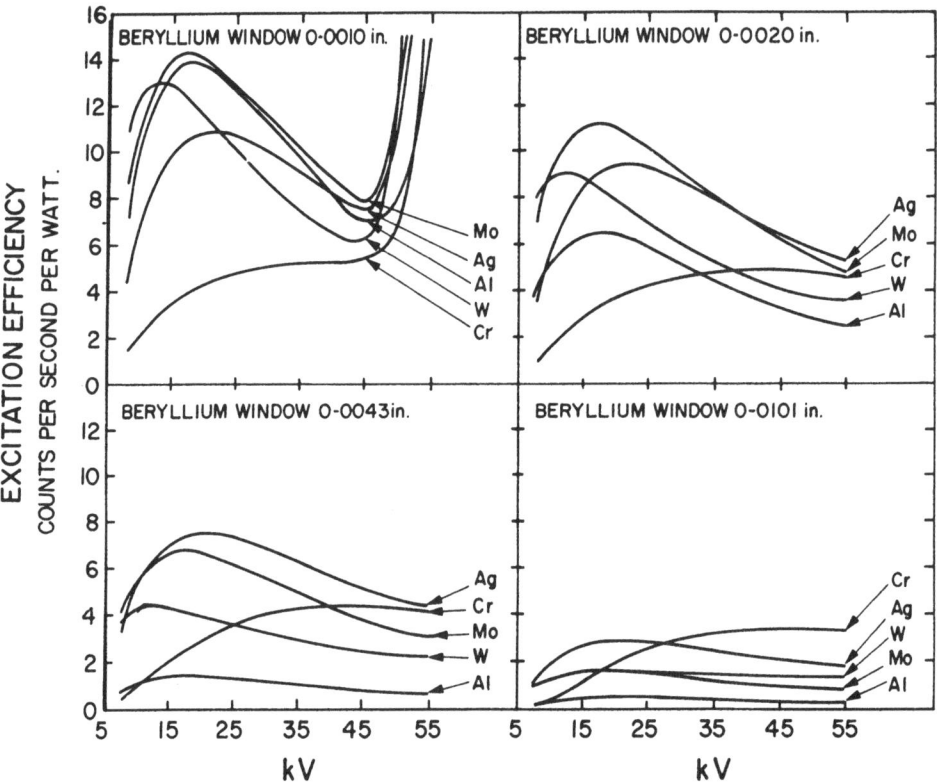

Figure 6. Experimental excitation efficiencies for aluminum, tungsten, molybdenum, silver and chromium targets obtained using a magnesium sample as a function of tube window thickness and tube voltage.

(using a target take-off angle of 20°) depends very much on the tube window thickness; in the case of a magnesium sample, for example, Mo L_α is the most efficient radiation from 0.001 in.–0.003 in. of beryllium, Ag L_α from 0.003 in.–0.009$_5$ in., and Cr K_α above

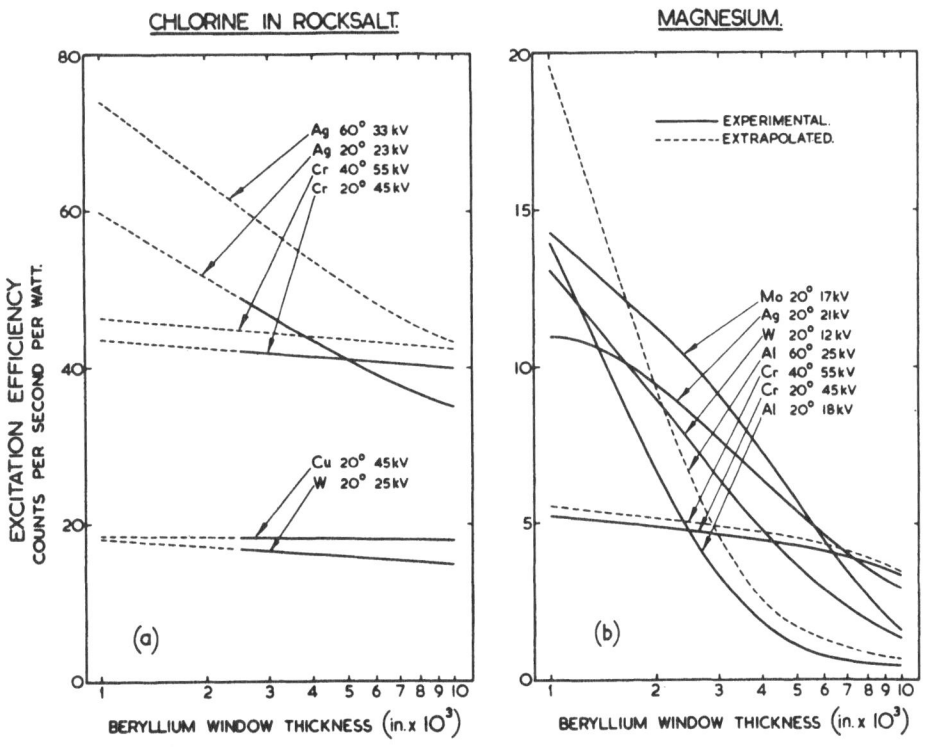

Figure 7. Experimental efficiencies for a variety of targets obtained for chlorine (NaCl) and magnesium at the optimum tube voltage as a function of tube window thickness.

0.009_5 in. Included in Figure 7a, which is for chlorine, are extrapolated values for Ag (60°) and Cr (40°) targets, and in Figure 7b, which is for magnesium, are extrapolated values for Al (60°) and Cr (40°) targets. It is now possible to make a comparison between the excitation efficiency to be expected using a sealed-tube chromium target and the best so far obtained using the demountable tube; this is given in Table II. It is expected that the intensities obtained using the demountable tube could be further improved by using an electron beam having a low angle of incidence and, in the case of the magnesium sample, by using a Mo (60°) target.

Table II. Comparison of Demountable and Sealed Tube Conditions for Magnesium and Chlorine

	Magnesium		Chlorine	
Type of tube	Sealed	Demountable	Sealed	Demountable
Target of angle	Cr, 40°	Al, 60°	Cr, 40°	Ag, 60°
Be window	0.010 in.	0.001 in.	0.010 in.	0.001 in.
cps/W	3.9	19.1	45	57
Optimum kV	55	25	55	33

Table III. Optimum Tube Voltages and Absorption Data

Radiation	Optimum voltage (this paper) with 20° target	Optimum voltage (Thatcher and Campbell[2]).	Self-absorption $\dfrac{\mu}{\rho}$
W M_α	12	10	1466
Mo L_α	17	15	728
Al K_α	18	—	386
Si K_α	—	16	328
Ag L_α	21	17	522
Cr K_α	45	—	88

The particularly low efficiencies of Cu K_α and W L_α for exciting Cl K_α shown in Figure 5 may be noted; this serves to emphasise the importance of using a primary wavelength that is close to the fluorescence absorption edge.

The optimum tube voltage is, of course, dependent on target take-off angle; in Table III we compare the optimum potentials established in the present work with those obtained by Thatcher and Campbell.[2] Their slightly lower values are probably due to a lower electron angle of incidence. Thatcher and Campbell conclude that this optimum tube voltage is inversely related to the mass absorption coefficient (also given in Table III) of the primary radiation in the target, with the exception—they say—of silicon. We find that aluminum, another light element target is also anomalous in this respect. This is not altogether surprising in view of the complexity of the parameters that it has been seen combine to produce the excitation efficiency maxima. It is noteworthy that the low optimum voltage found experimentally for Al K_α is confirmed by the theoretical calculations.

EXCITATION BY ELECTRON BACKSCATTER

If the increased excitation efficiency at high tube voltages and with very thin tube windows is indeed due to direct electron excitation, it would be expected that, for a given sample, excitation efficiency would increase with increasing atomic number of the target. Figure 8 shows the experimental intensities at 55 kV obtained using the magnesium sample and a 0.001-in. beryllium window; also shown are scaled values of the backscattering coefficient given by Burkhalter.[12] There is reasonable agreement between the two curves.

EXCITATION BY WHITE RADIATION

The elements present in the sample will be excited not only by the characteristic X-rays produced by the target but also to some extent by the white radiation. An attempt was made by Thatcher and Campbell[2] to take this into account on a quantitative basis in their work on excitation efficiencies of long wavelength targets in light element analysis; in so doing, they omitted to correct for the increased electron backscattering of higher atomic weight targets, thus casting doubt on the validity of their conclusions. We have preferred to investigate qualitatively but more rigorously the part played by the white radiation by examining the diminution in intensity caused by the increased thickness of the beryllium windows.

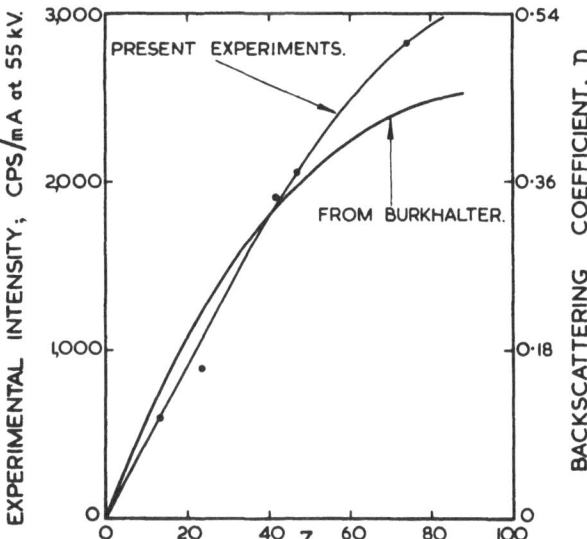

Figure 8. Intensity due to electron backscattering and electron backscattering coefficients as functions of atomic number Z.

If we consider initially the primary radiation as consisting solely of characteristic radiation then we may calculate precisely the decrease in intensity due to absorption in the tube window from the expression

$$I_{obs} = I_{inc} \exp(-\mu t)$$

where I_{obs} is the experimentally observed intensity, I_{inc} is the intensity of the radiation incident on the tube window, μ is the linear absorption coefficient of the beryllium, and t its thickness; $\exp(-\mu t)$ for the different radiations is given in Table I. Dividing I_{obs} by $\exp(-\mu t)$ should, therefore, give for each radiation a constant value of I_{inc} regardless of window thickness. If the values of $I_{obs}/\exp(-\mu t)$ increase with increasing window thickness then this implies that white radiation of a wavelength shorter than the characteristic wavelength is contributing to the production of fluorescence X-rays. If the values of $I_{obs}/\exp(-\mu t)$ decrease, then white radiation of a wavelength longer than the characteristic is contributing to the production of fluorescence. As the wavelength intensity distribution of the white radiation is dependent on the tube voltage then values of $I_{obs}/\exp(-\mu t)$ obtained at low and high voltages should be considered separately.

Values of $I_{obs}/\exp(-\mu t)$ have been calculated using the intensities obtained with aluminum, tungsten, molybdenum, silver, and chromium targets and the magnesium sample at 10 kV and 40 kV; these are shown as a function of tube window thickness in Figure 9. It can be seen that short-wavelength white radiation contributes to the fluorescence due nominally to Al K_α and W M_α, whose wavelengths are close to the Mg K absorption edge; wavelengths longer than the characteristic radiation contribute to the stimulation of fluorescence in the cases of Mo L_α, Ag L_α, and Cr K_α. It also may be seen that increased tube voltage has no qualitative effect on the effective wavelength of the white radiation despite the fact that the white radiation peak decreases from about 1.88 Å at 10 kV to about 0.46 Å at 40 kV. It is, therefore, only the very long wavelength white radiation that appears to stimulate fluorescence to an appreciable degree in light elements.

CONCLUSIONS

The experiments described above have shown that a high target take-off angle is desirable for longer wavelength characteristic radiations; this increases both the excitation

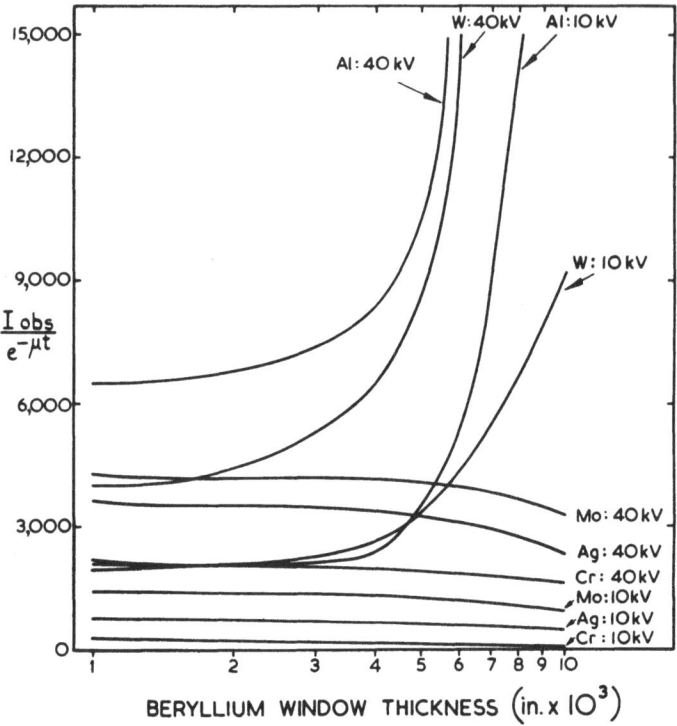

Figure 9. $I_{obs}/\exp(-\mu t)$ vs. thickness of beryllium for aluminum, tungsten, molybdenum, silver, and chromium targets at 10 kV and 40 kV.

efficiency and the optimum tube voltage. A high optimum tube voltage reduces the current output required from the generator for a given power input to the target. On the other hand, the relatively small increases in intensity at the higher take-off angles and the theoretical calculations on electron backscattering suggest that, especially for heavy element targets, a low angle of incidence is required for the electron beam on the target.

It also has been shown that a much higher excitation efficiency can be obtained for light elements by using a long-wavelength characteristic radiation with a high-transparency tube window when compared with the customary Cr K_α and a thick beryllium window. The most suitable target depends on many factors and must be determined experimentally. It is most important with a long-wavelength target to operate at the optimum tube voltage, as the excitation efficiency is particularly dependent on the voltage at these wavelengths. At this optimum, voltage fluctuations have less effect than at other voltages.

Two factors have not been brought out in the above presentation. Firstly, it is well known that the sensitivity of the fluorescence method of analysis is directly related to the amount of background radiation, and this has not been considered. It was, however, noted that the background increased with both atomic number of the target and also with tube voltage. This further emphasises the advantage to be gained in using long-wavelength primary radiation; on the other hand the optimum voltage with respect to sensitivity will be at a lower voltage than for the maximum excitation efficiency, thus imposing additional demands on the current capacity of the generator.

Secondly, the excitation efficiency in counts per second per watt does not take into account the differing power maxima that can be put into different targets; nor does it take into account the current which may or may not be available from the generator. Either of these factors may be sufficient to alter, in practice, our choice of target.

ACKNOWLEDGMENTS

We would like to thank our colleagues Dr. R. Witty and Mr. I. K. Openshaw for many helpful discussions; we would also like to thank the Management of the A.E.I. Electronics Group for permission to publish this paper.

REFERENCES

1. N. Spielberg, "Tube Fluorescence and Inherent Filtration as Factors in the Fluorescence Excitation of X-Rays," *J. Appl. Phys.* **33**: 2033, 1962.
2. J. W. Thatcher and W. J. Campbell, "Instrumentation for Primary and Secondary Excitation of Low-Energy X-Ray Spectral Lines," U.S. Bureau Mines, *Report Invest. 6689*, 1965.
3. N. Spielberg, "Intensities of Radiation from X-Ray Tubes and the Excitation of Fluorescence X-Rays," *Philips Res. Rept.* **14**: 215, 1959.
4. J. A. Bearden, "X-Ray Wavelengths," U.S. Atomic Energy Commission, *NYO-10586*, 1964.
5. H. A. Liebhafsky, H. G. Pfeiffer, E. H. Winslow, and P. D. Zemany, *X-Ray Absorption and Emission in Analytical Chemistry*, John Wiley and Sons, New York, 1960, p. 307.
6. K. F. J. Heinrich, "X-Ray Absorption Uncertainty," in: T. D. McKinley, K. F. J. Heinrich, and D. B. Wittry (eds.), *The Electron Microprobe*, John Wiley and Sons, New York, 1966, p. 296.
7. M. A. Short, "Detection and Correction of Nonlinearity in X-Ray Proportional Counters," *Rev. Sci. Instr.* **31**: 618, 1960.
8. M. Green and V. E. Cosslett, "The Efficiency of Production of Characteristic X-Radiation in Thick Targets of a Pure Element," *Proc. Phys. Soc. (London)* **78**: 1206, 1961.
9. J. Philibert, "A Method for Calculating the Absorption Correction in Electron Probe Microanalysis," in: H. H. Pattee, V. E. Cosslett, and A. Engstrom (eds.), *X-Ray Optics and X-Ray Microanalysis*, Academic Press, New York, 1963, p. 379.
10. P. Duncumb and P. K. Shields, "Effect of Critical Excitation Potential on the Absorption Correction in X-Ray Microanalysis," *Tube Investments Research Laboratories Technical Report No. 181*, 1964.
11. H. E. Bishop, private communication from Ph.D. thesis, University of Cambridge, 1966.
12. P. G. Burkhalter, "Measurements of Backscattered Electrons in an Electron Probe Microanalyser," U.S. Bureau Mines, *Report Invest. 6681*, 1965.

CHARACTERISTICS OF FLOW PROPORTIONAL COUNTERS FOR X-RAYS

N. Spielberg

Philips Laboratories
Briarcliff Manor, New York

ABSTRACT

Gas flow proportional counters for the detection of soft X-rays were introduced about ten years ago. These detectors offered the advantages of high sensitivity, good energy discrimination qualities and the ability to handle high counting rates. Since that time they have been used for ultra-soft and harder X-rays as well, both as detectors in standard spectrographic instruments and as energy discriminating instruments themselves in various so-called nondispersive applications. Depending upon the particular instrumental application, however, their use has led to considerable complication of the associated electronic circuitry in order to realize their advantages. For the most effective use of these counters (and of sealed proportional counters as well) it is necessary to have a clear understanding of the effect of various design parameters and operating conditions on their performance. The dependence of the shape of the pulse height distribution on the operating voltage, pressure and counting rate is described as a function of the energy of the radiation detected and the nature of the gas. Stability requirements on counter tube high voltage supplies and operating pressure are discussed. Shifts of pulse height distributions toward smaller pulse sizes with increasing counting rate are described and the dependence of these shifts on the various parameters and on wavelength are discussed. Techniques for eliminating the shifts and the implications of these techniques for the associated electronics are described.

INTRODUCTION

The use of side-window gas-flow proportional counters for the detection of X-rays is not a new technique.[1,2] They have been applied in a wavelength range from about 1–70 Å.[3,4] As in the case of permanently sealed gas-filled proportional counters, which are often used down to wavelengths of 0.5 Å, these detectors are reputed to offer high sensitivity, the ability to handle high counting rates, and the possibility of discriminating against the detection of unwanted wavelengths.[5] Usually in X-ray spectroscopy and diffraction applications, only relatively crude discrimination is required to suppress the effects of scattered background radiations and of higher order reflections of harder radiations, primary reliance being placed on crystal diffraction for finer resolution. Nevertheless, applications are reported from time to time of the use of the proportional counter alone, although coupled with fairly complex electronic circuitry, for the analysis of X-ray spectra, based on a knowledge of the pulse height distribution as a function of wavelength.[6,7]

There have been some reports recently which indicate that the proportional counter cannot handle as high counting rates as usually believed and still maintain its ability to discriminate against unwanted wavelengths.[8,9] The effect of the high counting rate seems

534

to result primarily in a counting rate dependent shift of the pulse height distribution toward smaller pulse heights. The effect seems to occur at surprisingly low counting rates. Such a shift, while serious enough for the usual use of this detector, plays havoc with analysis schemes based on a stable and definable pulse height distribution.

In the work reported here, it was found that the counting rate dependent shift of the pulse height distribution could be eliminated by increasing the diameter of the anode wire of the counter tube. Obviously such a change in counter tube design results in new requirements for the high voltage supply or the pulse amplifier circuits, or both. It is not clear, however, what effect, if any, there might be on such factors as the resolution and the voltage and pressure dependent stability of the counter tube. As a result, a study was undertaken of these matters also, and some preliminary results are reported here.

APPARATUS

Figure 1 shows a schematic drawing of the experimental setup. Radiation from a demountable X-ray tube was introduced into a vacuum spectrograph where, after collimation by a pair of slits and filtration, it was incident upon a potassium acid phthalate crystal cleaved parallel to the 100 planes ($2d = 26.6$ Å) and reflected into the flow proportional counter. The window between X-ray tube and vacuum spectrograph was 6 μ thick beryllium, and the entrance window on the detector was 12 μ thick beryllium. Electrical and gas connections were introduced through appropriate ports in the wall of the spectrometer chamber. Appropriate valving, not shown in the figure, was used to permit pumping the apparatus down and admitting counter gas without inducing undue stresses on the delicate beryllium windows.

The gas control system for the counter tube is shown in Figure 2. Gas from the supply was admitted to a pilot controlled absolute pressure regulator capable of maintaining set pressure within 0.05 mm Hg absolute. The regulated gas then flowed directly to the counter tube and was exhausted to vacuum through a very fine throttle valve. An absolute manometer of the aneroid type was used to monitor the pressure of the gas. This system worked very well, but for light gas mixtures, such as neon–helium–methane, the consumption of gas by the pilot flow was prohibitively high, necessitating the use of a parallel manually controlled throttle system, as shown in the lower part of the figure. Various other

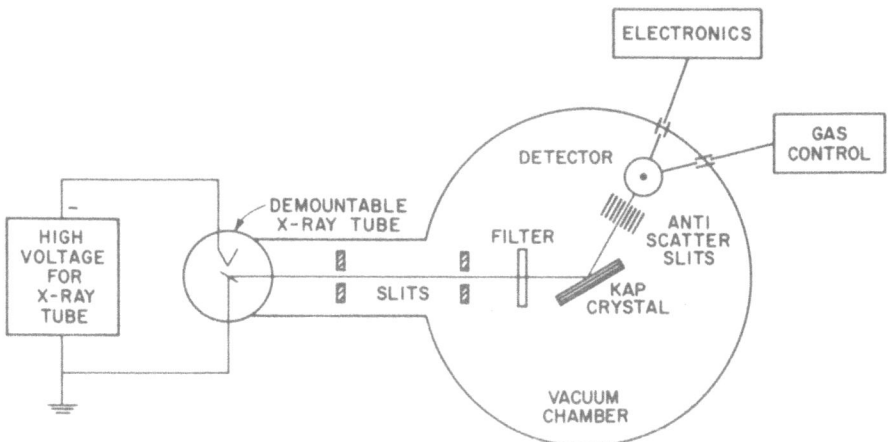

Figure 1. Schematic diagram of experimental arrangement.

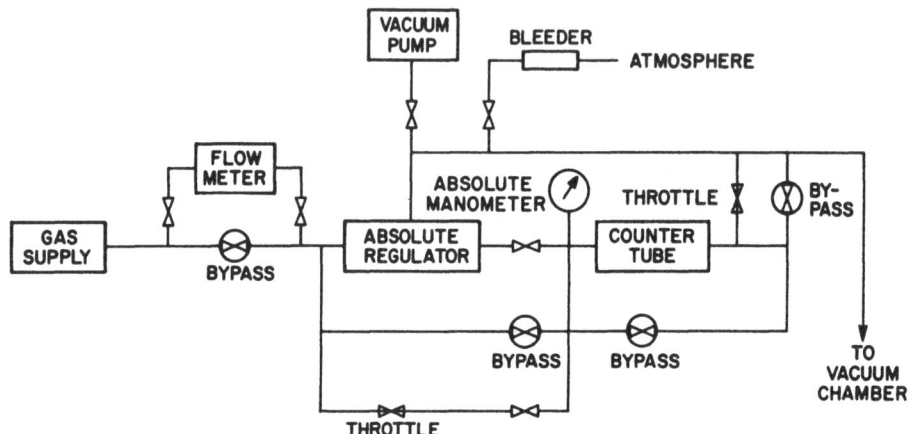

Figure 2. Gas control system for flow proportional counter.

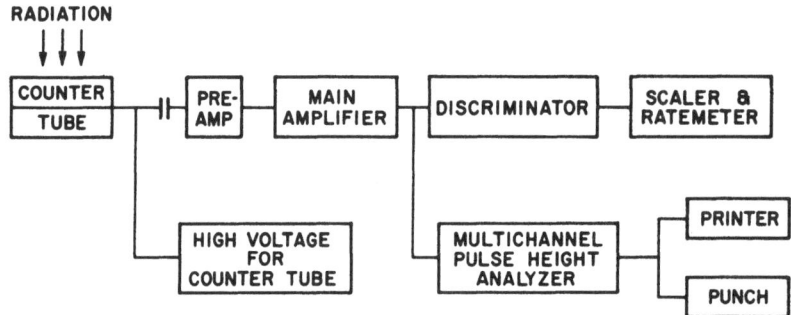

Figure 3. Block diagram of electronic circuitry for counter.

bypass valves were provided to facilitate the rapid evacuation of the counter tube. The direct connection to the spectrometer vacuum chamber through a valve (not shown) was used only in initial evacuation or final letdown to atmosphere of the spectrometer. In ordinary operation, the spectrometer was separately pumped.

A block diagram of the electronic circuitry is shown in Figure 3. The transistorized preamplifier was mounted inside the spectrometer adjacent to the counter tube, and had a voltage gain of approximately 20. The main amplifier was a rather-old tube design, but was adequate for the purposes of this work. Pulse height analysis was done with a 100 channel analyzer. The high voltage supply for the counter tube was stable to 0.01%. Checks over a period of time showed the overall system gain to exhibit a long term stability of one or two channels. Tests with a pulser showed that counting rates up to about 50,000 counts per second could be handled satisfactorily.

Gas gain of the counter tube was determined in the following manner: A wide 10–100 μsec voltage pulse of magnitude V from a pulse generator was introduced through a capacitance C (approximately 1 pF) onto the anode wire of the counter. This simulated the collection of a charge pulse of magnitude $Q = CV$ by the counter. With a low repetition rate of about 100 sec it was possible to observe the resulting pulse height distribution in the pulse height analyzer, thereby calibrating the pulse amplifier system in terms of charge delivered to the anode of the counter tube. By dividing the energy of the radiation being detected by the energy required to create an ion pair in the gas (approximately

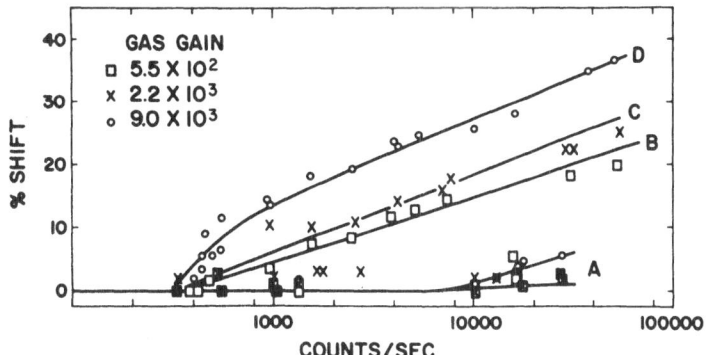

Figure 4. Percent shift of peak of pulse height distribution as a function of counting rate. Curves B, C, D with 50 μ center wire; curve A with 125 μ center wire. Counter gas, 90% argon–10% methane at 200 mm Hg pressure. Wavelength detected, 6.98 Å. Reprinted from *Rev. Sci. Instr.*[10] by permission of American Institute of Physics.

30 eV), the number of ion pairs and hence of primary electrons n per incident photon is determined. The gas gain is given then by

$$A_0 = Q/(ne)$$

where e is the charge on the electron.

The counter tubes were of conventional side window design, with both entrance and exit windows mounted by means of epoxy seals on demountable curved window frames to permit easy interchange and replacement. Gaskets of silicone rubber were used under the window frames. Beryllium as thin as 6 μ has been used successfully, but for this work the entrance window was 12 μ thick and the exit window 25 μ thick. The body and window frames were made from stainless steel, and the insulators from Teflon. The anode wire was spring-loaded to prevent sag, and was easily replaceable. Any desired wire diameter could be used; for this work the diameters ranged from 50 to 250 μ. The inside diameter of the counter tube shell was 22.2 mm and its length was 49.8 mm.

With this system the following parameters could be varied: wavelength and intensity of incident radiation, counter tube gas composition and pressure, counter tube voltage, anode wire diameter and composition.

SHIFT OF PULSE DISTRIBUTION WITH COUNTING RATE

Measurements of the shift of the pulse height distribution with counting rate have been made at 1.54 and 6.98 Å, at several different values of voltage and pressure (hence, of gas gain), and for three different gas mixtures: 90% argon–10% methane, 50% neon–49% helium–1% butane, and 46% nitrogen–46% helium–8% methane (the percentages quoted are partial pressures). Figure 4 shows typical results obtained for the argon-methane mixture at 200 mm Hg for 6.98 Å photons where curves B, C, and D were taken for increasing values of gas gain for a counter tube having a 50 μ diameter anode wire.

Since the shift increases with increasing gas gain, it might be thought that the shift should be dependent upon the amount of charge per unit time delivered to the anode wire, and would, thus, be independent of wavelength if the gas gain were adjusted to keep the size of the output pulse constant. In fact, this is not so; the shift for 1.54 Å photons is smaller than for 6.98 Å photons when the gas gain is adjusted to keep the output pulse size constant. Tentative evidence has also been found to indicate that the shift is greater for high absorption counter tube fillings than low absorption fillings, for constant size output pulse. Curve A of Figure 4 shows that the shifts are virtually completely eliminated by changing to a 125 μ diameter anode wire.[10] The elimination of the shift is essentially maintained for this diameter wire for all the gas fillings and pressures tested, up to the highest counting rates tested (30,000 counts per second).

In view of the results above and the work of Bender and Rapperport,[8] it may be generalized that the following geometric factors have significant effects on the magnitude of the shifts:

1. The density of ionizing radiation per unit length of anode wire; i.e., a fine beam of radiation will cause a greater shift than a broad beam having the same total photon flux.
2. The nonuniformity of absorption of radiation within the counter tube; i.e., if most of the absorption takes place between the entrance window and the anode wire, the shift will be considerably greater than if the absorption takes place throughout the path from entrance window to exit window.
3. The diameter of the anode wire, with this last factor being much more important than the first two factors, possibly because of its effect on the mechanisms of gas gain.

ENERGY RESOLUTION

One of the principal advantages of the proportional counter for detecting X-rays is the proportionality of the size of the output pulse to the energy of the incident photons. One is, therefore, interested in the energy resolution of the counter, and, hence, in the characteristics of the pulse height distribution obtained for monoenergetic incident photons. The shape of the distribution is determined primarily by the statistics of primary ionization, of gas multiplication, and of ion recombination, with the major contribution coming from the statistics of primary ionization, i.e., the number of primary ion pairs formed per detected photon. Assuming Gaussian statistics, one would expect that the full width at half maximum height of the pulse height distribution generated by 1.54 Å photons would have a minimum value of the order of 15% of the average or maximum height, whereas for 6.98 Å photons the corresponding number would be of the order of 32% (assuming 30 eV per ion pair in argon).

The implicit assumption has been made that the distribution is unimodal, but the escape peak phenomenon[11,12] leads to at least bimodal distributions which can cause serious interferences in the detection of soft X-rays, when harder radiations are also present, even when electronic pulse height discrimination techniques are employed. Figure 5 compares pulse height distributions obtained with gas mixtures of 90% argon–10% methane and 90% neon–10% methane respectively for incident Cu K_α photons. The suppression of the escape peak for the neon mixture is evident. The potassium K peak arises from fluorescence of the potassium acid phthalate crystal.

Table I presents the full width at half maximum ordinate of the pulse height distributions obtained with a 50 μ diameter anode wire at varying values of gas gain for the argon–methane, neon–helium–butane, and nitrogen–helium–methane mixtures at pressures of 200 mm and 760 mm Hg, for 1.54 Å and 6.98 Å X-rays. There is a slight increase in width for the higher values of gas gain for the low pressure argon–methane mixture, while for the other mixtures there is a significant increase in width for increasing gas gain. When these widths are quite large, as in the case of the neon–helium–butane mixture, or for the argon–methane mixture at quite low pressures, there is increasing asymmetry of the pulse height distribution with increasing gas gain, as exemplified in

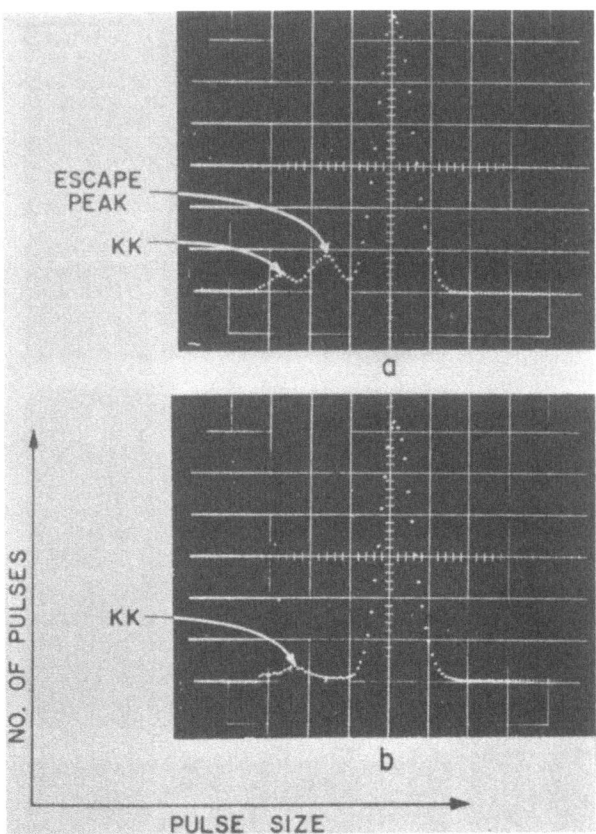

Figure 5. Comparison of pulse height distributions for 1.54 Å radiation for argon and neon based counter gases: (a) 90% argon–10% methane at 760 mm Hg pressure; (b) 90% neon–10% methane at 760 mm Hg pressure.

Table I. Width of PHD for Different Gases and Pressures for Counter with 50 μ Diameter Anode Wire

	Gas and pressure				
	Argon–methane		Neon–helium–butane		Nitrogen–helium–methane
Gas gain	200 mm Hg	760 mm Hg	200 mm Hg	760 mm Hg	200 mm Hg
1.54 Å					
1×10^2	0.18	0.20		0.23	
4×10^2	0.19	0.19		0.24	
16×10^2	0.24	0.21		0.35	
6.98 Å					
4.5×10^2	0.31	0.31	0.42	0.42	0.40
18×10^2	0.33	0.31	0.66	0.53	0.56
72×10^2	0.38		1.20	0.81	

Figure 6, which shows the changing shape as gas gain is increased for the helium–neon–butane mixture at 760 mm Hg when detecting 6.98 Å radiation. Possible evidence of this effect may be seen in some of the pulse height distributions for very soft X-rays reported by Henke.[3] Table II shows the effect of the anode wire diameter on the width of the distributions. It is clear that the larger diameter wire, in addition to its effect on

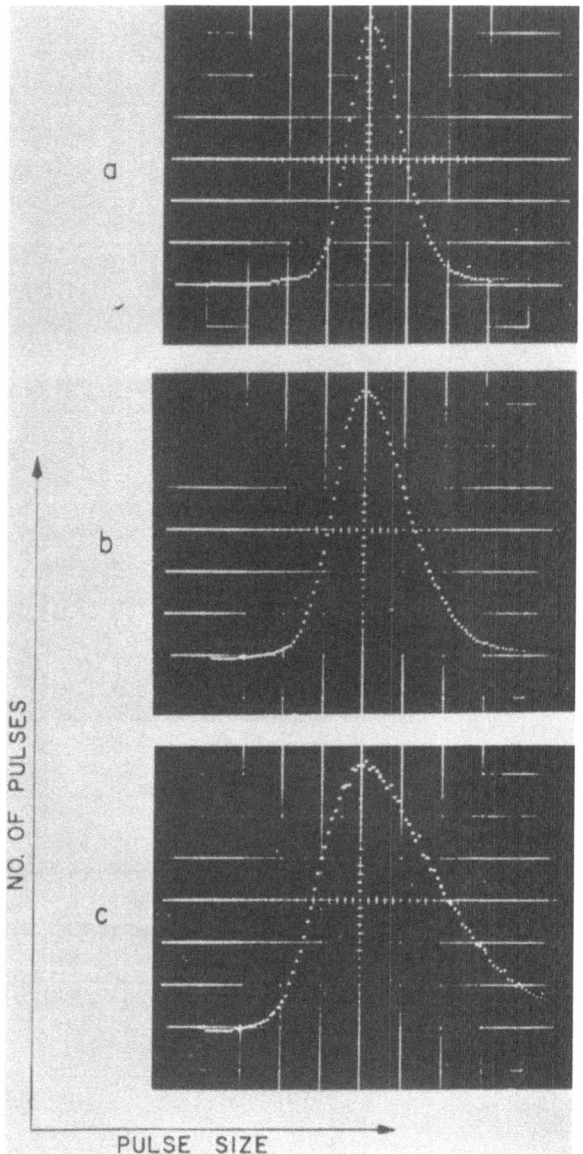

NO. OF PULSES

PULSE SIZE

Figure 6. Effect of excessive gas gain (overvoltage) on shape of pulse height distribution. Counter gas, 50% neon–49% helium–1% butane at 760 mm Hg pressure. Wavelength detected, 6.98 Å. (a) Gas gain, 4.5×10^2; 800 V. (b) Gas gain, 18×10^2; 915 V. (c) Gas gain, 72×10^2; 965 V.

the peak shift, as already discussed, does lead to some improvement in the width of the distributions. However, this point needs to be investigated further.

STABILITY

It is desirable to know the effect of small changes in operating parameters upon the pulse height distributions. It is often assumed that a given change in operating voltage will result in a ten-fold greater change in gas gain, and counter tube power supplies are stabilized accordingly. On the other hand, changes in gas density in the counter tube are

Table II. Width of PHD for Different Anode Wire Diameters

	Gas, pressure, and wire diameter							
	Argon–methane				Neon–helium–butane			
	200 mm Hg		760 mm Hg		200 mm Hg		760 mm Hg	
Gas gain	50 μ	125 μ	50 μ	125 μ	50 μ	125 μ	50 μ	125 μ
1.54 Å								
1×10^2	0.18	0.16	0.20	0.17	0.23	0.19		
4×10^2	0.19	0.16			0.24	0.18		
16×10^2	0.24	0.17			0.35	0.22		
6.98 Å								
4.5×10^2	0.31	0.31			0.42	0.36	0.42	0.42
18×10^2	0.33	0.33			0.53	0.42	0.66	0.66
72×10^2	0.38	0.36			0.81	0.59	1.20	1.20

Table III. Coefficient of Voltage Stability for Argon–Methane Mixture

	Pressure and wire diameter			
	50 μ anode wire		125 μ anode wire	
Gas gain	200 mm Hg	760 mm Hg	200 mm Hg	760 mm Hg
1.54 Å				
1×10^2	9.6	11.3	10.5	10.4
4×10^2	11.9	14.3	13.5	
16×10^2	15.1	15.3	16.0	
6.98 Å				
4.5×10^2	12.5	13.2	13.0	
18×10^2	15.4	14.6	15.3	
72×10^2	24.8		20.4	

frequently neglected. When it is realized that usually flow counters are vented to atmosphere, it is, therefore, clear that changes in atmospheric pressure may be of concern. In the New York area, for example, barometric pressure may vary by as much as 4% over a day or 7% over a year. A coefficient of voltage stability may be defined as the percent shift of the peak of the pulse height distribution for a one percent shift of voltage, and a similar definition may be made of the coefficient of density stability. In the measurements reported here, density was varied by varying the pressure of the gas mixture, and, hence, the results are reported in terms of a coefficient of pressure stability. The sign of the voltage coefficient is positive, while the sign of the pressure coefficient is negative.

Table III presents values of the coefficient of voltage stability for the argon–methane mixture for two different pressures and two different wire sizes for 1.54 and 6.98 Å

Table IV. Coefficient of Voltage Stability for Neon–Helium–Butane and Nitrogen–Helium–Methane Mixtures

	Gas, pressure, and wire diameter				
	Neon–helium–butane				Nitrogen–helium–methane
	50 μ anode wire		125 μ anode wire		50 μ anode wire
Gas gain	200 mm Hg	760 mm Hg	200 mm Hg	760 mm Hg	200 mm Hg
	1.54 Å				
1 × 10²		10.0		10.5	
4 × 10²				14.1	
16 × 10²				19.1	
	6.98 Å				
4.5 × 10²	13.2	12.5	14.7	14.6	14.0
18 × 10²				19.6	25.2
72 × 10²		32.0		32.0	

Table V. Coefficient of Pressure Stability for Argon–Methane and Nitrogen–Helium–Methane Mixtures

	Gas, pressure, and wire diameter				
	Argon–methane				Nitrogen–helium–methane
	50 μ anode wire		125 μ anode wire		50 μ anode wire
Gas gain	200 mm Hg	760 mm Hg	200 mm Hg	760 mm Hg	200 mm Hg
	1.54 Å				
1 × 10²	3.8	4.3	4.5	7.1	
4 × 10²	3.7	6.0	4.9		
16 × 10²	5.8	5.5	6.0		
	6.98 Å				
4.5 × 10²	3.7	5.8	6.1		4.8
18 × 10²	5.8	6.1	5.9		
72 × 10²	7.7		8.3		

radiations. The primary discernible effect is due to increasing gas gain. Table IV shows similar data for the neon–helium–butane and the nitrogen–helium–methane mixtures, with similar conclusions. It is clear that the voltage coefficient can be quite large for large values of gas gain.

Table V shows values of the coefficient of pressure stability for the argon–methane and the nitrogen–helium–methane mixtures. The pressure coefficients are about one-third to one-half of the corresponding voltage coefficients, and also increase with gas gain. There are also indications of a dependence upon gas pressure and anode wire size.

With regard to stability of operation, a *caveat* is in order when using neon based mixtures for the flow counter. During the course of the work reported here, it was found that rather long "warm-up" times, of the order of two or three hours, were required when studying these mixtures. A fairly substantial flow rate, of the order of 0.1 liters/min, was necessary in order to obtain any meaningful results at all. In view of the fact that a cylinder containing 200 liters of the neon–helium–butane mixture costs about $200, such a flow rate may be considered to be rather expensive. It is, therefore, desirable to take extra pains to remove sources of contamination from flow counter systems when using neon based mixtures.

CONCLUSIONS

It is quite apparent that the only effective way to eliminate the counting rate dependent shift of the pulse height distributions is to increase the diameter of the anode wire. Such an increase in diameter of the wire will necessitate the use of higher operating voltages for the counter tube in order to maintain a given value of gas gain, or alternatively higher gain in the pulse amplification electronics, or some combination of higher counter tube voltage and higher electronic gain. Care must be taken, of course, that the higher gain electronics does not introduce excess noise into the system. The use of the larger diameter anode wire has only a slight effect on the resolution and stability of the pulse height distribution with, if anything, a slight improvement in resolution, and a slightly greater coefficient of pressure stability.

For the detection of soft X-rays of wavelength greater than about 4 Å, the use of neon based gas mixtures offers the advantages of lower operating voltages, increased discrimination against short wavelengths, and the suppression of the escape peak. On the other hand, the resolution seems to be not as good, and stability of operation is at present somewhat more difficult to obtain than with the usual argon based gas mixtures.

Stabilization requirements on counter tube voltage and gas density are much higher than is commonly assumed, particularly if any type of electronic analysis based on the shape of the pulse height distributions of the observed spectra is to be carried out. While voltage stabilization is generally included in most instrumental arrangements, density stabilization is not. While the most desirable technique is to stabilize against density fluctuations,[13] usually the flow counter is in sufficiently good thermal contact with its environment and the gas flow rate is sufficiently low, that in an air conditioned laboratory it may be sufficient to stabilize pressure alone.

Although some of the factors affecting the counting rate dependent shift of the pulse height distributions have been uncovered, there is not as yet a clear understanding of the mechanism involved. One is tempted to regard the effect as being entirely due to a space charge of positive ions which effectively lower the counter tube voltage, and which would screen a smaller diameter wire more effectively than a larger wire; alternatively, a condition of saturation of microscopic discharges which would occur more readily with smaller diameter wires or a gas heating effect due to gas multiplication in the vicinity of the anode wire may be postulated. Further investigation is required.

NOTE ADDED IN PROOF

It has recently been shown that the anode wire material and its treatment have a significant effect on the shift of the pulse height distribution with counting rate (N. Spielberg, "Effect of Anode Material on Intensity Dependent Shifts in Proportional Counter Pulse Height Distributions," *Rev. Sci. Instr.* **38**: 291, 1967).

ACKNOWLEDGMENTS

I am indebted to A. J. Kirschner for help with the experimental apparatus, to Dr. P. G. Cath for help with the electronics, and to Dr. P. G. Cath and Dr. F. K. du Pré for discussion of some of the factors involved.

REFERENCES

1. U. W. Arndt, W. A. Coates, and A. R. Crathorn, "A Gas-Flow X-Ray Diffraction Counter," *Proc. Phys. Soc. (London)* **B67**: 357, 1954.
2. C. F. Hendee, S. Fine, and W. B. Brown, "Gas-Flow Proportional Counter for Soft X-Ray Detection," *Rev. Sci. Instr.* **27**: 531, 1956.
3. Burton L. Henke, "X-Ray Fluorescence Analysis for Sodium, Fluorine, Oxygen, Nitrogen, Carbon, and Boron," *Advances in X-Ray Analysis*, Vol. 7, Plenum Press, New York, 1964, pp. 460–488.
4. P. Lublin and W. J. Sutkowski, "Application of the Electron Probe to Electronic Materials," in: T. D. McKinley, K. F. J. Heinrich, and D. B. Wittry (eds.), *The Electron Microprobe*, The Electrochemical Society, John Wiley & Sons, Inc., New York, 1966, pp. 677–690.
5. W. Parrish and T. R. Kohler, "Use of Counter Tubes in X-Ray Analysis," *Rev. Sci. Instr.* **27**: 795, 1956.
6. R. M. Dolby, "Some Methods for Analyzing Unresolved Proportional Counter Curves of X-Ray Line Spectra," *Proc. Phys. Soc. (London)* **73**: 81, 1959.
7. L. S. Birks and A. P. Batt, "Use of a Multichannel Analyzer for Electron Probe Microanalysis," *Anal. Chem.* **35**: 778, 1963.
8. S. L. Bender and E. J. Rapperport, "Nonproportional Behavior of the Flow Proportional Detector," in: T. D. McKinley, K. F. J. Heinrich, and D. B. Wittry (eds.), *The Electron Microprobe*, The Electrochemical Society, John Wiley & Sons, Inc., New York, 1966, pp. 405–414.
9. D. R. Beaman, "Effect of Pulse Amplitude Shifts on Electron Probe Intensity Radios," *Anal. Chem.* **38**: 599, 1966.
10. N. Spielberg, "Elimination of Intensity Dependent Shifts in Proportional Counter Pulse Height Distributions," *Rev. Sci. Instr.* **37**: 1268, 1966.
11. S. C. Curran, "The Proportional Counter as Detector and Spectrometer," in: S. Flugge (ed.), *Handbuch der Physik*, Vol. XLV, *Nuclear Instrumentation II*, E. Creutz (coed.), Springer–Verlag, Berlin, 1958, pp. 174–221.
12. William Parrish, "Escape Peaks in X-Ray Diffractometry," *Advances in X-Ray Analysis*, Vol. 8, Plenum Press, New York, 1965, pp. 118–138.
13. Richard D. Deslattes, Bert G. Simpson, and Robert E. LaVilla, "Gas Density Stabilizer for Flow Proportional Counters," *Rev. Sci. Instr.* **37**: 596, 1966.

DISCUSSION

T. F. Swank (Cabot Corporation): I have a question on a subject that is slightly different than the one you have been talking about. What is the best way to retard the contamination of these flow counters?

N. Spielberg: Construct the counters from as clean materials as possible and use as clean a gas as possible and for the various components, use clean materials. In a good bit of our work we tried to use tygon tubings and things of this type. I don't think that is quite satisfactory. I think basically the idea is to try to keep things as clean as possible.

T. F. Swank: Do you have any idea what the contamination is inside the counter?

N. Spielberg: I might refer some remarks on this to Charles Hendee. He has a lot of experience in counter tubes. Perhaps he can comment on that.

C. F. Hendee (Bendix Corporation): I think that one of the worst offenders in gain stability, resolution, etc., is electronegative gases, consisting of oxygen or halogen, some carbon tetrachloride cleaner or something like that. As a rule of thumb, one should avoid chlorinated solvents for cleaning. You might use alcohol. I think it is common experience that in vacuum systems one should avoid the use of chlorinated solvents if at all possible. This is one suggestion.

K. F. J. Heinrich (National Bureau of Standards): In some of the applications of scanning microprobe analysis, we would like, for statistical reasons, to use counting rates on the order of 100,000–

200,000 cps. Would you care to extrapolate your experiences and indicate if work without unreasonable shrinkage can be done at these speeds?

N. Spielberg: I think (I have nothing but extrapolation on this) that maybe a 500 μ diameter wire would be helpful. The only problem with a larger diameter wire is, as you have seen, one of increased operating voltages. So, it may be desirable then to lower the pressure of the counter tube so that you can run it at lower voltages.

K. F. J. Heinrich: You say that as long as you have the dispersive system you basically discriminate against second order.

N. Spielberg: Yes, and if the shifts then are small, then it is not too serious for you. You may recall that the plot we have of the shifts was logarithmic with count rate. It is quite possible that the 125 μ wire may be satisfactory and may work fairly well with the usual sort of electronics at the usual operating voltages. For low-energy radiations, you must either increase the voltage to increase the gain, or alternatively lower the gas pressure, which would be all right for soft radiations.

K. F. J. Heinrich: It has also been commented sometimes that similar losses of pulse height might be due to the characteristics of the first stages of amplification rather than the detector.

N. Spielberg: This is certainly true. We were very careful with the electronics to pretest it with pulsers and with scintillation counters so that we stayed well within the range of the electronics there. We knew that our circuits were stable.

AUTHOR INDEX

Bold numbers refer to papers in this volume

A

Adler, I., 489, 493
Agar, J. N., 325, 327
Ageev, N. V., 213–215, 218, 219
Ahlers, M., **265–272**
Alexander, L. E., 255–257, 263, 285, 294, 316, 327
Alexandrov, K. S., 29, 31
Alford, W. J., 161, 171
Algie, S. H., 213, 214, 219
Almen, J. O., 295, 309
Amelinckx, S., 134, 151
Amick, J., 252, 263
Ananthanarayanan, N. I., **240–249**
Ando, Y., 27, 31, 46, 51, 60, 99, 106
Andrus, J., 40, 87, 90
Aoki, K., **342–353**
Appleton, A., 374, 387
Aqua, E. N., 257, 263
Arata, H., 126, 132
Archard, G. D., 419, 420
Archard, J. F., 311, 312, 326
Armstrong, R. W., 174, 184
Armstrong, W. J., 120, 132
Arndt, U. W., 534, 543
Ashby, M. F., 153,157, 158
Ashby, W. D., 508, 509, 519
Asimow, R., 255, 263
Asonson, J. R., 252, 263
Asp, E. T., 343, 346, 352
Atwood, J. G., 159, 171
Austerman, S. B., **134–152**
Authier, A., **9–31,** 92, 98, 106, 153, 158
Averbach, B. L., 313, 318, 326, 328, 340
Avery, D. H., 329, 340
Azaroff, L. V., 7, 8, 378, 388

B

Backofen, W. A., 329, 340
Baggerly, R. G., **328–341**
Baldrey, J. A., 120, 132
Ballman, A. A., 160, 171
Balluffi, R. W., 251, 262, 264
Bando, Y., 246, 248
Bardolle, J., 325, 327
Barkow, A. G., 251, 262
Barns, R. L., 159, 171
Baroody, E. M., 449, 460
Barrett, C. S., 2, 6, 7, 80, 90, 134, 135, 151, 285, 294
Barth, H., 2, 6, 7, 10, 30, **80–90**, 138, 151

Bartsch, G., 186, 187, 202
Batchelder, D. N., 358, 364
Batt, A. P., 534, 544
Batterman, B. W., 69, 78, 134, 151, 157, 158
Bauer, S. H., 259, 260, 263
Bauerle, J. E., 251, 262–264
Baun, W. L., **374–388,** 389, 397, 494, 505
Beaman, D. R., 534, 544
Beard, D. W., **506–519**
Bearden, J. A., 357, 358, 364, 522, 533
Beck, P. A., **213–220,** 345, 352
Beeley, P. R., 462, 472
Belanger, J. R., 213, 219
Belt, R. F., **159–172**
Bender, S. L., 534, 538, 544
Berg, W., 2, 6, 7, 80, 81, 90
Berghuis, J., 78
Bergin, R., 254, 263
Bergman, B. G., 213, 219
Bernstein, F., **494–505**
Bertin, E. P., **462–473,** 506, 519
Beton, R. H., 210, 211
Birks, L. S., 169, 170, 172, 416, 420, 432, 437, 444, 495, 505, 518, 519, 534, 544
Bishop, H. E., 454, 457, 461, 525, 533
Black, P. H., 295, 309
Blackman, M., 42, 45
Blech, I. A., 120, 126, 132
Blodgett, K. B., 389, 391, 397
Bloom, D. A., 213, 219
Boas, 289, 294
Bollmann, W., 46, 60
Bond, H. E., 163, 172
Bond, W. L., 40, 87, 90, 355, 360, 364
Bonfiglioli, A. F., 186, 202
Bonse, U. K., **1–8,** 9, 10, 30, 41, 87, 90, 104, 107, 131, 132, 134, 151
Borie, B., 235, 239
Borrmann, G., 3, 4, 8, 9, 20, 23, 30, 31, 73, 74, 78
Boswell, F. W. C., 251, 259, 263
Bowden, F. P., 314, 326
Bradinstein, M., 515, 518, 519
Bradley, C. S., 409, 420
Braski, D. N., **295–310**
Brauer, W., 448, 449, 460
Braun, I., 130, 132
Briggs, G., 425, 430
Brown, D. B., 457, 461
Brown, L. M., 153, 157, 158
Brown, W. B., 534, 544
Bruce, L., 343, 352

Bruining, H., 449, 460
Bubakova, R., 37, 38, 40
Buckrey, R. R., **250–264**
Burkhalter, P. G., 460, 461, 525, 530, 533
Burwell, J. T., 311, 326
Buteux, R. H., 111, 117
Butler, C. C., 260, 264

C
Cahn, R. W., 217, 265
Campbell, W. J., 509, 519, 520, 523, 530, 533
Castaing, R., 431, 441, 444, 456, 461
Cathcart, J. V., 235, 239
Chace, W. G., 251, 262
Chamberod, A., 247, 248
Chandrasekhar, S., 336, 341
Chanley, J. D., 391, 397
Chase, A. B., 163, 171
Chaudron, G., 186, 202
Chernock, W. P., 345, 352
Chikawa, J., 18, 30, 98, 104, 106, 107, **153–158**
Chipman, D. R., 42, 45
Christensen, A. L., 299, 300, 309
Claisse, F., 475, 483, 489, 493
Clarenbrough, L. M., 262, 264
Clarke, J. S., 108, 116
Claus, H., **213–220**
Coates, W. A., 534, 543
Cochran, W., 211, 212
Cohen, J. B., 265, 271, 329, 333, 340
Cohen, M., 313, 318, 326
Colby, J. W., **447–461**
Cole, H., 69, 78, 134, 151, 157, 158
Conley, D. K., **447–461**
Conrad, H., 160, 171
Cook, M. I., 161, 171
Cooper, H. W., 120, 132
Cooper, L., **422–430**
Cosslett, V. E., 524, 533
Cotterill, R. M. J., 262, 264
Craik, D. J., 102, 106
Crathorn, A. R., 534, 543
Cullington, E. H., 251, 262
Cullity, B. D., 257, 263, 281–283, 285, 294
Curran, S. C., 538, 544
Czanderna, A. W., 238, 239

D
Daane, R. A., 275, 283
Das, B. N., 378, 388
Da Veiga, L. M. D'A., 213, 214, 219
Davey, J. E., 259, 260, 263
Davies, D. E., 325, 327
Davies, R. G., 265, 271
De Barr, A. E., 343, 352
Decker, B. F., 213, 219, 343, 346, 352
Deiter, R. H., 259, 260, 263

Dekker, A. J., 449, 460
Delf, B. W., 355, 364
Derick, L., 120, 132
Deslattes, R. D., 543, 544
Dess, H. M., 160, 169–172
Dexter, H. B., 295, 309
Dickens, G. J., 213, 214, 219
Dickerman, C. E., 430
Dobrott, R. D., 118, 119, 123, 131
Doi, K., 205, 211
Dolby, R. M., 534, 544
Donachie, M. J., Jr., 273, 281, 283
Douglas, A. M. B., 213, 214, 219
Doyama, M., 262, 264
Dueker, G. W., 159, 171
Duffy, M. C., 120, 132
Dumesnil, F., 21, 31
DuMond, J. W. M., 256, 263
Duncumb, P., 416, 420, 441, 445, 525, 533
Dunn, C. G., 260, 264
Dunne, J. A., 462, 472, 475, 483
Duwez, P., 251, 262, 432, 444

E
Earley, D., **221–233**
Ehlert, R. C., **389–398,** 494, 505
Ehrenberg, W., 108, 116
Elias, J. A., 348, 349, 352
Elliot, S. B., 113, 117
Emslie, A. G., 252, 263
Eshelby, J. D., 18, 30, 156, 158
Evans, E. H., 161, 171
Evans, U. R., 325, 327
Evans, W. P., **273–283**, 285, 290, 294, 306, 309
Evitts, H. C., 120, 132
Ewald, P. P., 58, 60

F
Fairfield, J. M., 128, 130–132, 153, 158
Farineau, J., 380, 388
Faulring, G. M., **409–421**
Faust, J. W., Jr., 120, 132
Federighi, T., 185, 186, 191, 195–198, 201, 202
Feltner, C. E., 337, 341
Fetroff, J. F., 153, 158
Finch, R. H., 130–132
Fine, S., 534, 544
Fink, W. L., 186, 202
Finkel'shtein, L. D., 378, 388
Fischer, D. W., **374–388,** 389, 397, 494, 505
Fish, B. R., 251–253, 262, 263
Fisher, J. C., 260, 264
Flanagan, F. J., 489, 493
Flanagan, W. F., **328–341**
Flaschen, S. S., 120, 132

Fordemwalt, J. N., 120, 132
Forsyth, J. B., 213, 214, 219
Frank, F. C., 92, 97, 103, 105, 106
Frankenburg, W. G., 230, 232
Franks, A., 355, 364
Fridel, J., 381, 388
Frohnsdorf, G. J. C., 367, 372
Frondel, C., 93, 106
Frosch, C. J., 120, 132
Furnas, T. C., Jr., 389, 397
Furois, P. C., 130, 132
Furusho, K., 153, 158

G
Garrod, R. I., 282, 283
Gatto, F., 186, 202
Geisler, A. H., 241, 248
Geiss, R. H., 255, 263
Gerold, V., 10, 30, 87, 90, 185–190, 194, 196, 198, 200, 202, 204, 211
Gibbs, P., 163, 171
Gilfrich, J. V., 432, 437, 444
Glang, R., 121, 130, 132
Goetzberger, A., 130–132
Golay, M. J. E., 370, 372
Goldberg, E. D., 475, 483
Goldschmidt, H. J., 326, 327
Goldstein, J. I., 248, **431–446**
Goodrich, R. S., **284–294**
Gould, R. W., **185–203**
Grant, N. J., 213, 219
Green, M., 524, 533
Greenfield, P., 213, 219
Grewen, J., 343, 352
Grosskreutz, J. C., 329, 336, 340
Guentert, O. J., 108, 116
Guffy, J. C., **489–493**
Guinier, A., 1, 6, 7, 80, 90, 185, 186, 201, 202, 204, 211, 212, 224, 232, 255, 263
Gunn, E. L., 475, 483
Gwathmey, A. T., 235, 239

H
Haanappel, G. M., 78
Haberkorn, H., 186, 202
Hachenberg, O., 448, 449, 460
Hall, E. O., 213, 214, 219
Hanneman, R. E., 432, 444
Hansen, M., 240–242, 245–248, 434, 444
Hardy, H. K., 198, 202
Hargreaves, M. E., 262, 264
Harker, D., 343, 346, 352
Harris, G. T., 326, 327
Harris, P. H., 367, 372
Harrison, L. G., 260, 264
Hart, M., **1–8**, 10, 21, 30, 31, 97, 99, 104, 106, 107, 134, 151

Hartmann, R. J., 328, 339, 340
Hartwig, W., 3, 4, 8, 9, 20, 30, 73, 74, 78
Harvey, K. B., 163, 172
Hashimoto, F., 186, 187, 196, 198, 202
Haughton, J. L., 240, 248
Hawkes, G. A., 282, 283
Hayami, S., **342–353**
Haygood, J. D., 260, 264
Heal, T. J., 213, 219
Heimerl, G. J., 295, 309
Heinrich, K. F. J., 415, 419, 420, 432, 441, 442, 444, 445, 450, 451, 453, 454, 456, 460–462, 467, 472, 522, 533
Heiser, H. W., 230, 232
Helion, J. C., 265, 271
Hendee, C. F., 534, 544
Henke, B. L., 389, 397, 494, 505, 514, 519, 534, 539, 544
Herenguel, J., 186, 202
Herring, C., 259, 263
Hesser, D. R., 109, 116
Heyer, R. H., 348, 349, 352
Higuchi, S., 246, 248
Hilley, M. E., **284–294**
Hirsch, P. B., 46, 47, 49, 60, 98, 105, 106, 329, 336, 340
Hirst, W., 312, 326
Holmwood, R. A., 121, 130, 132
Homma, S., 46, 51, 59, 60
Honeycombe, R. W. K., 2, 6, 7
Horne, R. W., 46, 60
Hosemann, R., 2, 6, 7, 10, 30, 87, 90
Hough, R. L., **221–233**
Houseknecht, T. M., 477, 483
Howard, J. K., **118–133**
Hurley, J. W., 169, 170, 172
Hyler, W. S., 273, 283

I
Ibuka, S., 153, 158
Intrater, J., 3, 7, 235, 239
Irmler, H., 3, 4, 8, 9, 20, 30
Isaacs, T., 161, 171
Isherwood, B. J., 358, 364
Ishii, Z., 20, 30

J
Jackson, J. J., 251, 262
Jackson, L. R., 273, 283
James, D. R., 210–212
James, R. W., 69, 71, 78, 134, 151, 182, 184
Jan, J. P., 185, 201
Janowski, K. R., 160, 163, 171
Jenkinson, A. F., 118–120, 132
Johnson, M. P., 462, 472
Jopson, R. C., 399, 407
Joshi, M. L., 126, 132

Josso, E., 241, 248
Joyce, B. D., 120, 132
Juleff, E. M., **173–184**

K
Kachi, S., 246, 248
Kaczer, J., 103, 106
Kambe, K., 21, 31
Kamen, E. L., 213, 219
Kamiya, Y., 92, 105, 106
Kanter, H., 450, 460
Kappler, E., 2, 6, 7, 9, 30, 40, 41
Karioris, F. G., **250–264**
Kasper, J. S., 213–215, 218, 219
Katagawa, T., 23, 31, **46–66,** 92, 99, 106
Kato, H., 348, 349, 352
Kato, N., 23, 27, 31, **46–66,** 92, 98, 99, 106
Katzmann, M., 159, 171
Kaufman, L., 246, 248
Kehl, W. L., 371, 372
Kellen, P. F., 374, 387
Kellington, C. M., 159, 171
Kelly, A., 153, 158
Kelman, L. R., 429, 430
Khan, J. M., 399, 400, 402, 407
King, H. W., 257, 263, **354–365**
Kirkpatrick, H. A., 256, 263
Klement, W., Jr., 251, 262, 432, 437, 444, 445
Klug, H. P., 255–257, 263, 285, 294, 316, 327
Knapp, K. T., 389, 397
Koehler, J. S., 251, 262–264
Kohler, T. R., 534, 544
Kohra, K., 10, 30, 37, 40, 46, 60, 98, 106
Kohra, M., 2, 6, 7
Koistinen, D. P., 273, 275, 276, 283, 285, 286, 290, 294
Komarewsky, V. I., 230, 232
Kronberg, M. L., 161, 171

L
Laberrigue-Frolow, J., 441, 445
Lamla, E., 69, 71, 78
Lang, A. R., 3–6, 8, 9, 12, 16, 20, 23, 27, 30, 31, 46, 60, 87, 90, **91–107**, 118–120, 131, 132, 142, 151, 153, 157, 158, 161, 171, 174, 184, 336, 341
Langmuir, I., 391, 397
Lapierre, A. G., III, **173–184**
Larson, F. R., 308, 309
Larson, J. O., **489–493**
Laudise, R. A., 160, 171
von Laue, M., 58–60, 69, 78
Lauriente, M., 174, 184
LaVilla, R. E., 543, 544
Lawless, K. R., **234–239**
Lawley, A., 329, 336, 340
Lawn, B., 92, 106

Lee, F. S., 509, 519
Leeser, D. O., 275, 283
Lement, B. S., 213, 219
Lemons, K. E., 175, 184
Letner, H. R., 307, 309
Levins, A., 355, 364
Lewis, G. J., Jr., 475, 483
Libowitz, G. G., 259, 260, 263
Liebhafsky, H. A., 415, 419, 420, 522, 533
Liedl, G. L., **204–212**
Liefeld, R. J., 375, 387
Lihl, F., 241, 247, 248
Linares, R. C., 160, 171
Lipsitt, H. A., 223, 230–232
Liu, Y. H., 241, 242, 245, 246, 248
Livingston, J. D., 96, 106, 153, 158
Longobucco, R. J., 462, 472
Lonsdale, K., 357, 364
Loomis, T. C., 462, 472
Loopstra, B. O., 78
Loretto, M. H., 262, 264
Lublin, P., 534, 544
Lucas-Tooth, J., 475, 483
Luke, C. L., 489, 493
Lye, R. G., 449, 460
Lynch, C. T., 230, 232
Lynch, J. T., 462, 472

M
Mabis, A. J., 389, 397
MacGillavry, C. H., 78
Macherauch, E., 273, 281–283, 285, 294, 328, 339, 340
Madlem, K. W., 494, 505
Majeske, F. J., **431–446**
Malgrange, C., 20, 21, 31, 153, 158
Malizie, E. S., **409–421**
Manly, W. D., 213, 219
Marburger, R. E., 273, 275, 276, 283, 285, 286, 290, 294
Marchand, A., 247, 248
Mark, H., 399, 407
Marks, C. L., **399–408**
Marples, J. A. C., 425, 430
Martin, D. E., 273, 283
Massalski, T. B., 265, 271
Matsuo, M., **342–353**
Mattson, R. A., **389–398, 494–505**
Mayer, G., 69, 71, 78
Mazdiyasni, K. S., 230, 232
McCaleb, S. B., 367, 372
McDuffie, R., 463, 472
McGee, J. F., **108–117**
McKay, K. G., 448, 460
McLinden, H. C., 252, 263
Meakin, J. D., 329, 336, 340
Meier, F., 10, 21, 30, 31, 87, 90, 97, 106
Meieran, E. S., 120, 126, 132, 175, 184

Michaelis, R. E., 432, 444
Michaels, A. S., 260, 264
Mikkola, D. E., 265, 271
Miller, D. C., 462, 472
Miller, D. P., 178, 184
Miller, J., 308, 309
Miller, P. H., Jr., 260, 264
Miuscov, V. F., 104, 107
Moll, A. J., 252, 263
Moll, S. H., 328, 339, 340, 414, 420
Moore, C. R., 178, 184
Moore, H. K., 251, 262
Moore, J. E., 178, 184
Moore, M. G., 306, 309
Moreau, J., 325, 327
Morrison, J. A., 260, 264
Mulvey, T., 419, 420

N
Nagashima, S., 348, 349, 352
Nakayam, T., 153, 158
Nelson, D. F., 160, 171
Nemnonov, S. A., 378, 388
Nestor, O. H., 160, 169–172
Neuberger, M. C., 225, 232
Nevitt, M. V., 214, 219, 220
Newkirk, J. B., **1–8,** 9, 10, 30, 87, 90, 118, 119, 131, **134–152,** 174, 184, 241, 246, 248, 343, 352
Newsome, J. W., 230, 232
Nichols, J. A., 425, 430
Nicholson, R. B., 153, 158
Nicolson, M. M., 259, 260, 263
Nigam, A. N., 108, 116
Nilsson, N., 42, 45
Norton, J. T., 273, 281, 283
Nutting, J., 462, 472

O
Ogilvie, R. E., 248, 285, 294, 416, 420, 432, 441, 444, 445
Ohta, M., 186, 187, 196, 198, 202
O'Keefe, M., 238, 239
Olli, V. I., **108–117**
Otte, H. M., 223, 230, 232, 257, 263, 265, 271
Owen, E. A., 241, 242, 245, 246, 248

P
Panseri, C., 185, 186, 191, 195–198, 201, 202
Parrish, W., 355–357, 364, 534, 538, 544
Partridge, P. G., 329, 336, 340
Paskin, A., 42, 45
Paterson, M. S., 265, 271
Patterson, W., 477, 483
Pease, R. S., 256, 263
Peavler, R. J., **240–249**

Pelloux, R. M. N., **328–341**
Penning, P., 4, 8, 9, 20, 23, 30, 31, **67–79,** 131, 132
Perry, A. J., 186, 202
Petroff, J. F., 16, 27, 30, 31
Pfeiffer, H. G., 415, 419, 420, 522, 533
Philibert, J., 416, 420, 456, 460, 461, 525, 533
Phillips, V. A., 96, 106, 153, 158
Pike, E. R., 266, 271, 357, 364
Plooster, M. N., 160, 169–172
Polcarova, M., 16, 20, 30, 92, 97, 103, 106, 107
Polder, D., 4, 8, 20, 30, 131, 132
Polmear, I. J., 186, 198, 202
Poole, D. M., 456, 461
Potter, D. L., 399, 400, 402, 407
Potters, M., 78
Pratt, J. E., 337, 341
Preece, C. M., **354–365**
Preston, G. D., 204, 211
Prickett, R. L., **221–233**
Pride, R. A., 295, 309
Proctor, E. M., **506–519**
Prussin, S., 126, 132
Puri, O. P., 282, 283
Pyne, L., 475, 483

Q
Queisser, H. J., 126, 132, 174, 184
Quinn, T. F. J., **311–327**

R
Rachinger, W. A., 255, 263
Radvanyi, P., 441, 445
Ramachandran, G. N., 46, 51, 60
Rapperport, E. J., 534, 538, 544
Read, W. T., 18, 30
Remeika, J. P., 160, 171
Renninger, M., 9, 30, **32–41, 42–45,** 74, 78, 87, 90
Rex, R. W., **366–373**
Rhines, F. N., 241, 246, 248
Ricklefs, R. E., **273–283**
Rideal, E. K., 230, 232
Rideout, S. P., 213, 219
Rindner, W., 130, 132
Ringwood, A. E., 246, 248
Rizzo, F. E., 462, 463, 471, 472
Robertson, J. A., 329, 340
Rocktäschel, V. G., 409, 420
Rogers, C. B., 92, 106
Rollason, E. C., 210, 211
Rose, G. S., 260, 264
Rose, H. J., Jr., 489, 493
Rosenbaum, H. S., 185, 201
Rosenfeld, R. L., 121, 130, 132
Royster, D. M., **295–310**

Royster, G. W., Jr., 251, 252, 262
Rudman, P. S., 214, 220
Ruscoe, M. J. H., **520–533**
Russell, A. S., 230, 232
Russell, B. R., 260, 264
Russell, C. M., 354, 356, 361, 364
Ruzhova, T. V., 29, 31
Rymer, T. B., 260, 264

S
Saccocio, E. J., 72, 78
Salow, H., 449, 460
Sanderson, B. S., **474–488**
Sato, Y., 126, 132
Sauvage, M., 23, 31, 98, 106
Savitzky, A., 370, 372
Saylor, W. P., **399–408**
Scheuplein, R., 163, 171
Schlangenotto, H., 18, 30
Schlössin, H. H., 92, 106
Schmalzreid, H., 186, 202
Schmidt, 289, 294
Schossberger, F., 257, 263
Schuff, W., 409, 420
Schulz, L. G., 1, 2, 6, 7, 344, 346, 352
Schumacher, D., 251, 263
Schwartz, M., 343, 352
Schweizer, W., 185, 187, 189, 194, 200, 202
Schwuttke, G. H., 98, 106, **118–133**, 153, 158, 174, 184
Seeger, A., 251, 263
Segall, R. L., 262, 264, 329, 336, 337, 340, 341
Segmüller, A., 343, 352
Seitz, F., 185, 198, 201, 262, 264
Sello, H., 126, 132
Semchyshen, M., 273, 275, 276, 283, 285, 290, 294
Shaler, A. J., 260, 264
Shaw, G. G., 329, 336, 340
Shekhtman, V. Sh., 213–215, 218, 219
Sher, I. H., 391, 397
Shields, P. K., 416, 420, 441, 445, 525, 533
Shimizu, J., 2, 6, 7, 10, 30, 98, 106
Shockley, W., 18, 30
Shoemaker, D. P., 213, 219
Short, M. A., **520–533**
Shuttleworth, R., 259, 263
Siems, R., 18, 29, 30
Silberberg, M., 423, 430
Sils, V., 98, 106, 174, 184
Simmons, R. O., 251, 262, 264, 358, 364
Simpson, B. G., 543, 544
Sloan, R. D., 462, 472
Smallman, R. E., 336, 341
Smith, D. K., 147, 151
Smith, J. H., 348, 349, 352
Smith, J. S., 230, 232
Snowden, K. V., 329, 336, 340

Sparks, C. J., 235, 239
Spielberg, N., 515, 518–521, 533, **534–545**
Spooner, F. J., 213–215, 218, 219
Spor, R. W., **213–220**
Stachura, S. J., **422–430**
Stephens, D. L., 161, 171
Sterk, A. A., **399–408**
Sternglass, E. J., 448, 449, 460
Stickler, R., 174, 184
Stofel, E. J., 163, 171
Stoffels, J. J., 251, 262
Stokes, A. R., 266, 271, 329, 340
Stone, F. S., 238, 239
Storks, K. H., 462, 472
Strang, C. D., 311, 326
Straumanis, M. E., 355, 364
Stumpf, H. C., 230, 232
Sully, A. H., 213, 219
Sutkowski, W. J., 534, 544
Swalin, R. A., 238, 239
Swank, T. F., **234–239**
Swanson, H. E., 161, 171
Sweeney, W. E., 169, 170, 172
Swift, C. D., 399, 407

T
Tabor, D., 314, 326
Taira, S., 273, 281–283
Takagi, M., 92, 105
Takagi, S., 21, 31
Takamura, J., 251, 262
Takechi, H., 348, 349, 352
Taupin, D., 21, 31
Taylor, A., 225, 232, 285, 294
Taylor, W. H., 213, 214, 219
Tebble, R. S., 102, 106
Tennevin, J., 1, 6, 7, 80, 90
Thatcher, J. W., 520, 523, 530, 533
Theisen, P., 415, 419, 420
Theisen, R., 495, 505
Thomas, A. D., Jr., **204–212**
Thomas, G., 126, 132, 191, 202, 260, 264
Thomas, P. M., 456, 461
Thompson, B. J., 374, 387
Thompson, N., 329, 340
Togel, K., 462, 472
Tolansky, S., 111, 117
Toman, K., 204, 211
Toussaint, C. J., 494, 505
Townsend, J. E., 489, 493
Treaftis, H. N., 185, 201
Turkevich, J., 252, 263
Turnbull, D., 185, 201
Tylecote, R. F., 238, 239

U
Usami, K., 23, 31, **46–66**, 92, 99, 106

V

Vassamillet, L. F., 257, 263, **265–272,** 356, 357, 359, 360, 364
Veenendaal, A. L., 78
Vieth, D. L., 432, 444
Vos, G., 494, 505

W

Wadewitz, H., 85, 90
Wadsworth, N. J., 329, 337, 340, 341
Wagner, C. N. J., 257, 263, 265, 271
Wagner, J. B., 235, 239
Wallace, C. A., 358, 364
Walter, C. M., 429, 430
Warekois, E. P., 265, 271
Warner, R. M., 120, 132
Warren, B. E., 257, 263, 265, 268, 271, 328, 330, 340
Warren, B. S., 108, 116
Washburn, J., 126, 132, 191, 202, 260, 264
Wassermann, G., 343, 352
Waterstrat, R. M., 213–215, 218, 219
Watson, E. M., 251, 262
Weinryb, E., 456, 457, 460, 461
Weiss, A., 409, 420
Weissmann, S., 1, 3, 7
Wells, A. F., 225, 232
Wernick, J. H., 7, 8
Wert, J. J., **284–294**
Westerman, F., 463, 472
Weyl, R., 462, 472
Whelan, M. J., 46, 47, 49, 60, 98, 105, 106
White, E. W., 389, 397
Wieder, H., 238, 239
Wilcock, W. L., 111, 117
Wilhelm, F., 126, 132
Wilkens, M., 97, 106
Willaime, C., 16, 18, 30

Willens, R. H., 251, 262, 432, 444
Willey, L. A., 186, 202
Williamson, G. K., 336, 341
Wilson, A. J. C., 257, 263, 266, 271, 357, 359, 364
Wilson, C. G., 213–215, 218, 219
Winkler, R. A., **489–493**
Winslow, E. H., 415, 419, 420, 522, 533
Wise, W. N., **447–461**
Witte, H., 42, 45
Wittig, W. J., 462, 472
Wittry, D. B., 453, 454, 461
Wöhler, L., 409, 420
Wölfel, E., 42, 45
Wolfson, R. G., **173–184**
Wood, W., 337, 341
Woodard, J. M., 295, 309
Worley, R. D., 399, 407
Woyci, J. J., **250–264**

Y

Yakowitz, H., **431–446**
Yeck, J. A., **474–488**
Yoneda, Y., 108, 116
Yoshida, S., 380, 388
Yoshimatsu, M., 2, 6, 7, 10, 30, 46, 59, 60, 92, 98, 106
Yoshioka, Y., 273, 281–283
Young, F., 20, 21, 31
Young, F. W., Jr., 153, 158

Z

Zachariasen, W. H., 35, 40, 60, 62
Zajac, A., 72, 78
Zemany, P. D., 415, 419, 420, 522, 533
Ziebold, T. O., 416, 420, 432, 441, 444, 445
Zimmerman, R. H., 462, 472

SUBJECT INDEX

A

Absorption
  correction, 441
  effects, 409, 415–419
ADP analyzing crystal, 436, 439, 475, 521
Age-hardening, 204
Al_2O_3, 230, 377, 379, 382–384, 387
Alloys (*see constituents*)
Aluminum
  alloys, 284–294, 374–388
  analysis, 374–380, 392–395, 397, 439–442,
    451*f*, 457, 474–488, 512, 514*f*, 518
  –antimony system, 381
  –arsenic system, 381
  –cobalt system, 381, 387
  –copper system, 204–212, 379*ff*, 384*f*, 432
  –gold system, 380*f*, 384
  –hafnium system, 381
  –iron system, 381, 387
  –magnesium system, 379, 384*f*, 431–446
  –nickel system, 374, 377*f*, 381, 383*ff*, 387
  oxide, 159–172
  –phosphorus system, 381
  –silver system, 380*f*, 384
  target fluorescent tube, 521, 523–526, 529*ff*
  –titanium system, 381
  –zinc alloys, 185–203
  –zirconium system, 381
Amalgams, 240–249
Anisotropy, 348*f*
Anomalous
  images, 157
  reflections, 108–117
  stresses, 273–283
  transmission method, 3, 4, 10
Asymmetrical "Bragg case," 32–41
Atomic number effect, 409, 416, 441
Automation, 366–373

B

BBr_3, 183
B_4C, 223*ff*
BN, 432, 434
Backscattered electrons, 447–461, 495
Barth–Hosemann method, 5, 7
Berg–Barrett method, 2, 4, 5, 7, 9, 80, 134–
    152, 173–184
Beryllium
  crystal, 87
  –gold system, 381
Block function, 51, 58*f*
Blodgett–Langmuir technique, 389, 391, 396*f*

Bonding, 374–388
Borie thin film mechanism, 238
Boron carbide fibers, 221–233
Borrmann effect, 3, 5, 7, 9*f*, 12, 21, 24, 67, 74
Bragg
  geometry, 119
  reflection, 34, 156*f*
Brazil twins, 96, 99, 101
Bremsstrahlung radiation, 399
Burgers vector, 2, 5, 14, 16, 92, 97, 153, 163,
    166, 329, 338*f*

C

Calcium analysis, 414, 474–488, 509
Carbon analysis, 404, 425
CdS, 153–158
Centroids, 265–272
Ceratic acid, 391
Characteristic temperature, 42, 44
Chlorine analysis, 513, 521, 527
Chromium
  analysis, 425*f*, 429, 492, 526, 531
  –iron sigma, 213–220
  radiation, 137, 213, 215, 234*ff*, 276, 281,
    286
  target fluorescent tube, 395, 483, 495, 499,
    509, 512, 520*f*, 526–529, 532
Cobalt
  analysis, 491
  radiation, 243, 276, 286, 315, 342, 345
Concave reflectors, 108, 113*f*
Copper, 324–341
  analysis, 403, 406*f*, 450*f*, 457, 538
  –germanium system, 265–272
  –gold system, 432
  –magnesium system, 380
  radiation, 67, 69, 71, 73, 83, 85, 123, 137,
    148, 159, 161, 174*ff*, 182, 234*ff*, 254,
    286*f*, 299, 331, 357, 368, 412, 435, 525,
    530
  –silicon system, 380
  single crystals, 234*ff*
  –zinc system, 432
Cr_2O_3, 163
Creep, 348*f*
Cryogenics, 354, 365
Crystal defects, 82, 91–107
Crystallite size, 250, 264
CuO, 234–239
CuO_2, 234–239
Curved crystals, 231
Czochralski method, 159*f*, 168*ff*

D

Dauphin twins, 99*f*
Defect structure, 159–172
Diffraction
  topography, 1–31, 46–66, 91–107, 118–158, 160
  vector, 5
Diffractometry, 366–373
Diffuse scattering, 204–212
Dislocations, 7
  edge, 97
  line, 9
  screw, 97, 174, 180
Double-crystal method, 2–5, 7, 9, 32–41, 160, 168*ff*
Duwez method, 431–443
Dynamical X-ray effects, 4*f*, 9*f*, 20–23, 32*ff*, 67–79, 91–107, 131, 134, 153–158

E

EDDT analyzing crystal, 375, 389, 436, 512
Electron
  excitation, 494–505
  microprobe analysis, 409–442, 447–461
Emission analysis, 399–408
Energy of stacking faults, 340
Epitaxial
  films, 234
  forces, 236*ff*
Ewald sphere, 104

F

F-values, 42
Fatigue, 328–341
Fault vectors, 101
Faults, 92, 98–102, 125*f*, 174
FeNi, 247
FeSi$_2$, 409–413, 419
Fermi level, 238
Ferrotitanium ores, 474–488
Flow-proportional counters, 534, 545
Fluorescence analysis, 462–533
Fluorescence tube (*see target material*)
Flux method, 159–166, 169*f*
Formate process, 240–249
Fourier analysis, 205, 268–271
Fraunhofer effect, 20
Fresnel effect, 9, 20
Friedel's law, 100*f*

G

GaCl$_3$, 153*f*
Gaussian distribution, 538
Germanium, 37*f*, 69*ff*, 73*f*

Gold
  aerosols, 250–264
  analysis, 436, 449*ff*, 457
  –silicon system, 431–442
Guinier–Preston zones, 185–205, 207*f*
Guinier–Tennevin method, 1, 7
Gypsum analyzing crystal, 389

H

Half widths, 266*ff*
Hydrothermal method, 159*ff*, 166*f*, 169*f*

I

Identification, 366–373
Images, anomalous, 157
Intensities, 52–58
Interferometer, 104
Inverse pole figure, 342, 349
Iron, 230
  analysis, 422–430, 474–488
  –nickel system, 240–249, 432, 434
  radiation, 254, 412

K

KAP analyzing crystal, 389, 396, 474*f*
Kinematical effects, 4*f*, 9, 134

L

Lang method, 3, 5, 9*f*, 12, 16, 21, 46, 134–153, 159–172, 174, 180
Laser, 109, 111, 114
Lattice
  constants, 250–264
  defects, 174
  parameters, 354–365
Laue
  geometry, 119
  sphere, 34
  spots, 136
Lead
  caprate, 389, 392, 396
  laurate, 389, 396*f*
  lignocerate, 391
  melissate analyzing crystal, 391*f*, 397
  myristate, 389, 392*ff*, 396
  stearate analyzing crystal, 393*f*, 396
Li$_2$CO$_3$, 501
LiF, 254*f*, 500
  analyzing crystal, 437, 463
Light elements, 520–533
Lignoceric acid, 389
Line
  broadening, 221–223, 250, 328–341
  profiles, 265–272
  widths, 169*f*, 254

Liquid analysis, 508–521
Lomer–Cottrell dislocations, 178

M

Macrostresses, 273–283, 295–310
Magnesium
analysis, 439–442, 474–488, 494–505, 521, 523, 526f, 529
–silver system, 380
Magnetic domains, 7, 91f, 102f
Matrix effect, 449, 475, 494, 501f, 504
Melissic acid, 389, 391
MgO, 260
Mica, 1ff
analyzing crystal, 231, 389
Moiré patterns, 91f, 103ff
Molybdenum
films, 130
radiation, 11f, 16–19, 21, 42ff, 121, 148, 153, 159f, 345, 348f, 434, 496, 498f, 528f, 531
Montanic acid, 391
Mosaic structure, 42, 80f
Myristic acid, 389, 391

N

NaCl, 42–45, 498, 527
analyzing crystal, 512, 521, 523
Nelson–Riley function, 257
Ni$_3$Fe, 241, 246f
NiO, 260
Nickel, 339
analysis, 425f, 465–471, 491

O

Optical flats, 108
Ordering, 186, 213–220
Orientation study, 342–353

P

Palmitic acid, 389
PAP analyzing crystal, 535, 538
Particle size effect, 494
Paterson theory, 265
PbF$_2$, 161, 163
PbO, 162
Peak locations, 265–272
Pendellösung fringes, 5, 9, 21, 23f, 27, 46, 54f, 99, 105, 160, 165f
PET analyzing crystal, 512
Pimaric acid, 396
Pinhole camera, 115f
POCl$_3$, 183
Póisson's ratio, 16, 275f, 284, 289
Pole figures, 342–353

Precipitation, 153–158
hardening, 185–204
Proton excitation, 399–408
Pyrolytic deposition, 221–233

Q

Quantitative analysis, 431–442, 447–461
Quartz, 4, 83, 91–107

R

Random samples, 342
Rayleigh criterion, 108
Residual
strain, 170
stresses, 273–310
Resistance measurements, 191
Rhenium–iron sigma, 213–220
Rocking curve, 9, 20, 37, 42, 145, 156, 159, 161
Ruby, 159–172

S

Sapphire, 159f, 169f
Satellite lines, 374, 382–385, 387
Schulz method, 1f, 7, 342, 344, 348
Screw dislocation, 97, 174, 180
Secondary electrons, 447
Semiconductors, 118–133
Sherrer equation, 256
SiO$_2$, 409f
Sigma phase, 213–220
Silicon, 12, 16, 36f, 39f, 43–45, 409f, 413
analysis, 391, 414f, 437, 474–488, 512, 514f, 518
single crystals, 118–133, 159f, 173–184
Silver
analysis, 450f
radiation, 19, 148, 254, 521, 523–528, 531
Slip plane, 5
Slit apertures, 462–473
Small-angle scattering, 185–223
Soap film analyzing crystals, 389–398
Soft X-rays, 374–388
SOT topographs, 118–133
Splat cooling, 431–442
SrO$_2$, 409f
SrSi, 409
SrSi$_2$, 409–421
Sr$_2$Si, 409
Stacking faults, 7, 23, 46–66, 104f, 257, 265
Standards, 431–442
Stearic acid, 370
Steels, 248, 273–283, 311–327, 343
Strain
analysis, 328–341
residual, 170

Strontium analysis, 414*f*
Substructure, 92, 159–172
Sulfur, 512–515, 518
Symmetrical "Bragg case," 32–41

T
Take-off angle, 521–526, 531
Temperature factor, 42
9, 10, 12, 13,-Tetrabromostearic acid, 396
Thermal diffuse scattering, 42–45
Thermal expansion coefficients, 237
Thin films, 234
TiO$_2$, 482
Titanium
    alloy, 295–310
    analysis, 474–488
Topography (*see Diffraction topography*)
Transmission topography, 10–20
Triglycine sulfate, 16, 18
Tungsten target tube, 463, 474, 477, 526, 530*f*
Twin boundaries, 7
    fault probabilities, 265–272
Twins, 91, 98–102

U
UC, 422–430
UFe$_2$, 429

U$_6$Fe, 429
Uranium analysis, 422–430, 457

V
Verneuil method, 159*f*, 164

W
Wave field, 47–51
Wear, 311–327
Wurzite structure, 153

X
X-ray excitation, 494–505
X-ray tube (*see target material*)

Y
Young's modulus, 276*f*, 284, 288, 299, 349–
    352

Z
ZrO$_2$, 162, 230